S. Rempel · B.-W. Schulze

Index Theory of Elliptic Boundary Problems

Mathematische Lehrbücher und Monographien
Herausgegeben von der Akademie der Wissenschaften der DDR
Institut für Mathematik

II. Abteilung
Mathematische Monographien
Band 55

Index Theory of Elliptic Boundary Problems

von S. Rempel und B.-W. Schulze

Index Theory of Elliptic Boundary Problems

by Stephan Rempel and Bert-Wolfgang Schulze

Akademie-Verlag Berlin
1982

Verfasser:
Dr. Stephan Rempel
Prof. Dr. Bert-Wolfgang Schulze
Institut für Mathematik
der Akademie der Wissenschaften der DDR

Erschienen im Akademie-Verlag, DDR-1086 Berlin,
Leipziger Straße 3—4
Lektoren: Dr. Reinhard Höppner, Dipl.-Math. Gesine Reiher

Lizenznummer: 202 · 100/404/82
Gestaltung: Dietmar Kunz
Gesamtherstellung: VEB Druckerei „Thomas Müntzer“, 5820 Bad Langensalza
Bestellnummer: 762 597 2 (6502) · LSV 1065
Printed in GDR
DDR 75,— M
ISSN 0076-5430

Preface

Subject of this monograph are essential parts of the theory of boundary value problems for elliptic partial differential equations and pseudo-differential equations. The theory presented here is arranged with the consequent use of pseudo-differential operators and symbolic calculus.

The elliptic partial differential equations form a large region of the analysis. A great number of important applications mainly of theoretical physics gives rise to problems for such types of equations. Moreover its theory of solvability is used for the solution of many types of non-elliptic problems in partial differential equations.

The theory of pseudo-differential operators has been developed and first checked in connection with elliptic equations and includes many traditional aspects such as methods of potential theory, singular integral operators or functional analysis. A culminating point of the development was the index theory connecting analytical properties of the index of Fredholm operators, represented as elliptic pseudo-differential operators, and deep geometrical and topological facts. One of the main results is well-known as the Atiyah-Singer index theorem. Besides the original papers (cf. Atiyah/Singer [2]) there is a number of monographs (cf. Palais [1]). Recently there have been discovered interesting applications of the index theory for the solvability of fundamental equations in gauge theory.

Independently of geometrical and topological relations elliptic partial differential equations have been studied in the analysis already for a long time. Essential problems are the elliptic regularity in various function and distribution spaces (e.g. Sobolev and Hölder spaces) and in this connection a-priori estimates and parametrix constructions, further properties of the ellipticity of the boundary conditions, calculation of the index, investigation of the spectrum of the considered problems and the study of operators with various types of degeneracies. There are some basic monographs treating the various aspects of elliptic boundary problems (cf. Miranda [1], Lions/Magenes [1], Èskin [3]) and a large number of original papers.

As far as the authors know, until now there is no monograph in the international literature treating the theory of elliptic boundary value problems under the aspect of pseudo-differential operators including the topological aspects of the index theory. It is our aim to fill this gap by the present book. Here we refer to the papers of Višik/Èskin [2, 4, 6], Èskin [3] and of Boutet de Monvel [1, 3, 4].

The starting object is the class $\mathfrak{G}$ of operators introduced in Boutet de Monvel [4] containing besides the classical elliptic boundary problems their parametrices, too. Thus all assertions about elliptic operators in the class $\mathfrak{G}$ are valid especially for classical elliptic boundary problems. An essential point of view in the whole theory is the symbolic calculus for interior and boundary symbols. $\mathfrak{G}$ is closed under composition and other natural operations which are compatible with corresponding operations on symbolic level. The new feature in connection with boundary problems are the boundary symbols whereas the interior symbols correspond to the usual (principal) symbols of operators on manifolds without boundary. Despite of the fact that the boundary effects require deep and rather difficult analytical tools, the supporting idea in the majority of global constructions is nothing else than to make systematical use of the formal correspondence between the symbolic and operator level and to copy the usual methods of the case without boundary. This philosophy is of course a program to generalize much more results about elliptic pseudo-differential operators to elliptic boundary problems. That there are some unexpected hurdles is not only due to the transmission property of symbols, a property which refers to the boundary. There are topological obstructions for the existence of elliptic boundary problems and the analytical apparatus has to be bloated in an unpleasant way. Nevertheless, the idea of symbolic calculus in connection with boundary problems gives an excellent insight into the nature of elliptic operators in 𝔊 and a survey on the whole set of such operators. This is due to the homotopy classification in connection with the index theorem which has been proved by BOUTET DE MONVEL [4]. An essential part of our exposition is devoted to the analytical and topological foundation of the theory of operators in 𝔊 and to the usual Fredholm and regularity theory of elliptic operators in 𝔊. Moreover results of the authors are presented, among other things about elliptic complexes on manifolds with boundary, clutching construction for boundary problems, analytical index formulas for elliptic operators in 𝔊, non-elliptic boundary problems and a priori estimates in Sobolev spaces $H^{s,p}$ with $p \neq 2$ and Hölder spaces. In the last chapter a collection of some further problems and working directions is given. In section 4.3.7 there are given further comments to the contents of this monograph and to the literature. Finally let us mention that also in GRUBB [3]—[5] many details of the calculus of the class 𝔊 are carried out.

For reading this book some knowledge is supposed about pseudo-differential operators, functional analysis and K-theory. The basic facts of these tools are contained in the first chapter.

Acknowledgement:

The authors should like to thank Dr. W. HÖPPNER who contributed section 4.3.1 and gave some hints concerning section 3.1.2.1.

We also would like to thank Dr. G. ALBINUS for the improvement of some details and the Akademie-Verlag without its care and assistance the book would have been impossible.

Contents

Symbol Index

Standard notations

Notations from topology and *K*-theory

Spaces of functions and distributions

Spaces of operators

Spaces of symbols

Special operators or mappings

Special symbols of operators

The Fourier transform

Other notations

1. Introduction

1.1. Basic Facts from Functional Analysis and Algebraic Topology

1.1.1. Fredholm Operators

1.1.1.1. Fredholm Operators in Topological Vector Spaces

In this section we collect standard facts from the theory of Fredholm operators. Besides Fredholm operators in Banach spaces our applications will also require Fredholm operators in Fréchet spaces. Therefore we start with the case of general topological vector spaces. For the theory of topological vector spaces cf. ROBERTSON/ROBERTSON [1], SCHAEFER [1], TREVES [1].

Let E, F be topological vector spaces (over $\mathbb{C}$). Denote by $\mathcal{L}(E, F)$ the vector space of all linear continuous operators $A: E \to F$.

A linear operator $A: E \to F$ is called *bounded* if it sends any bounded set in E into a bounded one in F.

We assume in the following all topological vector spaces to be *Fréchet spaces*, i.e. complete metrizable locally convex vector spaces. Note that most of the statements in this section hold under more general assumptions (cf. TREVES [2]).

Proposition 1. *A linear operator $A: E \to F$ is continuous iff it is bounded.*

To any $A \in \mathcal{L}(E, F)$ define the vector spaces

$$\ker A = \{u \in E: Au = 0\},$$

the *kernel* of A, and

$$\operatorname{im} A = \{f \in F: Au = f \quad \text{for some} \quad u \in E\},$$

the *image* of A. Define by

$$\operatorname{coker} A = F/\operatorname{im} A$$

the *cokernel* of A. For a vector space V denote by $\dim V$ the *dimension* of V (which can be infinite).

Definition 2. An operator $A \in \mathcal{L}(E, F)$ is called *Fredholm operator* if

$$\dim \ker A < \infty \quad \text{and} \quad \dim \operatorname{coker} A < \infty .$$

The number $\dim \ker A - \dim \operatorname{coker} A$ is called the *index* of the Fredholm operator A and denoted by $\operatorname{ind} A$. The set of all Fredholm operators in $\mathcal{L}(E, F)$ is denoted by $\mathcal{F}(E, F)$.

Proposition 3. *Let* $A \in \mathcal{L}(E, F)$ *and* $\dim \operatorname{coker} A < \infty$. *Then* $\operatorname{im} A \subseteq F$ *is closed and there is a finite-dimensional subspace* $W \subset F$ *such that* $F \cong W \oplus \operatorname{im} A$ *(topological sum and isomorphism). Especially every* $A \in \mathcal{F}(E, F)$ *has these properties.*

Proposition 4. *Let* $A \in \mathcal{F}(F, G)$, $B \in \mathcal{F}(E, F)$. *Then* $AB \in \mathcal{F}(E, G)$ *and*

$$\operatorname{ind} AB = \operatorname{ind} A + \operatorname{ind} B .$$

A linear operator $C: E \to F$ is called *compact* if it sends some open neighbourhood of 0 in E into a relatively compact set in F. Denote by $\mathcal{K}(E, F)$ the vector space of all compact operators from E to F. Each compact operator is continuous, i.e. $\mathcal{K}(E, F) \subseteq \mathcal{L}(E, F)$. Let $A \in \mathcal{L}(E, F)$, $A' \in \mathcal{L}(G, E)$ and $B \in \mathcal{K}(F, G)$. Then $BA \in \mathcal{K}(E, G)$ and $A'B \in \mathcal{K}(F, E)$. Especially $\mathcal{K}(E, E)$ is a (two-sided) ideal in $\mathcal{L}(E, E)$.

Proposition 5. *Let* $C \in \mathcal{K}(E, E)$. *Then* $1 + C \in \mathcal{F}(E, E)$ *and* $\operatorname{ind}(1 + C) = 0$.

$A \in \mathcal{L}(E, F)$ is called *operator of finite rank* if $\dim \operatorname{im} A < \infty$. All continuous operators with finite rank are compact.

Proposition 6. *Let* E, F *be Fréchet spaces. Then for* $A \in \mathcal{L}(E, F)$ *the following conditions are equivalent:*

(i) $A \in \mathcal{F}(E, F)$;

(ii) *there exist operators* $L, R \in \mathcal{L}(F, E)$ *such that* $LA - 1_E \in \mathcal{K}(E, E)$, $AR - 1_F \in \mathcal{K}(F, F)$;

(iii) *there exists an operator* $P \in \mathcal{L}(F, E)$ *such that* $PA - 1_E$ *and* $AP - 1_F$ *are operators of finite rank.* (1_E and 1_F denote the identity operators in E and F, respectively).

The operators L and R in Proposition 6 (ii) are called *left* and *right parametrices*, of A, respectively. If $A \in \mathcal{L}(E, F)$ has a left and a right parametrix, then any right parametrix is also a left one and conversely.

Instead of '*parametrix*' sometimes the notation '*regularizer*' is used in the literature. We prefer the first notation, since in our applications we mainly deal with pseudodifferential operators.

Denote by $\pi: \mathcal{L}(E, F) \to \mathcal{L}(E, F)/\mathcal{K}(E, F)$ the natural projection onto the factor space and let $\pi A \in \mathcal{L}(E, F)/\mathcal{K}(E, F)$ be the image of $A \in \mathcal{L}(E, F)$ under π.

Corollary 7. *Under the assumptions of Proposition* 6 $A \in \mathcal{L}(E, F)$ *belongs to* $\mathcal{F}(E, F)$ *iff* $\pi A \in \mathcal{L}(E, F)/\mathcal{K}(E, F)$ *is invertible, i.e. has an inverse* $(\pi A)^{-1}$ *in* $\mathcal{L}(F, E)/\mathcal{K}(F, E)$. *If* $R \in \mathcal{F}(F, E)$ *is a parametrix of* A, *then* $(\pi A)^{-1} = \pi R$, *i.e., for any* $C \in \mathcal{K}(F, E)$ $R + C$ *is a parametrix of* $A + C'$, $C' \in \mathcal{K}(E, F)$ *arbitrary. If* R, R' *are parametrices of* A, *then we have* $R - R' \in \mathcal{K}(F, E)$.

1.1.1.2. Fredholm Operators in Banach Spaces

In this section let all vector spaces be *Banach spaces*, i.e. complete normed vector spaces. The theory of Fredholm operators in Banach spaces is presented in many monographs, cf. ATIYAH [2], ATKINSON [1], GOHBERG/FEL'DMAN [1], GOHBERG/KRUPNIK [2], PALAIS [1]. The assertions of the previous section are true in this special case, too. Point out that particularly the conditions of 1.1.1.1, Proposition 6 are satisfied and Fredholm operators are characterized by the existence of a parametrix.

An essential new point in the case of Banach spaces is the homotopy invariance of the index of Fredholm operators to be described later.

Let E, F be Banach spaces. Then $\mathcal{L}(E, F)$ is a Banach space with the norm

$$||A|| = \sup_{||u||_E \leq 1} \{||Au||_F\}, \quad A \in \mathcal{L}(E, F)$$

($||\cdot||_E$ and $||\cdot||_F$ denote the norms in E and F, respectively). Then $\mathcal{L}(E, E)$ is a Banach algebra and $\mathcal{K}(E, E)$ is a closed two-sided ideal in $\mathcal{L}(E, E)$.

Let E', F' be the dual spaces of the Banach spaces E, F equipped with the corresponding norms and $\langle\cdot,\cdot\rangle_E$, $\langle\cdot,\cdot\rangle_F$ the bilinear pairings between E, E' and F, F', respectively. Given $A \in \mathcal{L}(E, F)$, the transposed operator ${}^tA \in \mathcal{L}(F', E')$ is defined by $\langle Au, f\rangle_F = \langle u, {}^tAf\rangle_E$, $u \in E$, $f \in F$.

Proposition 1. *For $A \in \mathcal{L}(E, F)$ the following conditions are equivalent:*

(i) $A \in \mathcal{F}(E, F)$;

(ii) ${}^tA \in \mathcal{F}(F', E')$.

$A \in \mathcal{F}(E, F)$ *implies* $\dim \ker {}^tA = \dim \operatorname{coker} A$, $\dim \operatorname{coker} {}^tA = \dim \ker A$, *i.e.*,

$$\operatorname{ind} A = -\operatorname{ind} {}^tA .$$

An analogous proposition holds for the anti-dual spaces E^*, F^* and the adjoint operator A^* (if the spaces are over $\mathbb{C}$ and the pairing is linear in the first, antilinear in the second argument). We use the adjoint operators in the case of Hilbert spaces and the duality is given by Hermitean scalar products.

Proposition 2. *Let $A \in \mathcal{F}(E, F)$. Then there exists an $\varepsilon > 0$ such that $A + B \in \mathcal{F}(E, F)$ for any $B \in \mathcal{L}(E, F)$ with $||B|| < \varepsilon$ and*

$$\operatorname{ind}(A + B) = \operatorname{ind} A .$$

This means that $\mathcal{F}(E, F) \subset \mathcal{L}(E, F)$ is an open set if $\mathcal{F}(E, F)$ is considered in the topology induced by $\mathcal{L}(E, F)$.

Definition 3. Two operators A_0, $A_1 \in \mathcal{F}(E, F)$ are called *homotopic* (through Fredholm operators) if there is a continuous mapping

$$f\colon [0, 1] \to \mathcal{F}(E, F)$$

with $f(0) = A_0$, $f(1) = A_1$. Then we write $A_0 \simeq A_1$.

Let for instance $A \in \mathcal{F}(E, F)$, $C \in \mathcal{K}(E, F)$. Then $f(t) = A + tC$, $t \in [0, 1]$ is a homotopy of Fredholm operators (cf. 1.1.1.1, Corollary 7), i.e., $A + C \simeq A$. For two parametrices R_0, R_1 of $A \in \mathcal{F}(E, F)$ we have $R_0 \simeq R_1$ because of $R_0 - R_1 \in \mathcal{K}(F, E)$.

Proposition 4. A_0, $A_1 \in \mathcal{F}(E, F)$, $A_0 \simeq A_1$ *implies* $\operatorname{ind} A_0 = \operatorname{ind} A_1$.
For Hilbert spaces E, F the converse is also true, i.e., $A_0 \simeq A_1 \Leftrightarrow \operatorname{ind} A_0 = \operatorname{ind} A_1$.

The dimensions $\dim \ker A$ and $\dim \operatorname{coker} A$, $A \in \mathcal{F}(E, F)$, are semi-continuous in the following sense: For given $A_0 \in \mathcal{F}(E, F)$ there exists an open neighbourhood U

of A_0 in $\mathcal{F}(E, F)$ such that

$$\dim \ker A_0 \leqq \dim \ker A, \quad \dim \operatorname{coker} A_0 \leqq \dim \operatorname{coker} A$$

for all $A \in U$.

Example. The following example shows that the index of Fredholm operators in Fréchet spaces is no homotopy invariant. For the notations cf. 2.1.2.1. Let σ be a smooth function on S^1 with values in $\mathbb{C}$. Then $\pi^+\sigma: C_+^\infty(S^1) \to C_+^\infty(S^1)$ is a Fredholm operator iff σ has not more than a finite number of zeros of finite order. $\pi^+ z^k$, $k \in \mathbb{Z}$ arbitrary, is homotopic to the identity through Fredholm operators in $C_+^\infty(S^1)$. The homotopy follows from $\sigma_t = z^k - 2t$, $t \in [0, 1]$, since $z^k - 2$ is homotopic to a constant function.

Next we consider the case of Hilbert spaces.

Definition 5. Let H_1, H_2 be Hilbert spaces. An operator $A \in \mathcal{L}(H_1, H_2)$ is called *Hilbert-Schmidt operator* if $\sum_j \|Ae_j\|^2 < \infty$ for a complete orthonormal system $\{e_j\}$ in H_1. The space of all Hilbert-Schmidt operators from H_1 to H_2 is denoted by $\mathcal{H}(H_1, H_2)$.

If A^* denotes the adjoint operator of A (with respect to fixed Hermitean scalar products), we have $A \in \mathcal{H}(H_1, H_2) \Leftrightarrow A^* \in \mathcal{H}(H_2, H_1)$. All Hilbert-Schmidt operators are compact, i.e., $\mathcal{H}(H_1, H_2) \subseteqq \mathcal{K}(H_1, H_2)$.

Definition 6. An operator $A \in \mathcal{L}(H_1, H_2)$ is called *trace class operator* if $\sum_j |(Ae_j, f_j)| < \infty$ for all complete orthonormal systems $\{e_j\}$ in H_1 and $\{f_j\}$ in H_2. The space of all trace class operators from H_1 to H_2 is denoted by $\mathcal{T}(H_1, H_2)$.

It is easily checked that $A_j \in \mathcal{H}(H_j, H_0)$, $j = 1, 2$, and $A = A_2^* A_1$ implies $A \in \mathcal{T}(H_1, H_2)$. Conversely any trace class operator A has a representation as a composition of two Hilbert-Schmidt operators. All trace class operators are Hilbert-Schmidt operators, i.e., $\mathcal{T}(H_1, H_2) \subseteqq \mathcal{H}(H_1, H_2)$. For $H_1 = H_2 = H$ and $A \in \mathcal{T}(H, H)$ the finite number $\operatorname{tr} A = \sum_j (Ae_j, e_j)$ is called *trace* of the operator A. This definition is correct, i.e. independent of the choice of the complete orthonormal system $\{e_j\}$ in H.

Proposition 7. *Let $A \in \mathcal{T}(H, H)$. Then the spectrum of A consists of a countable number of eigenvalues λ_j and* $\operatorname{tr} A = \sum_j \lambda_j$ (the λ_j's are taken with their multiplicities).

This is the generalization of a well-known result from the theory of matrices.

Proposition 8. *Let $A \in \mathcal{L}(H_1, H_2)$ and suppose that there exists an operator $R \in \mathcal{L}(H_2, H_1)$ such that for some $N \in \mathbb{Z}_+$*

$$(1_{H_1} - RA)^N \in \mathcal{T}(H_1, H_1), \ (1_{H_2} - AR)^N \in \mathcal{T}(H_2, H_2).$$

Then $A \in \mathcal{F}(H_1, H_2)$ and

$$\operatorname{ind} A = \operatorname{tr} (1_{H_1} - RA)^N - \operatorname{tr} (1_{H_2} - AR)^N.$$

Definition 9. Let X be a topological space. A continuous map $A: X \to \mathcal{F}(H_1, H_2)$ is called *Fredholm family* (over X). Two Fredholm families A_0, A_1 are called *homotopic* if there is a continuous map $f: [0, 1] \times X \to \mathcal{F}(H_1, H_2)$ with $f(0, \cdot) = A_0$, $f(1, \cdot) = A_1$.

Proposition 10. *Let A be a Fredholm family over a compact space X. Then there exists a subspace $V \subsetneqq H_1$ of finite codimension with*

$$V \cap \ker A(x) = \{0\} \quad \textit{for all } x \in X$$

and a finite-dimensional subspace $W \subset H_2$ with

$$W + \operatorname{im} A(x) = H_2 \quad \textit{for all } \; x \in X .$$

Particularly, we have

$$\sup_{x \in X} \{\dim \ker A(x)\} < \infty , \qquad \sup_{x \in X} \{\dim \operatorname{coker} A(x)\} < \infty .$$

It is not hard to show that any Fredholm family $A: X \to \mathcal{F}(H_1, H_2)$ over a compact parameter space X has a parametrix $R: X \to \mathcal{F}(H_2, H_1)$, i.e., per def. $R(x)$ is a parametrix of $A(x)$ for all $x \in X$ and R is a Fredholm family. The *'parameter-depending index'* of a Fredholm family is called the *index element* and turns out to be an element of the K-group of the space X. For details cf. 1.1.3.4.

Later we shall have to consider families of Fredholm operators acting between Fréchet spaces H_1, H_2. In our special situation they have the same properties as for Hilbert spaces, since they have closures as Fredholm families in Hilbert spaces preserving kernels and cokernels.

The notion of Fredholm operators can be generalized in several directions. For instance, let H be a Hilbert space and M, N closed subspaces. The pair (M, N) is called a *Fredholm pair* if $\dim (M \cap N) < \infty$, $\operatorname{codim} (M + N) < \infty$ (cf. Kato [1], Cordes [1]). Then

$$\operatorname{ind} (M, N) = \dim (M \cap N) = \operatorname{codim} (M + N)$$

is called the *index of the Fredholm pair*. If for instance $A: H_1 \to H_2$ is a Fredholm operator and $H = H_1 \times H_2$, $M = \{(x, Ax): x \in H_1\}$, $N = \{(x, 0): x \in H_1\}$, then (M, N) is a Fredholm pair with $\operatorname{ind} A = \operatorname{ind} (M, N)$.

1.1.2. Vector Bundles

1.1.2.1. Vector Bundles over Topological Spaces

Certain aspects of the theory of differential and pseudo-differential operators are connected with real and complex vector bundles in a natural way. In this section we give (without proofs) some basic facts on vector bundles. They are also needed for the K-theoretic considerations in the next section. Concerning references with more detailed information cf. Atiyah [3], Atiyah/Singer [2], Jänich [1], Kuiper [1], Hirzebruch [1], Schwartz [1], Palais [1], Breuer [1], Friedrich [1].

Definition 1. Let E, X be Hausdorff spaces and $\pi: E \to X$ a continuous map. E is called a *vector bundle over X* if

(i) $E_x = \pi^{-1}(\{x\})$ is a vector space (over a field $\mathbb{K}$ equal to $\mathbb{R}$ or $\mathbb{C}$) for all $x \in X$,

(ii) for all $x \in X$ there exists an open neighbourhood U of x and a homeomorphism

$$\varphi: \pi^{-1}(U) \to U \times \mathbb{K}^k, \quad k = k(U) , \tag{1}$$

with $\varphi(E_x) = \{x\} \times \mathbb{K}^k$ so that the map $E_x \to \mathbb{K}^k$ (defined as composition of φ and the projection $\{x\} \times \mathbb{K}^k \to \mathbb{K}^k$) is an isomorphism between the vector spaces over $\mathbb{K}$.

X is called the *base* of E, π the *projection*. E_x is called the *fibre* over x. (1) is called a *(local) trivialization* and shall be denoted by (U, φ). E is called *trivial* if there exists a trivialization with $U = X$. E is called *complex (real)* if $\mathbb{K} = \mathbb{C}$ ($\mathbb{K} = \mathbb{R}$).

Vector bundles E with the base X and the projection π are sometimes denoted by the triple (E, π, X).

For simplicity we suppose in this section that the base space X is compact. It is left to the reader to check which definitions and assertions are to be modified for non-compact X. Later vector bundles occur as base spaces. They are locally compact.

We shall mainly consider the case $\mathbb{K} = \mathbb{C}$. The considerations for $\mathbb{K} = \mathbb{R}$ are similar. Let E be a (complex) vector bundle and take trivializations (U, φ), (V, ψ) of E over U and V, respectively. Then we have a mapping

$$\varphi \circ \psi^{-1}: (U \cap V) \times \mathbb{C}^k \to (U \cap V) \times \mathbb{C}^k \tag{2}$$

which is linear in the fibres over corresponding points so that there is defined a continuous function

$$g_{UV}: U \cap V \to \mathrm{GL}\,(k, \mathbb{C}) \tag{3}$$

(as usual GL $(k, \mathbb{C})$ denotes the group of complex $(k \times k)$-matrices with non-vanishing determinant). The functions g_{UV} are called *transition functions* of the considered vector bundle. If (U, φ), (V, ψ), (W, ω) are trivializations of E, we obiously have

$$g_{UV} \circ g_{VW} = g_{UW} \quad \text{over} \quad U \cap V \cap W\,. \tag{4}$$

The functions g_{UV} are also called the *cocycle* belonging to E.

If (E, π, X) is a vector bundle and $X_1 \subseteqq X$ a subspace, (E_1, π_1, X_1) with $E_1 = \pi^{-1}(X_1)$ is a vector bundle over X_1. We also set $E_1 = E|_{X_1}$, and $E|_{X_1}$ is called the *restriction* of E to X_1.

A continuous map $s: X \to E$ is called a *(continuous) section* if $\pi \circ s$ is equal to the identical mapping. The set of all (continuous) sections in E is denoted by $C(X, E)$. This set is obviously a vector space if we define $(\lambda s)(x) = \lambda s(x), \lambda \in \mathbb{C}$, and $(s_1 + s_2)(x) = s_1(x) + s_2(x)$, $x \in X$.

Let (U, φ) be a trivialization of E and $s \in C(X, E)$. Then $s_U = \varphi \circ s|_U: U \to U \times \mathbb{C}^k$ is a section in the trivial bundle $U \times \mathbb{C}^k$. If (V, ψ) is a second trivialization we have $s_U = g_{UV} s_V$ over $U \cap V$.

Definition 2. Let (E, π, X), (E', π', X') be vector bundles. If $a: E \to E'$, $a_0: X \to X'$ are continuous mappings with $a_0\pi = \pi' a$ and $a|_{E_x}: E_x \to E'_{a_0(x)}$ linear for all $x \in X$, (a, a_0) is called a *bundle morphism from* (E, π, X) *to* (E', π', X'). If a, a_0 are one-to-one and onto (a, a_0) is called an *isomorphism.*

Instead of bundle morphisms we also speak of *homomorphisms* between the corresponding bundles. If there exists an isomorphism between E, E', we write $E \cong E'$. This defines an equivalence relation for vector bundles. By Vect (X) we denote the set of classes of isomorphic (complex) vector bundles over X. Sometimes we do not distinguish between a bundle and the class in Vect (X). For abbreviation bundle morphisms are also denoted by $a: E \to E'$ (this is justified since the base space can be considered as the 0-section).

We denote by Hom (E, F) the set of all bundle morphisms $E \to F$ for $X = X'$ and $a_0 = \mathrm{id}_X$. In Hom (E, F) we have canonically a vector space structure (over $\mathbb{C}$).

A vector bundle (E, π, X) with the fibre $\mathbb{C}^k$ is trivial iff it is isomorphic to $(X \times \mathbb{C}^k, \pi_1, X)$ ($\pi_1: (x, e) \to x$ denotes the corresponding canonical projection). If we consider a trivial bundle with fibre $\mathbb{C}^k$ over some fixed space X we often write $\mathbb{C}^k$ instead of $X \times \mathbb{C}^k$ or sometimes simply k. A vector bundle (E, π, X) with fibre $\mathbb{C}^l$ is trivial iff there exist sections $s_1, \dots, s_l$ which are linearly independent over each $x \in X$.

If X is a contractible space every vector bundle over X is trivial.

A vector bundle E over X gives rise to a system $\{g_{UV}\}$ of transition functions. Here $\{U, V, \dots\}$ denotes an open covering $\mathfrak{U}$ of X which is supposed to be fixed. Conversely, knowing the system $\{g_{UV}\}$, the vector bundle E can be reproduced by the following procedure. Let $\tilde{E}$ be the disjoint union of all sets $U \times \mathbb{C}^k$ $(U \in \mathfrak{U})$ and $\tilde{X}$ the disjoint union of all $U \in \mathfrak{U}$. We consider $\tilde{E}$ and $\tilde{X}$ in the corresponding product topologies. Then the canonical projection $\tilde{\pi}: \tilde{E} \to \tilde{X}$ (defined by $U \times \mathbb{C}^k \to U$) is continuous. In $\tilde{E}$ we define an equivalence relation by $(x, e) \sim (y, f)$ for $(x, e) \in U \times \mathbb{C}^k$, $(y, f) \in V \times \mathbb{C}^k \Leftrightarrow x = y$, $e = g_{UV}(x) f$ (cf. (4)). Put $E = \tilde{E}/\sim$ and consider E with the factor topology. Then $\tilde{\pi}$ induces a continuous projection $\pi: E \to X$ and it is easily seen that E is a vector bundle over X in the sense of Definition 1.

In Definition 1 the number k may depend on U, but k is locally constant, i.e. constant on connected components of X. If k is the same for all $x \in X$, E is called a *vector bundle* over X with fibre $\mathbb{C}^k$. Then GL $(k, \mathbb{C})$ is called the *structure group* of the bundle. One can show that there always exists a system of trivializations for which the unitary group $U(k)$ is the structure group.

The concept of bundles over a space X with the fibre L can be easily generalized to the case if L is a more general topological space. Let Aut (L) be the group of homeomorphisms of L onto itself (equipped with a suitable topology) and G some topological group. Let $\mathfrak{U} = \{U, V, \dots\}$ be an open covering of X, and $g_{UV}: U \cap V \to G$ a cocycle of continuous functions in G. Moreover let $\alpha: G \to \text{Aut}(L)$ be a continuous homomorphism. Then $\alpha(g_{UV})$ is a cocycle of continuous functions $U \cap V \to \text{Aut}(L)$. Similarly as for vector bundles one defines, by using the disjoint union of the $U \times L$ $(U \in \mathfrak{U})$ and an equivalence relation given by $\alpha(g_{UV})$, an associated bundle over X with fibre L. Then $\alpha(G)$ is called the *structure group* of this bundle.

An important example is the case $L = G$ for a Lie group G and $\alpha = \text{id}_G$ (G acts on itself by multiplication from the left). The resulting bundles are called principal fibre bundles.

One can consider *Hilbert space bundles*, too, i.e. bundles over X the fibre of which is a (real or complex) Hilbert space and the structure group of linear automorphisms of H onto itself (cf. 1.1.3.4).

From now on it is useful to suppose that the open sets U of the covering $\mathfrak{U}$ are all contractible.

Starting from given bundles one can define further bundles. Let e.g. E, F be vector bundles over X with the fibre $\mathbb{C}^k$ and $\mathbb{C}^l$, respectively, and g_{UV} and h_{UV}, respectively, the corresponding cocycles of transition functions, with respect to the covering $\mathfrak{U}$ of X. Then $g_{UV} \oplus h_{UV}$ is a cocycle of isomorphisms in $\mathbb{C}^k \oplus \mathbb{C}^l$. The corresponding associated vector bundle over X is denoted by $E \oplus F$ (and is called *direct sum* of E and F).

Analogously $g_{UV} \otimes h_{UV}$ (defined by $g_{UV} \otimes h_{UV}(e \otimes f) = (g_{UV}e) \otimes (h_{UV}f)$) is a cocycle of isomorphisms in $\mathbb{C}^k \otimes \mathbb{C}^l$. The corresponding associated bundle is denoted by $E \otimes F$ (and called the *tensor product* of E and F). Moreover $g_{UV} \wedge \dots \wedge g_{UV}$ (p factors) (defined by $(g_{UV} \wedge \dots \wedge g_{UV})(e_1 \wedge \dots \wedge e_p) = (g_{UV}e_1) \wedge \dots \wedge (g_{UV}e_p)$, $e_1, \dots, e_p \in \mathbb{C}^k$) is a cocycle of isomorphisms in $\bigwedge^p \mathbb{C}^k$. Thus there follows an associated vector bundle $\bigwedge^p E$ (the *exterior product* of E). Let $e_1, \dots, e_k$ be a base in $\mathbb{C}^k$ inducing a bilinear pairing in $\mathbb{C}^k$. Then g_{UV} can be considered as matrix function over $U \cap V$. Then the inverse of the transpose ${}^t g_{UV}^{-1}$ is also a cocycle of isomorphisms of $\mathbb{C}^k$. The corresponding associated bundle, denoted by E^*, is called *dual bundle* of E. The complex conjugated ${}^t\bar{g}_{UV}^{-1}$ is also a cocycle of isomorphisms in $\mathbb{C}^k$. One can easily show that the corresponding associated vector bundle is isomorphic to E. The bundle E^* is isomorphic to that associated by $\bar{g}_{UV}$. If $h_{UV}: U \cap V \to \mathrm{GL}(k, \mathbb{R})$ $(U, V \in \mathfrak{U})$ is a real cocycle, one can consider h_{UV}, too, as cocycle of isomorphisms of $\mathbb{C}^k$. The associated vector bundle is called *complexification* of the corresponding real bundle H and denoted by $H_{\mathbb{C}}$.

There exists a canonical identification between the set of homorphisms of a vector bundle E in a vector bundle F and the space $C(X, E^* \otimes F)$. For the operations $\oplus$ and $\otimes$ between vector bundles we have the obvious rules $E \oplus F \cong F \oplus E$, $E \otimes F \cong F \otimes E$, $(E \oplus F) \otimes G \cong (E \otimes G) \oplus (F \otimes G)$.

Let $f: Y \to X$ be a continuous map and E a vector bundle over X with the fibre $\mathbb{C}^k$. We define a vector bundle f^*E over Y. Suppose that E is associated to a cocycle g_{UV} (corresponding to an open covering $\mathfrak{U} = \{U, V, \ldots\}$ of X). Then the sets $\{f^{-1}(U), f^{-1}(V), \ldots\}$ form an open covering $\mathfrak{V}$ of Y and $f^{-1}(U) \cap f^{-1}(V) = f^{-1}(U \cap V)$ for $U, V \in \mathfrak{U}$. By $(f^*g_{UV})(y) = g_{UV}(f(y))$ we obtain a cocycle f^*g_{UV} of isomorphisms of $\mathbb{C}^k$ with respect to $\mathfrak{V}$. The resulting vector bundle over Y is denoted by f^*E and called the *pull back* of E with respect to f.

If the vector bundle E has different fibre dimensions over different components of X, one can define the procedure in an analogous way. Thus we obtain a mapping $f^*: \mathrm{Vect}(X) \to \mathrm{Vect}(Y)$, and f^* is obviously compatible with the bundle operations $\oplus$, $\otimes$, Λ^p, $*$ (e.g. $f^*(E \oplus F) \cong (f^*E) \oplus (f^*F)$). For $f_0 \simeq f_1$ (homotopy between continuous mappings) we have $f_0^*E \cong f_1^*E$ for all vector bundles E over X.

$\mathrm{Vect}(X)$ is an Abelian semi-group with respect to $\oplus$ and a semi-ring with respect to $\oplus$, $\otimes$ with unit element. One can consider $X \to \mathrm{Vect}(X)$ as a contravariant functor into the category of Abelian semi-groups (or semi-rings). If X, Y are homotopy equivalent we have $\mathrm{Vect}(X) \cong \mathrm{Vect}(Y)$.

One can easily show the following. If X is a compact space and E a (complex) vector bundle over X there exists a vector bundle $E^\perp$ over X with $E \oplus E^\perp \cong X \times \mathbb{C}^N$ for some N sufficiently large.

Next we define the so called *clutching construction* between vector bundles. Let X be a compact topological Hausdorff space, X_1, X_2 compact subspaces with $X = X_1 \cup X_2$, $Y = X_1 \cap X_2$. Suppose that E_i is a vector bundle over X_i $(i = 1, 2)$ and $\varphi: E_1|_Y \to E_2|_Y$ an isomorphism. Denote by $E_1 \cup E_2$ the disjoint union of E_1, E_2. For $e_i \in E_i$ $(i = 1, 2)$ we write $e_1 \sim e_2$ iff $\pi_i(e_i) \in Y$ and $\varphi(e_1) = e_2$ (π_i denotes the projection in E_i). The factor space of $E_1 \cup E_2$ with respect to this equivalence relation is denoted by $E_1 \bigcup_{\varphi} E_2$ and called the *clutching of* E_1, E_2 *over* Y *with respect to* φ. The clutching operation has various natural properties.

(i) if E is a vector bundle over X and $E_i = E|_{X_i}$ $(i = 1, 2)$, φ the identical mapping over Y, we have $E \cong E_1 \bigcup_{\varphi} E_2$;

(ii) if $\lambda_i: E_i \to F_i$ are isomorphisms over X_i and $\varphi: E_1|_Y \to E_2|_Y$, $\psi: F_1|_Y \to F_2|_Y$ isomorphisms with $\psi \circ \lambda_1 = \lambda_2 \circ \varphi$, then $E_1 \bigcup_{\varphi} E_2 \cong F_1 \bigcup_{\psi} F_2$;

(iii) if E_i, F_i are vector bundles over X_i ($i = 1, 2$) and $\varphi: E_1|_Y \to E_2|_Y$, $\psi: F_1|_Y \to F_2|_Y$ isomorphisms, then

$$(E_1 \underset{\varphi}{\cup} E_2) \oplus (F_1 \underset{\psi}{\cup} F_2) \cong (E_1 \oplus F_1) \underset{\varphi\oplus\psi}{\cup} (E_2 \oplus F_2),$$

$$(E_1 \underset{\varphi}{\cup} E_2) \otimes (F_1 \underset{\psi}{\cup} F_2) \cong (E_1 \otimes F_1) \underset{\varphi\otimes\psi}{\cup} (E_2 \otimes F_2), \tag{5}$$

$$(E_1 \underset{\varphi}{\cup} E_2)^* \cong E_1^* \underset{(\varphi^*)^{-1}}{\cup} E_2^*.$$

Using the clutching construction one can define new vector bundles. Let e.g. $E_j = X_j \times \mathbb{C}^k$ ($j = 1, 2$). Then each isomorphism $E_1|_Y \to E_2|_Y$ can be represented by a continuous map

$$\varphi: Y \to \mathrm{GL}\,(k, \mathbb{C}). \tag{6}$$

Conversely (6) represents an isomorphism between trivial bundles. Thus φ defines a vector bundle E over X (the clutching of E_1, E_2 by means of φ). If $\psi: Y \to \mathrm{GL}\,(k, \mathbb{C})$ is a second map and F the corresponding bundle over X, we have $\varphi \simeq \psi \Rightarrow E \cong F$.

Let e.g. X_1, X_2 be two exemplars of the unit disc $|z| \leq 1$ in the complex plane $\mathbb{C}$ and $Y = S^1 = \{z: |z| = 1\}$. Topologically the sphere $X = S^2$ can be considered as the clutching of X_1, X_2 along Y identifying corresponding points of Y, where X_1 corresponds to the upper, X_2 to the lower hemisphere and Y to the equator. Thus any continuous matrix function

$$\varphi: S^1 \to \mathrm{GL}\,(k, \mathbb{C}) \tag{7}$$

generates some element in Vect (S^2). For $k = 1$ and $\varphi(z) = z^{-1}$ the corresponding bundle over S^2 is called the *Hopf bundle*. Similarly $\varphi(z) = z^l$ ($l \in \mathbb{Z}$) generates a one-dimensional bundle over S^2. The mappings φ are simply continuous functions $\varphi: S^1 \to \mathbb{C} \setminus \{0\}$. To each such φ there exists some l with $\varphi \cong z^l$. The number $-l$ is called the *mapping degree* of φ (denoted by $\deg \varphi$).

To each matrix function $\varphi: S^1 \to \mathrm{GL}\,(k, \mathbb{C})$ one can assign the function $\det \varphi: S^1 \to \mathbb{C} \setminus \{0\}$ and it is easily seen that φ is homotopic to the map given by the following $k \times k$-matrix

$$\varphi_1(z) = \begin{pmatrix} z^l & 0 & \dots & 0 \\ 0 & 1 & \dots & 0 \\ \vdots & \vdots & \ddots & \vdots \\ 0 & 0 & \dots & 1 \end{pmatrix}$$

where $-l = \deg\,(\det \varphi)$.

If X, Y are topological spaces, $Y \subseteq X$, we denote the quotient space as usual by X/Y. X is called a *space with base point* if some point $x_0 \in X$ is fixed. X/Y can be regarded as a space with base point $\{Y\}$ (the space Y contracted to a point). Mappings between spaces with base points are supposed to preserve the base points.

If X_1, X_2 are spaces with base points, we denote by $X_1 \vee X_2$ the space with a base point, defined as disjoint union of X_1, X_2 and then identification of the base points in X_1, X_2. Moreover set $X_1 \wedge X_2 = X_1 \times X_2/X_1 \vee X_2$. We regard S^1 as a space with a base point which is arbitrary and fixed. Let X be a compact space with base point. Then we put $SX = S^1 \wedge X$. If we define

$$C^+(X) = ([0, \tfrac{1}{2}] \times X)/(\{0\} \times X), \qquad C^-(X) = ([\tfrac{1}{2}, 1] \times X)/(\{1\} \times X)$$

and denote by $C^+(X) \cup C^-(X)$ the space defined by identification of corresponding points of $\{\frac{1}{2}\} \times X$ in $C^{\pm}(X)$, SX is homotopy equivalent to $C^+(X) \cup C^-(X)$. Note also that $X = C^+(X) \cap C^-(X)$. The spaces $C^{\pm}(X)$ are contractible. Thus every vector bundle over $C^{\pm}(X)$ is trivial. Each vector bundle over $C^+(X) \cup C^-(X)$ can be generated by clutching of trivial bundles over $C^{\pm}(X)$ by means of a continuous mapping $\varphi: X \to \mathrm{GL}(k, \mathbb{C})$. The isomorphy class of the bundle depends only on the homotopy class of φ.

We denote by $\mathrm{Vect}_k(SX)$ the set of all (complex) vector bundles over SX with fibre dimension k. Then

$$\mathrm{Vect}_k(SX) \cong [X, \mathrm{GL}(k, \mathbb{C})] \tag{8}$$

($[\cdot, \cdot]$ denotes the set of homotopy classes of continuous mappings between the spaces in the brackets). Thus a vector bundle over SX uniquely determines the homotopy class of some $\varphi: X \to \mathrm{GL}(k, \mathbb{C})$. This correspondence can be applied to the case $X = S^{n-1}$ and $SS^{n-1} = S^n$ $(n \in \mathbb{Z}_+)$.

1.1.2.2. Vector Bundles over Manifolds

Let X be a C^∞ manifold of dimension n in this section (we simply speak of a manifold). Consider some fixed open covering $\mathfrak{U}$ of X with the property that $U, V \in \mathfrak{U}$ implies $U \cap V \in \mathfrak{U}$ and that each $U \in \mathfrak{U}$ is contractible. The latter condition ensures that every vector bundle over any $U \in \mathfrak{U}$ is trivial. Moreover fix charts $\alpha: U \to \Omega$ $(U \in \mathfrak{U}, \Omega \subseteq \mathbb{R}^n$ open).

$\mathrm{GL}(k, \mathbb{C})$ can be considered as a C^∞ manifold. A vector bundle E over X is called *differentiable* or *smooth* if all transition functions g_{UV} are differentiable. In this section we only consider differentiable vector bundles over X. Then every vector bundle E over X is a C^∞ manifold with a system of local coordinates of the form $\Omega \times \mathbb{C}^k$. A section s in E is called *differentiable* (or *smooth*) if $s: X \to E$ is a differentiable map. The set of all differentiable sections in E is denoted by $C^\infty(X, E)$.

Each continuous vector bundle E over X in the sense of 1.1.2.1 is isomorphic to some smooth vector bundle over X. Thus we can identify $\mathrm{Vect}(X)$ with the set of all classes of isomorphic (complex) vector bundles over X. We suppose that all bundle homomorphisms under consideration are smooth.

The contravariant functor $X \to \mathrm{Vect}(X)$ from the category of manifolds into the category of semi-rings is needed with respect to complex vector bundles only. This and all the other notions to be introduced are analogous for real bundles (i.e., if the fibre is $\mathbb{R}^k$ and $\mathrm{GL}(k, \mathbb{R})$ the structure group).

Real bundles in our applications are mainly realized as *tangent bundles* TX or *cotangent bundles* T^*X of the manifold X and derived bundles, e.g. exterior products $\bigwedge^p T^*X$. One can define TX and T^*X, respectively, as follows. Consider two sets $U_i, U_j \in \mathfrak{U}$ and corresponding local coordinates $\alpha_i: U_i \to \Omega_i$, $\alpha_j: U_j \to \Omega_j$. Then

$$\chi_{ij} = \alpha_i \circ \alpha_j^{-1}: \alpha_j(U_i \cap U_j) \to \alpha_i(U_i \cap U_j) \tag{1}$$

defines a diffeomorphism.

Denote by $d\chi_{ij}(y)$ the Jacobi matrix of (1) in $y \in \alpha_j(U_i \cap U_j)$ and $l_{U_iU_j}(p) = d\chi_{ij}(\alpha_j(p))$, $p \in U_i \cap U_j$. Then we get a system of smooth mappings $l_{U_iU_j}: U_i \cap U_j \to \mathrm{GL}(n, \mathbb{R})$ $(U_i, U_j \in \mathfrak{U})$ being a cocycle in the sense of 1.1.2.1. The associated vector bundle over X with the fibre $\mathbb{R}^n$ is then per def. TX. The vector bundle with the fibre $\mathbb{R}^n$ associated with ${}^t l_{U_iU_j}^{-1}$ is per def. T^*X.

In connection with pseudo-differential operators we have to consider local coordinates $\Omega_j \times \mathbb{C}^k$ of a given vector bundle E. If $g_{U_iU_j}\colon U_i \cap U_j \to \mathrm{GL}\,(k, \mathbb{C})$ is the cocycle belonging to E, we get smooth mappings $g_{ij}\colon \alpha_j\,(U_i \cap U_j) \to \mathrm{GL}\,(k, \mathbb{C})$. The transition diffeomorphisms $(\chi_{ij}, g_{ij})\colon \alpha_j\,(U_i \cap U_j) \times \mathbb{C}^k \to \alpha_i\,(U_i \cap U_j) \times \mathbb{C}^k$ are coordinate diffeomorphisms of the manifold E and satisfy the usual condition $\chi_{li} \circ \chi_{ij} = \chi_{lj}$, $g_{li} \circ g_{ij} = g_{lj}$ (the transition functions of the fibres are defined after suitable substitution in $\alpha_j(U_i \cap U_j \cap U_k)$). Conversely the system of the open sets $\Omega_j \times \mathbb{C}^k$ and the (χ_{ij}, g_{ij}) determine the bundle E in a unique way.

Sections in bundles, morphisms between bundles and other notions can be defined as corresponding objects in the local coordinate systems with the property of invariance with respect to the substitutions (χ_{ij}, g_{ij}).

For instance, a system of sections $s_j\colon \Omega_j \to \Omega_j \times \mathbb{C}^k$ defines a section s in E if for each coordinate diffeomorphism $\chi_{ij}\colon \omega_j \to \omega_i\big(\omega_j = \alpha_j(U_i \cap U_j),\ \omega_i = \alpha_i(U_i \cap U_j)\big)$ the following diagram commutes

$$\begin{array}{ccc} \omega_j \times \mathbb{C}^k & \xrightarrow{(\chi_{ij},\, g_{ij})} & \omega_i \times \mathbb{C}^k \\ \uparrow{\scriptstyle s_j} & & \uparrow{\scriptstyle s_i} \\ \omega_j & \xrightarrow{\chi_{ij}} & \omega_i \end{array}$$

$(U_i, U_j \in \mathfrak{U})$. Moreover let E and F be vector bundles with fibres $\mathbb{C}^k$ and $\mathbb{C}^l$, and cocycles g_{UV} and h_{UV}, respectively $(U, V \in \mathfrak{U})$. Then a bundle morphism $A\colon E \to F$ (say with fixed base space) is defined by a system of bundle morphisms $A_j\colon \Omega_j \times \mathbb{C}^k \to \Omega_j \times \mathbb{C}^l$ (i.e., A_j are smooth matrix functions over Ω_j) and for each coordinate diffeomorphism $\chi_{ij}\colon \omega_j \to \omega_i$ the following diagram commutes

$$\begin{array}{ccc} \omega_i \times \mathbb{C}^k & \xrightarrow{A_i} & \omega_i \times \mathbb{C}^l \\ \uparrow{\scriptstyle (\chi_{ij},\, g_{ij})} & & \uparrow{\scriptstyle (\chi_{ij},\, g_{ij})} \\ \omega_j \times \mathbb{C}^k & \xrightarrow{A_j} & \omega_j \times \mathbb{C}^l \end{array} \tag{2}$$

Here $A\colon E \to F$ is an isomorphism if $k = l$ and the A_j are invertible.

In the following we suppose that X is paracompact and that $\mathfrak{U}$ is locally finite. If E is a (complex) vector bundle over X with the local coordinate systems $\Omega_j \times \mathbb{C}^k$ and the transition functions $g_{ij}\colon \omega_j \to \mathrm{GL}\,(k, \mathbb{C})$ (belonging to the coordinate diffeomorphisms $\chi_{ij}\colon \omega_j \to \omega_i$) we say that E is given by the system $\{\Omega_j \times \mathbb{C}^k, g_{ij}\}$. Here $i, j, \ldots$ correspond to open sets $U, V, \ldots \in \mathfrak{U}$.

In $\mathbb{C}^k$ we fix a base and a Hermitean scalar product $(\cdot, \cdot)$, linear in the first, antilinear in the second argument. Suppose that for each j a smooth matrix function $A_j\colon \Omega_j \to \mathrm{GL}\,(k, \mathbb{C})$ is given, A_j Hermitean (i.e., ${}^tA_j = \bar{A}_j$) with the property

$$A_j = {}^t\bar{g}_{ij} A_i g_{ij} \tag{3}$$

over ω_j for all i, j. Then the system $A = \{A_j\}$ is called a *Hermitean metric* of the vector bundle E.

The property (3) means that in each fibre of $\Omega_j \times \mathbb{C}^k$ a Hermitean scalar product $(A_j e_j, f_j)$ is defined $(e_j, f_j \in \mathbb{C}^k)$ smoothly depending on the base point $x \in \Omega_j$ and invariant in the sense

$$(A_j e_j, f_j) = (A_i e_i, f_i) \quad \text{for} \quad e_i = g_{ij} e_j,\ f_i = g_{ij} f_j\,.$$

If we put in (2) $l = k$ and $h_{ij} = {}^t\bar{g}^{-1}$ we see that a Hermitean metric in E induces an isomorphism of E to a bundle given by $\{\Omega_j \times \mathbb{C}^k, {}^t\bar{g}^{-1}\}$. The latter bundle is isomorphic to E.

Using a partition of unity one can easily show that there always exists a Hermitean metric A for each E.

Let now L be a real vector bundle over X given by $\{\Omega_j \times \mathbb{R}^q, l_{ij}\}$ and $\langle\cdot,\cdot\rangle$ a fixed scalar product in $\mathbb{R}^q$. A system $B = \{B_j\}$ of smooth functions $B_j\colon \Omega_j \to \mathrm{GL}(q, \mathbb{R})$ with ${}^tB_j = B_j$ and

$$B_j = {}^tl_{ij}B_il_{ij} \tag{4}$$

over ω_j for all i, j is called *a Riemannian metric* in L. In the fibres of L a Riemannian metric induces a scalar product smoothly depending on the base point. Using a partition of unity one can show that a Riemannian metric always exists. Similarly as in the complex case one can consider B as an isomorphism between the bundles represented by $\{\Omega_j \times \mathbb{R}^q, l_{ij}\}$ and $\{\Omega_j \times \mathbb{R}^q, {}^tl_{ij}\}$, respectively. In the case $L = TX$ this means that B induces an isomorphism $TX \to T^*X$. One also denotes B as a Riemannian metric on X.

Note that in a similar way Hermitean and Riemannian metrics can be defined in continuous vector bundles without C^∞ structure, too. This shall be used in 1.1.3.

Now we suppose that X is orientable and that the system $\{\Omega_j\}$ of local coordinate systems represents an orientation (i.e., the Jacobian of $\chi_{ij}\colon \omega_j \to \omega_i$ is positive for all i, j). Let $x^{(j)} = (x_1^{(j)}, \dots, x_n^{(j)}) \in \Omega_j$ and $B = \{B_j\}$ be a Riemannian metric in TX. Then the system $(\det B_j)^{1/2}\, dx_1^{(j)} \wedge \dots \wedge dx_n^{(j)}$ represents a never vanishing section in $\bigwedge^n T^*X$ $\big($since $(\det B_j)^{1/2} = (\det l_{ij})\,(\det B_i)^{1/2}$ over $\omega_j\big)$. This is a volume form on X which we denote by dx and we have an integration $C_0^\infty(X) \ni \varphi \to \int \varphi(x)\, dx$.

Let E be a complex vector bundle over X equipped with a Hermitean metric A. Let dx be a volume form over X. For given $u, v \in C_0^\infty(X, E)$ $\big(C_0^\infty(X, E)$ denotes the space of sections with compact support$\big)$ we get by scalar multiplication in the fibres a function $\big(u(x), v(x)\big)_A \in C_0^\infty(X)$. Then $(u, v) = \int \big(u(x), v(x)\big)_A\, dx$ defines a Hermitean scalar product over $C_0^\infty(X, E)$.

Let X be compact $\big(\Rightarrow C_0^\infty(X, E) = C^\infty(X, E)\big)$. The closure of $C^\infty(X, E)$ with respect to the scalar product $(\cdot,\cdot)$ is denoted by $L^2(X, E)$ $\big($later also by $\mathrm{H}^0(X, E)\big)$. This space is independent of the choice of the Hermitean metric in E and the Riemannian metric on X. For non-compact X one defines $L^2_{\mathrm{loc}}(X, E)$ (the space of all equivalence classes of measurable functions being square integrable over open relatively compact sets) and the space $L^2_{\mathrm{comp}}(X, E)$ (the subset in $L^2_{\mathrm{loc}}(X, E)$ with compact support).

In connection with boundary value problems we have to consider manifolds with boundary, too. They are defined as follows. Let X^0 be a manifold, $\Omega \subset X^0$ an open subset and $Y = \partial\Omega$ a (smooth) manifold of codimension 1. We set $X = \Omega \cup Y$. Then X is called a *manifold with boundary* if for each $x \in Y$ there exists a neighbourhood U^0 in X^0 and a chart $\alpha^0\colon U^0 \to \mathbb{R}^n$ with $\alpha^0(Y \cap U^0) \subseteq \mathbb{R}^{n-1} = \{x = (x', x_n) \in \mathbb{R}^n\colon x_n = 0\}$ and $\alpha^0(X \cap U^0) \subseteq \overline{\mathbb{R}}^n_+ = \{x \in \mathbb{R}^n\colon x_n \geqq 0\}$ $\big(x = (x', x_n),\ x' = (x_1, \dots, x_{n-1})\big)$. X^0 is called a *neighbouring manifold* and Y is called the *boundary* of X. Except the half space $\overline{\mathbb{R}}^n_+$ (which is a manifold with boundary) we consider mainly compact manifolds with boundary.

We often use the well known fact that there exists a (relatively open) neighbourhood U of Y in X and a diffeomorphism $\varrho\colon U \to Y \times [0, 1)$ $([0, 1) = \{t\colon 0 \leqq t < 1\})$. Then if $\mathfrak{B}$ is a covering of Y with coordinate neighbourhoods and $\beta\colon V \to \mathbb{R}^{n-1}$ a local chart ($V \in \mathfrak{B}$) we find a neighbourhood W in X so that

$$(\beta \times \mathrm{id}_{[0,1)}) \circ \varrho\colon W \to \mathbb{R}^{n-1} \times [0, 1) \tag{5}$$

is a local chart for W. The sets W of this form can be chosen in such a way that they form a covering of U with relatively open sets and that the coordinate diffeomorphisms are independent of the variable in $[0, 1)$. This variable shall be denoted by x_n. By (5) local coordinates are induced in $TX|_U$ and $T^*X|_U$, too. If, for instance, $(x, \xi) \in \mathbb{R}^n \times \mathbb{R}^n$ are local coordinates for T^*X over W, we can set $x = (x', x_n)$, $x' \in \mathbb{R}^{n-1}$ and $\xi = (\xi', \nu)$ with $\xi' \in \mathbb{R}^{n-1}$, $\nu \in \mathbb{R}^1$. The variables x_n and ν are invariant under our special coordinate diffeomorphisms. ν is called *conormal* variable. The above considerations show $T^*X|_Y \cong T^*Y \oplus (Y \times \mathbb{R})$, where $Y \times \mathbb{R}$ is the *conormal bundle* to Y (which is trivial). Local coordinates near the boundary of a mainfold X with boundary shall always be chosen in the described way. Moreover we suppose that over U the Riemannian metric on X under consideration is the same as the product metric of $Y \times [0, 1)$ for a fixed Riemannian metric on Y. It is always possible to find such a Riemannian metric on X.

1.1.3. Elements of K-Theory

1.1.3.1. Definitions and General Facts

The semi-group Vect (X) of classes of isomorphic complex vector bundles over X defined in 1.1.2.1 shall be completed by a standard construction to a group. In this section we suppose X to be compact.

Let $P(X) = \text{Vect}(X) \times \text{Vect}(X)$ and let $P_0(X)$ be the subset of all pairs (G, G), $G \in \text{Vect}(X)$. In $P(X)$ we have a semi-group structure with the addition $(E_1, F_1) \oplus (E_2, F_2) = (E_1 \oplus E_2,\ F_1 \oplus F_2)$. Let $K(X)$ be the set of all cosets $P(X)/P_0(X)$. Then the elements of $K(X)$ are equivalence classes of pairs (E, F) $\big(E, F \in \text{Vect}(X)\big)$ with $(E, F) \sim (E', F')$ iff there exist elements $G, G' \in \text{Vect}(X)$ with $(E, F) \oplus (G, G) = (E', F') \oplus (G', G')$ $(\Leftrightarrow E \oplus G \cong E' \oplus G',\ F \oplus G \cong F' \oplus G')$.

Denote by $[E] - [F]$ the equivalence class represented by (E, F) $\big([E]$ and $-[F]$ are represented by $(E, 0)$ and $(0, F)$, respectively$\big)$. $K(X)$ is an Abelian group with respect to the addition. The negative of $[E] - [F]$ is given by $[F] - [E]$.

Since for any $E \in \text{Vect}(X)$ there exists some $N \in \mathbb{Z}_+$ and some $E^\perp \in \text{Vect}(X)$ with $E \oplus E^\perp \cong \mathbb{C}^N$, every element in $K(X)$ can be represented by a pair of the form $(E, \mathbb{C}^M)$, $M \in \mathbb{Z}_+$.

In $P(X)$ a natural multiplication is defined by $(E, F) \cdot (V, W) = \big((E \otimes V) \oplus (F \otimes W),\ (F \otimes V) \oplus (E \otimes W)\big)$. Thus $K(X)$ becomes a ring. The unit is represented by $(X \times \mathbb{C}^1, 0)$. For $X = \{p\}$ (a single point) we obviously have $K(\{p\}) \cong \mathbb{Z}$.

Proposition 1. *Let $f\colon Y \to X$ be a continuous map. Then the pull back $f^*\colon$ Vect$(X) \to$ Vect (Y) induces a ring homomorphism*

$$f^*\colon K(X) \to K(Y)\,. \tag{1}$$

From $f_0 \cong f_1$ follows $f_0^* = f_1^*$. If $f\colon Y \to X$ is a homotopy equivalence (1) *is an isomorphism.*

The correspondence $X \to K(X)$ is called *K-functor*. (1) says that the K-functor is contravariant.

If X is a space with base point p the embedding $j\colon p \to X$ induces a homomorphism $j^*\colon K(X) \to K(\{p\})$. Denote by $\tilde{K}(X)$ the kernel of j^*. Then $K(X)$ is represented by the set of all pairs (E, F) with vector bundles E, F having the same fibre dimension over p.

Definition 2. Let X be an arcwise connected space. Then $E, F \in \mathrm{Vect}(X)$ are called *stable equivalent* if there exist integers M, N so that $E \oplus \mathbb{C}^M \cong F \oplus \mathbb{C}^N$.

An analogous definition can be formulated for real vector bundles.

Proposition 3. *Let X be an arcwise connected space. Then the map $\alpha: \mathrm{Vect}(X) \to \tilde{K}(X)$ induced by $E \to (E, \mathbb{C}^l)$ (l = fibre dimension of E) is surjective and $\alpha(E) = \alpha(F)$ is satisfied iff E, F are stable equivalent.*

Let (X, Y) be a pair of compact spaces, $Y \subseteq X$ and $K(X, Y) = \tilde{K}(X/Y)$. Here $\{Y\}$ is considered as the base point in X/Y. The group $K(X, Y)$ is called *relative K-group* for the pair (X, Y).

Let $j: X \to X/Y$ be the canonical mapping onto the quotient space. Every element in $K(X, Y)$ is represented by a pair (E', F'), where $E', F' \in \mathrm{Vect}\,(X/Y))$ have the same fibre dimension over $\{Y\}$. Then the pair $(E, F) = (j^*E', j^*F')$ represents an element in $K(X)$. By $(E', F') \to (E, F)$ a homomorphism $j^*: K(X, Y) \to K(X)$ is induced. Moreover by the restriction $(E, F) \to (E|_Y, F|_Y)$ $(E, F \in \mathrm{Vect}\,(X))$ a homomorphism $i^*: K(X) \to K(Y)$ is induced (this is the pull back with respect to the identical embedding $i: Y \to X$). One can easily show that the sequence

$$K(X, Y) \xrightarrow{j^*} K(X) \xrightarrow{i^*} K(Y)$$

is exact. If moreover there exists a retraction $r: X \to Y$ we have even an exact sequence

$$0 \to K(X, Y) \xrightarrow{j^*} K(X) \xrightarrow{i^*} K(Y) \to 0\,. \tag{2}$$

One obtains in this case

$$K(X) \cong K(X, Y) \oplus K(Y) \tag{3}$$

and for spaces with a base point (as usual contained in Y)

$$\tilde{K}(X) \cong K(X, Y) \oplus \tilde{K}(Y)\,. \tag{4}$$

Remark 4. Let (X, Y) be a pair of compact spaces and $CY = (Y \times I)/(Y \times \{1\})$; identify $Y \times \{0\}$ with Y. Then we have a canonical isomorphism $\gamma: \tilde{K}(X \cup CY) \to K(X, Y)$ (since CY is contractible; the base point of Y is also considered as base point of CY). Moreover there is a canonical isomorphism $\lambda: \tilde{K}(SY) \to \tilde{K}(CY/Y)$ (since SY is obtained from CY/Y by contracting an interval).

If the pairs (X, Y), (X', Y') are homotopy equivalent we have $K(X, Y) \cong K(X', Y')$.

It is useful to have further definitions of $K(X, Y)$. Consider the set $R(X, Y)$ of all complexes

$$0 \to E_0 \xrightarrow{\alpha} E_1 \to 0 \tag{5}$$

where E_0, E_1 are vector bundles over X and α a bundle morphism, $\alpha|_Y: E|_Y \to F|_Y$ an isomorphism. (5) shall be abbreviated by (E_0, E_1, α). Tripels of this type are called *difference bundles*.

In $R(X, Y)$ we can define an addition by

$$(E_0, E_1, \alpha) \oplus (F_0, F_1, \beta) = (E_0 \oplus F_0, E_1 \oplus F_1, \alpha \oplus \beta)$$

so that $R(X, Y)$ becomes a semi-group. Two elements $(E_0, E_1, \alpha), (F_0, F_1, \beta) \in R(X, Y)$ are called isomorphic ($\cong$) if there exist isomorphisms $u: E_0 \to F_0$, $v: E_1 \to F_1$ so that the restrictions to Y of the compositions $\beta \circ u$ and $v \circ \alpha$ are equal. Two elements $(E_0, E_1, \alpha), (E_0', E_1', \alpha') \in R(X, Y)$ are called equivalent if there exist elements of the form $(G, G, 1), (G', G', 1) \in R(X, Y)$ with

$$(E_0, E_1, \alpha) \oplus (G, G, 1) \cong (E_0', E_1', \alpha') \oplus (G', G', 1) .$$

Denote by $[E_0, E_1, \alpha]$ the equivalence class represented by (E_0, E_1, α). Let $L(X, Y)$ be the set of equivalence classes. Then $L(X, Y)$ is in a natural way an Abelian group. The element $-[E_0, E_1, \alpha]$ can be represented by (E_1, E_0, α^{-1}) if α^{-1} denotes a bundle morphism $E_1 \to E_0$ being inverse to α over Y.

Remark 5. *The following conditions are equivalent*

(i) $[E_0, E_1, \alpha] = 0$ *in* $L(X, Y)$;

(ii) *there exists an integer N so that the isomorphism*

$$E_0 \oplus \mathbb{C}^N|_Y \xrightarrow{\alpha \oplus 1_N} E_1 \oplus \mathbb{C}^N|_Y$$

can be extended to an isomorphism between $E_0 \oplus \mathbb{C}^N$ and $E_1 \oplus \mathbb{C}^N$ over X.

Proposition 6. *There is a canonical isomorphism $\chi: L(X, Y) \to K(X, Y)$.*

The correspondence χ between $L(X, Y)$ and $K(X, Y)$ is called *difference construction*. It can be realized by the following procedure. Set

$$Z_0 = (X \times \{0\}) \cup (Y \times I), \qquad Z_1 = (X \times \{1\}) \cup (Y \times I)$$

(I the unit interval), and $Z = Z_0 \cup Z_1$. Then we have natural projections $p_0: Z_0 \to X \times \{0\}$, $p_1: Z_1 \to X \times \{1\}$, $p: Z \to X \times \{1\}$. Since p is a retraction we have the following exact sequence

$$0 \to K(Z, X \times \{1\}) \xrightarrow{j^*} K(Z) \xrightarrow{i^*} K(X \times \{1\}) \to 0 \tag{6}$$

$\big(i: X \times \{1\} \to Z, j: Z \to Z/(X \times \{1\})\big)$. Moreover we have an obvious canonical isomorphism

$$q: K(Z, X \times \{1\}) \to K(X, Y) = \tilde{K}(X/Y) . \tag{7}$$

Now let (E_0, E_1, α) be a difference bundle over (X, Y). Then α induces an isomorphism $\beta: p_0^*E_0|_{Y \times I} \hookrightarrow p_1^* E_1|_{Y \times I}$. The element

$$[p_0^*E_0 \bigcup_\beta p_1^*E_1, p_1^*E_1] \in K(Z) \tag{8}$$

(cf. the clutching construction in 1.1.2.1) belongs to the kernel of i^*. Since (6) is exact one can identify (8) with an element of $K(Z, X \times \{1\})$. Denote by $[E_0', E_1']$ the image of the element in (8) with respect to q. The definition of χ is then

$$\chi([E_0, E_1, \alpha]) = [E_0', E_1'] .$$

$K(X, Y)$ is a ring with respect to the multiplication defined above. Thus by χ^{-1} a multiplication is given in $L(X, Y)$, too. One can show that this multiplication can be defined in the following way. Let (E, F, α), (V, W, β) be difference bundles over (X, Y). By $\alpha^{-1}: F \to E$, $\beta^{-1}: W \to V$ we denote bundle morphisms being inverse to α and β, respectively, over Y. Then

$$\lambda = \begin{pmatrix} \alpha \otimes 1 & -1 \otimes \beta^{-1} \\ 1 \otimes \beta & \alpha^{-1} \otimes 1 \end{pmatrix} : \begin{matrix} E \otimes V \\ \oplus \\ F \otimes W \end{matrix} \to \begin{matrix} F \otimes V \\ \oplus \\ E \otimes W \end{matrix}$$

is an isomorphism over Y. Over Y we obviously have

$$\lambda^{-1} = \frac{1}{2}\begin{pmatrix} \alpha^{-1} \otimes 1 & 1 \otimes \beta^{-1} \\ -1 \otimes \beta & \alpha \otimes 1 \end{pmatrix}.$$

If we put

$$[E, F, \alpha] \cdot [V, W, \beta] = \left[\begin{matrix} E \otimes V \\ \oplus \\ F \otimes W \end{matrix}, \begin{matrix} F \otimes V \\ \oplus \\ E \otimes W \end{matrix}, \lambda\right],$$

we obtain just the image with respect to χ^{-1} of the multiplication in $K(X, Y)$. It can be proved that equivalently

$$[E, F, \alpha] \cdot [V, W, \beta] = \left[\begin{matrix} E \otimes V \\ \oplus \\ F \otimes W \end{matrix}, \begin{matrix} F \otimes V \\ \oplus \\ E \otimes W \end{matrix}, \begin{pmatrix} \alpha \otimes 1 & -1 \otimes \beta^* \\ 1 \otimes \beta & \alpha^* \otimes 1 \end{pmatrix}\right]$$

with the adjoint morphisms α^*, β^* defined by fixed Hermitean metrics in the corresponding bundles.

Let $[Y, \mathrm{GL}(N, \mathbb{C})] \to \mathrm{Vect}_N(SY)$ be the isomorphism mentioned in 1.1.2.1 (8). For given $\varphi: Y \to \mathrm{GL}(N, \mathbb{C})$ we denote the corresponding bundle by $E_\varphi \in \mathrm{Vect}_N(SY)$. We have

Theorem 7. *By $\varphi \to [E_\varphi] - [N]$ the following isomorphism of groups is induced*

$$\lim_{N\to\infty} [Y, \mathrm{GL}(N, \mathbb{C})] \cong \tilde{K}(SY)$$

(the group structure on the left hand side is induced by the group structure in GL $(N, \mathbb{C})$ *for all N).*

Lemma 8. *Let A, $B: Y \to \mathrm{GL}(N, \mathbb{C})$ be continuous mappings. Then there exists a homotopy*

$$\begin{pmatrix} A & 0 \\ 0 & B \end{pmatrix} \simeq \begin{pmatrix} AB & 0 \\ 0 & 1_N \end{pmatrix}$$

in the class of continuous mappings $Y \to \mathrm{GL}(2N, \mathbb{C})$ (1_N denotes the unit $N \times N$-matrix).

Such a homotopy is realized by

$$\begin{pmatrix} A & 0 \\ 0 & 1_N \end{pmatrix} \begin{pmatrix} \cos t & \sin t \\ -\sin t & \cos t \end{pmatrix} \begin{pmatrix} 1_N & 0 \\ 0 & B \end{pmatrix} \begin{pmatrix} \cos t & -\sin t \\ \sin t & \cos t \end{pmatrix},$$

$(0 \leqq t \leqq \pi/2)$.

Note also the following fact (cf. HUSEMOLLER [1]). For $k \geqq 1$ there exists a group isomorphism

$$\tilde{K}(S^k) \cong \pi_{k-1}(U_{\mathbb{C}}) \tag{9}$$

(cf. also 1.1.3.3). Here $U_{\mathbb{C}}$ denotes the group of infinite matrices of the form $U \oplus$ (infinite diagonal matrix with elements 1) and U a unitary $N \times N$-matrix, N arbitrary. $\pi_{k-1}(\cdot)$ is the $(k-1)^{\text{th}}$ homotopy group of the space in the brackets.

Next we define a homomorphism

$$\delta: \tilde{K}(SY) \to \tilde{K}(X, Y).$$

Consider the isomorphisms $\lambda: \tilde{K}(SY) \to K(CY, Y)$ (cf. Remark 4), $\nu: K(CY, Y) \to K(X \cup CY, X)$, $(X \cup CY/X \cong CY/Y)$, $\eta = \nu \circ \lambda: \tilde{K}(SY) \to K(X \cup CY, X)$. By $X \cup CY \to X \cup CY/X$ is induced a homomorphism $\delta': K(X \cup CY, X) \to \tilde{K}(X \cup CY)$. We set $\delta = \gamma \circ \delta' \circ \eta$ (γ is defined in Remark 4). δ is called *coboundary operator*.

For $k \in Z_+$ we define

$$\tilde{K}^{-k}(X) - \tilde{K}(S^kX), \quad K^{-k}(X, Y) = \tilde{K}\big(S^k(X/Y)\big).$$

Here (X, Y) is a pair of compact spaces, Y a space with a base point and $S^kX = S(S^{k-1}X)$.

Theorem 9. *The following sequence is exact*

$$\dots \to \tilde{K}^{-k}(Y) \xrightarrow{\delta} K^{-k+1}(X, Y) \xrightarrow{j^*} \tilde{K}^{-k+1}(X) \xrightarrow{i^*} \tilde{K}^{-k+1}(Y) \to \dots$$
$$\dots \to \tilde{K}^{-2}(Y) \xrightarrow{\delta} K^{-1}(X, Y) \xrightarrow{j^*} \tilde{K}^{-1}(X) \xrightarrow{i^*} \tilde{K}^{-1}(Y)$$
$$\xrightarrow{\delta} K(X, Y) \xrightarrow{j^*} \tilde{K}(X) \xrightarrow{i^*} \tilde{K}(Y).$$

The homomorphisms j^* and i^*, respectively, are induced by $i: Y \hookrightarrow X$, $j: X \hookrightarrow X/Y$.

The meaning of a pair of compact spaces (X, Y) is clear without comments in the case $Y \neq \emptyset$. In the case $Y = \emptyset$ we define

$$K(X, \emptyset) = \tilde{K}(X^+),$$

here X^+ denotes the disjoint union of X with an isolated point $\{+\}$ (this is the base point of X^+). Then obviously $K(X) \cong K(X, \emptyset)$. Now we can define the functors K^{-k} also for spaces without base points by putting

$$K^{-k}(X) = K^{-k}(X, \emptyset) \; \big(= K(S^kX^+) = \tilde{K}^{-k}(X^+)\big).$$

Substituting X^+, Y^+ for X, Y in Theorem 9 we obtain an exact sequence

$$\dots \to K^{-2}(Y) \xrightarrow{\delta} K^{-1}(X, Y) \xrightarrow{j^*} K^{-1}(X) \xrightarrow{i^*} K^{-1}(Y)$$
$$\xrightarrow{\delta} K(X, Y) \xrightarrow{j^*} K(X) \xrightarrow{i^*} K(Y). \tag{10}$$

j^* and i^* are induced by $i: Y \to X$ and $j: (X, \emptyset) \to (X, Y)$, respectively.

1.1.3.2. The K-Functor for Locally Compact Spaces

In this section we study the K-functor for locally compact spaces. Let X be locally compact. A triple (E_0, E_1, α) is called *difference bundle* over X if E, F are complex vector bundles over X and $\alpha: E \to F$ is a bundle morphism for which there exists a relatively compact subset $K = K(\alpha) \subset X$ so that

$$\alpha|_{CK}: E_0|_{CK} \to E_1|_{CK} \tag{1}$$

is an isomorphism ($CK = X \setminus K$). The set of those points of X where α is no isomorphism is called the *support* of α. Denote by $R_1(X)$ the set of difference bundles over X. With respect to the addition $R_1(x)$ has an obvious semi-group structure. Two elements (E_0, E_1, α), $(F_0, F_1, \beta) \in R_1(X)$ are called isomorphic ($\cong$) if there exist isomorphisms $u: E_0 \to F_0$, $v: E_1 \to F_1$ so that the restrictions to $X \setminus \big(K(\alpha) \cup K(\beta)\big)$ of the compositions $\beta \circ u$ and $v \circ \alpha$ are equal. Two elements (E_0, E_1, α), $(E_0', E_1', \alpha') \in R_1(X)$ are called equivalent if there exist elements of the form $(G, G, 1)$, $(G', G', 1) \in R_1(X)$ with

$$(E_0, E_1, \alpha) \oplus (G, G, 1) \cong (E_0', E_1', \alpha') \oplus (G', G', 1) .$$

The equivalence class represented by (E_0, E_1, α) is denoted by $[E_0, E_1, \alpha]$. Let $K(X)$ be the set of equivalence classes. $X \to K(X)$ is per def. the K-functor for locally compact spaces (proper mappings with respect to X). $K(X)$ is an Abelian group and also a ring with respect to a multiplication similarly defined as in $L(X, Y)$ in 1.1.3.1. The functor K for locally compact spaces has analogous properties as the K-functor for compact spaces with respect to pull back and homotopy equivalence of the base space.

$K(X, Y)$ ($Y \subseteq X$ locally closed) is defined as the set of equivalence classes of those elements $(E_0, E_1, \alpha) \in R_1(X)$ for which $K(\alpha) \subset X \setminus Y$.

Remark 1. Let X be an open subset of a compact space A and $B = A \setminus X$. Then there is an isomorphism $\gamma: K(X) \to K(A, B)$.

The construction of γ is as follows. Let $(E, F, \alpha) \in R_1(X)$, $K \subset X$ relatively compact so that $\alpha|_{CK}$ is an isomorphism (cf. fig. 1, where A is the rectangle). Let K_0 be a compact set with $\bar{K} \subset \operatorname{int} K_0$ and H a vector bundle over K_0 with $F|_{K_0} \oplus H = K_0 \times \mathbb{C}^N$ for certain N.

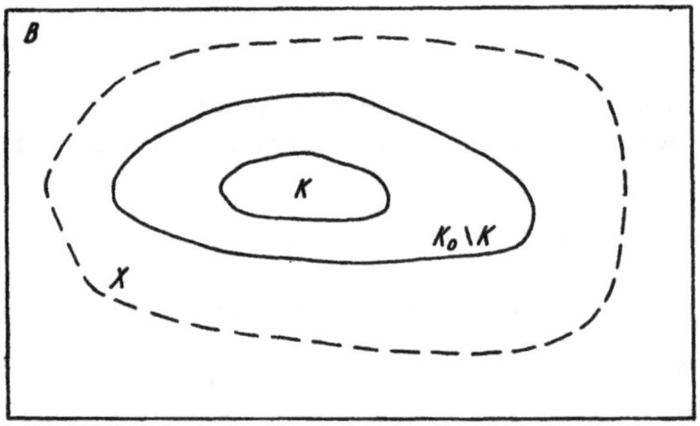

Fig. 1

Define by clutching the bundle $G = (E|_{K_0} \oplus H) \underset{\alpha_0 \oplus 1_H}{\cup} ((A \setminus K) \times \mathbb{C}^N)$ with $\alpha_0 = \alpha|_{K_0 \setminus K}$. This is a bundle over A. Then the triple $(G, \mathbb{C}^N, 1)$ (1 identity over B) belongs to $R(A, B)$ and γ is induced by $(E, F, \alpha) \to (G, \mathbb{C}^N, 1)$.

Conversely let $(G_0, G_1, \varrho) \in R(A, B)$. Then $\varrho: G_0 \to G_1$ is an isomorphism not only over B but also in an open neighbourhood B_0 of B. If $K = A \setminus B_0$, $E = G_0|_X$, $F = G_1|_X$, $\alpha = \varrho|_X$, then we have an element $(E, F, \alpha) \in R_1(X)$ and it can be proved that γ^{-1} is induced by $(G_0, G_1, \varrho) \to (E, F, \alpha)$.

Corollary 2. *Let X^+ be the one-point compactification of X. Then*

$$\tilde{K}(X^+) \cong K(X) . \tag{2}$$

$\{+\}$ is considered as the base point of X^+. Thus the K-functor for locally compact spaces X can be defined equivalently as K-functor for the one-point compactification X^+. Note that the proper mappings $Y \to X$ are those which have a continuous extension to Y^+.

Let U be an open subset in X and $i\colon U \hookrightarrow X$ the identical embedding. Then there is induced a homomorphism

$$i_*\colon K(U) \to K(X)\,. \tag{3}$$

In order to define i^* we consider the continuous map $X^+ \to U^+$ which is over U the identity and over the complement of U constant $= \{+\} \in U^+$. Then the pull back gives $\tilde{K}(U^+) \to \tilde{K}(X^+)$ and (3) follows from (2). For the definition of $K(X)$ it is sometimes convenient to start from *complexes*

$$\mathfrak{A}\colon 0 \to E_0 \xrightarrow{\alpha_0} E_1 \xrightarrow{\alpha_1} E_2 \longrightarrow \cdots \xrightarrow{\alpha_{N-1}} E_N \to 0\,. \tag{4}$$

The E_j are vector bundles over X and $\alpha_j\colon E_j \to E_{j+1}$ bundle morphisms. The number N is called the *length* of $\mathfrak{A}$. We consider only complexes for which there exists a relatively compact set $K = K(\alpha) \subsetneqq X$ $\big(\alpha = (\alpha_0, \dots, \alpha_{N-1})\big)$ so that the restriction of (4) to $\complement K$ is exact. The compact subset of X over which (4) is not exact is called *support* of $\mathfrak{A}$. Denote by $R_N(X)$ the set of all complexes with compact support and length N and set $R(X) = \bigcup_{N \in Z_+} R_N(X)$. A complex in $R_N(X)$ can also be regarded as a complex in $R_{N+1}(X)$ if one defines $\alpha_N = 0$, $E_{N+1} = X \times \{0\}$. Thus $R_N(X) \subsetneqq R_{N+1}(X)$. More general, two different complexes can always be considered as complexes of the same length by adding of zeros. $R(X)$ is in a natural way an additive semi-group. Set for abbreviation $\mathfrak{A} = \{E_j, \alpha_j;\ j = 0, \dots, N\}$ $(\alpha_N = 0)$. If $\mathfrak{A}' = \{E_j', \alpha_j';\ j = 0, \dots, N\}$ is a second complex $\big(\text{exact over } \complement K(\alpha')\big)$, $\mathfrak{A}$ is called *isomorphic* to $\mathfrak{A}'(\cong)$ if there exist isomorphisms $u_j\colon E_j \to E_j'$ $(j = 0, \dots, N)$ so that the restriction to $\complement\big(K(\alpha) \cup K(\alpha')\big)$ of $\alpha_j' \circ u_j$ and $u_{j+1} \circ \alpha_j$ are equal $(j = 0, \dots, N-1)$. $\mathfrak{A}_0$, $\mathfrak{A}_1$, are called *homotopic* $(\simeq)$ if there exists some $\mathfrak{A} \in R(X \times I)$ $(I = [0, 1])$ with $\mathfrak{A}|_{X \times \{0\}} \cong \mathfrak{A}_0$, $\mathfrak{A}|_{X \times \{1\}} \cong \mathfrak{A}_1$. Two complexes $\mathfrak{A}$, $\mathfrak{A}' \in R(X)$ are called *equivalent* if there exist elements $\mathfrak{C}, \mathfrak{C}' \in R(X)$ being exact over X with $\mathfrak{A} \oplus \mathfrak{C} \simeq \mathfrak{A}' \oplus \mathfrak{C}'$. If we denote by $K(X)$ the set of equivalence classes, we obtain an equivalent definition of the K-functor for locally compact spaces. In this definition it is sufficient to consider a fixed length. But for the multiplication it is convenient to take complexes with arbitrary length.

By multiplication of complexes in a standard way (cf. SPANIER [1]) an *external multiplication* $K(X) \otimes K(Y) \to K(X \times Y)$ is induced (for simplicity we denoted it again by the usual sign $\otimes$). Let e.g.

$$\big(0 \to E \xrightarrow{\alpha} F \to 0\big) \in R_1(X)\,, \qquad \big(0 \to V \xrightarrow{\beta} W \to 0\big) \in R_1(Y)\,.$$

The external product is then the following element in $R_2(X \times Y)$

$$0 \to \begin{matrix} E \otimes V \\ \oplus \\ 0 \end{matrix} \xrightarrow{\begin{pmatrix} \alpha\otimes 1 & 0 \\ 1\otimes\beta & 0 \end{pmatrix}} \begin{matrix} F \otimes V \\ \oplus \\ E \otimes W \end{matrix} \xrightarrow{\begin{pmatrix} 0 & 0 \\ -1\otimes\beta & \alpha\otimes 1 \end{pmatrix}} \begin{matrix} 0 \\ \oplus \\ F \otimes W \end{matrix} \to 0\,. \tag{5}$$

Here, for instance, $E \otimes V$ is an abbreviation of $p^*E \otimes q^*V$ $(p\colon X \times Y \to X$, $q\colon X \times Y \to Y$ the canonical projection). The morphisms $\alpha \otimes 1, \dots$ have an analogous meaning. (5) is equivalent (in the sense of the above equivalence relation) to

$$0 \to \begin{matrix} E \otimes V \\ \oplus \\ F \otimes W \end{matrix} \xrightarrow{\begin{pmatrix} \alpha\otimes 1 & -1\otimes\beta^* \\ 1\otimes\beta & \alpha^*\otimes 1 \end{pmatrix}} \begin{matrix} F \otimes V \\ \oplus \\ E \otimes W \end{matrix} \to 0\,.$$

This can be proved as follows. Let Z be a locally compact space and

$$\mathfrak{A}: 0 \to E \xrightarrow{\lambda} F \xrightarrow{\mu} G \to 0$$

be an element in $R_2(Z)$ being exact over $\complement K$. Addition of $\mathfrak{C}: 0 \to G \xrightarrow{1} G \to 0 \to 0$ gives

$$\mathfrak{A} \oplus \mathfrak{C}: 0 \to \begin{matrix} E \\ \oplus \\ G \end{matrix} \xrightarrow{\begin{pmatrix} \lambda & 0 \\ 0 & 1 \end{pmatrix}} \begin{matrix} F \\ \oplus \\ G \end{matrix} \xrightarrow{\begin{pmatrix} \mu & 0 \\ 0 & 0 \end{pmatrix}} \begin{matrix} G \\ \oplus \\ 0 \end{matrix} \to 0 .$$

Fix Hermitean metrics in the bundles. It is sufficient to prove that $\mathfrak{A} \oplus \mathfrak{C}$ is homotopic to the complex

$$\mathfrak{B}: 0 \to \begin{matrix} E \\ \oplus \\ G \end{matrix} \xrightarrow{\begin{pmatrix} \lambda & \mu^* \\ 0 & 0 \end{pmatrix}} \begin{matrix} F \\ \oplus \\ G \end{matrix} \xrightarrow{\begin{pmatrix} 0 & -1 \\ 0 & 0 \end{pmatrix}} \begin{matrix} G \\ \oplus \\ 0 \end{matrix} \to 0$$

$$= \left(0 \to \begin{matrix} E \\ \oplus \\ G \end{matrix} \xrightarrow{(\lambda, \mu^*)} F \to 0 \to 0 \right) \oplus \left(0 \to 0 \to G \xrightarrow{-1} G \to 0 \right).$$

Consider the following homotopy

$$\mathfrak{B}_t = \left(0 \to \begin{matrix} E \\ \oplus \\ G \end{matrix} \xrightarrow{\begin{pmatrix} \lambda & t\mu^* \\ 0 & (1-t)\mu\mu^* \end{pmatrix}} \begin{matrix} F \\ \oplus \\ G \end{matrix} \xrightarrow{\begin{pmatrix} (1-t)\mu & -t \\ 0 & 0 \end{pmatrix}} \begin{matrix} G \\ \oplus \\ 0 \end{matrix} \to 0 \right)$$

($0 \leqq t \leqq 1$). Then $\mathfrak{B} = \mathfrak{B}_1$ and $\mathfrak{A} \oplus \mathfrak{C} \simeq \mathfrak{B}_2$ because of $\mu\mu^* \simeq 1$. It is easily checked that the $\mathfrak{B}_t$ are exact over $\complement K$.

More generally one can easily show the following. If $\mathfrak{A} \in R(Z)$ is given by (4) with $N = 2k + 1$, $\mathfrak{A}$ is equivalent to the difference bundle $(H_0, H_1, \eta) \in R_1(Z)$ with $H_0 = E_0 \oplus E_2 \oplus \dots \oplus E_{2k}$, $H_1 = E_1 \oplus E_3 \oplus \dots \oplus E_{2k+1}$,

$$\eta = \begin{bmatrix} \alpha_0 & \alpha_1^* & 0 & \dots & 0 & 0 \\ 0 & \alpha_2 & \alpha_3^* & \dots & 0 & 0 \\ \cdot & \cdot & \cdot & \cdot & \cdot & \cdot \\ 0 & 0 & 0 & \dots & \alpha_{2k-2} & \alpha_{2k-1}^* \\ 0 & 0 & 0 & \dots & 0 & \alpha_{2k} \end{bmatrix}$$

(after addition of zeros the length $2k + 1$ can be assumed without loss of generality).

For compact X the complexes are simply sequences $(E_0, E_1, \dots, E_N)$ of vector bundles over X and $[E_0] - [E_1] + \dots + (-1)^N [E_N]$ is the image in $K(X)$.

Let us finally make some further remarks on the *external multiplication*. To begin with, let X, A be compact spaces. The external multiplication

$$K(X) \otimes K(A) \to K(X \times A) \tag{6}$$

is induced simply by

$$E \otimes V \overset{\text{def}}{=} (p^*E) \otimes (p^*V) \in \operatorname{Vect}(X \times A)$$

for $E \in \operatorname{Vect}(X)$, $V \in \operatorname{Vect}(A)$ and $p: X \times A \to X$, $q: X \times A \to A$ the canonical projections.

For compact spaces X and A with base points we also obtain a homomorphism $\mu_1: \tilde{K}(X) \otimes \tilde{K}(A) \to \tilde{K}(X \times A)$. Since X is a retract of $X \times A$ and A a retract of

$(X \times A)/X$ (cf. ATIYAH [2]), we have an isomorphism

$$\tilde{K}(X \times A) \to \tilde{K}(X \wedge A) \oplus \{p^*\tilde{K}(X) \oplus q^*\tilde{K}(A)\}\,.$$

In view of $p^*\tilde{K}(X) \oplus q^*\tilde{K}(A) \cong \tilde{K}(X \vee A)$ there follows an exact sequence

$$0 \to \tilde{K}(X \wedge A) \xrightarrow{\gamma} \tilde{K}(X \times A) \xrightarrow{\lambda} \tilde{K}(X \vee A) \to 0\,.$$

Since im $\mu_1 \subseteq \ker \lambda$, μ_1 induces a homomorphism

$$\mu\colon \tilde{K}(X) \otimes \tilde{K}(A) \to \tilde{K}(X \wedge A)$$

considered as external multiplication for the $\tilde{K}$-groups.

This yields a multiplication also for pairs of spaces. Let (X, Y), (A, B) be pairs of compact spaces. Then, in view of $K(X/Y) = K(X, Y)$ and $(X/Y) \wedge (A/B) \cong X \times A/(X \times B) \cup (Y \times A)$, we obtain an external multiplication

$$\mu\colon K(X, Y) \otimes K(A, B) \to K\big(X \times A, (X \times B) \cup (Y \times A)\big)\,. \tag{7}$$

Now let X, A be locally compact spaces. Using $K(X) = \tilde{K}(X^+)$ and $(X \times A)^+ \cong X^+ \wedge A^+$ we get an external multiplication

$$\mu\colon K(X) \otimes K(A) \to K(X \times A)\,. \tag{8}$$

This multiplication is equivalent to that defined above using complexes. For pairs of locally compact spaces the situation is analogous.

Put in (7) specially $A = X$. Let $\Delta\colon (X, Y \cup B) \to (X \times X, (X \times B) \cup (Y \times X))$ be the diagonal map. Then

$$\Delta^*\mu\colon K(X, Y) \otimes K(X, B) \to K(X, Y \cup B)\,.$$

This means particularly for $B = \emptyset$ that $K(X, Y)$ is a $K(X)$-module.

1.1.3.3. The Bott Periodicity

The Bott periodicity theorem is one of the most essential results of the K-theory. The Bott map to be described plays a role in the calculation of indices of elliptic operators on closed compact manifolds. We are interested in certain constructions connected with the Bott periodicity mainly because of its meaning for elliptic boundary value problems (cf. 3.2.2.2).

Consider the trivial bundle over the complex plane $\mathbb{C}$ with the fibre $\mathbb{C}$ and a bundle morphism $l\colon \mathbb{C} \times \mathbb{C} \to \mathbb{C} \times \mathbb{C}$ with $l(z, e) = (z, z^{-1} e)$ for $|z| \geqq 1$. The difference bundle $(\mathbb{C} \times \mathbb{C}, \mathbb{C} \times \mathbb{C}, l)$ represents an element $b \in K(\mathbb{C})$. Since $\mathbb{C} \cong \mathbb{R}^2$, we have a natural isomorphism $K(\mathbb{C}) \cong K(\mathbb{R}^2)$. The image of the element b is again denoted by b.

Let X be a locally compact space. By the multiplication $K(\mathbb{R}^2) \times K(X) \to K(\mathbb{R}^2 \times X)$ a homomorphism

$$\beta\colon K(X) \to K(\mathbb{R}^2 \times X)\,, \tag{1}$$

is induced if we set $\beta(a) = ba$, $a \in K(X)$. The following theorem is called *Bott periodicity theorem.*

Theorem 1. (1) *is an isomorphism.*

The map β is called *Bott isomorphism,* b is called the *Bott element.* Because of $\tilde{K}(\mathbb{R}^2) \cong \tilde{K}(S^2)$ $\big((\mathbb{R}^2)^+ \cong S^2\big)$ there are other equivalent formulations of Theorem 1, e.g.

Theorem 2. *For every space X the external multiplication*

$$K(S^2) \otimes K(X) \to K(S^2 \times X) \tag{2}$$

defines a ring isomorphism.

The proof of Theorem 2 is connected with some technical constructions which are of independent interest. First we recall the definition of the *Hopf bundle* H. Consider S^2 as unit sphere intersecting $\mathbb{C}$ in $S^1 \{z\colon |z| = 1\}$ so that the upper closed hemisphere B^+ corresponds to $B_\infty = \{z\colon |z| \geqq 1\}$ and the lower closed hemisphere B^- to $B_0 = \{z\colon |z| \leqq 1\}$ under stereographic projection. In the sense of the notations in 1.1.2.1 we can define $H = (B^+ \times \mathbb{C}) \underset{z^{-1}}{\cup} (B^- \times \mathbb{C})$ (this means the clutching of the trivial bundles $B^+ \times \mathbb{C}$, $B^- \times \mathbb{C}$ along S^1 by the isomorphism $z^{-1}\colon (z, e) \to (z, z^{-1} e)$, $z \in S^1$, $e \in \mathbb{C}$). Applying 1.1.2.1. (8) to $S^2 = S^1 \wedge S^1$ we see that every vector bundle F over S^2 with fibre dimension k can be defined by an isomorphism $\sigma\colon S^1 \times \mathbb{C}^k \to S^1 \times \mathbb{C}^k$, i. e.

$$F \cong (B^+ \times \mathbb{C}^k) \underset{\sigma}{\cup} (B^- \times \mathbb{C}^k)$$

with the clutching isomorphism $\sigma\colon (z, e) \to \big(z, \sigma(z)\, e\big)$ $(z \in S^1,\ e \in \mathbb{C}^k)$ and $F_1 \cong F_2$ iff $\sigma_1 \simeq \sigma_2$.

Now let F be an arbitrary vector bundle over $S^2 \times X$ (X compact). Then one shows similarly $F \cong F^+ \underset{\sigma}{\cup} F^-$ with $F^\pm = F|_{B^\pm}$ and some clutching isomorphism $\sigma(z, x)\colon F|_{S^1 \times X} \to F|_{S^1 \times X}$. The homotopy class of σ is uniquely determined by the isomorphy class of F. Since $B^\pm$ are contractible spaces, we have $F^\pm = (\pi^\pm)^* E$ with the canonical projections $\pi^\pm\colon B^\pm \times X \to X$ and some bundle E over X. Thus $F|_{S^1 \times X}$ can be replaced by s^*E with the canonical projection $s\colon S^1 \times X \to X$. If we regard $z \in S^1$ as parameter in σ, we can omit s^*.

The family $\sigma(z, x)$ of isomorphisms $E \to E$ can be approximated by a family of isomorphisms of the form

$$p(z, x) = \sum_{j=-M}^{M} a_j(x)\, z^j, \quad z \in S^1,$$

($a_j\colon E \to E$). Since the homotopy classes are open, we get $p \simeq \sigma$ for M sufficiently large. Thus replacing σ by p we obtain a bundle isomorphic to F by clutching.

Define q by $p = q z^{-M}$ so that $q(z, x) = \sum_{j=0}^{2M} c_j(x)\, z^j$ $(c_j = a_{j-M})$. Then

$$F \cong (F^+ \otimes \mathbb{C}) \underset{q \otimes z^{-M}}{\cup} (F^- \otimes \mathbb{C})$$

$$\cong (F^+ \underset{q}{\cup} F^-) \otimes (\mathbb{C} \underset{z^{-M}}{\cup} \mathbb{C}) \cong (F^+ \underset{q}{\cup} F^-) \otimes H^M$$

(with $H^M = H \otimes \ldots \otimes H$, M factors). This gives a good survey over the set of isomorphy classes of vector bundles over $S^2 \times X$.

Let now $q(z, x) = \sum_{j=0}^{n} c_j(x)\, z^j$ be a clutching polynomial for bundles $(\pi^\pm)^* E$ $\big(E \in \operatorname{Vect}(X)\big)$ and $F = (\pi^+)^* E \underset{q}{\cup} (\pi^-)^* E$. Moreover set $F_0 = \pi^* E$ ($\pi\colon S^2 \times X \to X$ the canonical projection). Then the bundle $F \oplus F_0 \oplus \ldots \oplus F_0$ is described by the clutching isomorphism $q \oplus 1_k \oplus \ldots \oplus 1_k$ (k is the fibre dimension of F and 1_k the $k \times k$-unit matrix).

Lemma 3. *There exists a homotopy*

$$\begin{pmatrix} q(z,x) & 0 & \dots & 0 \\ 0 & 1_k & \dots & 0 \\ \vdots & \vdots & & \vdots \\ 0 & 0 & \dots & 1_k \end{pmatrix} \simeq \begin{bmatrix} c_0(x) & c_1(x) & c_2(x) & \dots & c_{n-1}(x) & c_n(x) \\ -z\cdot 1_k & 1_k & 0 & \dots & 0 & 0 \\ 0 & -z\cdot 1_k & 1_k & & \vdots & \vdots \\ \vdots & \vdots & \vdots & & 1_k & 0 \\ 0 & 0 & 0 & \dots & -z\cdot 1_k & 1_k \end{bmatrix} \tag{3}$$

through isomorphisms of $F \oplus F_0 \oplus \dots \oplus F_0|_{S^1\times X}$ *onto itself.*

A proof of this lemma is given in HUSEMOLLER [1]. It is based on the fact that the matrix on the right of (3) can be obtained step by step by elementary transformations of the matrix on the left of (3). These can be represented as homotopies.

The structure of vector bundles over SX (e.g. $X = S^1$) gives information about $K(SX)$ (cf. 1.1.3.1, Theorem 7). For instance $K(S^2)$ is a free Abelian group with two generators [1], [H] (H the Hopf bundle over S^2 and 1 the one-dimensional trivial bundle over S^2).

For arbitrary $k = 0, 1, 2, \dots$ the ring $\tilde{K}(S^{2k+1})$ consists only of 0. With respect to addition $\tilde{K}(S^{2k})$ is an infinite cyclic group and the product of any two elements vanishes. Especially $\tilde{K}(S^2)$ is generated by $[H] - [1]$. The equation $([H] - [1])^2 = 0$ follows immediately from

$$\begin{pmatrix} z^{-1} & 0 \\ 0 & z^{-1} \end{pmatrix} \simeq \begin{pmatrix} z^{-2} & 0 \\ 0 & 1 \end{pmatrix}$$

(homotopy through clutching homomorphisms, cf. also 1.1.3.1, Proposition 8), so that $[H \oplus H] = [(H \otimes H) \oplus 1]$.

Denote by $\pi_k(\cdot)$ the kth homotopy group of the space in the brackets.

Theorem 4. *We have for* $1 \leqq i \leqq 2m$ *the following isomorphisms* $\pi_{i-1}(\mathrm{GL}(m, \mathbb{C})) \cong \pi_{i+1}(\mathrm{GL}\,(m, \mathbb{C}))$, *i. e. because of* $\pi_0\,(\mathrm{GL}\,(1, \mathbb{C})) = 0$, $\pi_1(\mathrm{GL}\,(1, \mathbb{C})) \cong Z$ *(infinite cyclic group) for* $1 \leqq k \leqq 2m$

$$\pi_k(\mathrm{GL}\,(m, \mathbb{C})) \cong \begin{cases} 0 & \text{for } k \text{ even}, \\ Z & \text{for } k \text{ odd}. \end{cases}$$

Theorem 4 can be considered as a formulation of the Bott periodicity theorem. A proof in this form is given in MILNOR [1]. The dimension condition can be dropped by passing to the group GL $(\infty, \mathbb{C})$ (cf. HUSEMOLLER [1]). Theorem 4 corresponds to 1.1.2.1. (8) for $X = S^{k+1}$.

In 1.1.3.1 the groups $K^{-j}(X)$ and $K^{-j}(X, Y)$ were defined. This definition can be extended to locally compact spaces by setting $K^{-j}(X) = \tilde{K}^{-j}(X^+)$, $K^{-j}(X, Y) = K^{-j}(X^+, Y^+)$. We have the following

Corollary 5. *For each integer* $k \geqq 0$ *the multiplication*

$$\tilde{K}(S^2) \otimes K^{-k}(X) \to K^{-k-2}(X)$$

induces an isomorphism

$$\beta\colon K^{-k}(X) \to K^{-k-2}(X) \tag{4}$$

if β *is defined by* $\beta(a) = ([H] - [1])\,a$, $a \in K^{-k}(X)$. *Similarly there is an isomorphism* $\beta\colon K^{-k}(X, Y) \to K^{-k-2}(X, Y)$.

Because of $(\mathbb{R}^k \times X)^+ = S^k \wedge X^+$ one can also set

$$K^{-k}(X) = K(\mathbb{R}^k \times X), \quad K^{-k}(X, Y) = K(\mathbb{R}^k \times (X/Y))$$

($k = 0, 1, 2, \ldots$). Similarly as in 1.1.3.1 we have the exact cohomology sequence of K-theory in the form 1.1.3.1 (10) for pairs of locally compact spaces (X, Y). Theorem 5 shows that this sequence is periodic. This gives an exact sequence consisting of 6 groups

$$\begin{array}{ccccc} K(X, Y) & \xrightarrow{j^*} & K(X) & \xrightarrow{i^*} & K(Y) \\ \uparrow{\scriptstyle\delta} & & & & \downarrow{\scriptstyle\delta\circ\beta} \\ K^{-1}(Y) & \xleftarrow{i^*} & K^{-1}(X) & \xleftarrow{j^*} & K^{-1}(X, Y) \end{array} . \tag{5}$$

Applying the Bott isomorphism β to the exact cohomology sequence of K-theory we obtain a commutative diagram of the form

$$\begin{array}{ccccccccc} \ldots \to K^{-k}(Y) & \xrightarrow{\delta} & K^{-k+1}(X, Y) & \xrightarrow{j^*} & K^{-k+1}(X) & \xrightarrow{i^*} & K^{-k+1}(Y) & \xrightarrow{\delta} & \ldots \\ \downarrow{\scriptstyle\beta} & & \downarrow{\scriptstyle\beta} & & \downarrow{\scriptstyle\beta} & & \downarrow{\scriptstyle\beta} & & \\ \ldots \to K^{-k-2}(Y) & \xrightarrow{\delta} & K^{-k-1}(X, Y) & \xrightarrow{j^*} & K^{-k-1}(X) & \xrightarrow{i^*} & K^{-k-1}(Y) & \xrightarrow{\delta} & \ldots \end{array} \tag{6}$$

Theorem 1 is a special case of a more general fact, namely the *Thom isomorphism*.

Let E, F be complex vector bundles over X. By the diagonal map $X \to X \times X$ $(x \to (x, x))$ an embedding $E \oplus E \to E \times E$ is induced. Then the composition $K(E) \otimes K(F) \to K(E \times F) \to K(E \oplus F)$ yields a multiplication $K(E) \otimes K(F) \to K(E \oplus F)$. Especially for $F = X$ (fibre dimension of F equal to zero) we have $K(E) \otimes K(X) \to K(E)$.

The bundle π^*E over E $(\pi\colon E \to X)$ has a section s with $s \neq 0$ over $E \setminus X$ (X is considered as zero section in E). s is defined by $s(e) = e$, $e \in E$. Then we can define $s \wedge\colon \bigwedge^k(\pi^*E) \to \bigwedge^{k+1}(\pi^*E)$ by pointwise exterior multiplication ($k = 0, 1, 2, \ldots$) and get a complex λ_E of vector bundles over E being exact over $E \setminus X$. External multiplication of λ_E by a difference bundle (E_1, E_2, α) over X (α is outside a compact subset of X an isomorphism, cf. 1.1.3.2) gives a complex over E being exact outside a compact set in E. Thus there is induced a homomorphism

$$\lambda_E\colon K(X) \to K(E) . \tag{7}$$

Theorem 6. (7) *is an isomorphism.*

(7) is called the *Thom isomorphism*. In the case of $E = X \times \mathbb{C}$ (trivial bundle over X with fibre $\mathbb{C}$) (7) is identical with the Bott isomorphism.

1.1.3.4. Fredholm Operators and K-Functor

Let X be a compact topological space and $K(X)$ the corresponding K-group. Moreover let H be an infinite dimensional complex separable Hilbert space. For a given Fredholm family $A\colon X \to \mathcal{F}(H, H)$ (cf. 1.1.1.2, Definition 9) we define an *index element*

$$\operatorname{ind}_X A \in K(X) . \tag{1}$$

According to 1.1.1.2, Proposition 10 there exists a finite dimensional subspace $W \subset H$ with $W + \operatorname{im} A(x) = H$ for all $x \in X$. Let $l = \dim W$. We choose an injective linear map $L\colon \mathbb{C}^l \to H$ with $\operatorname{im} L = W$. Then

$$(A, L)\colon \begin{array}{c} H \\ \oplus \\ \mathbb{C}^l \end{array} \to H \tag{2}$$

(defined by $f = Ah + Lv$, $h \in H$, $v \in \mathbb{C}^l$) is again a Fredholm family between Hilbert spaces. We use the following

Lemma 1. *Let $A: X \to \mathcal{F}(H_1, H_2)$ be a Fredholm family (H_j infinite dimensional complex separable Hilbert spaces, $j = 1, 2$) and suppose that $A(x)$ is surjective for all $x \in X$. Then*

$$\ker_X A = \bigcup_{x \in X} \ker A(x) \tag{3}$$

is a vector bundle over X.

Thus for the Fredholm family (2) we find a vector bundle $\ker_X (A, L)$. Let $\mathbb{C}^l$ be the trivial bundle over X with the fibre $\mathbb{C}^l$. Then we can define the element

$$\operatorname{ind}_X A = [\ker_X (A, L)] - [\mathbb{C}^l] \in K(X) . \tag{4}$$

It is easily seen that (4) is independent of the choice of W and L with the mentioned properties and that (4) depends only on the homotopy class of $A: X \to \mathcal{F}(H, H)$. Denote by $[X, \mathcal{F}]$ the set of homotopy classes of continuous maps $X \to \mathcal{F}(H, H)$. Then the above construction gives rise to a map

$$\operatorname{ind}_X : [X, \mathcal{F}] \to K(X) . \tag{5}$$

By the composition of Fredholm operators $\mathcal{F} \times \mathcal{F} \to \mathcal{F}$ in $[X, \mathcal{F}]$ a semi-group structure is defined. $[X, \mathcal{F}]$ is even a group. The inverse of an element represented by $A: X \to \mathcal{F}(H, H)$ is represented by a continuous family of parametrices of $A(x)$.

Theorem 2. (5) *is an isomorphism (X an arbitrary compact space).*

This theorem (cf. Jänich [1], Atiyah [2]) gives a connection between the K-functor and the space of Fredholm operators. It is functorial in the following sense. Let $f: Y \to X$ be a continuous map. To any Fredholm family $A: X \to \mathcal{F}(H, H)$ a Fredholm family $f^*A = A \circ f: Y \to \mathcal{F}(H, H)$ can be assigned. Then

$$\operatorname{ind}_Y (f^*A) = f^* \operatorname{ind}_X A . \tag{6}$$

Theorem 2 shall be used for various constructions connected with boundary value problems. In 3.2.2.2 we give an "analytical proof" of the Bott periodicity theorem based on Fredholm families (cf. also Atiyah [3]). In 3.1.1.1 the map (5) plays a role in understanding the so called boundary symbols. The following theorem of Kuiper shows another aspect of the nature of boundary symbols.

Let H be as above the Hilbert space under consideration and GL (H) the group of isomorphisms of H.

Theorem 3. *All homotopy groups of* GL(H) *are trivial*:

$$\pi_k (\mathrm{GL}\,(H)) = 0 \qquad (k = 0, 1, 2, \ldots) .$$

Corollary 4. *The topological group* GL(H) *is contractible.*

Since the unitary group $U(H)$ of H is a retract of GL (H), $U(H)$ is contractible, too.

Let X be a compact topological space (of homotopy type of a CW-complex, cf. Spanier [1]). Then we have

Corollary 5. *Every bundle over X with fibre H and structure group* GL(H) *is trivial.*

1.2. Pseudo-Differential Operators

1.2.1. Distributions and Fourier Transform

1.2.1.1. Distributions

In the theory of pseudo-differential operators an essential tool is the classical distribution theory and the Fourier transform. There are many books in which these tools are developed, e.g. SCHWARTZ [2], HÖRMANDER [2], GELFAND/SCHILOW [1], PALAIS [1], BRYČKOV/PRUDNIKOV [1].

Here we restrict ourselves to a collection of some definitions and basic results without any completeness. The proofs can be found in the mentioned literature. Concerning notions on topological vector spaces cf. ROBERTSON/ROBERTSON [1], SCHAEFER [1], TREVES [1].

By $x = (x_1, \dots, x_n)$ we denote a point in $\mathbb{R}^n$. If $\alpha = (\alpha_1, \dots, \alpha_n)$ is a multi-index, set $x^\alpha = x_1^{\alpha_1} \cdot \dots \cdot x_n^{\alpha_n}$,

$$D^\alpha = D_x^\alpha = \left(\frac{1}{i}\frac{\partial}{\partial x_1}\right)^{\alpha_1} \cdot \dots \cdot \left(\frac{1}{i}\frac{\partial}{\partial x_n}\right)^{\alpha_n},$$

$(i = \sqrt{-1})$, $\partial^\alpha = \partial_x^\alpha = (\partial/\partial_{x_1})^{\alpha_1} \cdot \dots \cdot (\partial/\partial_{x_n})^{\alpha_n}$. Moreover put $|\alpha| = \alpha_1 + \dots + \alpha_n$, $\alpha! = \alpha_1! \cdot \dots \cdot \alpha_n!$.

With these notations we have for arbitrary $x, y \in \mathbb{R}^n$ and any multi-index α

$$(x + y)^\alpha = \sum_{\beta+\gamma=\alpha} \frac{\alpha!}{\beta!\,\gamma!} x^\beta y^\gamma . \tag{1}$$

The Leibniz rule has the form

$$\partial^\alpha(fg) = \sum_{\beta+\gamma=\alpha} \frac{\alpha!}{\beta!\,\gamma!} (\partial^\beta f)\,(\partial^\gamma g), \tag{2}$$

f, g smooth in an open domain $\Omega \subseteq \mathbb{R}^n$. The formula (2) is a consequence of (1) in view of $\partial^\alpha\big(f(x)\,g(x)\big) = (\partial_x + \partial_y)^\alpha \big(f(x)\,g(y)\big)_{y=x}$.

For functions $a(x, \xi)$ $(x, \xi \in \mathbb{R}^n)$ we shall write

$$a^{(\alpha)}(x, \xi) = \partial_\xi^\alpha a(x, \xi) . \tag{3}$$

Then we have the following generalization of the Leibniz rule. Let $A(x, D)$ be a differential operator with the characteristic polynomial $a(x, \xi)$ $\big(\text{i.e. } A(x, D) = a(x, D)\big)$. Then

$$A(x, D)\,\big(f(x)\,g(x)\big) = \sum_\alpha \frac{1}{\alpha!} \big(a^{(\alpha)}(x, D)\,f(x)\big)\,D^\alpha g(x) \tag{4}$$

$\big(a^{(\alpha)}(x, D)$ denotes the differential operator obtained by substituting $D = (D_{x_1}, \dots, D_{x_n})$ for $\xi = (\xi_1, \dots, \xi_n)$ in the polynomial $a^{(\alpha)}(x, \xi)\big)$.

The n-th variable in $\mathbb{R}^n$ plays sometimes a special role. Then we put $x = (x', x_n)$, $x' = (x_1, \dots, x_{n-1})$, $\xi = (\xi', \xi_n)$, $\xi' = (\xi_1, \dots, \xi_{n-1})$ and $\nu = \xi_n$. Moreover set $\langle \xi \rangle = (1 + |\xi|^2)^{1/2}$.

The set of all integers is denoted by $\mathbb{Z}$. By $\mathbb{Z}_+$ denote the set of all $k \in \mathbb{Z}$, $k \geqq 0$.

Let E be a locally convex topological vector space (over $\mathbb{C}$) and E' the topological dual of E (the bilinear pairing is denoted by $\langle \cdot, \cdot \rangle$). A subset $B \subset E$ is called *weakly bounded* if

$$p_B(e') \overset{\text{def}}{=} \sup_{e \in B} |\langle e', e\rangle| < \infty \quad \text{for} \quad e' \in E'\,.$$

On E' a system of seminorms generating the so-called *w^*-topology* on E' is given by $e' \to |\langle e', e\rangle|$ $(e \in E)$. The system $\{p_B\colon B \subset E$ weakly bounded$\}$ of seminorms on E' generates a topology denoted by *s^*-topology.*

The topological vector spaces defined in the following as duals of given spaces are mainly considered in the w^*-topology. Otherwise the topology shall be explicitly indicated (various assertions are true for the s^*- and the w^*-topology; it is left to the reader to check this in details).

Let $\Omega \subseteq \mathbb{R}^n$ be an open set. Denote by $C^\infty(\Omega)$ the space of all (complex valued) infinitely differentiable functions in Ω with the topology given by the system of seminorms $p_{K,\alpha}\colon u \to \sup_{x \in K} |D^\alpha u(x)|$. Here α is an arbitrary multi-index and K an arbitrary compact subset of Ω (it is sufficient to take a countable set of compact subsets K_l, $l = 1, 2, \ldots$, with $\Omega = \bigcup_l K_l$). The closure in Ω of the set $\{x \in \Omega\colon u(x) \neq 0\}$ is called *support* of u and denoted by supp u.

Let $C_0^\infty(\Omega)$ be the space of all $u \in C^\infty(\Omega)$ with supp u compact. The topology in $C_0^\infty(\Omega)$ is described as follows. A sequence u_k $(k = 1, 2, \ldots)$ is convergent in $C_0^\infty(\Omega)$ if there exists a compact subset $K_0 \subset \Omega$ with supp $u_k \subseteq K_0$ for all k and $p_{K_0,\alpha}(u_k)$ convergent for all multi-indices α. The embedding $C_0^\infty(\Omega) \to C^\infty(\Omega)$ is then continuous.

The elements of $\mathcal{D}'(\Omega) \overset{\text{def}}{=} (C_0^\infty(\Omega))'$ are called *distributions* in Ω. For any $f \in \mathcal{D}'(\Omega)$ there exists a maximal open subset $\Omega_f \subseteq \Omega$ with $\langle f, u\rangle = 0$ for all $u \in C_0^\infty(\Omega_f)$. The relatively closed set supp $f \overset{\text{def}}{=} \Omega \setminus \Omega_f$ is called *support of* f (supp f). An example is the so-called *Dirac distribution* $\delta_{x_0}\colon u \to u(x_0)$, $\{x_0\} = \operatorname{supp} \delta_{x_0}$.

The space $\mathcal{E}'(\Omega) \overset{\text{def}}{=} \{f \in \mathcal{D}'(\Omega)\colon \operatorname{supp} f$ compact$\}$ can be identified with $(C^\infty(\Omega))'$. Thus for $\Omega = \mathbb{R}^n$ any $f \in \mathcal{E}'(\Omega)$ can be applied to $u = e^{-i\langle x, \xi\rangle}$ $(\xi \in \mathbb{R}^n$, $\langle x, \xi\rangle = \sum_{j=1}^n x_j \xi_j)$. The resulting function $\hat{f}(\xi) \overset{\text{def}}{=} \langle f, e^{-i\langle \cdot, \xi\rangle}\rangle \in C^\infty(\mathbb{R}^n_\xi)$ is called Fourier transform of f (cf. 1.2.1.2, Theorem 7).

Denote by $\mathcal{S}(\mathbb{R}^n)$ the space of all $u \in C^\infty(\mathbb{R}^n)$ with

$$p_{\alpha,\beta}(u) = \sup_{x \in \mathbb{R}^n} |x^\beta D^\alpha u(x)| < \infty \tag{5}$$

for all multi-indices α, β. The space $\mathcal{S}(\mathbb{R}^n)$ is considered in the Fréchet space topology generated by the system of semi-norms (5). $\mathcal{S}(\mathbb{R}^n)$ is called *Schwartz space.* $\mathcal{S}'(\mathbb{R}^n) \overset{\text{def}}{=} (\mathcal{S}(\mathbb{R}^n))'$ is called *space of temperate distributions* in $\mathbb{R}^n$. There are continuous embeddings $C_0^\infty(\mathbb{R}^n) \subset \mathcal{S}(\mathbb{R}^n) \subset C^\infty(\mathbb{R}^n)$, $\mathcal{E}'(\mathbb{R}^n) \subset \mathcal{S}'(\mathbb{R}^n) \subset \mathcal{D}'(\mathbb{R}^n)$.

Any locally integrable function f in Ω corresponds to a distribution

$$u \to \langle f, u\rangle = \int_\Omega f(x)\, u(x)\, dx\,, \quad u \in C_0^\infty(\Omega)\,. \tag{6}$$

This distribution is usually denoted again by f, since $\langle f, u\rangle = 0$ for all $u \in C_0^\infty(\Omega)$ implies $f = 0$ Lebesgue-almost everywhere. Thus one obtains continuous embeddings of $L^p(\Omega)$ $(1 \leq p \leq \infty)$, $C^k(\Omega)$, $C_0^k(\Omega)$ $(0 \leq k \leq \infty)$, ... in $\mathcal{D}'(\Omega)$ (as usual $L^p(\Omega)$ is the space of all functions f for which $|f|^p$ is integrable, $C^k(\Omega)$ is the space of all k-times continuously differentiable functions, and $C_0^k(\Omega) = \{f \in C^k(\Omega)\colon \operatorname{supp} f$ compact$\}$

with standard topologies). For any $f \in \mathscr{E}'(\Omega)$ there exists a sequence $f_k \in C_0^\infty(\Omega)$ with $f_k \to f$ with respect to the w^*-convergence.

Let U be an open subset of Ω. Then the restriction $f|_U \in \mathscr{D}'(U)$ for $f \in \mathscr{D}'(\Omega)$ is canonically defined. There is obviously a maximal open subset $U_0 \subset \Omega$ so that $f|_{U_0}$ belongs to $C^\infty(U_0)$. The relatively closed set $\Omega \setminus U_0$ is called *singular support* of f (denoted by sing supp f).

The multiplication of a distribution $f \in \mathscr{D}'(\Omega)$ by a function $w \in C^\infty(\Omega)$ is defined by $\langle wf, u\rangle = \langle f, wu\rangle$, $u \in C_0^\infty(\Omega)$. Then $wf \in \mathscr{D}'(\Omega)$. If $f \in \mathscr{D}'(\mathbb{R}^n)$, $w \in C_0^\infty(\Omega)$ and supp $w \cap$ sing supp $f = \emptyset$,

$$\langle wf, e^{-i\langle\cdot,\xi\rangle}\rangle \in \mathscr{S}(\mathbb{R}^n_\xi)\,. \tag{7}$$

Given $f \in \mathscr{D}'(\Omega)$ we can define $D^\alpha f \in \mathscr{D}'(\Omega)$ as the functional

$$u \to \langle f, (-1)^{|\alpha|} D^\alpha u\rangle, \qquad u \in C_0^\infty(\Omega)\,.$$

This gives a continuous map

$$D^\alpha\colon \mathscr{D}'(\Omega) \to \mathscr{D}'(\Omega) \tag{8}$$

extending the ordinary differentiation from $C^\infty(\Omega)$ to $\mathscr{D}'(\Omega)$.

Denote by

$$\Delta = -\sum_{i=1}^{n} \frac{\partial^2}{\partial x_i^2}$$

as usually the Laplace operator. For any given $f \in \mathscr{D}'(\Omega)$, $U \subset \Omega$ open with $\overline{U} \subset \Omega$ compact and any $k \in \mathbb{Z}_+$ there exists an $m \in \mathbb{Z}_+$ and a function $h \in C^k(U)$ with $\langle h, \Delta^m u\rangle = \langle f, u\rangle$ for all $u \in C_0^\infty(U)$, i.e. $\Delta^m h = f|_U$.

Next define WF(f), the so-called *wave front set* of $f \in \mathscr{D}'(\Omega)$. A point $(x_0, \xi_0) \in \mathbb{R}^n \times (\mathbb{R}^n \setminus \{0\})$ does not belong to WF (f) iff there exists some $w \in C_0^\infty(\mathbb{R}^n)$ with $w(x_0) \neq 0$ so that for each $N \in \mathbb{Z}_+$ there exists a constant c_N with

$$|\langle wf, e^{-i\langle\cdot,t\xi\rangle}\rangle| \leqq c_N t^{-N} \tag{9}$$

for all ξ belonging to a small neighbourhood of ξ_0 and all $t \geqq 1$ (the distribution wf in (9) is applied with respect to the variable $\cdot$).

WF (f) $\subset \mathbb{R}^n \times (\mathbb{R}^n \setminus \{0\})$ is a closed conic subset, i.e. $(x, \xi) \in$ WF (f) implies $(x, t\xi) \in$ WF (f) for all $t > 0$. If $\pi\colon \mathbb{R}^n \times (\mathbb{R}^n \setminus \{0\}) \to \mathbb{R}^n$ denotes the projection $(x, \xi) \to x$, we have $\pi(\text{WF}(f)) = $ sing supp f for each $f \in \mathscr{D}'(\Omega)$.

The wave front set of a distribution is studied in details in Hörmander [6] (cf. also Duistermaat [1]).

Theorem 1. *Let $U \subseteqq \mathbb{R}^n$, $V \subseteqq \mathbb{R}^m$ be open sets. Then there exists for any linear continuous operator*

$$A\colon C_0^\infty(U) \to \mathscr{D}'(V) \tag{10}$$

(continuous with respect to the s^*-topology in $\mathscr{D}'(V)$) a

$$K \in \mathscr{D}'(V \times U) \tag{11}$$

so that

$$\langle v, Au\rangle = \langle K, v \otimes u\rangle \tag{12}$$

for all $u \in C_0^\infty(U)$, $v \in C_0^\infty(V)$ $(v \otimes u \overset{\text{def}}{=} vu)$. Conversely a distribution (11) *corresponds to a uniquely determined continuous operator* (10) *for which* (12) *holds.*

K is called *distributional kernel* of A.

A proof of Theorem 1 is given in MAURIN [1], TREVES [1].

An element $K \in \mathcal{D}'(V \times U)$ is called *properly supported* if the two projections π_U: supp $K \to U$, π_V: supp $K \to V$ are proper (i.e. the pre-image of compact sets are compact; π_U and π_V denote the canonical projections of $V \times U$ to U and V, respectively).

An operator (10) with properly supported distributional kernel K is called *properly supported.*

1.2.1.2. The Fourier Transform and Sobolev Spaces

The *Fourier transform* of a function $u \in \mathcal{S}(\mathbb{R}^n)$ is defined by

$$(Fu)(\xi) = \int e^{-ix\xi} u(x) \, dx \,. \tag{1}$$

Here $\xi = (\xi_1, \dots, \xi_n)$, $x\xi \overset{\text{def}}{=} \langle x, \xi \rangle \overset{\text{def}}{=} x_1\xi_1 + \dots + x_n\xi_n$. We also use the abbreviation $\hat{u}(\xi) = (Fu)(\xi)$.

Theorem 1. *The Fourier transform defines an isomorphism*

$$F: \mathcal{S}(\mathbb{R}^n_x) \to \mathcal{S}(\mathbb{R}^n_\xi) \tag{2}$$

and

$$F(Fu)(x) = (2\pi)^n u(-x) \,, \tag{3}$$

$$(F^{-1}v)(x) = (2\pi)^{-n} \int e^{ix\xi} v(\xi) \, d\xi \,. \tag{4}$$

The definition obviously implies the following formulas

$$\frac{1}{i} \frac{\partial}{\partial x_k} = F^{-1} \xi_k F \,, \tag{5}$$

$$\frac{1}{i} x_k = F^{-1} \frac{\partial}{\partial \xi_k} F \tag{6}$$

($k = 1, \dots, n$). The Fourier transform F shall be extended to other function and distribution spaces. The corresponding linear operators will be simply denoted by F again.

F can be defined by (1) immediately for $u \in L^1(\mathbb{R}^n)$. Then $|\hat{u}(\xi)| \leqq \|u\|_{L^1}$ for all $\xi \in \mathbb{R}^n$ so that $F: L^1(\mathbb{R}^n) \to C(\mathbb{R}^n)$ is continuous.

For the *convolution product* $(u * v)(x) = \int u(y) v(x - y) \, dy$,

$$\|f * g\|_{L^p} \leqq \|f\|_{L^1} \|g\|_{L^p} \qquad (1 \leqq p \leqq \infty) \tag{7}$$

follows for all $f \in L^1(\mathbb{R}^n)$, $g \in L^p(\mathbb{R}^n)$. Especially $L^1(\mathbb{R}^n)$ is a commutative Banach algebra with respect to convolution.

Theorem 2.

$$F(u * v) = (Fu)(Fv) \tag{8}$$

for all $u, v \in L^1(\mathbb{R}^n)$.

Set $(u, v) = \int u(x) \overline{v(x)} \, dx$ (below the Hermitean scalar product shall be modified by some factor, cf. 2.1.1.1).

Theorem 3. *The operator* (2) *has a continuous extension as an isomorphism*

$$F: L^2(\mathbb{R}^n_x) \to L^2(\mathbb{R}^n_\xi) \tag{9}$$

and

$$(u, v) = (2\pi)^{-n} (\hat{u}, v) \tag{10}$$

for all $u, v \in L^2(\mathbb{R}^n)$.

Corollary 4.

$$(F^{-1}w, v) = (2\pi)^{-n} (w, Fv)\,, \tag{11}$$

$$(u, F^{-1}w) = (2\pi)^{-n} (Fu, w) \tag{12}$$

$\big(u, v, w \in L^2(\mathbb{R}^n)\big)$.

Remark 5. With respect to the bilinear pairing $\langle u, v\rangle = \int u(x)\, v(x)\, dx$ we have

$$\langle Fu, v\rangle = \langle u, Fv\rangle\,, \tag{13}$$

$$\langle w, u\rangle = \langle F^{-1}w, Fv\rangle \tag{14}$$

$\big(u, v, w \in L^2(\mathbb{R}^n)\big)$.

Let $v \in \mathcal{S}'(\mathbb{R}^n_\xi)$. Then the composition

$$\mathcal{S}(\mathbb{R}^n_x) \xrightarrow{F} \mathcal{S}(\mathbb{R}^n_\xi) \xrightarrow{v} \mathbb{C}$$

defines an element $Fv \in \mathcal{S}'(\mathbb{R}^n_x)$. Since $\langle Fu, v\rangle = \langle u, Fv\rangle$ for all $u, v \in \mathcal{S}(\mathbb{R}^n)$, we obtain an extension of the Fourier transform from $\mathcal{S}(\mathbb{R}^n_\xi)$ to $\mathcal{S}'(\mathbb{R}^n_\xi)$.

Theorem 6. *The Fourier transform* (2) *has an extension as an isomorphism*

$$F\colon \mathcal{S}'(\mathbb{R}^n_x) \to \mathcal{S}'(\mathbb{R}^n_\xi)\,. \tag{15}$$

For $u \in \mathcal{S}'(\mathbb{R}^{n-1}_{x'})$, $v \in \mathcal{S}'(\mathbb{R}_{x_n})$ $\big(x = (x', x_n),\ x' = (x_1, \dots, x_{n-1})\big)$ define $w = u \otimes v \in \mathcal{S}'(\mathbb{R}^n)$ by $\langle w, f(x')\, g(x_n)\rangle = \langle u, f\rangle\, \langle v, g\rangle$ $\big(f \in \mathcal{S}(\mathbb{R}^{n-1}_{x'}),\ g \in \mathcal{S}(\mathbb{R}_{x_n})\big)$. Then

$$F(u \otimes v) = F(u) \otimes F(v)\,. \tag{16}$$

Clearly the definition of F on the right hand side of (16) corresponds to the dimension of the underlying space.

For $v \in \mathcal{E}'(\mathbb{R}^n)$ we have $\hat{v}(\xi) = \langle v, e^{-i\langle\cdot,\xi\rangle}\rangle$. It can be proved that $\hat{v}(\xi)$ $\big(v \in \mathcal{E}'(\mathbb{R}^n)\big)$ has an extension as an entire analytic function $\hat{v}(\zeta)$ in $\mathbb{C}^n_\zeta$, $\zeta = \xi + i\eta$. This extension is called *Fourier-Laplace transform* of $v \in \mathcal{E}'(\mathbb{R}^n)$.

Theorem 7 (Paley-Wiener). *Let* $w(\zeta)$ *be an entire analytic function* $(\zeta \in \mathbb{C}^n)$ *and* $K_r = \{x \in \mathbb{R}^n\colon |x| \leqq r\}$. *There exists an* $u \in \mathcal{E}'(\mathbb{R}^n)$ *with* supp $u \subseteq K_r$ *and* $\hat{u}(\zeta) = w(\zeta)$ *iff there are constants* c, m *so that*

$$|w(\zeta)| \leqq c(1 + |\zeta|)^m\, e^{r|\eta|} \tag{17}$$

$(\eta = \operatorname{Im} \zeta)$. *The function* $w(\zeta)$ *is the Fourier-Laplace transform of some* $u \in C_0^\infty(\mathbb{R}^n)$ *with* supp $u \subseteq K_r$ *iff for each* $k \in \mathbb{Z}$ *there is a constant* c_k *with*

$$|w(\zeta)| \leqq c_k(1 + |\xi|)^{-k}\, e^{r|\eta|}\,. \tag{18}$$

Next recall the definition of *Sobolev spaces*. Set

$$H^s(\mathbb{R}^n) = \{u \in \mathcal{S}'(\mathbb{R}^n)\colon \|u\|_s < \infty\}$$

with

$$\|u\|_s^2 = \int (1 + |\xi|^2)^s\, |\hat{u}(\xi)|^2\, d\xi$$

$(s \in \mathbb{R})$. Then $H^s(\mathbb{R}^n)$ with the norm $\|\cdot\|_s$ is called *Sobolev space* with index s. Replacing $(1 + |\xi|^2)^s$ by $(1 + |\xi|)^{2s}$ gives an equivalent norm. Sometimes it shall be convenient to use another equivalent norm. $H^s(\mathbb{R}^n)$ is a Hilbert space with the scalar product

$$(u, v)_{H^s} = \int\limits_{\mathbb{R}^n} (1 + |\xi|^2)^s \, \hat{u}(\xi) \, \overline{\hat{v}(\xi)} \, d\xi \, .$$

With respect to the bilinear form $\langle u, v\rangle = \int \hat{u}(\xi) \, \hat{v}(\xi) \, d\xi$ a non-degenerate pairing $H^s(\mathbb{R}^n) \times H^{-s}(\mathbb{R}^n) \to \mathbb{C}$ is induced. Then

$$||u||_s = \sup_{w \in H^{-s}} \frac{|\langle u, w\rangle|}{||w||_{-s}} \, . \tag{19}$$

Similarly $H^{-s}(\mathbb{R}^n)$ can be identified with the anti-dual of $H^s(\mathbb{R}^n)$ with respect to the Hermitean form $(u, v)_{H^0}$, extended to $u \in H^s(\mathbb{R}^n)$, $v \in H^{-s}(\mathbb{R}^n)$.

Let Ω be an open bounded set in $\mathbb{R}^n$ with smooth boundary. Define

$$H^s(\Omega) = \{u|_\Omega : u \in H^s(\mathbb{R}^n)\}, \quad H_0^s(\overline{\Omega}) = \{u \in H^s(\mathbb{R}^n) : \operatorname{supp} u \subseteqq \overline{\Omega}\} \, .$$

In the case $s > -\frac{1}{2}$ write also $H_0^s(\Omega)$ instead of $H_0^s(\overline{\Omega})$. Analogously define $H^s(\complement\overline{\Omega})$, $H_0^s(\complement\Omega)$. Then $H^s(\Omega) = H^s(\mathbb{R}^n)/H_0^s(\complement\Omega)$, and $H^s(\Omega)$ is considered with the quotient norm. Similar definitions are used in the case $\Omega = \mathbb{R}_+^n = \{x \in \mathbb{R}^n : x_n > 0\}$ or $\Omega = \mathbb{R}_-^n = \{x \in \mathbb{R}^n : x_n < 0\}$. Then $H^s(\Omega)$ and $H_0^{-s}(\overline{\Omega})$ can be considered as anti-duals of each other for all $s \in \mathbb{R}$. Note that $H^s(\Omega) = H_0^s(\overline{\Omega})$ for $|s| < \frac{1}{2}$ and

$$H_0^s(\overline{\Omega}) \subseteqq H^s(\Omega) \text{ for } s \geqq 0 \, , \quad H_0^s(\overline{\Omega}) \supseteqq H^s(\Omega) \quad \text{for} \quad s \leqq 0 \, .$$

$H_0^s(\overline{\Omega})$ $(s \in \mathbb{R})$ can be obtained as closure of $C_0^\infty(\Omega)$ with respect to the topology induced by $H^s(\mathbb{R}^n)$. Per def. the elements $f \in H_0^s(\overline{\mathbb{R}_+^n})$ are distributions in $\mathcal{S}'(\mathbb{R}^n)$. Since f vanishes for $x_n < 0$, $f \, e^{x_n \mu}$ belongs to $\mathcal{S}'(\mathbb{R}^n)$ for all real $\chi < 0$. Thus the Fourier transform F can be applied to $f \, e^{x_n \mu}$, $\mu < 0$.

Proposition 8. *Let* $f \in H_0^s(\overline{\mathbb{R}_+^n})$ $(s \in \mathbb{R})$. *Then* $F(f \, e^{-x_n \mu}) = \hat{f}(\xi', \xi_n + i\mu)$ *is an analytic function of* $\zeta = \xi_n + i\mu$ *in* $\operatorname{Im} \zeta < 0$ *for almost all* $\xi' \in \mathbb{R}^{n-1}$ *and*

$$\int |\hat{f}(\xi', \zeta)|^2 \, (1 + |\xi'| + |\zeta|)^{2s} \, d\xi' \, d\xi_n \leqq c$$

with a constant c independent of $\operatorname{Im} \zeta$.

Conversely, let a locally integrable function $\hat{f}(\xi', \zeta)$ $(\xi' \in \mathbb{R}^{n-1}, \operatorname{Im} \zeta < 0)$ *satisfying the above estimate be given and being analytic in* $\operatorname{Im} \zeta < 0$ *for almost all* ξ'. *Then there is an* $f \in H_0^s(\overline{\mathbb{R}_+^n})$ *with* $\hat{f} = F(f \, e^{-x_n \mu})$.

Let M be a closed compact manifold and E a (smooth complex) vector bundle over M with fibre dimension k. Define the Sobolev space $H^s(M, E)$ of "sections in E belonging to H^s" in the following way. Let $\mathfrak{U} = (U_1, \ldots, U_N)$ be an open finite covering of M and $\{\varphi_j\}$ a corresponding partition of unity. Let $\alpha_j : U_j \to \Omega_j$ be a system of local coordinates, $\Omega_j \subset \mathbb{R}^n$ open sets and $\chi_j : E|_{U_j} \to \Omega_j \times \mathbb{C}^k$ trivializations of E. For any $u \in C^\infty(M, E)$ define functions $v_j = (\chi_j^{-1})^* \, (\varphi_j u) \in C^\infty(\Omega_j, \mathbb{C}^k)$ $(1 \leqq j \leqq N)$. Then $v_j = (v_j^1, \ldots, v_j^k)$ is a vector of functions in $C_0^\infty(\Omega_j)$. Thus the number $||v_j||_s^2 \stackrel{\text{def}}{=} \sum_{l=1}^{k} ||v_j^l||_s^2$ $(s \in \mathbb{R})$ is well-defined. Now

$$||u||_s \stackrel{\text{def}}{=} \left(\sum_{j=1}^{N} ||v_j||_s^2 \right)^{1/2}$$

is a norm on $C^\infty(M, E)$. The completion of $C^\infty(M, E)$ with respect to this norm is per def. $H^s(M, E)$. Choosing another open covering, other local coordinate systems and another partition of unity an equivalent norm is obtained, i.e. $H^s(M, E)$ is correctly defined.

Let $l_{U_iU_j}$ be the cocycle belonging to the tangent bundle TM and the local coordinate systems (cf. 1.1.2.2). Then $d_{U_iU_j} = |\det l_{U_iU_j}|$ is a one-dimensional cocycle. Denote by D the associated complex one-dimensional vector bundle over M, the so-called *density bundle* on M. If E is a complex vector bundle with its dual E^*, we have a natural bilinear pairing $E_x \times (E^* \otimes D)_x \to D_x$. This gives a bilinear map $C^\infty(M, E) \times C^\infty(M, E^* \otimes D) \to C^\infty(M, D)$. Composition with the integration $C^\infty(M, D) \to \mathbb{C}$ gives rise to a bilinear (separately continuous) map $C^\infty(M, E) \times C^\infty(M, E^* \otimes D) \to \mathbb{C}$ and hence a linear map

$$C^\infty(M, E) \to \mathcal{D}'(M, E) \overset{\text{def}}{=} \left(C^\infty(M, E^* \otimes D)\right)' . \tag{20}$$

$\mathcal{D}'(M, E)$ is called space of *distributional sections* of E (or E-valued distributions). Since, as in the usual distribution theory, the elements of $\mathcal{D}'(M, E)$ have locally finite order and if M is compact every $f \in \mathcal{D}'(M, E)$ belongs to a space $H^s(M, E)$ for certain $s \in \mathbb{R}$. Conversely, (20) can be extended to an injective linear map $H^s(M, E) \to \mathcal{D}'(M, E)$ for all $s \in \mathbb{R}$. Thus the elements in the Sobolev spaces $H^s(M, E)$ can be regarded as E-valued distributions on M. For non-compact M define $\mathcal{D}'(M, E) = \left(C_0^\infty(M, E^* \otimes D)\right)'$.

Let X be a smooth manifold with boundary Y (cf. 1.1.2.2) embedded into the compact closed manifold $M = 2X$ (the double of X obtained by gluing together two exemplars of X along Y). Obviously, any smooth vector bundle V over X can be considered as the restriction to X of a smooth vector bundle E over M. This gives rise to the following definitions $H^s(X, V) = \{u|_\Omega : u \in H^s(M, E)\}$ ($\Omega = \operatorname{int} X$) and $H_0^s(X, V) = \{u \in H^s(M, E) : \operatorname{supp} u \subseteqq X\}$. It would be more consequent to use the notation $H^s(\Omega, V)$ instead of $H^s(X, V)$ but our convention refers to X.

If M is a paracompact manifold and E a vector bundle over M, define $H^s_{\text{comp}}(M, E)$ ($s \in \mathbb{R}$) as the space of those $u \in \mathcal{D}'(M, E)$ for which a relatively compact submanifold $X = X(u)$ with smooth boundary exists so that $u \in H_0^s(X, E|_X)$. Similarly define $H^s_{\text{loc}}(M, E)$ ($s \in \mathbb{R}$) as the space of those $u \in \mathcal{D}'(M, E)$ for which $u|_\Omega \in H^s(X, E|_X)$ ($\Omega = \operatorname{int} X$) for every relatively compact submanifold X with smooth boundary. A sequence u_k ($k \in \mathbb{Z}_+$) in $H_0^s(M, E)$ is called convergent to $u \in H_0^s(M, E)$ iff $u_k \in H_0^s(X, E|_X)$ for some X and all $k \in \mathbb{Z}_+$ and $u_k \to u$ in $H_0^s(X, E|_X)$. A sequence u_k ($k \in \mathbb{Z}_+$) in $H^s_{\text{loc}}(M, E)$ is called convergent to $u \in H^s_{\text{loc}}(M, E)$ iff $u_k|_\Omega \to u|_\Omega$ in $H^s(X, E|_X)$ for all $X \subset M$ as above. Thus $H^s_{\text{comp}}(M, E)$, $H^s_{\text{loc}}(M, E)$ are complete locally convex spaces.

A scalar product in $H^s(X, E)$ for a manifold X with boundary and an antilinear duality between $H^s(X, E)$ and $H_0^{-s}(X, E)$ shall be fixed in 3.1.2.1.

An important standard fact to be used systematically in the theory of boundary problems is the following

Proposition 9. *Let X be a smooth manifold with boundary Y. Then the operator $r' : C^\infty(X, E) \to C^\infty(Y, E')$ ($E' = E|_Y$) restricting sections to the boundary has a continuous extension*

$$r' : H^s(X, E) \to H^{s-1/2}(Y, E') \tag{21}$$

for all $s \in \mathbb{R}$, $s > \frac{1}{2}$.

There are differential operators R_j, $j = 0, 1, \dots, [s - \frac{1}{2}]$ ($[t]$ denotes the largest integer smaller than $t \in \mathbb{R}$) such that the mapping

$$H^s(X, E) \ni u \to \{r' R_j u\}_{j=0}^{[s-1/2]} \in \bigoplus_{j=0}^{[s-1/2]} H^{s-1/2-j}(Y, E'), \quad s - \tfrac{1}{2} \notin \mathbb{Z}_+,$$

is surjective and there is a continuous linear right inverse. In the case of trivial bundles for R_j one can take the jth normal derivative.

Recall the definition of general Sobolev spaces $H^{s,p}$ and Besov spaces $B^{s,p}$ ($1 < p < \infty$) (more information is given in TRIEBEL [1]). Set

$$H^{s,p}(\mathbb{R}^n) = \{u \in \mathcal{S}'(\mathbb{R}^n) : \|u\|_{H^{s,p}} < \infty\}$$

where for $s \in \mathbb{Z}_+$

$$\|u\|_{H^{s,p}} = \Big(\sum_{|\alpha| \leq s} \|D^\alpha u\|_{L_p}^p\Big)^{1/p}.$$

For $0 < s < 1$ the norm is defined by

$$\|u\|_{H^{s,p}} = \left(\|u\|_{L_p}^p + \int_{\mathbb{R}^{2n}} \frac{|u(x) - u(y)|^p}{|x - y|^{n+ps}}\, dx\, dy\right)^{1/p}$$

and for arbitrary $s > 0$, $s = s_1 + s_2$, $s_1 \in \mathbb{Z}_+$, $0 < s_2 < 1$ by

$$\Big(\|u\|_{H^{s_1,p}}^p + \sum_{|\alpha| = s_1} \|D^\alpha u\|_{H^{s_2,p}}^p\Big)^{1/p}.$$

For $s < 0$ define $H^{s,p}(\mathbb{R}^n) = \big(H^{-s,p'}(\mathbb{R}^n)\big)'$, $1/p + 1/p' = 1$. For $s \notin \mathbb{Z}_+$, $p > 1$ set

$$B^{s,p}(\mathbb{R}^n) = \{u \in H^{s-1,p}(\mathbb{R}^n) : \|u\|_{B^{s,p}} < \infty\}$$

where

$$\|u\|_{B^{s,p}} = \left(\|u\|_{H^{s-1,p}}^p + \sum_{|\alpha| = s} \int_{\mathbb{R}^{2n}} \frac{\left|D^\alpha u(x) - 2D^\alpha u\left(\frac{x+y}{2}\right) + D^\alpha u(y)\right|^p}{|x - y|^{n+p}}\, dx\, dy\right)^{1/p}.$$

For $s < 0$ define $B^{s,p}(\mathbb{R}^n) = \big(B^{-s,p'}(\mathbb{R}^n)\big)'$. For non-integer $s > 0$ we have $B^{s,p}(\mathbb{R}^n) = H^{s,p}(\mathbb{R}^n)$. Note that for $p = 2$ we have $H^{s,p}(\mathbb{R}^n) = B^{s,p}(\mathbb{R}^n)$ for all $s \in \mathbb{R}$.

For a bounded domain Ω in $\mathbb{R}^n$ with smooth boundary and $s \geqq 0$ define

$$H^{s,p}(\Omega) = \{u|_\Omega : u \in H^{s,p}(\mathbb{R}^n)\}, \qquad B^{s,p}(\Omega) = \{u|_\Omega : u \in B^{s,p}(\mathbb{R}^n)\}$$

which are subspaces of $\mathcal{D}'(\Omega)$. The spaces

$$H_0^{s,p}(\overline{\Omega}) = \{u \in H^{s,p}(\mathbb{R}^n) : \operatorname{supp} u \subseteq \overline{\Omega}\},$$

$$B_0^{s,p}(\overline{\Omega}) = \{u \in B^{s,p}(\mathbb{R}^n),\ \operatorname{supp} u \subseteq \overline{\Omega}\}$$

are closed subspaces in $H^{s,p}(\mathbb{R}^n)$ and $B^{s,p}(\mathbb{R}^n)$, respectively. Then

$$H^{s,p}(\Omega) = H^{s,p}(\mathbb{R}^n)/H_0^{s,p}(\complement\Omega), \qquad B^{s,p}(\Omega) = B^{s,p}(\mathbb{R}^n)/B_0^{s,p}(\complement\Omega)$$

and $H^{s,p}(\Omega)$ and $B^{s,p}(\Omega)$ are considered in the factor norm. $C^\infty(\overline{\Omega})$ is a dense subspace both in $H^{s,p}(\Omega)$ and $B^{s,p}(\Omega)$.

For $s < 0$ set

$$H^{s,p}(\Omega) = \big(H_0^{-s,p'}(\overline{\Omega})\big)', \qquad B^{s,p}(\Omega) = \big(B_0^{-s,p'}(\overline{\Omega})\big)'.$$

For $0 \leqq s < 1/p$ the subspace $C_0^\infty(\Omega) \subset H^{s,p}(\Omega)$ $(\subset B^{s,p}(\Omega))$ is dense. Therefore,

$$H^{s,p}(\Omega) = H_0^{s,p}(\bar{\Omega}), \quad B^{s,p}(\Omega) = B_0^{s,p}(\bar{\Omega})$$

and

$$H^{s,p}(\Omega) = \left(H^{-s,p'}(\bar{\Omega})\right)', \quad B^{s,p}(\Omega) = \left(B^{-s,p'}(\bar{\Omega})\right)' \quad (0 \leqq s < 1/p) .$$

Let M be a closed compact manifold, E a vector bundle over M and φ_j a partition of unity corresponding to a (sufficiently fine) covering of M where local coordinates and local bases of sections of E are fixed. Define $H^{s,p}(X, E)$ $(B^{s,p}(X, E))$ as a subspace of $\mathcal{D}'(M, E)$ of those $u \in \mathcal{D}'(M, E)$ that $\varphi_j u$ in local coordinates belongs to $H^{s,p}(\mathbb{R}^n) \otimes \mathbb{C}^k((B^{s,p}(\mathbb{R}^n) \otimes \mathbb{C}^k)$ (k — the fibre dimension of E). It is easily seen that $H^{s,p}(X, E)$ and $B^{s,p}(X, E)$ are Banach spaces. Similar to Proposition 9 we have

Proposition 10. *Let X be a smooth manifold with boundary Y. Then the restriction operator $r' : C^\infty(X, E) \to C^\infty(Y, E')$ ($E' = E|_Y$) has a continuous extension*

$$r' : H^{s,p}(X, E) \to B^{s-1/p,p}(Y, E')$$

and

$$r' : B^{s,p}(X, E) \to B^{s-1/p,p}(Y, E') ,$$

respectively, for all $s \in \mathbb{R}$, $s > 1/p$. There are differential operators R_j, $j = 0, 1, \ldots, [s - 1/p]$ such that the mappings

$$H^{s,p}(X, E) \ni u \to \{r' R_j u\}_{j=0}^{[s-1/p]} \in \bigoplus_{j=0}^{[s-1/p]} B^{s,p}(Y, E')$$

and

$$B^{s,p}(X, E) \ni u \to \{r' R_j u\}_{j=0}^{[s-1/p]} \in \bigoplus_{j=0}^{[s-1/p]} B^{s,p}(Y, E') ,$$

respectively, are surjective and have continuous linear right inverses.

Below we shall have to use some facts from interpolation theory. Recall the terminology. Suppose that we are given Banach spaces A_0, A_1 continuously embedded in a linear Hausdorff space $\tilde{A}$, $A_0 \subset \tilde{A}$, $A_1 \subset \tilde{A}$. Then $\{A_0, A_1\}$ is called *interpolation pair*. Let $\{B_0, B_1\}$ be another one, $B_0 \subset \tilde{B}$, $B_1 \subset \tilde{B}$. Consider an arbitrary linear operator $T: \tilde{A} \to \tilde{B}$ such that the restriction to A_i yields a continuous operator $T: A_i \to B_i$ $(i = 0, 1)$. The Banach spaces $A \subset \tilde{A}$, $B \subset \tilde{B}$ are called *interpolation spaces* of the pairs $\{A_0, A_1\}$ and $\{B_0, B_1\}$ if the restriction of an arbitrary linear operator T with the above properties yields a continuous operator $T: A \to B$.

We shall refer to a special real interpolation method, i.e. a rule how to construct new spaces $A_\theta = (A_0, A_1)_\theta$, $0 < \theta < 1$ and $B_\theta = (B_0, B_1)_\theta$ out of the given ones. This is also called interpolation functor.

For details cf. TRIEBEL [1].

Proposition 11. *There is a real interpolation functor $(\cdot, \cdot)_\theta$, $0 < \theta < 1$, such that for any $-\infty < s_0 < s_1 < \infty$, $1 < p < \infty$ and any bounded domain Ω with smooth boundary (or $\Omega = \mathbb{R}^n$ or $\Omega = \mathbb{R}_+^n$) we have*

$$\left(H^{s_0,p}(\Omega), H^{s_1,p}(\Omega)\right)_\theta = B^{s,p}(\Omega) ,$$

$s = (1 - \theta) s_0 + \theta s_1$. For $s_0, s_1 \geqq 0$ with $s_0 - 1/p$, $s_1 - 1/p \notin \mathbb{Z}_+$ we have

$$\left(H_0^{s_0,p}(\bar{\Omega}), H^{s_1,p}(\Omega)\right)_\theta = B^{s,p}(\Omega) .$$

Recall two important properties of Sobolev spaces formulated in so-called *embedding theorems*.

Proposition 12. *Let X be compact and E be a vector bundle over X. Then the embedding*

$$H^{s,p}(X, E) \subset H^{s',p}(X, E)$$

is compact for $s > s'$.

Proposition 13. *For an arbitrary $k \in \mathbb{Z}_+$, $0 \leqq k < s - n/p$ there is an embedding*

$$H^{s,p}(X, E) \subset C^k(X, E)$$

in the sense of the choice of a representative in the equivalence class. The embedding is compact.

Note the following relation between compactly embedded spaces and norm estimates.

Proposition 14. *Let B_j be Banach spaces, with norms $\|\cdot\|_j$, $j = 1, 2, 3$, $B_1 \subset B_2 \subset B_3$. Assume that the embedding $B_1 \subset B_2$ is compact and the embedding $B_2 \subset B_3$ is continuous. Then, for any $\varepsilon > 0$, there is a constant $c(\varepsilon)$ such that*

$$\|u\|_2 \leqq \varepsilon \|u\|_1 + c(\varepsilon) \|u\|_3 \quad \textit{for all} \quad u \in B_1 .$$

1.2.2. Oscillatory Integrals

1.2.2.1. Amplitude Functions

In this section will be discussed a special class of distributions, the so called oscillatory integrals (or Fourier distributions). They have to be used in the theory of pseudo-differential operators, a special class of Fourier integral operators (cf. 1.2.2.4). For further literature cf. Hörmander [4, 6], Guillemin/Sternberg [1], Šubin [1].

Definition 1. Let $\Omega \subseteqq \mathbb{R}^n$ be an open set and $N \in \mathbb{Z}_+$. By $S^m(\Omega \times \mathbb{R}^N)$, $m \in \mathbb{R}$, we denote the set of all functions $a \in C^\infty(\Omega \times \mathbb{R}^N)$ so that for every compact set $K \subset \Omega$ and arbitrary multi-indices $\alpha \in \mathbb{Z}_+^N$, $\beta \in \mathbb{Z}_+^n$ there is a constant $c_{\alpha,\beta,K}$ with

$$|D_x^\beta D_\theta^\alpha a(x, \theta)| \leqq c_{\alpha,\beta,K}(1 + |\theta|)^{m-|\alpha|} . \tag{1}$$

The elements in $S^m(\Omega \times \mathbb{R}^N)$ are called *amplitude functions of order m* ($=$ ord a).

For abbreviation we sometimes write S^m instead of $S^m(\Omega \times \mathbb{R}^N)$. Put

$$S^\infty = \bigcup_{m \in \mathbb{R}} S^m, \qquad S^{-\infty} = \bigcap_{m \in \mathbb{R}} S^m .$$

$S^m(\Omega \times \mathbb{R}^N)$ is a Fréchet space with the system of semi-norms

$$p_{\alpha,\beta,K}^{(m)}(a) = \sup_{\substack{x \in K \\ \theta \in \mathbb{R}^N}} (1 + |\theta|)^{|\alpha|-m} |D_x^\beta D_\theta^\alpha a(x, \theta)| ,$$

α, β arbitrary multi-indices and K an arbitrary compact subset of Ω. Obviously

$$m \leqq m' \Rightarrow S^m \subseteqq S^{m'} , \tag{2}$$

$$a \in S^m \quad \Rightarrow D_x^\beta D_\theta^\alpha a \in S^{m-|\alpha|} \tag{3}$$

for arbitrary multi-indices α, β. From the Leibniz rule one obtains

$$a \in S^m, \ a' \in S^{m'} \Rightarrow aa' \in S^{m+m'} . \tag{4}$$

The embedding on the right hand side in (2) is continuous.

On bounded sets in $S^m(\Omega \times \mathbb{R}^N)$ the topology of pointwise convergence, the C^∞ topology and the topology induced by $S^{m'}(\Omega \times \mathbb{R}^N)$, $m' > m$ are equivalent. In fact, each of the mentioned topologies is not stronger than the following. From the estimate

$$\sup_{x\in K} |D_x^\beta D_\theta^\alpha a(x, \theta)| (1 + |\theta|)^{-m+|\alpha|}$$
$$\leqq \sup_{x\in K} |D_x^\beta D_\theta^\alpha a(x, \theta)| (1 + |\theta|)^{-m+|\alpha|} (1 + |\theta|)^{m-m'}$$

for $a \in S^m(\Omega \times \mathbb{R}^N)$, $K \subset \Omega$ a compact set, and Ascoli's theorem follows that on bounded sets in $S^m(\Omega \times \mathbb{R}^N)$ the $S^{m'}$ topology is not stronger than the topology of pointwise convergence.

A function $\varphi \in C^\infty(\mathbb{R}^N)$ is called *excision function* if φ is real valued and if there exist constants r, R $(0 < r < R)$ for which

$$\varphi(\theta) = \begin{cases} 0 & \text{if} \quad |\theta| \leqq r\,, \\ 1 & \text{if} \quad |\theta| \geqq R\,. \end{cases}$$

Usually we assume $R < 1$.

A function $a(x, \theta) \in C^\infty(\Omega \times (\mathbb{R}^N \setminus \{0\}))$ is called *positively homogeneous of degree* $m \in \mathbb{R}$ *(in* θ*)* if

$$a(x, t\theta) = t^m a(x, \theta) \quad \text{for all} \quad t > 0\,. \tag{5}$$

If (5) holds for $t \geqq 1$, $|\theta| \geqq c$ with a constant $c > 0$, the function a is called *positively homogeneous of degree* m *for large* $|\theta|$.

If $a(x, \theta) \in C^\infty(\Omega \times (\mathbb{R}^N \setminus \{0\}))$ is positively homogeneous of degree m and $\varphi(\theta)$ an excision function, φa is in S^m.

Theorem 2. *Let* $a_j \in S^{m_j}(\Omega \times \mathbb{R}^N)$ $(j = 0, 1, 2, \ldots)$ *and* $m_j \to -\infty$ *for* $j \to \infty$. *Then there exists an* $a \in S^m(\Omega \times \mathbb{R}^N)$, $m = \max\{m_i : j \geqq 0\}$ *with*

$$\operatorname{ord}\left(a - \sum_{j=0}^{k} a_j\right) \to -\infty \quad \text{for} \quad k \to \infty\,; \tag{6}$$

a is uniquely determined mod $S^{-\infty}(\Omega \times \mathbb{R}^N)$ *by the given sequence* a_j. *This remains true after an arbitrary rearranging of the sequence preserving the property "order of symbols* $\to -\infty$".

Write

$$a \sim \sum_{j=0}^{\infty} a_j\,. \tag{7}$$

The formal sum on the right is called *asymptotic expansion* of a.

Proof: Let t_j $(j = 0, 1, 2, \ldots)$ be a sequence of positive numbers, $t_j \to \infty$ for $j \to \infty$. Let φ be an excision function, $\varphi(\theta) = 0$ for $|\theta| \leqq \frac{1}{2}$ and $\varphi(\theta) = 1$ for $|\theta| \geqq 1$. Put

$$a(x, \theta) = \sum_{j=0}^{\infty} \varphi\left(\frac{\theta}{t_j}\right) a_j(x, \theta)\,. \tag{8}$$

The following considerations show that one can choose the t_j so fast increasing that (8) belongs to the space S^m.

Note first that $\varphi(\theta/t) \in S^0(\Omega \times \mathbb{R}^N)$ uniformly for $t \geqq 1$ (i.e. the estimates corresponding to S^0 hold uniformly in t). This follows from

$$\partial_\theta^\alpha \varphi\left(\frac{\theta}{t}\right) = (\partial_\theta^\alpha \varphi)\left(\frac{\theta}{t}\right) t^{-|\alpha|}$$

and $|\theta| \leqq t \leqq 2|\theta|$ for $\theta \in \operatorname{supp} \partial_\theta^\alpha \varphi$ $(\alpha \neq 0)$, i.e. $|\partial_\theta^\alpha \varphi(\theta/t)| \leqq c_\alpha(1 + |\theta|)^{-|\alpha|}$ with constants c_α.

Let Ω_k $(k = 0, 1, 2, \ldots)$ be a sequence of subdomains of Ω with $\overline{\Omega}_k \subset \Omega_{k+1}$ compact for all k and $\Omega = \bigcup_{k=0}^{\infty} \Omega_k$. Then for certain $c_j > 0$

$$\left|\partial_\theta^\alpha \partial_x^\beta \left(\varphi\left(\frac{\theta}{t}\right) a_j(x, \theta)\right)\right| \leqq c_j(1 + |\theta|)^{m_j - |\alpha|}$$

for all $x \in \overline{\Omega}_l$ and $|\alpha| + |\beta| + l \leqq j$, $t \geqq 1$. Without loss of generality suppose $m_{j+1} < m_j$ for all j. In view of

$$(1 + |\theta|)^{m_j} = (1 + |\theta|)^{m_{j-1}} (1 + |\theta|)^{m_j - m_{j-1}} \quad (j > 0)$$

it follows that

$$c_j(1 + |\theta|)^{m_j} < 2^{-j}(1 + |\theta|)^{m_{j-1}}$$

for $|\theta|$ sufficiently large. Since $\partial_\theta^\alpha \partial_x^\beta \varphi(\theta/t)\, a_j(x, \theta) = 0$ for $|\theta| \leqq t/2$, we obtain for $t = t_j$ sufficiently large

$$|\partial_\theta^\alpha \partial_\beta^\beta \varphi(\theta/t_j)\, a_j(x, \theta)| \leqq 2^{-j}(1 + |\theta|)^{m_{j-1} - |\alpha|}$$

for all $x \in \overline{\Omega}_j$, $|\alpha| + |\beta| + l \leqq j$. Thus (8) is convergent including all partial derivatives and

$$\left|\partial_\theta^\alpha \partial_x^\beta \left(\sum_{j=r+1}^{\infty} \varphi(\theta/t_j)\, a_j(x, \theta)\right)\right| \leqq 2^{-r}(1 + |\theta|)^{m_r - |\alpha|},$$

$x \in \overline{\Omega}_l$. Then obviously $a - \sum_{j=0}^{r} a_j \in S^{m_r}$, i.e. $a \sim \sum_j a_j$. The uniqueness of a mod $S^{-\infty}$ is an immediate consequence of the definition of asymptotic sums. □

Because of the relation (7) the definition of S^m contains all partial derivatives of $a - \sum_{j=0}^{r} a_j$ with respect to x, θ $(r \in \mathbb{Z}_+)$. The following theorem yields conditions under which consideration of partial derivatives can be avoided.

Theorem 3. *Let $a_j \in S^{m_j}(\Omega \times \mathbb{R}^N)$ be a given sequence with $m_j \to -\infty$ for $j \to \infty$ and $m = \max\{m_j : j \in \mathbb{Z}_+\}$. Let $a \in C^\infty(\Omega \times \mathbb{R}^N)$ be a function with the following properties:*

(i) *for all multi-indices α, β and any compact set $K \subset \Omega$ there exist constants $c = c_{\alpha,\beta,K}$, $\mu = \mu_{\alpha,\beta,K}$ with*

$$|\partial_\theta^\alpha \partial_x^\beta a(x, \theta)| \leqq c(1 + |\theta|)^\mu \quad \text{for all} \quad x \in K,$$

(ii) *for any compact set $K \subset \Omega$ there exist constants $\mu_l = \mu_l(K)$ $(l = 1, 2, \ldots)$ with $\mu_l \to -\infty$ for $l \to \infty$ and constants $c_l = c_l(K)$ with*

$$\left|a(x, \theta) - \sum_{j=0}^{l-1} a_j(x, \theta)\right| \leqq c_l(1 + |\theta|)^{\mu_l} \quad \textit{for all} \quad x \in K.$$

Then $a \in S^m(\Omega \times \mathbb{R}^N)$ and $a \sim \sum_{j=0}^{\infty} a_j$.

For the proof we need the following

Lemma 4. *Let $G \subseteq \mathbb{R}^M_y$ be an open set and $K_1, K_2 \subset G$ compact subsets with* int $K_1 \subset K_2$. *Then for all $f \in C^2(G)$ there exists a constant $c > 0$ so that*

$$\Big(\sup_{\substack{|\alpha|=1 \\ y\in K_1}} |\partial^\alpha f(y)|\Big)^2 \leqq c\Big(\sup_{y\in K_2} |f(y)|\Big)\Big(\sup_{\substack{|\alpha|=2 \\ y\in K_2}} |\partial^\alpha f(y)|\Big). \tag{9}$$

Proof: It is easily seen that the assertion follows from the following special result. Let $M = 1$ and $f \in C^2(\mathbb{R})$. Then $|f(t)| \leqq \lambda$, $|f''(t)| \leqq \gamma$ for all $t \in \mathbb{R}$ implies $|f'(t)|^2 \leqq c\lambda\gamma$ for all $t \in \mathbb{R}$ with some constant $c > 0$. In order to prove this, use that

$$f(t+\varepsilon) = f(t) + \varepsilon f'(t) + \tfrac{1}{2}\varepsilon^2 f''(\tau_1)$$

for $\varepsilon > 0$ with some τ_1, $t \leqq \tau_1 \leqq t + \varepsilon$, i.e.

$$f(t+\varepsilon) - f(t) = \varepsilon f'(t) + \tfrac{1}{2}\varepsilon^2 f''(\tau_1)\,.$$

Similarly

$$f(t-\varepsilon) - f(t) = -\varepsilon f'(t) + \tfrac{1}{2}\varepsilon^2 f''(\tau_2)$$

with some τ_2, $t - \varepsilon \leqq \tau_2 \leqq t$. Thus

$$f(t+\varepsilon) - f(t-\varepsilon) = 2\varepsilon f'(t) + \tfrac{1}{2}\varepsilon^2\big(f''(\tau_1) - f''(\tau_2)\big)\,,$$

i.e.

$$f'(t) = \frac{1}{2\varepsilon}\big(f(t+\varepsilon) - f(t-\varepsilon)\big) - \frac{1}{4}\varepsilon\big(f''(\tau_1) - f''(\tau_2)\big)\,,$$

hence $|f'(t)| \leqq \lambda/\varepsilon + \varepsilon\gamma/2$. The right hand side of the last inequality takes its minimum for $\varepsilon = (2\lambda/\gamma)^{1/2}$.

Proof of Theorem 3: In view of Theorem 2 there exists some $b \in S^m$ with $b \sim \sum_{j=0}^{\infty} a_j$. Put $g(x, \theta) = a(x, \theta) - b(x, \theta)$. Then for any compact set $K \subset X$ we have estimates

$$|\partial_\theta^\alpha \partial_x^\beta g(x, \theta)| \leqq c(1 + |\theta|)^\mu\,, \quad x \in K \tag{10}$$

with constants c, μ depending on α, β, K. Moreover

$$|g(x, \theta)| \leqq c_r(1 + |\theta|)^{-r}\,, \quad x \in K \tag{11}$$

for arbitrary $r \in \mathbb{Z}_+$ with a constant $c_r = c_r(K)$. The function $h(x, \theta, \eta) = g(x, \theta + \eta)$ satisfies

$$\partial_\eta^\alpha \partial_x^\beta h(x, \theta, \eta)|_{\eta=0} = \partial_\theta^\alpha \partial_x^\beta g(x, \theta)\,.$$

Apply now Lemma 4 with $K_1 = K \times \{0\}$ (0 is the origin in $\mathbb{R}^N_\eta$), $K_2 = K' \times \{|\eta| \leqq 1\}$, K' a compact set in Ω with $K' \subset$ int K. Using (10), (11) it follows that

$$\sup_{x\in K} \sum_{|\alpha|+|\beta|\leqq 1} |\partial_\theta^\alpha \partial_x^\beta g(x, \theta)| \leqq c(1 + |\theta|)^{-r}(1 + |\theta|)^\mu\,,$$

$r \in \mathbb{Z}_+$ arbitrary, $c = c(\alpha, \beta, K, r)$, $\mu = \mu(\alpha, \beta, K)$ constant. Thus the function $\partial_\theta^\alpha \partial_x^\beta g(x, \theta)$ is faster decreasing than any power of $1 + |\theta|$ for $|\theta| \to \infty$ for all $x \in K$, $|\alpha| + |\beta| \leqq 1$. Induction gives analogous properties for all α, β, i. e. $g \in S^{-\infty}$. □

Let $S_0^{-\infty}(\Omega \times \mathbb{R}^N)$ be the space of all $a \in C^\infty(\Omega \times \mathbb{R}^N)$ with $a(x, \theta) = 0$ for all $x \in K$, $|\theta| > c_K$ for each compact $K \subset \Omega$ ($c_K > 0$ a constant). Obviously $S_0^{-\infty} \subset S^{-\infty}$.

Proposition 5. *Let F be a Fréchet space and $l_0 \colon S_0^{-\infty}(\Omega \times \mathbb{R}^N) \to F$ a linear mapping which is continuous in the topology induced by $S^m(\Omega \times \mathbb{R}^N)$ for all $m \in \mathbb{R}$. Then there* *exists a uniquely determined extension of l_0 to a linear mapping $l: S^\infty(\Omega \times \mathbb{R}^N) \to F$ the restriction of which to $S^m(\Omega \times \mathbb{R}^N)$ is continuous for all $m \in \mathbb{R}$ (i.e. a continuous linear operator $S^m \to F$).*

Proof: A subset $A \subset S^m$ is bounded if $p^{(m)}_{\alpha,\beta,K}(a) \leqq c^{(m)}_{\alpha,\beta,K}$ for all $a \in A$ with constants $c^{(m)}_{\alpha,\beta,K}$ for all α, β, K as in (1). According to Ascoli's theorem on bounded sets in S^m the following topologies are equivalent: The topology of pointwise convergence, the topology induced by $C^\infty(\Omega \times \mathbb{R}^N)$ and the topology induced by $S^{m'}(\Omega \times \mathbb{R}^N)$ for arbitrary $m' > m$. Let $\chi(\theta) \in C_0^\infty(\mathbb{R}^N)$, $\chi(\theta) = 1$ in a neighbourhood of the origin in $\mathbb{R}^N$. Then $|\partial_\theta^\beta \chi(\varepsilon\theta)| \leqq c_\beta(1 + |\theta|)^{-|\beta|}$ with constants c_β independent of ε, $0 < \varepsilon \leqq 1$. Thus for any fixed $a \in S^m$ the functions $\chi(\varepsilon\theta)\, a(x, \theta)$, $0 < \varepsilon \leqq 1$ form a bounded set in S^m and

$$\chi(\varepsilon\theta)\, a(x, \theta) \to a(x, \theta) \quad \text{for} \quad \varepsilon \to 0 \tag{12}$$

with respect to pointwise convergence. This means that (12) is also valid with respect to $S^{m'}$-convergence for any $m' > m$. Thus l_0 has a uniquely determined extension $l_m: S^m \to F$ which is continuous in the topology induced by $S^{m'}$ in S^m, $m' > m$. Since the embedding $S^m \to S^{m'}$ is continuous for $m' > m$, l_m is continuous with respect to the S^m-topology, too. This proves Proposition 5. □

A subset Γ of $\Omega \times \mathbb{R}^N$ is called *conic* if $(x, \theta) \in \Gamma$ implies $(x, t\theta) \in \Gamma$ for all $t > 0$. Let $\Gamma \subseteqq \Omega \times \mathbb{R}^N$ be an open conic set. Denote by $S^m(\Gamma)$ the set of all $a \in S^\infty(\Gamma)$ satisfying the estimates of the form (1) for all $(x, \theta) \in K^c$ and all compact $K \subset \Gamma$; here $K^c = \{(x, t\theta): (x, \theta) \in K, t \geqq 1\}$. In the case of $\Gamma = \Omega \times \mathbb{R}^N$ this is equivalent to Definition 1.

For $\Gamma \subseteqq \Omega \times (\mathbb{R}^N \setminus \{0\})$ functions in $S^m(\Gamma)$ are in general not bounded for small $|\theta|$ ($\mathbb{R}^N \setminus \{0\}$ means as usual $\{\theta \in \mathbb{R}^N: \theta \neq 0\}$). We are interested in the behaviour of $a \in S^m(\Gamma)$ only for large $|\theta|$. Therefore we suppose sometimes $a(x, \theta) = 0$ for $|\theta| \leqq$ const. With respect to the space $S^m(\Gamma)$ analogous propositions can be proved as for $S^m(\Omega \times \mathbb{R}^N)$. Their formulation is left to the reader. Therefore, in the following the case $\Gamma = \Omega \times \mathbb{R}^N$ shall be mainly considered.

If $\Gamma \subseteqq \Omega \times \mathbb{R}^N$ is a conic set, we put $t(x, \theta) = (x, t\theta)$ for $(x, \theta) \in \Gamma$, $t > 0$. Thus $\mathbb{R}_+$ acts on Γ as a group of mappings. By $(x, \theta) \to x$ a projection $\Gamma \to \Omega$ is defined.

For the sake of completeness mention the following result without proof (cf. HÖRMANDER [6]).

Proposition 6. *Let $\Gamma_i \subseteqq \Omega_i \times (\mathbb{R}^N \setminus \{0\})$ be open conic sets ($\Omega_i \subseteqq \mathbb{R}^n$ open, $i = 1, 2$) and $\chi: \Gamma_1 \to \Gamma_2$ a C^∞ map commuting with the action of $\mathbb{R}_+$ (i.e. $\chi(x, t\theta) = t\chi(x, \theta)$ for all $t > 0$) and χ fibre preserving (i.e. the projection of $\chi(x, \theta)$ onto Ω_2 depends only on $x \in \Omega_1$). Then*

$$\chi^*: S^m(\Gamma_2) \to S^m(\Gamma_1) \tag{13}$$

for $m \in \mathbb{R}$.

If $\Gamma \subset \Omega \times (\mathbb{R}^N \setminus \{0\})$ is an open conic set and $U = (\Omega \times \{\theta: |\theta| = 1\}) \cap \Gamma$, then $\Gamma \to U \times \mathbb{R}_+$ defined by $(x, \theta) \to (x, \theta/|\theta|, |\theta|)$ is a diffeomorphism inducing an isomorphism $S^m(U, \mathbb{R}_+) \cong S^m(\Gamma)$ ($\mathbb{R}_+$ is considered as an open conic subset in $\mathbb{R}$).

1.2.2.2. Phase Functions

Definition 1. Let $\Omega \subseteqq \mathbb{R}^n$ be open and $\Gamma \subseteqq \Omega \times (\mathbb{R}^N \setminus \{0\})$ an open conic set. A function $\varphi \in C^\infty(\Gamma)$ is called *(homogeneous) phase function* if φ has the following properties:

(i) φ is real valued;

(ii) $\varphi(x, t\theta) = t\varphi(x, \theta)$ for all $t > 0$, $(x, \theta) \in \Gamma$;

(iii) $\sum_{i=1}^{n} \left|\frac{\partial\varphi}{\partial x_i}(x, \theta)\right|^2 + \sum_{j=1}^{N} \left|\frac{\partial\varphi}{\partial\theta_j}(x, \theta)\right|^2 \neq 0$ for all $(x, \theta) \in \Gamma$.

The variable θ is called *frequency variable*.

Set for abbreviation

$$\varphi_x' = \left(\frac{\partial\varphi}{\partial x_1}, \dots, \frac{\partial\varphi}{\partial x_n}\right), \qquad \varphi_\theta' = \left(\frac{\partial\varphi}{\partial\theta_1}, \dots, \frac{\partial\varphi}{\partial\theta_N}\right)$$

and similarly φ_{xx}'', $\varphi_{x\theta}''$, ... for matrices of corresponding second derivatives. For simplicity consider mainly the case $\Gamma = \Omega \times (\mathbb{R}^N \setminus \{0\})$. The general case corresponds to the assumption that the amplitude functions have support in Γ. If L is a differential operator, denote by ${}^tL$ the formal adjoint of L with respect to a standard bilinear pairing $\langle\cdot, \cdot\rangle$.

Lemma 2. *Let φ be a phase function in $\Gamma = \Omega \times (\mathbb{R}^N \setminus \{0\})$. Then there exists a differential operator of first order*

$$L = \sum_{j=1}^{N} a_j(x, \theta) \frac{\partial}{\partial\theta_j} + \sum_{k=1}^{n} b_k(x, \theta) \frac{\partial}{\partial x_k} + c(x, \theta) \tag{1}$$

with

$${}^tL\, e^{i\varphi} = e^{i\varphi} \tag{2}$$

and

$$a_j \in S^0(\Omega \times \mathbb{R}^N) \quad (j = 1, \dots, N), \qquad b_k, c \in S^{-1}(\Omega \times \mathbb{R}^N) \quad (k = 1, \dots, n). \tag{3}$$

Proof: First construct an operator

$$M_0 = \sum_{j=1}^{N} \alpha_j \frac{\partial}{\partial\theta_j} + \sum_{k=1}^{n} \beta_k \frac{\partial}{\partial x_k}$$

with $M_0\, e^{i\varphi} = e^{i\varphi}$ in Γ. Obviously

$$M_0\, e^{i\varphi} = \left(i \sum_{j=1}^{N} \alpha_j \frac{\partial\varphi}{\partial\theta_j} + i \sum_{k=1}^{n} \beta_k \frac{\partial\varphi}{\partial x_k}\right) e^{i\varphi}.$$

The sum $\omega = \sum_j |\theta|^2 (\partial\varphi/\partial\theta_j)^2 + \sum_k (\partial\varphi/\partial x_k)^2$ is homogeneous in θ of degree 2. In view of (iii) in Definition 1 $\omega \neq 0$. Thus we can set $\alpha_j = (i\omega)^{-1} |\theta|^2 \partial\varphi/\partial\theta_j$, $\beta_k = (i\omega)^{-1} \partial\varphi/\partial x_k$ and obviously $\alpha_j \in S^0$, $\beta_k \in S^{-1}$ for large $|\theta|$. Let now $\chi(\theta)$ be an excision function. Then the operator $M = \chi M_0 + (1 - \chi)$ has the property $M\, e^{i\varphi} = e^{i\varphi}$, too, and $L = {}^tM$ has the desired properties, since

$$a_j = -\chi\alpha_j \in S^0(\Omega \times \mathbb{R}^N), \qquad b_k = -\chi\beta_k \in S^{-1}(\Omega \times \mathbb{R}^N),$$

$$c = (1 - \chi) - \sum_{j=1}^{N} \frac{\partial(\chi\alpha_j)}{\partial\theta_j} - \sum_{k=1}^{n} \frac{\partial(\chi\beta_k)}{\partial x_k} \in S^{-1}(\Omega \times \mathbb{R}^N). \quad \square$$

Define the set

$$C_\varphi = \{(x, \theta) \in \Omega \times (\mathbb{R}^N \setminus \{0\}) : \varphi_\theta'(x, \theta) = 0\}. \tag{4}$$

Using Euler's homogeneity relation $\varphi(x, \theta) = \sum_{j=1}^{N} \theta_j \frac{\partial\varphi}{\partial\theta_j}(x, \theta)$ one obtains

$$\varphi(x, \theta) = 0 \quad \text{for} \quad (x, \theta) \in C_\varphi . \tag{5}$$

Definition 3. A phase function φ is called *non-degenerate* if the vectors $d_{x,\theta}(\partial\varphi/\partial\theta_j)$ ($j = 1, \dots, N$) are linearly independent for all $(x, \theta) \in C_\varphi$ ($d_{x,\theta}\psi$ means the vector of partial derivatives $(\partial\psi/\partial x_1, \dots, \partial\psi/\partial x_n, \partial\psi/\partial\theta_1, \dots, \partial\psi/\partial\theta_N)$).

Lemma 4. *Let φ be a non-degenerate phase function on $\Omega \times (\mathbb{R}^N \setminus \{0\})$. Then C_φ is a conic C^∞ manifold of dimension n.*

Proof: Since $d_{x,\theta}(\partial\varphi/\partial\theta_j) \neq 0$ in a neighbourhood of C_φ ($j = 1, \dots, N$), $C_\varphi^j = \{(x, \theta): \partial\varphi/\partial\theta_j \neq 0\}$ is a C^∞ manifold of dimension $N + n - 1$. The assumption on φ implies that the C_φ^j ($j = 1, \dots, N$) intersect each other transversally, i.e. $C_\varphi = C_\varphi^1 \cap \dots \cap C_\varphi^N$ is a C^∞ manifold of dimension $(N + n) - N = n$. That C_φ is conic is an immediate consequence of the homogeneity of φ. □

Given a phase function define the conic set

$$\Lambda_\varphi = \{(x, \xi) \in T^*\Omega \setminus \{0\} : \xi = \varphi_x'(x, \theta), (x, \theta) \in C_\varphi\} .$$

Thus we have a map

$$t_\varphi: C_\varphi \to \Lambda_\varphi, \quad t_\varphi(x, \theta) = (x, \varphi_x'(x, \theta)) , \tag{6}$$

homogeneous in θ of degree 1, i.e. $\varphi_x'(x, t\theta) = t\varphi_x'(x, \theta)$ for $t > 0$.

Theorem 5. *Let φ be a non-degenerate phase function. Then Λ_φ is locally a conic C^∞ submanifold of $T^*\Omega \setminus \{0\}$ and (6) is locally a diffeomorphism.*

Proof: It is sufficient to show that the map t_φ has an injective differential at all points $(x, \theta) \in C_\varphi$. Denote by $(\delta x, \delta\theta)$ and $(\delta x, \delta\xi)$ tangent vectors of $\Omega \times \mathbb{R}^N$ and $T^*\Omega = \Omega \times \mathbb{R}^N$, respectively. The condition $v = (\delta x, \delta\theta) \in T_{(x,\theta)}C_\varphi$ is equivalent to $\varphi_{\theta x}''\delta x + \varphi_{\theta\theta}''\delta\theta = 0$ in (x, θ).

The differential of t_φ is $dt_\varphi: (\delta x, \delta\theta) \to (\delta x, \varphi_{xx}''\delta x + \varphi_{x\theta}''\delta\theta)$. Thus $dtv = 0$ is equivalent to $\delta x = 0$, $\varphi_{\theta\theta}''\delta\theta = \varphi_{x\theta}''\delta\theta = 0$. Since φ is non-degenerate, the last property implies $\delta\theta = 0$, i e. $v = 0$. □

The bilinear form

$$\sigma(w_1, w_2) = \langle\delta\xi^1, \delta x^2\rangle - \langle\delta x^1, \delta\xi^2\rangle \tag{7}$$

($w_j = (\delta x^j, \delta\xi^j) \in T(T^*\Omega)$, $j = 1, 2$) is called *symplectic form* on $T^*\Omega$ (σ is obviously antisymmetric and non-degenerate). An n-dimensional submanifold $\Lambda \subset T^*\Omega$ is called *Lagrangean manifold* if (7) vanishes for arbitrary vectors $w_1, w_2 \in T_\varrho\Lambda$ and all $\varrho \in \Lambda$. If e.g. $Y \subseteqq \Omega$ is a smooth submanifold of dimension $k \leqq n$, then

$$N^*Y = \{(y, \eta) \in T^*\Omega: y \in Y, \eta|_{T_yY} = 0\}$$

is a (conic) Lagrangean manifold. N^*Y is called *conormal bundle* of Y.

Concerning further notions of the symplectic geometry cf. Duistermaat [1], Guillemin/Sternberg [1], Maslov/Fedorjuk [1].

Theorem 6. *Let $\Gamma \subseteqq \Omega \times (\mathbb{R}^N \setminus \{0\})$ be an open conic set and φ a non-degenerate phase function in Γ. Then $\Lambda_\varphi \subset T^*\Omega \setminus \{0\}$ is a conic Lagrangean manifold (if necessary Γ is assumed to be sufficiently small).*

Proof: Let $\varrho \in \Lambda$ be fixed and $w_k = (\delta x^k, \delta\xi^k) \in T_\varrho\Lambda$ $(k = 1, 2)$. Then we have representations $\delta\xi^k = \varphi''_{xx}\delta x^k + \varphi''_{x\theta}\delta\theta^k$, where

$$\varphi''_{\theta x}\delta x^k + \varphi''_{\theta\theta}\delta\theta^k = 0 \tag{8}$$

$(k = 1,2)$. Using (8) and the symmetry of the Hessian we obtain

$$\begin{aligned}\sigma(w_1, w_2) &= \langle\varphi''_{xx}\,\delta x^1 + \varphi''_{x\theta}\,\delta\theta^1, \delta x^2\rangle - \langle\delta x^1, \varphi''_{xx}\,\delta x^2 + \varphi''_{x\theta}\,\delta\theta^2\rangle \\ &= \langle\delta\theta^1, \varphi''_{\theta x}\,\delta x^2\rangle - \langle\delta\theta^2, \varphi''_{\theta x}\,\delta x^1\rangle \\ &= -\langle\delta\theta^1, \varphi''_{\theta\theta}\,\delta\theta^2\rangle - \langle\delta\theta^2, \varphi''_{\theta\theta}\,\delta\theta^1\rangle = 0\,. \quad \square\end{aligned}$$

Remark 7. Conversely it can be proved that every conic Lagrangean manifold $\Lambda \subset T^*\Omega \setminus \{0\}$ in an open conic neighbourhood of any point $\varrho \in \Lambda$ can be described in the form Λ_φ with a non-degenerate phase function φ (the number of frequency variables N and the open conic set $\Gamma \subset \Omega \times (\mathbb{R}^N \setminus \{0\})$ are to be chosen in a suitable way). One can always find local coordinates in Ω so that φ has the form $\varphi(x, \xi) = \langle x,\xi\rangle - H(\xi)$, where $H(\xi) \in C^\infty(\mathbb{R}^n_\xi \setminus \{0\})$ is real and positively homogeneous of degree 1 (the frequency variable is here $\xi = (\xi_1, \dots, \xi_n)$). One can prove that if $n - k$ is the rank of the projection $\pi: \Lambda \to \Omega$ at $\varrho \in \Lambda$ (i.e. rank of the differential), one can find a phase function $\varphi_1(x, \eta)$ in an open conic set $\Gamma \subset U \times (\mathbb{R}^k \setminus \{0\})$, $U \subset \Omega$ open, so that $V \cap \Lambda = \Lambda_{\varphi_1}$ for an open conic neighbourhood V of ϱ (cf. HÖRMANDER [6]).

Remark 7 shows that a phase function φ with $\Lambda = \Lambda_\varphi$ in a conic neighbourhood of a point $\varrho \in \Lambda$ is not uniquely determined. This gives rise to the following

Definition 8. Let $\Gamma_\theta \subset \Omega \times (\mathbb{R}^N_\theta \setminus \{0\})$ be an open conic set, $\varphi(x, \theta)$ a non-degenerate phase function in Γ_θ. Let $(x_0, \theta_0) \in C_\varphi$, $\varrho_0 = (x_0, \xi_0) = t_\varphi(x_0, \theta_0)$. Moreover let $\Gamma_\Sigma \subset \Omega \times (\mathbb{R}^M_\Sigma \setminus \{0\})$ be open, conic, and $\chi(x, \Sigma)$ a non-degenerate phase function in Γ_Σ, $(x_0, \Sigma_0) \in C_\chi$, $\varrho_0 = t_\chi(x_0, \Sigma_0)$. Then φ, χ are called *equivalent* at the corresponding points (x_0, θ_0) and (x_0, Σ_0), respectively, if there are open neighbourhoods C'_φ and C'_χ of these points in C_φ and C_χ, respectively, for which $t_\varphi(C'_\varphi) = t_\chi(C'_\chi)$.

For instance, let $\varphi(x, \theta)$ be a non-degenerate phase function in an open conic set $\Gamma_\theta \subset \Omega \times (\mathbb{R}^N \setminus \{0\})$. If $Q(\eta, \eta)$ is a non-degenerate quadratic form in $\mathbb{R}^k$, then by $\varphi_1(x, \theta, \eta) = \varphi(x, \theta) + |\theta|^{-1} Q(\eta, \eta)$ a non-degenerate phase function with respect to the frequency variable (θ, η) is defined and obviously $\Lambda_\varphi = \Lambda_{\varphi_1}$ in a conic neighbourhood of the considered point $\varrho \in \Lambda$.

We are mainly interested in linear phase functions (i.e. linear in the frequency variable). Let $Y \subset \Omega$ be a submanifold of dimension k. For any $y_0 \in Y$ we find a neighbourhood $U \in y_0$ in Ω and functions $\psi_j \in C^\infty(U)$ with linearly independent differentials $d_x\psi_j$ over U $(j = 1, \dots, n - k)$ so that $Y \cap U = \{x \in U: \psi_j = 0, j = 1, \dots, n - k\}$. Then obviously the conormal bundle to Y is described by

$$N^*(Y \cap U) = \{(y, (\psi_1)'_x(y), \dots, (\psi_{n-k})'_x(y)) : y \in Y \cap U\}\,.$$

Putting $\Gamma = U \times (\mathbb{R}^{n-k} \setminus \{0\})$, $\varphi(x, \theta) = \sum_{j=1}^{n-k} \psi_j(x)\,\theta_j$ is a non-degenerate phase function in Γ and $N^*(Y \cap U) = \Lambda_\varphi$.

Conversely, one can prove the following (cf. DUISTERMAAT [1]). If Λ is an arbitrary conic Lagrangean submanifold of $T^*\Omega \setminus \{0\}$ and $W \subset \Lambda$ an open conic neighbourhood so that the projection $\pi: W \to \Omega$ has constant rank (say $n - k$), then W is part of the conormal bundle of a corresponding submanifold $Y \subset \Omega$ of dimension k.

Example 9. Let $\Omega = \mathbb{R}^n \times \mathbb{R}^n$ and

$$\varphi(x, y, \xi) = \langle x - y, \xi\rangle\,. \tag{9}$$

Then φ is a non-degenerate linear phase function in $(\mathbb{R}^n \times \mathbb{R}^n) \times (\mathbb{R}^n \setminus \{0\})$ and $\Lambda_\varphi = N^*(\operatorname{diag} \mathbb{R}^n)$, $\operatorname{diag} \mathbb{R}^n = \{(x, x): x \in \mathbb{R}^n\}$; $N^*(\ldots)$ denotes as above the conormal bundle of the manifold in the brackets.

1.2.2.3. Oscillatory Integrals

Let $\varphi(x, \theta)$ be a phase function in $\Omega \times (\mathbb{R}^N \setminus \{0\})$. Consider the integral

$$\langle A, u\rangle = \int_\Omega \int_{\mathbb{R}^N} e^{i\varphi(x,\theta)} a(x, \theta)\, u(x)\, d\theta\, dx\,. \tag{1}$$

(1) exists in the usual sense for all $a \in S^m(\Omega \times \mathbb{R}^N)$ with $m + N < 0$ and all $u \in C_0^\infty(\Omega)$.

The following considerations are valid in the more general case if φ is given in an open conic set $\Gamma \subset \Omega \times (\mathbb{R}^N \setminus \{0\})$ and if a is supported by K^c for some compact set $K \subset \Gamma$. For simplicity we restrict ourselves to

Theorem 1. *Let φ be a phase function. Then* (1) *considered for $a \in S_0^{-\infty}(\Omega \times \mathbb{R}^N)$ has a uniquely determined extension for all $a \in S^\infty(\Omega \times \mathbb{R}^N)$, $u \in C_0^\infty(\Omega)$, continuously depending on $a \in S^m(\Omega \times \mathbb{R}^N)$ for any fixed $a \in S^m(\Omega \times \mathbb{R}^N)$. For given fixed $a \in S^m(\Omega \times \mathbb{R}^N)$ with $m + N < k \in \mathbb{Z}_+$ the linear form $u \to \langle A, u\rangle$ represents a distribution over Ω of order $\leqq k$.*

Proof: In order to construct the extension of (1) to S^∞ check the conditions of 1.2.2.1, Proposition 5 ($F = \mathbb{C}$). Let $u \in C_0^\infty(\Omega)$, $a \in S_0^{-\infty}$ be fixed. In view of 1.2.2.2, Lemma 2 we obtain by partial integration

$$\iint e^{i\varphi(x,\theta)} a(x, \theta)\, u(x)\, dx\, d\theta = \iint ({}^tL\, e^{i\varphi(x,\theta)})\, a(x, \theta)\, u(x)\, dx\, d\theta$$
$$= \iint e^{i\varphi(x,\theta)} L\big(a(x, \theta)\, u(x)\big)\, dx\, d\theta = \ldots = \iint e^{i\varphi(x,\theta)} L\big(a(x, \theta)\, u(x)\big)\, dx\, d\theta\,. \tag{2}$$

L^k regarded as operator $L^k\colon S^m(\Omega \times \mathbb{R}^N) \to S^{m-k}(\Omega \times \mathbb{R}^N)$ is continuous. For $k > m + N$ the right hand side of (2) depends continuously on a in the topology induced by S^m in $S_0^{-\infty}$ and exists for all $a \in S^m$ in the classical sense. Applying 1.2.2.1, Proposition 5 we obtain the first part of the assertion. Thus we have also proved

$$\langle A, u\rangle = \lim_{\varepsilon \to 0} \iint e^{i\varphi(x,\theta)} a(x, \theta)\, \chi(\varepsilon\theta)\, u(x)\, dx\, d\theta$$
$$= \iint e^{i\varphi(x,\theta)} L^k\big(a(x, \theta)\, u(x)\big)\, dx\, d\theta \tag{3}$$

(cf. 1.2.2.1.(12)) for arbitrary $a \in S^m(\Omega \times \mathbb{R}^N)$. This implies for $m + N < k$ that (1) defines a distribution of order $\leqq k$. □

The integral (1) shall always be interpreted in the sense of the extension (3). Then (1) is called *oscillatory integral* (or *Fourier distribution*) with the phase function φ and the amplitude a.

A consequence of the definition of oscillatory integrals is that the usual rules of integration theory (such as Fubini's theorem or partial integration) are valid for oscillatory integrals, too. Namely, they are valid for $a \in S_0^{-\infty}$ and follow for $a \in S^\infty$ because of the uniqueness of the extension.

If $a \in S^{-\infty}$, A is a distribution with C^∞ density over Ω. Replacing $a \in S^m$ by $a_1 \in S^m$ with $a - a_1 = 0$ for $|\theta| \geqq$ const. we thus obtain $A - A_1 \in C^\infty(\Omega)$ for the corresponding Fourier distributions. In applications we are mainly interested in equivalence classes of A mod C^∞. Therefore the values of amplitude functions for small $|\theta|$ are not essential.

Let $W \subseteq \mathbb{R}^q$ be an open set and $\varphi\colon W \times \Omega \times (\mathbb{R}^N \setminus \{0\}) \to \mathbb{R}$ be a C^∞ function which is for every fixed $z \in W$ a phase function with respect to $(x, \theta) \in \Omega \times (\mathbb{R}^N \setminus \{0\})$. We claim that then

$$\iint e^{i\varphi(z,x,\theta)}\, a(z, x, \theta)\, u(x)\, dx\, d\theta \in C^\infty(W) \tag{4}$$

$\left(a \in S^m(W \times \Omega \times \mathbb{R}^N)\right)$. Similarly as in the proof of Theorem 1 (4) can be written in the form

$$\iint e^{i\varphi(z,x,\theta)}\, L^k(z, x, \theta, D_x, D_\theta)\, \big(a(z, x, \theta)\, u(x)\big)\, dx\, d\theta = \iint e^{i\varphi(z,x,\theta)}\, b(z, x, \theta)\, dx\, d\theta \tag{5}$$

for some $b \in S^{m-k}(W \times \Omega \times \mathbb{R}^N)$, $m + N < k$. Now, under the sign of integration it can be differentiated with respect to z because of $D_z^\gamma b \in S^{m-k}$ for all multi-indices γ. It is easily checked that the equation (5) defines a continuous operator $C_0^\infty(\Omega) \to C^\infty(W)$.

Theorem 2. *Let R_φ be the image of C_φ in Ω under the projection $\Omega \times (\mathbb{R}^N \setminus \{0\})$. For a given Fourier distribution*

$$A\colon u \to \iint e^{i\varphi(x,\theta)}\, a(x, \theta)\, u(x)\, dx\, d\theta \tag{6}$$

$\left(a \in S^m(\Omega \times \mathbb{R}^N)\right)$ *we have*

$$\operatorname{sing\,supp} A \subseteq R_\varphi\,, \tag{7}$$

i.e. $A|_{\Omega \setminus R_\varphi} \in C^\infty(\Omega \setminus R_\varphi)$.

Proof: In $\Omega \setminus R_\varphi$ we can consider φ as a phase function with respect to θ and x as a parameter . Then the above remark shows

$$\int e^{i\varphi(x,\theta)}\, a(x, \theta)\, d\theta \in C^\infty(\Omega \setminus R_\varphi)\,. \quad \square$$

If a vanishes in an open conic neighbourhood of C_φ, the corresponding Fourier distribution A belongs to $C^\infty(\Omega)$. This is a consequence of Theorem 2.

Theorem 3. *Let φ be a non-degenerate phase function in $\Omega \times (\mathbb{R}^N \setminus \{0\})$, $a \in S^m(\Omega \times \mathbb{R}^N)$ and (1) the corresponding Fourier distribution. If a vanishes on C_φ, there exists a $b \in S^{m-1}(\Omega \times \Omega^N)$ with*

$$\iint e^{i\varphi(x,\theta)}\, a(x, \theta)\, u(x)\, dx\, d\theta = \iint e^{i\varphi(x,\theta)}\, b(x, \theta)\, u(x)\, dx\, d\theta$$

for all $u \in C_0^\infty(\Omega)$, i.e. ord $A \leq k - 1$ *for $m + N < k$. If a vanishes on C_φ of infinite order, then $A \in C^\infty(\Omega)$.*

For the proof we need the following

Lemma 4. *Let $\Gamma \subset \Omega \times (\mathbb{R}^N \setminus \{0\})$ be an open conic set and $\psi_j(x, \theta) \in C^\infty(\Gamma)$ $(j = 1, \dots, k)$ positively homogeneous functions in θ of degree 0. Suppose that the differentials $d\psi_j(x, \theta)$ $(j = 1, \dots, k)$ are linearly independent on the set $C = \{(x, \theta) \in \Gamma\colon \psi_j(x, \theta) = 0,\ j = 1, \dots, k\}$. If $a \in S^m(\Gamma)$, $a|_C = 0$ and $a = 0$ for small $|\theta|$, then there exist functions $a_j \in S^m(\Gamma)$ with $a = \sum_{j=1}^{k} a_j\psi_j$. If a vanishes on C of infinite order, then the a_j vanish on C of infinite order, too.*

Proof: Note first that it is sufficient to find an open conic neighbourhood of any (x_0, θ_0) for which the assertion holds. Such conic neighbourhoods then form an open covering of Γ and one can choose a locally finite subcovering $(\Gamma_l)_{l=1,2,\dots}$. Let

$(\delta_l(x, \theta))_{l=1,2,\dots}$ be a corresponding partition of unity and δ_l positively homogeneous of degree 0 in θ. It is left to the reader to check that the homogeneity condition for the δ_l can be ensured. If

$$a = \sum_{j=1}^{k} a_j^l \psi_j \tag{8}$$

is a representation with the asserted property over Γ_l, then

$$a = \sum_{j=1}^{k} \left\{ \sum_l \delta_l a_j^l \right\} \psi_l$$

is a representation over Γ as desired. For $(x_0, \theta_0) \notin C$ one can easily find an open conic neighbourhood so that a representation (8) holds (for some j_0 we have $\psi_{j_0}(x_0, \theta_0) \neq 0$ and it is sufficient to set $a_{j_0}^l = a/\psi_{j_0}$, $a_j^l = 0$ for $j \neq j_0$). Suppose now $(x_0, \theta_0) \in C$. The differentials of the functions $\psi_1, \dots, \psi_k$ are linearly independent in $(x_0, \theta_0/|\theta_0|)$. Thus there exist (in view of the implicit function theorem) homogeneous functions of degree zero $\psi_{k+1}, \dots, \psi_{n+N-1}$ vanishing in $(x_0, \theta_0/|\theta_0|)$, so that $\psi_1, \dots, \psi_{n+N-1}$ in a neighbourhood of $(x_0, \theta_0/|\theta_0|) \ni \Gamma \cap \{(x, \theta) : |\theta| = 1\}$ form a local coordinate system. The map

$$(x, \theta) \to (\varphi_1, \dots, \varphi_{n+N-1}, |\theta|) \tag{9}$$

is then a homogeneous diffeomorphism of an open conic neighbourhood $\Gamma_{l_0} \in (x_0\ \theta_0)$ onto a set of the form $V \times \mathbb{R}_+$ for an open ball V in $\mathbb{R}^{n+N-1}$ with the center at the origin. In view of 1.2.2.1, Proposition 6 and the following remark the space $S^m(\Gamma_{l_0})$ is transformed to $S^m(V \times \mathbb{R}_+)$. Thus the assertion is reduced to the formula $a = \sum_{j=1}^{k} a_j y_j$ if one substitutes (9). By construction a vanishes for $y_1 = \dots = y_k = 0$. Taylor's formula gives

$$a = \sum_{j=1}^{k} y_j \int_0^1 \frac{\partial a}{\partial y_j} (ty_1, \dots, ty_k, y_{k+1}, \dots, y_{n+N-1}, |\theta|)\, dt\,.$$

Since $\partial a/\partial y_j \in S^m$, Lemma 4 is proved. □

Proof of Theorem 3: Application of Lemma 4 for $\psi_j = \partial\varphi/\partial\theta_j$ $(j = 1, \dots, N)$ yields

$$\iint e^{i\varphi(x,\theta)}\, a(x, \theta)\, u(x)\, dx\, d\theta = \iint e^{i\varphi(x,\theta)} \left(\sum_{j=1}^{N} a_j \frac{\partial\varphi}{\partial\theta_j} \right) u(x)\, dx\, d\theta$$
$$= \iint e^{i\varphi(x,\theta)} \left(\sum_{j=1}^{N} i \frac{\partial a_j}{\partial\theta_j} \right) u(x)\, dx\, d\theta\,.$$

The last equation is understood in the sense of oscillatory integrals, i.e. as unique extension of corresponding identities for $a \in S_0^{-\infty}$ following from $\iint \partial/\partial\theta_j\, (ae^{i\varphi})\, u\, dx\, d\theta = 0$. This proves the first assertion. If now a vanishes on C_φ of infinite order, all a_j vanish on C_φ of infinite order. Therefore one can apply the consideration to $b = b_1 = i \sum_{j=1}^{N} \partial a_j/\partial\theta_j \in S^{m-1}$. The resulting $b_2 \in S^{m-2}$ satisfies again the conditions and so on. Thus for each $k \in \mathbb{Z}_+$ one can construct an amplitude function $b_k \in S^{m-k}$ so that A can be written as Fourier distribution with amplitude b_k and phase function φ. Thus $A \in C^l(\Omega)$ for all $l \in \mathbb{Z}_+$, i.e. $A \in C^\infty(\Omega)$. □

Remark 5. One can prove (cf. HÖRMANDER [6], DUISTERMAAT [1]) that a Fourier distribution A of the form (1) satisfies $WF(A) \subseteq \Lambda_\varphi$.

If $(x_0, \theta_0) \in C_\varphi$ and $(x_0, \xi_0) = t_\varphi(x_0, \theta_0)$, there exists an amplitude function $a(x, \theta)$ with $(x_0, \xi_0) \in \mathrm{WF}(A)$. It is sufficient to choose some a which is positively homogeneous of degree zero and does not vanish in a conic neighbourhood of (x_0, θ_0).

1.2.2.4. Fourier Integral Operators

Let $U \subseteq \mathbb{R}^p$, $V \subseteq \mathbb{R}^q$ be open sets. Given a phase function $\varphi(x, y, \theta)$ over $U \times V \times (\mathbb{R}^N \setminus \{0\})$ and an amplitude function $a(x, y, \theta) \in S^m(U \times V \times \mathbb{R}^N)$ we can consider the oscillatory integral

$$\langle A, w\rangle = \iiint e^{i\varphi(x, y, \theta)}\, a(x, y, \theta)\, w(x, y)\, dx\, dy\, d\theta\,, \tag{1}$$

$w \in C_0^\infty(U \times V)$. Then $A \in \mathcal{D}'(U \times V)$. According to 1.2.1.1, Theorem 1, there is defined a linear continuous operator

$$A\colon C_0^\infty(V) \to \mathcal{D}'(U) \tag{2}$$

the kernel of which is the given distribution A (for abbreviation the letter A is used again in (2)). (2) is called a *Fourier integral operator* (FIO) with phase function φ and amplitude function a. For $u \in C_0^\infty(V)$ write also

$$(Au)\,(x) = \iint e^{i\varphi(x, y, \theta)}\, a(x, y, \theta)\, u(y)\, dy\, d\theta\,, \tag{3}$$

interpreted in the sense of (2).

If A is a FIO with amplitude function in $S^{-\infty}(U \times V \times \mathbb{R}^N)$, A is an operator with C^∞ kernel.

Examples. (1) Let $\Omega = \mathbb{R}^n \times \mathbb{R}^n$ and $\varphi(x, y, \xi) = \langle x - y, \xi\rangle$. A Fourier integral operator a with kernel

$$\langle A, w\rangle = (2\pi)^{-n} \iiint e^{i\langle x-y, \xi\rangle}\, a(x, y, \xi)\, w(x, y)\, dx\, dy\, d\xi$$

$(w \in C_0^\infty(\mathbb{R}^n \times \mathbb{R}^n))$ is called a *pseudo-differential operator*. The basic properties of pseudo-differential operators shall be studied in 1.2.3. For instance, each differential operator $A = \sum_{|\alpha| \leq m} a_\alpha(x)\, D^\alpha$ with the characteristic polynomial $a(x, \xi) = \sum_{|\alpha| \leq m} a_\alpha(x)\, \xi^\alpha$ is a pseudo-differential operator and (3) has the form

$$(Au)\,(x) = (2\pi)^{-n} \iint e^{i\langle x-y, \xi\rangle}\, a(x, \xi)\, u(y)\, dy\, d\xi\,. \tag{4}$$

(2) Let $U \subseteq \mathbb{R}^p_x$, $V \subseteq \mathbb{R}^n_y$ be open sets, $h\colon U \to V$ a differentiable map. Then the pull back $h^*\colon C^\infty(V) \to C^\infty(U)$ is defined by $h^*u = u \circ h$. Putting $a \equiv 1$ in (4) and substituting $h(x)$ for x yields

$$(h^*u)\,(x) = (2\pi)^{-n} \int e^{i\langle h(x)-y, \xi\rangle}\, u(y)\, dy\, d\xi\,,$$

$u \in C_0^\infty(V)$. It is easily seen that $\varphi(x, y, \xi) = \langle h(x) - y, \xi\rangle$ is a non-degenerate phase function. Thus the pull back of functions can be represented as a FIO. Especially, let $y = (y', y_n)$, $U = \mathbb{R}^{n-1}_{x'}$, $V = \mathbb{R}^n_y$ and $h\colon x' \to (x', 0)$ the embedding as hypersurface $y_n = 0$. Then $h^*\colon C^\infty(\mathbb{R}^n) \to C^\infty(\mathbb{R}^{n-1})$ is simply the restriction. For $\xi = (\xi', \tau)$ $(\xi' = (\xi_1, \dots, \xi_{n-1}) \in \mathbb{R}^{n-1}, \tau \in \mathbb{R})$,

$$\varphi_T(x', y, \xi', \tau) = \langle x' - y', \xi'\rangle - y_n\tau$$

is the corresponding phase function. Note that φ_T is linear in the frequency variable ξ. Fourier integral operators of the form

$$(Tu)\,(x') = \iiint e^{i\varphi_T(x', y, \xi', \tau)}\, t(x', y, \xi', \tau)\, u(y)\, dy\, d\xi'\, d\tau$$

are called *trace operators*; $T\colon C_0^\infty(\mathbb{R}^n) \to \mathcal{D}'(\mathbb{R}^{n-1})$. Obviously,

$$C_{\varphi_T} = \{(x', y, \xi', \tau)\colon x' = y', y_n = 0, (\xi', \tau) \in \mathbb{R}^n \setminus \{0\}\},$$
$$\Lambda_{\varphi_T} = \{(x', \xi', y, \eta)\colon x' = y', y_n = 0, \xi' = -\eta', \eta \neq 0\}.$$

Λ_{φ_T} is the conormal bundle of $\{(x', y) \in \mathbb{R}^{n-1} \times \mathbb{R}^n\colon x' = y', y_n = 0\}$ in $T^*(\mathbb{R}^{n-1} \times \mathbb{R}^n)$.

(3) A FIO of the form

$$(Kv)(x) = \iiint e^{i\varphi_K(x, y', \xi', \nu)}\, k(x, y', \xi', \nu)\, v(y')\, dy'\, d\xi'\, d\nu$$

$(x, y') \in \mathbb{R}^n \times \mathbb{R}^{n-1}$, $\xi = (\xi', \nu) \in \mathbb{R}^{n-1} \times \mathbb{R}$ with the phase function

$$\varphi_K(x, y', \xi', \nu) = \langle x' - y', \xi' \rangle + x_n\nu$$

is called *potential operator*; $K\colon C_0^\infty(\mathbb{R}^{n-1}) \to \mathcal{D}'(\mathbb{R}^n)$. We have

$$C_{\varphi_K} = \{(x, y', \xi', \nu)\colon x' = y', x_n = 0, (\xi', \nu) \in \mathbb{R}^n \setminus \{0\}\},$$
$$\Lambda_{\varphi_K} = \{(x, y', \xi, \eta')\colon x' = y', x_n = 0, \xi' = -\eta', \xi \neq 0\}.$$

(4) A FIO of the form

$$(Bu)(x) = \iiiint e^{i\varphi_B(x, y, \xi', \nu, \tau)}\, b(x, y, \xi', \nu, \tau)\, u(y)\, dy\, d\xi'\, d\nu\, d\tau$$

$(x, y) \in \mathbb{R}^n \times \mathbb{R}^n$, $(\xi', \nu, \tau) \in \mathbb{R}^{n-1} \times \mathbb{R}^1 \times \mathbb{R}^1$ with the phase function

$$\varphi_B(x, y, \xi', \nu, \tau) = \langle x' - y', \xi' \rangle - y_n\tau + x_n\nu$$

is called *Green operator*; $B\colon C_0^\infty(\mathbb{R}^n) \to \mathcal{D}'(\mathbb{R}^n)$. In this case

$$C_{\varphi_B} = \{(x, y, \xi', \nu, \tau)\colon x' = y', x_n = y_n = 0, (\xi', \nu, \tau) \in \mathbb{R}^{n+1} \setminus \{0\}\},$$
$$\Lambda_{\varphi_B} = \{(x, y, \xi, \eta)\colon x' = y', x_n = y_n = 0, \xi' = -\eta', |\xi| + |\eta| \neq 0\}.$$

The notations trace, potential and Green operator in (2), (3), (4) for the classes of FIO have a formal meaning here. Later we shall define corresponding operators in the half space or on manifolds with boundary as an essential tool for studying boundary value problems.

A FIO is called *properly supported* if its distributional kernel has this property (cf. 1.2.1.1). If A is a FIO with amplitude $a \in S^m(U \times V \times \mathbb{R}^N)$ denote by $\operatorname{supp}_0 a \subseteq U \times V$ the image of $\operatorname{supp} a$ with respect to the projection $U \times V \times \mathbb{R}^N \to U \times V$. It is easily seen that A is *properly supported* if the projections of $\operatorname{supp}_0 a$ to U and V, respectively, are proper.

Example. Let $\varphi = \langle x - y, \xi \rangle$ (cf. Example 1 above) with $C_\varphi = (\operatorname{diag} \Omega) \times (\mathbb{R}^n \setminus \{0\})$ and $a(x, y, \xi) \in S^m(\Omega \times \Omega \times \mathbb{R}^n)$. If then $\chi(x, y) \in C^\infty(\Omega \times \Omega)$ is a function with proper support in $\Omega \times \Omega$ and $\chi = 1$ in a neighbourhood of $\operatorname{diag} \Omega$, the function $(1 - \chi)a$ vanishes in a neighbourhood of C_φ. The Fourier distribution belonging to $(1 - \chi)a$ lies in $C^\infty(\Omega \times \Omega)$ (cf. 1.2.2.3, Theorem 3). Let A be the FIO with amplitude a, A_0 the FIO with amplitude χa (and phase $\langle x - y, \xi \rangle$). Then $A - A_0 \in C^\infty(\Omega \times \Omega)$. Per def. A and A_0 are both pseudo-differential operators, A_0 is properly supported. Thus, for any pseudo-differential operator A, there is a properly supported pseudo-differential operator A_0 so that $A - A_0$ is an operator with C^∞ kernel.

Let $U \subseteq \mathbb{R}_x^p$, $V \subseteq \mathbb{R}_y^n$ be open sets and A a FIO considered as a mapping (2). Then it is natural question whether (2) induces a linear operator

$$A\colon C_0^\infty(V) \to C^\infty(U) \tag{5}$$

and when A has an extension

$$A\colon \mathcal{E}'(V) \to \mathcal{D}'(U). \tag{6}$$

For completeness we shall mention corresponding assertions. Theorems of this type for operators (2) with general distributional kernel in $\mathcal{D}'(U \times V)$ are proved in Hörmander [6]. The conditions are formulated in terms of wave front sets of the distributional kernels.

Theorem 1. *Let $\varphi(x, y, \theta)$ be a non-degenerate phase function on $U \times V \times (\mathbb{R}^N \setminus \{0\})$ with the property*

$$d_{y,\theta}\varphi \neq 0 \quad \text{for all} \quad x \in U . \tag{7}$$

Then (1) *corresponds to a linear continuous operator* (5). *If*

$$d_{x,\theta}\varphi \neq 0 \quad \text{for all} \quad y \in V , \tag{8}$$

(2) *has a continuous extension* (6).

Proof: (7) means that φ can be considered as a phase function smoothly depending on $x \in U$ (x considered as a parameter). Then the first assertion is a simple consequence of the remark in 1.2.2.3 about parameter depending oscillatory integrals. Now the condition (8) means that the transpose ${}^tA$ of A (defined by $\langle Au, v\rangle = \langle u, {}^tAv\rangle$, $u \in C_0^\infty(U)$, $v \in C_0^\infty(V)$) satisfies the condition (7) with respect to the other variable. Thus, applying ${}^tA: C_0^\infty(U) \to C^\infty(V)$, we get by duality the extension (6). □

Remark 2. A properly supported FIO A with non-degenerate phase function φ possessing the property (7) induces continuous operators

$$A: C_0^\infty(V) \to C_0^\infty(U) , \quad A: C^\infty(V) \to C^\infty(U) . \tag{9}$$

If φ has the property (8), A induces continuous operators

$$A: \mathcal{E}'(V) \to \mathcal{E}'(U), \quad A: \mathcal{D}'(V) \to \mathcal{D}'(U) . \tag{10}$$

A non-degenerate phase function φ on $U \times V \times (\mathbb{R}^N \setminus \{0\})$ with the properties (7), (8) is called *operator phase function.* Denote by $R_\varphi \subset U \times V$ the image of C_φ with respect to the projection $U \times V \times \mathbb{R}^N \to U \times V$. If $K \subset V$ is a subset, put

$$R_\varphi \circ K = \{x \in U: (x, y) \in R_\varphi \quad \text{for some} \quad y \in K\} .$$

Theorem 3. *Let A be a FIO with an operator phase function φ $\big($on $U \times V \times (\mathbb{R}^N \setminus \{0\})\big)$. Then*

$$\operatorname{sing\,supp} (Av) \subseteqq R_\varphi \circ \operatorname{sing\,supp} v \tag{11}$$

for all $v \in \mathcal{E}'(V)$.

Proof: Let V_0 be an open neighbourhood of sing supp v in V with $\overline{V}_0 \subset V$. Then we find a decomposition $v = v_0 + v_1$ with supp $v_0 \subset V_0$, $v_1 \in C_0^\infty(V)$. Since $Av_1 \in C^\infty(U)$, it is sufficient to prove the assertion for v_0. If $K \subset U$ is an arbitrary compact set with $K \cap R_\varphi \circ \overline{V}_0 = \emptyset$, there exists an open neighbourhood U_0 of K with $U_0 \cap R_\varphi \circ \overline{V}_0 = \emptyset$. Obviously, $A \in C^\infty(U_0 \times V_0)$, thus $Av_0|_{U_0} \in C^\infty(U_0)$. Since V_0 is arbitrary, the assertion is proved. □

There is an analogue of Theorem 3 for the wave front sets of the corresponding distributions (cf. Hörmander [6], Duistermaat [1]).

1.2.3. Pseudo-Differential Operators

1.2.3.1. Definitions and Basic Facts

In this section we start with a more classical definition of pseudo-differential operators (PDO) than in 1.2.2.4 and study the connection with the earlier definition in 1.2.3.2.

In order to give a motivation consider a linear differential operator in $\mathbb{R}^n$ with C^∞ coefficients

$$A(x, D) = \sum_{|\alpha| \leq m} a_\alpha(x) D^\alpha \tag{1}$$

and let

$$a(x, \xi) = \sum_{|\alpha| \leq m} a_\alpha(x) \xi^\alpha \tag{2}$$

be its characteristic polynomial. Since $D^\alpha = F^{-1} \xi^\alpha F$, we obtain

$$A(x, D) = F^{-1} a(x, \xi) F \tag{3}$$

(considered as operator, for instance, on $C_0^\infty(\mathbb{R}^n)$). Now it is useful to connect the operators defined by (3) with general functions $a(x, \xi) \in S^m(\mathbb{R}^n \times \mathbb{R}^n)$, $m \in \mathbb{R}$.

Applied to $u \in C_0^\infty(\mathbb{R}^n)$ for arbitrary $a \in S^m(\mathbb{R}^n \times \mathbb{R}^k)$ we obtain well-defined integrals

$$(Au)(x) = (2\pi)^{-n} \int e^{i\langle x, \xi\rangle} a(x, \xi) \hat{u}(\xi) \, d\xi \tag{4}$$

(use $\hat{u}(\xi) \in \mathcal{S}(\mathbb{R}^n_\xi)$). It is clear that (4) can be differentiated with respect to x under the integral sign so that $Au \in C^\infty(\mathbb{R}^n)$. By a similar definition we obtain for open $\Omega \subseteq \mathbb{R}^n$, $a(x, \xi) \in S^m(\Omega \times \mathbb{R}^n)$ an operator

$$A: C_0^\infty(\Omega) \to C^\infty(\Omega) \tag{5}$$

and (5) is continuous. For A we also write $a(x, D_x)$.

An operator $C: C_0^\infty(\Omega) \to C^\infty(\Omega)$ of the form $u \to Cu = \int C(x, y) u(y) \, dy$ with $C(x, y) \in C^\infty(\Omega \times \Omega)$ is called *operator with C^∞ kernel* (or *smoothing operator*). It is clear that such an operator has a continuous extension $C: \mathcal{E}'(\Omega) \to C^\infty(\Omega)$ (extensions and restrictions of operators we often denote by the same letter). For instance, (4) is a smoothing operator for $a \in S^{-\infty}$. On the other hand, not each smoothing operator has a representation (4) with some $a \in S^{-\infty}$ (cf. 1.2.3.2, Remark 3).

Definition 1. An operator of the form

$$A + C: C_0^\infty(\Omega) \to C^\infty(\Omega) \tag{6}$$

with A given by (4), $a \in S^m(\Omega \times \mathbb{R}^n)$, $m \in \mathbb{R}$ and C a smoothing operators is called *pseudo-differential operator* (PDO) in Ω of order m. The function $a(x, \xi)$ is called a (*complete*) *symbol* of (6). Denote by $L^m(\Omega)$ (or L^m for fixed domain) the set of all PDOs in Ω of order m. Moreover set $L^\infty = \bigcup_{m \in \mathbb{R}} L^m$, $L^{-\infty} = \bigcap_{m \in \mathbb{R}} L^m$.

By definition $L^{-\infty}(\Omega)$ is the set of smoothing operators in Ω. In 1.2.3.2, Theorem 6 it will be proved that the definition of a complete symbol is correct, i.e. a, a' are complete symbols of an operator in $L^\infty(\Omega)$ iff $a - a' \in S^{-\infty}(\Omega \times \mathbb{R}^n)$. In this section symbols always mean complete symbols (contrarily to principal symbols to be defined below).

If $A_1, A_2 \in L^m(\Omega)$ and $A_1 - A_2 \in L^{-\infty}(\Omega)$, A_1, A_2 will be often identified, because many properties of PDOs depend on the class mod $L^{-\infty}$ only.

An operator $A \in L^m(\Omega)$ is called *classical* if $a \in S^m_{\mathrm{cl}}(\Omega \times \mathbb{R}^n)$ ($\Leftrightarrow a \sim \sum_j a_j$, $a_j(x, \xi)$ homogeneous in ξ of order $m - j$ for $|\xi| \geqq$ const., $m = \max\{m - j\}$). The set of classical PDOs in $L^m(\Omega)$ shall be denoted by L^m_{cl}. There is a uniquely determined positively homogeneous function $\sigma_A(x, \xi)$ ($\xi \neq 0$) of order m with $\sigma_A(x, \xi) = a_m(x, \xi)$ when $|\xi| =$ const. σ_A is called *homogeneous principal symbol of A of order m.* For $m \geqq 0$ an operator can be defined by substituting a positively homogeneous function $a(x, \xi)$ into the formula (4). In the case $m = 0$ this yields a singular integral operator (cf. Mihlin [1]).

Let $a \in S^{-\infty}(\Omega \times \mathbb{R}^n)$. Then (4) can be written in the form

$$(Au)(x) = (2\pi)^{-n} \iint e^{i\langle x-y, \xi\rangle} a(x, \xi)\, u(y)\, dy\, d\xi\,. \tag{7}$$

Multiplication by $v \in C_0^\infty(\Omega)$ and integration gives

$$\langle Au, v\rangle = (2\pi)^{-n} \iiint e^{i\langle x-y, \xi\rangle} a(x, \xi)\, u(y)\, v(x)\, dx\, dy\, d\xi\,. \tag{8}$$

In view of 1.2.2.3, Theorem 1 the distribution (8) has a unique extension to S^m depending continuously on $a \in S^m$ for any fixed $m \in \mathbb{R}$. Such an extension is given by

$$\langle Au, v\rangle = (2\pi)^{-n} \iint e^{i\langle x, \xi\rangle} a(x, \xi)\, \hat{u}(\xi)\, v(x)\, dx\, d\xi\,, \tag{9}$$

too, so that (9) is equal to the oscillatory integral (8) for arbitrary $a \in S^m$. Thus (7) is correctly defined as a Fourier integral operator for $u \in S^m$ with the operator phase function $\varphi = \langle x - y, \xi\rangle$ (cf. 1.2.2.4). The set $R_\varphi \subset \Omega \times \Omega$ from 1.2.2.4, Theorem 3 is here $\operatorname{diag} \Omega = \{(x, x) : x \in \Omega\}$.

Theorem 2. *The distributional kernel K_A of a PDO $A \in L^m(\Omega)$ belongs to C^∞ over $\Omega \times \Omega \setminus \operatorname{diag} \Omega$. For $m + j < -n$ it follows that $K_A \in C^j(\Omega \times \Omega)$.*

The first assertion follows from the preceding remark (cf. 1.2.2.4, Theorem 3) and the second one by differentiation j times under the integral sign of

$$K_A(x, y) = (2\pi)^{-n} \int e^{i\langle x-y, \xi\rangle} a(x, \xi)\, d\xi\,.$$

Theorem 3. *Let $A \in L^m(\Omega)$ be of the form (4) with $a \in S^m$. Then, for each $j \in \mathbb{Z}_+$, .here exists a $k \in \mathbb{Z}_+$ and a function $g_k \in C^j(\Omega \times \Omega)$ with*

$$(Au)(x) = \int g_k(x, y)\, (1 - \Delta_y)^k\, u(y)\, dy\,, \quad u \in C_0^\infty(\Omega)\,. \tag{10}$$

Proof: Using $\hat{u}(\xi) = (1 + |\xi|^2)^{-k} \left((1 - \Delta)^k u\right)^\wedge(\xi)$ we obtain

$$(Au)(x) = (2\pi)^{-n} \int e^{-i\langle x, \xi\rangle} b_k(x, \xi) \left((1 - \Delta)^k u\right)^\wedge(\xi)\, d\xi$$

with $b_k(x, \xi) = (1 + |\xi|^2)^{-k} a(x, \xi) \in S^{m-2k}$. For $m - 2k < -n$ we get (10) with

$$g_k(x, y) = (2\pi)^{-n} \int e^{i\langle x-y, \xi\rangle} b_k(x, \xi)\, d\xi\,.$$

For $k > \frac{1}{2}(m + n + j)$ the last integral can be differentiated j times under the integral sign. □

Lemma 4. *Let $a \in S^m(\Omega \times \mathbb{R}^n)$, $v \in C_0^\infty(\Omega)$,*

$$b_v(\xi, \eta) = \int e^{-i\langle x, \eta\rangle} a(x, \xi)\, v(x)\, dx\,.$$

Then there exists for any $N \in \mathbb{Z}_+$ a constant c_N with

$$|b_v(\xi, \eta)| \leqq c_N(1 + |\eta|)^{-N} (1 + |\xi|)^m\,.$$

Proof: The assertion is an immediate consequence of

$$\eta^\alpha \int e^{-i\langle x, \eta\rangle} a(x, \xi)\, v(x)\, dx = \int e^{-i\langle x, \eta\rangle} D_x^\alpha\big(a(x, \xi)\, v(x)\big)\, dx$$

for any multi-index α. □

Proposition 5. *Let $A \in L^m(\Omega)$. Then there exists a continuous extension*

$$A: \mathscr{E}'(\Omega) \to \mathscr{D}'(\Omega)\,. \tag{11}$$

Moreover sing supp $Au \subseteq$ sing supp u *for all* $u \in \mathscr{E}'(\Omega)$.

Proof: Obviously it is sufficient to extend (9) to $u \in \mathscr{E}'(\Omega)$ with $v \in C_0^\infty(\Omega)$. For $u \in C_0^\infty(\Omega)$ (9) can be written in the form

$$\langle Au, v\rangle = \int b_v(\xi)\, \hat{u}(\xi)\, d\xi \tag{12}$$

with $b_v(\xi) = (2\pi)^{-n} \int e^{i\langle x, \xi\rangle} a(x, \xi)\, v(x)\, dx$. Because of Lemma 4 $b_v(\xi)$ is rapidly decreasing in ξ. Thus (12) remains convergent for $u \in \mathscr{E}'(\Omega)$ and (12) represents an element in $\mathscr{D}'(\Omega)$ as a functional acting on v. The assertion about the singular support of Au is a special case of 1.2.2.4, Theorem 3. □

It can be proved that $\mathrm{WF}(Au) \subseteq \mathrm{WF}\,(u)$ for each $u \in \mathscr{E}'(\Omega)$.

Remark 6. If $A \in L^m(\Omega)$ is properly supported, A induces continuous mappings $C_0^\infty(\Omega) \to C_0^\infty(\Omega)$, $C^\infty(\Omega) \to C^\infty(\Omega)$, $\mathscr{E}'(\Omega) \to \mathscr{E}'(\Omega)$, $\mathscr{D}'(\Omega) \to \mathscr{D}'(\Omega)$.

Let $A \in L^m(\Omega)$ be properly supported. Consider $u \in C_0^\infty(\Omega)$ and apply A to $u(x) = (2\pi)^{-n} \int e^{i\langle x, \xi\rangle} \hat{u}(\xi)\, d\xi$. Then A can be applied under the integral sign with respect to x and it follows that

$$(Au)\,(x) = (2\pi)^{-n} \int e^{i\langle x, \xi\rangle} \sigma(A)\,(x, \xi)\, \hat{u}(\xi)\, d\xi \tag{13}$$

with

$$\sigma(A)\,(x, \xi) = e^{-i\langle x, \xi\rangle} A\, e^{i\langle \cdot, \xi\rangle} \in S^m(\Omega \times \mathbb{R}^n)\,. \tag{14}$$

Thus (14) is a complete symbol of A and each properly supported $A \in L^m(\Omega)$ can be written in the form (13).

If A is a differential operator of the from (1), then (14) is identical with (2).

1.2.3.2. Definition by General Amplitude Functions

Let $a(x, y, \xi) \in S^m(\Omega \times \Omega \times \mathbb{R}^n)$, $u \in C_0^\infty(\Omega)$ and

$$(Au)\,(x) = \iint e^{i\langle x-y, \xi\rangle} a(x, y, \xi)\, u(y)\, dy\, d\xi \tag{1}$$

(cf. 1.2.2.4, Example 1). We shall show that (1) is a PDO $\in L^m(\Omega)$ in the sense of 1.2.3.1, Definition 1 and calculate its symbol in terms of a (for simplicity the factor $(2\pi)^{-n}$ is considered as part of a).

Theorem 1. *Suppose that the operator* (1) *is properly supported. Then* $A \in L^m(\Omega)$ *and*

$$\sigma(A)\,(x, \xi) \sim (2\pi)^n \sum_\alpha \frac{1}{\alpha!} \partial_\xi^\alpha D_y^\alpha a(x, y, \xi)\big|_{y=x} \tag{2}$$

($\sigma(A)$ is defined by 1.2.3.1, (14)*).*

Proof: Since (1) is properly supported, A can be applied to functions $u \in C^\infty(\Omega)$. Then for $u(y) = e^{i\langle y, \xi\rangle}$ it follows that

$$\begin{aligned}\sigma(A)\,(x,\xi) &= e^{-i\langle x,\xi\rangle} A\, e^{i\langle\cdot,\xi\rangle} = e^{-i\langle x,y\rangle} \iint e^{i\langle x-y,\eta\rangle} e^{i\langle y,z\rangle} a(x,y,\eta)\,dy\,d\eta\\ &= \iint e^{-i\langle y-x,\eta-\xi\rangle} a(x,y,\eta)\,dy\,d\eta\\ &= \iint e^{-i\langle y,\eta\rangle} a(x,x+y,\xi+\eta)\,dy\,d\eta\,.\end{aligned}$$

Set $b(x,y,\xi) = a(x,x+y,\xi)$ and consider the partial Fourier transform

$$\hat{b}(x,\eta,\xi) = \int e^{-i\langle y,\eta\rangle} b(x,y,\xi)\,dy\,.$$

From $a \in S^m$ follows

$$|D_x^\alpha \eta^\beta D_\xi^\gamma \hat{b}(x,\eta,\xi)| \leqq c(1+|\xi|)^{m-|\gamma|}$$

for arbitrary multi-indices α, β, γ and x in a compact set. Thus there exists for each $N \in Z_+$ a constant $c = c_N$ with

$$|D_x^\alpha D_\xi^\gamma \hat{b}(x,\eta,\xi)| \leqq c(1+|\xi|)^{m-|\gamma|}(1+|\eta|)^{-N}\,. \tag{3}$$

We obtain $\sigma(A)\,(x,\xi) = \int \hat{b}(x,\eta,\xi+\eta)\,d\eta$. Because of (3) each partial derivative of $\sigma(A)$ can be estimated by a power of $|\xi|$. In order to get the asymptotic expansion for $\sigma(A)$ use the Taylor expansion of $\hat{b}(x,\eta,\xi+\eta)$. Then (3) yields

$$\begin{aligned}&\left|\hat{b}(x,\eta,\xi+\eta) - \sum_{|\alpha|<N} \partial_\xi^\alpha \hat{b}(x,\eta,\xi)\frac{\eta^\alpha}{\alpha!}\right|\\ &\leqq c\,|\eta|^N \sup_{0<t<1}(1+|\xi+t\eta|)^{m-N}(1+|\eta|)^{-M}\end{aligned} \tag{4}$$

with arbitrary $N \in Z_+$ and a constant $c = c_M$. $c(1+|\xi|)^{m-N}$ is a bound for (4) for $|\eta| < |\xi|/2$, $M = N$. For $|\eta| > |\xi|/2$ and M sufficiently large an arbitrary power of $(1+|\eta|)^{-1}$ is a bound of (4). Thus the Fourier inversion formula can be applied. For this use $(D_y^\alpha f)\,(0) = (2\pi)^{-n}\int \eta^\alpha \hat{f}(\eta)\,d\eta$ and the inequalities

$$\int_{|\eta|<|\xi|/2} d\eta \leqq c(1+|\xi|)^n\,, \qquad \int_{|\eta|>|\xi|/2} (1+|\eta|)^{-M}\,d\eta \leqq c(1+|\xi|)^{-M_1}$$

for arbitrary $M_1 \in Z$ and suitable $M \in Z_+$. Then

$$|\sigma(A)\,(x,\xi) - (2\pi)^n \sum_{|\alpha|<N} \partial_\xi^\alpha D_y^\alpha b(x,y,\xi)|_{y=0}\,| \leqq c(1+|\xi|)^{m+n-N}$$

follows from (4). Using the definition of b and 1.2.2.1, Theorem 3 we obtain the asymptotic expansion (2). □

Corollary 2. *The distributional kernel $K_A(x,y)$ of the operator* (1) *is in C^∞ over $\Omega \times \Omega \setminus \operatorname{diag}\Omega$. Moreover* (1) *defines continuous mappings*

$$A: C_0^\infty(\Omega) \to C^\infty(\Omega)\,, \quad A: \mathscr{E}'(\Omega) \to \mathscr{D}'(\Omega)$$

and sing supp $Au \subseteqq$ sing supp u, $u \in \mathscr{E}'(\Omega)$. *Each operator $A \in L^m(\Omega)$ has a decomposition $A = A_0 + C$ with $A_0 \in L^m(\Omega)$ properly supported and an operator C with C^∞ kernel.*

Remark 3. Let $C: C_0^\infty(\Omega) \to C^\infty(\Omega)$ be an operator with C^∞ kernel $K(x,y)$. Then there exists an $a(x,y,\xi) \in S^{-\infty}(\Omega \times \Omega \times \mathbb{R}^n)$ so that C can be written in the form (1). It is sufficient to put $a(x,y,\xi) = e^{-i\langle x-y,\xi\rangle} K(x,y)\,\varphi(\xi)$ with a real function $\varphi(\xi) \in C_0^\infty(\mathbb{R}^n)$, $\varphi \geqq 0$, $\int \varphi(\xi)\,d\xi = 1$.

Remark 4. If $a \in S^m(\Omega \times \Omega \times \mathbb{R}^n)$ vanishes on diag Ω, there exists a $b \in S^{m-1}(\Omega \times \Omega \times \mathbb{R}^n)$ defining (1) with a replaced by b.

This is an immediate consequence of the considerations in 1.2.2.3.

Lemma 5. *For each $a(x, \xi) \in S^m(\Omega \times \mathbb{R}^n)$ there exists a properly supported PDO $A \in L^m(\Omega)$ with $a(x, \xi) - \sigma(A)(x, \xi) \in S^{-\infty}(\Omega \times \mathbb{R}^n)$ (cf. formula* 1.2.3.1. (14)).

Proof: Consider a function $\chi \in C^\infty(\Omega \times \Omega)$, $0 \leqq \chi(x, y) \leqq 1$, $\chi \equiv 1$ near diag Ω and supp $\chi \subset \Omega \times \Omega$ proper. Define with $b(x, y, \xi) = \chi(x, y)\, a(x, \xi) \in S^m(\Omega \times \Omega \times \mathbb{R}^n)$ by (1) a PDO $A \in L^m(\Omega)$. Then A is properly supported. Since $b(x, y, \xi) - a(x, \xi)$ vanishes near diag Ω, applying Remark 4 we obtain the assertion. □

For given $A \in L^m(\Omega)$ defined by (1) with $a(x, y, \xi) \in S^m(\Omega \times \Omega \times \mathbb{R}^n)$ we take some properly supported $A_0 \in L^m(\Omega)$ with $A - A_0 \in L^{-\infty}(\Omega)$ and consider the equivalence class of $\sigma(A_0)(x, \xi)$ in $S^m(\Omega \times \mathbb{R}^n)/S^{-\infty}(\Omega \times \mathbb{R}^n)$. Thus there is induced a map

$$L^m(\Omega)/L^{-\infty}(\Omega) \to S^m(\Omega \times \mathbb{R}^n)/S^{-\infty}(\Omega \times \mathbb{R}^n)\,. \tag{5}$$

Theorem 6. *The map* (5) *is an isomorphism.*

Proof: Applying Remark 3 we find an amplitude function $a \in S^{-\infty}(\Omega \times \Omega \times \mathbb{R}^n)$ for each operator A in $L^{-\infty}(\Omega)$ so that A is given by (1). Thus A transforms to zero in $S^m/S^{-\infty}$ by the map (5), i.e. the definition of (5) is correct. If $a(x, \xi) \in S^m(\Omega \times \mathbb{R}^n)$, the corresponding PDO A of the form 1.2.3.1.(4) goes into the class represented by a. Thus (5) is surjective. For $a \in S^{-\infty}(\Omega \times \mathbb{R}^n)$ we have for the corresponding PDO A of Lemma 5 $A \in L^{-\infty}(\Omega)$. This shows the injectivity of (5). □

1.2.3.3. Adjoints and Compositions

Suppose that we are given a PDO A written in the form 1.2.3.1.(7) with $a(x, \xi) \in S^m(\Omega \times \mathbb{R}^n)$. Then, with respect to the bilinear pairing $\langle Au, v\rangle = \langle u, {}^tAv\rangle$ $(u, v \in C_0^\infty(\Omega))$, we can define the uniquely defined transposed operator ${}^tA : C_0^\infty(\Omega) \to \mathcal{D}'(\Omega)$,

$$\begin{aligned}({}^tAv)(x) &= (2\pi)^{-n} \iint e^{-i\langle x-y, \xi\rangle} a(y, \xi)\, v(y)\, dy\, d\xi \\ &= (2\pi)^{-n} \iint e^{i\langle x-y, \xi\rangle} a(y, -\xi)\, v(y)\, dy\, d\xi\,.\end{aligned}$$

1.2.3.3, Theorem 1 shows the following

Theorem 1. *Let $A \in L^m(\Omega)$, then ${}^tA \in L^m(\Omega)$. If A is properly supported, ${}^tA$ is properly supported, too, and*

$$\sigma({}^tA)(x, \xi) \sim \sum_\alpha \frac{1}{\alpha!} (-1)^{|\alpha|} (\partial_\xi^\alpha D_x^\alpha q)(x, -\xi)\,. \tag{1}$$

Remark 2. Let $A \in L^m(\Omega)$ be of the form 1.2.3.1.(7). If one defines an adjoint A^* with respect to the Hermitean pairing $(u, v) = \int u\bar{v}\, dx$, i.e. $(Au, v) = (u, A^*v)$ $(u, r \in C_0^\infty(\Omega))$, we get $A^* \in L^m(\Omega)$. For properly supported A it follows that A^* is properly supported and there is an asymptotic expansion

$$\sigma(A^*)(x, \xi) \sim \sum_\alpha \frac{1}{\alpha!} \partial_\xi^\alpha D_x^\alpha \overline{a(x, \xi)}\,. \tag{2}$$

Let $A \in L^m(\Omega)$ be a properly supported PDO written in the form 1.2.3.1.(7), $a \in S^m(\Omega \times \mathbb{R}^n)$. Then there exists an $\tilde{a} \in S^m(\Omega \times \mathbb{R}^n)$ with

$$(Au)(x) = (2\pi)^{-n} \iint e^{i\langle x-y, \xi\rangle} \tilde{a}(y, \xi)\, u(y)\, dy\, d\xi\,. \tag{3}$$

In order to find $\tilde{a}$ take ${}^tA$,

$$({}^tAv)(y) = (2\pi)^{-n} \int e^{i\langle y, \xi\rangle} \sigma({}^tA)(y, \xi)\, \hat{v}(\xi)\, d\xi\,.$$

Using $\langle Au, v\rangle = \langle F^{-1}Au, Fv\rangle$ we obtain from

$$\langle Au, v\rangle = (2\pi)^{-n} \iint e^{i\langle y, \xi\rangle} \sigma({}^tA)(y, \xi)\, \hat{v}(\xi)\, u(y)\, d\xi\, dy$$

that Au is the Fourier transform of

$$\xi \to (2\pi)^{-n} \int e^{i\langle y, \xi\rangle} \sigma({}^tA)(y, \xi)\, u(y)\, dy\,.$$

Hence

$$\begin{aligned}(Au)(x) &= (2\pi)^{-n} \int e^{-i\langle x, \xi\rangle} \left\{\int e^{i\langle y, \xi\rangle} \sigma({}^tA)(y, \xi)\, u(y)\, dy\right\} d\xi \\ &= (2\pi)^{-n} \iint e^{i\langle x-y, \xi\rangle} \tilde{a}(y, \xi)\, u(y)\, dy\, d\xi\end{aligned} \tag{4}$$

with $\tilde{a}(y, \xi) = \sigma({}^tA)(y, -\xi)$. (4) is equivalent to

$$(Au)^\wedge(\xi) = \int e^{-i\langle y, \xi\rangle} \tilde{a}(y, \xi)\, u(y)\, dy\,. \tag{5}$$

If A is given in the form 1.2.3.2. (1) with $a(x, y, \xi) \in S^m(\Omega \times \Omega \times \mathbb{R}^n)$, we get by 1.2.3.2, Theorem 1

$$\tilde{a}(y, \xi) \sim (2\pi)^n \sum_\alpha \frac{1}{\alpha!} (-1)^{|\alpha|}\, \partial_\xi^\alpha D_x^\alpha a(x, y, \xi)|_{x=y}\,. \tag{6}$$

Next consider the composition of PDOs. If $A_i \in L^\infty(\Omega)$ $(i = 1, 2)$ and A_1 or A_2 properly supported, then the compositions $A_1 \circ A_2$, $A_2 \circ A_1$ are well defined as operators $C_0^\infty(\Omega) \to C^\infty(\Omega)$.

Theorem 3. *Let $A_i \in L^{m_i}(\Omega)$ $(i = 1, 2)$ and A_1 or A_2 properly supported. Then $A = A_2A_1 \in L^{m_1+m_2}(\Omega)$. If A_i is written in the form 1.2.3.1. (4) with $a_i \in S^{m_i}(\Omega \times \mathbb{R}^n)$ $(i = 1, 2)$, there is the following asymptotic expansion*

$$\sigma(A_2A_1)(x, \xi) \sim \sum_\alpha \frac{1}{\alpha!} \left(\partial_\xi^\alpha a_2(x, \xi)\right) D_x^\alpha a_1(x, \xi)\,. \tag{7}$$

For abbreviation we write $a_2 \circ a_1$.

Proof: In view of 1.2.3.2, Corollary 2 we can assume that A_1, A_2 are both properly supported. Write A_1 in the form (5) with a symbol $\tilde{a}_1(y, \xi)$. Application of A_2 gives then

$$(Au)(x) = (2\pi)^{-n} \iint e^{i\langle x-y, \xi\rangle} a_2(x, \xi)\, \tilde{a}_1(y, \xi)\, u(y)\, dy\, d\xi\,.$$

Thus because of 1.2.3.2, Theorem 1, A belongs to $L^{m_1+m_2}(\Omega)$. Obviously A is properly supported. Using 1.2.3.2. (2) we get

$$\begin{aligned}\sigma(A)(x, \xi) &\sim \sum_\alpha \frac{1}{\alpha!} \partial_\xi^\alpha D_y^\alpha \left(a_2(x, \xi)\, \tilde{a}_1(y, \xi)\right)_{y=x} \\ &= \sum_\alpha \frac{1}{\alpha!} \partial_\xi^\alpha \left(a_2(x, \xi)\, D_y^\alpha \tilde{a}_1(y, \xi)\right)_{y=x} \sim \sum_{\alpha, \beta} \frac{1}{\alpha!\,\beta!} \partial_\xi^\alpha \left(a_2(x, \xi)\, (-\partial_\xi)^\beta D_x^{\alpha+\beta} a_1(x, \xi)\right).\end{aligned}$$

Application of 1.2.1.1. (1) gives

$$\sigma(A)(x,\xi) \sim \sum_{\substack{\alpha,\beta,\gamma,\delta\\ \gamma+\delta=\alpha}} (\partial_\xi^\gamma a_2(x,\xi)) \left((-\partial_\xi)^\beta (\partial_\xi^\delta D_x^{\alpha+\beta} a_1(x,\xi))\right)/\beta!\,\gamma!\,\delta!$$

$$= \sum_{\beta,\gamma,\delta} (-1)^{|\beta|} (\partial_\xi^\gamma a_2(x,\xi)) \left(\partial_\xi^{\beta+\delta} D_x^{\beta+\gamma+\delta} a_1(x,\xi)\right)/\beta!\,\gamma!\,\delta!$$

$$= \sum_\gamma \sum_\delta \left(\sum_{\beta+\delta=\lambda} \frac{(-1)^{|\beta|}}{\beta!\,\delta!} \right) (\partial_\xi^\gamma a_2(x,\xi)) \left(\partial_\xi^\lambda D_x^{\lambda+\gamma} a_1(x,\xi)\right)/\gamma!\,. \tag{8}$$

Now apply 1.2.1.1. (1) for $x = -y = h = (1, 1, \dots, 1)$, i.e.

$$(h-h)^\lambda = \sum_{\beta+\delta=\lambda} \frac{\lambda!}{\beta!\,\delta!} h^\delta (-h)^\beta = \lambda! \sum_{\beta+\delta=\lambda} \frac{(-1)^{|\beta|}}{\beta!\,\delta!}\,.$$

This vanishes for $\lambda \neq 0$ and is equal to 1 for $\lambda = 0$. Thus (8) takes the form (7). □

The assertion of Theorem 3 corresponds to the Leibniz rule for differential operators..Let

$$A(x,D) = \sum_{|\alpha|\le m} a_\alpha(x) D^\alpha\,, \quad B(x,D) = \sum_{|\beta|\le l} b_\beta(x) D^\beta$$

be differential operators with C^∞ coefficients. Moreover let $a(x,\xi) = \sum_{|\alpha|\le m} a_\alpha(x)\,\xi^\alpha$, $a^{(\alpha)} = \partial_\xi^\alpha a$. Then the Leibniz rule (cf. 1.2.1.1. (2)) says that for arbitrary functions $f, b \in C^m(\mathbb{R}^n)$

$$A(x,D)\left(f(x)\,b(x)\right) = \sum_{|\alpha|\le m} \frac{1}{\alpha!} \left(a^{(\alpha)}(x,D)\,f(x)\right) D^\alpha b(x)\,.$$

For $f(x) = D_x^\beta h(x)$ and $b(x) = b_\beta(x)$ we have

$$A(x,D)\left(b_\beta(x)\,D_x^\beta h(x)\right) = \sum_{|\alpha|\le m} \frac{1}{\alpha!} \left(a^{(\alpha)}(x,D)\,D_x^\beta h(x)\right) D_x^\alpha b_\beta(x)\,.$$

Summation over all multi-indices β with $|\beta| \le l$ shows that the complete symbol of the composition $A(x,D)\,B(x,D)$ has the from

$$\sum_{|\alpha|\le m} \frac{1}{\alpha!} (\partial_\xi^\alpha a(x,\xi))\, D_x^\alpha b(x,\xi)$$

with $b(x,\xi) = \sum_{|\beta|\le l} b_\beta(x)\,\xi^\beta$.

Let in Theorem 3 $A_j \in L_{\mathrm{cl}}^{m_j}(\Omega)$ with homogeneous principal symbols σ_{A_j}, $j = 1, 2$. Then $A_2 A_1 \in L_{\mathrm{cl}}^{m_1+m_2}(\Omega)$ and

$$\sigma_{A_2A_1}(x,\xi) = \sigma_{A_2}(x,\xi)\,\sigma_{A_1}(x,\xi)\,, \tag{9}$$

i.e. the principal symbol of the composition of classical PDOs is equal to the product of the principal symbols.

1.2.3.4. Change of Coordinates

In this section we consider the behaviour of PDOs under coordinate transformations. Let $X, Y \subseteq \mathbb{R}^n$ be open sets and $\varkappa: X \to Y$ a diffeomorphism. Then there are induced mappings $\varkappa^*: C_0^\infty(Y) \to C_0^\infty(X)$, $\varkappa^*: C^\infty(Y) \to C^\infty(X)$ and any $A \in L^m(X)$ corresponds to an operator

$$\varkappa_* A \overset{\text{def}}{=} (\varkappa^*)^{-1} \circ A \circ \varkappa^*: C_0^\infty(Y) \to C^\infty(Y)\,. \tag{1}$$

Coordinates in T^*X and T^*Y are denoted by (x, ξ) and (y, η), respectively. By $\varkappa$ a map $(\varkappa, {}^t(\mathrm{d}\varkappa)^{-1}) : T^*X \to T^*Y$ is given and for $(x, \xi) \to (y, \eta)$ we have $y = \varkappa(x)$, $\xi = {}^t(\mathrm{d}\varkappa)\, \eta$.

Theorem 1. *If $\varkappa: X \to Y$ is a diffeomorphism, by* (1) *is induced a bijection*

$$\varkappa_*: L^m(X) \to L^m(Y) \tag{2}$$

($m \in \mathbb{R}$). *If $A \in L^m(X)$ is properly supported and written in the form* 1.2.3.1. (7) *with $a(x, \xi) \in S^m(\Omega \times \mathbb{R}^n)$, the following asymptotic expansion holds*

$$\sigma(\varkappa_* A)\,(y, \eta) \sim \sum_\alpha \frac{1}{\alpha!} a^{(\alpha)}\big(x, {}^t(\mathrm{d}\varkappa)\,(x)\,\eta\big)\, D_z^\alpha\, \mathrm{e}^{\mathrm{i}h(z,x)\eta}\big|_{z=x}, \tag{3}$$

where $x = \varkappa^{-1}(y)$, $a^{(\alpha)}(x, \xi) = \partial_\xi^\alpha a(x, \xi)$, $h(z, x) = \varkappa(z) - \varkappa(x) - \mathrm{d}\varkappa(x)\,(z - x)$ on the right hand side and

$$\tau_\alpha(x, \eta) = D\, \mathrm{e}^{\mathrm{i}h(z,x)\eta}\big|_{z=x} \tag{4}$$

is a polynomial in η of degree $\leqq |\alpha|/2$.

Instead of $\sigma(\varkappa_* A)$ we write sometimes $(\sigma(A))_\varkappa$.

Proof: Note first that the function $h(z, x)$ has a zero of second order at $z = x$. Thus $\tau_\alpha(x, \eta)$ is a polynomial of degree $\leqq |\alpha|/2$. It follows that

$$a^{(\alpha)}\big(x, {}^t\mathrm{d}\varkappa(x)\,\eta\big)\,\tau_\alpha(x, \eta) \in S^{m-|\alpha|/2}(X \times \mathbb{R}^n)$$

and the asymptotic sum (3) is defined. Without loss of generality we can assume that A is properly supported. Write A in the form

$$(Av)\,(x) = (2\pi)^{-n} \iint \mathrm{e}^{\mathrm{i}\langle x-z, \xi\rangle}\, a(x, \xi)\, v(z)\, \mathrm{d}z\, \mathrm{d}\xi$$

$(v \in C_0^\infty(X))$. Then, with $u = (\varkappa^*)^{-1} v \in C_0^\infty(Y)$ and $y = \varkappa(x)$, $w = \varkappa(z)$,

$$\begin{aligned}(\varkappa_* A)\, u(y) &= (2\pi)^{-n} \iint \mathrm{e}^{\mathrm{i}\langle \varkappa^{-1}(y)-z, \xi\rangle}\, a\big(\varkappa^{-1}(y), \xi\big)\, u\big(\varkappa(z)\big)\, \mathrm{d}z\, \mathrm{d}\xi \\ &= (2\pi)^{-n} \iint \mathrm{e}^{\mathrm{i}\langle \varkappa^{-1}(y)-\varkappa^{-1}(w), \xi\rangle}\, a\big(\varkappa^{-1}(y), \xi\big)\, u(w)\, |\det \mathrm{d}\varkappa^{-1}(w)|\, \mathrm{d}w\, \mathrm{d}\xi\end{aligned} \tag{5}$$

follows. The phase function

$$\varphi(y, w, \xi) = \langle \varkappa^{-1}(y) - \varkappa^{-1}(w), \xi\rangle$$

has the property

$$\varphi'_\xi(y, w, \xi) = 0 \Leftrightarrow y = w\,. \tag{6}$$

Before we finish the proof of Theorem 1 prove the following

Theorem 2. *Let $\varphi(y, w, \xi)$ be a non-degenerate phase function on $\Omega \times \Omega \times \mathbb{R}^n$, $\Omega \subseteqq \mathbb{R}^n$ open, linear in ξ with the property* (6). *If A is a Fourier integral operator with phase function φ and amplitude function $a(y, w, \xi) \in S^m(\Omega \times \Omega \times \mathbb{R}^n)$, then $A \in L^m(\Omega)$.*

For the proof we need

Lemma 3. *Let φ be as in Theorem 2. Then there exists a neighbourhood $U \subset \Omega \times \Omega$ of* diag Ω *and a C^∞ map $\psi: U \to \mathrm{GL}\,(n, \mathbb{R})$ so that*

$$\varphi\big(y, w, \psi(y, w)\,\eta\big) = \langle y - w, \eta\rangle\,, \tag{7}$$

$(y, w) \in U$ *and*

$$\det \psi(y, y)\, \det \varphi''_{y\xi}(y, w, \xi)\big|_{y=w} = 1\,. \tag{8}$$

Proof: By assumption

$$\varphi(y, w, \xi) = \sum_{j=1}^{n} \varphi_j(y, w)\, \xi_j \tag{9}$$

with $\varphi_j(y, y) = 0$ for all j and $\varphi_j(y, w) = 0$ for $j = 1, \dots, n$ implies $y = w$. We get

$$\varphi_j(y, w) = \varphi_j\big(w + t(y - w), w\big)_{t=1}$$

$$= \varphi_j(w, w) + \int_0^1 \frac{\mathrm{d}}{\mathrm{d}t}\varphi_j\big(w + t(y - w), w\big)\, \mathrm{d}t = \sum_{k=1}^{n} (y_k - w_k)\, \varphi_{kj}(y, w) \tag{10}$$

with functions $\varphi_{kj} \in C^\infty(U')$ (U' a neighbourhood of diag Ω) with

$$\varphi_{kj}(y, w) = \frac{\partial \varphi_j(y, w)}{\partial y_k}\bigg|_{y=w}. \tag{11}$$

Moreover

$$\varphi'_{y,\xi} = \left(\sum_{j=1}^{n} \frac{\partial \varphi_j}{\partial y_1}\xi_j, \dots, \sum_{j=1}^{n} \frac{\partial \varphi_j}{\partial y_n}\xi_j, \varphi_1, \dots, \varphi_n\right). \tag{12}$$

Since φ is a phase function, $\varphi'_{y,w,\xi} \neq 0$ for $\xi \neq 0$. Because of $\varphi'_\xi(y, y, \xi) = 0$ we consequently get $\varphi'_y(y, y, \xi) \neq 0$, i.e. for any $\xi \neq 0$ there exists a k, $1 \leq k \leq n$ with $\sum_{j=1}^{n} (\partial \varphi_j/\partial y_k)_{y=w}\, \xi_j \neq 0$ (cf. (11)). Hence

$$\det\left(\frac{\partial \varphi_j(y, w)}{\partial y_k}\right)\bigg|_{y=w} \neq 0. \tag{13}$$

Denote by μ the matrix $\varphi_{kj}(y, w)$ ($j, k = 1, \dots, n$). From (11), (13) it follows that $\mu(y, w)$ is non-singular in a neighbourhood of diag Ω. Put $\psi(y, w) = \mu^{-1}(y, w)$. From (9), (10) follows

$$\varphi(y, w, \xi) = \sum_{j,k=1}^{n} \varphi_{kj}(y, w)\, \xi_j(y_k - w_k) = \langle y - w, \mu(y, w)\, \xi\rangle .$$

Thus we get (7) and (8) is a consequence of (11) and the definition of ψ. □

Proof of Theorem 2: Because of 1.2.2.4, Corollary 2 A can be written in the form $A = A_0 + C$ with a smoothing operator C and A_0 properly supported, where the amplitude function $a(y, w, \xi)$ of A_0 vanishes outside some neighbourhood U of diag Ω. Substituting $\xi = \psi(y, w)\, \eta$ for A_0 we obtain

$$(A_0 u)(y) = \iint e^{i\langle y-w, \eta\rangle}\, a\big(y, w, \psi(y, w)\, \eta\big)\, |\det \psi(y, w)|\, u(w)\, \mathrm{d}w\, \mathrm{d}\eta .$$

In view of 1.2.2.1, Proposition 6, Theorem 2 is proved.

Continuation of the proof of Theorem 1: Apply the substitution of Lemma 3 in (5). Then

$$(\varkappa_* A)\, u(y) = (2\pi)^{-n} \iint e^{i\langle y-w, \eta\rangle}\, a\big(\varkappa^{-1}(y), \psi(y, w)\, \eta\big)\, D(y, w)\, u(w)\, \mathrm{d}w\, \mathrm{d}\eta$$

with $D(y, w) = |\det \mathrm{d}\varkappa^{-1}(w)|\, |\det \psi(y, w)|$, and 1.2.3.2, Theorem 1 shows $\varkappa_* A \in L^m(Y)$,

$$\sigma(\varkappa_* A)(y, \eta) \sim \sum_\alpha \frac{1}{\alpha!} \partial_\eta^\alpha D_w^\alpha \big[a\big(\varkappa^{-1}(y), \psi(y, w)\, \eta\big) D(y, w)\big]_{w=y} . \tag{14}$$

From (8) it follows that $\psi(y, y) = {}^t\mathrm{d}\varkappa(y)$. The αth summand in (14) is a sum of terms of the form

$$c(y)\,\eta^\gamma a^{(\beta)}\big(\varkappa^{-1}(y), {}^t d\varkappa(y)\,\eta\big) \tag{15}$$

where $c(y)$ only depends on the phase function (5) but not on the amplitude function. For the multi-indices in (15) we have

$$|\beta| \leqq 2\,|\alpha|, \qquad |\gamma| + |\alpha| \leqq |\beta|\,. \tag{16}$$

The first inequality is obvious. The second is true because the operator D_y, which gives rise to (15), does not change the difference $|\beta| - |\gamma|$, whereas ∂_η increases the difference $|\beta| - |\gamma|$ by 1. From (16) we obtain

$$|\gamma| \leqq |\beta| - |\alpha| \leqq |\beta| - |\beta|/2 = |\beta|/2\,. \tag{17}$$

For (14) it follows that

$$\sigma(\varkappa_* A)(y, \eta) \sim \sum_\beta \frac{1}{\beta!}\, a^{(\beta)}\big(\varkappa^{-1}(y), {}^t d\varkappa(y)\,\eta\big)\, \varrho_\beta(y, \eta) \tag{18}$$

with polynomials $\varrho_\beta(y, \eta)$ in η of degree $\leqq |\beta|/2$ and smooth coefficients with respect to y depending only on (5) and $\varrho_0 \equiv 1$. Substituting $y = \varkappa(x)$ yields

$$\sigma(\varkappa_* A)\big(\varkappa(x), \eta\big) \sim \sum_\beta \frac{1}{\beta!}\, a^{(\beta)}\big(x, {}^t\mathrm{d}\varkappa(x)\,\eta\big)\, \tau_\beta(x, \eta) \tag{19}$$

with polynomials $\tau_\beta(x, \eta)$ in η of degree $\leqq |\beta|/2$ depending only on $\varkappa$, but not on the operator A and $\tau_0 \equiv 1$. It remains to show that the τ_β have the form (4). The polynomials $\tau_\beta(x, \eta)$ can be calculated by substitution of differential operators. If A is a differential operator over X and $B = \varkappa_* A$,

$$\begin{aligned} \sigma(B)\,(y, \eta)|_{y=\varkappa(x)} &= \mathrm{e}^{-\mathrm{i}\langle y, \eta\rangle}\, B\, \mathrm{e}^{\mathrm{i}\langle y, \eta\rangle}|_{y=\varkappa(x)} \\ &= \mathrm{e}^{-\mathrm{i}\langle \varkappa(z), \eta\rangle}\, a(z, D_z)\, \mathrm{e}^{\mathrm{i}\langle \varkappa(z), \eta\rangle}|_{z=x}\,. \end{aligned} \tag{20}$$

With $\varkappa(z) = \varkappa(x) + \mathrm{d}\varkappa(x)\,(z - x) + h(z, x)$ it follows that

$$\varkappa(z)\,\eta = \varkappa(x)\,\eta + z\,{}^t\mathrm{d}\varkappa(x)\,\eta + h(z, x)\,\eta - x\,{}^t\mathrm{d}\varkappa(x)\,\eta\,.$$

Then, from (20) we obtain

$$\sigma(B)\,(y, \eta)|_{y=\varkappa(x)} = \mathrm{e}^{-\mathrm{i}\langle x, {}^t\mathrm{d}\varkappa(x)\,\eta\rangle}\, \big(a(z, D_z)\,\{\mathrm{e}^{\mathrm{i}\langle z, {}^t\mathrm{d}\varkappa(x)\,\eta\rangle}\, \mathrm{e}^{\mathrm{i}h(z, x)\,\eta}\}\big)_{z=x}\,. \tag{21}$$

Application of the Leibniz rule gives

$$\sigma(B)\,(y, \eta)|_{y=\varkappa(x)} = \sum_\alpha \frac{1}{\alpha!}\, a^{(\alpha)}\big(x, {}^t\mathrm{d}\varkappa(x)\,\eta\big)\, D_z^\alpha\, \mathrm{e}^{\mathrm{i}h(z, x)\,\eta}|_{z=x}\,. \quad \square$$

Let $A \in L^m_{\mathrm{cl}}(X)$ be with the homogeneous principal symbol $\sigma_A(x, \xi)$. From (3) we obtain that the principal symbol $\sigma_{\varkappa_* A}(y, \eta)$ of the PDO $\varkappa_* A$ is given by

$$\sigma_{\varkappa_* A}(y, \eta) = \sigma_A\big(x, {}^t\mathrm{d}\varkappa(x)\,\eta\big)|_{x=\varkappa^{-1}(y)}\,. \tag{22}$$

1.2.3.5. Continuity in Sobolev Spaces

Theorem 1. *Every PDO $A \in L^m(\Omega)$, $\Omega \subseteq \mathbb{R}^n$ open, $m \in \mathbb{R}$, has a continuous extension*

$$A : H^s_{\mathrm{comp}}(\Omega) \to H^{s-m}_{\mathrm{loc}}(\Omega) \tag{1}$$

for all $s \in \mathbb{R}$. If A is given by 1.2.3.1. (4) *and $\{x : a(x, \xi) \neq 0$ for some $\xi \in \mathbb{R}^n\}$ is relatively compact in Ω, A has a continuous extension*

$$A : H^s(\Omega) \to H^{s-m}(\Omega)\,, \tag{1'}$$

$s \in \mathbb{R}$. The correspondence $a \to A$ is continuous with respect to the S^m topology and the norm topology in $\mathcal{L}(H^s(\Omega), H^{s-m}(\Omega))$.

Proof: The first assertion follows from the second one. Without loss of generality we can assume that $\Omega = \mathbb{R}^n$. From

$$(Au)(x) = (2\pi)^{-n} \int e^{ix\xi} a(x, \xi) \hat{u}(\xi) \, d\xi, \quad u \in C_0^\infty(\mathbb{R}^n)$$

it follows that

$$(\widehat{Au})(\eta) = (2\pi)^{-n} \iint e^{i\langle x, \xi-\eta\rangle} a(x, \xi) \hat{u}(\xi) \, d\xi \, dx .$$

The integrations can be changed. With the function $b(\zeta, \xi) = \int e^{-ix\zeta} a(x, \xi) \, dx$ we get

$$(\widehat{Au})(\eta) = (2\pi)^{-n} \int b(\eta - \xi, \xi) \hat{u}(\xi) \, d\xi .$$

Let $w \in H^{m-s}(\mathbb{R}^n)$. Then

$$\begin{aligned}\int (\widehat{Au})(\eta) \hat{w}(\eta) \, d\eta &= (2\pi)^{-n} \iint b(\eta - \xi, \xi) \hat{w}(\eta) \hat{u}(\xi) \, d\xi \, d\eta \\ &= (2\pi)^{-n} \iint b(\eta - \xi, \xi) (1 + |\xi|)^{-s} (1 + |\eta|)^{s-m} \\ &\quad \times \hat{w}(\eta) (1 + |\eta|)^{m-s} \hat{u}(\xi) (1 + |\xi|)^s \, d\xi \, d\eta .\end{aligned}$$

We shall show that the kernel

$$K(\xi, \eta) = b(\eta - \xi, \xi) (1 + |\xi|)^{-s} (1 + |\eta|)^{s-m}$$

satisfies the inequalities

$$\int |K(\xi, \eta)| \, d\xi \le c, \qquad \int |K(\xi, \eta)| \, d\eta \le c \tag{2}$$

with a constant c independent of ξ and η. Then

$$\begin{aligned}\left|\int (\widehat{Au})(\eta) \hat{w}(\eta) \, d\eta\right| &\le (2\pi)^{-n} \iint |K(\xi, \eta) (1 + |\eta|)^{m-s} \\ &\quad \times \hat{w}(\eta) (1 + |\xi|)^s \hat{u}(\xi)| \, d\xi \, d\eta \\ &\le (2\pi)^{-n} \left(\iint |K(\xi, \eta)| \{(1 + |\eta|)^{m-s} \hat{w}(\eta)|\}^2 \, d\xi \, d\eta\right)^{1/2} \\ &\quad \times \left(\iint |K(\xi, \eta)| \{(1 + |\xi|)^s |\hat{u}(\xi)|\}^2 \, d\xi \, d\eta\right)^{1/2} \\ &\le c \, \|w\|_{m-s} \|u\|_s .\end{aligned}$$

Since $\|v\|_{s-m} = \sup\limits_{w \in H^{m-s}} \dfrac{|\langle v, w\rangle|}{\|w\|_{m-s}}$ (cf. 1.2.1.2) for $v = Au$ we get $\|Au\|_{s-m} \le c \, \|u\|_s$. It remains to show (2). Since a has compact support with respect to x for each $N \in \mathbb{Z}_+$, there exists a constant c_N with

$$|b(\zeta, \xi)| \le c_N (1 + |\zeta|)^{-N} (1 + |\xi|)^m$$

(cf. 1.2.3.1, Lemma 4). Hence we have the estimate

$$|K(\xi, \eta)| = |b(\eta - \xi, \xi) (1 + |\xi|)^{-s} (1 + |\eta|)^{s-m}| \le c \frac{(1 + |\xi|)^{m-s}}{(1 + |\eta|)^{m-s}} (1 + |\xi - \eta|)^{-N} .$$

Using the inequality $(1 + |a|)(1 + |b|)^{-1} \le 1 + |a - b|$ it follows that

$$|K(\xi, \eta)| \le c \, (1 + |\xi - \eta|)^{|m-s|-N} .$$

Thus for N sufficiently large, (2) follows. The continuity $a \to A$ results follows since the constant c in (2) is bounded by $Cc_N(b) \le C' \sup\limits_{\xi \in \mathbb{R}^n, |\alpha| \le N} (1 + |\xi|) \int |D_x^\alpha a(x, \xi)| \, dx$ with C, C' independent of b. □

1.2.4. Elliptic Pseudo-Differential Operators on Manifolds

1.2.4.1. Pseudo-Differential Operators on Manifolds

The general calculus of PDOs developed in 1.2.3 can be generalized in a natural way to systems of PDOs. If $X \subseteq \mathbb{R}^n$ is an open set, denote by $L^m(X; \mathbb{C}^k, \mathbb{C}^l)$, $m \in \mathbb{R}$, the space of linear operators $A: C_0^\infty(X, \mathbb{C}^k) \to C^\infty(X, \mathbb{C}^l)$ which can be represented by an $l \times k$-matrix of operators in $L^m(X)$. Similarly use the notation $L_{\mathrm{cl}}^m(X; \mathbb{C}^k, \mathbb{C}^l)$ for matrices of operators in $L_{\mathrm{cl}}^m(X)$. In this section we consider PDOs on manifolds. Most of the assertions are simple consequences of the definitions and the corresponding local facts. Therefore we only sketch the proofs. Details are left to the reader (concerning literature cf. ATIYAH/SINGER [2], PALAIS [1], ŠUBIN [1]).

Let M be a paracompact C^∞ manifold of dimension n and $E, F \in \mathrm{Vect}\,(M)$. A linear operator

$$A: C_0^\infty(M, E) \to C^\infty(M, F) \tag{1}$$

is called *pseudo-differential operator* on M (between the spaces of sections in the corresponding bundles) if for arbitrary trivializations

$$\chi_E: E|_V \to \Omega \times \mathbb{C}^k, \qquad \chi_F: F|_V \to \Omega \times \mathbb{C}^l \tag{2}$$

($V \subseteq M$, $\Omega \subseteq \mathbb{R}^n$ open) the operator A_Ω defined by the commutative diagram

$$\begin{array}{ccc} C_0^\infty(M, E) & \xrightarrow{\ A\ } & C^\infty(M, F) \\ \uparrow{\scriptstyle \chi_E^*} & & \downarrow{\scriptstyle (\chi_F)_*} \\ C_0^\infty(\Omega, \mathbb{C}^k) & \xrightarrow{\ A_\Omega\ } & C^\infty(\Omega, \mathbb{C}^l) \end{array}$$

belongs to $L^m(\Omega; \mathbb{C}^k, \mathbb{C}^l)$ for some $m \in \mathbb{R}$ $\left((\chi_F)_* u = (\chi_F^*)^{-1} (u|_V)\right)$. Let $L^m(M; E, F)$ be the set of all PDOs on M possessing the order $\leq m$ in this sense. The notations $L^\infty(M; E, F)$, $L^{-\infty}(M; E, F)$ are used in a similar meaning as in the local situation. Denote by $L_{\mathrm{cl}}^m(M; E, F)$ the subset of those PDOs in $L^m(M; E, F)$ for which $A_\Omega \in L_{\mathrm{cl}}^m(\Omega; \mathbb{C}^k, \mathbb{C}^l)$ for all trivializations (2). Operators in $L_{\mathrm{cl}}^m(M; E, F)$ are called classical PDOs. Note that in (2) the coordinate neighbourhoods may be non-connected.

Let $\Omega, \Omega' \subseteq \mathbb{R}^n$ be open sets and $\chi_E: E|_U \to \Omega \times \mathbb{C}^k$, $\chi_E': E|_U \to \Omega' \times \mathbb{C}^k$ ($U \subseteq M$ open) two trivializations of $E|_U$. Then $\chi_E' \circ \chi_E^{-1}$ has the form $(\varkappa_{\Omega'\Omega}, g_{\Omega'\Omega}^E): \Omega \times \mathbb{C}^k \to \Omega' \times \mathbb{C}^k$ with the coordinate diffeomorphism $\varkappa_{\Omega'\Omega}$ and the cocycle $g_{\Omega'\Omega}^E: \Omega \to \mathrm{GL}\,(k, \mathbb{C})$ belonging to E. Let $g_{\Omega'\Omega}^F: \Omega \to \mathrm{GL}\,(l, \mathbb{C})$ be the corresponding cocycle belonging to F. Then the following diagram commutes

$$h_{\Omega'\Omega}^E \left| \begin{array}{ccc} C_0^\infty(\Omega', \mathbb{C}^k) & \xrightarrow{\ A_{\Omega'}\ } & C^\infty(\Omega, \mathbb{C}^l) \\ \downarrow{\scriptstyle \chi_E'^*} & & \uparrow{\scriptstyle (\chi_F')_*} \\ C_0^\infty(M, E) & \xrightarrow{\ A\ } & C^\infty(M, F) \\ \uparrow{\scriptstyle \chi_E^*} & & \downarrow{\scriptstyle (\chi_F)_*} \\ C_0^\infty(\Omega, \mathbb{C}^k) & \xrightarrow{\ A_\Omega\ } & C^\infty(\Omega, \mathbb{C}^l) \end{array} \right| h_{\Omega'\Omega}^F,$$

with $h_{\Omega'\Omega}^E = (\chi_E'^*)^{-1} \circ \chi_E^*: C_0^\infty(\Omega, \mathbb{C}^k) \to C_0^\infty(\Omega', \mathbb{C}^k)$ and similarly $h_{\Omega'\Omega}^F$. Write $h_{\Omega'\Omega}^E$ as a composition $g_{\Omega'\Omega}^E \varkappa_{\Omega'\Omega}^*$ (i.e. substitution in the base coordinates and multiplication of vectors by the matrix function $g_{\Omega'\Omega}^E$). Then $A_{\Omega'} = h_{\Omega'\Omega}^F A_\Omega (h_{\Omega'\Omega}^E)^{-1}$ can be written in the form

$$A_{\Omega'} = g^F_{\Omega'\Omega} \varkappa^*_{\Omega'\Omega} A_\Omega (\varkappa^*_{\Omega'\Omega})^{-1} (g^E_{\Omega'\Omega})^{-1} . \tag{3}$$

Denote the coordinates in $T^*\Omega$ and $T^*\Omega'$ by (x, ξ) and (y, η), respectively. Let $\sigma(A_\Omega)(x, \xi)$ and $\sigma(A_{\Omega'})(y, \eta)$ be the complete symbols of A_Ω and $A_{\Omega'}$, respectively. Then

$$\sigma(A_{\Omega'})(y, \eta) \sim g^F_{\Omega'\Omega}(y) \left(\sigma(A_\Omega)\right)_{\varkappa_{\Omega'\Omega}} (y, \eta) \circ (g^E_{\Omega'\Omega})^{-1} (y) . \tag{4}$$

In (4) $(\ldots)_{\varkappa_{\Omega'\Omega}}$ denotes the matrix symbol obtained element-wise by applying the rule of transformation of complete symbols under the coordinate transformation $\varkappa_{\Omega'\Omega}\colon \Omega \to \Omega'$ given in 1.2.3.4 and $\circ$ denotes the composition of complete symbols according to 1.2.3.3, Theorem 3 with matrix multiplication before.

For $A \in L^m_{\mathrm{cl}}(M; E, F)$ we have for each A_Ω a homogeneous principal symbol σ_{A_Ω} of order m. From 1.2.3.3. (9) and 1.2.3.4. (22) it follows that

$$\sigma_{A_{\Omega'}}(y, \eta) = g^F_{\Omega'\Omega}(y)\, \sigma_{A_\Omega}(x, \xi)\, \left(g^E_{\Omega'\Omega}(y)\right)^{-1} \tag{5}$$

with $x = \varkappa^{-1}_{\Omega'\Omega}(y)$, $\xi = {}^t\left(\mathrm{d}\varkappa_{\Omega'\Omega}(x)\right) \eta$. Thus the following diagram commutes

$$\begin{array}{ccc} \Omega' \times \mathbb{R}^n \times \mathbb{C}^k & \xrightarrow{\sigma_{A_{\Omega'}}} & \Omega' \times \mathbb{R}^n \times \mathbb{C}^l \\ \Big\uparrow (\varkappa_{\Omega'\Omega}, {}^t\mathrm{d}\varkappa^{-1}_{\Omega'\Omega}, g^E_{\Omega'\Omega}) & & \Big\uparrow (\varkappa_{\Omega'\Omega}, {}^t\mathrm{d}\varkappa^{-1}_{\Omega'\Omega}, g^F_{\Omega'\Omega}) \\ \Omega \times \mathbb{R}^n \times \mathbb{C}^k & \xrightarrow{\sigma_{A_\Omega}} & \Omega \times \mathbb{R}^n \times \mathbb{C}^l \end{array} . \tag{6}$$

Since $(\varkappa_{\Omega'\Omega}, {}^t\mathrm{d}\varkappa^{-1}_{\Omega'\Omega})$ is just the cocycle of T^*M, the cotangent bundle of M, (6) shows that the system of σ_{A_Ω} can be interpreted as a bundle morphism

$$\sigma_A\colon \pi^*E \to \pi^*F , \tag{7}$$

where $\pi\colon T^*M \setminus \{0\} \to M$ denotes the canonical projection. σ_A is called homogeneous principal symbol (of order m) of $A \in L^m_{\mathrm{cl}}(M; E, F)$. Note that the homogeneous principal symbol of $A \in L^m_{\mathrm{cl}}$ may vanish for $A \neq 0$. Then a lower order principal symbol of order $m' < m$ is invariantly defined. For simplicity we speak of the principal symbol of order m for $A \in L^m_{\mathrm{cl}}$. Otherwise (for $A \in L^{m'}_{\mathrm{cl}}$, $m' < m$) the order is explicitly indicated.

Denote by $\mathrm{Hom}^m(\pi^*E, \pi^*F)$ the set of all bundle morphisms $\sigma\colon \pi^*E \to \pi^*F$ for which the base points $\varrho \in T^*M \setminus \{0\}$ remain fixed and which are positively homogeneous of degree m, i.e. $\sigma(x, t\xi) = t^m\sigma(x, \xi)$, $t > 0$ for each local coordinate system. Here $\sigma(x, \xi)$ represents a linear map $E_x \to F_x$.

Let S^*M be the cosphere bundle induced by T^*M with respect to some fixed Riemannian metric on M. Then each $\sigma \in \mathrm{Hom}^m(\pi^*E, \pi^*F)$ is uniquely determined by $\sigma|_{S^*M}$ and m, since the extension of $\sigma|_{S^*M}$ by homogeneity m reproduces σ. Denote by $\pi_1\colon S^*M \to M$ the canonical projection. Then $\sigma_1 = \sigma|_{S^*M}$ can be considered as a smooth section in the bundle $\mathrm{hom}\,(\pi_1^*E, \pi_1^*F) = (\pi_1^*E)^* \otimes \pi_1^*F$ over S^*M, i.e. $\sigma_1 \in C^\infty\left(S^*M, \mathrm{hom}\,(\pi_1^*E, \pi_1^*F)\right)$. This leads to a natural C^∞ topology in $\mathrm{Hom}^m(\pi^*E, \pi^*F)$.

The notion of the principal symbol can be generalized to general PDOs $A \in L^m$ as the system of equivalence classes in $S^m\left(\Omega \times \mathbb{R}^n, (\mathbb{C}^k)^* \otimes \mathbb{C}^l\right)/S^{m-1}\left(\Omega \times \mathbb{R}^n, (\mathbb{C}^k)^* \otimes \mathbb{C}^l\right)$ for the local coordinate systems $\Omega \times \mathbb{R}^n$ of T^*M. The correctness of such a definition follows from (4). We omit the details (cf. Šubin [3]) here, since on manifolds we consider mainly classical PDOs.

Denote by σ^m the map which assigns to an operator $A \in L^m(M; E, F)$ its homogeneous principal symbol σ_A of order m.

Theorem 1. *The following sequence is exact*

$$0 \to L^{m-1}_{\text{cl}}(M; E, F) \xrightarrow{i} L^m_{\text{cl}}(M; E, F) \xrightarrow{\sigma^m} \text{Hom}^m(\pi^*E, \pi^*F) \to 0 \tag{8}$$

(*i denotes the natural inclusion*).

Proof: First it is easily seen that $\ker \sigma^m = L^{m-1}_{\text{cl}}(M; E, F)$, because $\sigma^m(A) = 0$ for all $A \in L^{m-1}_{\text{cl}}$ and $\sigma^m(A) = 0$ implies $A \in L^{m-1}_{\text{cl}}$ (the last assertion follows easily from the corresponding local representations). Thus we have to show the surjectivity of σ^m. For doing this it is sufficient to construct a right inverse of σ^m. Let $\mathfrak{U} = \{U_j\}$ be a covering of M by coordinate neighbourhoods so that for all i, j with $U_i \cap U_j \neq \emptyset$ the bundles E and F are trivial over $U_i \cup U_j$. Let $\{\varphi_j\}$ be a partition of unity belonging to $\mathfrak{U}$. Consider arbitrary indices i, j with $U_i \cap U_j \neq \emptyset$ and denote by $(\sigma_A)_{ij}$ the local representation of a given $\sigma_A \in \text{Hom}^m(\pi^*E, \pi^*F)$ in a local coordinate system Ω_{ij} belonging to $U_i \cup U_j$ and fixed trivializations of E, F over $U_i \cup U_j$. Let A_{ij} be the PDO on Ω_{ij} defined by $A_{ij}u = (2\pi)^{-n} \int e^{ix\xi} (\sigma_A)_{ij}(x, \xi)\, \hat{u}(\xi)\, d\xi$ ((x, ξ) coordinates in $T^*\Omega_{ij}$). Then, using symbolic calculus (i.e. behaviour under coordinate transformations and composition) it is clear that $A \overset{\text{def}}{=} \sum_{i,j} \varphi_i A_{ij} \varphi_j$ is an operator in L^m_{cl} possessing the prescribed principal symbol. □

Remark 2. The construction of a right inverse of the symbol map σ^m in the proof of Theorem 1 depends on the choice of $\mathfrak{U}$, $\{\varphi_j\}$ and the trivializations of the bundles. If these objects are fixed, denote by Op the resulting right inverse, i.e. $A = \text{Op}(\sigma_A)$, $\sigma_A \in \text{Hom}^m(\pi^*E, \pi^*F)$.

Theorem 3. *Let E, F, $G \in \text{Vect}(M)$, $B \in L^m_{\text{cl}}(M; F, G)$, $C \in L^{m'}_{\text{cl}}(M; E, F)$ and B or C properly supported. Then $A = BC \in L^{m+m'}_{\text{cl}}(M; E, G)$ and*

$$\sigma_A = \sigma_B \sigma_C \tag{9}$$

(*this means the composition of the corresponding bundle morphisms*).

Proof: Let χ_E, χ_F and χ_G be trivializations of E, F and G, respectively, over $U \subset M$ open and let e.g. C be properly supported. Then corresponding to the notations at the beginning the following diagram commutes

$$\begin{array}{ccccc}
C_0^\infty(M, E) & \xrightarrow{C} & C_0^\infty(M, F) & \xrightarrow{B} & C^\infty(M, G) \\
\uparrow \chi_E^* & & \uparrow \chi_F^* & & \uparrow (\chi_G)_* \\
C_0^\infty(\Omega, \mathbb{C}^k) & \xrightarrow{C_\Omega} & C_0^\infty(\Omega, \mathbb{C}^l) & \xrightarrow{B_\Omega} & C^\infty(\Omega, \mathbb{C}^j)
\end{array}$$

From the corresponding local theorem on composition $A_\Omega = B_\Omega C_\Omega \in L^m_{\text{cl}}(\Omega; \mathbb{C}^k, \mathbb{C}^j)$ the assertion about the principal symbols follows, since locally the principal symbol of the composition is equal to the composition of the principal symbols (cf. 1.2.3.3). □

Theorem 4. *Let E, $F \in \text{Vect}(M)$ be equipped with Hermitean metrics, M with a Riemannian metric and let $A \in L^m_{\text{cl}}(M; E, F)$ with the homogeneous principal symbol $\sigma_A \in \text{Hom}^m(\pi^*E, \pi^*F)$. Let A^* be the formal adjoint of A defined by $(Au, v)_F = (u, A^*v)_E$ for all $u \in C_0^\infty(M, E)$, $v \in C_0^\infty(M, F)$ ($(\cdot, \cdot)_E$ the scalar product in $L^2(M, E)$,* *similarly* $(\cdot,\cdot)_F$). *Then* $A^* \in L^m_{\mathrm{cl}}(M; F, E)$ *and the homogeneous principal symbol* $\sigma_{A^*} \in \mathrm{Hom}^m(\pi^*F, \pi^*E)$ *is the Hermitean adjoint of* σ_A.

This assertion is an immediate consequence of the corresponding local theorems on adjoints and compositions (cf. 1.2.3.3).

Theorem 5. *Let* $A \in L^m(M; E, F)$. *Then* $A: C_0^\infty(M, E) \to C^\infty(M, F)$ *has a continuous extension*

$$A: H^s_{\mathrm{comp}}(M, E) \to H^{s-m}_{\mathrm{loc}}(M, F) \tag{10}$$

for arbitrary $s \in \mathbb{R}$.

The simple reduction to the local theorem (cf. 1.2.3.5) is left to the reader. Note that for compact closed M we have the continuity

$$A: H^s(M, E) \to H^{s-m}(M, F), \quad s \in \mathbb{R}. \tag{11}$$

In particular, any $A \in L^m_{\mathrm{cl}}(M; E, F)$ can be regarded as an element of $\mathcal{L}\big(H^s(M, E), H^{s-m}(M, F)\big)$ for each $s \in \mathbb{R}$. Consider $L^m_{\mathrm{cl}}(M; E, F)$ in the topology induced by the norm topology in $\mathcal{L}\big(H^s(M, E), H^{s-m}(M, F)\big)$.

Let us mention also asymptotic sums of PDOs. For sequence $A_j \in L^{m_j}(M; E, F)$, $j = 0, 1, 2, \ldots$ with $m_j \to -\infty$ when $j \to \infty$ there exists an operator $A \in L^m(M; E, F)$, $m = \max\{m_j\}$, with

$$A - \sum_{j=0}^{N} A_j \in L^{m(N)}(M; E, F),$$

$m(N) \to -\infty$ when $N \to \infty$ and A is uniquely determined mod $L^{-\infty}(M; E, F)$. Write $A \sim \sum_j A_j$ and denote A as asymptotic sum of the operators A_j.

As noted at the beginning we have, for fixed $m \in \mathbb{R}$, an injective map from $\mathrm{Hom}^m(\pi^*E, \pi^*F)$ into $C^\infty\big(S^*M, \mathrm{hom}(\pi_1^*E, \pi_1^*F)\big)$. Thus we can speak of the topology in $\mathrm{Hom}^m(\pi^*E, \pi^*F)$ induced by $C^t\big(S^*M, \mathrm{hom}(\pi_1^*E, \pi_1^*F)\big)$, $t \in \mathbb{Z}_+$.

Theorem 6. *Let* $\mathrm{Op}: \mathrm{Hom}^m(\pi^*E, \pi^*F) \to L^m_{\mathrm{cl}}(M; E, F)$ *be a right inverse of* σ^m *mentioned in Remark 2. Then there exists* $t = t(s) \in \mathbb{Z}_+$ *for each* $s \in \mathbb{R}$ *so that* Op *is continuous with respect to the topologies induced by* $C^{t(s)}\big(S^*M, \mathrm{hom}(\pi_1^*E, \pi_1^*F)\big)$ *and* $\mathcal{L}\big(H^s(M, E), H^{s-m}(M, F)\big)$, *respectively.*

This theorem is a consequence of 1.2.3.5, Theorem 1. Note that $t(s) > (|s| + n)/2$, $n = \dim M$ and that Op defines a continuous map from $\mathrm{Hom}^m(\pi^*E, \pi^*F)$ equipped with the $C^\infty\big(S^*M, \mathrm{hom}(\pi_1^*E, \pi_1^*F)\big)$ topology into $L^m_{\mathrm{cl}}(M; E, F)$ with the induced topology by the projective limit $\lim\limits_{t\to\infty} \bigcap\limits_{|s|<t} \mathcal{L}\big(H^s(M, E), H^{s-m}(M, F)\big)$ (i.e. the weakest topology in L^m_{cl} for which the embeddings $L^m_{\mathrm{cl}} \to \mathcal{L}(H^s(\ldots), H^{s-m}(\ldots))$ are continuous for all $s \in \mathbb{R}$).

Theorem 1 yields that the principal symbol map σ^m defines a map $L^m_{\mathrm{cl}}/L^{m-1}_{\mathrm{cl}} \to \mathrm{Hom}^m(\pi^*E, \pi^*F)$. Suppose that M is a closed compact C^∞ manifold. Then $L^{m-1}_{\mathrm{cl}}(M; E, F)$ is a subspace of $\mathcal{K} = \mathcal{K}\big(H^s(M, E), H^{s-m}(M, F)\big)$, the set of compact operators $H^s(\ldots) \to H^{s-m}(\ldots)$. Denote by $\|\cdot\|_s$ the norm in $\mathcal{L}\big(H^s(\ldots), H^{s-m}(\ldots)\big)$.

Theorem 7. *For* $A \in L^m_{\mathrm{cl}}(M; E, F)$ *we have*

$$\inf_{C \in \mathcal{K}} \|A + C\|_s = \sup_{\varrho \in S^*M} \|\sigma_A(\varrho)\|_{\mathrm{hom}(E_x, F_x)}, \tag{12}$$

$(\pi\varrho = x)$.

A proof of Theorem 7 is included in the considerations of 2.3.4.

Corollary 8. *The principal symbol map*

$$\sigma^m\colon L^m_{\rm cl}(M; E, F) \to \operatorname{Hom}^m(\pi^*E, \pi^*F)$$

is continuous, if $L^m_{\rm cl}$ *is considered in the induced topology of* $\mathcal{L}(H^s(\ldots), H^{s-m}(\ldots))$ *and* $\operatorname{Hom}^m(\pi^*E, \pi^*F)$ *in the topology defined by* $C(S^*M;\ \hom(\pi_1^*E,\ \pi_1^*F))$. *Denote by* $\overline{L}^m_{\rm cl}$ *the closure of* $L^m_{\rm cl}$ *in* $\mathcal{L}(H^s(\ldots), H^{s-m}(\ldots))$. *Let* $\sigma_1^m(A) = \sigma^m(A)|_{S^*M}$, $A \in L^m_{\rm cl}$. *Then* σ_1^m *has a continuous extension* $\sigma_1^m\colon \overline{L}^m_{\rm cl} \to C(S^*M, \hom(\pi_1^*E, \pi_1^*F))$ *and* σ_1^m *is surjective and* $\mathcal{K} = \ker \sigma_1^m$.

Remark 9. The above considerations show that $\overline{L}^0_{\rm cl}(M; E, F)/\overline{L}^{-1}_{\rm cl}(M; E, F)$ is a C^* algebra and that $\sigma_1^0\colon \overline{L}^0_{\rm cl}(\ldots)/\overline{L}^{-1}_{\rm cl}(\ldots) \to C(S^*M, \hom(\pi_1^*E, \pi_1^*F))$ is a $*$ isomorphism (the latter space is considered in the uniform convergence topology).

1.2.4.2. Elliptic Operators and Their Index

Let M be a closed compact C^∞ manifold, $E, F \in \operatorname{Vect}(M)$. A PDO $A \in L^m_{\rm cl}(M; E, F)$ is called *elliptic* if its homogeneous principal symbol $\sigma_A\colon \pi^*E \to \pi^*F$ ($\pi\colon T^*M \setminus \{0\} \to M$) is an isomorphism.

In this section we collect the most essential assertions about elliptic PDOs on M and sketch only some proofs. In chapter 3 an analogue of the Fredholm theory presented here will be discussed in all details for elliptic boundary problems. So the case of manifolds without boundary can be considered as an exercise (cf. Palais [1]).

Theorem 1. *Let* $A \in L^m_{\rm cl}(M; E, F)$ *be elliptic. Then there exists an elliptic* PDO $P \in L^{-m}_{\rm cl}(M; F, E)$ *being a* C^∞ *parametrix of* A (*i.e.* $AP - I \in L^{-\infty}$, $PA - I \in L^{-\infty}$).

The construction of P is as follows. Starting with $\sigma_P \overset{\rm def}{=} \sigma_A^{-1}\colon \pi^*F \to \pi^*E$ define $P_0 \in \operatorname{Op}(\sigma_P) \in L^{-m}_{\rm cl}(M; F, E)$. Then $\sigma_{P_0A} = \sigma_{P_0}\sigma_A = 1_{\pi^*E}$ implies $K \overset{\rm def}{=} P_0A - I \in L^{-1}_{\rm cl}(M; E, E)$. Now use the fact that, if we are given a sequence $A_j \in L^{m_j}$ with $m_j \to -\infty$ for $j \to \infty$, we have an asymptotic sum $\sum_j A_j \in L^{\max\{m_j\}}_{\rm cl}$ uniquely determined $\bmod L^{-\infty}$. Especially, $\sum_{j=0}^{\infty} (-1)^j K^j \in L^0_{\rm cl}(M; E, E)$ is well-defined $\bmod L^{-\infty}$ and obviously $P \overset{\rm def}{=} \left(\sum_{j=0}^{\infty} (-1)^j K^j\right) P_0$ is a left parametrix of A in the sense of $PA - I \in L^{-\infty}$. Similarly a right parametrix $P' \in L^{-m}_{\rm cl}(M; F, E)$ can be constructed and it is a simple exercise to check that then any left parametrix is a right parametrix, too, and conversely.

Another simple remark is that if P, P' are two parametrices of A, then $P - P' \in L^{-\infty}$.

Theorem 2. *Let* $A \in L^m_{\rm cl}(M; E, F)$ *be elliptic. Then* A *defines a Fredholm operator*

$$A\colon H^s(M, E) \to H^{s-m}(M, F) \tag{1}$$

for all $s \in \mathbb{R}$. *For every fixed* $s, t \in \mathbb{R}$, $t < s$, *there exists a constant* $c > 0$ *so that*

$$\|u\|_s \le c(\|Au\|_{s-m} + \|u\|_t) \tag{2}$$

for all $u \in H^s(M, E)$. *The kernel of* (1) *is a finite dimensional subspace of* $C^\infty(M, E)$ *and independently of* s *there exists a finite dimensional subspace* $L \subset C^\infty(M, F)$ *with* $L \oplus \operatorname{im} A = H^{s-m}(M, F)$. *Thus the index of* (1) *is independent of* s. *By* (1) *a Fred-*

holm operator

$$A: C^\infty(M, E) \to C^\infty(M, F) \tag{3}$$

with $L \oplus \operatorname{im} A = C^\infty(M, F)$ *is induced, i.e.* (3) *has the same index as* (1).

The Fredholm property of (1) is an immediate consequence of the fact that $P \in L_{\mathrm{cl}}^{-m}(M; F, E)$ mentioned above defines a parametrix of A in Sobolev spaces. Already $P_0 = \operatorname{Op}(\sigma_A^{-1})$ has this property. Now (2) follows from $\|u\|_s = \|(PA - K)u\|_s \leqq \|PAu\|_s + \|Ku\|_s \leqq c(\|Au\|_{s-m} + \|u\|_t)$ because of ord $P = -m$, ord $K = -\infty$. Note that in the latter estimate P was only used as a left parametrix of A. (2) easily yields $\ker A \subset C^\infty(M, E)$. Suppose now $m = \operatorname{ord} A = 0$. Let $(\cdot, \cdot)_E$ and $(\cdot, \cdot)_F$ be Hermitean scalar products of $L^2(M, E)$ and $L^2(M, F)$, respectively, induced by corresponding fixed Hermitean metrics in the bundles and a fixed Riemannian metric on M. Then $A^* \in L_{\mathrm{cl}}^0(M; F, E)$ defined by $(Au, v)_F = (u, A^*v)_E$ is again elliptic, i.e. $\ker A^* \subset C^\infty(M, F)$ is finite-dimensional and $(\operatorname{im} A) \oplus \ker A^* = L^2(M, F)$. Thus $L = \ker A^*$ is a space as asserted in Theorem 2. The case of arbitrary $m \in \mathbb{R}$ can be treated by reducing the order to $m = 0$ (cf. Theorem 5). That ind A is independent of s is obvious.

Of course, L can be constructed directly by using a parametrix of A and without reference to the reduction of orders (cf. 3.1.1.1, Theorem 5).

Note that the *Hermitean adjoint* A^* of an elliptic $A \in L_{\mathrm{cl}}^m(M; E, F)$ defined by $(Au, v)_F = (u, A^*v)_E$, $u \in C^\infty(M, E)$, $v \in C^\infty(M, F)$ is again elliptic, since $(\sigma_A)^* = \sigma_{A^*}$ and

$$\operatorname{ind} A = -\operatorname{ind} A^*.$$

Here $\ker A = \ker(A^*A)$, $\ker A^* = \ker(AA^*)$, $(\ker A^*) \oplus \operatorname{im} A = C^\infty(M, F)$ and one can set $L = \ker A^*$ in the case $m \neq 0$, too.

Note that if ind $A = 0$, one can always find an operator $K \in L^{-\infty}$ so that $A + K$ is an isomorphism in Sobolev spaces as well as in C^∞. Similarly, if ind $A > 0$ (ind $A < 0$), there is a $K \in L^{-\infty}$ so that $A + K$ is surjective (injective). In order to give an idea how to find K, consider for instance ind $A = 0$. Let $K_1: L^2(M, E) \to \ker A$ be the orthogonal projection onto $\ker A$. Choose an isomorphism $B: \ker A \to L \cong \operatorname{coker} A$ (cf. Theorem 2). Then $A + BK_1$ has the desired property (verify $BK_1 \in L^{-\infty}$ as an exercise).

In connection with the general index theory the index of A will be denoted as *analytical index* ($\operatorname{ind}_a A$). Another so-called *topological index* to be used later is denoted be $\operatorname{ind}_t A$.

If $A_1, A_2 \in L_{\mathrm{cl}}^m(M; E, F)$ are elliptic operators with $\sigma_{A_1} = \sigma_{A_2}$, we get in view of 1.2.4.1, Theorem 1 $A_1 - A_2 \in L_{\mathrm{cl}}^{m-1}(M; E, F) \subset \mathcal{K}(H^s(M, E), H^{s-m}(M, F))$, hence

$$\sigma_{A_1} = \sigma_{A_2} \Rightarrow \operatorname{ind} A_1 = \operatorname{ind} A_2 .$$

The assertion (2) is called *elliptic regularity* and has the following interpretation. If $u \in H^t(M, E)$ for some $t \in \mathbb{R}$ and $Au \in H^{s-m}(M, F)$, then $u \in H^s(M, E)$. Note that each distributional section in E belongs to $H^t(M, E)$ for a certain $t \in \mathbb{R}$ (M is compact). The elliptic regularity has a microlocal version, too, and can be used to give an equivalent definition of the wave front set of a distribution. For instance, if $\Omega \subseteqq \mathbb{R}^n$ is open, $u \in \mathcal{E}'(\Omega)$, WF (u) is the intersection of all sets $\{\varrho \in T^*\Omega \setminus \{0\}: \sigma(A)(\varrho) = 0\}$ for which $Au \in C^\infty(\Omega)$, $A \in L_{\mathrm{cl}}^0(\Omega)$. Elliptic PDOs have the property WF (Au) = WF (u) for all $u \in \mathcal{E}'(\Omega)$.

Theorem 3. *Let $A \in L^m_{\text{cl}}(M; E, F)$ and (1) a Fredholm operator for a certain fixed $s \in \mathbb{R}$. Then A is elliptic.*

Consider $\text{Hom}^m(\pi^*E, \pi^*F)$ ($m \in \mathbb{R}$ fixed) in the topology induced by $C^\infty(S^*M, \text{hom}(\pi_1^*E, \pi_1^*F))$ (cf. 1.2.4.1). A continuous map $\sigma_{(\cdot)}: [0, 1] \to \text{Hom}^m(\pi^*E, \pi^*F)$ is called (*continuous*) *homotopy* of homogeneous symbols (homotopy between σ_0 and σ_1). If this map is C^∞, the homotopy is called C^∞ *homotopy* (if only C^∞ homotopies are considered we simply speak of homotopies). Homotopies of symbols are denoted by the sign $\simeq$. Two (continuous) homotopies σ_t^0, σ_t^1 $(0 \leq t \leq 1)$ through elliptic symbols are called *homotopic* if there exists a continuous map $\sigma_t^s: [0, 1]_t \times [0, 1]_s \to \text{Hom}^m(S^*M, \text{hom}(\pi_1^*E, \pi_1^*F))$ into the set of elliptic symbols equal to the given homotopies at $s = 0$ and $s = 1$, respectively. One can easily show that each continuous homotopy σ_t between elliptic symbols σ_0, σ_1 is homotopic to a C^∞ homotopy between σ_0, σ_1 and that in an arbitrary neighbourhood of σ_t one can find a C^∞ homotopy homotopic to σ_t.

Now consider the map

$$\text{Op}: \text{Hom}^m(\pi^*E, \pi^*F) \to \mathcal{L}(H^s(M, E), H^{s-m}(M, F)) \tag{4}$$

defined in 1.2.4.1. In view of 1.2.4.1, Theorem 6, the map Op is continuous. Since Op (σ) is Fredholm for elliptic σ, we get from a homotopy $\sigma_0 \simeq \sigma_1$ through elliptic symbols a homotopy Op $(\sigma_0) \simeq$ Op (σ_1) of the corresponding Fredholm operators. Thus, according to 1.1.1.2, Proposition 4, we get the following

Remark 4. The index of an elliptic operator $A \in L^m_{\text{cl}}(M; E, F)$ depends only on the homotopy class of its homogeneous principal symbol.

Theorem 5. *For each $E \in \text{Vect}(M)$ and any $m \in \mathbb{R}$ there exists an elliptic operator $\Lambda_E^m \in L^m_{\text{cl}}(M; E, E)$ inducing an isomorphism*

$$\Lambda_E^m: H^s(M, E) \to H^{s-m}(M, E) \tag{5}$$

for each $s \in \mathbb{R}$. Then Λ_E^m induces an isomorphism $C^\infty(M, E) \to C^\infty(M, E)$, too.

For the construction of Λ_E^m it is sufficient to find an elliptic PDO $A: C^\infty(M, E) \to C^\infty(M, E)$ of order m with index zero. Adding a suitable operator with C^∞ kernel we then obtain an isomorphism $C^\infty(M, E) \to C^\infty(M, F)$ and the closure in Sobolev spaces gives isomorphisms $H^s(M, E) \to H^{s-m}(M, E)$, $s \in \mathbb{R}$. Let $\sigma_A: \pi^*E \to \pi^*F$ be an elliptic symbol positively homogeneous of order m and $\sigma_A = (\sigma_A)^*$. Then $A = \text{Op}(\sigma_A)$ has the property $A - A^* \in L^{m-1}_{\text{cl}}(M; E, F)$ and hence $\text{ind}\, A = \text{ind}\, A^* = -\,\text{ind}\, A = 0$.

A special Hermitean self-adjoint homogeneous symbol is $|\xi|^m 1_{\pi^*E}$ (in local coordinates (x, ξ) of $T^*M \setminus \{0\}$). Thus one can choose Λ_E^m in such a way that $\sigma_{\Lambda_E^m}(x, \xi) = |\xi|^m 1_{\pi^*E}$.

Now, if $A \in L^m_{\text{cl}}(M; E, F)$ is an arbitrary elliptic PDO, the operator $\Lambda_F^{s-m} A \Lambda_E^{-s} \in L^0_{\text{cl}}(M; E, F)$ is elliptic, too, for each $s \in \mathbb{R}$ and

$$\text{ind}\, A = \text{ind}\,(\Lambda_F^{s-m} A \Lambda_E^{-s}) \,. \tag{6}$$

Using the scalar product in $L^2(M, E)$ introduced in 1.1.2.2 a Hermitean scalar product in $H^s(M, E)$ can be defined by

$$(u, v)_{H^s(M, E)} = (\Lambda_E^s u, \Lambda_E^s v)_{L^2(M, E)} \,. \tag{7}$$

The correspondence $A \to \Lambda_F^{s-m} A \Lambda_E^{-s} = A_0$ is called *reduction of the order*. Since $\sigma_A|_{S^*M} = \sigma_{A_0}|_{S^*M}$, we get from (6)

Remark 6. Let $A \in L_{\text{cl}}^m(M; E, F)$ be elliptic. Then $\operatorname{ind} A$ only depends on the homotopy class of $\sigma_A|_{S^*M}$.

Let $A \in L_{\text{cl}}^0(M; E, F)$ be elliptic and $\sigma_A = \sigma^0(A)$ the homogeneous principal symbol. Suppose that σ_A depends only on the base point in T^*M. Then $\sigma_A : \pi^*E \to \pi^*F$ has a natural extension to an isomorphism of the pull backs of E and F to T^*M (including the zero section) and restriction to the zero section gives an isomorphism $\sigma_A : E \to F$ depending only on the base point. Denote the induced map of corresponding sections by $(\sigma_A)_* : C^\infty(M, E) \to C^\infty(M, F)$. This is obviously an isomorphism and $A = (\sigma_A)_* + A_{-1}$ with some $A_{-1} \in L_{\text{cl}}^{-1}(M; E, F)$. Thus $\operatorname{ind} A = 0$.

Denote the identical operator $C^\infty(M, E) \to C^\infty(M, E)$ by 1_E.

Between PDOs on M a direct addition $\oplus$ is defined in a natural way. For $A \in L_{\text{cl}}^0(M; E, F)$, $B \in L_{\text{cl}}^0(M; E', F')$ we get $A \oplus B \in L_{\text{cl}}^0(M; E \oplus E', F \oplus F')$. If A, B are elliptic, we have obviously $\operatorname{ind}(A \oplus B) = \operatorname{ind} A + \operatorname{ind} B$. Especially for $G \in \operatorname{Vect}(M)$ follows $\operatorname{ind}(A \oplus 1_G) = \operatorname{ind} A$.

An elliptic symbol $\sigma_A \in \operatorname{Hom}^m(\pi^*E, \pi^*F)$ regarded as an isomorphism outside the compact subset $M \subset T^*M$ (M identified with the zero section) between bundles over T^*M, namely π^*E, π^*F, represents an element $d(\sigma_A)$ of $K(T^*M)$ (cf. 1.1.3.2). The preceding remarks show that $\operatorname{ind} A$ only depends on the class $d(\sigma_A)$ and after reduction of the order it is sufficient to consider zero order symbols. The definition of $d(\sigma_A)$ yields $d(\sigma_A \oplus \sigma_{A'}) = d(\sigma_A) + d(\sigma_{A'})$, $d(\sigma_A \circ \sigma_B) = d(\sigma_A) + d(\sigma_B)$, $d(\sigma_A^{-1}) = -d(\sigma_A)$, $d(\sigma_A \oplus 1_{\pi^*E}) = d(\sigma_A)$ $(E \in \operatorname{Vect}(M))$. Moreover $\sigma_{A_0} \simeq \sigma_{A_1}$ through elliptic symbols implies $d(\sigma_{A_0}) = d(\sigma_{A_1})$.

The following theorem is called *Atiyah-Singer index theorem.*

Theorem 7. *There exists a homomorphism* $\operatorname{ind}_t : K(T^*M) \to \mathbb{Z}$ *(the so-called topological index) so that*

$$\operatorname{ind}_a A = \operatorname{ind}_t \big(d(\sigma_A)\big) \tag{8}$$

for each classical PDO A on M.

The homomorphism ind_t can be expressed in topological terms without any reference to pseudo-differential operators or Fredholm property in functional spaces (cf. 3.2.2.3). In the K-theoretic construction only the difference element $d(\sigma_A)$ plays a role.

There are other formulations of the index theorem with a lot of geometrical consequences, cf. Atiyah/Singer [2], Palais [1], Gilkey [1]. We do not use them explicitly and recall only examples (cf. Atiyah/Singer [2, III]).

Example. (1) Let $\dim M$ be odd and $A : C^\infty(M, E) \to C^\infty(M, F)$ be an elliptic differential operator on M. Then $\operatorname{ind} A = 0$.

Example. (2) Let $M = S^1$ so that S^*M can be identified with two exemplars $S_\pm^1$ of the unit circle $S^1 = \{z \in \mathbb{C} : |z| = 1\}$. Consider a scalar symbol $\sigma_A : \pi^*\mathbb{C}^1 \to \pi^*\mathbb{C}^1$ of order zero defined by $\sigma_A(z, 1) = z^k$, $k \in \mathbb{Z}$, $\sigma_A(z, -1) = 1$. Then σ_A is an elliptic symbol on S^1 and $\operatorname{ind}\big(\operatorname{Op}(\sigma_A)\big) = -k$.

Now let

$$A : C^\infty(M, \mathbb{C}^N) \to C^\infty(M, \mathbb{C}^N) \tag{9}$$

be an elliptic PDO between the corresponding trivial bundles of dimension N and $n = \dim M$. Then σ_A can be interpreted as a continuous map

$$\sigma_A\colon S^*M \to \mathrm{GL}\,(N, \mathbb{C})\,. \tag{10}$$

The unitary group $U(N)$ is the deformation retract of $\mathrm{GL}\,(N, \mathbb{C})$. Denote by ϱ_1: $\mathrm{GL}\,(N, \mathbb{C}) \to U(N)$ a retraction. Moreover let $\tau\colon U(N) \to U(N)/U(N-1) \cong S^{2N-1}$ be the canonical projection.

The cohomology $H^*(U(N))$ is an exterior algebra generated by elements $h_j^N \in H^{2j-1}(U(N))$ ($i = 1, \dots, N$) and h_j^N transforms into h_j^{N-1} under the restriction homomorphism corresponding to $U(N-1) \to U(N)$ ($j = 1, \dots, N-1$). Therefore, we write h_j instead of h_j^N.

Example. (3) Let M be a hypersurface in $\mathbb{R}^{n+1}$ ($n = \dim M$). Then the elliptic operator (9) has the index

$$\operatorname{ind} A = \begin{cases} \dfrac{(-1)^{N+n-1}}{(n-1)!}\,\sigma_A^* h_n[S^*M] & \text{if} \quad N \geqq n\,, \\ 0 & \text{if} \quad N < n\,. \end{cases} \tag{11}$$

Here σ_A^* denotes the pull back of the cohomology with the identification $H^*(\mathrm{GL}(N, \mathbb{C})) \cong H^*(U(N))$. For $n = N$ the above expression is equal to

$$\operatorname{ind} A = -\frac{1}{(n-1)!}\deg\,(\tau \circ \varrho_1 \circ \sigma_A) \tag{12}$$

(deg denotes the degree of the mapping, cf. Example 2).

Example. (4) Let A be an elliptic operator of the form (9) on a compact manifold M of dimension n, where $N \leqq n/2$. Then $\operatorname{ind} A = 0$ except for $n = 2N = 4k$, X is orientable and has Euler number zero. In particular

$$\operatorname{ind} A = 0 \quad \text{if} \quad n = \dim M > 1,\ N = 1\,. \tag{13}$$

Finally discuss external tensor products of operators. Let $\sigma_A\colon \pi^*E \to \pi^*F$ be a continuous map (not necessarily C^∞), $\pi\colon T^*M \setminus \{0\} \to M$ homogeneous of degree m. Then, with σ one can connect an operator $A\colon H^s(M, E) \to H^{s-m}(M, F)$ ($s \in \mathbb{R}$ fixed) obtained as the limit of PDOs A_k for $k \to \infty$ ($m = \operatorname{ord} A_k$) in the operator topology with homogeneous principal symbols σ_{A_k} converging to σ_A uniformly over S^*M (cf. 1.2.4.1, Corollary 8). This will be discussed in 2.3.4. in the more general case of manifolds with boundary. A is called a PDO with *continuous homogeneous principal* symbol σ_A. Ellipticity is defined in an obvious way.

Let $M = M_1 \times M_2$ and $A_i\colon C^\infty(M_i, E_i) \to C^\infty(M_i, F_i)$ be elliptic PDOs with homogeneous principal symbols $\sigma_{A_i}\colon \pi_i^*E \to \pi_i^*F$ ($\pi_i\colon T^*M_i \to M_i$) and $\operatorname{ord} \sigma_{A_i} = r > 0$. Then

$$\sigma_{A_1} \# \sigma_{A_2} = \begin{pmatrix} \sigma_{A_1} \otimes 1 & -1 \otimes \sigma_{A_2}^* \\ 1 \otimes \sigma_{A_2} & \sigma_{A_1}^* \otimes 1 \end{pmatrix} : \begin{matrix} \pi_1^*E_1 \otimes \pi_2^*E_2 \\ \oplus \\ \pi_1^*F_1 \otimes \pi_2^*F_2 \end{matrix} \to \begin{matrix} \pi_1^*F_1 \otimes \pi_2^*E_2 \\ \oplus \\ \pi_1^*E_1 \otimes \pi_2^*F_2 \end{matrix}$$

is an isomorphism over $T^*(M_1 \times M_2) \setminus \{0\}$ (cf. the external multiplication defined in 1.1.3.2). Hermitean metrics in the bundles are fixed. $\sigma_{A_1} \# \sigma_{A_2}$ is homogeneous of degree r but in general only a continuous symbol over $M_1 \times M_2$. Then

$$A_1 \# A_2 = \begin{pmatrix} A_1 \otimes 1 & -1 \otimes A_2^* \\ 1 \otimes A_2 & A_1^* \otimes 1 \end{pmatrix} : \begin{matrix} C^\infty(M, E_1 \otimes E_2) \\ \oplus \\ C^\infty(M, F_1 \otimes F_2) \end{matrix} \to \begin{matrix} C^\infty(M, F_1 \otimes E_2) \\ \oplus \\ C^\infty(M, E_1 \otimes F_2) \end{matrix}$$

(with the corresponding external products in the bundles) is an elliptic PDO over $M = M_1 \times M_2$ with the continuous symbol $\sigma_{A_1} \# \sigma_{A_2}$ and it can be proved that

$$\operatorname{ind} (A_1 \# A_2) = \operatorname{ind} A_1 \operatorname{ind} A_2 .$$

2. Operators in the Half-Space and on Manifolds

2.1. Operators on the Half-Axis

2.1.1. Certain Function Spaces

2.1.1.1. The Spaces H, H^+, H^-

Denote by $\mathscr{S}(\mathbb{R})$ the Schwartz space of all rapidly decreasing (complex valued) C^∞ functions on the real t-axis with the usual Fréchet space topology given by the set of semi-norms

$$u \to \sup_{t \in \mathbb{R}} |t^k D_t^l u(t)|\,, \quad k, l \in \mathbb{Z}_+$$

$\left(D_t = -i\dfrac{\mathrm{d}}{\mathrm{d}t}\right)$. Let r^+ denote the restriction operator of a function to $\overline{\mathbb{R}}_+ = \{t \in \mathbb{R}: t \geqq 0\}$ and set

$$\mathscr{S}(\overline{\mathbb{R}}_+) = \{r^+u : u \in \mathscr{S}(\mathbb{R})\}\,.$$

$\mathscr{S}(\overline{\mathbb{R}}_+)$ with the set of semi-norms

$$u \to \sup_{t \in \overline{\mathbb{R}}_+} |t^k D_t^l\, u(t)|\,, \quad k, l \in \mathbb{Z}_+$$

is a Fréchet space. The subspace $\mathscr{S}_0(\overline{\mathbb{R}}_-) = \{u \in \mathscr{S}(\mathbb{R}): \operatorname{supp} u \subseteqq \overline{\mathbb{R}}_-\} \subset \mathscr{S}(\mathbb{R})$ is considered with the induced topology. Then there is a topological isomorphism $\mathscr{S}(\overline{\mathbb{R}}_+) \cong \mathscr{S}(\mathbb{R})/\mathscr{S}_0(\overline{\mathbb{R}}_-)$. Let θ^+ be the characteristic function of $\overline{\mathbb{R}}_+$, i. e. $\theta^+(t) = 1$ for $t \in \overline{\mathbb{R}}_+$ and $\theta^+(t) = 0$ for $t \in \mathbb{R}_-$. θ^+u $\big(u \in \mathscr{S}(\mathbb{R})\big)$ is the extension of r^+u by zero to $t < 0$. We shall usually identify θ^+u and r^+u because of the obvious one-to-one correspondence between them. If a distinction is necessary, denote by j^+v the extension of a function $v \in \mathscr{S}(\overline{\mathbb{R}}_+)$ by zero for $t < 0$. By $v \to j^+v$ we define a continuous embedding

$$j^+ : \mathscr{S}(\overline{\mathbb{R}}_+) \to L^2(\mathbb{R})\,.$$

Similar definitions and notations are used for $\overline{\mathbb{R}}_- = \{t \in \mathbb{R}: t \leqq 0\}$, e.g. r^-, $\mathscr{S}(\overline{\mathbb{R}}_-)$ and there is a continuous embedding $j^- : \mathscr{S}(\overline{\mathbb{R}}_-) \to L^2(\mathbb{R})$.

We identify $\mathscr{S}(\overline{\mathbb{R}}_\pm)$ with $j^\pm\big(\mathscr{S}(\overline{\mathbb{R}}_\pm)\big) \subset L^2(\mathbb{R})$. Then $\mathscr{S}(\overline{\mathbb{R}}_+)$ and $\mathscr{S}(\overline{\mathbb{R}}_-)$ are orthogonal. Set $\mathscr{S}_\theta(\mathbb{R}) = \mathscr{S}(\overline{\mathbb{R}}_+) \oplus \mathscr{S}(\overline{\mathbb{R}}_-)$. The extension mappings $j^\pm$ have continuous extensions $j^\pm : L^2(\mathbb{R}_\pm) \to L^2(\mathbb{R})$ (denoted in the same way). If we consider $L^2(\mathbb{R}_\pm)$

as subspaces of $L^2(\mathbb{R})$, we have $L^2(\mathbb{R}) = L^2(\mathbb{R}_+) \oplus L^2(\mathbb{R}_-)$. The restriction mappings $r^\pm : \mathscr{S}(\mathbb{R}) \to \mathscr{S}(\overline{\mathbb{R}}_\pm)$ have continuous extensions $r^\pm : L^2(\mathbb{R}) \to L^2(\mathbb{R}_\pm)$. These are complementary projections parallel to $L^2(\mathbb{R}_-)$ and $L^2(\mathbb{R}_+)$, respectively.

Denote by P' the vector space of all distributions ϱ of the form $\varrho = \sum_{j=0}^{q-1} a_j D_t^j \delta_0$ for some $q > 0$, $a_j \in \mathbb{C}$ (δ_0 is the Dirac distribution in the origin). Let P'_q be the subspace of P' of distributions containing at most $q - 1$ derivatives of the Dirac distribution. Obviously $P'_q \cong \mathbb{C}^q$ and we consider P'_q with the topology of $\mathbb{C}^q$. Let P' be equipped with the direct limit topology $\ldots \subset P'_q \subset P'_{q+1} \subset \ldots \subset P'$, i.e. a sequence ϱ_k, $k \in \mathbb{Z}_+$ converges iff $\varrho_k \in P'_r$ for some r and all k and it converges in P'_r.

Let $u \to Fu$, $Fu(\nu) = \int e^{-it\nu} u(t)\, dt$, $u \in L^2(\mathbb{R})$, be the Fourier transform on the axis. Instead of Fu we also write $\hat{u}$. The Fourier transform induces an isomorphism

$$F : L^2(\mathbb{R}_t) \to L^2(\mathbb{R}_\nu)\,.$$

The Fourier inversion formula yields

$$u(t) = F^{-1}\hat{u}(t) = (2\pi)^{-1} \int e^{it\nu}\, h(\nu)\, d\nu\,.$$

Consider $L^2(\mathbb{R})$ with the Hermitean scalar product $(u, v) = (2\pi)^{-1} \int u(t)\, \overline{v(t)}\, dt$. Then $(u, v) = (2\pi)^{-1}(\hat{u}, \hat{v})$, i.e. $(2\pi)^{-1/2} F$ defines an isometry of $L^2(\mathbb{R})$.

The image of P' with respect to F is denoted by H'. This is the space of all polynomials in ν. Via the one-to-one mapping $F: P' \to H'$ we obtain a topology in H'. We set $H'_q = F(P'_q)$.

Let $H^+(H_0^-)$ denote the image of $\mathscr{S}(\overline{\mathbb{R}}_+)$ $(\mathscr{S}(\overline{\mathbb{R}}_-))$ with respect to F. This asymmetric notation is motivated by the application to symbol spaces. It is convenient to set $H^- = H_0^- \oplus H'$. Moreover set $H_0 = H^+ \oplus H_0^-$, $H = H^+ \oplus H^-$, $H_q^- = H_0^- \oplus H'_q$, $H_q = H^+ \oplus H_q^-$. We consider H^+ and H_q^- in the image topology of $\mathscr{S}(\overline{\mathbb{R}}_+)$ and $\mathscr{S}(\overline{\mathbb{R}}_-)$, respectively. Then $F: \mathscr{S}_\theta(\mathbb{R}) \to H_0$ is an isomorphism of Fréchet spaces and the Fourier transforms $\Pi^\pm$ of $r^\pm$. i.e. $r^\pm = F^{-1}\Pi^\pm F$, define continuous projections

$$\Pi^+ : H_0 \to H^+, \quad \Pi^- : H_0 \to H_0^-\,.$$

Now extend the projections $\Pi^\pm$ to H by setting $\Pi^+ = 0$ over H', $\Pi^- = 1_{H'}$ over H'.

Let $V^\pm$ be the image of $L^2(\mathbb{R}_\pm) \subset L^2(\mathbb{R})$ with respect to F. Then V^+ and V^- are orthogonal and span $L^2(\mathbb{R})$, $V^+ \oplus V^- = L^2(\mathbb{R})$. The projections $\Pi^\pm$ extend continuously to $L^2(\mathbb{R}_\nu)$ as orthogonal projections onto $V^\pm$ parallel to $V^\mp$.

We introduce the complex ζ-plane with $\mathbb{R}_\nu$ as real axis, $\zeta = \nu + i\mu$.

Proposition 1. *The following conditions are equivalent:*

(i) $h \in H^+$;

(ii) $h \in C^\infty(\mathbb{R})$ *has an analytic extension into the lower complex half plane* $\operatorname{Im} \zeta < 0$, *i.e. there exists an analytic function in the half plane* $\operatorname{Im} \zeta < 0$ *continuous in* $\operatorname{Im} \zeta \leq 0$ *with the given function h as boundary values. There is an asymptotic expansion*

$$h(\zeta) \sim \sum_{k \leq -1} a_k \zeta^k \quad \textit{for } |\zeta| \to \infty \textit{ in } \operatorname{Im} \zeta \leq 0 \tag{1}$$

and all derivatives $D_\nu^l h$ $(l \in \mathbb{Z}_+)$ *have asymptotic expansions obtained by formal differentiation of* (1).

Note that the coefficients in the expansion (1) are related to the derivatives of the Fourier preimage of h at $t = +0$. For $h = \hat{u}$, $u \in \mathscr{S}(\overline{\mathbb{R}}_+)$ we have for instance

$$a_{-k-1} = -i \lim_{t \to +0} D_t^k u(t)\,, \quad k = 0, 1, 2, \dots\,. \tag{2}$$

Proof: (i) $\Rightarrow$ (ii): Let $h = \hat{u}$, $u \in \mathscr{S}(\overline{\mathbb{R}}_+)$, i.e.

$$h(\nu) = \int_0^\infty e^{-it\nu}\, u(t)\, dt\,.$$

Partial integration yields for $\nu \neq 0$

$$h(\nu) = \left[-\frac{1}{i\nu} e^{-it\nu}\, u(t)\right]_0^\infty + \frac{1}{\nu}\int_0^\infty e^{-it\nu}\, D_t u(t)\, dt = \dots =$$

$$= \frac{1}{i\nu} u(0) + \frac{1}{i\nu^2} D_t u(0) + \dots + \frac{1}{i\nu^{k+1}} D_t^k u(0) + \frac{1}{i\nu^{k+1}} \int_0^\infty e^{-it\nu}\, D_t^{k+1} u(t)\, dt\,,$$

therefore

$$h(\nu) \sim \sum_{k \leq -1} \frac{1}{i\nu^{-k}} D_t^{-k-1} u(0)\,. \tag{3}$$

Similarly

$$h'(\nu) = \int_0^\infty e^{-it\nu}\, (-it)\, u(t)\, dt \sim -i \sum_{k \leq -1} \frac{1}{i\nu^{-k}} D_t^{-k-1} (tu)\,(0)\,.$$

From $D_t^l(tu)\,(0) = l D_t^{l-1} u(0)$, $l \in \mathbb{Z}_+$, it follows that

$$h'(\nu) \sim \sum_{k \leq -1} \frac{k}{i\nu^{-k+1}} D_t^{-k-1} u(0)$$

and this is the result of formal differentiation of (3). The higher derivatives can be considered similarly. Obviously, the asymptotic expansions remain valid for $\operatorname{Im} \zeta < 0$. Thus (i) $\Rightarrow$ (ii) is proved.

(ii) $\Rightarrow$ (i): Let u be the inverse Fourier transform of the function h satisfying (ii). $h \in L^2(\mathbb{R})$ implies $u \in L^2(\mathbb{R})$ and by 1.2.1.2, Proposition 8 $u(t)$ vanishes for almost all $t < 0$. By (ii) the function $\nu^k D_\nu^l h(\nu) \in C^\infty(\mathbb{R})$ is the sum of a polynomial and a function $h_{kl}(\nu)$ satisfying (ii) for any $k, l \in \mathbb{Z}_+$. Thus $F^{-1}(\nu^k D_\nu^l h(\nu))$ is the sum of derivatives of the Dirac distribution at the origin and an L^2 function with support in $\overline{\mathbb{R}}_+$, i.e. $D_t^k t^l u(t)|_{t>0} \in L^2(\mathbb{R}_+)$. Since k and l are arbitrary, we obtain $u \in \mathscr{S}(\overline{\mathbb{R}}_+)$. □

Functions in H_0^- can be similarly characterized.

Corollary 2. *The following conditions are equivalent:*

(i) $h \in H_0^-$,

(ii) *$h \in C^\infty(\mathbb{R})$ has an analytic extension into the upper complex half plane $\operatorname{Im} \zeta > 0$ and processes an asymptotic expansion (1) in $\operatorname{Im} \zeta \geq 0$ which can be formally differentiated.*

For $h = \hat{u}$, $u \in \mathscr{S}(\overline{\mathbb{R}})_-$ we have (2) with $t \to -0$ instead of $+0$ and $+i$ instead of $-i$.

Corollary 2 follows from Proposition 1 and the observation that the complex conjugation $h \to \bar{h}$ defines a topological isomorphism between H^+ and H_0^-.

From the definition of H and Proposition 1 with Corollary 2 we get

Corollary 3. *The following conditions are equivalent:*

(i) $h \in H$;

(ii) $h \in C^\infty(\mathbb{R})$ *has an asymptotic expansion*

$$h(\nu) \sim \sum_{j=0}^{q-1} b_j \nu^j + \sum_{k \leq -1} a_k \nu^k \quad \textit{for} \quad |\nu| \to \infty \tag{4}$$

which can be formally differentiated.

Proof. (i) $\Rightarrow$ (ii) is obvious. To prove (ii) $\Rightarrow$ (i) note first that after subtraction of a polynomial it can be assumed that $q = 0$. Then $h \in L^2(\mathbb{R})$. To finish the proof we can apply Proposition 1 and Corollary 2 (ii) $\Rightarrow$ (i) to $\Pi^+ h$ and $\Pi^- h$, respectively. □

Note that H^+, H_0^-, H^-, and H are algebras, i.e. closed under multiplication.

Define the space H_q for $q \in \mathbb{Z}$ as the subspace of those functions $h \in H$ for which $\nu^{-q+1} h(\nu) \leq$ const for all $\nu \in \mathbb{R}$. Similarly define H_q^+ and H_q^- for $q \in \mathbb{Z}$.

Any excision function φ (i.e. $\varphi \in C^\infty(\mathbb{R})$ and φ vanishes near the origin and is equal to 1 outside a bounded set) belongs to H. Thus we have $\varphi h \in H$ for any $h \in H$. Note that $(1 \pm i\nu)^m \in H$ for any $m \in \mathbb{Z}$. Therefore $(1 \pm i\nu)^m h \in H_{q+m}$ for $h \in H_q$.

Now we consider the conformal mapping of the complex ζ-plane into a complex z-plane defined by

$$z = \frac{1 - i\zeta}{1 + i\zeta} \quad \text{or} \quad \zeta = \frac{1}{i} \frac{1 - z}{1 + z}.$$

Set $S^1 = \{z \in \mathbb{C}: |z| = 1\}$, $B_0 = \{z \in \mathbb{C}: |z| \leq 1\}$, $B_\infty = \{z \in \mathbb{C}: |z| \geq 1\}$. Then $\mathbb{R}_\nu$ transforms into S^1, $\{\operatorname{Im} \zeta \leq 0\}$ into B_0, $\{\operatorname{Im} \zeta \geq 0\}$ into B_∞ and $\nu = \pm\infty$ corresponds to $z = -1$. By the substitution $z = (1 - i\nu)(1 + i\nu)^{-1}$ functions on S^1 are transformed into functions on the real axis. Denote this transformation by $\varkappa^*$ (the pull back of functions with respect to the mapping $\varkappa: \mathbb{R}_\nu \to S^1$). By $[-\pi, \pi] \in \varphi \to z = e^{i\varphi} \in S^1$ is given a one-to-one correspondence between the interval $[-\pi, \pi]$ with identified end points and S^1. Functions on S^1 are then considered as functions on $[-\pi, \pi]$, too.

Proposition 4. *The following conditions are equivalent:*

(i) $h \in H^+$ ($h \in H_0^-$);

(ii) $g \stackrel{\text{def}}{=} (\varkappa^*)^{-1} h \in C^\infty(S^1)$ *vanishes at* $\varphi = \pm\pi$ ($z = -1$) *and has an analytic extension into* B_0 (B_∞).

Note that the assertion for H_0^- is an immediate consequence of that for H^+ taking complex conjugation of functions. .

Proof: One has to show that the asymptotic expansion (4) near $\nu = \pm\infty$ with $q = 0$ holds iff g is smooth at $\varphi = \pm\pi$. The other assertions are obvious.

(i) $\Rightarrow$ (ii): Let $h \in H^+$, $g \stackrel{\text{def}}{=} (\varkappa^*)^{-1} h$. We have $g'(\varphi) = -h'(\nu)(1 + \nu^2)/2$. The right hand side is a sum of a constant equal to $\lim\limits_{\varphi \to +\pi} g'(\varphi)$ and a function possessing an expansion like (4) ($q = 0$). By induction we get that all derivatives $g^{(k)}(\varphi)$ are continuous at $\varphi = \pm\pi$.

(ii) $\Rightarrow$ (i): Let $g \in C^\infty(S^1)$, $h \overset{\text{def}}{=} \varkappa^* g$. The Taylor expansion of $g(\varphi)$ near $\varphi = \pm\pi$ yields the asymptotic expansion of $h(\nu)$ for $|\nu| \to \infty$. It is easily checked that this expansion can be differentiated formally. □

The proof of Proposition 4 gives another characterization of functions in H in terms of its transform to S^1.

Corollary 5. *The following conditions are equivalent:*

(i) $h \in H_q$;

(ii) $g \overset{\text{def}}{=} (\varkappa^*)^{-1} h$ *has the property that* $(1 + e^{i\varphi})^{q-1} g(\varphi) \in C^\infty(S^1)$.

We now pass to a characterization of $C^\infty(S^1)$ in terms of Fourier coefficients. First we show that the standard topology of $C^\infty(S^1)$ given by the semi-norms

$$\sup_{-\pi \leq \varphi \leq \pi} |D_\varphi^k g(\varphi)|\,, \qquad k \in \mathbb{Z}_+\,, \quad g \in C^\infty(S^1)$$

can be equivalently described by the countable set of semi-norms

$$||D_\varphi^k g(\varphi)||_{L^2(S^1)}\,, \qquad k = 0, 1, 2, \ldots, \quad g \in C^\infty(S^1)\,. \tag{5}$$

It is easily seen that the vector space $C^\infty(S^1)$ is complete with respect to this topology. Hence, in view of the closed graph theorem, it is sufficient to show that the standard topology is weaker (or stronger) than the topology given by (5). But this follows from the estimates

$$||D_\varphi^k g(\varphi)||_{L^2(S^1)}^2 = (2\pi)^{-1} \int_{-\pi}^{\pi} |D_\varphi^k g(\varphi)|^2 \, d\varphi \leq \text{const} \sup |D_\varphi^k g(\varphi)|^2\,.$$

We shall use the following well-known

Lemma 6. *Any real (complex) separable Hilbert space $\mathcal{H}$ is isometrically isomorphic to the space $l_2(\mathbb{R})$ ($l_2(\mathbb{C})$) of square summable sequences of real (complex) numbers. If $\{e_j\} \subset \mathcal{H}$ is an orthonormal base of $\mathcal{H}$, the mentioned isometric isomorphism is given by*

$$\mathcal{H} \ni f \to \{(f, e_j)\} \in l_2\,,$$

i.e. $||f||^2 = \sum_j |(f, e_j)|^2$.

An orthonormal base in $L^2(S^1)$ with the scalar product

$$(u, v) = (2\pi)^{-1} \int_{-\pi}^{\pi} u(\varphi)\, \overline{v(\varphi)} \, d\varphi\,, \quad u, v \in L^2(S^1) \tag{6}$$

is given by $e_j(\varphi) = e^{ij\varphi}$, $j \in \mathbb{Z}$. Then $g \in L^2(S^1)$ is equivalent to

$$\sum_j |(g, e_j)|^2 < \infty\,,$$

i.e. the sequence of Fourier coefficients is square summable, and this sum equals the square of the norm of g. Let

$$g(\varphi) = \sum_j g_j\, e^{ij\varphi}\,, \quad g_j = (g, e_j)$$

be the Fourier decomposition of $g \in L^2(S^1)$ and assume that $D_\varphi g \in L^2(S^1)$. Then

$$D_\varphi g(\varphi) = \sum_j j g_j e^{ij\varphi}$$

implies that $\sum_j j^2 |g_j|^2 < \infty$. By induction we get

Corollary 7. *A function g belongs to $C^\infty(S^1)$ iff the sequence of Fourier coefficients g_j of g is rapidly decreasing, i.e. for any $k \in \mathbb{Z}_+$ we have*

$$\sum_j j^{2k} |g_j|^2 < \infty .$$

For $g \in C^\infty(S^1)$ the Fourier series of g converges.

We use now the complex variable $z = e^{i\varphi}$. For $g \in L^2(S^1)$ the Fourier decomposition is

$$g(z) = \sum_{j=-\infty}^{\infty} g_j z^j , \qquad |z| = 1 .$$

Then $g_+(z) = \sum_{j=0}^{\infty} g_j z^j$ and $g_-(z) = \sum_{j=-\infty}^{-1} g_j z^j$ are analytic functions in B_0 (B_∞) and $g = g_+ + g_-$. Denote the projections

$$L^2(S^1) \ni g \to g_\pm \in L^2(S^1)$$

by $\pi^\pm$. Let $L^2_+(S^1)$ $\left(L^2_-(S^1)\right)$ be the image of $\pi^+(\pi^-)$. Similarly define the spaces $C^\infty_\pm(S^1) = \pi^\pm\left(C^\infty(S^1)\right)$. Obviously $C^\infty_\pm(S^1)$ are subspaces of $C^\infty(S^1)$ and $C^\infty(S^1) = C^\infty_+(S^1) \oplus C^\infty_-(S^1)$.

It is useful to discuss some relations between functions on the real axis and on S^1. Define an operator

$$A : L^2(S^1) \to L^2(\mathbb{R})$$

by $A = \varkappa^* \circ (1 + z)$, i.e. $Ag(\nu) = \varkappa^*[(1 + z)\, g(z)]\,(\nu) = \frac{2}{1 + i\nu} g\left((1 - i\nu)\,(1 + i\nu)^{-1}\right)$. $Ag \in L^2(\mathbb{R})$ for $g \in L^2(S^1)$ follows from

$$(2\pi)^{-1} \int |Ag(\nu)|^2 \, d\nu = (2\pi)^{-1} \int |g(e^{i\varphi})\,(1 + e^{i\varphi})|^2 \frac{2}{(|1 + e^{i\varphi}|)^2} \, d\varphi = 2\, ||g||_{L^2(S^1)} .$$

Moreover $A^{-1} = (1 + z)^{-1} \circ (\varkappa^*)^{-1}$, hence A is an isomorphism.

Note that

$$Az^j = 2 \frac{(1 - i\nu)^j}{(1 + i\nu)^{j+1}} \qquad \text{for} \quad j \geqq 0 ,$$

$$Az^{-j-1} = 2 \frac{(1 + i\nu)^j}{(1 - i\nu)^{j+1}} \qquad \text{for} \quad j \geqq 0 .$$

The functions $\{z^j\}_{j=-\infty}^{\infty}$ form an orthonormal base in $L^2(S^1)$ and $\{z^j\}_{j=0}^{\infty}$ span the subspace $L^2_+(S^1)$ in $L^2(S^1)$. Therefore the functions $\{(1 - i\nu)^j\,(1 + i\nu)^{-j-1}, (1 + i\nu)^j\,(1 - i\nu)^{-j-1}\}_{j=0}^{\infty}$ form an orthogonal base in $L^2(\mathbb{R}_\nu)$ and V^+ is spanned by $\{(1 - i\nu)^j\,(1 + i\nu)^{-j-1}\}_{j=0}^{\infty}$.

Thus the following diagram commutes

$$\begin{array}{ccc} L^2(S^1) & \xrightarrow{\pi^+} & L^2_+(S^1) \\ \downarrow A & & \downarrow A \\ L^2(\mathbb{R}) & \xrightarrow{\Pi^+} & V^+ \end{array}$$

(the analogous diagram with minus signs commutes, too). Restriction to C^∞ gives the commutative diagram

$$\begin{array}{ccc} C^\infty(S^1) & \xrightarrow{\pi^+} & C^\infty_+(S^1) \\ \downarrow A & & \downarrow A \\ H_0 & \xrightarrow{\Pi^+} & H^+ . \end{array}$$

Corollary 8. *A function $h(\nu)$ belongs to $L^2(\mathbb{R})$ iff*

$$h(\nu) = \sum_{j=-\infty}^{\infty} a_j(1 - i\nu)^j (1 + i\nu)^{-j-1} \tag{7}$$

with $\sum_{j=-\infty}^{\infty} |a_j|^2 < \infty$. *The projections Π^+h and Π^-h are given by*

$$\Pi^+h(\nu) = \sum_{j=0}^{\infty} a_j(1 - i\nu)^j (1 + i\nu)^{-j-1} , \tag{8}$$

$$\Pi^-h(\nu) = \sum_{j=-\infty}^{-1} a_j(1 - i\nu)^j (1 + i\nu)^{-j-1} . \tag{9}$$

h belongs to H_0 iff there is an expansion (7) *with rapidly decreasing coefficients a_j, i.e.* $\sum_{j=-\infty}^{\infty} |j|^{2k} |a_j|^2 < \infty$ *for any $k \in \mathbb{Z}_+$ and* (8), (9) *are the projections to H^+ and H_0^-, respectively, introduced at the beginning. A function h belongs to H_q iff there is an expansion*

$$h(\nu) = \sum_{k=0}^{q-1} c_k\nu^k + \sum_{j=-\infty}^{\infty} a_j(1 - i\nu)^j (1 + i\nu)^{-j-1} \tag{10}$$

with rapidly decreasing sequence $\{a_j\}$.

Note that the sequence of coefficients $\{a_j\}$ in (7) is uniquely determined by the function h, since $\{(1 - i\nu)^j (1 + i\nu)^{-j-1}\}$ is an orthogonal system of functions. Similarly the coefficients in (10) are uniquely determined by h.

Let

$$h(\nu) = \sum_{j=0}^{\infty} a_j(1 - i\nu)^j (1 + i\nu)^{-j-1} \in H^+$$

and $h(\nu) \sim \sum_{k \le -1} b_k\nu^k$ for $|\nu| \to \infty$. Then

$$b_{-1} = -i \sum_{j=0}^{\infty} (-1)^j a_j .$$

As a consequence of (2) $u(+0) = \sum_{j=0}^{\infty} (-1)^j a_j$ follows for $u = F^{-1}h \in \mathscr{S}(\overline{\mathbb{R}}_+)$.

Note that the above assertions about expansions of functions in H^+, H_0^-, H remain valid, if we replace the coordinate z by $(\lambda - i\nu)(\lambda + i\nu)^{-1}$ for an arbitrary $\lambda > 0$. Instead of (10) we get the expansion

$$h(\nu) = \sum_{k=0}^{q-1} c_k\nu^k + \sum_{j=-\infty}^{\infty} b_j(\lambda - i\nu)^j (\lambda + i\nu)^{-j-1} , \quad h \in H_q .$$

In Proposition 1, Proposition 4 and in the Corollaries 2, 3, 5, and 8 we obtained several characterizations of the spaces H^+, H_0^-, H. Now we shall discuss the topology of these spaces. We shall mainly consider H^+ (H_0^- is analogous). Then the assertions about H are simple corollaries.

The following countable systems of semi-norms are defined on the space H^+:

$$p_l^{(1)}(h) = \|\Pi^+(\nu^{l_1} D_\nu^{l_2} h(\nu))\|_{L^2(\mathbb{R})}, \quad l = (l_1, l_2) \in \mathbb{Z}_+ \times \mathbb{Z}_+,$$

$$p_l^{(2)}(h) = \sup_{-\pi \leq \varphi \leq \pi} |D_\varphi^l g(\varphi)|, \quad l \in \mathbb{Z}_+, g = (\varkappa^*)^{-1} h,$$

$$p_l^{(3)}(h) = \left(\sum_{j=0}^{\infty} |a_j|^2 j^{2l}\right)^{1/2}, \quad l \in \mathbb{Z}_+, \quad h(\nu) = \sum_{j=0}^{\infty} a_j (1 - i\nu)^j (1 + i\nu)^{-j-1},$$

$h \in H^+$.

The semi-norms $p_l^{(1)}$ define the topology on H^+, which is carried over from $\mathscr{S}(\overline{\mathbb{R}}_+)$ by the Fourier transform. This can be proved by a similar argument as for the space $C^\infty(S^1)$ with the semi-norms (5).

Proposition 9. $p_l^{(1)}, p_l^{(2)}, p_l^{(3)}$ *define the same topology on* H^+.

Proof: The equivalence of $p_l^{(2)}$ and $p_l^{(3)}$ has already been proved. Now we show that convergence with respect to $p_l^{(3)}$ implies $p_l^{(1)}$-convergence. Therefore H^+ is also complete with respect to $p_l^{(3)}$ and in view of the Banach Theorem the equivalence of $p_l^{(1)}$ and $p_l^{(3)}$ follows.

First we note that the numbers $\|\Pi^+(\nu^{l_1} D_\nu^{l_2} e_j(\nu))\|_{L^2(\mathbb{R})}$, $e_j(\nu) = (1 - i\nu)^j (1 + i\nu)^{-j-1}$ can be estimated by $c_{l_1 l_2} |j|^{l_2}$ with constants $c_{l_1 l_2}$ depending on l_1, l_2, but not on j. With constants c_{ml_1} we have $\nu^{l_1} = \sum_{m=0}^{l_1} c_{ml_1} (1 + i\nu)^m$.

From $D_\nu[(1 - i\nu)^k (1 + i\nu)^{k'}] = -k(1 - i\nu)^{k-1} (1 + i\nu)^{k'} + k'(1 - i\nu)^k (1 + i\nu)^{k'-1}$ we obtain by induction that the coefficients $\alpha_{kk'}(j)$ in $D_\nu^{l_2}[(1 - i\nu)^j (1 + i\nu)^{-j-1}] = \sum_k \alpha_{kk'}(j) (1 - i\nu)^k (1 + i\nu)^{k'}$ can be estimated by const $|j|^{l_2}$. Therefore

$$\|\Pi^+(\nu^{l_1} D_\nu^{l_2} e_j(\nu)\|_{L^2(\mathbb{R})} \leq \sum_{m=0}^{l_1} |c_m| \left\|\Pi^+\left(\sum_k \alpha_{kk'}(j) (1 - i\nu)^k (1 + i\nu)^{-k'+m}\right)\right\|$$

$$= \sum_{m=0}^{l_1} |c_m| \left\| \sum_{\substack{k' < m \\ k < m - k'}} \alpha_{kk'}(j) (1 - i\nu)^k (1 + i\nu)^{-k'+m} \right\| \leq \text{const } |j|^{l_2},$$

and the assertion then follows from

$$\|\Pi^+(\nu^{l_1} D_\nu^{l_2} h(\nu))\|_{L^2(\mathbb{R})}^2 = \left\| \sum_{j=0}^{\infty} a_j \Pi^+(\nu^{l_1} D_\nu^{l_2} e_j(\nu)) \right\|_{L^2(\mathbb{R})}^2$$

$$\leq \left(\sum_{j=0}^{\infty} |a_j|^2 j^{2m}\right) \sum_{j=0}^{\infty} j^{-2m} \|\Pi^+(\nu^{l_1} D_\nu^{l_2} e_j(\nu))\|_{L^2(\mathbb{R})}^2 \leq \text{const} \sum_{j=0}^{\infty} |a_j|^2 j^{2m}. \quad \square$$

The obvious modification of the assertion for H_0^- is left to the reader.

Corollary 10. *The topology of* H_q *is defined by the set of semi-norms*

$$p_l(h) = \sup_{-\pi \leq \varphi \leq \pi} |D_\varphi^l[(1 + e^{i\varphi})^{q-1} (\varkappa^*)^{-1} h(\varphi)]|.$$

For $H_1 = \varkappa^*(C^\infty(S^1))$ *this is the topology for which* $\varkappa^*$ *is an isomorphism.*

Note that by Corollary 10 $(\varkappa^*)^{-1} H^+$ and $(\varkappa^*)^{-1} H_0^-$ are closed subspaces of $C^\infty(S^1)$. They can be characterized as the subspaces of $C_+^\infty(S^1)$ $(C_-^\infty(S^1))$ vanishing at $\varphi = \pm\pi$ (or equivalently in $z = -1$).

Denote by $\tilde{H} \subset H$ the subspace of functions for which the expansion (10) is a finite sum, i.e. any $h \in \tilde{H}$ has the form

$$h(\nu) = \sum_{k=0}^{q-1} c_k \nu^k + \sum_{j=-N}^{N} a_j (1 - i\nu)^j (1 + i\nu)^{-j-1}$$

for some $N \in Z_+$ depending on h. $\tilde{H} \subset H$ is dense, since the Fourier expansion converges in $C^\infty(S^1)$.

The projections $\Pi^\pm$ restricted to $\tilde{H}$ can be defined by integration over a contour in the upper complex ζ half plane. Then $\Pi^\pm$ are the extensions by continuity.

Proposition 11. *Let $h \in \tilde{H}$. Then, for any closed Jordan curve Γ in the half plane* $\operatorname{Im} \zeta > 0$ *containing the point* i *and moreover the point ζ_0 for the second formula, we have*

$$\Pi^+ h(\zeta_0) = \frac{1}{2\pi i} \int_\Gamma \frac{f(\zeta)}{\zeta_0 - \zeta} d\zeta \qquad \text{for} \quad \operatorname{Im} \zeta_0 < 0,$$

$$\Pi^- h(\zeta_0) = -\frac{1}{2\pi i} \int_\Gamma \frac{f(\zeta)}{\zeta_0 - \zeta} d\zeta \quad \text{for} \quad \operatorname{Im} \zeta_0 > 0 .$$

The proof based on the Cauchy formula is left to the reader.

2.1.1.2. Further Properties of the Projectors Π^+

In this section we consider *integrals of Cauchy type*

$$\frac{1}{2\pi i} \int_R \frac{f(\lambda)}{\zeta - \lambda} d\lambda . \tag{1}$$

If we assume $|f(\lambda)| \leq \text{const}\, |\lambda|^{-1}$ for $|\lambda| \geq 1$ and $f \in C^\infty(\mathbb{R})$, then (1) is an analytic function in $\operatorname{Im} \zeta > 0$ and $\operatorname{Im} \zeta < 0$, $\zeta = \nu + i\mu$.

Proposition 1. *For $f \in \mathcal{S}(\mathbb{R})$ we have*

$$\Pi^+ f(\nu) = \lim_{\substack{\mu<0 \\ \mu\to 0}} \int \frac{1}{2\pi i} \frac{f(\lambda)\, d\lambda}{(\nu + i\mu) - \lambda} = \frac{1}{2} f(\nu) + \frac{1}{2\pi i} \int_{\text{p.v.}} \frac{f(\lambda)\, d\lambda}{\nu - \lambda} \tag{2}$$

and

$$\Pi^- f(\nu) = \lim_{\substack{\mu>0 \\ \mu\to 0}} -\frac{1}{2\pi i} \int \frac{f(\lambda)\, d\lambda}{(\nu + i\mu) - \lambda} = \frac{1}{2} f(\nu) - \frac{1}{2\pi i} \int_{\text{p.v.}} \frac{f(\lambda)\, d\lambda}{\nu - \lambda} . \tag{3}$$

Proof: We deal with Π^+, the case Π^- is similar. For the first equality of (2) we have to show

$$F(\theta^+ u) = \lim_{\substack{\mu<0 \\ \mu\to 0}} \frac{1}{2\pi i} \int \frac{f(\lambda)\, d\lambda}{(\nu + i\mu) - \lambda} \qquad \text{for} \quad u \in \mathcal{S}(\mathbb{R}) ,\ f = \hat{u} .$$

From

$$F(\theta^+ e^{\mu t}) = \int_0^\infty e^{-i\nu t + \mu t}\, dt = \frac{-i}{\nu + i\mu} , \quad \mu < 0$$

it follows that

$$F\left(\theta^+ e^{\mu t} u(t)\right) = \frac{1}{2\pi i} \int \frac{\hat{u}(\lambda)}{\nu + i\mu - \lambda} d\lambda$$

and therefore in view of $F(wu) = (2\pi)^{-1} \int \hat{w}(\nu - \lambda)\, \hat{u}(\lambda)\, d\lambda$, $u, w \in L^1(\mathbb{R})$,

$$F\left(\theta^+ u(t)\right) = \lim_{\substack{\mu<0 \\ \mu\to 0}} F\left((\theta^+ e^{\mu t} u(t)\right) = \lim_{\substack{\mu<0 \\ \mu\to 0}} \frac{1}{2\pi i} \int \frac{\hat{u}(\lambda)\, d\lambda}{\nu + i\mu - \lambda}.$$

Now we pass to the second equality in (2), i.e. the relation with the Hilbert transform. Let

$$g(\nu + i\mu) = \int \frac{f(\lambda)\, d\lambda}{\nu + i\mu - \lambda}, \quad \mu < 0,$$

and set

$$g_1(\nu, \mu) = \int_{|\nu-\lambda|\le 1} \frac{f(\lambda) - f(\nu)}{\nu + i\mu - \lambda} d\lambda + \int_{|\nu-\lambda|>1} \frac{f(\lambda)}{\nu + i\mu - \lambda} d\lambda,$$

$$g_2(\nu, \mu) = f(\nu) \int_{|\nu-\lambda|<1} \frac{d\lambda}{\nu + i\mu - \lambda},$$

thus $g = g_1 + g_2$. We have

$$g_2(\nu, \mu) = f(\nu) \int_{|\varrho|<1} \frac{d\varrho}{\varrho + i\mu} = f(\nu) \left\{ \int_{-1}^{1} \frac{\varrho\, d\varrho}{\varrho^2 + \mu^2} - i\mu \int_{-1}^{1} \frac{d\varrho}{\varrho^2 + \mu^2} \right\} =$$

$$= \frac{2}{i} f(\nu) \arctan \frac{1}{\mu} \to i\pi f(\nu) \quad \text{for} \quad \mu \to -0.$$

In $g_1(\nu, \mu)$ one can take the limit $\mu \to -0$ under the integrals. Furthermore

$$\int_{\varepsilon<|\nu-\lambda|<1} \frac{f(\nu)}{\nu - \lambda} d\lambda = 0,$$

since the integrand is an odd function. Therefore

$$g_1(\nu, 0) = \lim_{\varepsilon\to 0} \int_{|\nu-\lambda|>\varepsilon} \frac{f(\lambda)\, d\lambda}{\nu - \lambda} = \int_{\text{p.v.}} \frac{f(\lambda)\, d\lambda}{\nu - \lambda}. \quad \square$$

Proposition 2. *The projector $\Pi^+ : \mathcal{S}(\mathbb{R}) \to V^+$ has a continuous extension to an operator in $\mathcal{L}\left(F(H^s(\mathbb{R})), F(H^s(\overline{\mathbb{R}}_+))\right)$ for any $s \in (-\frac{1}{2}, \frac{1}{2})$.*

For $s = 0$ this result is contained in 2.1.1.1. Note that (2) and (3) remain valid for $f \in H^s(\mathbb{R})$, $s \in (-\frac{1}{2}, \frac{1}{2})$. A proof of Proposition 2 is given in ÈSKIN [3]. A similar result is true for Π^-.

Corollary 3. *Any function $f \in F\left(H^s(\mathbb{R})\right)$, $s \in (-\frac{1}{2}, \frac{1}{2})$ has a unique decomposition $f = f^+ + f^-$ where $f^+ \in F\left(H^s(\overline{\mathbb{R}}_+)\right)$, $f^- \in F\left(H^s(\overline{\mathbb{R}}_-)\right)$ and $f^\pm = \Pi^\pm f$.*

In our applications in the sections 2.1.2.1, 2.1.2.3 and 2.1.2.4 it is useful to have explicit expressions for $\Pi^+[(1 - i\nu)^p (1 + i\nu)^q]$, $p, q \in \mathbb{Z}$. Now it is obvious that $(1 - i\nu)^p (1 + i\nu)^q \in H^+$ for $-q > p \geqq 0$ and $(1 - i\nu)^p (1 + i\nu)^q \in H^-$ for $q \geqq 0$, $p \in \mathbb{Z}$.

The case $p \geqq -q > 0$ still remains. By the binomial formula we have

$$\left(\frac{1-i\nu}{1+i\nu}\right)^p = \left(\frac{2}{1+i\nu} - 1\right)^p = \sum_{k=0}^{p} 2^k \binom{p}{k} (1+i\nu)^{-k} (-1)^{p-k}.$$

With the notation $j = p + q$ it follows that

$$(1-i\nu)^p (1+i\nu)^q = (1+i\nu)^j \left(\frac{1-i\nu}{1+i\nu}\right)^p = \sum_{k=0}^{\infty} 2^k \binom{p}{k} (1+i\nu)^{-k+j} (-1)^{p-k},$$

and therefore

$$\Pi^+ [(1-i\nu)^p (1+i\nu)^q] = \sum_{k=j+1}^{p} 2^k \binom{p}{k} (1-i\nu)^{-k+p+q} (-1)^{p-k}.$$

2.1.1.3. The Operator Π'

Consider the space $\mathscr{S}(\mathbb{R})$ on the real t-axis and denote by

$$r' : \mathscr{S}(\mathbb{R}) \to \mathbb{C}$$

the restriction operator $r'u \overset{\text{def}}{=} u(0)$, $u \in \mathscr{S}(\mathbb{R})$. We set

$$r'u = \lim_{\substack{t>0 \\ t\to 0}} u(t)$$

for $u \in \mathscr{S}_\theta(\mathbb{R})$ and $r'p = 0$ for any $p \in P'$. Then

$$r' : \mathscr{S}_\theta(\mathbb{R}) \oplus P' \to \mathbb{C}$$

is a continuous linear functional. The composition

$$H \xrightarrow{F^{-1}} \mathscr{S}_\theta(\mathbb{R}) \oplus P' \xrightarrow{r'} \mathbb{C}$$

defines the continuous linear functional

$$\Pi' : H \to \mathbb{C}, \quad \Pi' = r' \circ F^{-1}.$$

We shall give some equivalent expressions for Π'. Let $h \in H$ be absolutely integrable, i.e. $h \in H_{-1}$, then

$$\Pi' h = (2\pi)^{-1} \int h(\nu)\, d\nu. \tag{1}$$

This follows from

$$u(t) = (2\pi)^{-1} \int e^{it\nu} h(\nu)\, d\nu$$

putting $t = 0$. If moreover h belongs to H^+ (in other words it has an analytic extension into the lower complex half plane), the Cauchy Theorem shows that the integral vanishes, i.e.

$$\Pi' h = 0 \quad \text{for} \quad h \in H^+ \quad \text{and} \quad \int |h(\nu)|\, d\nu < \infty. \tag{2}$$

Proposition 1. *Let Γ be a closed Jordan curve in* $\operatorname{Im} \zeta > 0$ *around the point* $\zeta = i$ *and counter-clockwise oriented. Then for any* $h \in \tilde{H}$

$$\Pi' h = (2\pi)^{-1} \int_\Gamma f(\zeta)\, d\zeta \tag{3}$$

(the integrand is the meromorphic extension of f to the whole complex plane). Since $\tilde{H} \subset H$ is dense, the operator Π' on H is uniquely defined by (3) *as continuous extension.*

Proof: The integral over $\Pi^- f$ vanishes, since $\Pi^- f$ is analytic in $\operatorname{Im} \zeta > 0$. Set $u = F^{-1}\Pi^+ f$. The assertion follows from

$$u(0) = \sum_{j=0}^{N} (-1)^j a_j \quad \text{for} \quad f(\nu) = \sum_{j=0}^{N} a_j (1 - i\nu)^j (1 + i\nu)^{-j-1},$$

in view of

$$(2\pi)^{-1} \int_{\Gamma} a_j \frac{(1 - i\zeta)^j}{(1 + i\zeta)^{j+1}} \, d\zeta = (-1)^j a_j .$$

The last equation is a consequence of the well-known formula

$$g^{(j)}(\zeta_0) = \frac{j!}{2\pi i} \int_{\Gamma_0} \frac{g(\zeta) \, d\zeta}{(\zeta - \zeta_0)^{j+1}}$$

for analytic functions g and a closed Jordan curve Γ_0 around ζ_0 putting $g(\zeta) = (1 - i\zeta)^j$, $\zeta_0 = i$. □

2.1.2. Boundary Symbols in the One-Dimensional Case

2.1.2.1. Fredholm Property of Wiener-Hopf Operators

In this section we consider so-called Wiener-Hopf operators on the real axis and on S^1. After the classical work of Wiener/Hopf [1] this theory was enriched by new results and generalizations (cf. Mushelišvili [2], Gohberg/Fel'dman [1], Prössdorf [1], Kreĭn [1], Mihlin [1]).

Recall some basic facts of this theory in a form adapted for our applications.

Let $\sigma \in H$. Then there is a linear continuous operator

$$\Pi^+\sigma: H^+ \to H^+$$

defined by

$$H^+ \ni h \to \Pi^+(\sigma h) \in H^+ .$$

Operators of the form $\Pi^+\sigma$ ($\sigma \in H$) are called *Wiener-Hopf operators*. This notation comes from an obvious relation between $\Pi^+\sigma$ and classical Wiener-Hopf operators via the Fourier transform. (One should not mix up the operator $\Pi^+\sigma: H^+ \to H^+$ with the image of $\sigma \in H$ under the projection $\Pi^+: H \to H^+$.)

In Wiener/Hopf [1] the following integral equation is studied

$$f(x) - \int_0^\infty K(x - y) \, f(y) \, dy = 0 \quad \text{for} \quad x > 0 . \tag{1}$$

Consider the left hand side of (1) as an operator applied to $f \in \mathcal{S}(\overline{\mathbb{R}}_+)$. Assume $K(t) \in \mathcal{S}_0(\mathbb{R})$. The Fourier transform of (1) is

$$\Pi^+ (\hat{f}(\nu) - \hat{K}(\nu) \, f(\nu)) = 0 ,$$

i.e. we get an operator of the form $\Pi^+\sigma$ with $\sigma(\nu) = 1 - \hat{K}(\nu) \in H_1$. Solutions f of (1) correspond to functions in the kernel of $\Pi^+\sigma$ in H^+ and conversely.

Let $\sigma \in H_1$ and $\sigma^1 \overset{\text{def}}{=} (\varkappa^*)^{-1}\sigma \in C^\infty(S^1)$ be the transformed function on the sphere S^1 (cf. 2.1.1.1). The following commuting diagram

$$\begin{array}{ccc} C^\infty_+(S^1) & \xrightarrow{\pi^+\sigma^1} & C^\infty_+(S^1) \\ \downarrow A & & \downarrow A \\ H^+ & \xrightarrow{\Pi^+\sigma} & H^+ \end{array}$$

gives a natural one-to-one correspondence of Wiener-Hopf operators on the axis $\mathbb{R}$ and induced operators on S^1. Here $\pi^+\sigma^1$ is defined in an analogous way as $\Pi^+\sigma$.

Note that the Fréchet space H^+ is a *countable-normed space*, i.e. its topology can be given by a countable set of norms $|\cdot|_j$, $j \in \mathbb{Z}_+$, and we can assume that $||h||_j \leqq ||h||_k$, $h \in H^+$, for arbitrary $j \leqq k$.

A linear operator $A: E \to F$, E, F countable-normed spaces, is continuous iff for any $j \in \mathbb{Z}_+$ there exists a $k = k(j) \in \mathbb{Z}_+$ such that

$$||Au||_j^F \leqq c_j \, ||u||_j^E$$

with suitable constants c_j ($||\cdot||_j^E$, and $||\cdot||_j^F$ denote the jth norm in E and F, respectively). The best constants c_j define a countable set of norms on $\mathscr{L}(E, F)$. In particular we consider $\mathscr{L}(H^+, H^+)$ with this topology.

Lemma 1. *The mapping* $H \ni \sigma \to \Pi^+\sigma \in \mathscr{L}(H^+, H^+)$ *is continuous.*

Proof: Obviously $\Pi^+\sigma \in \mathscr{L}(H^+, H^+)$. Further, the assertion is clear for $\sigma(\nu) \in H'$. For $\sigma \in H_1$ it follows via transform to σ^1 from the remark that the multiplication operator $C^\infty_+(S^1) \ni g(\varphi) \to \sigma^1(\varphi)\, g(\varphi) \in C^\infty(S^1)$ depends continuously on σ^1. □

Similarly we define systems of Wiener-Hopf operators. Let $\sigma \in H \otimes \hom(\mathbb{C}^k, \mathbb{C}^l)$, $k, l \in \mathbb{Z}_+$, i.e. a matrix $\sigma_{ij}(\nu)$ of functions in H, $h \in H^+ \otimes \mathbb{C}^k$, $h = (h_1, \ldots, h_k)$, $h_j(\nu)$ functions in H^+. Write $\sigma(\nu)\, h(\nu)$ for $\left(\sum_j \sigma_{ij}(\nu)\, h_j(\nu)\right)_{i=1,\ldots,l}$, and define

$$(\Pi^+\sigma)\, h = \left\{\sum_j \Pi^+\left(\sigma_{ij}(\nu)\, h_j(\nu)\right)\right\}_{i=1,\ldots,l} .$$

For the right hand side write also $\Pi^+\sigma(\nu)\, h(\nu)$, where the projection Π^+ acts on each component of the vector $\sigma(\nu)\, h(\nu)$ of functions in H.

Proposition 2. *Let* $\sigma(\nu) \in H_1 \otimes \hom(\mathbb{C}^k, \mathbb{C}^k)$, $k \in \mathbb{Z}_+$ *and* $\det \sigma(\nu) \neq 0$ *for all* $\nu \in \overline{\mathbb{R}}$, *i.e. in particular* $\det\left(\lim_{\nu \to \pm\infty} \sigma(\nu)\right) \neq 0$. *Then*

$$\Pi^+\sigma: H^+ \otimes \mathbb{C}^k \to H^+ \otimes \mathbb{C}^k$$

is a Fredholm operator with the index $\operatorname{ind} \Pi^+\sigma = -\deg(\det \sigma)$ (deg *denotes the mapping degree). A parametrix is given by* $\Pi^+\sigma^{-1}$.

A proof shall be given in 2.1.2.3.

Note that the condition of a non-vanishing determinant is not necessary. In 2.1.2.3 will be proved (i) ⇒ (ii) of the following

Proposition 3. *Let* $\sigma \in H \otimes \hom(\mathbb{C}^k, \mathbb{C}^k)$. *Then the following conditions are equivalent:*

(i) $\det \sigma(\nu)$ *has a finite number of zeros of finite orders on* $\overline{\mathbb{R}}$ (*i.e. including* $\pm\infty$);

(ii) $\Pi^+\sigma: H^+ \otimes \mathbb{C}^k \to H^+ \otimes \mathbb{C}^k$

is a Fredholm operator.

(ii) ⇒ (i) can be proved as in **Prössdorf** [1]. Illustrate the situation by the following example. Take $\sigma \in H^+$ such that $\sigma(\nu) \neq 0$ for all ν, but the limits of all derivatives vanish at ∞. A function with this property can be obtained by restriction of $e^{-|\zeta - i|}$, $\zeta \in \mathbb{C}$, to the real axis. Then $\Pi^+\sigma$ is not Fredholm. In fact, we have $\Pi^+(\sigma h) = \sigma h$, $h \in H^+$ and the equation $\Pi^+(\sigma h) = g$, $g \in H^+$ has no solution for infinitely many linearly independent right hand sides, since $\sigma h = g$ implies that g and all its derivatives vanish at ∞.

Sometimes we consider $\Pi^+\sigma$ not on H^+ but its extension to the L^2-closure V^+ of H^+.

Lemma 4. *Let $\sigma \in H_1$. Then $\Pi^+\sigma$ has a unique extension to an operator in $\mathcal{L}(V^+, V^+)$.*

For the proof note that $\sigma h \in L^2(\mathbb{R})$ for $h \in V^+$ and Π^+ has a continuous extension to $L^2(\mathbb{R})$ (cf. 2.1.1.1).

Note that $\Pi^+\sigma \in \mathcal{L}(V^+, V^+)$ depends continuously on $\sigma \in H_1$ (cf. Lemma 1).

Non-vanishing symbols or determinants imply the Fredholm property of the V^+-extension and this condition is also necessary:

Proposition 5. *Let $\sigma \in H_1 \otimes \hom(\mathbb{C}^k, \mathbb{C}^k)$. Then*

$$\Pi^+\sigma: V^+ \otimes \mathbb{C}^k \to V^+ \otimes \mathbb{C}^k$$

is Fredholm iff

$$\det \sigma(\nu) \neq 0 \quad \textit{for all} \quad \nu \in \overline{\mathbb{R}}.$$

The sufficiency follows as in the proof of Proposition 2, since $1 - \Pi^+\sigma\,\Pi^+\sigma^{-1}$ is compact in $V^+ \otimes \mathbb{C}^k$ (cf. 2.1.2.2). The converse is a classical result (cf. **Gohberg/Fel'dman** [1])

2.1.2.2. Definition of Boundary Symbols

Besides the operators

$$\Pi^+\sigma_A: H^+ \otimes \mathbb{C}^k \to H^+ \otimes \mathbb{C}^j, \qquad \sigma_A \in H \otimes \hom(\mathbb{C}^k, \mathbb{C}^j)$$

considered in 2.1.2.1, we now study further types of operators occuring in the so-called *boundary symbols.*

A function $\sigma_K \in H^+ \otimes \hom(\mathbb{C}^{l_-}, \mathbb{C}^j)$ is called *potential symbol.* Multiplication of vectors in $\mathbb{C}^l$ by σ_K yields a mapping

$$\sigma_K: \mathbb{C}^{l_-} \to H^+ \otimes \mathbb{C}^j. \tag{1}$$

It continuously depends on the matrix function σ_K. More generally, for a given $\sigma'_K \in H \otimes \hom(\mathbb{C}^{l_-}, \mathbb{C}^j)$, we can define

$$\Pi^+\sigma'_K: \mathbb{C}^{l_-} \to H^+ \otimes \mathbb{C}^j$$

by $v \to \Pi^+(\sigma'_K v) = (\Pi^+\sigma'_K)\, v$ and get a mapping (1) with $\sigma_K = \Pi^+\sigma'_K$.

A matrix function $\sigma_T: H^- \otimes \hom(\mathbb{C}^k, \mathbb{C}^{l_+})$ is called *trace symbol.* We define a mapping

$$\Pi'\sigma_T: H^+ \otimes \mathbb{C}^k \to \mathbb{C}^{l_+} \tag{2}$$

by $h \to \Pi'_\nu(\sigma_T(\nu)\, h(\nu))$. Obviously it continuously depends on the matrix function σ_T. For $\sigma'_T \in H \otimes \hom(\mathbb{C}^k, \mathbb{C}^{l_+})$ the mapping (2) depends on $\Pi^-\sigma_T$ only, since for

$h \in H^+ \otimes \mathbb{C}^k$

$$\Pi'(\sigma_T h) = \Pi'\big((\Pi^+\sigma_T)\,h\big) + \Pi'(\Pi^-\sigma_T)\,h = \Pi'\big((\Pi^-\sigma_T)\,h\big)\,.$$

(cf. 2.1.1.3.(2)).

The smallest $q \in \mathbb{Z}_+$ with $\sigma_T \in H_q^- \otimes \hom(\mathbb{C}^k, \mathbb{C}^{l_+})$ is called the *type* of the trace symbol σ_T.

Next we define the so-called *Green symbols*. Consider the topological tensor product $H^+ \otimes H_d^-$ of Fréchet spaces (the definition and elementary properties of the topological tensor product of Fréchet spaces are given in 2.2.1.1). Let $(p_l^+)_{l \in \mathbb{Z}_+}$ and $(p_m^-)_{m \in \mathbb{Z}_+}$ be sets of semi-norms on H^+ and H_d^-, respectively, defining the topology of these spaces. Then on the algebraic tensor product of H^+ and H_d^-, i.e. on the space of functions of the form

$$g(\nu, \varrho) = \sum_{j=1}^{N} h_j(\nu)\, f_j(\varrho) \tag{3}$$

$h_j \in H^+$, $f_j \in H_d^-$, $N \in \mathbb{Z}_+$ arbitrary, semi-norms

$$(p_l^+ \otimes p_m^-)\,(g) = \inf \left\{ \sum_{j=1}^{N} p_l^+(h_j)\, p_m^-(f_j) \right\}$$

are defined, where the infimum is taken over all decompositions (3) of g. This countable set of semi-norms defines a topology on the algebraic tensor product and the closure is called topological tensor product denoted by $H^+ \otimes H_d^-$. The topological tensor product is a Fréchet space.

Note that H^+ and H_d^- are nuclear spaces as closed subspaces of the nuclear space $\varkappa^*(C^\infty(S^1))$ (cf. 2.1.1.1, remark following Corollary 10). Then $H^+ \otimes H_d^-$ is nuclear, too.

Define $H^+ \otimes H^-$ as the projective limit of $H^+ \otimes H_d^-$ for $d \to \infty$. In a similar way the topological tensor product $H_\nu \otimes H_\varrho$ is defined (the indices denote the variable, confusion with the type should not be possible). By the multiplication $(f, h) \to fh$ a continuous mapping $(H_\nu \otimes H_\varrho) \times H_\varrho \to H_\nu \otimes H_\varrho$, $\big(f(\nu) \otimes g(\varrho), h(\varrho)\big) \to f(\nu) \otimes g(\varrho)\, h(\varrho)$ is induced. By $\Pi' : H_\varrho \to \mathbb{C}$ is induced a continuous mapping

$$\Pi'_\varrho : H_\nu \otimes H_\varrho \to H_\nu\,, \tag{4}$$

$\Pi'_\varrho\big(f(\nu) \otimes g(\varrho)\big) = f(\nu)\, \Pi'(g)$, $f, g \in H$. The same is true for $\Pi'_\varrho : H_\nu^+ \otimes H_\varrho \to H_\nu^+$.

From 2.2.1.1, Proposition 2 follows

Corollary 1. *Any function $g \in H^+ \otimes H_q^-$ has a decomposition*

$$g(\nu, \varrho) = \sum_{j=0}^{q-1} g_j(\nu)\, \varrho^j + \sum_{l,m=0}^{\infty} c_{lm} \frac{(1 - i\nu)^l}{(1 + i\nu)^{l+1}} \frac{(1 + i\varrho)^m}{(1 - i\varrho)^{m+1}} \tag{5}$$

with rapidly decreasing c_{lm}, i.e. for arbitrary M, N there exists a constant $C = C(M, N)$ such that $l^M m^N c_{lm} \leq C$. Any right hand side in (5) with rapidly decreasing coefficients belongs to $H^+ \otimes H_q^-$.

Denote by Π_0^- the projection $\Pi_0^- : H \to H_0^-$ defined by

$$\Pi_0^-|_{H_0} = \Pi^-|_{H_0}\,, \quad \Pi_0^-|_{H'} = 0\,.$$

Proposition 2. *The following conditions are equivalent:*

(i) $g \in H^+ \otimes H_0^-$.

(ii) *$g(\nu, \varrho) \in C^\infty(\mathbb{R} \times \mathbb{R})$ and for any fixed ν there is an analytic extension into $\operatorname{Im} \varrho > 0$ and for any fixed ϱ there is an analytic extension into $\operatorname{Im} \nu < 0$. Moreover the* *following semi-norms are defined and finite*

$$r^{(1)}_{k,l}(g) = \|\Pi^+_\nu \Pi^-_{0\varrho}(\nu^{k_1} D^{k_2}_\nu \varrho^{l_1} D^{l_2}_\varrho g(\nu, \varrho))\|_{L^2(R^2)},$$

$$k = (k_1, k_2) \in Z^2_+, \; l = (l_1, l_2) \in Z^2_+.$$

(iii) $(1+z)^{-1}(1+w)^{-1} g(\mathrm{i}(z-1)(z+1)^{-1}, \mathrm{i}(w-1)(w+1)^{-1}) \in C^\infty(S^1 \times S^1)$ *and for any fixed z there is an analytic extension into $|w| > 1$ and for any fixed w an analytic extension into $|z| < 1$ and the following semi-norms are finite*

$$r^{(2)}_{k,l}(g) = \sup_{\substack{-\pi \le \varphi \le \pi \\ -\pi \le \psi \le \pi}} |D^k_\varphi D^l_\psi((1+\mathrm{e}^{\mathrm{i}\varphi})^{-1}(1+\mathrm{e}^{\mathrm{i}\psi})^{-1}) \times$$
$$\times\, g(\mathrm{i}(\mathrm{e}^{\mathrm{i}\varphi}-1)(\mathrm{e}^{\mathrm{i}\varphi}+1)^{-1}, \mathrm{i}(\mathrm{e}^{\mathrm{i}\psi}-1)(\mathrm{e}^{\mathrm{i}\psi}+1)^{-1}),$$

$k, l \in Z_+$, $z = \mathrm{e}^{\mathrm{i}\varphi}$, $w = \mathrm{e}^{\mathrm{i}\psi}$.

Moreover the given sets of semi-norms are equivalent and define the topology of $H^+ \otimes H_0^-$.

The simple **proof** based on Corollary 1 is left to the reader.

A matrix function $\sigma_B(\nu, \varrho) \in H^+ \otimes H_q^- \otimes \hom(C^k, C^j)$ is called *Green symbol.* Any Green symbol has a decomposition (5) with rapidly decreasing matrix coefficients. The smallest number $q \in Z_+$ with $\sigma_B \in H^+ \otimes H_q^- \otimes \hom(C^k, C^j)$ is called *type* of the Green symbol σ_B. By $h(\varrho) \to \Pi'_\varrho(\sigma_B(\nu, \varrho) h(\varrho))$, $h \in H^+ \otimes C^k$ a continuous mapping

$$\Pi'\sigma_B \colon H^+ \otimes C^k \to H^+ \otimes C^j \tag{6}$$

is defined. It depends continuously on the matrix function σ_B.

For $\sigma_B \in H \otimes H \otimes \hom(C^k, C^j)$ we define mapping $H^+ \otimes C^k \to H^+ \otimes C^j$ by

$$h \to \Pi^+_\nu \Pi'_\varrho(\sigma_B(\nu, \varrho) h(\varrho)) \tag{6'}$$

This mapping depends only on $\Pi^+_\nu \Pi^-_\varrho(\sigma_B(\nu, \varrho))$.

If the coefficients c_{lm} in the decomposition (5) of a Green symbol σ_B are different from zero for a finite number of indices l only, then $\Pi'\sigma_B$ is an operator of finite rank. More generally we have

Proposition 3. *Let* $\sigma_B \in H^+ \otimes H_d^- \otimes \hom(C^k, C^j)$. *Then* $\Pi'\sigma_B \colon H^+ \otimes C^k \to H^+ \otimes C^j$ *is compact. For* $\sigma_B \in H^+ \otimes H_0^- \otimes \hom(C^k, C^j)$, *i.e.* σ_B *has type* 0, *the extension*

$$\Pi'\sigma_B \colon V^+ \otimes C^k \to V^+ \otimes C^j$$

is compact and $\mathrm{im}(\Pi'\sigma_B) \subset H^+$.

Proof: We may consider the scalar case and assume $d = 0$, since $\Pi'\sigma_B$, $\sigma_B \in H^+ \otimes H'_d$, is finite-dimensional. Then

$$\Pi'\sigma_B h(\nu) = (2\pi)^{-1} \int \sigma_B(\nu, \varrho) h(\varrho) \, \mathrm{d}\varrho.$$

Assume first that $\sigma_B(\nu, \varrho)$ is a finite sum $\sum_j \sigma^j_K(\nu) \sigma^j_T(\varrho)$, $\sigma^j_K \in H^+$, $\sigma^j_T \in H_0^-$.

For any continuous semi-norm p on H^+ we have

$$p(\Pi'\sigma_B h) = p\Big(\sum_j \sigma^j_K(\nu) (2\pi)^{-1} \int \sigma^j_T(\varrho) h(\varrho) \, \mathrm{d}\varrho\Big) \le$$
$$\le \sum_j p(\sigma^j_K) \|\sigma^j_T\|_{L^2(R)} \|h\|_{L^2(R)}.$$

An analogous estimate follows by continuity for arbitrary $\sigma_B \in H^+ \otimes H_0^-$. Thus the compactness of $\Pi'\sigma_B$ in H^+ follows and, moreover, continuity in V^+ with an image in H^+. For $\sigma_B \in H^+ \otimes H_0^-$ the operator $\Pi'\sigma_B$ is an integral operator with $L^2(\mathbb{R} \otimes \mathbb{R})$-kernel, hence a Hilbert-Schmidt operator which is compact. □

Note that for $\sigma_T \in H_0^-$ the operator $\Pi'\sigma_T : H^+ \to \mathbb{C}$ has a continuous extension to an operator in $\mathcal{L}(V^+, \mathbb{C})$. This is shown in the above proof. The space $H^+ \otimes \mathbb{C}^k$ is always considered as a subspace of $V^+ \otimes \mathbb{C}^k$ with a fixed scalar product, which is given by scalar products in V^+ and $\mathbb{C}^k$, respectively.

Lemma 4. *Let $V \subset H^+ \otimes \mathbb{C}^{k_1}$, $W \subset H^+ \otimes \mathbb{C}^{k_2}$ be finite dimensional subspaces.*

(i) *Let $\beta : V \to W$ be an isomorphism. Then there exists a Green operator $\Pi'\sigma_B : H^+ \otimes \mathbb{C}^{k_1} \to W \otimes \mathbb{C}^{k_2}$ with $\Pi'\sigma_B|_V = \beta$. Moreover a projection $\lambda : H^+ \otimes \mathbb{C}^{k_1} \to V$ can be represented as Green operator $\Pi'\sigma_L$.*

(ii) *Any isomorphism $\varkappa : \mathbb{C}^j \to W$ can be represented as a potential operator σ_K, i.e. $\varkappa = \sigma_K$.*

(iii) *Any isomorphism $\tau : V \to \mathbb{C}^l$ is induced by some trace operator $\Pi'\sigma_T$, i.e. $\tau = \Pi'\sigma_T|_V$. $\sigma_B, \sigma_L, \sigma_T$ can be chosen in such a way that their type is zero.*

Proof: (ii) follows if one represents $\varkappa$ as element of $(\mathbb{C}^j)^* \otimes W$. (iii) follows by representing τ as an element in $V^* \otimes \mathbb{C}^l$. Note that V^* can be identified with $\overline{V}$, the space of complex conjugate elements consisting component-wise of functions in H_0^- and that the duality $(t, h)_{V^* \times V}$ is realized by $\Pi'(th)$. Now it is obvious that β in (i) corresponds to an element in $V^* \otimes W$ which can be regarded as a matrix of products $t(\varrho)\, h(\nu)$ for suitable functions $t \in H_0^-$, $h \in H^+$. If we apply this to the identity in V, the resulting Green symbol gives rise to a projection $H^+ \otimes \mathbb{C}^{k_1} \to V$. □

Any matrix of functions as above

$$\mathfrak{c} = \begin{pmatrix} \sigma_A, \sigma_B & \sigma_K \\ \sigma_T & \sigma_Q \end{pmatrix}$$

defines a continuous operator

$$\mathrm{op}'(\mathfrak{c}) \overset{\mathrm{def}}{=} \begin{pmatrix} \Pi^+\sigma_A + \Pi'\sigma_B & \sigma_K \\ \Pi'\sigma_T & \sigma_Q \end{pmatrix} : \begin{matrix} H^+ \otimes \mathbb{C}^k \\ \oplus \\ \mathbb{C}^{l_-} \end{matrix} \to \begin{matrix} H^+ \otimes \mathbb{C}^j \\ \oplus \\ \mathbb{C}^{l_+}. \end{matrix} \tag{7}$$

Definition 5. By $\mathfrak{R}^{m,d}(k, j; l_-, l_+)$, $m \in \mathbb{Z}$, $d \in \mathbb{Z}_+$ we denote the space of all operators of the form (7) with $\sigma_A \in H_{m+1} \otimes \mathrm{hom}\,(\mathbb{C}^k, \mathbb{C}^j)$, $\sigma_B \in H^+ \otimes H_d^- \otimes \mathrm{hom}\,(\mathbb{C}^k, \mathbb{C}^j)$, $\sigma_K \in H^+ \otimes \mathrm{hom}\,(\mathbb{C}^{l_-}, \mathbb{C}^j)$, $\sigma_T \in H_d^- \otimes \mathrm{hom}\,(\mathbb{C}^k, \mathbb{C}^{l_+})$, $\sigma_Q \in \mathrm{hom}\,(\mathbb{C}^{l_-}, \mathbb{C}^{l_+})$.

For $h \in H^+ \otimes \mathbb{C}^k$, $v \in \mathbb{C}^{l_-}$ we have per def.

$$\begin{pmatrix} \Pi^+\sigma_A + \Pi'\sigma_B & \sigma_K \\ \Pi'\sigma_T & \sigma_Q \end{pmatrix} \begin{pmatrix} h \\ v \end{pmatrix} = \begin{pmatrix} \Pi_\nu^+(\sigma_A(\nu)\, h(\nu)) + \Pi_\varrho'(\sigma_B(\nu, \varrho)\, h(\varrho)) + \sigma_K(\nu)\, v \\ \Pi_\nu'(\sigma_T(\nu)\, h(\nu)) + \sigma_Q v \end{pmatrix}.$$

Denote by $\mathfrak{R}^{m,d}$ the union of $\mathfrak{R}^{m,d}(k, j; l_-, l_+)$ over all k, j, l_-, l_+. The elements of $\mathfrak{R}^{m,d}$ are called *boundary symbols of order m and type d* (more precisely boundary symbols on the line, since the proper boundary symbols in 2.2.5.1 are families of boundary symbols on the line). Set $\mathfrak{R} = \bigcup_{m,d} \mathfrak{R}^{m,d}$, $\mathfrak{R}^{\infty,d} = \bigcup_m \mathfrak{R}^{m,d}$, $\mathfrak{R}^m = \bigcup_d \mathfrak{R}^{m,d}$, $\mathfrak{R}^{-\infty,d} = \bigcap_m \mathfrak{R}^{m,d}$. The matrix function $\sigma_A \in H_{m+1} \otimes \mathrm{hom}\,(\mathbb{C}^k, \mathbb{C}^j)$ is called the *operator symbol of the boundary symbol* (7).

We have checked earlier that the components of a boundary symbol define continuous operators, hence $\mathfrak{R}^{m,d}(k, j; l_-, l_+) \subset \mathcal{L}(H^+ \otimes \mathbb{C}^k \otimes \mathbb{C}^{l_-}, H^+ \otimes \mathbb{C}^j \otimes \mathbb{C}^{l_+})$. We consider $\mathfrak{R}^{m,d}$ in the topology induced by the mapping op′ from the direct sum of function spaces $H_{m+1} \otimes \mathrm{hom}\,(\mathbb{C}^k, \mathbb{C}^j) \oplus H^+ \otimes H_d^- \otimes \mathrm{hom}\,(\mathbb{C}^k, \mathbb{C}^j) \oplus H^+ \otimes \mathrm{hom}\,(\mathbb{C}^k, \mathbb{C}^{l_+}) \oplus H_d^- \otimes \mathrm{hom}\,(\mathbb{C}^{l_-}, \mathbb{C}^j) \otimes \mathrm{hom}\,(\mathbb{C}^{l_-}, \mathbb{C}^{l_+})$. With this topology $\mathfrak{R}^{m,d}$ is a Fréchet space. Later on we will show that the mapping $\mathfrak{c} \to \mathrm{op}'(\mathfrak{c})$ is one-to-one. As we observed above for the components of a boundary symbol the embedding $\mathfrak{R}^{m,d} \subset \mathcal{L}(H^+ \otimes \mathbb{C}^k \oplus \mathbb{C}^{l_-}, H^+ \otimes \mathbb{C}^j \oplus \mathbb{C}^{l_+})$ is continuous. Note that $\mathfrak{R}^{m,d}$ is not closed in $\mathcal{L}(H^+ \otimes \mathbb{C}^k \oplus \mathbb{C}^{l_-}, H^+ \otimes \mathbb{C}^j \oplus \mathbb{C}^{l_+})$. The closure of $\mathfrak{R}^{m,d}$ with the induced topology contains, for instance, trace symbols $\Pi'\sigma_T$ with $\sigma_T \in V^-$ (V^- is the L^2-closure of H_0^-) and Green symbols $\Pi'\sigma_B$ with $\sigma_B \in H^+ \otimes V^-$. This follows from the estimates in the proof of Proposition 3.

2.1.2.3. Compositions and Adjoints

The operators in $\mathfrak{R}$ play a basic role in the theory of elliptic boundary value problems. An essential point to be discussed in this section is that $\mathfrak{R}$ is closed under composition.

Proposition 1. *Let $\mathfrak{a} \in \mathfrak{R}^{m,d}(k, j; l_1, l_2)$ and $\mathfrak{b} \in \mathfrak{R}^{m',d'}(j, l; l_2, l_3)$. Then $\mathfrak{ba} \in \mathfrak{R}^{m'',d''}(k, l; l_1, l_3)$, where $m'' = m + m'$, $d'' = \max(m + d', d)$. In particular, for $m = m' = d = d' = 0$, we have $m'' = d'' = 0$, i.e. $\mathfrak{R}^{0,0}(k, k; l, l)$ is an algebra.*

The proof follows from the lemmata below. It can be assumed that all symbols are scalar ones. Set

$$\mathfrak{a} = \begin{pmatrix} \Pi^+\sigma_{A_1} + \Pi'\sigma_{B_1} & \sigma_{K_1} \\ \Pi'\sigma_{T_1} & \sigma_{Q_1} \end{pmatrix}, \qquad \mathfrak{b} = \begin{pmatrix} \Pi^+\sigma_{A_2} + \Pi'\sigma_{B_2} & \sigma_{K_2} \\ \Pi'\sigma_{T_2} & \sigma_{Q_2} \end{pmatrix}.$$

Then

$$\mathfrak{ba} = \begin{pmatrix} \alpha & \varkappa \\ \tau & \delta \end{pmatrix}$$

where

$$\alpha = \Pi^+\sigma_{A_2}\Pi^+\sigma_{A_1} + \Pi^+\sigma_{A_2}\Pi'\sigma_{B_1} + \Pi'\sigma_{B_2}\Pi^+\sigma_{A_1} + \Pi'\sigma_{B_2}\Pi'\sigma_{B_1} + \sigma_{K_2}\Pi'\sigma_{T_1},$$

$$\varkappa = \Pi^+\sigma_{A_2}\sigma_{K_1} + \Pi'\sigma_{B_2}\sigma_{K_1} + \sigma_{K_2}\sigma_{Q_1},$$

$$\tau = \Pi'\sigma_{T_2}\Pi^+\sigma_{A_1} + \Pi'\sigma_{T_2}\Pi'\sigma_{B_1} + \sigma_{Q_2}\Pi'\sigma_{T_1},$$

$$\delta = \Pi'\sigma_{T_2}\sigma_{K_1} + \sigma_{Q_2}\sigma_{Q_1}.$$

We have to show that there are functions $\sigma_A \in H_{m+m'+1}$, $\sigma_B \in H^+ \otimes H_{d''}^-$, $\sigma_K \in H^+$, $\sigma_T \in H_{d''}^-$ and a constant $\sigma_Q \in \mathbb{C}$ such that

$$\alpha = \Pi^+\sigma_A + \Pi'\sigma_B, \quad \varkappa = \sigma_K, \quad \tau = \Pi'\sigma_T, \quad \delta = \sigma_Q.$$

The following Lemma is a simple consequence of 2.1.1.3.(2).

Lemma 2.

$$\Pi'\sigma_{B_2}\Pi^+\sigma_{A_1} = \Pi'_\varrho(\Pi_\varrho^-(\sigma_{B_2}(\nu, \varrho)\,\sigma_{A_1}(\varrho))),$$

$$\Pi^+\sigma_{A_2}\Pi'\sigma_{B_1} = \Pi'_\varrho(\Pi_\nu^+(\sigma_{A_2}(\nu)\,\sigma_{B_1}(\nu, \varrho))),$$

$$\Pi'\sigma_{B_2}\Pi'\sigma_{B_1} = \Pi'_\lambda(\Pi'_\varrho(\sigma_{B_2}(\nu, \varrho)\,\sigma_{B_1}(\varrho, \lambda))),$$

$$\sigma_{K_2}\Pi'\sigma_{T_1} = \Pi'_\varrho(\sigma_{K_2}(\nu)\,\sigma_{T_1}(\varrho))$$

and the right hand sides contain Green symbols in $H^+ \otimes H_{d''}^-$, $d'' = \max(m + d', d)$.

Lemma 3.

$$\Pi^+\sigma_{A_2}\sigma_{K_1} = \Pi_\nu^+\left(\sigma_{A_2}(\nu)\,\sigma_{K_1}(\nu)\right),$$
$$\Pi'\sigma_{B_2}\sigma_{K_1} = \Pi'_\varrho\left(\sigma_{B_2}(\nu,\varrho)\,\sigma_{K_1}(\varrho)\right)$$

and the right hand sides are potential symbols.

Lemma 4.

$$\Pi'\sigma_{T_2}\Pi^+\sigma_{A_1} = \Pi'_\nu\left(\Pi_\nu^-(\sigma_{T_2}(\nu)\,\sigma_{A_1}(\nu))\right),$$
$$\Pi'\sigma_{T_2}\Pi'\sigma_{B_1} = \Pi'_\varrho\left(\Pi'_\nu(\sigma_{T_2}(\nu)\,\sigma_{B_1}(\nu,\varrho))\right),$$
$$\sigma_{Q_2}\Pi'\sigma_{T_1} = \Pi'_\nu\left(\sigma_{Q_2}\sigma_{T_1}(\nu)\right),$$

and the right hand sides are trace symbols in $H^-_{d''}$, $d'' = \max(m + d', d)$.

The proofs of Lemma 3 and 4 are trivial.

Obviously $\sigma_Q = \Pi'\left(\sigma_{T_2}(\nu)\,\sigma_{K_1}(\nu)\right)$ is a complex factor. Now Proposition 1 follows from

Lemma 5.

$$\Pi^+\sigma_{A_2}\Pi^+\sigma_{A_1} - \Pi^+\sigma_{A_2}\sigma_{A_1} = \Pi'\sigma_B$$

with Green symbol

$$\sigma_B(\nu,\varrho) = \Pi_\nu^+\Pi_\varrho^-\left(\frac{\left(\sigma_2^+(\nu) - \sigma_2^+(\varrho)\right)\left(\sigma_1^-(\nu) - \sigma_1^-(\varrho)\right)}{\mathrm{i}(\nu-\varrho)}\right) \in H^+ \otimes H^-_m \tag{1}$$

where $\sigma_2^+(\nu) = \Pi_\nu^+\left(\sigma_2(\nu)\right)$, $\sigma_1^-(\nu) = \Pi_\nu^-\left(\sigma_1(\nu)\right)$, $\sigma_i = \sigma_{A_i}$, $i = 1, 2$.

Proof: For $h \in H^+$,

$$\begin{aligned} L(\sigma_2,\sigma_1)\,h &= \Pi^+\left(\sigma_2(\nu)\,\sigma_1(\nu)\,h(\nu)\right) - \Pi^+\left(\sigma_2(\nu)\,\Pi^+(\sigma_1(\nu)\,h(\nu))\right) \\ &= \Pi^+\left(\sigma_2(\nu)\,\Pi^-(\sigma_1(\nu)\,h(\nu))\right) = \Pi^+\left(\sigma_2^+(\nu)\,\Pi^-(\sigma_1^-(\nu)\,h(\nu))\right), \end{aligned} \tag{2}$$

since Π^+ vanishes on H^- and Π^- vanishes on H^+. Assume now that $\sigma_1^-(\nu) \in H_1^-$. Then

$$\Pi^-(\sigma_1^- h) = \Pi'_\tau\left(\frac{\sigma_1^-(\nu) - \sigma_1^-(\tau)}{\mathrm{i}(\nu-\tau)}\,h(\tau)\right).$$

In fact, we have $\Pi^-(\sigma_1^- h) = \sigma_1^-\Pi^+h - \Pi^+(\sigma_1^- h)$, and according to 2.1.1.2, Proposition 1

$$\sigma_1^-(\nu)\,\Pi^+h(\nu) = \frac{\sigma_1^-(\nu)\,h(\nu)}{2} + \sigma_1^-(\nu)\,(2\pi\mathrm{i})^{-1}\int_{\text{p.v.}}\frac{h(\tau)}{\nu-\tau}\,\mathrm{d}\tau,$$

$$\Pi^+(\sigma_1^- h)\,(\nu) = \frac{\sigma_1^-(\nu)\,h(\nu)}{2} + (2\pi\mathrm{i})^{-1}\int_{\text{p.v.}}\frac{\sigma_1^-(\tau)\,h(\tau)}{\nu-\tau}\,\mathrm{d}\tau,$$

therefore

$$\Pi^-(\sigma_1^- h)\,(\nu) = (2\pi)^{-1}\int\frac{\sigma_1^-(\nu) - \sigma_1^-(\tau)}{\mathrm{i}(\tau-\nu)}\,h(\tau)\,\mathrm{d}\tau = \Pi'_\tau\left(\frac{\sigma_1^-(\nu) - \sigma_1^-(\tau)}{\mathrm{i}(\nu-\tau)}\,h(\tau)\right).$$

From 2.1.1.3.(2) it follows that

$$\Pi^+\left(\sigma_2^+\Pi^-(\sigma_1^-h)\right) = \Pi^+\sigma_2^+\Pi'_\tau\left(\frac{\sigma_1^-(\nu)-\sigma_1^-(\tau)}{\mathrm{i}(\nu-\tau)}h(\tau)\right) =$$
$$= \Pi'_\tau\left(\Pi^+_\nu\left(\sigma_2^+(\nu)\frac{\sigma_1^-(\nu)-\sigma_1^-(\tau)}{\mathrm{i}(\nu-\tau)}h(\tau)\right)\right) =$$
$$= \Pi'_\tau\left(\Pi^+_\nu\Pi^-_\tau\left(\sigma_2^+(\nu)\frac{\sigma_1^-(\nu)-\sigma_1^-(\tau)}{\mathrm{i}(\nu-\tau)}\right)h(\tau)\right).$$

For $\sigma_1^-\in H_1^-$ we have $\left(\sigma_1^-(\nu)-\sigma_1^-(\tau)\right)/\mathrm{i}(\nu-\tau)\in H_0^-\otimes H_0^-$. In fact, take the Fourier expansion of $\sigma_1^-\in H_1^-$

$$\sigma_1^-(\nu)=\sum_{j=0}^\infty a_j\left(\frac{1+\mathrm{i}\nu}{1-\mathrm{i}\nu}\right)^j,$$

a_j rapidly decreasing. Then

$$\left(\sigma_1^-(\nu)-\sigma_1^-(\tau)\right)/\mathrm{i}(\nu-\tau)=\sum_{j=0}^\infty a_j\left(\left(\frac{1+\mathrm{i}\nu}{1-\mathrm{i}\nu}\right)^j-\left(\frac{1+\mathrm{i}\tau}{1-\mathrm{i}\tau}\right)^j\right)\Big/\mathrm{i}(\nu-\tau)$$
$$=2(1-\mathrm{i}\nu)^{-1}(1-\mathrm{i}\tau)^{-1}\sum_{j=0}^\infty a_j\left(\left(\frac{1+\mathrm{i}\nu}{1-\mathrm{i}\nu}\right)^j-\left(\frac{1+\mathrm{i}\tau}{1-\mathrm{i}\tau}\right)^j\right)\Big/\left(\frac{1+\mathrm{i}\nu}{1-\mathrm{i}\nu}-\frac{1+\mathrm{i}\tau}{1-\mathrm{i}\tau}\right)$$
$$=2(1-\mathrm{i}\nu)^{-1}(1-\mathrm{i}\tau)^{-1}\sum_{j=1}^\infty\sum_{k=0}^{j-1}a_j\left(\frac{1+\mathrm{i}\nu}{1-\mathrm{i}\nu}\right)^k\left(\frac{1+\mathrm{i}\tau}{1-\mathrm{i}\tau}\right)^{j-k}$$
$$=2(1-\mathrm{i}\nu)^{-1}(1-\mathrm{i}\tau)^{-1}\sum_{k,l=0}^\infty a'_{kl}\left(\frac{1+\mathrm{i}\nu}{1-\mathrm{i}\nu}\right)^k\left(\frac{1+\mathrm{i}\tau}{1-\mathrm{i}\tau}\right)^l$$

where $a'_{kl}=a_j$ for $k+l=j\geqq 1$ and zero otherwise. Since the a_j are rapidly decreasing, it follows that $\left(\sigma_1^-(\nu)-\sigma_1^-(\tau)\right)/\mathrm{i}(\nu-\tau)\in H_0^-\otimes H_0^-$. Then obviously

$$\Pi^+_\nu\Pi^-_\tau\left(\sigma_2^+(\nu)\left[\left(\sigma_1^-(\nu)-\sigma_1^-(\tau)\right)/\mathrm{i}(\nu-\tau)\right]\right)\in H^+\otimes H_0^-$$

and

$$\Pi^+_\nu\Pi^-_\tau\left(\sigma_2^+(\tau)\left[\left(\sigma_1^-(\nu)-\sigma_1^-(\tau)\right)/\mathrm{i}(\nu-\tau)\right]\right)=0,$$

since

$$\Pi^+_\nu\left[\left(\sigma_1^-(\nu)-\sigma_1^-(\tau)\right)/\mathrm{i}(\nu-\tau)\right]=0.$$

Next consider the case $\sigma_1^-(\nu)=\nu^q$. For $h\in H^+$ and

$$h(\nu)\sim\sum_{k=0}^\infty a_k\nu^{-k-1}\qquad\text{for}\qquad|\nu|\to\infty$$

we have $a_k=-\mathrm{i}\Pi'\left(\nu^k h(\nu)\right)$. In fact,

$$\nu^k h(\nu)=\sum_{l=0}^{k-1}a_l\nu^{-l-1+k}+h_0,\qquad h_0\in H^+$$

and h_0 has the asymptotic expansion

$$h_0(\nu)\sim\sum_{l=0}^\infty a_{l+k}\nu^{-l-1},$$

therefore there is an $h_1\in H^+\cap L^1(\mathbb{R})$ with

$$h_0(\nu)=\mathrm{i}a_k(1+\mathrm{i}\nu)^{-1}+h_1(\nu).$$

Because of 2.1.1.3. (2) we get

$$\Pi'\left(\nu^k h(\nu)\right)=\Pi' h_0=\Pi'\left(\mathrm{i}a_k(1+\mathrm{i}\nu)^{-1}\right)=\mathrm{i}a_k.$$

If $g \in H^+$ decreases for $\nu \to \pm\infty$ faster than ν^{-k}, then $\nu^k g(\nu) \in H^+$. Therefore

$$\Pi^-(\nu^q h(\nu)) = \sum_{0 \leqq k \leqq q-1} a_k \nu^{q-k-1} = -\mathrm{i} \sum_{0 \leqq k \leqq q-1} \nu^{q-k-1} \Pi'(\nu^k h(\nu)) .$$

Using the identity

$$\sum_{0 \leqq k \leqq q-1} \nu^{q-k-1} \tau^k = (\nu^q - \tau^q)(\nu - \tau)^{-1}$$

we get

$$\Pi^-(\sigma_1^- h) = \Pi^-(\nu^q h(\nu)) = \Pi'_\tau\left(\frac{\sigma_1^-(\nu) - \sigma_1^-(\tau)}{\mathrm{i}(\nu - \tau)} h(\tau)\right),$$

$$L(\sigma_2, \sigma_1) h(\nu) = \Pi^+ \sigma_2^+ \Pi^-(\sigma_1^- h) = \Pi^+ \sigma_2^+ \Pi'_\tau\left(\frac{\sigma_1^-(\nu) - \sigma_1^-(\tau)}{\mathrm{i}(\nu - \tau)} h(\tau)\right)$$

$$= \Pi'_\tau\left(\Pi^+_\nu\left(\sigma_2^+(\nu) \frac{\sigma_1^-(\nu) - \sigma_1^-(\tau)}{\mathrm{i}(\nu - \tau)}\right) h(\tau)\right).$$

Since $(\nu^q - \tau^q)(\nu - \tau)^{-1}$ is a polynomial in ν and τ of order $q - 1$, it follows that

$$\Pi^+_\nu\left(\sigma_2^+(\nu) \frac{\sigma_1^-(\nu) - \sigma_1^-(\tau)}{\mathrm{i}(\nu - \tau)}\right) = \Pi^+_\nu \Pi^-_\tau\left(\frac{(\sigma_2^+(\nu) - \sigma_2^+(\tau))(\sigma_1^-(\nu) - \sigma_1^-(\tau))}{\mathrm{i}(\tau - \nu)}\right)$$

$$\in H^+ \otimes H_q^- . \quad \square$$

Boundary symbols in $\mathfrak{R}^{0,0}$ have a continuous extension to operators in $\mathcal{L}(V^+ \otimes \mathbb{C}^k \oplus \mathbb{C}^l, V^+ \otimes \mathbb{C}^j \oplus \mathbb{C}^m)$. For $\Pi^+\sigma_A$ this was observed in 2.1.2.1. For σ_K and σ_Q the assertion is obvious and for $\Pi'\sigma_B$ it was proved in 2.1.2.2, Proposition 3. For $\sigma_T \in H_0^-$ we have

$$|\Pi'\sigma_T h| = (2\pi)^{-1} \left|\int \sigma_T(\nu) h(\nu)\, \mathrm{d}\nu\right| \leqq \text{const.} \|\sigma_T\|_{L^2(\mathbb{R})} \|h\|_{L^2(\mathbb{R})},$$

$h \in V^+$. It is obvious that for trace and Green symbols of type > 0 such extensions cannot exist.

Consider the Hilbert space $V^+ \otimes \mathbb{C}^k \oplus \mathbb{C}^l$ with the standard Hermitean scalar product

$$(h^1 \oplus w^1, h^2 \oplus w^2) = \sum_{j=1}^{k} (2\pi)^{-1} \int h_j^1(\nu) \overline{h_j^2(\nu)}\, \mathrm{d}\nu + \sum_{j=1}^{l} w_i^1 \overline{w_i^2},$$

$$h^r = (h_1^r, \dots, h_k^r) \in H^+ \otimes \mathbb{C}^k, \quad w^r = (w_1^r, \dots, w_l^r) \in \mathbb{C}^l, \quad r = 1, 2 .$$

If it is necessary to indicate the dimensions k, l, we write $(\cdot, \cdot)_{k,l}$.

Proposition 6. *Consider*

$$\mathfrak{a} = \begin{pmatrix} \Pi^+\sigma_A + \Pi'\sigma_B & \sigma_K \\ \Pi'\sigma_T & \sigma_Q \end{pmatrix} \in \mathfrak{R}^{0,0}(k, j; l, m)$$

as an operator in $\mathcal{L}(V^+ \otimes \mathbb{C}^k \oplus \mathbb{C}^l, V^+ \otimes \mathbb{C}^j \oplus \mathbb{C}^m)$. *Then we have* $\mathfrak{a}^* \in \mathfrak{R}^{0,0}(j, k; m, l)$ *for the adjoint* $\mathfrak{a}^*$. *More explicitly*

$$\mathfrak{a}^* = \begin{pmatrix} \Pi^+ {}^t\overline{\sigma}_A + \Pi' {}^t\overline{\sigma}_B & {}^t\overline{\sigma}_T \\ \Pi' {}^t\overline{\sigma}_K & {}^t\overline{\sigma}_Q \end{pmatrix} : \begin{matrix} V^+ \otimes \mathbb{C}^j \\ \oplus \\ \mathbb{C}^m \end{matrix} \to \begin{matrix} V^+ \otimes \mathbb{C}^k \\ \oplus \\ \mathbb{C}^l \end{matrix} \qquad (3)$$

where in ${}^t\overline{\sigma}_B$ *the role of the variables* ν *and* τ *is changed.*

Proof: It is sufficient to consider scalar symbols. Let $\sigma_K \in H^+$ be a potential symbol. For any $h \in V^+$, $w \in \mathbb{C}$ we have

$$(\sigma_K w, h) = (2\pi)^{-1} \int \sigma_K(\nu)\, w\overline{h(\nu)}\, d\nu = w\overline{\Pi'_\nu(\overline{\sigma}_K(\nu)\, h(\nu))}\,.$$

Since $\overline{\sigma}_K \in H_0^-$, the assertion is proved for potential and trace symbols.

Let $\sigma_B(\nu, \tau) \in H^+ \otimes H_0^-$ be a Green symbol, $\Pi'\sigma_B\colon V_\tau^+ \to V_\nu^+$. Then, for arbitrary $h, f \in V^+$

$$\left(\Pi'_\tau(\sigma_B(\nu, \tau)\, h(\tau)), f(\nu)\right)_{V_\nu^+} = (2\pi)^{-2} \int_{\mathbb{R}} \left\{ \int_{\mathbb{R}} \sigma_B(\nu, \tau)\, h(\tau)\, d\tau \right\} \overline{f(\nu)}\, d\nu$$

$$= (2\pi)^{-2} \int_{\mathbb{R}} h(\tau) \overline{\left\{ \int_{\mathbb{R}} \overline{\sigma_B(\nu, \tau)}\, f(\nu)\, d\nu \right\}}\, d\tau = \left(h(\tau), \Pi'_\nu(\overline{\sigma_B(\nu, \tau)}\, f(\nu))\right).$$

Finally note that $\sigma_B(\nu, \tau) \in H^+ \otimes H_0^-$ implies $\overline{\sigma_B(\nu, \tau)} \in H_0^- \otimes H^+$.

Let $\sigma_A \in H_1$, $f, h \in V^+$. Then

$$\begin{aligned}(\Pi^+\sigma_A h, f) &= \Pi'(\Pi^+\sigma_A h\bar{f}) = \Pi'(\sigma_A h\bar{f}) = \Pi'\left(h\overline{(\overline{\sigma}_A f)}\right)\\ &= \Pi'(h\overline{\Pi^+\overline{\sigma}_A f}) = (h, \Pi^+\overline{\sigma}_A f)\,.\end{aligned}$$

In the last equation 2.1.1.3.(2) is used. Since $\overline{\sigma}_A \in H_1$, Proposition 6 is proved. □

From 2.1.2.2, Proposition 3 and Proposition 1 we get

Proposition 7. *Let $\sigma_A \in H_1 \otimes \hom(\mathbb{C}^k, \mathbb{C}^k)$ and $\det \sigma_A(\nu) \neq 0$ for all $\nu \in \overline{\mathbb{R}}$ (including $\nu = \pm\infty$). Then for an arbitrary Green symbol $\sigma_B \in H^+ \otimes H_0^- \otimes \hom(\mathbb{C}^k, \mathbb{C}^k)$ the operator*

$$\Pi^+\sigma_A + \Pi'\sigma_B\colon V^+ \otimes \mathbb{C}^k \to V^+ \otimes \mathbb{C}^k$$

s Fredholm and $\Pi^+\sigma_A^{-1}$ is a parametrix. Moreover $\ker(\Pi^+\sigma_A + \Pi'\sigma_B) \subset H^+ \otimes \mathbb{C}^k$ and there exists a finite dimensional subspace $W \subset H^+ \otimes \mathbb{C}^k$ such that $W \oplus \operatorname{im}(\Pi^+\sigma_A + \Pi'\sigma_B) = V^+ \otimes \mathbb{C}^k$, i.e. $W \cap \operatorname{im}(\Pi^+\sigma_A + \Pi'\sigma_B) = \{0\}$ and $W + \operatorname{im}(\Pi^+\sigma_A + \Pi'\sigma_B) = V \otimes \mathbb{C}^k$. One can choose W in such a way that $W \oplus \operatorname{im}\left((\Pi^+\sigma_A + \Pi'\sigma_B)|_{H^+\otimes\mathbb{C}^k}\right) = H^+ \otimes \mathbb{C}^k$. In particular the index of $\Pi^+\sigma_A + \Pi'\sigma_B\colon V^+ \otimes \mathbb{C}^k \to V^+ \otimes \mathbb{C}^k$ is equal to the index of the closure $\Pi^+\sigma_A + \Pi'\sigma_B\colon H^+ \otimes \mathbb{C}^k \to H^+ \otimes \mathbb{C}^k$.

Proof: Because of Proposition 1 and the compactness of Green symbols (cf. 2.1.2.2, Proposition 3) it is clear that $\Pi^+\sigma_A^{-1}$ is a parametrix of $\Pi^+\sigma_A + \Pi'\sigma_B$, since $(\Pi^+\sigma_A + \Pi'\sigma_B)\,\Pi^+\sigma_A^{-1} = 1 + \Pi'\sigma_{B_1}$, $\Pi^+\sigma_A^{-1}(\Pi^+\sigma_A + \Pi'\sigma_B) = 1 + \Pi'\sigma_{B_2}$. Now let $h \in \ker(\Pi^+\sigma_A + \Pi'\sigma_B)$, i.e. $(\Pi^+\sigma_A + \Pi'\sigma_B)h = 0$. Applying $\Pi^+\sigma_A^{-1}$ we get $(1 + \Pi'\sigma_{B_2})h = 0$. Since $\operatorname{im}\Pi'\sigma_{B_2} \subset H^+ \otimes \mathbb{C}^k$, it follows that $h \in H^+ \otimes \mathbb{C}^k$. For the last assertion note that the cokernel W can be identified with the kernel of the adjoint which belongs to the same class in view of Proposition 6. □

Lemma 8. *For any $m \in \mathbb{Z}$ we have $l_m^\pm(\nu) = (1 \pm i\nu)^m \in H_{m+1}$. Moreover the operator*

$$\Pi^+ l_m^-\colon H^+ \to H^+ \tag{4}$$

is an isomorphism and $(\Pi^+ l_m^-)^{-1} = \Pi^+ l_{-m}^-$.

Proof: $l_m^\pm \in H_{m+1}$ follows from 2.1.1.1, Corollary 3. By Lemma 5 we have $\Pi^+ l_m^- \Pi^+ l_{-m}^- = \Pi^+(l_m^- l_{-m}^-) = 1$. □

Let $H^s(\mathbb{R}_+)$ be the Sobolev space on the half-axis, i.e. $H^s(\mathbb{R}_+) = \{r^+ f\colon f \in H^s(\mathbb{R})\}$, where r^+ is the restriction to $\mathbb{R}_+$. Any $f \in H^s(\mathbb{R}_+)$ has a unique extension to a function in $H^s(\mathbb{R})$ mod $H_0^s(\overline{\mathbb{R}}_-)$, i.e. modulo functions in $H^s(\mathbb{R})$ with support in $\overline{\mathbb{R}}_-$. So we have the isomorphism $H^s(\mathbb{R}_+) \cong H^s(\mathbb{R})/H_0^s(\overline{\mathbb{R}}_-)$ and $H^s(\mathbb{R}_+)$ is considered in the factor topology.

Lemma 9. *The space $F(j^+H^s(\mathbb{R}_+))$ of Fourier transforms of functions in $j^+H^s(\mathbb{R}^+)$ is the closure of H^+ with respect to the norm $\|\Pi^+(l_s^-h)\|_{L^2(\mathbb{R})}$, $h \in H^+$.*

Proof: For $f \in \mathscr{S}(\mathbb{R})$ and $h = F(\theta^+ f)$ (θ^+ denotes the characteristic function of $\overline{\mathbb{R}}_+$) we have

$$h(\nu) = \Pi^+\big(Ff(\nu)\big) = \Pi^+(1 - i\nu)^s\, \Pi^+\big((1 - i\nu)^{-s}\, Ff(\nu)\big)\,.$$

The assertion follows from the equivalence $f \in H^s(\mathbb{R}) \Leftrightarrow (1 - i\nu)^s\, Ff(\nu) \in L^2(\mathbb{R})$ and Lemma 8. □

In the following we write $(H^+)_s$ for $F\big(H^s(\mathbb{R}_+)\big)$, where obviously $V^+ = (H^+)_0$. Similarly $F\big(H^s(\mathbb{R}_-)\big)$ is defined and equal to $(H_0^-)_s$, the closure of H_0^- with respect to the norm $\|\Pi_0^-(l_s^+h)\|_{L^2(\mathbb{R})}$, $h \in H_s^-$. By Lemma 9 we have an isomorphism $\Pi^+l_s^-: (H^+)_t \to (H^+)_{t-s}$, $s, t \in \mathbb{Z}$ arbitrary.

In a similar way the space $F\big(H_0^s(\overline{\mathbb{R}}_+)\big)$, $s \in \mathbb{Z}$ can be characterized as the closure of H^+ with respect to the norm $\|l_s^+h\|_{L^2(\mathbb{R})}$, $h \in H^+$, and

$$l_{-s}^+ \colon V^+ \to F\big(H_0^s(\overline{\mathbb{R}}_+)\big) \tag{5}$$

defined by multiplication with l_{-s}^+ is an isomorphism.

By the scalar product on V^+

$$(u, v) = (2\pi)^{-1} \int u(\nu)\, \overline{v(\nu)}\, d\nu\,, \qquad u, v \in V^+$$

a pairing between $(H^+)_s$ and $F\big(H_0^{-s}(\overline{\mathbb{R}}_+)\big)$ is induced

$$(u_1, v_1) = \int \Pi^+l_s^-(\nu)\, u_1(\nu)\, \overline{l_{-s}^+(\nu)\, v_1(\nu)}\, d\nu = \int u_1(\nu)\, \overline{v_1(\nu)}\, d\nu\,.$$

Thus $F\big(H_0^{-s}(\overline{\mathbb{R}}_+)\big)$ can be identified with the dual space $(H^+)_s^*$ of $(H^+)_s$. The operators $\Pi^+l_s^-$ and l_s^+ are dual with respect to this pairing.

Using the isomorphism $\Pi^+l_m^-$ we find a one-to-one correspondence between $\mathfrak{R}^{m,d}$ and $\mathfrak{R}^{0,0}$. Let $\mathfrak{a} \in \mathfrak{R}^{m,d}(k, j; l_-, l_+)$. Set $\mathfrak{a}_0 = (\Pi^+l_{s-m}^- \oplus 1_{l_+})\, \mathfrak{a}(\Pi^+l_{-s}^- \oplus 1_{l_-})$ (for abbreviation $\Pi^+l_t^-$ is used instead of $\Pi^+l_t^- \otimes 1_k$). Then

$$\mathfrak{a} = (\Pi^+l_{-s+m}^- \oplus 1_{l_+})\, \mathfrak{a}_0(\Pi^+l_s^- \oplus 1_{l_-}) \tag{6}$$

and by Proposition 1 we have $\mathfrak{a}_0 \in \mathfrak{R}^{0,0}$.

The mapping $\mathfrak{a} \to \mathfrak{a}_0$ is called *reduction of the order* and the type of $\mathfrak{a}$ to zero. The following diagram commutes

$$\begin{array}{ccc}
(H^+)_s \otimes \mathbb{C}^k \oplus \mathbb{C}^{l_-} & \xrightarrow{\ \mathfrak{a}\ } & (H^+)_{s-m} \otimes \mathbb{C}^j \otimes \mathbb{C}^{l_+} \\
\Big\downarrow \Pi^+l_s^- \oplus 1 & & \Big\downarrow \Pi^+l_{s-m}^- \oplus 1 \\
(H^+)_0 \otimes \mathbb{C}^k \oplus \mathbb{C}^{l_-} & \xrightarrow{\ \mathfrak{a}_0\ } & (H^+)_0 \otimes \mathbb{C}^j \oplus \mathbb{C}^{l_+}\,.
\end{array} \tag{7}$$

This shows that every $\mathfrak{a} \in \mathfrak{R}^{m,d}$ has a continuous extension to an operator in $\mathscr{L}\big((H^+)_s \otimes \mathbb{C}^k \oplus \mathbb{C}^{l_-}, (H^+)_{s-m} \otimes \mathbb{C}^j \oplus \mathbb{C}^{l_+}\big)$ for any $s \geqq d$.

From Proposition 6 we obtain the following

Corollary 10. *Consider* $\mathfrak{a} \in \mathfrak{R}^{m,d}(k, j; l_-, l_+)$ *as an operator in* $\mathcal{L}((H^+)_s \otimes \mathbf{C}^k \oplus \mathbf{C}^{l_-}, (H^+)_{s-m} \otimes \mathbf{C}^j \oplus \mathbf{C}^{l_+})$, $s \geqq d$. *Then the adjoint* $\mathfrak{a}^* \in \mathcal{L}((H^+)^*_{s-m} \otimes \mathbf{C}^j \oplus \mathbf{C}^{l_+}, (H^+)^*_s \otimes \mathbf{C}^k \oplus \mathbf{C}^{l_-})$ *is a boundary symbol in* $\mathfrak{R}^{m,d}$ *(with a modified domain as indicated).*

Proof: With $\mathfrak{a} = (\Pi^+ l^-_{-s+m} \oplus 1)\, \mathfrak{a}_0 (\Pi^+ l^-_s \oplus 1)$ we have

$$\mathfrak{a}^* = (\Pi^+ l^-_s \oplus 1)^*\, \mathfrak{a}_0^* (\Pi^+ l^-_{-s+m} \oplus 1)^* = (l^+_{-s} \oplus 1)\, \mathfrak{a}_0^* (l^+_{s-m} \oplus 1). \quad \square$$

Proof of 2.1.2.1, Proposition 2: Let $\sigma(\nu)$ be as in 2.1.2.1, Proposition 2. The Fredholm property of $\Pi^+\sigma$ follows from Proposition 7. For the calculation of the index consider the V^+ extension, because we use homotopies. As a corollary of the Bott periodicity theorem (cf. 1.1.3.3, Theorem 4) we obtain that any matrix function $\sigma: S^1 \to \mathrm{GL}\,(k, \mathbf{C})$ is stable homotopic to $z^\varkappa \oplus 1_N$, $z \in S^1$, for some $\varkappa \in \mathbf{Z}$ and $N \in \mathbf{Z}_+$ sufficiently large. Since $\operatorname{ind} \Pi^+(z^\varkappa \oplus 1) = \operatorname{ind} \Pi^+ z^\varkappa$, it is sufficient to consider the scalar symbol $z^\varkappa$. If $\varkappa = -1$, we have $\operatorname{ind} \Pi^+ z^{-1} = 1$. In fact, using the expansion

$$g(\nu) = \sum_{j=0}^{\infty} g_j \frac{(1 - i\nu)^j}{(1 + i\nu)^{j+1}}, \quad \sum_j |g_j|^2 < \infty, \quad g \in V^+,$$

(cf. 2.1.1.1, Corollary 8), we get

$$\Pi^+ \frac{1 + i\nu}{1 - i\nu} \sum_{j=0}^{\infty} g_j \frac{(1 - i\nu)^j}{(1 + i\nu)^{j+1}} = \sum_{j=0}^{\infty} g_{j+1} \frac{(1 - i\nu)^j}{(1 + i\nu)^{j+1}},$$

i.e. the kernel of $\Pi^+ z^{-1}$ is spanned by $(1 + i\nu)^{-1}$ and the cokernel has dimension zero. For $\varkappa \in \mathbf{Z}$ arbitrary $\operatorname{ind} \Pi^+ z^\varkappa = -\varkappa$ follows, since $(\Pi^+ z^\varkappa)(\Pi^+ z^{\varkappa'}) = \Pi^+ z^{\varkappa + \varkappa'}$ modulo compact operators. Hence it is proved that

$$\operatorname{ind} \Pi^+ z^\varkappa = -\varkappa = -\deg (z^\varkappa).$$

The homotopy $\sigma(z) \oplus 1_M \simeq z^\varkappa \oplus 1_N$ mentioned above preserves $\operatorname{ind} \Pi^+\sigma$ and the mapping degree of the determinant. $\square$

Proof of 2.1.2.1, Proposition 3, (i) ⇒ (ii): First consider the case $\sigma(\nu) \neq 0$ for $|\nu| < \infty$. Let $\sigma(\nu) \sim \sum_{j=0}^{\infty} a_j \nu^{m-j}$, $m \in \mathbf{Z}$, be the asymptotic expansion of σ with $\det a_0 \neq 0$. Then $l^-_{-m}(\nu)\, \sigma(\nu)$ belongs to $H_1 \otimes \hom(\mathbf{C}^k, \mathbf{C}^k)$ and satisfies the conditions of 2.1.2.1, Proposition 2. Thus $\Pi^+ l^-_{-m}\sigma: H^+ \otimes \mathbf{C}^k \to H^+ \otimes \mathbf{C}^k$ is an isomorphism. In view of Lemma 8, $\Pi^+ l^-_{-m}: H^+ \otimes \mathbf{C}^k \to H^+ \otimes \mathbf{C}^k$ is an isomorphism, too. l^-_{-m} is a minus symbol. From (2)

$$\Pi^+(l^-_{-m}\sigma h) = \Pi^+\left(l^-_{-m}(\Pi^+\sigma h)\right),$$

$h \in H^+ \otimes \mathbf{C}^k$ follows, i.e. $\Pi^+ l^-_{-m}\sigma = \Pi^+ l^-_{-m} \circ \Pi^+\sigma$ and hence $\Pi^+\sigma: H^+ \otimes \mathbf{C}^k \to H^+ \otimes \mathbf{C}^k$ is an isomorphism.

Now consider the transformation $A: C^\infty(S^1) \to H_0$ introduced in 2.1.1.1 and the projector $\pi^+ = A^{-1} \circ \Pi^+ \circ A: C^\infty(S^1) \to C^\infty_+(S^1)$. Starting with the given symbol σ we can suppose without loss of generality that $|\sigma(\nu)| \leqq$ const., $\nu \in \mathbf{R}$. Otherwise multiply by l^-_{-m} for $m \in \mathbf{Z}$ sufficiently large and use that $\Pi^+ l^-_{-m}: H^+ \otimes \mathbf{C}^k \to H^+ \otimes \mathbf{C}^k$ is an isomorphism. Then $\sigma^1 = (\varkappa^*)^{-1}\sigma$ (with $\varkappa: \mathbf{R}_\nu \to S^1$ introduced in 2.1.1.1) belongs to $C^\infty(S^1)$ and $\det \sigma^1$ has a finite number of zeros of finite order on the circle iff $\det \sigma(\nu)$ satisfies the analogous condition on $\mathbf{R}$. The point $\nu = \pm\infty$ is included and corresponds to $z = -1$. In order to simplify the notations, consider scalar operators from now on.

We can write σ^1 as a product $\sigma_1^1 \cdot \ldots \cdot \sigma_l^1$, where σ_j^1 has a zero of finite order at a point z_j, and $\sigma_j^1(z) \neq 0$ when $z \neq z_j$. Operators on $C_+^\infty(S^1)$ of the form $\pi^+\sigma^1\colon C_+^\infty(S^1) \to C_+^\infty(S^1)$ have analogous properties with respect to composition. Therefore,

$$\pi^+\sigma^1 = \pi^+\sigma_1^1 \circ \pi^+\sigma_2^1 \circ \ldots \circ \pi^+\sigma_l^1 + \lambda\,,$$

where $\lambda\colon C_+^\infty(S^1) \to C_+^\infty(S^1)$ is a compact operator. If we show that $\pi^+\sigma_j^1\colon C_+^\infty(S^1) \to C_+^\infty(S^1)$ is Fredholm, we get in view of the commuting diagram

$$\begin{array}{ccc} C_+^\infty(S^1) & \xrightarrow{\pi^+\sigma_j^1} & C_+^\infty(S^1) \\ \big\downarrow A & & \big\downarrow A \\ H^+ & \xrightarrow{\Pi^+\sigma_j} & H^+ \end{array}$$

with $\sigma_j^1 = (\varkappa^*)^{-1}\sigma_j$ that $\Pi^+\sigma_j\colon H^+ \to H^+$ is Fredholm, too. Thus the assertion is reduced to $\pi^+\sigma_j^1$. By a rotation we can transform σ_j^1 into a function $\tilde{\sigma}_j^1$ with $\tilde{\sigma}_j^1(-1) = 0$ of finite order and $\tilde{\sigma}_j^1(z) \neq 0$ when $z \in S^1$, $z \neq -1$. Obviously, this transformation preserves the Fredholm property. Thus we are reduced to the case considered in the first step of the proof. □

In connection with the index of operators in this book we only use functions $\sigma(\nu) \in H \otimes \hom(\mathbb{C}^k, \mathbb{C}^k)$ with non-vanishing determinant.

Proposition 11. *The mapping* op *is injective, i.e.* $\mathrm{op}'(\mathfrak{c}) = 0$ *implies* $\mathfrak{c} = 0$.

Proof: Let $\mathrm{op}'(\mathfrak{c}) \in \mathfrak{R}^{m,d}$ and

$$\mathrm{op}'(\mathfrak{c})\begin{pmatrix} h \\ w \end{pmatrix} = \begin{pmatrix} \Pi^+\sigma_A + \Pi'\sigma_B & \sigma_K \\ \Pi'\sigma_T & \sigma_Q \end{pmatrix}\begin{pmatrix} h \\ w \end{pmatrix} = 0 \quad \text{for all } h \in H^+ \otimes \mathbb{C}^k,\ w \in \mathbb{C}^{l_-}. \tag{8}$$

(8) immediately implies $\sigma_K = 0$, $\sigma_Q = 0$. Since (8) implies the same for all components of the matrices $\Pi^+\sigma_A + \Pi'\sigma_B$ and $\Pi'\sigma_T$, we can assume that all symbols are scalar. Consider the equation

$$\Pi'\sigma_T h = 0 \quad \text{for all} \quad h \quad H^+\,. \tag{9}$$

The trace symbol σ_T has the decomposition

$$\sigma_T(\nu) = \sum_{k=0}^{d-1} c_k\nu^k + \sigma_T^0(\nu)\,,$$

$\sigma_T^0 \in H_0^-$. The inverse Fourier transform sends (9) into

$$\sum_{k=0}^{q-1} c_k D_t^k f(0) + \int_{\mathbb{R}_+} F^{-1}\sigma_T^0(-t)\, f(t)\, dt = 0 \quad \text{for all} \quad f \in \mathcal{S}(\overline{\mathbb{R}}_+)\,. \tag{10}$$

If $F^{-1}\sigma_T^0 \not\equiv 0$ on $\mathbb{R}_-$, one can find a function $f \in C_0^\infty(\mathbb{R}_+)$ such that

$$\int_{\mathbb{R}_+} F^{-1}\sigma_T^0(-t)\, f(t)\, dt \neq 0$$

in contradiction to (10). By substituting functions $f_k(t) = t^k\varphi(t)$, $\varphi \in C_0^\infty(\overline{\mathbb{R}}_+)$, $\varphi = 1$ near 0 into (10), it follows that $c_k = 0$ for all k.

Now it remains to show that

$$\Pi^+\sigma_A h + \Pi'\sigma_B h = 0 \quad \text{for all} \quad h \in H^+$$

implies $\sigma_A = 0$ and $\sigma_B = 0$. The map $\Pi^+(1 - i\nu)^q\colon H^+ \to H^+$, $q \in \mathbb{Z}$ defines an isomorphism with the inverse $\Pi^+(1 - i\nu)^{-q}$. Proposition 1 shows that $(\Pi^+\sigma_A + \Pi'\sigma_B)$

$\times \Pi^+(1-i\nu)^q = \Pi^+\sigma'_A + \Pi'\sigma'_B$, $\sigma'_A \in H_{m+q+1}$, $\sigma'_B \in H^+ \otimes H^-_{d+q}$ and $\Pi^+\sigma'_A h + \Pi'\sigma'_B h = 0$ for all $h \in H^+$ follows from the assumption. For $-q \geqq \max(m, d)$ there is a continuous extension to V^+. The inverse Fourier transform gives

$$\int_{\mathbb{R}_+} F^{-1}\sigma'_A(t-x)\, f(x)\, dx + \int_{\mathbb{R}_+} F_\nu^{-1} F_\varrho^{-1} \sigma'_B(t-x)\, f(x)\, dx = 0$$

for all $f \in L^2(\mathbb{R}_+)$. Then the assertion follows from

Lemma 12. *Let $A: L^2(\mathbb{R}_+) \to L^2(\mathbb{R}_+)$ be a convolution operator,*

$$Af(t) = \int_{\mathbb{R}_+} A(t-t')\, f(t')\, dt', \quad f \in L^2(\mathbb{R}_+),$$

$A(t) \in L^1(\mathbb{R})$ and $C: L^2(\mathbb{R}_+) \to L^2(\mathbb{R}_+)$ a compact operator. Then, for any $\nu_0 \in \mathbb{R}$, there is a sequence $\{f_\lambda\} \subset L^2(\mathbb{R}_+)$, $\lambda \in \mathbb{Z}_+$, $\|f_\lambda\|_{L^2(\mathbb{R}_+)} = 1$, weakly converging to zero for $\lambda \to \infty$ such that

$$\|(A+C)\, f_\lambda - \mathrm{FA}\,(\nu_0)\, f_\lambda\|_{L^2(\mathbb{R}_+)} \to 0 \quad \textit{for} \quad \lambda \to \infty$$

(FA denotes the Fourier transform of the function $A(t)$).

Proof: Take $f_\lambda(t) = \lambda^{-1/2} e^{i(t+\lambda)\nu_0} f(t\lambda^{-1} + 1)$ for $f \in C_0^\infty(\mathbb{R}_+)$, $\operatorname{supp} f \subset [1, \infty)$. The Fourier transform $\hat{f}_\lambda$ of f_λ is given by

$$\hat{f}_\lambda(\nu) = \lambda^{-1/2} \int e^{-i\nu t}\, e^{i(t+\lambda)\nu_0} f(t\lambda^{-1} + 1)\, dt$$
$$= \lambda^{1/2} \int e^{-i\lambda t'(\nu-\nu_0)}\, e^{i\nu\lambda} f(t')\, dt' = \lambda^{1/2} e^{i\nu\lambda} \hat{f}(\lambda(\nu-\nu_0)),$$

where $\hat{f} = Ff$. Obviously $\|\hat{f}_\lambda\|_{L^2(\mathbb{R})} = 1$ for all $\lambda > 0$. For $\lambda \to \infty$ the sequence $\hat{f}_\lambda$ weakly converges to zero. Since $\operatorname{supp} f \subset [1, \infty)$, for the sequence f_λ we have $\operatorname{supp} f_\lambda \subset \mathbb{R}_+$. Now it remains to show that $\|Af_\lambda - \mathrm{FA}(\nu_0)\, f_\lambda\|_{L^2(\mathbb{R}_+)} \to 0$. We have

$$\int |\lambda^{1/2}(\mathrm{FA}(\nu)\, \hat{f}(\lambda(\nu-\nu_0)) - \mathrm{FA}\,(\nu_0)\, \hat{f}(\lambda(\nu-\nu_0)))|^2\, d\nu$$
$$= \int |\mathrm{FA}\,(\nu_0 + \lambda^{-1}\tau) - \mathrm{FA}(\nu_0)|^2\, |\hat{f}(\tau)|^2\, d\tau.$$

The last integral tends to zero for $\lambda \to \infty$, since FA is a continuous function. □

Note that a consequence of Proposition 11 is that

$$\Pi^+\sigma_A + \Pi'\sigma_B = \Pi^+\sigma'_A + \Pi'\sigma'_B$$

implies $\sigma_A = \sigma'_A$, $\sigma_B = \sigma'_B$.

2.1.2.4. Calculation of Inverses

In this section we show that $\mathfrak{R}$ and $\mathfrak{R}^{0,0}$ are closed with respect to taking inverse operators.

Definition 1. A symbol $\sigma \in H_{m+1} \otimes \hom(\mathbb{C}^k, \mathbb{C}^k)$ is called *elliptic* (*of order m*) if for $\sigma_0(\nu) = (1-i\nu)^{-m}\sigma(\nu) \in H_1 \otimes \hom(\mathbb{C}^k, \mathbb{C}^k)$ we have $\det \sigma_0(\nu) \neq 0$ for all $\nu \in \overline{\mathbb{R}}$, i.e. including $\nu = \pm\infty$. A symbol $\sigma \in H_{m+1} \otimes \hom(\mathbb{C}^k, \mathbb{C}^l)$, $l \geqq k$, is called *overdetermined elliptic* if $\sigma_0(\nu)$ is an injective matrix for all $\nu \in \overline{\mathbb{R}}$. Similarly one defines *underdetermined elliptic* symbols (for $l \leqq k$).

If σ is elliptic of order m, then σ^{-1} is elliptic of order $-m$. If σ is overdetermined elliptic, the adjoint matrix function $\sigma^*(\nu)$ is underdetermined elliptic (of the same order) and $\sigma^*(\nu)\, \sigma(\nu)$ is elliptic of order m. It is easily seen that any overdetermined elliptic symbol of order m has a left inverse (which is underdetermined elliptic of order $-m$), namely $(\sigma^*(\nu)\,\sigma(\nu))^{-1}\,\sigma^*(\nu)$. An analogous assertion is true for underdetermined elliptic systems.

If the order is given and fixed, we simply speak of elliptic symbols.

Lemma 2. *Let $\sigma_P \in H_{m+1} \otimes \hom(\mathbb{C}^k, \mathbb{C}^k)$ be elliptic. If* $\operatorname{ind} \Pi^+\sigma_P = j - l$, $j, l \in \mathbb{Z}_+$, *there is a boundary symbol*

$$\mathfrak{b} = \begin{pmatrix} \Pi^+\sigma_P + \Pi'\sigma_{B_0} & \sigma_{K_0} \\ \Pi'\sigma_{T_0} & \sigma_{Q_0} \end{pmatrix} : \begin{matrix} H^+ \otimes \mathbb{C}^k \\ \otimes \\ \mathbb{C}^l \end{matrix} \to \begin{matrix} H^+ \otimes \mathbb{C}^k \\ \oplus \\ \mathbb{C}^j \end{matrix}$$

$\mathfrak{b} \in \mathfrak{R}^{m,0}$ *which is bijective.*

Proof: We shall construct a σ_{B_0} such that $\dim\ker(\Pi^+\sigma_P + \Pi'\sigma_{B_0}) = j$ and $\dim\operatorname{coker}(\Pi^+\sigma_P + \Pi'\sigma_{B_0}) = l$. Let $W \subset H^+ \otimes \mathbb{C}^k$ be an l-dimensional subspace which is complementary to $\operatorname{im}(\Pi^+\sigma_P + \Pi'\sigma_{B_0})$. Any isomorphism $\mathbb{C}^l \to W$ can be given by a potential symbol $\sigma_{K_0}: \mathbb{C}^l \to H^+ \otimes \mathbb{C}^k$, $\operatorname{im}\sigma_{K_0} = W$ (cf. 2.1.2.2, Lemma 4). Then

$$(\Pi^+\sigma_P + \Pi'\sigma_{B_0}, \sigma_{K_0}): \begin{matrix} H^+ \otimes \mathbb{C}^k \\ \oplus \\ \mathbb{C}^l \end{matrix} \to H^+ \otimes \mathbb{C}^k$$

is surjective and its kernel V has the dimension j. Now we can find an isomorphism $V \to \mathbb{C}^j$ which is induced from

$$(\Pi'\sigma_{T_0}, \sigma_{Q_0}): \begin{matrix} H^+ \otimes \mathbb{C}^k \\ \oplus \\ \mathbb{C}^l \end{matrix} \to \mathbb{C}^j$$

with the trace symbol $\sigma_{T_0} \in H_0^- \otimes \hom(\mathbb{C}^k, \mathbb{C}^j)$ and a matrix $\sigma_{Q_0} \in \hom(\mathbb{C}^l, \mathbb{C}^j)$. In fact, the inverse isomorphism $\mathbb{C}^j \to V$ can be given by ${}^t(\sigma_{K'}, \sigma_{Q'})$ with the potential symbol $\sigma_{K'}$. Then we get an isomorphism in the desired form.

In order to construct σ_{B_0} we choose $j_0, l_0 \in \mathbb{Z}_+$ such that $j - l = j_0 - l_0$, $j_0 \geqq \dim\ker\Pi^+\sigma_P$ and $j_0 \geqq j$. If $V \subset H^+ \otimes \mathbb{C}^k$ is an arbitrary subspace with $V \supseteqq \ker\Pi^+\sigma_P$ and $j_0 = \dim V$, we find by 2.1.2.2, Lemma 4 a Green symbol $\sigma_B^{(V)} \in H^+ \otimes H_0 \otimes \hom(\mathbb{C}^k, \mathbb{C}^k)$ such that

$$\Pi'\sigma_B^{(V)}: H^+ \otimes \mathbb{C}^k \to H^+ \otimes \mathbb{C}^k$$

is a projection onto V. Then $V = \ker\left(\Pi^+\sigma_P(1 - \Pi'\sigma_B^{(V)})\right)$ and, since $j_0 = \dim V$, we get $l_0 = \dim\operatorname{coker}\left(\Pi^+\sigma_P(1 - \Pi'\sigma_B^{(V)})\right)$. For $j_0 = j$ we can take $\Pi'\sigma_{B_0} = -\Pi^+\sigma_P \Pi'\sigma_B^{(V)}$. If $j_0 > j$, let $V_0 \subset \ker\left(\Pi^+\sigma_P(1 - \Pi'\sigma_B^{(V)})\right)$ be a $j_0 - j$ dimensional subspace and W_0 a $l_0 - l$ dimensional subspace in a complement of $\operatorname{im}\left(\Pi^+\sigma_P(1 - \Pi'\sigma_B^{(V)})\right)$, $j_0 - j = l_0 - l$. Take an isomorphism $\beta: V_0 \to W_0$ in the form of a Green symbol of type 0. Then the composition $H^+ \otimes \mathbb{C}^k \xrightarrow{\Pi'\sigma_B^{(V_0)}} V_0 \xrightarrow{\beta} W_0$ is equal to $\Pi'\sigma_{B_1}$ with a Green symbol σ_{B_1} of type 0. Then $\Pi'\sigma_{B_0} = -\Pi^+\sigma_P\Pi'\sigma_B^{(V)} + \Pi'\sigma_{B_1}$ has the desired properties. □

Let $\sigma_B(\nu, \varrho) \in H^+ \otimes H_d^-$. Then there is a decomposition

$$\sigma_B(\nu, \varrho) = \sum_{l=0}^{d-1} \sigma_{K_l}(\nu)\,\varrho^l + \sum_{m,n=0}^{\infty} a_{mn} \frac{(1 - i\nu)^m}{(1 + i\nu)^{m+1}} \frac{(1 + i\varrho)^n}{(1 - i\varrho)^{n+1}}$$

(cf. 2.1.2.2, Corollary 1). Denote by $(H^+ \otimes H_d^-)^{MN}$ the subspace of all Green symbols of type 0 for which $a_{mn} = 0$ for $m \leqq M$, $n \leqq N$. There is the canonical projection $p_{MN}: H^+ \otimes H_d^- \to (H^+ \otimes H_d^-)^{MN}$ defined by putting zero all coefficients a_{mn}, $m \leqq M$, $n \leqq N$.

Lemma 3. *For any operator*

$$\mathfrak{g} = \begin{pmatrix} \Pi'\sigma_B & \sigma_K \\ \Pi'\sigma_T & \sigma_Q \end{pmatrix} : \begin{matrix} H^+ \otimes \mathbb{C}^k \\ \oplus \\ \mathbb{C}^l \end{matrix} \to \begin{matrix} H^+ \otimes \mathbb{C}^j \\ \oplus \\ \mathbb{C}^{l'} \end{matrix}$$

and arbitrary $M, N \in \mathbb{Z}_+$ there exists a decomposition

$$\mathfrak{g} = \begin{pmatrix} 0 & \sigma_{K_1} \\ 0 & 0 \end{pmatrix} \begin{pmatrix} 0 & 0 \\ \Pi'\sigma_{T_1} & 0 \end{pmatrix} + \begin{pmatrix} 0 & \sigma_K \\ \Pi'\sigma_T & \sigma_Q \end{pmatrix} \begin{pmatrix} \Pi'\sigma_{B_1} & 0 \\ 0 & 0 \end{pmatrix},$$

where $\sigma_{B_1} = p_{NM}\sigma_B$ and suitable potential and trace symbols σ_{K_1} and σ_{T_1}.

Note that the first operator in the above decomposition is of finite rank.

Proof: Obviously we can write the decomposition 2.1.2.2.(5) of the Green symbol σ_B as

$$\sigma_B(\nu, \varrho) = \sum_{i=1}^{i_0} \sigma_{K_1,i}(\nu)\, \sigma_{T_1,i}(\varrho) + p_{MN}\sigma_B(\nu, \varrho) . \quad \square$$

Lemma 4. *Let $\sigma_B \in H^+ \otimes H_0^- \otimes \hom(\mathbb{C}^k, \mathbb{C}^k)$ and*

$$(2\pi)^{-1} \int_{\mathbb{R}^2} |\sigma_B(\nu, \varrho)|^2 \, d\nu \, d\varrho < 1 .$$

Then $1 + \Pi'\sigma_B: H^+ \otimes \mathbb{C}^k \to H^+ \otimes \mathbb{C}^k$ is invertible and there is a Green symbol $\sigma_G \in H^+ \otimes H_0^- \otimes \hom(\mathbb{C}^k, \mathbb{C}^k)$ such that $(1 + \Pi'\sigma_B)^{-1} = 1 + \Pi'\sigma_G$.

Proof: Consider the Neumann series $\sum_{k=0}^{\infty} (-1)^k (\Pi'\sigma_B)^k$. According to 2.1.2.3, Lemma 2 we have

$$(\Pi'\sigma_B)^k = \Pi'_\tau (2\pi)^{-k+1} \int_{\mathbb{R}^{k-1}} \sigma_B(\nu, \tau_1)\, \sigma_B(\tau_1, \tau_2) \ldots \sigma_B(\tau_{k-1}, \tau)\, d\tau_1 \ldots d\tau_{k-1} .$$

Denote the function on the right hand side by $\sigma_B^{(k)}(\nu, \tau)$. If we consider $H^+ \otimes H_0^-$ with the semi-norms

$$||\sigma_B(\tau, \nu)||_{j_1 j_2 j_3 j_4} = ||\Pi_\nu^+ \Pi_{0\tau}^- (\nu^{j_1} D_\nu^{j_2} \tau^{j_3} D_\tau^{j_4} \sigma_B(\nu, \tau))||_{L^2(\mathbb{R}^2)} ,$$

by the Schwartz inequality, we obtain

$$||\sigma_B^{(k)}(\nu, \tau)||_{j_1 j_2 j_3 j_4} \leqq (2\pi)^{-k}\, ||\sigma_B||_{j_1 j_2 00}\, ||\sigma_B||_{L^2(\mathbb{R}^2)}^{k-2}\, ||\sigma_B||_{00 j_3 j_4} .$$

Hence the Neumann series $\sum_k (-1)^k \sigma_B^{(k)}$ converges in $H^+ \otimes H_0^-$. Since $\Pi'\sigma_B$ continuously depends on σ_B, the same follows for $\sum_k (-1)^k (\Pi'\sigma_B)^k$. $\square$

Lemma 5. *Let V be an (infinite dimensional) vector space and $\beta: V \to V$ a linear operator of finite rank. If $1 + \beta: V \to V$ is bijective, there exist constants $c_1, \ldots, c_N$ depending on β such that*

$$(1 + \beta)^{-1} = 1 + \sum_{k=1}^{N} c_k \beta^k .$$

Proof: Let V_1 be a complementary subspace to $\ker\beta$, $W = V_1 + \operatorname{im}\beta$ and V_0 a complement of W in $\ker\beta$. Then $V = V_0 \oplus W$ and for $\alpha = \beta|_W$ we have $\beta = 0 \oplus \alpha$, and $1 + \alpha\colon W \to W$ is bijective. Note that W is finite dimensional. Since $1 + \alpha$ is invertible, -1 does not belong the spectrum $\operatorname{sp}(\alpha)$ of α and the function $f(z) = (1 + z)^{-1}$ is analytic in a neighbourhood of $\operatorname{sp}(\alpha)$. The spectrum consists of a finite number of eigenvalues $\lambda_j \in \mathbb{C}$ with finite multiplicity m_j $(j = 1, \dots, k)$. Thus there is a polynomial $g(z) = \sum_{j=0}^{N} c_j z^j$ with

$$f^{(m)}(\lambda_j) = g^{(m)}(\lambda_j) \quad \text{for all } m \leqq m_j\,.$$

By a classical result from the theory of operator functions (cf. DUNFORD/SCHWARTZ [1]) we then have $f(\alpha) = g(\alpha)$, i.e. $(1 + \alpha)^{-1} = \sum_{j=0}^{N} c_j\alpha^j$ where obviously $c_0 = 1$. □

Proposition 6. *Let* $\sigma_A \in H_{m+1} \otimes \hom(\mathbb{C}^k, \mathbb{C}^k)$ *be elliptic and* $\mathfrak{a} \in \mathfrak{R}^{m,d}$, $d \geqq 0$, *bijective,*

$$\mathfrak{a} = \begin{pmatrix} \Pi^+\sigma_A + \Pi'\sigma_B & \sigma_K \\ \Pi'\sigma_T & \sigma_Q \end{pmatrix} : \begin{matrix} H^+ \otimes \mathbb{C}^k \\ \oplus \\ \mathbb{C}^j \end{matrix} \to \begin{matrix} H^+ \otimes \mathbb{C}^k \\ \oplus \\ \mathbb{C}^l \end{matrix}.$$

Then $\mathfrak{a}^{-1} \in \mathfrak{R}^{-m,d'}$, $d' = \max(d - m, 0)$.

Proof: Since σ_A is elliptic (of order m), there exists a $\sigma_P = \sigma_A^{-1}$ and this is elliptic (of order $-m$). By Lemma 2 there exists a bijective boundary symbol $\mathfrak{b} \in \mathfrak{R}^{-m,0}$ containing σ_P. Therefore

$$\mathfrak{c} = \mathfrak{a}\mathfrak{b} = \begin{pmatrix} 1 + \Pi'\sigma_{G_0} & \sigma_C \\ \Pi'\sigma_S & \sigma_Q \end{pmatrix} : \begin{matrix} H^+ \otimes \mathbb{C}^k \\ \oplus \\ \mathbb{C}^l \end{matrix} \to \begin{matrix} H^+ \otimes \mathbb{C}^k \\ \oplus \\ \mathbb{C}^l \end{matrix} \in \mathfrak{R}^{0,d'},$$

$d' = \max(d - m, 0)$, is an isomorphism.

Now it is sufficient to show that $(\mathfrak{a}\mathfrak{b})^{-1} \in \mathfrak{R}^{0,d'}$. Then by 2.1.2.3, Proposition 1, we obtain $\mathfrak{a}^{-1} = \mathfrak{b}(\mathfrak{a}\mathfrak{b})^{-1} \in \mathfrak{R}^{-m,d'}$. Set

$$\mathfrak{g} = \begin{pmatrix} \Pi'\sigma_{G_0} & \sigma_C \\ \Pi'\sigma_S & \sigma_R \end{pmatrix}.$$

By Lemma 3 there is a decomposition $\mathfrak{g} = \mathfrak{d} + \mathfrak{e}$ with

$$\mathfrak{e} = \begin{pmatrix} \Pi'\sigma_G & 0 \\ 0 & 0 \end{pmatrix},$$

where $\mathfrak{d}$ is an operator of finite rank and the coefficients in the decomposition of σ_G are so small that the condition of Lemma 4 is satisfied, i.e. $(1 + \Pi'\sigma_G)^{-1} = 1 + \Pi'\sigma_L$ with a suitable $\sigma_L \in H^+ \otimes H_0^- \otimes \hom(\mathbb{C}^k, \mathbb{C}^k)$. Then

$$\mathfrak{c} = \begin{pmatrix} 1 + \Pi'\sigma_G & 0 \\ 0 & 1 \end{pmatrix} \left\{ \begin{pmatrix} 1 & 0 \\ 0 & 0 \end{pmatrix} + \begin{pmatrix} 1 + \Pi'\sigma_L & 0 \\ 0 & 1 \end{pmatrix} \mathfrak{d} \right\}.$$

Since the first factor is invertible, the second one is invertible, too. It has the form

$$\begin{pmatrix} 1 & 0 \\ 0 & 1 \end{pmatrix} + \beta = \begin{pmatrix} 1 & 0 \\ 0 & 1 \end{pmatrix} + \left\{ \begin{pmatrix} 1 + \Pi'\sigma_L & 0 \\ 0 & 1 \end{pmatrix} \mathfrak{d} - \begin{pmatrix} 0 & 0 \\ 0 & 1 \end{pmatrix} \right\}$$

and β is an operator of finite rank. From Lemma 5 we conclude

$$\mathfrak{c}^{-1} = \left\{\begin{pmatrix}1 & 0\\ 0 & 1\end{pmatrix} + \sum_{j=1}^{N} c_j \beta^j\right\}\begin{pmatrix}1 + \Pi' \sigma_L & 0\\ 0 & 1\end{pmatrix}.$$

Using again 2.1.2.3, Proposition 1 we obtain the assertion. □

Proposition 7. *Let* $\mathfrak{a} \in \mathfrak{R}^{m,d}$ *be a boundary symbol with the operator symbol* $\sigma_A \in H_{m+1} \otimes \hom(\mathbb{C}^k, \mathbb{C}^m)$ *underdetermined elliptic and* $\mathfrak{a}$ *surjective. Then there is a right inverse* $\mathfrak{a}^{-1} \in \mathfrak{R}^{-m,d'}$ *of* $\mathfrak{a}$ *with the operator symbol* σ_A^{-1} *where* σ_A^{-1} *is a right inverse of* σ_A. *For an injective boundary symbol* $\mathfrak{a}' \in \mathfrak{R}^{m,d}$ *with an overdetermined elliptic operator symbol* σ_A' *there is a left inverse in* $\mathfrak{R}^{-m,d'}$ *with a left inverse of* σ_A' *as operator symbol.*

Proof: Let $\mathfrak{a} \in \mathfrak{R}^{m,d}$ be a surjective boundary symbol with an underdetermined elliptic operator symbol σ_A. Let $\mathfrak{a}_0 = (\Pi^+ l_{d-m}^- \oplus 1)\, \mathfrak{a}(\Pi^+ l_{-d}^- \oplus 1)$ be the reduced boundary symbol of order and type zero, $\mathfrak{a}_0 \in \mathfrak{R}^{0,0}$. It is obviously underdetermined elliptic. Then $\mathfrak{a}_0^* \in \mathfrak{R}^{0,0}$ (cf. 2.1.2.3, Proposition 6) is an injective boundary symbol with the overdetermined operator symbol ${}^t\bar{\sigma}_{A_0}$. Then $\mathfrak{a}\mathfrak{a}_0^* \in \mathfrak{R}^{0,0}$ is a bijective boundary symbol for the elliptic symbol $\sigma_{A_0} {}^t\bar{\sigma}_{A_0}$. According to Proposition 6 there exists an inverse $(\mathfrak{a}_0\mathfrak{a}_0^*)^{-1} \in \mathfrak{R}^{0,0}$ and $\mathfrak{a}_0^*(\mathfrak{a}_0\mathfrak{a}_0^*)^{-1} \in \mathfrak{R}^{0,0}$ is a right inverse of $\mathfrak{a}_0$. Consequently $(\Pi^+ l_d^- \oplus 1)\, \mathfrak{a}_0^*(\mathfrak{a}_0\mathfrak{a}_0^*)^{-1}\, (\Pi^+ l_{m-d}^- \oplus 1)$ is the desired right inverse. The assertion in the overdetermined case can be similarly proved. □

Note that if $\mathfrak{a} \in \mathfrak{R}$ is a surjective boundary symbol with an underdetermined elliptic operator symbol σ_A, there exists a boundary symbol $\mathfrak{p} \in \mathfrak{R}$ inducing a projection onto $\ker \mathfrak{a}$, namely $\mathfrak{p} = 1 - \mathfrak{a}_r^{-1}\mathfrak{a}$ with a right inverse $\mathfrak{a}_r^{-1} \in \mathfrak{R}$ of $\mathfrak{a}$. A similar assertion holds for injective boundary symbols with overdetermined elliptic operator symbols.

In Section 3.1.1.2 we give another proof of Proposition 6.

2.2. Symbols in the Half Space

2.2.1. The Spaces $H \otimes S^m$, $H^+ \otimes S^m$, $H^- \otimes S^m$

2.2.1.1. The Space $H^+ \otimes S^m$

In the following we shall consider topological tensor products of Fréchet spaces. Recall the definitions and simple facts. For more detailed information cf. [S8], [R2].

Let E, F be vector spaces over a field $\mathbb{K}$ ($\mathbb{K}$ may be $\mathbb{R}$ or $\mathbb{C}$) and denote the space of all bilinear forms on $E \times F$ by $B(E \times F, \mathbb{K})$. For a given couple $(x, y) \in E \times F$ by $h \mapsto h(x, y)$, $h \in B(E \times F, \mathbb{K})$ a linear form on $B(E \times F, \mathbb{K})$ is defined, i.e. an element of $B(E \times F, \mathbb{K})'$. This mapping

$$\varphi: E \times F \to B(E \times F, \mathbb{K})'$$

is bilinear. The linear hull of $\varphi(E \times F)$ in $B(E \times F, \mathbb{K})'$ is denoted by $E \otimes F$ and called *tensor product* of E and F. Instead of $\varphi(x, y)$ we write $x \otimes y$. Any element of $E \otimes F$ can be given as a finite sum $\sum_{i=1}^{N} \lambda_i(x_i \otimes y_i)$, $\lambda_i \in \mathbb{K}$, $N \in \mathbb{Z}_+$.

Let $\mathcal{L}(E, G)$ be the vector space of all linear mappings from the vector space E to the vector space G and $B(E \times F, G)$ the vector space of all bilinear mappings from $E \times F$ to G. The following proposition yields a connection between linear mappings on the tensor product $E \otimes F$ and bilinear mappings on $E \times F$.

Proposition 1. *Let E, F be vector spaces over K and φ as above. Then, for an arbitrary vector space G over $\mathbb{K}$, the mapping $\mathcal{L}(E \otimes F, G) \ni u \mapsto u \circ \varphi \in B(E \times F, G)$ is an isomorphism.*

The property of the tensor product stated in Proposition 1 is called *universality*.

Assume in the following that E, F, ... are locally convex topological vector spaces. Consider all locally convex topologies on $E \otimes F$ such that $\varphi: E \times F \to E \otimes F$ is continuous. The strongest topology with this property is called *projective topology* of the tensor product $E \otimes F$.

We consider $E \otimes F$ with the projective topology. Then the mapping in Proposition 1 is an isomorphism from the space of all linear continuous mappings $\mathcal{L}(E \otimes F, G)$ onto the space of all bilinear continuous mappings $B(E \times F, G)$.

Let $\{p_i\}_{i\in I}$, $\{q_j\}_{j\in J}$ be systems of semi-norms on E, F defining the topologies Then the projective topology on $E \otimes F$ is given by the set of semi-norms

$$r_{ij}(u) = \inf\left\{\sum_{k=1}^{N} p_i(x_k)\, q_i(y_k)\right\}, \qquad u \in E \otimes F\,,$$

where the infimum is taken over all possible decompositions $u = \sum_{k=1}^{N} x_k \otimes y_k$ ($N \in \mathbb{Z}_+$ may depend on the decomposition). The semi-norm r_{ij} is called tensor product of the semi-norms p_i and q_j, $r_{ij} = p_i \otimes q_j$.

Even for complete E and F the tensor product $E \otimes F$ does not have this property in general. Denote by $E \hat{\otimes} F$ the (unique up to isomorphism) completion of $E \otimes F$. For metrizable E and F the topological tensor product $E \hat{\otimes} F$ is a Fréchet space.

Note that there are other natural topologies in $E \otimes F$. In the case that E or F are nuclear, all these topologies are equivalent (cf. SCHAEFER [1]). In our applications this condition will be satisfied. We shall write $E \otimes F$ for the topological tensor product (instead of $E \hat{\otimes} F$), since no confusion should be possible.

Proposition 2. *Let E and F be Fréchet spaces. Then any element $u \in E \otimes F$ admits a decomposition*

$$u = \sum_{k=1}^{\infty} \lambda_k x_k \otimes y_k\,, \quad \lambda_k \in \mathbb{K}\,, \quad x_k \in E, \quad y_k \in F \tag{1}$$

such that $\sum_{k=1}^{\infty} |\lambda_k| \leqq 1$ and the sequences x_k and y_k tend to zero in E and F, respectively. The series in (1) *converges absolutely, i.e. for an arbitrary semi-norm $r_{ij} = p_i \otimes q_j$ on $E \otimes F$ holds*

$$\sum_{k=1}^{\infty} r_{ij}(\lambda_k x_k \otimes y_k) < \infty\,.$$

This implies

$$r_{ij}(u) \leqq \sum_{k=1}^{\infty} |\lambda_k|\, p_i(x_k)\, q_j(y_k)\,.$$

For a proof cf. SCHAEFER [1].

Obviously, for $E = E_1 \oplus E_2$, we have $E \otimes F = (E_1 \otimes F) \oplus (E_2 \otimes F)$.

Now we give several characterizations of functions in $H^+ \otimes S^m(\Omega' \times \mathbb{R}^{n-1})$, $\Omega' \subseteq \mathbb{R}^{n-1}$ an open set (the definition of $S^m(\Omega' \times \mathbb{R}^{n-1})$ is given in 1.2.2.1).

Let (x', ξ') be coordinates in $\Omega' \times \mathbb{R}^{n-1}$ and ν the coordinate in $\mathbb{R}$.

Proposition 3. *The following assertions are equivalent:*

(i) $a \in H^+ \otimes S^m(\Omega' \times \mathbb{R}^{n-1})$;

(ii) $a \in C^\infty(\Omega' \times \mathbb{R}^{n-1} \times \mathbb{R})$ *has an analytic extension into the lower complex half plan* $\operatorname{Im} \zeta < 0$, $\zeta = \nu + i\mu$ *for any fixed* $(x', \xi') \in \Omega' \times \mathbb{R}^{n-1}$, *and there is an asymptotic expansion*

$$a(x', \xi', \nu) \sim \sum_{k \leq -1} c_k(x', \xi')\,(1 + i\nu)^k \quad \text{for } |\nu| \to \infty ,$$

which can be differentiated formally, i.e. for arbitrary multi-indices $\alpha, \beta \in \mathbb{Z}_+^{n-1}$, $l \in \mathbb{Z}_+^3$, $l = (l_1, l_2, l_3)$, *and any compact set* $K \subset \Omega'$ *there is a constant* $c = c_{\alpha\beta l K}$ *such that*

$$\Big\| D_{x'}^\alpha D_{\xi'}^\beta \Pi_\nu^+ \Big(\nu^{l_1} D_\nu^{l_2} \Big(a(x', \xi', \nu) - \sum_{-l_3 \leq k \leq -1} c_k(x', \xi')\,(1 + i\nu)^k\Big)\Big)\Big\|_{L^2(\mathbb{R}_\nu)}$$
$$\leq c\langle \xi' \rangle^{m-|\beta|}, \quad x' \in K,\ \xi' \in \mathbb{R}^{n-1};$$

(iii) $g(x', \xi', z) \overset{\text{def}}{=} (1 + z)^{-1}\, a\big(x', \xi', i(z-1)/(z+1)\big)$ *is a* C^∞ *function on* $\Omega' \times \mathbb{R}^{n-1} \times S^1$ *having an analytic extension into* $|z| < 1$ *for any fixed* $(x', \xi') \in \Omega' \times \mathbb{R}^{n-1}$ *and for an arbitrary* $\gamma \in \mathbb{Z}_+$, *arbitrary multi-indices* $\alpha, \beta \in \mathbb{Z}_+^{n-1}$ *and any compact set* $K \subset \Omega'$ *there is a constant* $c = c_{\alpha\beta\gamma K}$ *such that*

$$\sup_{-\pi \leq \varphi \leq \pi} |D_{x'}^\alpha D_{\xi'}^\beta D_\varphi^\gamma g(x', \xi', e^{i\varphi})| \leq C\langle \xi' \rangle^{m-|\beta|}, \quad x' \in K, \xi' \in \mathbb{R}^{n-1} \quad (z = e^{i\varphi});$$

(iv) *a admits an expansion*

$$a(x', \xi', \nu) = \sum_{k=0}^{\infty} a_k(x', \xi') \frac{(1 - i\nu)^k}{(1 + i\nu)^{k+1}},$$

where the a_k *form a rapidly decreasing sequence in* $S^m(\Omega' \times \mathbb{R}^{n-1})$, *i.e. for any continuous semi-norm* p *on* $S^m(\Omega' \times \mathbb{R}^{n-1})$ *and any* $N \in \mathbb{Z}_+$ *there exists a constant* $c = c_{N,p}$ *such that* $p(a_k) \leq c(1 + k)^{-N}$.

Note that the rational functions $(1 + i\nu)^k$ in (ii) are chosen in such a way that they have analytic extensions into $\operatorname{Im} \zeta < 0$ and therefore this property has the difference $a(x', \xi', \nu) - \sum_{-l \leq k \leq -1} c_k(x', \xi')\,(1 + i\nu)^k$. Of course the assertion (ii) remains true for other rational functions with the same behaviour near ∞, e.g. ν^k or $(1 - i\nu)^k$.

Proof: (i) ⇒ (iv): According to Proposition 2 any function $a \in H^+ \otimes S^m(\Omega' \times \mathbb{R}^{n-1})$ possesses an expansion

$$a(x', \xi', \nu) = \sum_{j=1}^{\infty} \lambda_j b_j(x', \xi')\, h_j(\nu),$$

where $\sum_j |\lambda_j| \leq 1$, b_j tends to 0 in $S^m(\Omega' \times \mathbb{R}^{n-1})$ and h_j tends to 0 in H^+. The functions $h_j \in H^+$ admit expansions

$$h_j(\nu) = \sum_{k=0}^{\infty} h_{jk} e_k(\nu), \quad h_{jk} \in \mathbb{C}, \quad e_k(\nu) = (1 - i\nu)^k/(1 + i\nu)^{k+1}$$

with rapidly decreasing h_{jk} with respect to k (cf. 2.1.1.1, Corollary 8). Set

$$a_k(x', \xi') = \sum_{j=1}^{\infty} \lambda_j b_j(x', \xi')\, h_{jk} .$$

Obviously the a_k form a rapidly decreasing sequence of symbols in $S^m(\Omega' \times \mathbb{R}^{n-1})$. Changing the order of summation we get

$$a(x', \xi', \nu) = \sum_{j=1}^{\infty} \sum_{k=0}^{\infty} \lambda_j b_j(x', \xi')\, h_{jk} e_k(\nu) = \sum_{k=0}^{\infty} a_k(x', \xi')\, e_k(\nu)\,.$$

(iv) $\Rightarrow$ (i): Condition (iv) yields that the series $\sum_{k=0}^{\infty} a_k(x', \xi')\, e_k(\nu)$ converges in $H^+ \otimes S^m(\Omega' \times \mathbb{R}^{n-1})$, therefore $a \in H^+ \otimes S^m(\Omega' \times \mathbb{R}^{n-1})$. Now the equivalence of all assertions follows from the implications (iv) $\Rightarrow$ (ii) $\Rightarrow$ (iii) $\Rightarrow$ (iv). We use the characterizations of H^+ functions given in 2.1.1.1. In the following let $K \subset \Omega'$ be an arbitrary compact set, $\alpha, \beta \in \mathbb{Z}_+^{n-1}$ and $l \in \mathbb{Z}_+^3$ multi-indices, $\gamma \in \mathbb{Z}_+$; c denotes a suitable constant depending on some of these objects, not necessarily the same in different connections.

(iv) $\Rightarrow$ (ii): Let (iv) be satisfied for $a(x', \xi', \nu)$. Then $a \in C^\infty(\Omega' \times \mathbb{R}^{n-1} \times \mathbb{R})$ and for arbitrary fixed $(x', \xi') \in \Omega' \times \mathbb{R}^{n-1}$ there is an analytic extension into $\operatorname{Im} \zeta < 0$. Moreover we have the desired asymptotic expansion for fixed (x', ξ'). For $e \in H^+$ from 2.1.1.1, Proposition 9 it follows that

$$\left\| \Pi^+ \nu^{l_1} D_\nu^{l_2} \Big(e(\nu) - \sum_{-l_3 \leq k \leq -1} c_k (1 + i\nu)^k \Big) \right\|_{L^2(\mathbb{R}_\nu)} \leq c p_{l'}^{(1)}(e) \leq c p_{l''}^{(3)}(e)$$

for suitable $l', l'' \in \mathbb{Z}_+$ depending only on $l \in \mathbb{Z}_+^3$. Therefore

$$|D_{x'}^\alpha D_{\xi'}^\beta a(x', \xi')| \left\| \Pi^+ \nu^{l_1} D_\nu^{l_2} \Big(e(\nu) - \sum_{-l_3 \leq k \leq -1} c_k (1 + i\nu)^k \Big) \right\|_{L^2(\mathbb{R}_\nu)} \leq c \langle \xi' \rangle^{m - |\beta|}$$

for $x' \in K$. The same estimate we get for finite sums in the expansion in (iv). For $a(x', \xi', \nu)$, i.e. the whole series in (iv), this follows, too, since by Proposition 2 the convergence is absolute.

(ii) $\Rightarrow$ (iii): For fixed $(x', \xi') \in \Omega' \times \mathbb{R}^{n-1}$ the function $g(x', \xi, z)$ is C^∞ on S^1 and has an analytic extension into $|z| < 1$ if (ii) is satisfied. By 2.1.1.1, Proposition 9 we have

$$|D_{x'}^\alpha D_{\xi'}^\beta D_\varphi^\gamma (a(x', \xi')\, f(e^{i\varphi}))| \leq c \langle \xi' \rangle^{m - |\beta|}\, p_{l'}^{(2)}(f) \leq c \langle \xi' \rangle^{m - |\beta|}\, p_{l''}^{(1)}(f)$$

with suitable $l', l'' \in \mathbb{Z}_+$ depending on γ. Then the assertion follows by the same argument as above.

(iii) $\Rightarrow$ (iv): The functions $a_k(x', \xi')$ are Fourier coefficients of $g(x', \xi', z)$ with respect to the orthonormal base z^j, $j \in \mathbb{Z}_+$, in $L_+^2(S^1)$, i.e.

$$a_k(x', \xi') = (2\pi)^{-1} \int_{-\pi}^{\pi} g(x', \xi', e^{i\varphi})\, e^{ik\varphi}\, d\varphi\,.$$

Therefore the a_k are smooth functions. From

$$\frac{k!}{(k - \gamma)!} a_k(x', \xi') = (2\pi)^{-1} \int_{-\pi}^{\pi} (-D_\varphi)^\gamma g(x', \xi', e^{i\varphi})\, e^{ik\varphi}\, d\varphi$$

$(k \geq \gamma)$ and the estimate in (iii) we get $k^\gamma q(a_k(x', \xi')) \leq C$ for an arbitrary semi-norm q on $S^m(\Omega' \times \mathbb{R}^{n-1})$. □

As observed in 2.1.1.1 the space H^+ is nuclear. Denote by q_l, $l \in \mathbb{Z}_+$, a countable set of continuous semi-norms on $S^m(\Omega' \times \mathbb{R}^{n-1})$ defining the Fréchet space topology of

S^m. The following sets of semi-norms are defined on $H^+ \otimes S^m(\Omega' \times \mathbb{R}^{n-1})$:

$$r^{(1)}_{k,l}(a) = \{\int q_l(\Pi^+(\nu^{k_1} D^{k_2}_\nu a(x', \xi', \nu))^2 \, d\nu)\}^{1/2}, \quad k = (k_1, k_2) \in \mathbb{Z}^2_+,$$

$$r^{(2)}_{\gamma,l}(a) = \sup_{-\pi \leq \varphi \leq \pi} q_l(D^\gamma_\varphi g(x', \xi', e^{i\varphi})), \quad \gamma \in \mathbb{Z}_+,$$

$$r^{(3)}_{\delta,l}(a) = \left\{\sum_{j=0}^{\infty} q_l(a_j)^2 \, j^{2\delta}\right\}^{1/2}, \quad \delta \in \mathbb{Z}_+.$$

The proof of Proposition 3 contains the proof of the following

Proposition 4. *Each of the systems of semi-norms $r^{(1)}_{k,l}$, $r^{(2)}_{\gamma,l}$, $r^{(3)}_{\delta,l}$ defines the topology of the tensor product on $H^+ \otimes S^m$.*

2.2.1.2. The Spaces $H^- \otimes S^m$ and $H \otimes S^m$

Note that the mapping

$$H^+ \otimes S^m(\Omega' \times \mathbb{R}^{n-1}) \ni a \mapsto \bar{a} \in H_0^- \otimes S^m(\Omega' \times \mathbb{R}^{n-1})$$

defined by complex conjugation of the function $a(x', \xi', \nu)$ is a topological isomorphism, since complex conjugation defines a topological isomorphism between H^+ and H_0^-. Then 2.2.1.1, Proposition 3 immediately yields characterizations of functions in $H_0^- \otimes S^m(\Omega' \times \mathbb{R}^{n-1})$ (replace Π^+ by Π_0^- and $\operatorname{Im} \zeta < 0$ ($|z| < 1$) by $\operatorname{Im} \zeta > 0$ ($|z| > 1$)).

A function $a(x', \xi', \nu)$ belongs to $H_d' \otimes S^m(\Omega' \times \mathbb{R}^{n-1})$ iff there is a decomposition

$$a(x', \xi', \nu) = \sum_{j=0}^{d-1} b_j(x', \xi') \, \nu^j,$$

$b_j \in S^m(\Omega' \times \mathbb{R}^{n-1})$. A function a belongs to $H' \otimes S^m(\Omega' \times \mathbb{R}^{n-1})$ if there is a number $d > 0$ and $a \in H_d' \otimes S^m(\Omega' \times \mathbb{R}^{n-1})$.

Let q_l, $l \in \mathbb{Z}_+$, be a countable set of semi-norms on $S^m(\Omega' \times \mathbb{R}^{n-1})$ defining the topology. Then the topology on $H_d' \otimes S^m(\Omega' \times \mathbb{R}^{n-1})$ is given by the set of semi-norms $q_l'(a) = \sum_{j=0}^{d-1} q_l(b_j)$. Define the topology in $H' \otimes S^m(\Omega' \times \mathbb{R}^{n-1})$ as the inductive limit topology with respect to the inclusions $H_d' \otimes S^m \subset H' \otimes S^m$. Since $H_d^- = H_0^- \oplus H_d'$, we have

$$H_d^- \otimes S^m = (H_0^- \otimes S^m) \oplus (H_d' \otimes S^m)$$

and

$$H^- \otimes S^m = (H_0^- \otimes S^m) \oplus (H' \otimes S^m).$$

From 2.2.1.1, Proposition 3 we get

Proposition 1. *The following assertions are equivalent:*

(i) $a \in H_d^- \otimes S^m(\Omega' \times \mathbb{R}^{n-1})$;

(ii) $a \in C^\infty(\Omega' \times \mathbb{R}^{n-1} \times \mathbb{R})$ *has an analytic extension into the upper complex half plane* $\operatorname{Im} \zeta > 0$, $\zeta = \nu + i\mu$ *for any fixed* $(x', \xi') \in \Omega' \times \mathbb{R}^{n-1}$ *and there is an asymptotic expansion*

$$a(x', \xi', \nu) \sim \sum_{k \leq d-1} c_k(x', \xi') (1 - i\nu)^k, \quad \text{for} \quad |\nu| \to \infty,$$

which can be formally differentiated, i.e. there are similar estimates as in 2.2.1.1, *Proposition* 3 (ii);

(iii) $g(x', \xi', z) \overset{\text{def}}{=} (1+z)^{d-1} a(x', \xi', \mathrm{i}(z-1)/(z+1))$ *is a* C^∞ *function on* $\Omega' \times \mathbb{R}^{n-1} \times S^1$ *having an analytic extension into* $|z| > 1$ *(including* ∞*) for any fixed* x', ξ' *and there are similar estimates as in* 2.1.1.1, *Proposition* 3 (iii);

(iv) *a admits an expansion*

$$a(x', \xi', \nu) = \sum_{j=0}^{d-1} b_j(x', \xi') \nu^j + \sum_{k=0}^{\infty} a_k(x', \xi') \frac{(1+\mathrm{i}\nu)^k}{(1-\mathrm{i}\nu)^{k+1}},$$

where $b_j \in S^m(\Omega' \times \mathbb{R}^{n-1})$ *and the* a_k *form a rapidly decreasing sequence in* $S^m(\Omega \times \mathbb{R}^{n-1})$ (*cf.* 2.1.1.1, *Proposition* 3).

Note that an asymptotic expansion like that in (ii) is valid for other rational functions with the same asymptotic behaviour near ∞. A simple conclusion of 2.1.1.1, Proposition 4 is that the best constants in the estimates indicated under (ii) — (iv) form sets of continuous semi-norms on $H_d^- \otimes S^m(\Omega' \times \mathbb{R}^{n-1})$ defining the topology.

The Fréchet space $H_d \otimes S^m(\Omega' \times \mathbb{R}^{n-1})$ is the direct sum $H^+ \otimes S^m(\Omega' \times \mathbb{R}^{n-1}) \oplus H_d^- \otimes S^m(\Omega' \times \mathbb{R}^{n-1})$. For convenience of references we formulate

Proposition 2. *The following assertions are equivalent:*

(i) $a \in H_d \otimes S^m(\Omega' \times \mathbb{R}^{n-1})$;

(ii) $a \in C^\infty(\Omega' \times \mathbb{R}^{n-1} \times \mathbb{R})$ *admits an asymptotic expansion*

$$a(x', \xi', \nu) \sim \sum_{k \le d-1} c_k(x', \xi') \nu^k, \quad \text{for } |\nu| \to \infty,$$

which can be formally differentiated, i.e. for arbitrary multi-indices $\alpha, \beta \in \mathbb{Z}_+^{n-1}$ *and* $l \in \mathbb{Z}_+^3$, $l = (l_1, l_2, l_3)$ *and any compact set* $K \subset \Omega'$ *there is a constant* $c = c_{\alpha\beta l K}$ *such that*

$$\Big\| D_{x'}^\alpha D_{\xi'}^\beta \Pi_{0,\nu} \Big(\nu^{l_1} D_\nu^{l_2} \Big(a(x', \xi', \nu) - \sum_{-l_3 \le k \le 1} c_k(x', \xi') \chi(\nu) \nu^k\Big)\Big)\Big\|_{L^2(\mathbb{R}_\nu)}$$
$$\le C \langle \xi' \rangle^{m-|\beta|}, \quad x' \in K, \quad \xi' \in \mathbb{R}^{n-1},$$

where $\Pi_0 = \Pi^+ + \Pi_0^-$ *and* χ *an excision function;*

(iii) $g(x', \xi', z) \overset{\text{def}}{=} (1+z)^{d-1} a(x', \xi', \mathrm{i}(z-1)/(z+1))$ *is a* C^∞*-function on* $\Omega' \times \mathbb{R}^{n-1} \times S^1$ *and for an arbitrary* $\gamma \in \mathbb{Z}_+$, *arbitrary multi-indices* $\alpha, \beta \in \mathbb{Z}_+^{n-1}$ *and any compact set* $K \subset \Omega'$ *there is a constant* $c = c_{\alpha\beta\gamma K}$ *such that*

$$\sup_{-\pi \le \varphi \le \pi} |D_{x'}^\alpha D_{\xi'}^\beta D_\varphi^\gamma g(x', \xi', e^{\mathrm{i}\varphi})| \le c \langle \xi' \rangle^{m-|\beta|}, \quad x' \in K, \quad \xi' \in \mathbb{R}^{n-1} \quad (z = e^{\mathrm{i}\varphi}),$$

(iv) *a admits an expansion*

$$a(x', \xi', \nu) = \sum_{j=0}^{d-1} b_j(x', \xi') \nu^j + \sum_{k=-\infty} a_k(x', \xi') \frac{(1-\mathrm{i}\nu)^k}{(1+\mathrm{i}\nu)^{k+1}}$$

where $b_j \in S^m(\Omega' \times \mathbb{R}^{n-1})$ *and the* a_k *form a rapidly decreasing sequence in* $S^m(\Omega' \times \mathbb{R}^{n-1})$.

Note that the properties (ii) — (iv) with non-fixed d characterize functions in $H \otimes S^m(\Omega' \times \mathbb{R}^{n-1})$.

Proposition 2 follows from 2.2.1.1, Proposition 3 and Proposition 1. In fact, the projections $\Pi^+ : H_d \to H^+$ and $\Pi^- : H_d \to H_d^-$ induce surjective projections $H_d \otimes S^m \to H^+ \otimes S^m$ and $H_d \otimes S^m \to H_d^- \otimes S^m$ (also denoted by Π^+ and Π^-, respectively). Then $a \in H_d \otimes S^m$ is equivalent to $\Pi^+ a \in H^+ \otimes S^m$ and $\Pi^- a \in H_d^- \otimes S^m$. An independent proof can be given using similar arguments as in the proof of 2.2.1.1, Proposition 3. It is left to the reader.

Similarly as in 2.2.1.1, Proposition 4 we have that the best constants in the estimates in (ii) and (iii), respectively, can be considered as semi-norms on $H_d \otimes S^m(\Omega' \times \mathbb{R}^{n-1})$ defining the topology.

2.2.1.3. The Space $H^+ \otimes H^- \otimes S^m$

Again we mainly consider the Fréchet space $H^+ \otimes H_d^- \otimes S^m(\Omega' \times \mathbb{R}^{n-1})$. The space $H^+ \otimes H^- \otimes S^m(\Omega' \times \mathbb{R}^{n-1})$ is the inductive limit with respect to the inclusions $H^+ \otimes H_d^- \otimes S^m(\Omega' \times \mathbb{R}^{n-1}) \subset H^+ \otimes H^- \otimes S^m(\Omega' \times \mathbb{R}^{n-1})$ for arbitrary $d \geqq 0$.

As in 2.1.2.2 denote the space of H^+ functions in the variable $\nu \in \mathbb{R}$ by $H^+(\nu)$.

Proposition 1. *The following assertions are equivalent:*

(i) $a(x', \xi', \nu, \tau) \in H^+(\nu) \otimes H_d^-(\tau) \otimes S^m(\Omega' \times \mathbb{R}^{n-1})$;

(ii) *a admits an expansion of the form*

$$a(x', \xi', \nu, \tau) = \sum_{j=0}^{d-1} b_j(x', \xi', \nu)\, \tau^j + \sum_{k=0}^{\infty} c_k(x', \xi', \nu)\, (1 + i\tau)^k/(1 - i\tau)^{k+1},$$

where $b_j \in H^+ \otimes S^m(\Omega' \times \mathbb{R}^{n-1})$, $j = 0, \dots, d-1$, and the c_k, $k \in \mathbb{Z}_+$, form a rapidly decreasing sequence in $H^+ \otimes S^m(\Omega' \times \mathbb{R}^{n-1})$, i.e. for any continuous semi-norm p on $H^+ \otimes S^m$ and any $N \in \mathbb{Z}_+$ there is a constant $C = C(N)$ such that $p(c_k) \leqq C(1 + k)^{-N}$;

(iii) *a admits an expansion of the form*

$$a(x', \xi', \nu, \tau) = \sum_{l=0}^{\infty} e_l(x', \xi', \tau)\, (1 - i\nu)^l/(1 + i\nu)^{l+1},$$

where the e_l form a rapidly decreasing sequence in $H_d^- \otimes S^m(\Omega' \times \mathbb{R}^{n-1})$;

(iv) *a admits an expansion of the form*

$$a(x', \xi', \nu, \tau) = \sum_{\substack{0 \leq l < \infty \\ 0 \leq j < d}} \alpha_{lj}(x', \xi') \frac{(1 - i\nu)^l}{(1 + i\nu)^{l+1}} \tau^j + \sum_{0 \leq l, k < \infty} \beta_{lk}(x', \xi') \frac{(1 - i\nu)^l}{(1 + i\nu)^{l+1}} \frac{(1 + i\tau)^k}{(1 - i\tau)^{k+1}},$$

where α_{lj} in $S^m(\Omega' \times \mathbb{R}^{n-1})$ are rapidly decreasing with respect to l and β_{lk} in $S^m(\Omega' \times \mathbb{R}^{n-1})$ are rapidly decreasing with respect to $l + k$.

Note that the coefficients in the expansions are uniquely defined by the function a. Similar expansions are valid with $\lambda + i\nu$ and $\lambda - i\nu$, $\lambda \in \mathbb{R}_+$ instead of $1 + i\nu$ and $1 - i\nu$, respectively. The proof is similar to that of 2.2.1.1, Proposition 3 and left to the reader.

From (ii) — (iv) it is easy to deduce several countable sets of semi-norms on $H^+ \otimes H_d^- \otimes S^m$ defining the topology. In particular, for a countable set q_γ, $\gamma \in \mathbb{Z}_+$ of semi-norms on $S^m(\Omega' \times \mathbb{R}^{n-1})$ defining the topology the following semi-norms

$$\left\{ \sum_{\substack{0 \leq l < \infty \\ 0 \leq j < d}} q_\gamma(\alpha_{lj})^2\, l^{2N} + \sum_{l,k=0}^{\infty} q_\gamma(\beta_{lk})^2\, (l + k)^{2N'} \right\}^{1/2}, \quad N, N' \in \mathbb{Z}_+,$$

yield the topology of $H^+ \otimes H_d^- \otimes S^m$. Moreover the expansions in (ii) — (iv) converge absolutely in $H^+ \otimes H_d^- \otimes S^m$. Note that for $(x', \xi') \in \Omega' \times \mathbb{R}^{n-1}$ there are analytic extensions of $a(x', \xi', \nu, \tau) \in H^+ \otimes H_d^-$ into the complex half plane $\operatorname{Im} \zeta < 0, \zeta = \nu + i\mu$, and $\operatorname{Im} \eta > 0, \eta = \tau + i\varrho$. For $\zeta \to \infty$ in $\operatorname{Im} \zeta < 0$ and $\eta \to \infty$ in $\operatorname{Im} \eta > 0$ there are asymptotic expansions as mentioned in 2.1.2.2 for functions in $H^+ \otimes H_d^-$. The formulation of further characterizations of functions in $H^+ \otimes H_d^- \otimes S^m$ is left to the reader.

2.2.2. Symbols with the Transmission Property

2.2.2.1. Definition and Various Characterizations of the Space $\mathfrak{A}^m$

In the investigation of boundary value problems for pseudo-differential operators in the half space and on manifolds with boundary the so-called transmission property of the symbol of the pseudo-differential operator is needed. This property will be defined and studied in this section.

Let $\Omega' \subseteq \mathbb{R}^{n-1}$ be open. Denote by $S^m((\Omega' \times \overline{\mathbb{R}}_+) \times \mathbb{R}^n)$ the subspace of $S^m((\Omega' \times \mathbb{R}_+) \times \mathbb{R}^n)$ of all symbols having an extension as symbols in $S^m((\Omega' \times \mathbb{R}) \times \mathbb{R}^n)$. Obviously, a symbol in $S^m((\Omega' \times \overline{\mathbb{R}}_+) \times \mathbb{R}^n)$ belongs to $S^m((\Omega' \times \overline{\mathbb{R}}_+) \times \mathbb{R}^n)$ if it admits an extension as a symbol in $S^m((\Omega' \times \mathbb{R}_\varepsilon) \times \mathbb{R}^n)$, $\varepsilon > 0$, $\mathbb{R}_\varepsilon = \{x_n \in \mathbb{R}, x_n > -\varepsilon\}$.

Lemma 1. *A symbol $a \in S^m((\Omega' \times \mathbb{R}_+) \times \mathbb{R}^n)$ belongs to $S^m((\Omega' \times \overline{\mathbb{R}}_+) \times \mathbb{R}^n)$ iff all derivatives $a_{(\alpha)}^{(\beta)}(x', x_n, \xi) = D_x^\alpha D_\xi^\beta a(x', x_n, \xi)$, $\alpha, \beta \in \mathbb{Z}_+^n$, have limits for $x_n \to +0$ and for an arbitrary compact set $K \subset \Omega' \times \overline{\mathbb{R}}_+$, there is a constant $c = c_{\alpha\beta K}$ such that*

$$|D_x^\alpha D_\xi^\beta a(x, \xi)| \leq c\langle\xi\rangle^{m-|\beta|} \quad \text{for } x \in K, \xi \in \mathbb{R}^n .$$

Proof: It is well-known that there exists a rapidly decreasing sequence of constants c_p, $p \in \mathbb{Z}_+$, such that $\sum_{p=0}^{\infty} p^k c_p = (-1)^k$ for any $k \in \mathbb{Z}_+$. Then define the extension of $a(x', x_n, \xi)$ for $x_n < 0$ by

$$a(x', x_n, \xi) = \sum_{p=0}^{\infty} c_p a(x', -p x_n, \xi) .$$

Obviously this yields a C^∞ function and the estimates in the definition of $S^m((\Omega' \times \overline{\mathbb{R}}_+) \times \mathbb{R}^n)$ hold for x in compact sets in $\Omega' \times \mathbb{R}$. □

Note that for $a \in S^m((\Omega' \times \overline{\mathbb{R}}_+) \times \mathbb{R}^n)$ the functions

$$a_{[\gamma]}(x', \xi', \nu) = D_{x_n}^\gamma a(x', +0, \xi', \nu\langle\xi'\rangle) , \tag{1}$$

$\gamma \in \mathbb{Z}_+$ arbitrary, are well defined and smooth. Point out that in $a_{[\gamma]}$, besides the γth derivative in normal direction to the boundary and restriction to the boundary, there is a transformation in the covariable ν which is dual to the normal variable x_n.

Definition 2. A symbol $a \in S^m((\Omega' \times \overline{\mathbb{R}}_+) \times \mathbb{R}^n)$, $m \geq -\infty$, has the *transmission property* if for any $\gamma \in \mathbb{Z}_+$ we have $a_{[\gamma]}(x', \xi', \nu) \in H(\gamma) \otimes S^m(\Omega' \times \mathbb{R}^{n-1})$. The space of all symbols in $S^m((\Omega' \times \overline{\mathbb{R}}_+) \times \mathbb{R}^n)$ with the transmission property is denoted by $\mathfrak{A}^m((\Omega' \times \overline{\mathbb{R}}_+) \times \mathbb{R}^n)$ or $\mathfrak{A}^m$. Set $\mathfrak{A}^\infty = \bigcup_m \mathfrak{A}^m$, $\mathfrak{A}^{-\infty} = \bigcap_m \mathfrak{A}^m$.

Proposition 3. *For $a \in S^m((\Omega' \times \overline{\mathbb{R}}_+) \times \mathbb{R}^n)$ the following assertions are equivalent:*

(i) $a \in \mathfrak{A}^m((\Omega' \times \overline{\mathbb{R}}_+) \times \mathbb{R}^n)$;

(ii) *for any $\gamma \in \mathbb{Z}_+$ the function $a_{[\gamma]}(x', \xi', \nu)$ admits an asymptotic expansion*

$$a_{[\gamma]}(x', \xi', \nu) \sim \sum_{k \leq m} c_k(x', \xi')\, \nu^k$$

for $|\nu| \to \infty$ which can be formally differentiated, i.e. for arbitrary multi-indices $\alpha, \beta \in \mathbb{Z}_+^{n-1}$, $l \in \mathbb{Z}_+^3$, $l = (l_1, l_2, l_3)$ and any compact set $K \subset \Omega'$ there is a constant $c = c_{\alpha\beta l K}$ such that

$$\left\| D_{x'}^\alpha D_{\xi'}^\beta \Pi_{0,\nu} \left(\nu^{l_1} D_\nu^{l_2} \left(a_{[\gamma]}(x', \xi', \nu) - \sum_{-l_3 \leq k \leq m} c_k(x', \xi')\, \chi(\nu)\, \nu^k\right)\right)\right\|_{L^2(\mathbb{R}_\nu)}$$

$$\leq c\langle \xi' \rangle^{m-|\beta|}, \quad x' \in K,\ \xi' \in \mathbb{R}^{n-1}$$

(χ denotes an excision function);

(iii) *for any $\gamma \in \mathbb{Z}_+$ the function $g(x', \xi', z) \stackrel{\text{def}}{=} (1+z)^{d-1}\, a_{[\gamma]}\left((x', \xi', \mathrm{i}\,\frac{z-1}{z+1}\right)$ belongs to $C^\infty(\Omega' \times \mathbb{R}^{n-1} \times S^1)$ and for an arbitrary $l \in \mathbb{Z}_+$, multi-indices $\alpha, \beta \in \mathbb{Z}_+^{n-1}$ and any compact set $K \subset \Omega'$ there is a constant $C = C_{\alpha\beta l K}$ such that*

$$\sup_{-\pi \leq \varphi \leq \pi} |D_{x'}^\alpha D_{\xi'}^\beta D_\varphi^l g(x', \xi', \mathrm{e}^{\mathrm{i}\varphi})| \leq C\langle \xi' \rangle^{m-|\beta|}, \quad x' \in K,\ \xi' \in \mathbb{R}^{n-1}\ (z = \mathrm{e}^{\mathrm{i}\varphi});$$

(iv) *for any $\gamma \in \mathbb{Z}_+$ the function $a_{[\gamma]}(x', \xi', \nu)$ admits an expansion of the form*

$$a_{[\gamma]}(x', \xi', \nu) = \sum_{j=0}^{m} b_{\gamma j}(x', \xi')\nu^j + \sum_{k=-\infty}^{\infty} a_{[\gamma],k}(x', \xi')\, \frac{(1 - \mathrm{i}\nu)^k}{(1 + \mathrm{i}\nu)^{k+1}}$$

where $b_{\gamma j} \in S^m(\Omega' \times \mathbb{R}^{n-1})$ and the $a_{[\gamma],k}$ form a rapidly decreasing sequence in $S^m(\Omega' \times \mathbb{R}^{n-1})$ with respect to k.

Note that (ii) remains valid if we replace $\chi(\nu)\,\nu^k$ by other rational functions with the same behaviour at infinity (without poles on the real axis). Moreover instead of L^2 norms on $\mathbb{R}_\nu$ we can use the supremum norm. For $l_3 \geq l_1 - l_2$ one can drop Π_0. It is easily seen that the condition for all those l_1, l_2, l_3 is equivalent to (ii).

Obviously one can formulate the conditions (ii) — (iv) in terms of $a_{(\gamma)}(x', 0, \xi', \nu) = \partial_{x_n}^\gamma a(x', 0, \xi', \nu)$ by replacing ν by $\nu\langle \xi' \rangle^{-1}$. In particular, we have instead of (iv)

(iv)' $$a_{(\gamma)}(x', 0, \xi', \nu) = \sum_{j=0}^{m} b_{\gamma j}(x', \xi')\, (\nu\langle \xi' \rangle^{-1})^j + \sum_{k=-\infty}^{\infty} a_{[\gamma],k}(x', \xi')\, \frac{(1 - \mathrm{i}\nu\langle \xi' \rangle^{-1})^k}{(1 + \mathrm{i}\nu\langle \xi' \rangle^{-1})^{k+1}}$$

where $b_{\gamma j} \in S^m(\Omega' \times \mathbb{R}^{n-1})$ and the $a_{[\gamma],k}$ form a rapidly decreasing sequence in $S^m(\Omega' \times \mathbb{R}^{n-1})$ with respect to k.

Remark. The expansion (iv)' for a function $a(x', \xi', \nu)$ does not imply $a \in S^m(\Omega' \times \mathbb{R}^n)$ in general. For $m > 0$ this is easily seen by the example $a(x', \xi', \nu) = b(x', \xi')\nu^j$, $b \in S^{m-j}(\Omega' \times \mathbb{R}^{n-1})$ since, for x' in a compact set $K \subset \mathbb{R}'$, we have

$$|D_{\xi'}^\alpha D_\nu^j a(x', \xi', \nu)| = j!\, |D_{\xi'}^\alpha b(x', \xi')| \leq c\langle \xi' \rangle^{m-j-|\alpha|}$$

with a suitable constant c instead of an estimate by $c\langle\xi\rangle^{m-j-|\alpha|}$ ($|\alpha| > m - j$). If a admits an expansion

$$a(x', \xi', \nu) = \sum_{k=-\infty}^{\infty} a_k(x', \xi') \frac{(1 - i\nu\langle\xi'\rangle^{-1})^k}{(1 + i\nu\langle\xi'\rangle^{-1})^{k+1}},$$

$a_k \in S^m(\Omega' \times \mathbb{R}^{n-1})$ rapidly decreasing, we get only

$$|D^\alpha_{\xi'} a(x', \xi', \nu)| \leq c\langle\xi'\rangle^{m-|\alpha|}, \quad x \in K,$$

instead of an estimate by $c\langle\xi\rangle^{m-|\alpha|}$ ($|\alpha| > m$).

Consider the space $\mathfrak{A}^m((\Omega' \times \overline{\mathbb{R}}_+) \times \mathbb{R}^n)$ in the topology given by the set of all continuous semi-norms on $S^m((\Omega' \times \overline{\mathbb{R}}_+) \times \mathbb{R}^n)$ and a (countable) set of semi-norms given by the best constants of the estimates in Proposition 3. This set of semi-norms is induced from $H \otimes S^m(\Omega' \times \mathbb{R}^{n-1})$ by the mapping

$$\mathfrak{A}^m((\Omega' \times \overline{\mathbb{R}}_+) \times \mathbb{R}^n) \ni a \mapsto \{a_{[\gamma]}(x', \xi', \nu)\}_{\gamma \in \mathbb{Z}_+} \subset H \otimes S^m(\Omega' \times \mathbb{R}^{n-1}).$$

Lemma 4. *We have* $S^{-\infty}((\Omega' \times \overline{\mathbb{R}}_+) \times \mathbb{R}^n) = \mathfrak{A}^{-\infty}((\Omega' \times \overline{\mathbb{R}}_+) \times \mathbb{R}^n)$ *and moreover* $D^\alpha_x D^\beta_\xi a(x, \xi) \in \mathfrak{A}^{m-|\beta|}$ *for* $a \in \mathfrak{A}^m$ *and* $aa' \in \mathfrak{A}^{m+m'}$ *for* $a' \in \mathfrak{A}^{m'}$.

Proof: Let $a(x, \xi) \in S^{-\infty}((\Omega' \times \overline{\mathbb{R}}_+) \times \mathbb{R}^n) = \bigcap_m S^m((\Omega' \times \overline{\mathbb{R}}_+) \times \mathbb{R}^n)$. Then $\nu^{l_1} D^{l_2}_\nu (\partial^\gamma_{x_n} a(x', 0, \xi', \langle\xi'\rangle \nu)) = \nu^{l_1} D^{l_2}_\nu \partial^\gamma_{x_n} a(x', 0, \xi', \langle\xi'\rangle \nu) \langle\xi'\rangle^{l_2}$ implies for an arbitrary $N \in \mathbb{Z}_+$ and any compact set $K \subset \Omega'$

$$\|\nu^{l_1} D^{l_2}_\nu a_{[\gamma]}(x', \xi', \nu)\|_{L^2(\mathbb{R}_\nu)} \leq c_N \langle\xi'\rangle^{-N} \quad \text{for} \quad x' \in K$$

with a suitable constant c_N. The other assertions are obvious. □

Example. Let $a(x', x_n, \xi', \nu) \in S^m((\Omega' \times \overline{\mathbb{R}}_+) \times \mathbb{R}^n)$ be a polynomial with respect to ν near $x_n = 0$. Then $a \in \mathfrak{A}^m((\Omega' \times \overline{\mathbb{R}}_+) \times \mathbb{R}^n)$.

In fact, the condition implies (iv) in Proposition 3.

In particular, any function $a(x', x_n) \in C^\infty(\Omega' \times \overline{\mathbb{R}}_+)$ can be considered as a symbol (of order 0) with the transmission property.

Example. Let $b(x', \xi') \in S^m(\Omega' \times \mathbb{R}^{n-1})$ and $\varphi \in \mathscr{S}(\mathbb{R})$. Then, for

$$a(x', \xi, \nu) = b(x', \xi') \varphi(\nu\langle\xi'\rangle^{-1}),$$

we have $a \in \mathfrak{A}^m((\Omega' \times \overline{\mathbb{R}}_+) \times \mathbb{R}^n)$.

First we show that for any $N \in \mathbb{Z}_+$, $\beta \in \mathbb{Z}^{n-1}_+$ there is a constant $c = c_{N\beta}$ such that

$$|D^\beta_{\xi'}(\varphi(\nu\langle\xi'\rangle^{-1}))| \leq c\langle\xi'\rangle^{-|\beta|} \langle\nu\langle\xi'\rangle^{-1}\rangle^{-N}. \tag{2}$$

In fact, we have

$$D_{\xi_j}(\varphi(\nu\langle\xi'\rangle^{-1})) = \varphi'(\nu\langle\xi'\rangle^{-1}) \nu\langle\xi'\rangle^{-1} b_1(\xi'), \quad 1 \leq j \leq n-1,$$

where $b_1 \in S^{-1}(\Omega' \times \mathbb{R}^{n-1})$ and by induction over $|\beta|$

$$D^\beta_{\xi'} \varphi(\nu\langle\xi'\rangle^{-1}) = \sum_k \varphi_k(\nu\langle\xi'\rangle^{-1}) b_k(\xi')$$

with functions $\varphi_k \in \mathscr{S}(\mathbb{R})$, $b_k \in S^{-|\beta|}(\Omega' \times \mathbb{R}^{n-1})$. Then (2) is a consequence of the estimates

$$|b_k(\xi')| \leq c'\langle\xi'\rangle^{-|\beta|}, \quad |\varphi_k(\nu\langle\xi'\rangle^{-1})| \leq c''\langle\nu\langle\xi'\rangle^{-1}\rangle^{-N}$$

with suitable constants c', c''. Since $b \in S^m(\Omega' \times R^{n-1})$ for any multi-index α and an arbitrary compact set $K \subset \Omega'$, there is a constant $c = c_{\alpha K}$ such that

$$|D^\alpha_{\xi'} b(x', \xi')| \leq c\langle\xi'\rangle^{m-|\alpha|}, \quad x' \in K. \tag{3}$$

(2) and (3) together imply

$$|D^\alpha_{\xi'} b(x', \xi')\, D^\beta_{\xi'}(\varphi(\nu\langle\xi'\rangle^{-1}))| \leq c\langle\xi'\rangle^{m-|\alpha|-|\beta|}\, \langle\nu\langle\xi'\rangle^{-1}\rangle^{-N} \leq c\langle\xi\rangle^{m-|\alpha|-|\beta|}$$

for $N > |\alpha| + |\beta| - m$, $x' \in K$, since $\langle\xi\rangle = \langle\xi'\rangle\, \langle\nu\langle\xi'\rangle^{-1}\rangle$.

Similar arguments can be applied to x' derivatives. Therefore, $a(x', \xi', \nu) \in S^m((\Omega' \times \overline{R}_+) \times R^n)$. Moreover

$$\|D^\alpha_{x'} D^\beta_{\xi'} b(x', \xi')\, \nu^{l_1} D^{l_2}_\nu \varphi(\nu)\|_{L^2(R_\nu)} \leq c\langle\xi'\rangle^{m-|\beta|}$$

for $x' \in K$, i.e. (ii) of Proposition 3 is satisfied and thus $a \in \mathfrak{A}^m((\Omega' \times \overline{R}_+) \times R^n)$.

2.2.2.2. Further properties

In connection with the symbolic calculus we need the fact that $\mathfrak{A}^m$ is closed under asymptotic sums in the sense of the following

Proposition 1. *Let $a_j \in \mathfrak{A}^{m_j}((\Omega' \times \overline{R}_+) \times R^n)$, $j \in Z_+$, and $m_j \to -\infty$ for $j \to \infty$. Then we have $a \in \mathfrak{A}^m((\Omega' \times \overline{R}_+) \times R^n)$, $m = \max\limits_j m_j$ for any a, $a \sim \sum\limits_j a_j$ ($\sim$ in the sense of symbols in S^m).*

Proof: By definition for any $N \in Z_+$ there is an $N' \in Z_+$ such that for any compact set $K \subset \Omega' \times \overline{R}_+$ and $l_2, \gamma \in Z_+$ there exists a constant $c = c_{l_2\gamma K}$ such that

$$\left| D^{l_2}_\nu D^\gamma_{x_n}\left(a(x', x_n, \xi', \nu) - \sum_{j=0}^{N'} a_j(x', x_n, \xi', \nu)\right)\right| \leq c\langle\xi\rangle^{-N-l_2} \leq c\langle\xi'\rangle^{-N-l_2}\, \langle\nu_1\rangle^{-N-l_2}$$

for $x \in K$ where $\nu_1 = \nu\langle\xi'\rangle^{-1}$. Then

$$\left|\nu_1^{l_1} D^{l_2}_{\nu_1}\left(a_{[\gamma]}(x', \xi', \nu_1) - \sum_{j=0}^{N'} a_{j[\gamma]}(x', \xi', \nu_1)\right)\right| \leq c\langle\xi'\rangle^{-N}\, \langle\nu_1\rangle^{-N-l_2+l_1}.$$

According to 2.2.2.1, Proposition 3, we have for $a_j \in \mathfrak{A}^{m_j}$

$$\left\|\nu_1^{l_1} D^{l_2}_{\nu_1}\left(a_{j[\gamma]}(x', \xi', \nu_1) - \sum_{-l_3 \leq k \leq m_j} c_{jk}(x', \xi')\, \chi(\nu_1)\, \nu_1^k\right)\right\|_{L^2(R_\nu)}$$
$$\leq c\langle\xi'\rangle^m, \quad x' \in K,\ l_3 > l_1 - l_2,$$

and therefore

$$\left\|\nu_1^{l_1} D^{l_2}_{\nu_1}\left(a_{[\gamma]}(x', \xi', \nu_1) - \sum_{j=0}^{N'} \sum_{-l_3 \leq k \leq m_j} c_{jk}(x', \xi')\, \chi(\nu_1)\, \nu_1^k\right)\right\|_{L^2(R_\nu)}$$
$$\leq c\langle\xi'\rangle^{-N}\, \langle\nu_1\rangle^{-N-l_2+l_1} + c\langle\xi'\rangle^m \leq c\langle\xi'\rangle^m$$

for $x' \in K$, $N \geq \max(-m, l_1 - l_2)$. Similar estimates are valid for the derivatives. Thus all $a_{[\gamma]}(x', \xi', \nu_1)$ satisfy the condition (ii) in 2.2.2.1, Proposition 3. □

The symbolic calculus for pseudo-differential operators, 2.2.2.1, Lemma 4 and Proposition 1 immediately imply

Corollary 2. (i) $a_1 \in \mathfrak{A}^{m_1}, a_2 \in \mathfrak{A}^{m_2} \Rightarrow a_1 \circ a_2 \sim \sum_\alpha (1/\alpha!)\, \partial_\xi^\alpha a_1\, D_x^\alpha a_2 \in \mathfrak{A}^{m_1+m_2}$.

(ii) $a \in \mathfrak{A}^m \Rightarrow a^* \sim \sum_\alpha (1/\alpha!)\, \partial_\xi^\alpha D_x^\alpha \bar{a} \in \mathfrak{A}^m$ *and*

$$a^t(x, \xi) \sim \sum_\alpha (1/\alpha!)\, (-1)^{|\alpha|}\, \partial_\xi^\alpha D_x^\alpha a(x, -\xi) \in \mathfrak{A}^m .$$

(iii) *Let* $\varkappa: \Omega' \to \Omega_1'$ *be a diffeomorphism and* $\varkappa \times 1: \Omega' \times \mathbb{R}_+ \to \Omega_1' \times \mathbb{R}_+$ *the induced diffeomorphism. Then* $a \in \mathfrak{A}^m\big((\Omega' \times \overline{\mathbb{R}}_+) \times \mathbb{R}^n\big)$ *implies that*

$$a_{\varkappa\times 1}(y', y_n, \eta', \eta_n) \sim \sum_\alpha (1/\alpha!)\, \partial_{\xi'}^\alpha a(x', x_n, \xi', \xi_n)\, \big(D_{z'}^\alpha\, e^{ih(z', x')\eta'}|_{z'=0}\big)$$

belongs to $\mathfrak{A}^m\big((\Omega_1' \times \overline{\mathbb{R}}_+) \times \mathbb{R}^n\big)$ *(on the right hand side take* $x' = \varkappa^{-1}(y')$, $x_n = y_n$, $\xi' = {}^t d\varkappa(x')\, \eta'$, $\xi_n = \eta_n$, *cf.* 1.2.3.4.(3)).

The definition of the transmission property of a symbol with respect to $\{x_n = 0\}$ makes sense for symbols given in $\mathbb{R}^n$. Then there is no difference between $\overline{\mathbb{R}}^n_+$ and $\overline{\mathbb{R}}^n_-$, i.e. the transformation in $\mathbb{R}^n$ given by $x_n \mapsto -x_n$ maps symbols with transmission property into symbols with transmission property. We shall speak about symbols with the transmission property with respect to the hyperplane $\{x_n = 0\}$.

Sometimes it is useful to have the transmission property with respect to several hyperplanes. We say that a symbol $a \in S^m$ has the transmission property with respect to the hyperplane $\{x_n = \varepsilon\}$, if for $T_\varepsilon: \mathbb{R}^n \to \mathbb{R}^n$, $T_\varepsilon(x', x_n) = (x', x_n - \varepsilon)$ the symbol $T_\varepsilon^* a(x, \xi) = a\big(T_\varepsilon(x), \xi\big)$ has the transmission property with respect to the hyperplane $\{x_n = 0\}$.

Lemma 3. *Let* $a \in \mathfrak{A}^m\big((\Omega' \times \overline{\mathbb{R}}_+) \times \mathbb{R}^n\big)$, *i.e. a has the transmission property with respect to* $\{x_n = 0\}$. *Then there exists a symbol* $a_1 \in S^m\big((\Omega' \times \overline{\mathbb{R}}_+) \times \mathbb{R}^n\big)$ *having the transmission property with respect to any hyperplane* $\{x_n = \varepsilon\}$, $\varepsilon > 0$ *and such that*

$$D_{x_n}^k\big(a(x', x_n, \xi', \nu) - a_1(x', x_n, \xi', \nu)\big)\big|_{x_n=+0} = 0$$

for any $k \in \mathbb{Z}_+$ *and arbitrary* x', ξ', ν.

Proof: Construct a_1 in the form $a_1(x', x_n, \xi) = \sum_{j=0}^\infty (x_n^j/j!)\, \partial_{x_n}^j a(x', 0, \xi)\, \varphi(t_j x_1)$, $\varphi \in C_0^\infty(\overline{\mathbb{R}}_+)$, $\varphi \equiv 1$ near 0. We show that one can choose t_j increasing so fast that the right hand side defines a symbol in $S^m\big((\Omega' \times \overline{\mathbb{R}}_+) \times \mathbb{R}^n\big)$. Let K_l, $l \in \mathbb{Z}_+$, be an increasing sequence of compact subsets in Ω', $K_l \subset \operatorname{int} K_{l+1}$, $\alpha, \beta \in \mathbb{Z}_+^n$. For a fixed $p \in \mathbb{Z}_+$ and $j > p$ there is an estimate

$$|D_x^\alpha D_\xi^\beta((x_n^j/j!)\, \partial_{x_n}^j a(x', 0, \xi)\, \varphi(t_j x_n))| \leqq c_p t_j^{-j+p}, \quad x' \in K_l,\ x_n \geqq 0$$

for all $|\alpha| + |\beta| + l \leqq p$ with suitable constant c_p. Then the considered sequence converges in $S^m\big((\Omega' \times \overline{\mathbb{R}}_+) \times \mathbb{R}^n\big)$ if we choose t_j for $j > p$ so that $t_j > c_p 2^j$. Similar arguments can be applied to $(x_n^j/j!)\, \partial_{x_n}^j a_{[\gamma]}(x', 0, \xi', \nu)$, $\gamma \in \mathbb{Z}_+$ and we obtain convergence even in $\mathfrak{A}^m\big((\Omega' \times \overline{\mathbb{R}}_+) \times \mathbb{R}^n\big)$. Now it is obvious that $a(x', x_n, \xi) - a_1(x', x_n, \xi)$ vanishes at $x_n = 0$ with all derivatives and that a_1 has the transmission property for any hyperplane $x_n = \varepsilon$, $\varepsilon > 0$, since the sequence is finite for fixed $x_n > 0$. □

Denote by $\mathfrak{A}_1^m\big((\Omega' \times \overline{\mathbb{R}}_+) \times \mathbb{R}^n\big)$ the subspace of $\mathfrak{A}^m\big((\Omega' \times \overline{\mathbb{R}}_+) \times \mathbb{R}^n\big)$ of such symbols being independent of x_n in a neighbourhood of $x_n = 0$. The restriction r' to $x_n = 0$ defines a mapping $\mathfrak{A}^m\big((\Omega' \times \overline{\mathbb{R}}_+) \times \mathbb{R}^n\big) \to \mathfrak{A}_1^m\big((\Omega' \times \overline{\mathbb{R}}_+) \times \mathbb{R}^n\big)$ to be used in the definition of boundary symbols.

2.2.2.3. Homogeneous Symbols with the Transmission Property

First we give a characterization of symbols with the transmission property being homogeneous for large $|\xi|$, i.e. $a(x', x_n, \xi', \nu) \in C^\infty((\Omega' \times \overline{\mathbb{R}}_+) \times \mathbb{R}^n)$,

$$a(x', x_n, \lambda\xi', \lambda\nu) = \lambda^m a(x', x_n, \xi', \nu), \lambda \geqq 1, |\xi'|^2 + \nu^2 \geqq c$$

with a suitable constant c. The degree of homogeneity m is assumed to be real. Denote by $\mathfrak{A}^{(m)}$ the subspace of $\mathfrak{A}^m$ of positively homogeneous symbols of order m for large $|\xi|$. We assume in the following that $c = 1$.

Proposition 1. *The following assertions are equivalent for homogeneous symbols for* $|\xi| \geqq 1$:

(i) $a(x', x_n, \xi', \nu) \in \mathfrak{A}^m((\Omega' \times \overline{\mathbb{R}}_+) \times \mathbb{R}^n)$,

(ii) $m \in \mathbb{Z}$ *and for any* $k \in \mathbb{Z}_+$ *and an arbitrary multi-index* $\alpha \in \mathbb{Z}_+^{n-1}$ *we have*

$$D_{x_n}^k D_{\xi'}^\alpha a(x', 0, 0, +1) = e^{i\pi(m-|\alpha|)} D_{x_n}^k D_{\xi'}^\alpha a(x', 0, 0, -1).$$

The condition (ii) is called *symmetry condition*. If $a(x', x_n, \xi)$ near $x_n = 0$ is homogeneous for $|\xi| \geqq c$, take the points $(x', 0, 0, +c)$ and $(x', 0, 0, -c)$ in (ii).

Proof: Consider the Taylor expansion of $a_{[k]}(x', \xi', \nu) = D_{x_n}^k a(x', 0, \xi', \langle\xi'\rangle \nu)$ with respect to ξ' near $\xi' = 0$. Using the homogeneity for $\langle\xi'\rangle \nu \geqq 1$ we get

$$\begin{aligned} a_{[k]}(x', \xi', \nu) &= (\langle\xi'\rangle |\nu|)^m D_{x_n}^k a(x', 0, \xi'/\langle\xi'\rangle |\nu|, \pm 1) \\ &= \sum_{|\alpha|=0}^{N-1} \frac{1}{\alpha!} \partial_{\xi'}^\alpha D_{x_n}^k a(x', 0, 0, \pm 1)\, \xi'^\alpha (\langle\xi'\rangle |\nu|)^{m-|\alpha|} \\ &\quad + r_N(x', \xi', \nu), \quad N \in \mathbb{Z}_+, \end{aligned}$$

where for any compact set $K \subset \Omega'$ there is a constant $c = c_{KN}$ with

$$|r_N(x', \xi', \nu)| \leqq c\langle\xi'\rangle^m |\nu|^{m-N}, \quad x' \in K.$$

Using (ii) we get

$$a_{[k]}(x', \xi', \nu) = \sum_{|\alpha|=0}^{N-1} \frac{1}{\alpha!} \partial_{\xi'}^\alpha D_{x_n}^k a(x', 0, 0, +1)\, \xi'^\alpha (\langle\xi'\rangle \nu)^{m-|\alpha|} + r_N(x', \xi', \nu).$$

Similarly, for $l_1, l_2 \in \mathbb{Z}_+$,

$$\begin{aligned} \nu^{l_1} D_\nu^{l_2} a_{[k]}(x', \xi', \nu) &= \sum_{|\alpha|=0}^{N-1} \frac{1}{\alpha!} \partial_{\xi'}^\alpha D_{x_n}^k a(x', 0, 0, +1)\, \xi'^\alpha (\langle\xi'\rangle \nu)^{m-|\alpha|-l_2} \langle\xi'\rangle^{l_2} \nu^{l_1} \\ &\quad + r_{Nl_1l_2}(x', \xi', \nu) \end{aligned}$$

where

$$|r_{Nl_1l_2}(x', \xi', \nu)| \leqq c\langle\xi'\rangle^m |\nu|^{m-N+l_1-l_2}, \quad x' \in K.$$

Concluding in the same way for the derivatives $D_{x'}^\beta D_{\xi'}^\gamma a_{[k]}$ we obtain (ii) in 2.2.2.1, Proposition 3.

Conversely, if (ii) in 2.2.2.1, Proposition 3 is satisfied, we must have symmetry, since the coefficients in the asymptotic expansion are uniquely determined and all occur. Moreover the asymptotic expansion shows that the degree of homogeneity m has to be integer. □

Let $a \in C^\infty((\Omega' \times \overline{R}_+) \times (R^n \setminus \{0\}))$ be positively homogeneous of order m for all $\xi \neq 0$, i.e.

$$a(x', x_n, \lambda\xi) = \lambda^m a(x', x_n, \xi)\,, \quad \lambda > 0\,, \ \xi \neq 0\,.$$

Denote by $\chi(\xi)$ an arbitrary excision function (set $\chi \equiv 1$ for $|\xi| \geqq 1$). Then we have

Corollary 2. *$\chi(\xi)\, a(x', x_n, \xi)$ defines a symbol in $\mathfrak{A}^{(m)}((\Omega' \times \overline{R}_+) \times R^n)$ iff a satisfies the symmetry condition. In particular, for different excision functions $\chi_1(\xi)$, $\chi_2(\xi)$ we have*

$$\big(\chi_1(\xi) - \chi_2(\xi)\big)\, a(x', x_n, \xi) \in S^{-\infty}((\Omega' \times \overline{R}_+) \times R^n)\,.$$

This is an immediate consequence of Proposition 1.

Proposition 3. *Let $a(x', x_n, \xi) \in C^\infty((\Omega' \times \overline{R}_+) \times (R^n \setminus \{0\}))$ be positively homogeneous of order $m \in Z$ for all $\xi \neq 0$ and assume that for any $\gamma \in Z_+$*

$$a_{[\gamma]}(x', \xi', \nu) = D^\gamma_{x_n} a(x', 0, \xi', \langle\xi'\rangle\, \nu)$$

is a function in $C^\infty(\Omega' \times S^{n-2}, H_{m+1})$ (S^{n-2} is given by $|\xi'| = 1$). Then for any excision function $\chi(\xi)$ we have $\chi(\xi)\, a(x', x_n, \xi) \in \mathfrak{A}^{(m)}$.

Proof: By assumption, for $x' \in K$ (K a compact set in Ω') and $|\xi'| = 1$, we have

$$\Big\| D^\alpha_{x'} D^\beta_{\xi'} \Pi_0 \nu^{l_1} D^{l_2}_\nu \Big(a_{[\gamma]}(x', \xi', \nu) - \sum_{-l_2 \leqq k \leqq m} c_k(x', \xi')\, \chi_1(\nu)\, \nu^k\Big)\Big\|_{L^2(R_\nu)} \leqq c$$

where $\chi_1 \in C^\infty(R)$ is an excision function (cf. 2.1.1.1.(4)). Extending $c_k(x', \xi')$ for $|\xi'| \geqq 1$ with homogeneity $m - |\beta|$ and using that

$$\begin{aligned} D^\alpha_{x'} D^\beta_{\xi'} a_{[\gamma]}(x', \xi', \nu) &= D^\alpha_{x'} D^\beta_{\xi'} D^\gamma_{x_n} a(x', 0, \xi', \langle\xi'\rangle\, \nu) \\ &= |\xi'|^{m-|\beta|}\, D^\alpha_{x'} D^\beta_{\xi'} D^\gamma_{x_n} a\left(x', 0, \xi'/|\xi'|, \frac{\langle\xi'\rangle}{|\xi'|}\, \nu\right) \end{aligned}$$

differs from $|\xi'|^{m-|\beta|}\, D^\alpha_{x'} D^\beta_{\xi'} \tilde{a}_{[\gamma]}(x', \xi'/|\xi'|, \nu)$ by $c\,|\xi'|^{m-|\beta|}$ we obtain (ii) in 2.2.2.1, Proposition 3. Therefore $\chi a \in \mathfrak{A}^{(m)}((\Omega' \times \overline{R}_+) \times R^n)$. □

Denote by $\mathcal{A}^m$ the subspace of $\mathfrak{A}^m$ of all symbols possessing an asymptotic expansion in homogeneous ones (for large $|\xi|$), i.e. $a \in \mathcal{A}^m$ iff there is a sequence $a_j \in \mathcal{A}^{(m_j)}$, $j \in Z_+$, $m = m_0$, $m_j \to -\infty$ for $j \to \infty$ such that $a \sim \sum_j a_j$. It is natural to assume that m_j tend to $-\infty$ strongly monotone. Obviously, $\mathcal{A}^m = \mathfrak{A}^m \cap S^m_{\mathrm{cl}}$.

Since S^m_{cl} and $\mathcal{A}^m$ are closed under all operations of the symbolic calculus (cf. 1.2.3.3, (4) and 2.2.2.2, Corollary 2), we get

Corollary 4. *The assertions of 2.2.2.2, Corollary 2 remain valid for $\mathcal{A}^m$ instead of $\mathfrak{A}^m$.*

Proposition 5. *$a \in \mathcal{A}^m$ iff any $a_j \in S^{m_j}((\Omega' \times \overline{R}_+) \times R^n)$ in the asymptotic expansion $a \sim \sum_j a_j$ with respect to homogeneous symbols satisfies the symmetry condition* (ii) *in Proposition* 1 *(the orders m_j are assumed to be strongly monotonely decreasing).*

Proof: If the condition is satisfied, the assertion follows from Proposition 1 and 2.2.2.2, Proposition 1. Conversely, if the symmetry condition is not valid for some a_j, then $a \in \mathcal{A}^m$ is impossible. □

Note as a simple consequence of 1.2.3.3, 1.2.4.2 and 2.2.2.2, Proposition 1 the following

Proposition 6. *Let $a \in \mathfrak{A}^m((\Omega' \times \overline{\mathbb{R}}_+) \times \mathbb{R}^n)$ be elliptic. Then, for the complete symbol b of a parametrix given in 1.2.4.2, we have $b \in \mathfrak{A}^{-m}((\Omega' \times \overline{\mathbb{R}}_+) \times \mathbb{R}^n)$.*

Mention without proof the following result (cf. DUISTERMAAT/SINGER [1]).

Proposition 7. *Let $a(x, \xi) \in S^m_{\mathrm{cl}}(\mathbb{R}^n \times \mathbb{R}^n)$ and $a \sim \sum_j a_j$ the asymptotic expansion in homogeneous symbols (for large $|\xi|$). Then a has the transmission property with respect to any hyperplane in $\mathbb{R}^n$ if*

$$a_j(x, \xi) = e^{i\pi(m-j)} a_j(x, -\xi) \quad \text{for large } |\xi|$$

($m - j$ the degree of homogeneity of a_j).

2.2.3. Potential and Trace Symbols

2.2.3.1. Definition and Various Characterizations of the Space $\mathfrak{K}^m$

In this section we define potential and trace symbols in the half space. In 2.1.2.2 potential and trace symbols on the half axis were defined and a one-to-one correspondence between these symbols and functions in H^+ and H^-, respectively, was used. In the half space potential and trace symbols are defined using $H^+ \otimes S^m$ and $H^- \otimes S^m$.

Definition 1. A function $k(x', \xi', \nu) \in C^\infty(\Omega' \times \mathbb{R}^{n-1} \times \mathbb{R})$ is called *potential symbol of order $m \geq -\infty$* if

$$k_{[0]}(x', \xi', \nu) \overset{\text{def}}{=} k(x', \xi', \langle\xi'\rangle \nu) \in H^+_\nu \otimes S^m(\Omega' \times \mathbb{R}^{n-1})$$

((x', ξ') denote coordinates in $\Omega' \times \mathbb{R}^{n-1}$). The space of all potential symbols of order m is denoted by $\mathfrak{K}^m(\Omega' \times \mathbb{R}^n)$ or $\mathfrak{K}^m$.

2.2.1.1, Proposition 3 yields several characterizations of potential symbols collected in the following

Proposition 2. *The following assertions are equivalent:*

(i) $k(x', \xi', \nu) \in \mathfrak{K}^m(\Omega' \times \mathbb{R}^{n-1} \times \mathbb{R})$;

(ii) $k_{[0]} \in C^\infty(\Omega' \times \mathbb{R}^{n-1} \times \mathbb{R})$ *has for any fixed $(x', \xi') \in \Omega' \times \mathbb{R}^{n-1}$ an analytic extension into the lower complex half plane $\operatorname{Im} \zeta < 0$, $\zeta = \nu + i\mu$, and there is an asymptotic expansion*

$$k_{[0]}(x', \xi', \nu) \sim \sum_{j \leq -1} b_j(x', \xi') (1 + i\nu)^j \quad \text{for} \quad |\nu| \to \infty$$

which can be formally differentiated, i.e. for arbitrary multi-indices $\alpha, \beta \in \mathbb{Z}^{n-1}_+$ and $l \in \mathbb{Z}^3_+$, $l = (l_1, l_2, l_3)$ and any compact set $K \subset \Omega'$ there is a constant $c = c_{\alpha\beta l K}$ such that

$$\Big\| D^\alpha_{x'} D^\beta_{\xi'} \Pi^+_\nu \Big(\nu^{l_1} D^{l_2}_\nu \Big(k_{[0]}(x', \xi', \nu) - \sum_{-l_3 \leq j < 0} b_j(x', \xi') (1 + i\nu)^j\Big)\Big)\Big\|_{L^2(\mathbb{R}_\nu)}$$
$$\leq c\langle\xi'\rangle^{m-|\beta|}, \quad x' \in K, \xi' \in \mathbb{R}^{n-1};$$

(iii) $g_k(x', \xi', z) \overset{\text{def}}{=} (1 + z)^{-1} k\big(x', \xi', i\langle\xi'\rangle (z - 1)/(z + 1)\big)$ *is a C^∞ function on $\Omega' \times \mathbb{R}^{n-1} \times S^1$ having for any fixed $(x', \xi') \in \Omega' \times \mathbb{R}^{n-1}$ an analytic extension* *into* $|z| < 1$ *and for an arbitrary* $\gamma \in Z_+$, *arbitrary multi-indices* $\alpha, \beta \in Z_+^{n-1}$ *and any compact set* $K \subset \Omega'$ *there is a constant* $c = c_{\alpha\beta\gamma K}$ *such that*

$$\sup_{-\pi \leq \varphi \leq \pi} |D_{x'}^\alpha D_{\xi'}^\beta D_\varphi^\gamma g(x', \xi', e^{i\varphi})| \leqq c\langle\xi'\rangle^{m-|\beta|}, \quad x' \in K, \ \xi' \in R^{n-1} \quad (z = e^{i\varphi});$$

(iv) *k admits an expansion*

$$k(x', \xi', \nu) = \sum_{j=0}^{\infty} k_j(x', \xi')\,(1 - i\nu\langle\xi'\rangle^{-1})^j\,(1 + i\nu\langle\xi'\rangle^{-1})^{-j-1}$$

where the k_j *form a rapidly decreasing sequence in* $S^m(\Omega' \times R^{n-1})$.

Suppose that $k(x', \xi', \nu) \in \mathfrak{R}^m$ vanishes for small $|\xi'|$. Then k admits an expansion

$$k(x', \xi', \nu) = \sum_{j=0}^{\infty} k_j(x', \xi')\,(1 - i\nu\,|\xi'|^{-1})^j\,(1 + i\nu\,|\xi'|^{-1})^{-j-1},$$

where $k_j \in S^m(\Omega' \times R^{n-1})$ is rapidly decreasing and all $k_j(x', \xi')$ vanish for $x' \in K$, $|\xi'| \leqq c = c_K$.

Consider $\mathfrak{R}^m(\Omega' \times R^n)$ with the topology induced from $H^+ \otimes S^m(\Omega' \times R^{n-1})$ by the mapping $H^+ \otimes S^m(\Omega' \times R^{n-1}) \ni k_{[0]}(x', \xi, \nu) \mapsto k(x', \xi', \nu) \in \mathfrak{R}^m(\Omega' \times R^n)$, i.e. the strongest locally convex topology which makes this mapping continuous. Then $\mathfrak{R}^m$ is a Fréchet space and the best constants in the estimates in Proposition (ii) and (iii), respectively yield equivalent systems of semi-norms defining the topology. Another such system of semi-norms is given by

$$\left\{\sum_{j=0}^{\infty} q_l(k_j)^2\, j^{2l'}\right\}^{1/2}, \quad l, l' \in Z_+$$

(q_l denotes a system of continuous semi-norms on S^m). The equivalence of the sesystems of semi-norms follows from 2.2.1.1, Proposition 4.

As observed in 2.2.2.1 $k \in \mathfrak{R}^m$ does not imply $k \in S^m(\Omega' \times R^n)$ in general. Any potential symbol $k(x', \xi', \nu)$ with the property that for any compact set $K \subset \Omega'$ there is a constant $c = c_K$ such that $k(x', \xi', \nu) = 0$ for $x' \in K$ and $|\xi'| \geqq c$ belongs to $\mathfrak{R}^{-\infty}(\Omega' \times R^n)$. In fact,

$$k_j(x', \xi') = (2\pi)^{-1}\,\langle\xi'\rangle^{-1} \int k(x', \xi', \nu)\,\frac{(1 - i\nu\langle\xi'\rangle^{-1})^j}{(1 + i\nu\langle\xi'\rangle^{-1})^{j+1}}\,d\nu$$

implies $k_j \in S^{-\infty}(\Omega' \times R^{n-1})$ rapidly decreasing. Simple examples of potential symbols in $\mathfrak{R}^m(\Omega' \times R^n)$ are the functions

$$k(x', \xi,' \nu) = k_j(x', \xi')\,\frac{(1 - i\nu\langle\xi'\rangle^{-1})^j}{(1 + i\nu\langle\xi'\rangle^{-1})^{j+1}}, \quad j \in Z_+,$$

where $k_j \in S^m(\Omega' \times R^{n-1})$. Since the expansion in Proposition 2 (iv) converges in $\mathfrak{R}^m$, the linear hull of these symbols is dense in $\mathfrak{R}^m$.

2.2.3.2. Definition and Various Characterizations of the Space $\mathfrak{T}^{m,d}$

Definition 1. A function $t(x', \xi', \nu) \in C^\infty(\Omega' \times R^{n-1} \times R)$ is called *trace symbol of order* $m \geqq -\infty$ *and type* d if

$$t_{[0]}(x', \xi', \nu) \overset{\text{def}}{=} t(x', \xi', \langle\xi'\rangle\,\nu) \in H_d^- \otimes S^m(\Omega' \times R^{n-1}).$$

The space of all trace symbols of order m and type d is denoted by $\mathfrak{T}^{m,d}(\Omega' \times \mathbb{R}^n)$ or $\mathfrak{T}^{m,d}$. Set $\mathfrak{T}^m = \bigcup_{d \geqq 0} \mathfrak{T}^{m,d}$.

Note that complex conjugation yields an isomorphism $\mathfrak{R}^m \to \mathfrak{T}^{m,0}$. Then 2.2.3.1, Proposition 2 yields several characterizations of trace symbols in $\mathfrak{T}^{m,d}$.

Proposition 2. *The following assertions are equivalent:*

(i) $t(x', \xi', \nu) \in \mathfrak{T}^{m,d}(\Omega' \times \mathbb{R}^{n-1} \times \mathbb{R})$;

(ii) $t_{[0]}(x' \xi', \nu) \in C^\infty(\Omega' \times \mathbb{R}^{n-1} \times \mathbb{R})$ *has for any fixed* $(x', \xi') \in \Omega' \times \mathbb{R}^{n-1}$ *an analytic extension into the upper complex half plane* $\operatorname{Im} \zeta > 0$, $\zeta = \nu + i\mu$, *and there is an asymptotic expansion*

$$t_{[0]}(x', \xi', \nu) \sim \sum_{j<d} b_j(x', \xi')\,(1 - i\nu)^j \quad \text{for} \quad |\nu| \to \infty$$

which can be formally differentiated, i.e. there are estimates as in 2.2.3.1, Proposition 2 (ii) with Π_0^- *instead of* Π^+;

(iii) $g_t(x', \xi', z) \overset{\text{def}}{=} (1 + z)^{d-1}\, t\big(x', \xi', i\langle\xi'\rangle\,(z - 1)/(z + 1)\big)$ *is a* C^∞ *function on* $\Omega' \times \mathbb{R}^{n-1} \times S^1$ *having for any fixed* $(x', \xi') \in \Omega' \times \mathbb{R}^{n-1}$ *an analytic extension into* $|z| > 1$ *(including* ∞*) and there are estimates as in* 2.2.3.1, *Proposition* 2 (iii);

(iv) *t admits an expansion*

$$t(x', \xi', \nu) = \sum_{k=0}^{d-1} b_k(x', \xi')\,(\nu\langle\xi'\rangle^{-1})^k + \sum_{j=0}^{\infty} t_j(x', \xi')\, \frac{(1 + i\nu\langle\xi'\rangle^{-1})^j}{(1 - i\nu\langle\xi'\rangle^{-1})^{j+1}}$$

where $b_k \in S^m(\Omega' \times \mathbb{R}^{n-1})$ *and* t_j *in* $S^m(\Omega' \times \mathbb{R}^{n-1})$ *rapidly decreasing.*

Consider $\mathfrak{T}^{m,d}$ in the topology induced from $H_d^- \otimes S^m$. Then the isomorphism $\mathfrak{R}^m \xrightarrow{\cong} \mathfrak{T}^{m,0}$ mentioned above is continuous. Moreover, the expansion in Proposition 2 (iv) converges in $\mathfrak{T}^{m,d}$ and the linear hull of the trace symbols $a(x', \xi')\,(\nu\langle\xi'\rangle^{-1})^k$, $k \in \mathbb{Z}_+$, $b(x', \xi') \dfrac{(1 + i\nu\langle\xi'\rangle^{-1})^j}{(1 - i\nu\langle\xi'\rangle^{-1})^{j+1}}$, $j \in \mathbb{Z}_+$, $a, b \in S^m(\Omega' \times \mathbb{R}^{n-1})$ is dense in $\mathfrak{T}^{m,d}$.

Any trace symbol $t(x', \xi', \nu)$ with the property that for any compact set $K \in \Omega'$ there is a constant $c = c_K$ such that

$$t(x', \xi', \nu) = 0 \quad \text{for} \quad x' \in K,\ |\xi'| \geqq c$$

belongs to $\mathfrak{T}^{-\infty,d}(\Omega' \times \mathbb{R}^n)$ for some d.

2.2.3.3. Homogeneous Symbols

Denote by $\mathfrak{R}^{(m)}(\Omega' \times \mathbb{R}^n)$ the subspace of $\mathfrak{R}^m(\Omega' \times \mathbb{R}^n)$ of potential symbols being positively homogeneous of order m for large $|\xi'|$, i.e. $k \in \mathfrak{R}^m(\Omega' \times \mathbb{R}^n)$ belongs to $\mathfrak{R}^{(m)}(\Omega' \times \mathbb{R}^n)$ if

$$k(x', \lambda\xi', \lambda\nu) = \lambda^m k(x', \xi', \nu), \quad \lambda \geqq 1,\ |\xi'| \geqq c$$

for a suitable constant c.

Similarly let $\mathfrak{T}^{(m),d}(\Omega' \times \mathbb{R}^n)$ be the subspace of symbols in $\mathfrak{T}^{m,d}(\Omega' \times \mathbb{R}^n)$ being positively homogeneous of order m for large $|\xi'|$.

Proposition 1. *Let* $\chi \in C^\infty(\mathbb{R}^{n-1})$ *be an arbitrary excision function, i.e.* $\chi \equiv 0$ *near the origin and* $\chi \equiv 1$ *near* ∞. *If* $k(x', \xi', \nu) \in C^\infty\big(\Omega' \times (\mathbb{R}^{n-1} \setminus \{0\}) \times \mathbb{R}\big)$ *is positively* *homogeneous of oder m (with respect to* (ξ', ν)*) and* $k(x', \xi', \cdot) \in C^\infty(\Omega' \times S^{n-2}, H^+)$, *then* $\chi(\xi')\, k(x', \xi', \nu) \in \mathfrak{K}^m(\Omega' \times \mathbb{R}^n)$. *For* $t(x', \xi', \nu) \in C^\infty(\Omega' \times (\mathbb{R}^{n-1} \setminus \{0\}) \times \mathbb{R})$ *positively homogeneous of order m and* $t(x', \xi', .) \in C^\infty(\Omega' \times S^{n-2}, H_{\bar d}^-)$, *follows* $\chi(\xi')\, t(x', \xi', \nu) \in \mathfrak{T}^{(m), d}(\Omega' \times \mathbb{R}^n)$. *Different excision functions yield the same potential (trace) symbol modulo* $\mathfrak{K}^{-\infty}(\mathfrak{T}^{-\infty, d})$.

Proof: In view of the isomorphism $\mathfrak{K}^m \xrightarrow{\cong} \mathfrak{T}^{m, 0}$ the first assertion follows from the second. The latter one is obvious. Let $t(x', \xi', \nu) = a(x', \xi')\, \nu^k$, where a is positively homogeneous with respect to ξ' of order $m - k$. By assumption $a(x', \xi') \in C^\infty(\Omega' \times S^{n-2})$, therefore $\chi(\xi')\, a(x', \xi') \in S^{m-k}(\Omega' \times \mathbb{R}^{n-1})$. Now, let $t(x', \xi', \nu) \in C^\infty(\Omega' \times S^{n-2}, H_{\bar 0}^-)$. Then there is an expansion

$$t(x', \xi', \nu) = \sum_{j=0}^{\infty} t_j(x', \xi') \frac{(1 + i\nu\, |\xi'|^{-1})^j}{(1 - i\nu\, |\xi'|^{-1})^{j+1}}, \quad |\xi'| \neq 0$$

with a rapidly decreasing sequence $t_j \in C^\infty(\Omega' \times S^{n-2})$. Extension to all ξ' by homogeneity of order m and multiplication with the excision function $\chi(\xi')$ yield a rapidly decreasing sequence of symbols in $S^m(\Omega' \times \mathbb{R}^{n-1})$. Then the assertion follows from 2.2.3.2, Proposition 2 (iv). □

Corollary 2. *Let* $m \in \mathbb{R}$ *be fixed. Then there are bijections*

$$C^\infty(\Omega' \times S^{n-2}, H^+) \to \mathfrak{K}^{(m)}/\mathfrak{K}^{-\infty},$$

$$C^\infty(\Omega' \times S^{n-2}, H_{\bar d}^-) \to \mathfrak{T}^{(m), d}/\mathfrak{T}^{-\infty, d}$$

defined by extension with homogeneity m with respect to (ξ', ν) *and multiplication by an arbitrary excision function* $\chi(\xi')$.

Proposition 3. *Let* $k_j \in \mathfrak{K}^{m_j}(\Omega' \times \mathbb{R}^n)$, $j \in \mathbb{Z}_+$, $m_j \to -\infty$ *for* $j \to \infty$. *Then there exists a unique modulo* $\mathfrak{K}^{-\infty}(\Omega' \times \mathbb{R}^n)$ *potential symbol* $k \in \mathfrak{K}^m(\Omega' \times \mathbb{R}^n)$, $m \doteq \sup_{j \in \mathbb{Z}_+} m_j$, *such that for any* $N \in \mathbb{Z}_+$ *there is a* $N' \in \mathbb{Z}_+$ *with* $k - \sum_{j=0}^{N} k_j \in \mathfrak{K}^{-N}(\Omega' \times \mathbb{R}^n)$. *Then we write* $k \sim \sum_j k_j$. *The same assertion holds for trace symbols, i.e. with* $\mathfrak{T}$ *instead of* $\mathfrak{K}$.

Proof: Consider the expansion from Proposition 2 (iv) for k_j

$$k_j(x', \xi', \nu) = \sum_{l=0}^{\infty} k_{jl}(x', \xi')\, (1 - i\nu\langle\xi'\rangle^{-1})^l\, (1 + i\nu\langle\xi'\rangle^{-1})^{-l-1}$$

with $k_{jl} \in S^{m_j}(\Omega' \times \mathbb{R}^{n-1})$ rapidly decreasing with respect to $l \in \mathbb{Z}_+$. Let $k_l \in S^m(\Omega' \times \mathbb{R}^{n-1})$ be given by the construction in 1.2.2.1, Theorem 2, $k_l \sim \sum_j k_{jl}$. Then k_l is rapidly decreasing in $S^m(\Omega' \times \mathbb{R}^{n-1})$ and

$$k(x', \xi', \nu) = \sum_{l=0}^{\infty} k_l(x', \xi')\, (1 - i\nu\langle\xi'\rangle^{-1})^l\, (1 + i\nu\langle\xi'\rangle^{-1})^{-l-1}$$

defines the potential symbol with the desired properties. Similar arguments apply to trace symbols. □

Denote by $\mathfrak{K}^m$ ($\mathfrak{T}^{m, d}$) the subspace of all symbols in $\mathfrak{K}^m$ ($\mathfrak{T}^{m, d}$) having all coefficients in S^m_{cl} in the expansion in 2.2.3.1, Proposition 2 (iv) (2.2.3.2, Proposition 2 (iv)). From Proposition 3 it is clear that $k \in \mathfrak{K}^m$ is equivalent to the existence of a sequence

$k_j \in \mathfrak{R}^{(m_j)}, j \in \mathbb{Z}_+, m_j \to -\infty$ for $j \to \infty$ with $k \sim \sum_j k_j$. $\mathfrak{R}^m$ $(\mathfrak{T}^{m,d})$ is closed with respect to asymptotic sums.

Note that for symbols in $\mathfrak{R}^m$ $(\mathfrak{T}^{m,d})$ a homogeneous principal symbol in $\mathfrak{R}^{(m)}$ $(\mathfrak{T}^{(m),d})$ is uniquely defined modulo $\boldsymbol{\mathfrak{R}}^{-\infty}$ $(\boldsymbol{\mathfrak{T}}^{-\infty,d})$.

2.2.4. Green Symbols

2.2.4.1. Definition and Various Characterizations of the Space $\mathfrak{B}^{m,d}$

Definition 1. A function $b(x', \xi', \nu, \tau) \in C^\infty(\Omega' \times \mathbb{R}^{n-1} \times \mathbb{R} \times \mathbb{R})$ is called *Green symbol of order $m \geqq -\infty$ and type d if* $b_{[0]}(x', \xi', \nu) \overset{\text{def}}{=} b(x', \xi', \langle\xi'\rangle \nu, \langle\xi'\rangle \tau) \in H_\nu^+ \otimes H_{d,\tau}^- \otimes S^m(\Omega' \times \mathbb{R}^{n-1})$. The space of all Green symbols of order m and type d is denoted by $\boldsymbol{\mathfrak{B}}^{m,d}(\Omega' \times \mathbb{R}^{n+1})$ or $\boldsymbol{\mathfrak{B}}^{m,d}$. Set $\boldsymbol{\mathfrak{B}}^m = \bigcup_{d \geqq 0} \boldsymbol{\mathfrak{B}}^{m,d}$.

Various characterizations of Green symbols follow immediately from 2.2.1.3, Proposition 1.

Proposition 2. *The following assertions are equivalent:*

(i) $b(x', \xi', \nu, \tau) \in \boldsymbol{\mathfrak{B}}^{m,d}(\Omega' \times \mathbb{R}^{n+1})$,

(ii) *b admits an expansion of the form*

$$b(x', \xi', \nu, \tau) = \sum_{j=0}^{d-1} c_j(x', \xi', \nu) (\tau\langle\xi'\rangle^{-1})^j + \sum_{k=0}^{\infty} b_k(x', \xi', \nu) (1 + i\tau\langle\xi'\rangle^{-1})^k (1 - i\tau\langle\xi'\rangle^{-1})^{-k-1}$$

where $c_j \in \boldsymbol{\mathfrak{R}}^m(\Omega' \times \mathbb{R}^n)$, $j = 0, 1, \ldots, d-1$, *and b_k form a rapidly decreasing sequence in* $\boldsymbol{\mathfrak{R}}^m(\Omega' \times \mathbb{R}^n)$,

(iii) *b admits an expansion of the form*

$$b(x', \xi', \nu, \tau) = \sum_{l=0}^{\infty} e_l(x', \xi', \tau) (1 - i\nu\langle\xi'\rangle^{-1})^l (1 + i\nu\langle\xi'\rangle^{-1})^{-l-1},$$

where e_l in $\boldsymbol{\mathfrak{T}}^{m,d}(\Omega' \times \mathbb{R}^n)$ *is rapidly decreasing,*

(iv) *b admits an expansion of the form*

$$b(x', \xi', \nu, \tau) = \sum_{\substack{0 \leqq j < \infty \\ 0 \leqq l < d}} c_{jl}(x', \xi') \frac{(1 - i\nu\langle\xi'\rangle^{-1})^j}{(1 + i\nu\langle\xi'\rangle^{-1})^{j+1}} (\tau\langle\xi'\rangle^{-1})^l + \sum_{0 \leqq j, l < \infty} b_{jl}(x', \xi') \frac{(1 - i\nu\langle\xi'\rangle^{-1})^j}{(1 + i\nu\langle\xi'\rangle^{-1})^{j+1}} \frac{(1 + i\tau\langle\xi'\rangle^{-1})^l}{(1 - i\tau\langle\xi'\rangle^{-1})^{l+1}},$$

where c_{jl} in $S^m(\Omega' \times \mathbb{R}^{n-1})$ with respect to j and b_{jl} in $S^m(\Omega' \times \mathbb{R}^{n-1})$ with respect to $j + l$ rapidly decreasing.

If b vanishes for small $|\xi'|$, one can replace $\langle\xi'\rangle$ by $|\xi'|$ in the above expansions.

Consider $\boldsymbol{\mathfrak{B}}^{m,d}(\Omega' \times \mathbb{R}^{n+1})$ with the topology induced from $H^+ \otimes H_d^- \otimes S^m(\Omega' \times \mathbb{R}^{n-1})$. Then the expansions in (ii) — (iv) converge in $\boldsymbol{\mathfrak{B}}^{m,d}$.

Any Green symbol $b(x', \xi', \nu, \tau)$ with the property that for any compact set $K \subset \Omega'$ there is a constant $c = c_K$ such that

$$b(x', \xi', \nu, \tau) = 0 \quad \text{for} \quad x' \in K\,, \ |\xi'| \geqq c\,,$$

belongs to $\mathfrak{B}^{-\infty, d}$.

Let $k(x', \xi', \nu) \in \mathfrak{K}^m(\Omega' \times \mathbb{R}^n)$ and $t(x', \xi', \nu) \in \mathfrak{T}^{m', d}(\Omega' \times \mathbb{R}^n)$. Then $k(x', \xi', \nu) \times t(x', \xi', \tau) \in \mathfrak{B}^{m+m', d}(\Omega' \times \mathbb{R}^{n+1})$ as follows from the expansion in Proposition 2. Moreover the linear hull of such symbols is dense in $\mathfrak{B}^{m, d}$.

Proposition 3. *Let $b_j \in \mathfrak{B}^{m_j, d}(\Omega' \times \mathbb{R}^{n+1})$, $j \in \mathbb{Z}_+$, $m_j \to -\infty$ for $j \to \infty$. Then there exists a Green symbol $b \in \mathfrak{B}^{m, d}(\Omega' \times \mathbb{R}^{n+1})$, $m = \max_{j \in \mathbb{Z}_+} m_j$, uniquely determined modulo $\mathfrak{B}^{-\infty, d}(\Omega' \times \mathbb{R}^{n+1})$ such that for any $N \in \mathbb{Z}_+$ there exists an $N' \in \mathbb{Z}_+$ with $b - \sum_{j=0}^{N'} b_j \in \mathfrak{B}^{-N, d}(\Omega' \times \mathbb{R}^{n+1})$. Then we write $b \sim \sum_j b_j$.*

The proof is similar to that of 2.3.3.2, Proposition 3 and is left to the reader.

2.2.4.2. Homogeneous Green Symbols

Denote by $\mathfrak{B}^{(m), d}(\Omega' \times \mathbb{R}^{n+1})$ the subspace of $\mathfrak{B}^{m, d}(\Omega' \times \mathbb{R}^{n+1})$ of Green symbols being positively homogeneous of order m for large $|\xi'|$, i.e. $b \in \mathfrak{B}^{m, d}(\Omega' \times \mathbb{R}^{n+1})$ belongs to $\mathfrak{B}^{(m), d}(\Omega' \times \mathbb{R}^{n+1})$ if

$$b(x', \lambda\xi', \lambda\nu, \lambda\tau) = \lambda^m b(x', \xi', \nu, \tau)\,, \quad \lambda \geqq 1\,, \ |\xi'| \geqq c\,,$$

for a suitable constant c.

Proposition 1. *Let $\chi \in C^\infty(\mathbb{R}^{n-1})$ be an arbitrary excision function, i.e. $\chi \equiv 0$ near the origin and $\chi \equiv 1$ near ∞. If $b(x', \xi', \nu, \tau) \in C^\infty(\Omega' \times (\mathbb{R}^{n-1} \setminus \{0\}) \times \mathbb{R}^2)$ is positively homogeneous of order m (with respect to (ξ', ν, τ)) and $b(x', \xi', \cdot, \cdot) \in C^\infty(\Omega' \times S^{n-2}, H^+ \hat{\otimes} H_{\bar{d}}^-)$, then $\chi(\xi')\, b(x', \xi', \nu, \tau) \in \mathfrak{B}^{(m), d}(\Omega' \times \mathbb{R}^{n+1})$. Different excision functions yield the same Green symbol modulo $\mathfrak{B}^{-\infty, d}$.*

The proof is completely analogous to the proof of 2.2.3.3, Proposition 1.

Corollary 2. *Let $m \in \mathbb{R}$ be fixed. Then there is a bijection*

$$C^\infty(\Omega' \times S^{n-2}, H^+ \hat{\otimes} H_{\bar{d}}^-) \to \mathfrak{B}^{(m), d}/\mathfrak{B}^{-\infty, d}$$

defined by extension with homogeneity m with respect to (ξ', ν, τ) and multiplication with an excision function $\chi(\xi')$.

Denote by $\mathfrak{B}^{m, d}$ the subspace of all Green symbols in $\mathfrak{B}^{m, d}$ having an asymptotic expansion with respect to homogeneous Green symbols (for large $|\xi'|$). $\mathfrak{B}^{m, d}$ is obviously closed under asymptotic sums. Note that for Green symbols in $\mathfrak{B}^{m, d}$ a homogeneous principal symbol is defined.

2.2.5. Boundary Symbols and Symbols in the Half Space

2.2.5.1. The Space $\mathfrak{R}^{m, d}$

In the same way as in 2.1.2.2 we assign to symbols in $\mathfrak{A}^m$ ($\mathfrak{B}^{m, d}$, $\mathfrak{K}^m$, $\mathfrak{T}^{m, d}$) boundary symbols being smooth families of boundary symbols on the axis with parameters (x', ξ'). Let $-\infty \leqq m \leqq \infty$, $d \geqq 0$ and $k, k', j, j' \in \mathbb{Z}_+$ be fixed.

Let $a \in \mathfrak{A}^m((\Omega' \times \overline{R}_+) \times R^n) \otimes \hom(C^k, C^{k'})$. For fixed $(x', \xi') \in \Omega' \times R^{n-1}$ there is a continuous mapping

$$\Pi^+ a_{[0]}(x', \xi', \nu): H^+ \otimes C^k \to H^+ \otimes C^{k'},$$

$a_{[0]}(x', \xi', \nu) \overset{\text{def}}{=} a(x', 0, \xi', \langle \xi' \rangle \nu)$ defined by

$$H^+ \otimes C^k \in h(\nu) \mapsto \Pi^+_\nu(a_{[0]}(x', \xi', \nu)\, h(\nu)) \in H^+ \otimes C^{k'}.$$

Since $a_{[0]}$ can be considered as smooth function on $\Omega' \times R^{n-1}$ with values in H and $\Pi^+\sigma_A \in \mathscr{L}(H^+, H^+)$ continuously depends on $\sigma_A \in H$, this is a smooth function on $\Omega' \times R^{n-1}$ with values in $\mathscr{L}(H^+ \otimes C^k, H^+ \otimes C^{k'})$ denoted by $\Pi^+ a_{[0]}(x', \xi')$.

Let $b \in \mathfrak{B}^{m,d}(\Omega' \times R^{n-1} \times R^2) \otimes \hom(C^k, C^{k'})$. For fixed $(x', \xi') \in \Omega' \times R^{n-1}$ there is a continuous mapping

$$\Pi' b_{[0]}(x', \xi', \nu, \tau): H^+ \otimes C^k \to H^+ \otimes C^{k'},$$

$b_{[0]}(x', \xi', \nu, \tau) \overset{\text{def}}{=} b(x', \xi', \langle \xi' \rangle \nu, \langle \xi' \rangle \tau)$ defined by $H^+ \otimes C^k \in h(\nu) \mapsto \Pi'_\tau(b_{[0]}(x', \xi', \nu, \tau) \times h(\tau)) \in H^+ \otimes C^{k'}$. This defines a smooth function $\Pi' b_{[0]}(x', \xi')$ on $\Omega' \times R^{n-1}$ with values in $\mathscr{L}(H^+ \otimes C^k, H^+ \otimes C^{k'})$.

Let $k \in \mathfrak{K}^m(\Omega' \times R^{n-1} \times R) \otimes \hom(C^j, C^k)$ and $k_{[0]}(x', \xi', \nu) \overset{\text{def}}{=} k(x', \xi', \langle \xi' \rangle \nu)$. For fixed $(x', \xi') \in \Omega' \times R^{n-1}$ there is a continuous mapping

$$k_{[0]}(x', \xi', \nu): C^j \to H^+ \otimes C^{k'}$$

defined by

$$C^j \in v \mapsto k_{[0]}(x', \xi', \nu)\, v \in H^+ \otimes C^{k'}.$$

It defines a smooth function $k_{[0]}(x', \xi')$ on $\Omega' \times R^{n-1}$ with values in $\mathscr{L}(C^j, H^+ \otimes C^{k'})$.

Finally let $t \in \mathfrak{T}^{m,d}(\Omega' \times R^{n-1} \times R) \otimes \hom(C^k, C^{j'})$. For fixed $(x', \xi') \in \Omega' \times R^{n-1}$ there is a continuous mapping

$$\Pi' t_{[0]}(x', \xi', \nu): H^+ \otimes C^k \to C^j,$$

$t_{[0]}(x', \xi', \nu) \overset{\text{def}}{=} t(x', \xi', \langle \xi' \rangle \nu)$ defined by

$$H^+ \otimes C^k \ni h(\nu) \mapsto \Pi'_\nu(t_{[0]}(x', \xi', \nu)\, h(\nu)) \in C^{j'}.$$

It defines a smooth function $\Pi' t_{[0]}(x', \xi')$ on $\Omega' \times R^{n-1}$ with values in $\mathscr{L}(H^+ \otimes C^k, C^{j'})$

Note that the convention to take $\langle \xi' \rangle \nu$ and $\langle \xi' \rangle \tau$ plays a role in connection with the behaviour of $\Pi^+ a_{[0]}(x', \xi')$, $\Pi' b_{[0]}(x', \xi')$, $k_{[0]}(x', \xi')$ and $\Pi' t_{[0]}(x', \xi')$ for large ξ'.

Given a tupel of functions $a_{[0]} \in H^+(\nu) \otimes S^m(\Omega' \times R^{n-1}) \otimes \hom(C^k, C^{k'})$, $b_{[0]} \in H^+(\nu) \otimes H^-_d(\tau) \otimes S^m(\Omega' \times R^{n-1}) \otimes \hom(C^k, C^{k'})$, $k_{[0]} \in H^+(\nu) \otimes S^m(\Omega' \times R^{n-1}) \otimes \hom(C^j, C^{k'})$, $t_{[0]} \in H^-_d(\nu) \otimes S^m(\Omega' \times R^{n-1}) \otimes \hom(C^k, C^{j'})$, $q \in S^m(\Omega' \times R^{n-1}) \otimes \hom(C^j, C^{j'})$, we get a family of continuous mappings

$$\mathfrak{a}_{[0]}(x', \xi') = \begin{pmatrix} \Pi^+ a_{[0]}(x', \xi') + \Pi' b_{[0]}(x', \xi') & k_{[0]}(x', \xi') \\ \Pi' t_{[0]}(x', \xi') & q(x', \xi') \end{pmatrix} : \begin{matrix} H^+ \otimes C^k \\ \oplus \\ C^j \end{matrix} \to \begin{matrix} H^+ \otimes C^{k'} \\ \oplus \\ C^{j'} \end{matrix}$$

moothly depending on $(x', \xi') \in \Omega' \times R^{n-1}$. Obviously

$$\mathfrak{a}_{[0]}(x', \xi') \in \mathfrak{R}^{m,d} \otimes S^m(\Omega' \times R^{n-1}).$$

Denote by $\varkappa_t^*: H \to H$, $t \in \mathbb{R}_+$ fixed, the mapping defined by $\varkappa_t^* h(\nu) = h(t\nu)$, $h \in H$. For arbitrary $l_1, l_2 \in \mathbb{Z}_+$, $h \in H^+$, we have

$$\|\Pi^+\nu^{l_1}D_\nu^{l_2}\varkappa_t^* h(\nu)\|_{L^2(\mathbb{R})} = \{\int |\Pi^+\nu^{l_1}D_\nu^{l_2}\varkappa_t^* h(\nu)|^2 \, d\nu\}^{1/2}$$
$$= t^{-l_1+l_2-1/2} \|\Pi^+\nu^{l_1}D_\nu^{l_2} h(\nu)\|_{L^2(\mathbb{R})},$$

since $\Pi^+\varkappa_t^* = \varkappa_t^*\Pi^+$.

Definition 1. Denote by $\mathfrak{R}^{m,d}(\Omega' \times \mathbb{R}^{n-1}; \mathbb{C}^k, \mathbb{C}^{k'}, \mathbb{C}^j, \mathbb{C}^{j'})$ the space of all mappings

$$\mathfrak{a}(x', \xi'): \begin{matrix} H^+ \otimes \mathbb{C}^k \\ \oplus \\ \mathbb{C}^j \end{matrix} \to \begin{matrix} H^+ \otimes \mathbb{C}^{k'} \\ \oplus \\ \mathbb{C}^{j'} \end{matrix}$$

of the form

$$\mathfrak{a}(x', \xi') = (\varkappa^*_{\langle\xi'\rangle^{-1}} \oplus 1_{j'})\, \mathfrak{a}_{[0]}(x', \xi')\, (\varkappa^*_{\langle\xi'\rangle} \oplus 1_j)$$

where $\mathfrak{a}_{[0]}(x', \xi') \in \mathfrak{R}^{m,d} \otimes S^m(\Omega' \times \mathbb{R}^{n-1})$ (1_j denotes the $j \times j$ identity matrix and $\varkappa^*_{\langle\xi'\rangle}$ acts on all components of the vectors). The elements of $\mathfrak{R}^{m,d}$ are called *boundary symbols of order m and type d*. Consider $\mathfrak{R}^{m,d}(\Omega' \times \mathbb{R}^{n-1}; \mathbb{C}^k, \mathbb{C}^{k'}, \mathbb{C}^j, \mathbb{C}^{j'})$ in the topology induced from

$$\mathfrak{R}^{m,d}(\mathbb{C}^k, \mathbb{C}^{k'}, \mathbb{C}^j, \mathbb{C}^{j'}) \otimes S^m(\Omega' \times \mathbb{R}^{n-1}).$$

For $a \in \mathfrak{A}^m((\Omega' \times \overline{\mathbb{R}}_+) \times \mathbb{R}^n)$ we set $\Pi^+ r' a(x', \xi') = \varkappa^*_{\langle\xi'\rangle^{-1}} \Pi^+ a_{[0]}(x', \xi')\, \varkappa^*_{\langle\xi'\rangle}$ (r' restriction to $x_n = 0$ and $a_{[0]}$ defined by 2.2.2.1.(1)), therefore

$$\Pi^+ r' a(x', \xi')\, h(\nu) = \Pi^+_{\nu_1}\big(a_{[0]}(x', \xi', \nu_1)\, h(\langle\xi'\rangle \nu_1)\big)\big|_{\nu=\nu_1\langle\xi'\rangle}$$
$$= \Pi^+_{\nu_1}\big(a(x', 0, \xi', \langle\xi'\rangle \nu_1)\, h(\langle\xi'\rangle \nu_1)\big)\big|_{\nu=\nu_1\langle\xi'\rangle}$$
$$= \Pi^+_\nu\big(a(x', 0, \xi', \nu)\, h(\nu)\big), \quad h \in H^+.$$

For $b \in \mathfrak{B}^{m-1,d}(\Omega' \times \mathbb{R}^{n+1})$ set

$$\Pi' b(x', \xi') \overset{\text{def}}{=} \langle\xi'\rangle\, \varkappa^*_{\langle\xi'\rangle^{-1}} \Pi' b_{[0]}(x', \xi')\, \varkappa^*_{\langle\xi'\rangle},$$

hence

$$\Pi' b(x', \xi')\, h(\nu) = \langle\xi'\rangle\, \Pi'_{\tau_1} b(x', \xi', \langle\xi'\rangle \nu_1, \langle\xi'\rangle \tau_1)\, h(\langle\xi'\rangle \tau_1)\big|_{\nu=\nu_1\langle\xi'\rangle}$$
$$= \Pi'_\tau\big(b(x', \xi', \nu, \tau)\, h(\tau)\big), \quad h \in H^+.$$

For $k \in \mathfrak{K}^m(\Omega' \times \mathbb{R}^n)$ set

$$k(x', \xi') \overset{\text{def}}{=} \varkappa^*_{\langle\xi'\rangle^{-1}} k_{[0]}(x', \xi').$$

Thus for $v \in \mathbb{C}$

$$k(x', \xi')\, v(\nu) = k(x', \xi', \langle\xi'\rangle \nu_1)\, v\big|_{\nu=\nu_1\langle\xi'\rangle} = k(x', \xi', \nu)\, v.$$

Finally, for $t \in \mathfrak{T}^{m-1,d}(\Omega' \times \mathbb{R}^n)$, set

$$\Pi' t(x', \xi') \overset{\text{def}}{=} \langle\xi'\rangle\, \Pi' t_{[0]}(x', \xi')\, \varkappa^*_{\langle\xi'\rangle}.$$

Thus for $h \in H^+$

$$\Pi' t(x', \xi')\, h(\nu) = \langle\xi'\rangle\, \Pi'_{\nu_1}\big(t(x', \xi', \langle\xi'\rangle \nu_1)\, h(\langle\xi'\rangle \nu_1)\big) = \Pi'_\nu\big(t(x', \xi', \nu)\, h(\nu)\big).$$

For $\mathfrak{a}(x', \xi') \in \mathfrak{R}^{m,d}(\Omega' \times \mathbb{R}^{n-1})$ constructed with the help of $a \in \mathfrak{A}^m((\Omega' \times \overline{\mathbb{R}}_+) \times \mathbb{R}^n)$, $b \in \mathfrak{B}^{m-1,d}(\Omega' \times \mathbb{R}^{n+1})$, $k \in \mathfrak{K}^m(\Omega' \times \mathbb{R}^n)$, $t \in \mathfrak{T}^{m-1,d}(\Omega' \times \mathbb{R}^n)$ and $q \in S^m(\Omega' \times \mathbb{R}^{n-1})$ we get

$$\mathfrak{a}(x', \xi') = \begin{pmatrix} \Pi^+ r' a(x', \xi') + \Pi' b(x', \xi') & k(x', \xi') \\ \Pi' t(x', \xi') & q(x', \xi') \end{pmatrix}.$$

$a \in \mathfrak{A}^m((\Omega' \times \overline{\mathbb{R}}_+) \times \mathbb{R}^n)$ is called the *operator symbol* of the boundary symbol $\mathfrak{a} \in \mathfrak{R}^{m,d}(\Omega' \times \mathbb{R}^{n-1})$.

The above definition is motivated by the homogeneous case. Let $a(x, \xi) \in C^\infty(\Omega \times \mathbb{R}^n)$ be positively homogeneous of order m for $|\xi| \geqq 1$. Moreover let $\mathfrak{a}(x', \xi') \in C^\infty(\Omega' \times S^{n-2}, \mathfrak{R}^{m,d})$ ($S^{n-2}$is defined by $|\xi'| = 1$), where the operator symbol of $\mathfrak{a}$ is $a(x', 0, \xi', \nu)$ for $|\xi'| = 1$. Extend $\mathfrak{a}$ by positive homogeneity of order m to all $\xi' \neq 0$, i.e.

$$\mathfrak{a}_{[0]}(x', \xi') = |\xi'|^m \, \mathfrak{a}(x', \xi'/|\xi'|) .$$

Then the operator symbol is equal to $a(x', 0, \xi', \nu)$ for all $\xi' \neq 0$, $\nu \geqq 1$. Let

$$\mathfrak{a}(x', \xi') = \begin{pmatrix} \Pi^+ a(x', 0, \xi', \nu) + \Pi' b(x', \xi', \nu, \tau) & k(x', \xi', \nu) \\ \Pi' t(x', \xi', \nu) & q(x', \xi') \end{pmatrix} \quad \text{for } |\xi'| = 1 .$$

Then

$$\mathfrak{a}(x', \xi') = (\varkappa^*_{|\xi'|^{-1}} \oplus 1)\, \mathfrak{a}_{(0)}(x', \xi')\, (\varkappa^*_{|\xi'|} \oplus 1)$$

has the form

$$\mathfrak{a}(x', \xi') = \begin{pmatrix} \Pi^+ a(x', 0, \xi', \nu) + \Pi' b(x', \xi', \nu, \tau) & k(x', \xi', \nu) \\ \Pi' t(x', \xi', \nu) & q(x', \xi') \end{pmatrix}, \quad \xi' \neq 0 ,$$

where $b(x', \xi', \nu, \tau)$ and $t(x', \xi', \nu)$ are positively homogeneous of order $m - 1$ in ξ', ν, τ and $k(x', \xi', \nu)$, $q(x', \xi')$ are positively homogeneous of order m in ξ', ν ($|\xi'| \neq 0$). In fact, for the Green symbol b we have, for instance,

$$\begin{aligned} &|\xi'|^m \, \Pi'_{\tau_1}\big(b(x', \xi'/|\xi'|, \nu_1, \tau_1)\, h(|\xi'|\, \tau_1)\big)\big|_{\nu = \nu_1|\xi'|} \\ &= |\xi'|^{m-1}\, \Pi'_\tau\big(b(x', \xi'/|\xi'|, \nu/|\xi'|, \tau/|\xi'|)\, h(\tau)\big) . \end{aligned}$$

Then $\chi(\xi')\, \mathfrak{a}(x', \xi') \in \mathfrak{R}^{m,d}(\Omega' \times \mathbb{R}^{n-1})$ for any excision function χ ($\chi = 0$ for $|\xi'| \leqq 1$). We call a boundary symbol homogeneous for large ξ' if $\mathfrak{a}_{(0)}(x', \xi') = (\varkappa^*_{|\xi'|} \oplus 1)\, \mathfrak{a}(x', \xi')\, (\varkappa^*_{|\xi'|^{-1}} \oplus 1)$ has this property. The space of all boundary symbols being positively homogeneous of order m for large $|\xi'|$ and of type d is denoted by $\mathfrak{R}^{(m),d}(\Omega' \times \mathbb{R}^{n-1})$. The above construction shows that we have the following

Lemma 2. *For fixed $m \in \mathbb{Z}$ there is an isomorphism*

$$C^\infty(\Omega' \times \mathbb{R}^{n-1}, \mathfrak{R}^{m,d}) \to \mathfrak{R}^{(m),d}(\Omega' \times \mathbb{R}^{n-1}) / \mathfrak{R}^{-\infty,d}(\Omega' \times \mathbb{R}^{n-1}) .$$

Proposition 3. *Let $\mathfrak{a}_j \in \mathfrak{R}^{m_j,d}(\Omega' \times \mathbb{R}^{n-1})$, $j \in \mathbb{Z}_+$, $m_j \to -\infty$ for $j \to \infty$. Then there exists a unique modulo $\mathfrak{R}^{-\infty,d}(\Omega' \times \mathbb{R}^{n-1})$ boundary symbol $\mathfrak{a} \in \mathfrak{R}^{m,d}(\Omega' \times \mathbb{R}^{n-1})$, $m = \max\limits_{j \in \mathbb{Z}_+} m_j$, such that for any $N \in \mathbb{Z}_+$ there is an $N' \in \mathbb{Z}_+$ with*

$$\mathfrak{a} - \sum_{j=0}^{N'} \mathfrak{a}_j \in \mathfrak{R}^{-N,d}(\Omega' \times \mathbb{R}^{n-1}) .$$

Then we write $\mathfrak{a} \sim \sum\limits_j \mathfrak{a}_j$.

This is an immediate consequence of the corresponding assertion for the symbol spaces $\mathfrak{B}^{m,d}$, $\mathfrak{K}^m$, $\mathfrak{T}^{m,d}$, $\mathfrak{A}^m$ and S^m.

Proposition 4. (i) *Let $\mathfrak{a} \in \mathfrak{R}^{m,d}(\Omega' \times \mathbb{R}^{n-1};\ \mathbb{C}^k, \mathbb{C}^{k'}, \mathbb{C}^j, \mathbb{C}^{j'})$, $\mathfrak{b} \in \mathfrak{R}^{m',d'}(\Omega' \times \mathbb{R}^{n-1};\ \mathbb{C}^{k'}, \mathbb{C}^{k''}, \mathbb{C}^{j'}, \mathbb{C}^{j''})$. Then $\mathfrak{b}(x', \xi')\, \mathfrak{a}(x', \xi') \in \mathfrak{R}^{m'',d''}(\Omega' \times \mathbb{R}^{n-1};\ \mathbb{C}^k, \mathbb{C}^{k''}, \mathbb{C}^j, \mathbb{C}^{j''})$, $m'' = m' + m$, $d'' = \max\,(m + d', d)$ with the composition of boundary symbols on the axis for fixed (x', ξ') (cf. 2.1.2.3).*

(ii) *Let* $\mathfrak{a} \in \mathfrak{R}^{m,d}(\Omega' \times \mathbb{R}^{n-1})$. *Then, for arbitrary multi-indices* $\alpha, \beta \in \mathbb{Z}_+^{n-1}$, $D_{x'}^\alpha D_{\xi'}^\beta \mathfrak{a}(x', \xi') \in \mathfrak{R}^{m-|\beta'|,d}(\Omega' \times \mathbb{R}^{n-1})$.

(iii) *For* $\mathfrak{a} \in \mathfrak{R}^{0,0}(\Omega' \times \mathbb{R}^{n-1})$ *define* $\mathfrak{a}^*(x', \xi') \overset{\text{def}}{=} (\mathfrak{a}(x', \xi'))^*$ *with the adjoint in the sense of boundary symbols (cf. 2.1.2.3). Then* $\mathfrak{a}^* \in \mathfrak{R}^{0,0}(\Omega' \times \mathbb{R}^{n-1})$.

Proof. We have per def. $\mathfrak{a}(x', \xi') = (\varkappa^*_{\langle\xi'\rangle^{-1}} \oplus 1)\, \mathfrak{a}_{[0]}(x', \xi')\, (\varkappa^*_{\langle\xi'\rangle} \oplus 1)$ and $\mathfrak{b}(x', \xi') = (\varkappa^*_{\langle\xi'\rangle} \oplus 1)\, \mathfrak{b}_{[0]}(x', \xi')\, (\varkappa^*_{\langle\xi'\rangle} \oplus 1)$, $\mathfrak{a}_{[0]} \in \mathfrak{R}^{m,d} \otimes S^m$, $\mathfrak{b}_{[0]} \in \mathfrak{R}^{m',d'} \otimes S^{m'}$. Therefore $\mathfrak{b}\mathfrak{a} = (\varkappa^*_{\langle\xi'\rangle^{-1}} \oplus 1)\, \mathfrak{b}_{[0]}\mathfrak{a}_{[0]}(\varkappa^*_{\langle\xi'\rangle} \oplus 1)$ and obviously $\mathfrak{b}_{[0]}\mathfrak{a}_{[0]} \in \mathfrak{R}^{m'',d''} \otimes S^{m''}$. Thus (i) is proved.

For the proof of (ii) represent the functions $a(x', \xi', \nu)$, $b(x', \xi', \nu, \tau)$, $k(x', \xi', \nu)$, $t(x', \xi', \nu)$ by expansions as in 2.2.3.1, 2.2.3.2 and 2.2.4.1. Consider, for instance, the Green symbol

$$b(x', \xi', \nu, \tau) = \sum_{i,k \in \mathbb{Z}_+} b_{jk}(x', \xi')\, e_j(\nu\langle\xi'\rangle^{-1})\, f_k(\tau\langle\xi'\rangle^{-1})$$

where $e_j \in H^+$, $f_k \in H^-$ and b_{jk} in S^m rapidly decreasing. Then it is sufficient to consider $b(x', \xi', \nu, \tau) = b_1(x', \xi')\, e(\nu\langle\xi'\rangle^{-1})\, f(\tau\langle\xi'\rangle^{-1})$. Derivatives with respect to x' are functions of the same type. Using the Leibniz rule we get the assertion from the formula

$$D_{\xi'}^\alpha\big(e(\nu\langle\xi'\rangle^{-1})\big) = \sum_k e_k(\nu\langle\xi'\rangle^{-1})\, b_k(\xi'),$$

$e_k \in H^+$, $b_k \in S^{-|\alpha|}$, which is easily checked by induction over $|\alpha|$. The latter assertion (iii) follows from 2.1.2.3, Proposition 6 and the expansions used above. □

Denote by $\mathfrak{R}^{m,d}(\Omega' \times \mathbb{R}^{n-1})$ the subspace of $\mathfrak{R}^{m,d}(\Omega' \times \mathbb{R}^{n-1})$ of boundary symbols with an asymptotic expansion with respect to homogeneous symbols.

2.2.5.2. Symbols in the Half Space

Symbols in the half space denote pairs of symbols with the transmission property and boundary symbols satisfying a natural compatibility condition. A pair of $(a, \mathfrak{a}) \in \mathfrak{A}^m\big((\Omega' \times \overline{\mathbb{R}}_+) \times \mathbb{R}^n\big) \times \mathfrak{R}^{m,d}(\Omega' \times \mathbb{R}^{n-1})$ is called *compatible* if

$$r'a(x', \xi', \nu) \overset{\text{def}}{=} a(x', 0, \xi', \nu)$$

is equal to the operator symbol in the boundary symbol $\mathfrak{a}(x', \xi')$ for $|\xi'| \geqq \text{const}$, i.e.

$$\mathfrak{a}(x', \xi') = \begin{pmatrix} \Pi^+ a(x', 0, \xi', \nu) + \Pi' b(x', \xi', \nu, \tau) & k(x', \xi', \nu) \\ \Pi' t(x', \xi', \nu) & q(x', \xi') \end{pmatrix}$$

with some $b \in \mathfrak{B}^{m-1,d}(\Omega' \times \mathbb{R}^{n+1})$, $k \in \mathfrak{K}^m(\Omega' \times \mathbb{R}^n)$, $t \in \mathfrak{T}^{m-1,d}(\Omega' \times \mathbb{R}^n)$, $q \in S^m(\Omega' \times \mathbb{R}^{n-1})$, $|\xi'| \geqq \text{const}$.

Denote by $\mathfrak{S}^{m,d}(\Omega' \times \overline{\mathbb{R}}_+)$ the space of all compatible pairs $(a, \mathfrak{a}) \in \mathfrak{A}^m\big((\Omega' \times \overline{\mathbb{R}}_+) \times \mathbb{R}^n\big) \times \mathfrak{R}^{m,d}(\Omega' \times \mathbb{R}^{n-1})$. The elements of $\mathfrak{S}^{m,d}(\Omega' \times \overline{\mathbb{R}}_+)$ are called *symbols of pseudo-differential boundary value problems of order m and type d*. For $\Omega' = \mathbb{R}^{n-1}$ we speak of *symbols of pseudo-differential boundary value problems in the half space*, or shorter, of *symbols in the half space*.

$\mathfrak{S}^{m,d}$ is a linear space with operations applied to the corresponding components. Moreover, define the multiplication of $\sigma_i = (a_i, \mathfrak{a}_i) \in \mathfrak{S}^{m_i,d_i}(\Omega' \times \overline{\mathbb{R}}_+)$, $i = 1, 2$ by

$$\sigma_1\sigma_2 = (a_1a_2, \mathfrak{a}_1\mathfrak{a}_2)$$

assuming that the compositions are defined.

From 2.2.2.2, Corollary 2 and 2.2.5.1, Proposition 4 we get

Proposition 1. (i) *Let $\sigma_i \in \mathfrak{S}^{m_i,d_i}$, $i = 1, 2$, and let $\sigma_1\sigma_2$ be defined. Then $\sigma_1\sigma_2 \in \mathfrak{S}^{m',d'}$, $m' = m_1 + m_2$, $d' = \max(m_2 + d_1, d_2)$.*

(ii) *Let $\sigma = (a, \mathfrak{a}) \in \mathfrak{S}^{0,0}$. Then $\sigma^* \overset{\text{def}}{=} (a^*, \mathfrak{a}^*) \in \mathfrak{S}^{0,0}$.*

It is easily checked that the composition and taking of the adjoint do not destroy compatibility.

For $\sigma = (a, \mathfrak{a}) \in \mathfrak{S}^{m,d}$ the component a is called *interior symbol* and $\mathfrak{a}$ is called *boundary symbol.*

Proposition 2. *Let $\sigma_i \in \mathfrak{S}^{m_i,d_i}$, $j \in \mathbb{Z}_+$, $m_j \to -\infty$ for $j - \infty$. Then there exists a $\sigma \in \mathfrak{S}^{m,d}$, $m = \max_{j \in \mathbb{Z}_+} m_j$, uniquely determined modulo $\mathfrak{S}^{-\infty,d}$ such that for any $N \in \mathbb{Z}_+$ there is an $N' \in \mathbb{Z}_+$ with*

$$\sigma - \sum_{j=0}^{N'} \sigma_j \in \mathfrak{S}^{-N,d}\,.$$

Then we write $\sigma \sim \sum_j \sigma_j$.

Denote by p_1 and p_2 the natural projections of $\mathfrak{S}^{m,d}$ to $\mathfrak{A}^m$ and $\mathfrak{R}^{m,d}$, respectively. Then $\sigma \sim \sum_j \sigma_j$ is equivalent to $p_1(\sigma) \sim \sum_j p_1(\sigma_j)$ in $\mathfrak{A}^m$ and $p_2(\sigma) \sim \sum_j p_2(\sigma_j)$ in $\mathfrak{R}^{m,d}$.

2.3. Classes of Operators in the Half Space and on Manifold

2.3.1. Pseudo-Differential Operators with the Transmission Property in the Half Space

2.3.1.1. Properties of Op ($\mathfrak{A}^m$)

Let $\Omega' \subseteq \mathbb{R}^{n-1}$ be an open set and $A \in L^m(\Omega' \times \overline{\mathbb{R}}_+)$, i.e. there is an open neighbourhood $\Omega \supset \Omega' \times \overline{\mathbb{R}}_+$ and A can be extended to a pseudo-differential operator $lA \in L^m(\Omega)$. Denote by $j^+ : C^\infty(\Omega' \times \overline{\mathbb{R}}_+) \to \mathcal{D}'(\Omega)$ the extension by zero for $x_n < 0$ ((x', x_n) are coordinates in $\Omega' \times \mathbb{R}$). Let $r^+ : \mathcal{D}'(\Omega) \to \mathcal{D}'(\Omega' \times \mathbb{R}_+)$ be the restriction of distributions on Ω to the open subset $\Omega' \times \mathbb{R}_+$. Consider the mapping

$$C_0^\infty(\Omega' \times \overline{\mathbb{R}}_+) \xrightarrow{j^+} \mathcal{E}'(\Omega) \xrightarrow{lA} \mathcal{D}'(\Omega) \xrightarrow{r^+} \mathcal{D}'(\Omega' \times \overline{\mathbb{R}}_+)\,.$$

For a differential operator A (defined in a neighbourhood of $\Omega' \times \overline{\mathbb{R}}_+$) it is clear that r^+Aj^+ maps $C_0^\infty(\Omega' \times \overline{\mathbb{R}}_+)$ continuously into $C_0^\infty(\Omega' \times \overline{\mathbb{R}}_+)$. For a pseudo-differential operator $A \in L^m(\Omega)$ the image of $C_0^\infty(\Omega' \times \overline{\mathbb{R}}_+)$ with respect to r^+Aj^+ does not belong to $C^\infty(\Omega' \times \overline{\mathbb{R}}_+)$ in general. Singularities are possible at $x_n = +0$. A pseudo-differential operator $A \in L^m(\Omega)$ is called *smooth* in $\Omega' \times \mathbb{R}_+$ (smooth in the half space for $\Omega' = \mathbb{R}^{n-1}$) if r^+Aj^+ defines a mapping $C_0^\infty(\Omega' \times \overline{\mathbb{R}}_+) \to C^\infty(\Omega' \times \overline{\mathbb{R}}_+)$.

Since smoothing operators, i.e. integral operators with C^∞ kernel, are obviously smooth in $\Omega' \times \overline{\mathbb{R}}_+$, it is reasonable to look for conditions for smoothness in $\Omega' \times \overline{\mathbb{R}}_+$ in terms of the symbol.

Definition 1. A pseudo-differential operator $A \in L^m(\Omega' \times \overline{\mathbb{R}}_+)$ has the transmission property if any complete symbol $a \in S^m((\Omega' \times \overline{\mathbb{R}}_+) \times \mathbb{R}^n)$ of A belongs to $\mathfrak{A}^m((\Omega' \times \overline{\mathbb{R}}_+) \times \mathbb{R}^n)$. Denote by $\mathrm{Op}\,(\mathfrak{A}^m)\,(\Omega' \times \overline{\mathbb{R}}_+)$, or shorter by $\mathrm{Op}\,(\mathfrak{A}^m)$, the subspace of $L^m(\Omega' \times \overline{\mathbb{R}}_+)$ of all pseudo-differential operators with the transmission property.

Note that $A \in \mathrm{Op}\,(\mathfrak{A}^m)\,(\Omega' \times \overline{\mathbb{R}}_+)$ if we have $a \in \mathfrak{A}^m((\Omega' \times \overline{\mathbb{R}}^+) \times \mathbb{R}^n)$ for some complete symbol a, since two complete symbols of A are different by a symbol in $S^{-\infty}((\Omega' \times \overline{\mathbb{R}}_+) \times \mathbb{R}^n) = \mathfrak{A}^{-\infty}((\Omega' \times \overline{\mathbb{R}}_+) \times \mathbb{R}^n)$. Similarly, the space $\mathrm{Op}\,(\mathfrak{A}^m)$ is defined as the subspace of $\mathrm{Op}\,(\mathfrak{A}^m)$ of all classical pseudo-differential operators.

In 2.3.1.2 we shall prove that the transmission property implies smoothness in the half space. It is useful to collect some properties of $\mathrm{Op}\,(\mathfrak{A}^m)$. Point out that the transmission property is a condition about the symbol near the boundary, more exactly a condition about all derivatives in normal direction at $x_n = 0$. As noted in 2.2.2.1 symbols of differential operators with C^∞ coefficients have the transmission property.

A properly supported pseudo-differential operator $A \in L^m(\Omega' \times \mathbb{R}_+)$ defines a continuous mapping

$$A: C_0^\infty(\Omega' \times \mathbb{R}_+) \to C_0^\infty(\Omega' \times \mathbb{R}_+)\,. \tag{1}$$

In particular we have such a mapping for $A \in \mathrm{Op}\,(\mathfrak{A}^m)\,(\Omega' \times \overline{\mathbb{R}}_+) \subset L^m(\Omega' \times \overline{\mathbb{R}}_+) \subset L^m(\Omega' \times \mathbb{R}_+)$. Note that we consider the pseudo-differential operators acting in an open domain.

Proposition 2. (i) *Let $A_i \in \mathrm{Op}(\mathfrak{A}^{m_i})$, $i = 1, 2$, be properly supported. Then $A_1A_2 \in \mathrm{Op}\,(\mathfrak{A}^{m_1+m_2})$ (composition in the sense of* (1)*).*

(ii) *Let $A \in \mathrm{Op}\,(\mathfrak{A}^m)$ be properly supported. Then $A^* \in \mathrm{Op}\,(\mathfrak{A}^m)$.*

(iii) *Let $\varkappa\colon \Omega' \to \Omega_1'$ be a diffeomorphism, $A \in \mathrm{Op}\,(\mathfrak{A}^m)(\Omega' \times \overline{\mathbb{R}}_+)$, and $\varkappa_* A$ the induced pseudo-differential operator* (*cf.* 1.2.3.4). *Then $\varkappa_*A \in \mathrm{Op}\,(\mathfrak{A}^m)\,(\Omega_1 \times \overline{\mathbb{R}}_+)$.*

The assertions immediately follow from the symbolic calculus for pseudo-differential operators in open domains (cf. 1.2.3.3 and 1.2.3.4) and 2.2.2.2, Corollary 2.

2.3.1.2. Smoothness in the Half Space

In this section we shall prove that the transmission property implies smoothness in the half space or in $\Omega' \times \overline{\mathbb{R}}_+$.

Theorem 1. *Let $A \in \mathrm{Op}\,(\mathfrak{A}^m)\,(\Omega' \times \overline{\mathbb{R}}_+)$, $m \in \mathbb{R}$ arbitrary. Then r^+A defines a continuous mapping*

$$r^+A: C_0^\infty(\Omega' \times \overline{\mathbb{R}}_+) \to C^\infty(\Omega' \times \overline{\mathbb{R}}_+)\,.$$

The proof is based on a reduction to the 1-dimensional case. Then we use the following

Proposition 2. *Let $a(x, \nu) \in \mathfrak{A}^m(\overline{\mathbb{R}}_+ \times \mathbb{R})$, $x \in \overline{\mathbb{R}}_+$, $\nu \in \mathbb{R}$, $m \in \mathbb{R}$ arbitrary. Then any pseudo-differential operator $A \in \mathrm{Op}(\mathfrak{A}^m)(\overline{\mathbb{R}}_+)$ with the symbol a has a decomposition $A = A_1 + A_2 + A_3$, where A_1 is defined with a symbol a_1 vanishing at $x = 0$ of infinite order, i.e. $D_x^\gamma a(0, \nu) = 0$ for all $\nu \in \mathbb{R}$, $\gamma \in \mathbb{Z}_+$, A_2 is a differential operator with C^∞ coefficients and the distributional kernel $K \in \mathcal{D}'(\overline{\mathbb{R}}_+ \times \overline{\mathbb{R}}_+)$ of A_3 is smooth up to the boundary in $\overline{\mathbb{R}}_+ \times \overline{\mathbb{R}}_+ \setminus \mathrm{diag}\, \overline{\mathbb{R}}_+$ and for an arbitrary x the function $K(x, y)$ has at most a jump at $y = x$.*

Proof: Define $a_1'(x, \nu) = \sum_{j=0}^{\infty} (x^j/j!)\, \partial_x^j a(0, \nu)\, \varphi(t_j x)$, $\varphi \in C_0^\infty(\overline{\mathbb{R}}_+)$, $\varphi = 1$ near 0 and t_j increasing sufficiently fast for $j \to \infty$ (cf. 2.2.2.2, Lemma 3) so that the series converges in $\mathfrak{A}^m(\overline{\mathbb{R}}_+)$. Then $a_1 = a - a_1'$ vanishes at $x = 0$ of infinite order and $a_1 \in S^m(\overline{\mathbb{R}}_+)$. Denote by $a_2(x, \nu)$ the polynomial part of a_1' with respect to ν. The corresponding operator A_2 is a differential operator with C^∞ coefficients. Set $a_3(x, \nu) = a_1'(x, \nu) - a_2(x, \nu)$. Then $a_3(x, \nu) \in C_0^\infty(\overline{\mathbb{R}}_+) \otimes H_0$, i.e.

$$a_3(x, \nu) = \sum_{k=0}^{\infty} \varphi_k(x)\, b_k(\nu)\,,$$

with $\varphi_k \in C_0^\infty(\overline{\mathbb{R}}_+)$, $b_k \in H_0$, and the series converges in $C_0^\infty(\overline{\mathbb{R}}_+) \otimes H_0$. The distributional kernel of the pseudo-differential operator defined with the symbol $b(\nu)$ is

$$(2\pi)^{-1} \int e^{i(x-y)\nu}\, b(\nu)\, d\nu = \check{b}(x - y)\,,$$

where $\check{b}$ denotes the inverse Fourier transform of b, $\check{b} \in \mathcal{S}(\overline{\mathbb{R}}_+) \otimes \mathcal{S}(\overline{\mathbb{R}}_-)$, if $b \in H_0$. For the kernel K of the pseudo-differential operator A_3 defined with the symbol a_3 we get

$$K(x, y) = \sum_{k=0}^{\infty} \varphi_k(x)\, \check{b}_k(x - y)\,,$$

$\check{b}_k \in \mathcal{S}(\overline{\mathbb{R}}_+) \otimes \mathcal{S}(\overline{\mathbb{R}}_-)$ and the assertion is proved. □

Corollary 3. *Let $A \in \mathrm{Op}(\mathfrak{A}^m)(\overline{\mathbb{R}}_+)$. Then A is smooth on the half axis, i.e.*

$$r^+ A : C_0^\infty(\overline{\mathbb{R}}_+) \to C^\infty(\overline{\mathbb{R}}_+)\,.$$

Moreover if A is defined by a symbol a smoothly depending on a parameter, then r^+A smoothly depends on the parameter.

Proof: We show the assertion for all components of the decomposition in Proposition 2. Note that the second assertion follows if we show that the mapping

$$\mathfrak{A}^m(\overline{\mathbb{R}}_+ \times \mathbb{R}) \ni a \mapsto A \in \mathcal{L}\big(C_0^\infty(\overline{\mathbb{R}}_+), C^\infty(\overline{\mathbb{R}}_+)\big)$$

is continuous,

$$Au(x) = (2\pi)^{-1} \int e^{ix\nu}\, a(x, \nu)\, \hat{u}(\nu)\, d\nu\,.$$

The assertion is obvious for a differential operator, i.e. $a(x, \nu) = \sum_{k=0}^{m} c_k(x)\, \nu^k$. Now let $a(x, \nu) \in S^m(\overline{\mathbb{R}}_+)$ be vanishing at $x_n = 0$ of infinite order. It is sufficient to consider a compact set $K \subset \overline{\mathbb{R}}_+$ containing a neighbourhood of 0 in $\overline{\mathbb{R}}_+$. We have for $x > 0$

$$|D_x^\gamma A j^+ u(x)| \leq c \sum_{\alpha+\beta \leq \gamma}' \left| \int e^{ix\nu}\, \nu^\alpha D_x^\beta a(x, \nu)\, \widehat{j^+ u}(\nu)\, d\nu \right|$$

($\alpha, \beta, \gamma \in \mathbb{Z}_+$, $c = c_\gamma$). By Taylor expansion of $D_x^\beta a(x, \nu)$ with respect to x at $x = 0$ we get for an arbitrary fixed $N \in \mathbb{Z}_+$

$$|D_x^\beta a(x, \nu)| \leqq c\langle\nu\rangle^m x^N\,, \quad x \in K$$

and

$$|D_\nu^\gamma D_x^\beta a(x, \nu)| \leqq c\langle\nu\rangle^{m-\gamma} x^N\,, \quad x \in K\,.$$

Then we have

$$\sup_{x\in K} |D_x^\gamma A j^+ u(x)| \leqq c \sum_{\alpha+\beta\leqq\gamma} \left|\int e^{ix\nu} D_\gamma^N\big(\nu^\alpha D_x^\beta a(x, \nu)\, \widehat{j^+ u}(\nu)\big)\, d\nu\right|$$

$$\leqq c \sum_{\substack{\alpha+\beta\leqq\gamma \\ N_1+N_2+N_3=N}} \int \langle\nu\rangle^{\alpha-N_1} \langle\nu\rangle^{m-N_2} \langle\nu\rangle^{-N_3-1}\, d\nu < \infty$$

for sufficiently large $N \in \mathbb{Z}_+$ with constants depending on γ and K which can be estimated by semi-norms of $a(x, \nu) \in S^m(\overline{\mathbb{R}}_+ \times \mathbb{R})$.

Assume now that $a(x, \nu)$ has the form of $a_3(x, \nu)$ in Proposition 2. It is sufficient to consider $a(x, \nu) = \varphi(x)\, b(\nu)$, $\varphi \in C_0^\infty(\overline{\mathbb{R}}_+)$, $b \in H_0$. Then, for a compact set $K \subset \overline{\mathbb{R}}_+$,

$$\sup_{x\in K} \left|D_x^\gamma \int e^{ix\nu}\, \varphi(x)\, b(\nu)\, \widehat{j^+ u}(\nu)\, d\nu\right|$$

$$\leqq c \sum_{\alpha+\beta=\gamma} \left|\int e^{ix\nu} D_x^\beta \varphi(x)\, \nu^\alpha b(\nu)\, \widehat{j^+ u}(\nu)\, d\nu\right|\,.$$

Obviously, $C_0^\infty(\overline{\mathbb{R}}_+) \ni \varphi \mapsto D^\beta\varphi \in C_0^\infty(\overline{\mathbb{R}}_+)$ and $H_0 \ni b \mapsto \nu^\alpha b(\nu) \in H_\alpha$ are continuous mappings and the assertion follows. □

Proof of Theorem 1: Obviously we have $\theta^+(x_n)\, u(x', x_n) = u(x', x_n)\, e^{x_n} e^{\langle x'\rangle} e^{-x_n}$ $\theta^+(x_n)\, e^{-\langle x'\rangle}$, θ^+ the characteristic function of $\overline{\mathbb{R}}_+$ and $u \in C_0^\infty(\Omega' \times \overline{\mathbb{R}}_+)$. Since the multiplication by the function $u(x', x_n)\, e^{x_n} e^{\langle x'\rangle} \in C_0^\infty(\Omega' \times \overline{\mathbb{R}}_+)$ is an operator (of order 0) with the transmission property, in view of 2.3.1.1, Proposition 2 (i), it is sufficient to show $r^+A\big(e^{-\langle x'\rangle} e^{-x_n}\, \theta^+(x_n)\big) \in C^\infty(\Omega' \times \overline{\mathbb{R}}_+)$ for an arbitrary operator $A \in \mathrm{Op}\,(\mathfrak{A}^m)\,(\Omega' \times \overline{\mathbb{R}}_+)$. Define

$$b(x', x_n, \nu) = (2\pi)^{-n+1} \int e^{i(x'-y')\xi'}\, a(x', x_n, \xi', \nu)\, e^{-\langle y'\rangle}\, dy'\, d\xi'$$

$$= (2\pi)^{-n+1} \int e^{ix'\xi'}\, a(x', x_n, \xi', \nu)\, g(\xi')\, d\xi'\,,$$

$g \in \mathcal{S}(\mathbb{R}^{n-1})$. Then $b(x', x_n, \nu) \in C^\infty\big(\Omega', \mathfrak{A}^m(\overline{\mathbb{R}}_+ \times \mathbb{R})\big)$. In fact, for arbitrary $\alpha \in \mathbb{Z}_+^{n-1}$, $\beta, \gamma, N \in \mathbb{Z}_+$ and an arbitrary compact set $K \subset \Omega' \times \overline{\mathbb{R}}_+$ there is a constant such that

$$|D_{x'}^\alpha D_{x_n}^\beta D_\nu^\gamma b(x', x_n, \nu)| \leqq c \int \langle\xi\rangle^{m-\gamma} \langle\xi'\rangle^{-N}\, d\xi' \leqq c\, \langle\nu\rangle^{m-\gamma}\,.$$

This implies $b(x', x_n, \nu) \in C^\infty\big(\Omega', S^m(\overline{\mathbb{R}}_+)\big)$. From $D_{x_n}^\gamma b(x', 0, \nu) = (2\pi)^{-n+1} \int e^{ix'\xi'} D_{x_n}^\gamma$ $\times\, a(x', 0, \xi', \nu)\, g(\xi')\, d\xi'$ and the asymptotic expansion for $D_{x_n}^\gamma a(x', 0, \xi', \nu)$ from 2.2.2.1, Proposition 3 (ii) we obtain $D_{x_n}^\gamma b(x', 0, \nu) \in C^\infty(\Omega', H)$ and therefore $b(x', x_n, \nu) \in C^\infty\big(\Omega',$ $\mathfrak{A}^m(\overline{\mathbb{R}}_+ \times \mathbb{R})\big)$. We have

$$r^+A\big(e^{-\langle x'\rangle} e^{-x_n}\, \theta^+(x_n)\big) = (2\pi)^{-n}\, r^+ \int e^{ix'\xi'+ix_n\nu}\, a(x', x_n, \xi', \nu)\, g(\xi')\,,$$

$$F\big(e^{-y_n}\, \theta(y_n)\big)(\nu)\, d\xi'\, d\nu = (2\pi)^{-1}\, r^+ \int e^{ix_n\nu}\, b(x', x_n, \nu)\, (1 + i\nu)^{-1}\, d\nu\,.$$

By Corollary 3 this is a C^∞ function on $\overline{\mathbb{R}}_+$ smoothly depending on the parameter $x' \in \Omega'$. □

Let us briefly consider the situation of pseudo-differential operators on manifolds with boundary. Let X be a manifold with boundary $Y = \partial X$, E, $F \in \text{Vect}(X)$. Denote by $S^m(X; E, F)$ the subspace of all such symbols in $S^m(\Omega; E|_\Omega, F|_\Omega)$, $\Omega = \text{int}\, X$, all local representations of which in local coordinates near Y can be extended, i.e. they belong to $S^m(\overline{\mathbb{R}}^n_+;\ \mathbb{C}^k, \mathbb{C}^{k'})$ ($n = \dim X$, k, k' the fibre dimensions of E and F, respectively). In view of 2.2.2.1, Lemma 1 symbols in $S^m(X; E, F)$ can be characterized as symbols in $S^m(\Omega;\ E|_\Omega,\ F|_\Omega)$ with the extension as a symbol in $S^m(X^0; E^0, F^0)$, $E^0, F^0 \in \text{Vect}(X^0)$ for an arbitrary neighbouring manifold $X^0 \supset X$, $E^0|_X = E$, $F^0|_X = F$.

Denote by $\mathfrak{A}^m(X; E, F)$ the subspace of $S^m(X; E, F)$ of all symbols, the local representations of which in local coordinates near Y have the transmission property, i.e. they belong to $\mathfrak{A}^m(\overline{\mathbb{R}}^n_+ \times \mathbb{R}^n) \otimes \hom(\mathbb{C}^k, \mathbb{C}^{k'})$. In view of 2.2.2.2, Corollary 2 (iii) it is sufficient to have this property for a fixed atlas. The space of pseudo-differential operators on X acting between sections of the bundles E and F defined with symbols in $S^m(X; E, F)$ is denoted by $L^m(X; E, F)$.

The subspace of $L^m(X; E, F)$ of all pseudo-differential operators with a symbol in $\mathfrak{A}^m(X; E, F)$ is denoted by $\text{Op}(\mathfrak{A}^m)(X; E, F)$ or shorter by $\text{Op}(\mathfrak{A}^m)$.

Theorem 4. *Let $r^+A \in \text{Op}(\mathfrak{A}^m)(X; E, F)$. Then r^+A defines a continuous mapping*

$$r^+A: C^\infty(X, E) \to C^\infty(X, F)\,.$$

Proof: Fix an open covering $\{U_j\}_{j \in J}$ of X with the property that for U_i, U_j: $U_i \cap U_j \neq \emptyset$ there is a coordinate neighbourhood $U_{ij} \subset X$, $U_{ij} \supset U_i \cup U_j$ with $E|_{U_{ij}}$, $F|_{U_{ij}}$ trivial. Then for $u \in C^\infty(X, E)$ we have $Au = \sum_{i,j} \varphi_i A \varphi_j u$ (cf. 1.2.4.1, Theorem 1). In local coordinates of U_{ij} using local trivializations of $E|_{U_{ij}}$ and $F|_{U_{ij}}$ we get a system of PDOs in $\mathbb{R}^n$ for U_{ij} away from the boundary Y and a system of PDOs with the transmission property in $\overline{\mathbb{R}}^n_+$ near the boundary. Then the assertion follows from Theorem 1 in 1.2.4.1 and Theorem 1. □

2.3.2. Potential, Trace and Green Operators in the Half Space

In this section we give definitions and basic properties of classes of trace, potential and Green operators. Pseudo-differential trace operators are a natural generalization of the usual differential trace operators from classical boundary value problems. Potential operators can be described as the adjoints of trace operators. They also appear as 'boundary conditions' for PDOs (cf. chapter 3). Green operators are introduced in order to get an algebra of operators. In fact, the simplest example of a Green operator is the composition of a trace operator and a potential operator.

In a similar way as PDOs the operators are described by their (boundary) symbols. It is the symbolic calculus which makes possible a lot of constructions as for ordinary PDOs (compositions, adjoints, parametrices for elliptic operators, and so on).

2.3.2.1. Definitions and Basic Properties

First we define smoothing potential, trace and Green operators. Let $k \in C^\infty(\Omega' \times \overline{\mathbb{R}}_+ \times \Omega')$, $\Omega' \subseteq \mathbb{R}^{n-1}$ open. Then the operator

$$K: C_0^\infty(\Omega') \to C^\infty(\Omega' \times \overline{\mathbb{R}}_+)$$

defined by

$$Kv(x', x_n) = \int_{\Omega'} k(x', x_n, y')\, v(y')\, dy', \quad v \in C_0^\infty(\Omega'),$$

is called *smoothing potential operator*. Denote by Op $(\mathfrak{K}^{-\infty})$ the space of all smoothing potential operators.

Every smoothing potential operator admits a continuous extension $\mathscr{E}'(\Omega') \to C^\infty(\Omega' \times \overline{\mathbb{R}}_+)$. This follows from the definition. Conversely, every continuous operator $K: C_0^\infty(\Omega') \to C^\infty(\Omega' \times \overline{\mathbb{R}}_+)$ admitting an extension $\mathscr{E}'(\Omega') \to C^\infty(\Omega' \times \overline{\mathbb{R}}_+)$ is a smoothing potential operator, since its distributional kernel can be represented by a function in $C^\infty(\Omega' \times \overline{\mathbb{R}}_+ \times \Omega')$.

Let B_k, $k = 0, \dots, d-1$, be smoothing operators in Ω' and $t(x', y', y_n) \in C^\infty(\Omega' \times \Omega' \times \overline{\mathbb{R}}_+)$. Then the operator

$$r'T: C_0^\infty(\Omega' \times \overline{\mathbb{R}}_+) \to C^\infty(\Omega')$$

defined by

$$r'T\, u(x') = \sum_{k=0}^{d-1} B_k(r' D_{x_n}^k u)(x') + \int_{\Omega' \times \mathbb{R}_+} t(x', y', y_n)\, u(y', y_n)\, dy'\, dy_n,$$

$u \in C_0^\infty(\Omega' \times \overline{\mathbb{R}}_+)$, is called *smoothing trace operator* of type d. Denote by Op $(\mathfrak{T}^{-\infty, d})$ the space of all smoothing trace operators of type d.

Every smoothing trace operator of type $d > 0$ admits a continuous extension $H^s_{\text{comp}}(\Omega' \times \mathbb{R}_+) \to C^\infty(\Omega')$ for any $s > d - 1/2$ (cf. 1.2.1.2, Proposition 9). For $d = 0$ there exists a continuous extension $\mathscr{E}'(\Omega' \times \overline{\mathbb{R}}_+) \to C^\infty(\Omega')$.

Let K_k, $k = 0, \dots, d-1$, be smoothing potential operators and $b(x, y) \in C^\infty(\Omega' \times \overline{\mathbb{R}}_+ \times \Omega' \times \overline{\mathbb{R}}_+)$. Then the operator

$$r'B: C_0^\infty(\Omega' \times \overline{\mathbb{R}}_+) \to C^\infty(\Omega' \times \overline{\mathbb{R}}_+)$$

defined by

$$r'Bu(x', x_n) = \sum_{k=0}^{d-1} K_k(r' D_{x_n}^k u)(x', x_n) + \int_{\Omega' \times \mathbb{R}_+} b(x, y)\, u(y)\, dy,$$

$$u \in C_0^\infty(\Omega' \times \overline{\mathbb{R}}_+),$$

is called *smoothing Green operator* of type d. Denote by Op $(\mathfrak{B}^{-\infty, d})$ the space of all smoothing Green operators of type d.

Every smoothing Green operator of type $d > 0$ admits a continuous extension $H^s_{\text{comp}}(\Omega' \times \overline{\mathbb{R}}_+) \to C^\infty(\Omega' \times \overline{\mathbb{R}}_+)$ for any $s > \mathrm{d} - 1/2$. For $d = 0$ there exists a continuous extension $\mathscr{E}'(\Omega' \times \overline{\mathbb{R}}_+) \to C^\infty(\Omega' \times \overline{\mathbb{R}}_+)$.

Using the classes of Green, trace and potential symbols defined in 2.2.3 and 2.2.4 we define corresponding operator classes.

Lemma 1. *Let* $k \in \mathfrak{K}^m(\Omega' \times \overline{\mathbb{R}}^n)$ *and define*

$$\check{k}(x', \xi', x_n) = (2\pi)^{-1} \int e^{ix_n\nu}\, k(x', \xi', \nu)\, d\nu, \quad x_n > 0.$$

For any $j \in \mathbb{Z}_+$ *the function* $\check{k}$ *can be considered as an element of* $C^j(\overline{\mathbb{R}}_+, S^{m+j+1}(\Omega' \times \mathbb{R}^{n-1}))$.

Proof: For $x_n > 0$ we have

$$D_{x_n}^j \check{k}(x', \xi', x_n) = (2\pi)^{-1} \int e^{ix_n\nu}\, \nu^j k(x', \xi', \nu)\, d\nu.$$

Let $K \subset \Omega'$ be a compact set and $\alpha, \beta \in \mathbb{Z}_+^{n-1}$ be arbitrary multi-indices. Then, after the substitution $\nu \mapsto \langle \xi' \rangle \nu$ we get

$$D_{x'}^\alpha D_{\xi'}^\beta D_{x_n}^j \check{k}(x', \xi', x_n) = (2\pi)^{-1} D_{x'}^\alpha D_{\xi'}^\beta \int e^{ix_n\nu\langle\xi'\rangle} \langle \xi' \rangle^{j+1} \nu^j k_{[0]}(x', \xi', \nu)\, d\nu,$$

where $k_{[0]}(x', \xi', \nu) \in H_\nu^+ \otimes S^m(\Omega' \times \mathbb{R}^{n-1})$. It is sufficient to consider the equation $\Pi_\nu^+ \langle \xi' \rangle^{j+1} \nu^j k_{[0]}(x', \xi' \nu) = a(x', \xi') f(\nu)$, $a \in S^{m+j+1}(\Omega' \times \mathbb{R}^{n-1})$, $f \in H^+$. Let $\check{f}$ be the inverse Fourier transform of f, $\check{f} \in \mathcal{S}(\overline{\mathbb{R}}_+)$. Then

$$D_{x'}^\alpha D_{\xi'}^\beta D_{x_n}^j \check{k}(x', \xi', x_n) = D_{\xi'}^\beta (D_{x'}^\alpha a(x', \xi') \check{f}(\langle \xi' \rangle x_n))$$

and for $x' \in K$ the desired estimate follows from

$$|D_{\xi'}^\beta \check{f}(\langle \xi' \rangle x_n)| \leqq c \langle \xi' \rangle^{-|\beta|} \quad \text{for all} \quad x_n > 0$$

which can be easily checked by induction using

$$D_{\xi_j}(\check{f}(\langle \xi' \rangle x_n)) = \check{f}'(\langle \xi' \rangle x_n) x_n \langle \xi' \rangle b(\xi'), \quad b \in S^{-1}(\mathbb{R}^{n-1}). \quad \square$$

Definition 2. Let $k(x', y', \xi', \nu) \in \mathfrak{R}^m(\Omega' \times \Omega' \times \mathbb{R}^{n-1} \times \mathbb{R})$. Then the oscillatory integral

$$(2\pi)^{-n+1} F_n^{-1} \int e^{i(x'-y')\xi'} k(x', y', \xi', \nu) v(y')\, dy'\, d\xi'$$
$$= (2\pi)^{-n+1} \int e^{i(x'-y')\xi'} \check{k}(x', y', \xi', x_n) v(y')\, dy'\, d\xi', \tag{1}$$

$v \in C_0^\infty(\Omega')$, $x_n \in \overline{\mathbb{R}}_+$, depending on the parameter x_n defines the *potential operator*

$$K = \operatorname{Op}(k): C_0^\infty(\Omega') \to C^\infty(\Omega' \times \overline{\mathbb{R}}_+).$$

The space of all sums of smoothing and potential operators with an amplitude function $k \in \mathfrak{R}^m$ is denoted by $\operatorname{Op}(\mathfrak{R}^m)$. We have $\operatorname{Op}(\mathfrak{R}^{-\infty}) = \bigcap_m \operatorname{Op}(\mathfrak{R}^m)$, since it is easily seen that $\operatorname{Op}(k)$ with $k \in \mathfrak{R}^{-\infty}$ is smoothing. Note that not every smoothing potential operator can be given as $\operatorname{Op}(k)$, $k \in \mathfrak{R}^{-\infty}$.

Take $\varphi \in C_0^\infty(\overline{\mathbb{R}}_+)$, $\varphi = 1$ near 0. Then $(1 - \varphi(x_n)) K$, $K \in \operatorname{Op}(\mathfrak{R}^m)$ is a smoothing potential operator. In fact, for $k(x', y', \xi', \nu) \in \mathfrak{R}^m$ we have $(1 - \varphi(x_n))\, \check{k}(x', y', \xi', x_n) \in C^\infty(\Omega' \times \Omega' \times \mathbb{R}^{n-1} \times \mathbb{R})$, since $\check{k}_{[0]}(x', y', \xi', x_n) = \check{k}(x', y', \xi', \langle \xi' \rangle^{-1} x_n) \in S^m(\Omega' \times \Omega' \times \mathbb{R}^{n-1}) \otimes \mathcal{S}(\overline{\mathbb{R}}_+)$ multiplied by $1 - \varphi(x_n)$ gives a function in $S^m(\Omega' \times \Omega' \times \mathbb{R}^{n-1}) \otimes \mathcal{S}(\mathbb{R})$. Thus the potential operator is defined modulo smoothing operators by its image functions restricted to an arbitrary neighbourhood of $x_n = 0$.

We shall need symbols with values in a Banach space B. Let $\Omega \subseteqq \mathbb{R}^n$ be open. Define $S^m(\Omega \times \mathbb{R}^n, B)$ as the subspace of smooth functions $a \in C^\infty(\Omega \times \mathbb{R}^n, B)$ such that for arbitrary multi-indices $\alpha, \beta \in \mathbb{Z}_+^n$ and any compact set $K \subset \Omega$ there is a constant $c = c_{\alpha\beta K}$ such that

$$\|D_x^\alpha D_\xi^\beta a(x, \xi)\| \leqq c \langle \xi \rangle^{m-|\beta|} \quad \text{for} \quad x \in K.$$

Obviously, $S^m(\Omega \times \mathbb{R}^n, B)$ is a Fréchet space, $S^m(\Omega \times \mathbb{R}^n, B) \subseteqq S^{m'}(\Omega \times \mathbb{R}^n, B)$ for $m \leqq m'$ and $a \in S^m(\Omega \times \mathbb{R}^n, B) \Rightarrow D_x^\alpha D_\xi^\beta a(x, \xi) \in S^{m-|\beta|}(\Omega \times \mathbb{R}^n, B)$. In our application B is often the Banach algebra $\mathcal{L}(B_1, B_2)$ for Banach spaces B_i. Then $a_1 \in S^{m_1}(\Omega \times \mathbb{R}^n, \mathcal{L}(B_1, B_2))$, $a_2 \in S^{m_2}(\Omega \times \mathbb{R}^n, \mathcal{L}(B_2, B_3))$ implies $a_2 a_1 \in S^{m_1+m_2}(\Omega \times \mathbb{R}^n, \mathcal{L}(B_1, B_3))$. For $u \in C_0^\infty(\Omega, B_1)$ and $a \in S^m(\Omega \times \mathbb{R}^n, \mathcal{L}(B_1, B_2))$ consider the oscillatory integral

$$(2\pi)^{-n} \int e^{i(x-y)\xi} a(x, \xi) u(y)\, dy\, d\xi. \tag{2}$$

Then we have

Proposition 3. *The operator A defined by* (2) *yields a continuous mapping*

$$A: C_0^\infty(\Omega, B_1) \to C^\infty(\Omega, B_2) .$$

The distributional kernel K_A of A, $K_A \in \mathcal{D}'(\Omega \times \Omega, \mathcal{L}(B_1, B_2))$ is C^∞ in $\Omega \times \Omega \setminus \operatorname{diag} \Omega$. For $m + j < -n$ we have $K_A \in C^j(\Omega \times \Omega, \mathcal{L}(B_1, B_2))$.

The proof repeats the arguments from the scalar case (cf. 1.2.3.1). As usual one defines properly supported operators. For A properly supported its symbol $\sigma(A)(x, \xi)$ is given by

$$\sigma(A)(x, \xi) = e^{-ix\xi} A(e^{i\cdot\xi}) .$$

Denote by $L^m(\Omega; B_1, B_2)$ the space of all PDOs given by (2). As in 1.2.3.2 we have

Proposition 4. *Any operator A*

$$Au(x) = (2\pi)^{-n} \int e^{i(x-y)\xi} a(x, y, \xi) u(y) \, dy \, d\xi , \quad u \in C_0^\infty(\Omega, B_1) \tag{3}$$

with $a \in S^m(\Omega \times \Omega \times \mathbb{R}^n, \mathcal{L}(B_1, B_2))$ belongs to $L^m(\Omega; B_1, B_2)$. For A properly supported its symbol evaluated by $\sigma(A) = e^{-ix\xi} A(e^{i\cdot\xi})$ has an asymptotic expansion given by 1.2.3.2.(2). *We have an isomorphism*

$$L^m(\Omega; B_1, B_2)/L^{-\infty}(\Omega; B_1, B_2) \to S^m(\Omega; \mathcal{L}(B_1, B_2))/S^{-\infty}(\Omega; \mathcal{L}(B_1, B_2)) .$$

Similarly as in 1.2.3.3 and 1.2.3.4 the symbolic calculus holds for PDOs in $L^m(\Omega; B_1, B_2)$ and the formulas change in an obvious way.

In the following we shall assume all spaces $B_1, B_2, \ldots$ to be Hilbert spaces. For $s \geqq 0$ define the Sobolev space of B valued functions $H^s(\mathbb{R}^n, B)$ as subspace of $L^2(\mathbb{R}^n, B)$ of all functions u with the finite norm

$$\|u\|_s = \{\int (1 + |\xi|^2)^s \|\hat{u}(\xi)\|_B^2 \, d\xi\}^{1/2}$$

where $\hat{u} \in L^2(\mathbb{R}^n, B)$ is the Fourier transform. Set $H^s(\mathbb{R}^n, B) = H^{-s}(\mathbb{R}^n, B)'$ for $s < 0$.

In the same way as in 1.2.3.5, Theorem 1 one can show

Proposition 5. *Any $A \in L^m(\Omega; B_1, B_2)$ admits a continuous extension*

$$A: H^s_{\text{comp}}(\Omega, B_1) \to H^{s-m}_{\text{loc}}(\Omega, B_2) .$$

If A is defined by (2) *with $a(x, \xi) \in S^m(\Omega \times \mathbb{R}^n; \mathcal{L}(B_1, B_2))$ having compact support with respect to x, there is a continuous extension*

$$A: H^s(\Omega, B_1) \to H^{s-m}(\Omega, B_2)$$

and its norm is bounded by a sum of semi-norms of $a \in S^m$.

Now we can give the definition of trace and Green operators. For $u \in C_0^\infty(\Omega' \times \overline{\mathbb{R}}_+)$ we have $j^+u(x', .) \in L^2(\mathbb{R})$ smoothly depending on $x' \in \Omega'$. Then

$$u(x', \nu) = \int e^{-ix_n\nu} j^+u(x', x_n) \, dx_n$$

can be considered as an element of $C^\infty(\Omega', V^+)$ (V^+ the L^2 closure of H^+).

Lemma 6. *By $(x', \xi') \mapsto \Pi' t(x', \xi', \nu) \in \mathcal{L}(V^+, V^+)$, $t \in \mathfrak{T}^{m,0}(\Omega' \times \mathbb{R}^{n-1} \times \mathbb{R})$, there is defined an element $\Pi' t \in S^{m+1/2}(\Omega' \times \mathbb{R}^{n-1}, \mathcal{L}(V^+, V^+))$.*

Proof: Since $\|\Pi' t\|_{\mathcal{L}(V^+, V^+)} \leqq \|t\|_{L^2(\mathbb{R})}$, we consider $\|D_{x'}^\alpha D_{\xi'}^\beta t(x', \xi', \nu)\|_{L^2(\mathbb{R})}$. The substitution $\nu \mapsto \nu\langle\xi'\rangle^{-1}$ yields an element of $H_0^- \otimes S^m(\Omega' \times \mathbb{R}^{n-1})$. Then it is sufficient to consider $t(x', \xi', \nu) = a(x', \xi') f(\langle\xi'\rangle^{-1}\nu)$, $a \in S^m(\Omega' \times \mathbb{R}^{n-1})$, $f \in H_0^-$ and the assertion follows from

$$||D_{\xi'}^{\beta} f(\langle\xi'\rangle^{-1}\nu)||_{L^2(\mathbb{R})} \leqq c\langle\xi'\rangle^{1/2-|\beta|}. \quad \square$$

Definition 7. Let $t(x', y', \xi', \nu) \in \mathfrak{T}^{m,d}(\Omega' \times \Omega' \times \mathbb{R}^{n-1} \times \mathbb{R})$. Then the oscillatory integral

$$\left.\begin{aligned} &(2\pi)^{-n+1} \int e^{i(x'-y')\xi'} \Pi'\big(t(x', y', \xi', \nu)\, \hat{u}(y', \nu)\big)\, dy'\, d\xi', \\ &u \in C_0^\infty(\Omega' \times \overline{\mathbb{R}}_+),\quad \hat{u}(y', \nu) = \int e^{-iy_n\nu} j^+u(y', y_n)\, dy_n \in C_0^\infty(\Omega', V^+) \end{aligned}\right\} \tag{4}$$

defines the *trace operator*

$$r'T = \mathrm{Op}\,(t) : C_0^\infty(\Omega' \times \overline{\mathbb{R}}_+) \to C^\infty(\Omega').$$

The space of all sums of smoothing trace operators of type d and trace operators defines with an amplitude function $t \in \mathfrak{T}^{m,d}$ is denoted by $\mathrm{Op}\,(\mathfrak{T}^{m,d})$. We have $\mathrm{Op}\,(\mathfrak{T}^{-\infty,d}) = \bigcap_m \mathrm{Op}\,(\mathfrak{T}^{m,d})$.

Definition 8. Let $b(x', y', \xi', \nu, \tau) \in \mathfrak{B}^{m,d}(\Omega' \times \Omega' \times \mathbb{R}^{n-1} \times \mathbb{R} \times \mathbb{R})$. Then the oscillatory integral

$$\begin{aligned} &(2\pi)^{-n+1} F_\nu^{-1} \int e^{i(x'-y')\xi'} \Pi_\tau'\big(b(x', y', \xi', \nu, \tau)\, \hat{u}(y', \tau)\big)\, dy'\, d\xi' \\ &= (2\pi)^{-n} \int e^{i(x'-y')\xi'} \Pi_\tau'\big(\check{b}(x', y', \xi'\, x_n, \tau)\, \hat{u}(y', \tau)\big)\, dy'\, d\xi', \end{aligned} \tag{5}$$

$$u \in C_0^\infty(\Omega' \times \overline{\mathbb{R}}_+),\ \hat{u}(y', \tau) = \int e^{-iy_n\tau} j^+u(y', y_n)\, dy_n \in C_0^\infty(\Omega', V^+),$$

$$\check{b}(x', y', \xi', x_n, \tau) = (2\pi)^{-1} \int e^{ix_n\nu}\, b(x', y', \xi', \nu, \tau)\, d\nu,$$

smoothly depending on the parameter $x_n \in \overline{\mathbb{R}}_+$ defines a *Green operator*

$$r'B = \mathrm{Op}\,(b) : C_0^\infty(\Omega' \times \overline{\mathbb{R}}_+) \to C^\infty(\Omega' \times \overline{\mathbb{R}}_+).$$

The space of all sums of smoothing Green operators of type d and Green operators defined with an amplitude function $b \in \mathfrak{B}^{m,d}$ is denoted by $\mathrm{Op}\,(\mathfrak{B}^{m,d})$. We have $\mathrm{Op}\,(\mathfrak{B}^{-\infty,d}) = \bigcap_m \mathrm{Op}\,(\mathfrak{B}^{m,d})$.

Similar as for potential operators consider $\varphi \in C_0^\infty(\overline{\mathbb{R}}_+)$, $\varphi = 1$ near 0. Then for any $r'T \in \mathrm{Op}\,(\mathfrak{T}^{m,d})$, $r'B \in \mathrm{Op}\,(\mathfrak{B}^{m,d})$ the operators $r'T(1 - \varphi)$, $r'B(1 - \varphi)$ and $(1 - \varphi)\, r'B$ are smoothing ones (by ψ we denote the multiplication operator with the function ψ).

In the next section we shall deal with compositions, adjoints and so on of Green, potential and trace operators. It will be convenient to consider matrices of operators. For abbreviation use again the notations $\mathrm{Op}\,(\mathfrak{A}_1^m)$, $\mathrm{Op}\,(\mathfrak{B}^{m-1,d})$, ... without explicitly indicating dimensions.

Definition 9. Denote by $\mathrm{Op}\,(\mathfrak{R}^{m,d})\,(\Omega' \times \overline{\mathbb{R}}_+)$ the space of all operators

$$\mathcal{A} = \begin{pmatrix} r^+A + r'B & K \\ & \\ r'T & Q \end{pmatrix} : \begin{matrix} C_0^\infty(\Omega' \times \overline{\mathbb{R}}_+, \mathbb{C}^k) \\ \oplus \\ C_0^\infty(\Omega', \mathbb{C}^j) \end{matrix} \to \begin{matrix} C^\infty(\Omega' \times \overline{\mathbb{R}}_+, \mathbb{C}^{k'}) \\ \oplus \\ C^\infty(\Omega', \mathbb{C}^{j'}) \end{matrix}$$

with $r^+A \in \mathrm{Op}\,(\mathfrak{A}_1^m)$, $r'B \in \mathrm{Op}\,(\mathfrak{B}^{m-1,d})$, $K \in \mathrm{Op}\,(\mathfrak{K}^m)$, $r'T \in \mathrm{Op}\,(\mathfrak{T}^{m-1,d})$, $Q \in L^{\mathring{m}}$.

We assume here that the operator symbol $a \in \mathfrak{A}_1^m$ is independent of $x_n \in \overline{\mathbb{R}}_+$. This is a special case of the more general situation if $r^+A \in \mathrm{Op}\,(\mathfrak{A}^m)$. The space of all such operators with $r^+A \in \mathrm{Op}\,(\mathfrak{A}^m)$ is denoted by $\mathrm{Op}\,(\mathfrak{S}^{m,d})$. For $\mathrm{Op}\,(\mathfrak{S}^{m,d})$ we shortly write $\mathfrak{G}^{m,d}$.

Obviously, modulo smoothing operators (in $\mathrm{Op}(\mathfrak{R}^{-\infty,d})$) any $\mathcal{A} \in \mathrm{Op}(\mathfrak{R}^{m,d})$ is defined by an amplitude function $\mathfrak{a}(x', \xi')$ with values in the space of one-dimensional boundary symbols.

2.3.2.2. Interpretation as Pseudo-Differential Operators with Operator Valued Symbols

In this section we shall systematically use the representation of Green, potential and trace operators as pseudo-differential operators with the operator valued symbol already mentioned in 2.3.2.1. We assume that the pseudo-differential symbol entering in the boundary symbol is independent of x_n near $x_n = 0$. Then we obtain the calculus of complete boundary symbols. The case of symbols depending on x_n will be discussed in 2.3.2.3.

Recall that a boundary symbol $\mathfrak{a} \in \mathfrak{R}^{m,d}$ on the axis

$$\mathfrak{a}: \begin{matrix} H^+ \otimes \mathbb{C}^k \\ \oplus \\ \mathbb{C}^j \end{matrix} \to \begin{matrix} H^+ \otimes \mathbb{C}^{k'} \\ \oplus \\ \mathbb{C}^{j'} \end{matrix}$$

via Fourier transform F defines a mapping

$$F_*\mathfrak{a} \overset{\text{def}}{=} (F^{-1} \oplus 1)\, \mathfrak{a}(F \oplus 1): \begin{matrix} \mathcal{S}(\overline{\mathbb{R}}_+) \otimes \mathbb{C}^k \\ \oplus \\ \mathbb{C}^j \end{matrix} \to \begin{matrix} \mathcal{S}(\overline{\mathbb{R}}_+) \otimes \mathbb{C}^{k'} \\ \oplus \\ \mathbb{C}^{j'} \end{matrix}$$

For $\mathfrak{a}(x', y', \xi') \in \mathfrak{R}^{m,d}(\Omega' \times \Omega' \times \mathbb{R}^{n-1};\ \mathbb{C}^k,\ \mathbb{C}^{k'},\ \mathbb{C}^j,\ \mathbb{C}^{j'})$ ($\Omega' \subseteq \mathbb{R}^{n-1}$ open) and $(u, v) \in C_0^\infty(\Omega' \times \overline{\mathbb{R}}_+, \mathbb{C}^k) \oplus C_0^\infty(\Omega', \mathbb{C}^j)$ we consider

$$(2\pi)^{-n+1}(F_n^{-1} \oplus 1) \int e^{i(x'-y')\xi'}\, \mathfrak{a}(x', y', \xi')\, (F_n \oplus 1) \begin{pmatrix} j^+ u(y', y_n) \\ v(y') \end{pmatrix} dy'\, d\xi'$$

$$= (2\pi)^{-n+1} \int e^{i(x'-y')\xi'}\, F_*\mathfrak{a}(x', y', \xi') \begin{pmatrix} j^+ u(y', y_n) \\ v(y') \end{pmatrix} dy'\, d\xi' \tag{1}$$

(here F_n denotes the partial Fourier transform with respect to y_n).

Obviously, since $F: \mathcal{S}(\overline{\mathbb{R}}_+) \to H^+$ is an isomorphism, (1) can be considered as pseudo-differential operator with a symbol taking values in $\mathfrak{R}^{m,d}$ or in $F_*\mathfrak{R}^{m,d}$.

In fact (1) defines a pseudo-differential operator with operator valued symbols if we show that the norms of all derivatives of $\mathfrak{a}$ for $|\xi'| \to \infty$ grow in the usual way if on H^+ we take the following norms

$$\|h\|_l = \|\Pi^+ ((|\xi'| - i\nu)^l\, h(\nu))\|_{L^2(\mathbb{R})}, \quad l \in \mathbb{Z}_+ .$$

The completion of H^+ with respect to $\|\cdot\|_l$ is denoted by $(H^+)_l$.

Proposition 1. *Let*

$$\mathfrak{a}(x', \xi') = \begin{pmatrix} \Pi^+ a(x', \xi') + \Pi' b(x', \xi') & k(x', \xi') \\ \Pi' t(x', \xi') & q(x', \xi') \end{pmatrix}$$

where $a \in \mathfrak{A}_1^m((\Omega' \times \overline{\mathbb{R}}_+) \times \mathbb{R}^n)$, $b \in \mathfrak{B}^{m-1,d}(\Omega' \times \mathbb{R}^{n+1})$, $k \in \mathfrak{K}^{m-1/2}(\Omega' \times \mathbb{R}^n)$, $t \in \mathfrak{T}^{m-1/2}(\Omega' \times \mathbb{R}^n)$, $q \in S^m(\Omega' \times \mathbb{R}^{n-1})$. *Then for every* $l \in \mathbb{Z}_+$, $l > d - 1/2$, *arbitrary multi-indices* $\alpha, \beta \in \mathbb{Z}_+^{n-1}$ *and an arbitrary compact set* $K \subset \Omega'$ *there is a constant* $c = c_{l\alpha\beta K}$ *such that*

$$\|D_{x'}^\alpha D_{\xi'}^\beta\, \mathfrak{a}(x', \xi')\|_{\mathcal{L}((H^+)_l \otimes \mathbb{C}^k \oplus \mathbb{C}^j,\, (H^+)_{l-m} \otimes \mathbb{C}^{k'} \oplus \mathbb{C}^{j'})} \leq c \langle \xi' \rangle^{m-|\beta|}$$

for $x' \in K$, $\xi' \in \mathbb{R}^{n-1}$.

Proof. First consider the case $l = m = d = 0$. We can assume all symbols to be scalars. We have $\|\Pi^+ a(x', \xi')\| \leqq \sup_{\nu \in R} |a(x', \xi', \nu)|$ and a decomposition $a(x', \xi', \nu) = \sum_{\alpha \in Z_+} c_\gamma(x', \xi') f_\gamma(\langle \xi' \rangle^{-1} \nu)$, $c_\gamma \in S^0(\Omega' \times R^{n-1})$, $f_\gamma \in H_0$ and the series converges in $\mathfrak{A}^0((\Omega' \times \overline{R}_+) \times R^n)$. Thus it is sufficient to consider $a(x', \xi', \nu) = c(x', \xi') f(\langle \xi' \rangle^{-1} \nu)$, $c \in S^0$, $f \in H_1$. Obviously

$$|D^\alpha_{x'} D^\beta_{\xi'} c(x', \xi')| \leqq c \langle \xi' \rangle^{-|\beta|} \quad \text{for } x' \in K .$$

By induction one easily proves

$$\sup |D^\beta_{\xi'} f(\langle \xi' \rangle^{-1} \nu)| \leqq c \langle \xi' \rangle^{-|\beta|} .$$

The best constants in the estimates continuously depend on a. Then the assertion for $\Pi^+ a(x', \xi')$ follows.

Similarly, for the potential symbol $k(x', \xi', \nu)$ we have

$$\|k(x', \xi')\|_{\mathcal{L}(C, (H^+)_0)} \leqq \|k(x', \xi', \nu)\|_{L^2(R)} .$$

Let $k(x', \xi', \nu) = c(x', \xi') f(\langle \xi' \rangle^{-1} \nu)$, $c \in S^{-1/2}(\Omega' \times R^{n-1})$, $f \in H^+$. Then it follows that

$$\|D^\alpha_{x'} D^\beta_{\xi'} k(x', \xi', \nu)\|_{L^2(R)} \leqq c \langle \xi' \rangle^{-|\beta|} \quad \text{for } x' \in K$$

and by continuity these estimates extend to arbitrary $k \in \mathfrak{K}^{-1/2}$. For the trace symbol $t(x', \xi', \nu) \in \mathfrak{T}^{-1/2, 0}$ we have

$$\|\Pi' t(x', \xi')\|_{\mathcal{L}((H^+)_0, C)} \leqq \|t(x', \xi', \nu)\|_{L^2(R)}$$

and we conclude as above.

For the Green symbol $b(x', \xi', \nu, \tau) \in \mathfrak{B}^{-1, 0}$ of the special form $b(x', \xi', \nu, \tau) = k(x', \xi', \nu)\, t(\xi', \tau)$, $k \in \mathfrak{K}^{-1/2}$, $t \in \mathfrak{T}^{-1/2, 0}$, we have

$$\|\Pi' b(x', \xi')\|_{\mathcal{L}((H^+)_0, (H^+)_0)} \leqq \|b(x', \xi', \nu, \tau)\|_{L^2(R^2)}$$

and the desired estimate follows as before.

For non-vanishing l, m, d reduction of order and type to zero yields

$$\begin{array}{ccc} (H^+)_l \oplus C & \xrightarrow{\ \mathfrak{a}\ } & (H^+)_{l-m} \oplus C \\ \downarrow & & \downarrow \\ (H^+)_0 \oplus C & \xrightarrow{\ \mathfrak{a}_0\ } & (H^+)_0 \quad \oplus C , \end{array}$$

where $\|\mathfrak{a}\|_{\mathcal{L}((H^+)_l \oplus C, (H^+)_{l-m} \oplus C)} = \|\mathfrak{a}_0\|_{\mathcal{L}((H^+)_0 \oplus C, (H^+)_0 \oplus C)}$ (cf. 2.1.2.3.(7)). □

Corollary 2. (i) $b \in \mathfrak{B}^{m-1, d}(\Omega' \times R^{n+1})$ *implies*

$$F_* b(x', \xi') \in S^m(\Omega' \times R^{n-1}, \mathcal{L}(H^l(R_+), H^{l-m}(R_+)))$$

for any $l \in Z_+$, $l > d - 1/2$.

(ii) $k \in \mathfrak{K}^m(\Omega' \times R^n)$ *implies*

$$F_* k(x', \xi') \in S^{m+1/2}(\Omega' \times R^{n-1}, \mathcal{L}(C, H^l(R_+)))$$

for any $l \in Z_+$.

(iii) $t \in \mathfrak{T}^{m, d}(\Omega' \times R^n)$ *implies*

$$F_* t(x', \xi') \in S^{m-1/2}(\Omega' \times R^{n-1}, \mathcal{L}(H^l(R_+), C))$$

for any $l \in Z_+$, $l > d - 1/2$.

The following assertions are immediate consequences of Proposition 1 and the results on operator-valued PDOs formulated in 2.3.2.1.

Corollary 3. *$\mathcal{A} \in \mathrm{Op}\,(\mathfrak{R}^{m,d})(\Omega' \times \mathbb{R}^{n-1};\ \mathbb{C}^k,\ \mathbb{C}^{k'},\ \mathbb{C}^j,\ \mathbb{C}^{j'})$ defines a continuous mapping*

$$\mathcal{A}: C_0^\infty(\Omega' \times \overline{\mathbb{R}}_+,\ \mathbb{C}^k) \oplus C_0^\infty(\Omega',\ \mathbb{C}^j) \to C^\infty(\Omega' \times \overline{\mathbb{R}}_+,\ \mathbb{C}^{k'}) \oplus C^\infty(\Omega',\ \mathbb{C}^{j'}),$$

which can be continuously extended to a mapping

$$\mathcal{A}: H^t_{\mathrm{comp}}(\Omega', H^s(\mathbb{R}_+) \otimes \mathbb{C}^k) \oplus H^{t+1/2}_{\mathrm{comp}}(\Omega', \mathbb{C}^j)$$
$$\to H^{t-m}_{\mathrm{loc}}(\Omega', H^{s-m}(\mathbb{R}_+) \otimes \mathbb{C}^{k'}) \oplus H^{t-m+1/2}_{\mathrm{loc}}(\Omega', \mathbb{C}^{j'})$$

for any $s > d - {}^1\!/_2$ and arbitrary t.

An operator $\mathcal{A} \in \mathrm{Op}\,(\mathfrak{R}^{m,d})$ is called *properly supported* if its distributional kernel, i.e. the matrix of distributional kernels $K_A(x, y)$, $K_B(x, y)$, $K_K(x, y')$, ... , is properly supported. Any $\mathcal{A} \in \mathrm{Op}\,(\mathfrak{R}^{m,d})$ is a sum of a properly supported operator and a smoothing one.

A potential operator $K \in \mathrm{Op}\,(\mathfrak{K}^m)$, for instance, is properly supported iff for any compact set $C \subset \Omega'$ there is a compact set $C_1 \subset \Omega' \times \overline{\mathbb{R}}_+$ such that $\mathrm{supp}\, Kv \subset C_1$ for $\mathrm{supp}\, v \subset C$, $v \in C_0^\infty(\Omega')$, and if an analogous condition is satisfied for the adjoint.

Any operator $\mathcal{A} \in \mathrm{Op}\,(\mathfrak{R}^{m,d})$ given by (1) with an amplitude function $\mathfrak{a}(x', y', \xi')$ with values in the boundary symbols on the axis is a sum of a properly supported and a smoothing operator. For a properly supported $\mathcal{A} \in \mathrm{Op}\,(\mathfrak{R}^{m,d})$ we get a uniquely defined boundary symbol $\mathfrak{a}(\mathcal{A})(x', \xi')$ from $\mathfrak{a}(\mathcal{A})(x', \xi') = e^{-ix'\xi'}\mathcal{A}(e^{i\cdot\xi'})(x')$. Obviously the right hand side is a well defined function with values in the Fourier transforms of the boundary symbols on the axis. Thus, for properly supported operators, we can assume that the amplitude function in (1) depends on x', ξ' only.

Proposition 4. *Let $\mathcal{A}_1 \in \mathrm{Op}\,(\mathfrak{R}^{m_1,d_1})(\Omega' \times \mathbb{R}^{n-1};\ \mathbb{C}^k,\ \mathbb{C}^{k'},\ \mathbb{C}^j,\ \mathbb{C}^{j'})$ and*

$$\mathcal{A}_2 \in \mathrm{Op}\,(\mathfrak{R}^{m_2,d_2})(\Omega' \times \mathbb{R}^{n-1};\ \mathbb{C}^{k'},\ \mathbb{C}^{k''},\ \mathbb{C}^{j'},\ \mathbb{C}^{j''}),$$

one of them properly supported. Then $\mathcal{A} = \mathcal{A}_2\mathcal{A}_1 \in \mathrm{Op}\,(\mathfrak{R}^{m_1+m_2,d})(\Omega' \times \mathbb{R}^{n-1};\ \mathbb{C}^k,\ \mathbb{C}^{k''},\ \mathbb{C}^j,\ \mathbb{C}^{j''})$, $d = \max\,(m_1 + d_2,\ d_1)$ and there is an asymptotic expansion

$$\mathfrak{a}(x', \xi') \sim \sum_\alpha (1/\alpha!)\, \partial^\alpha_{\xi'}\, \mathfrak{a}_2(x', \xi') \circ D^\alpha_{x'}\, \mathfrak{a}_1(x', \xi'),$$

where $\mathfrak{a}_i$ are boundary symbols of $\mathcal{A}_i$ and $\circ$ denotes the composition in the algebra of boundary symbols on the axis $\mathfrak{R}^{m,d}$.

In the following we drop $\circ$ if no confusion is possible.

Let Ω', Ω_1' be open sets in $\mathbb{R}^{n-1}$ and $\varkappa: \Omega' \to \Omega_1'$ a diffeomorphism. Then a mapping $\varkappa^*: C_0^\infty(\Omega_1' \times \overline{\mathbb{R}}_+) \oplus C_0^\infty(\Omega_1') \to C_0^\infty(\Omega' \times \overline{\mathbb{R}}_+) \oplus C_0^\infty(\Omega')$ is induced and the same for C^∞ and vector-valued functions (denoted again by $\varkappa^*$). Define for any $\mathcal{A} \in \mathrm{Op}\,(\mathfrak{R}^{m,d})(\Omega' \times \mathbb{R}^{n-1})$

$$\varkappa_*\mathcal{A} = (\varkappa^*)^{-1}\mathcal{A}\varkappa^*: \begin{matrix} C_0^\infty(\Omega_1 \times \overline{\mathbb{R}}_+,\ \mathbb{C}^k) \\ \oplus \\ C_0^\infty(\Omega_1,\ \mathbb{C}^j) \end{matrix} \to \begin{matrix} C^\infty(\Omega_1 \times \overline{\mathbb{R}}_+,\ \mathbb{C}^{k'}) \\ \oplus \\ C^\infty(\Omega_1,\ \mathbb{C}^{j'}). \end{matrix}$$

Let (x', ξ') and (y', η') be coordinates in $T^*\Omega'$ and $T^*\Omega_1'$, respectively.

Proposition 5. *Any diffeomorphism $\varkappa\colon \Omega' \to \Omega_1'$ induces a bijection*

$$\varkappa_*\colon \operatorname{Op}(\mathfrak{R}^{m,d})(\Omega' \times \mathbb{R}^{n-1}) \to \operatorname{Op}(\mathfrak{R}^{m,d})(\Omega_1' \times \mathbb{R}^{n-1}).$$

If $\mathcal{A} \in \operatorname{Op}(\mathfrak{R}^{m,d})(\Omega' \times \mathbb{R}^{n-1})$ is properly supported with the boundary symbol $\mathfrak{a}(\mathcal{A})(x', \xi')$ the boundary symbol $\mathfrak{a}(\varkappa_*\mathcal{A})(y', \eta')$ of $\varkappa_*\mathcal{A}$ is given by $\mathfrak{a}(\varkappa_*\mathcal{A})(y', \eta') \sim \sum_\alpha (1/\alpha!)\, \partial_{\xi'}^\alpha\, \mathfrak{a}\big(x', {}^t(\mathrm{d}\varkappa)(x')\,\eta'\big)\, D_{z'}^\alpha\, e^{ih(z',x')\eta'}\big|_{z'=x'}$, $x' = \varkappa^{-1}(y')$, $h(z', x') = \varkappa(z') - \varkappa(x') - \mathrm{d}\varkappa(x')(z' - x')$.

Note that all the above assertions remain true if we replace $\operatorname{Op}(\mathfrak{R}^{m,d})$ by $\operatorname{Op}(\mathfrak{R}^{m,d})$, since we remarked in 2.2.5.1 that $\mathfrak{R}^{m,d}$ is closed under the operations of the symbolic calculus.

The natural inclusions $\operatorname{Op}(\mathfrak{B}^{m-1,d})$, $\operatorname{Op}(\mathfrak{R}^m)$, $\operatorname{Op}(\mathfrak{T}^{m-1,d})$, and $\operatorname{Op}(S^m)$ in $\operatorname{Op}(\mathfrak{R}^{m,d})$ yield that similar facts as in Proposition 5 hold for these classes of operators. The precise formulation is left to the reader.

Until now we considered the subclass of operators in $\operatorname{Op}(\mathfrak{S}^{m,d})$ the interior symbols of which are independent of the normal variable x_n. In section 2.3.2.3 we shall completely drop this assumption. Observe that all assertions of this section remain true if the interior symbol is only independent of x_n near $x_n = 0$. In fact, any $a \in \mathfrak{A}_1^m$ is a sum $a(x', x_n, \zeta) = a_1(x', \xi) + a_2(x', x_n, \xi)$ where $a_1(x', \xi) = a(x', 0, \xi)$ and a_2 identically vanishes near $x_n = 0$. The boundary symbol $\Pi^+ a$ is equal to $\Pi^+ a_1$ and a_2 gives no contribution to compositions (they are all smoothing).

2.3.2.3. Relations Between the Various Operator Classes

First we consider relations between potential, trace and pseudo-differential operators with the transmission property.

Example 1. Let $A \in \operatorname{Op}(\mathfrak{A}_1^m)$ be a pseudo-differential operator with the transmission property (the symbol is independent of x_n near $x_n = 0$). For $v \in C_0^\infty(\Omega')$, $\Omega' \subseteq \mathbb{R}^{n-1}$ open,

$$\langle v, \varphi\rangle = \int_{\Omega'} v(x')\,\varphi(x', 0)\,\mathrm{d}x', \quad \varphi \in C^\infty(\Omega' \times \mathbb{R})$$

defines a distribution with compact support denoted by $v \otimes \delta(x_n)$. Define an operator by $C_0^\infty(\Omega') \ni v \mapsto r^+\big(A(v \otimes \delta(x_n))\big) \in C^\infty(\Omega' \times \mathbb{R}_+)$. This is a potential operator. In fact, let A be properly supported with symbol $\sigma(A)(x', \xi)$. Then

$$\begin{aligned} r^+A\big(v \otimes \delta(x_n)\big) &= (2\pi)^{-n}\, r^+ \int e^{ix\xi}\,\sigma(A)(x', \xi)\,\hat{v}(\xi')\,\mathrm{d}\xi \\ &= (2\pi)^{-n} \int e^{ix\xi}\,[\Pi_\nu^+\sigma(A)(x', \xi,' \nu)]\,\hat{v}(\xi')\,\mathrm{d}\xi'\,\mathrm{d}\nu\,. \end{aligned}$$

Since $\Pi^+\colon \mathfrak{A}^m \to \mathfrak{K}^m$, the last expression is a potential operator.

Note that the potential operator defined with a pseudo-differential operator A with the symbol $\sigma(A)$ depends only on $\Pi_\nu^+\sigma(A)(x', 0, \xi', \nu)$.

Proposition 2. *Every potential operator $K \in \operatorname{Op}(\mathfrak{K}^m)$ can be defined as in Example 1, i.e. there is a pseudo-differential operator $A \in \operatorname{Op}(\mathfrak{A}_1^m)$ such that $Kv = r^+A\big(v \otimes \delta(x_n)\big)$ for any $v \in C_0^\infty(\Omega')$.*

Proof: A smoothing potential operator with kernel $k(x', x_n, y') \in C^\infty(\Omega' \times \overline{\mathbb{R}}_+ \times \Omega')$ can be considered as smoothing pseudo-differential operator with this kernel. Now the assertion follows from

Lemma 3. *The projection $\Pi^+ : \mathfrak{A}_1^m \to \mathfrak{K}^m$ is surjective.*

Proof: Denote by $l: C_0^\infty(\overline{\mathbb{R}}_+) \to C_0^\infty(\mathbb{R})$ the Seeley extension operator (cf. the proof of 2.2.2.1, Lemma 1) and set $\hat{l}\hat{u} = F(lu)$. The operators l and $\hat{l}$ commute with $\varkappa_t^*$, $\varkappa_t$: $\mathbb{R} \to \mathbb{R}$, $\varkappa_t(s) = ts$, $t \in \mathbb{R}_+$. For $k(x', \xi', \nu) \in \mathfrak{K}^m$ define

$$a_{[0]}(x', \xi', \nu) = \hat{l}_\nu k_{[0]}(x', y', \nu)\,,$$

where

$$k_{[0]}(x', \xi', \nu) = k(x', \xi', \nu\langle\xi'\rangle) \in H_\nu^+ \otimes S^m(\Omega' \times \mathbb{R}^{n-1})\,.$$

Then $a_{[0]}(x', \xi', \nu) \in H_0 \otimes S^m(\Omega' \times \mathbb{R}^{n-1})$ and as in the second example 2.2.2.1 we see that $a(x', \xi', \nu) = a_{[0]}(x', \xi', \nu\langle\xi'\rangle^{-1})$ belongs to $S^m(\Omega' \times \mathbb{R}^n)$. □

In Example 1 we assumed for simplicity that the symbol of the pseudo-differential operator is independent of x_n near $x_n = 0$. We show now that for an arbitrary $r^+A \in \mathrm{Op}\,(\mathfrak{A}^m)$ the operator

$$C_0^\infty(\Omega') \ni v \mapsto r^+A\big(v \otimes \delta(x_n)\big) \in C^\infty(\Omega' \times \mathbb{R}_+)$$

is a potential operator and evaluate its symbol.

We have

$$\begin{aligned} r^+A\big(v \otimes \delta(x_n)\big) &= (2\pi)^{-n}\, r^+ \textstyle\int e^{i(x'\xi' + x_n\nu)}\, \sigma(A)\,(x', x_n, \xi', v)\, \hat{v}(\xi')\, d\xi' d\nu \\ &= (2\pi)^{-n}\, F_\tau^{-1}\Pi_\tau^+ \left(F_{x_n} \textstyle\int e^{i(x'\xi' + x_n\nu)}\, \sigma(A)\,(x', x_n, \xi', \nu)\, \hat{v}(\xi')\, d\xi'\, d\nu\right) \\ &= (2\pi)^{-n} \textstyle\int e^{i(x'\xi' + x_n\tau)}\, k_A(x', \xi', \tau)\, \hat{v}(\xi')\, d\xi'\, d\tau \end{aligned}$$

where

$$k_A(x', \xi', \tau) = \Pi_\tau^+ \textstyle\int e^{ix_n(\tau - \nu)}\, \sigma(A)|(x', x_n, \xi', \nu)\, d\nu\, dx_n\,.$$

Show $k_A \in \mathfrak{K}^m$. Set $\nu = \nu_1\langle\xi'\rangle$, $\tau = \tau_1\langle\xi'\rangle$. Then

$$k_A(x', \xi', \tau) = \Pi_\tau^+\langle\xi'\rangle \textstyle\int e^{i\langle\xi'\rangle x_n(\nu_1 - \tau_1)}\, \sigma(A)\,(x', x_n, \xi', \nu_1\langle\xi'\rangle)\, d\nu_1\, dx_n\,.$$

$\sigma(A)(x, \xi) \in \mathfrak{A}^m$ implies $\sigma(A)_{[k]}\,(x', \xi', \nu_1) = \partial_{x_n}^k \sigma(A)(x', 0, \xi', \nu_1\langle\xi'\rangle) \in H \otimes S^m(\Omega' \times \mathbb{R}^{n-1})$.
Taylor expansion with respect to $\eta = \nu_1 - \tau_1$ in $\eta = 0$ yields

$$\begin{aligned} k_A(x', \xi', \tau) &= \Pi_\tau^+\langle\xi'\rangle \textstyle\int e^{i\langle\xi'\rangle x_n\eta}\, \sigma(A)_{[0]}(x', x_n, \xi', \tau_1 + \eta)\, d\eta\, dx_n \\ &\sim \Pi_\tau^+\langle\xi'\rangle \sum_j (1/j!) \textstyle\int e^{i\langle\xi'\rangle x_n\eta}\, \eta^j\, \partial_{\tau_1}^j \sigma(A)_{[0]}\,(x', x_n, \xi', \tau_1)\, d\eta\, dx_n \\ &= \sum_j \big((-1)^j/j!\big)\, \langle\xi'\rangle^{-j}\, \Pi_{\tau_1}^+ D_{x_n}^j\, \partial_{\tau_1}^j \sigma(A)_{[0]}\,(x', x_n, \xi', \tau_1)|_{x_n=0,\, \tau_1=\tau\langle\xi'\rangle^{-1}} \\ &= \sum_j \big((-1)^j/j!\big)\, \Pi_\tau^+ D_{x_n}^j\, \partial_\tau^j \sigma(A)\,(x', 0, \xi', \tau)\,. \end{aligned}$$

The remainder estimate necessary for justification of $\sim$ is standard and left to the reader.

The above considerations prove the following

Proposition 4. *Any operator*

$$Kv(x', x_n) = (2\pi)^{-n}\, r^+ \textstyle\int e^{i(x'\xi' + x_n\nu)}\, k(x', x_n, \xi', \nu)\, \hat{v}(\xi')\, d\xi'\, d\nu\,,$$

$v \in C_0^\infty(\Omega')$, where

$$D_{x_n}^j k(x', x_n, \xi', \nu)|_{x_n=0} \in \mathfrak{K}^m \quad \textit{for all} \quad j \in \mathbb{Z}_+\,,$$

is a potential operator $\in \mathrm{Op}(\mathfrak{K}^m)$ with the symbol asymptotically equal to

$$\sum_j \big((-1)^j/j!\big)\, D_{x_n}^j\, \partial_\tau^j k(x', 0, \xi', \nu)\,. \tag{1}$$

Let $A \in \mathrm{Op}(\mathfrak{A}_1^m)$ with the symbol $\sigma(A)(x', \xi', \nu) \in \mathfrak{A}_1^m$. Then the composition $r' \circ r^+A$ defines a trace operator $r'T_A\colon C_0^\infty(\Omega' \times \overline{\mathbb{R}}_+) \to C^\infty(\Omega')$, $r'T_A \in \mathrm{Op}(\mathfrak{T}^{m, m+1})$ with the symbol $t(x', \xi', \nu) = \Pi_\nu^- \sigma(A)(x', \xi', \nu)$. In fact, we have

$$\begin{aligned} r' \circ r^+Au(x') &= (2\pi)^{-n}\, r'_{x_n} r^+ \int e^{i(x'\xi' + x_n\nu)}\, \sigma(A)(x', \xi', \nu)\, \widehat{j^+u}(\xi', \nu)\, d\xi'\, d\nu \\ &= (2\pi)^{-n+1} \int e^{ix'\xi'}\, \Pi'_\nu\big(\sigma(A)(x', \xi', \nu)\, \widehat{j^+u}(\xi', \nu)\big)\, d\xi' \\ &= (2\pi)^{-n+1} \int e^{ix'\xi'}\, \Pi'_\nu\big((\Pi_\nu^-\sigma(A)(x', \xi', \nu))\, \widehat{j^+u}(\xi', \nu)\big)\, d\xi'\,, \\ & u \in C_0^\infty(\Omega' \times \overline{\mathbb{R}}_+)\,. \end{aligned}$$

Obviously complex conjugation yields an isomorphism $\mathfrak{K}^m \to \mathfrak{T}^{m,0}$ and $\Pi^+\colon \mathfrak{A}_1^m \times \mathfrak{K}^m$ transforms into $\Pi_0^-\colon \mathfrak{A}_1^m \to \mathfrak{T}^{m,0}$. Thus, by Lemma 3, we have

Corollary 5. *Any trace operator* $r'T \in \mathrm{Op}(\mathfrak{T}^{m,0})$ *of type zero admits a representation* $r'T = r' \circ r^+A$, $r^+A \in \mathrm{Op}(\mathfrak{A}_1^m)$.

For trace operators of positive type Corollary 5 is not valid in general. Take, for instance, $t(x', \xi', \nu) \in \mathfrak{T}^{m,d}$, $d > 0$ such that $t \notin S^m(\Omega' \times \mathbb{R}^n)$.

Similarly as for potential operators it is natural to consider $r' \circ r^+A$ for $A \in \mathrm{Op}(\mathfrak{A}^m)$, i.e. for a pseudo-differential operator with a symbol depending on x_n. In the same way as above one can show that this is a trace operator with the symbol

$$t(x', \xi', \nu) \sim \sum_j \big((-1)^j/j!\big)\, \Pi_\nu^- D_{x_n}^j\, \partial_\nu^j \sigma(A)(x', 0, \xi', \nu) \tag{2}$$

if $\sigma(A) \in \mathfrak{A}^m\big((\Omega' \times \overline{\mathbb{R}}_+) \times \mathbb{R}^n\big)$ is a symbol of A. (2) follows from the case of potential operators via the adjoint operator (cf. calculations in this section below).

Thus we obtain

Proposition 6. *Any operator*

$$r'Tu(x') = (2\pi)^{-n}\, r'_{x_n} \int e^{i(x'\xi' + x_n\nu)}\, t(x', x_n, \xi', \nu)\, \widehat{j^+u}(\xi', \nu)\, d\xi'\, d\nu\,,$$

$u \in C_0^\infty(\Omega' \times \overline{\mathbb{R}}_+)$, *where*

$$D_{x_n}^j t(x', x_n, \xi', \nu)|_{x_n=0} \in \mathfrak{T}^{m,d} \quad \textit{for all} \quad j \in \mathbb{Z}_+\,,$$

is a trace operator $\in \mathrm{Op}(\mathfrak{T}^{m,d})$ *with the symbol asymptotically equal to*

$$\sum_j \big((-1)^j/j!\big)\, D_{x_n}^{j\,1}\, \partial_\nu^j t(x', x_n, \xi', \nu)|_{x_n=0}\,. \tag{3}$$

As observed in 2.2.4.1 any Green symbol $b(x', \xi', \nu, \tau) \in \mathfrak{B}^{m,d}(\Omega' \times \mathbb{R}^{n-1})$ has a decomposition

$$b(x', \xi', \nu, \tau) = \sum_{j=1}^\infty k_j(x', \xi', \nu)\, t_j(\xi', \tau)$$

where $k_j \in \mathfrak{K}^m$ and $t_j \in \mathfrak{T}^{0,d}$, and the series absolutely converges in $\mathfrak{B}^{m,d}$. Let K_j and $r'T_j$ be the potential and trace operators defined with k_j and t_j by 2.3.2.1. (1) and (4), respectively. The composition $K_j r'T_j$ obviously has the Green symbol $k_j(x', \xi', \nu)\, t_j(\xi', \tau)$, since t_j does not depend on x'. Thus we have

Lemma 7. *Each Green operator* $r'B \in \mathrm{Op}(\mathfrak{B}^{m,d})$ *has a decomposition*

$$r'B = \sum_j K_j r'T_j\,, \quad K_j \in \mathrm{Op}(\mathfrak{K}^m)\,,\ r'T_j \in \mathrm{Op}(\mathfrak{T}^{0,d})\,,$$

where the corresponding series of symbols $\sum_j k_j(x', \xi', \nu)\, t_j(x', \xi', \tau)$ *converges in* $\mathfrak{B}^{m,d}$.

Similar to Lemma 3 we have

Lemma 8. *The projections*

$$\Pi_\nu^+ : \mathfrak{A}_1^m \otimes \mathfrak{T}^{0,d} \to \mathfrak{B}^{m,d},$$

$$\Pi_{0\tau}^- : \mathfrak{K}^m \otimes \mathfrak{A}_1^0 \to \mathfrak{B}^{m,0},$$

and

$$\Pi_\nu^+ \Pi_{0\tau}^- : \mathfrak{A}_1^m \otimes \mathfrak{A}_1^0 \to \mathfrak{B}^{m,0}$$

are surjective.

From Lemma 7 and Proposition 2 we get a decomposition of any Green operator $r'B \in \mathrm{Op}\,(\mathfrak{B}^{m,d})$

$$r'Bu = \sum_j r^+A_j\big((r'T_ju) \otimes \delta(x_n)\big), \quad u \in C_0^\infty(\Omega' \times \overline{\boldsymbol{R}}_+),$$

where $r^+A_j \in \mathrm{Op}\,(\mathfrak{A}_1^m)$ and $r'T_j \in \mathrm{Op}\,(\mathfrak{T}^{m,d})$.

For $r^+A \in \mathrm{Op}\,(\mathfrak{A}^m)$ (the symbol may depend on x_n near $x_n = 0$) and $r'T \in \mathrm{Op}\,(\mathfrak{T}^{0,d})$ we have $r^+A\big(r'T \otimes \delta(x_n)\big) \in \mathrm{Op}\,(\mathfrak{B}^{0,d})$.

Proposition 9. *Any operator*

$$r'Bu(x', x_n) = (2\pi)^{-n}\, r_{x_n}^+ \int e^{i(x'\xi' + x_n\nu)}\, \Pi_\tau'\big(b(x', x_n, \xi', \nu, \tau)\, \widehat{j^+u}(\xi', \tau)\big)\, d\xi'\, d\nu,$$

$u \in C_0^\infty(\Omega' \times \overline{\boldsymbol{R}}_+)$, *where*

$$D_{x_n}^j b(x', x_n, \xi', \nu, \tau)|_{x_n=0} \in \mathfrak{B}^{m,d} \quad \textit{for all} \quad j \in \boldsymbol{Z}_+$$

is a Green operator $\in \mathrm{Op}\,(\mathfrak{B}^{m,d})$ *with a symbol asymptotically equal to*

$$\sum_j \big((-1)^j/j!\big)\, D_{x_n}^j \partial_\nu^j b(x', 0, \xi', \nu, \tau). \tag{4}$$

Denote by $(\cdot,\cdot)_{\boldsymbol{C}^n}$ the Hermitean scalar product on $\boldsymbol{C}^n$. Then by

$$(u, v)_{L^2(X, \boldsymbol{C}^n)} = \int_X \big(u(x), v(x)\big)_{\boldsymbol{C}^n}\, dx$$

there is defined a scalar product on $L^2(X, \boldsymbol{C}^n)$ (X a manifold with smooth positive density dx).

For $(\cdot,\cdot)_{L^2(\Omega' \times \boldsymbol{R}_+, \boldsymbol{C}^k)}$, $k \in \boldsymbol{Z}_+$, $\Omega' \subseteq \boldsymbol{R}^{n-1}$ open, write simply $(\cdot,\cdot)$ and for $(\cdot,\cdot)_{L^2(\Omega', \boldsymbol{C}^k)}$ shorter $(\cdot,\cdot)'$.

Consider a trace operator

$$r'T : C_0^\infty(\Omega' \times \overline{\boldsymbol{R}}_+, \boldsymbol{C}^k) \to C^\infty(\Omega', \boldsymbol{C}^j),$$

$r'T \in \mathrm{Op}\,(\mathfrak{T}^{m,d})$. The adjoint $(r'T)^*$ defined by

$$(r'Tu, v)' = \big(u, (r'T)^* v\big), \quad u \in C_0^\infty(\Omega' \times \boldsymbol{R}_+, \boldsymbol{C}^k), \quad v \in C_0^\infty(\Omega', \boldsymbol{C}^j)$$

turns out to be a potential operator of order m. Point out that the support of u is assumed to be in the open set $\Omega' \times \boldsymbol{R}_+ \in \Omega' \times \boldsymbol{R}$, i.e. u identically vanishes near $x_n = 0$.

In fact, the adjoint of a smoothing trace operator is a smoothing potential operator depending only on the part of type zero of $r'T$. Then we can assume that $r'T$ is defined by 2.3.2.1. (4) with symbol $t(x', \xi', \nu) \in \mathfrak{T}^{m,d}$ and $r'Tu = r'T_0u$ for any $u \in C_0^\infty(\Omega' \times \boldsymbol{R}_+, \boldsymbol{C}^k)$, where $r'T_0$ is defined with $t_0(x', \xi', \nu) = \Pi_0^-\big(t(x', \xi', \nu)\big)$.

Then we have

$$\begin{aligned}&(r'Tu, v)'\\ &= (2\pi)^{-n} \int e^{i(x'-y')\xi' - iy_n\nu} \left(t_0(x', \xi', \nu)\, j^+u(y', y_n), v(x')\right)_{C^k} d\xi'\, d\nu\, dy\, dx'\\ &= (2\pi)^{-n} \int \left(j^+u(y', y_n), \int e^{i(y'-x')\xi' + iy_n\nu}\, t_0^*(x', \xi', \nu)\, v(x')\, d\xi'\, d\nu\, dx'\right)_{C^k} dy,\end{aligned}$$

where $t_0^*(x', \xi', \nu)$ is the adjoint matrix of $t_0(x', \xi', \nu)$, i.e. $t_0^*(x', \xi', \nu) = {}^t\bar{t}_0(x', \xi', \nu)$ and therefore $t_0^* \in \mathfrak{K}^m$. Per def. $(r'T)^* \in \mathrm{Op}(\mathfrak{K}^m)$. In the same way we find that the adjoint K^* of a potential operator $K \in \mathrm{Op}(\mathfrak{K}^m)$ is a trace operator of type zero in $\mathrm{Op}(\mathfrak{T}^{m,0})$.

Consider now Green operators. For

$$r'B = Kr'T \in \mathrm{Op}(\mathfrak{B}^{m,d}), \quad K \in \mathrm{Op}(\mathfrak{K}^m), \quad r'T \in \mathrm{Op}(\mathfrak{T}^{0,d})$$

we have $(r'B)^* = (r'T)^* K^* \in \mathrm{Op}(\mathfrak{B}^{m,0})$ since $(r'T)^* \in \mathrm{Op}(\mathfrak{K}^0)$, $K^* \in \mathrm{Op}(\mathfrak{T}^{m,0})$. For arbitrary Green operators the same follows from Lemma 7.

Finally let $r^+A \in \mathrm{Op}(\mathfrak{A}_1^m)$ with the symbol $a(x', \xi', \nu)$. Then the adjoint r^+A^* defined by $(r^+Au_1, u_2) = (u_1, r^+A^*u_2)$, $u_1 \in C_0^\infty(\Omega' \times \mathbb{R}_+, \mathbb{C}^k)$, $u_2 \in C_0^\infty(\Omega' \times \mathbb{R}_+, \mathbb{C}^{k'})$ belongs to $\mathrm{Op}(\mathfrak{A}_1^m)$, since

$$\begin{aligned}&(r^+Au_1, u_2)\\ &= (2\pi)^{-n} \int e^{i(x'-y')\xi' + ix_n\nu} \left(\Pi_\nu^+(\sigma(A)(x', \xi', \nu)\, j^+u_1(y', \hat{\nu}), j^+u_2(x', x_n))\right)_{C^k} d\zeta'\, d\nu\, dx\, dy'\\ &= (2\pi)^{-n} \int \left(j^+u_1(y', y_n), \int e^{i(y'-x')\xi' + iy_n\nu}\, \Pi_\nu^+(\sigma(A)^*(x', \xi', \nu)\, j^+u_2(x, \hat{\nu}))\, d\xi'\, d\nu\, dx'\right)_{C^k} dy'\, dy_n,\end{aligned}$$

i.e. $(r^+A^*) = r^+A^*$ where A^* has the symbol $\sigma(A)^*(x', \xi', \nu) \in \mathfrak{A}_1^m$ (this is the formal, adjoint of A in the open domain $\Omega' \times \mathbb{R}_+$).

Thus, in view of 2.1.2.3, Proposition 6, we have proved

Proposition 10. *Let* $\mathcal{A} \in \mathrm{Op}(\mathfrak{R}^{m,d})$ *and let* $\mathcal{A}^*$ *be defined by*

$$(\mathcal{A}u_1, u_2) = (u_1, \mathcal{A}^*u_2),$$

$u_1 \in C_0^\infty(\Omega' \times \mathbb{R}_+, \mathbb{C}^k) \oplus C_0^\infty(\Omega', \mathbb{C}^j)$, $u_2 \in C_0^\infty(\Omega' \times \mathbb{R}_+, \mathbb{C}^{k'}) \oplus C_0^\infty(\Omega', \mathbb{C}^{j'})$ *and the scalar products on* $L^2(\Omega' \times \mathbb{R}_+, \mathbb{C}^{k'}) \oplus L^2(\Omega', \mathbb{C}^{j'})$ *and* $L^2(\Omega' \times \mathbb{R}_+, \mathbb{C}^k) \oplus L^2(\Omega', \mathbb{C}^j)$, *respectively. Then* $\mathcal{A}^* \in \mathrm{Op}(\mathfrak{R}^{m+1,0})$ *only depends on the part of type* 0 *of* $\mathcal{A}$ *and is given by*

$$\mathcal{A}^* = \begin{pmatrix} r^+A^* + r'B^* & (r'T)^* \\ K^* & Q^* \end{pmatrix}.$$

Corollary 11. *Let* $\mathcal{A} \in \mathrm{Op}(\mathfrak{S}^{m,d})$. *Then* $\mathcal{A}^*$ *defined as in Proposition* 10 *belongs to* $\mathrm{Op}(\mathfrak{S}^{m+1,0})$. *For* $\mathcal{A} \in \mathfrak{G}^{0,0}$ *we have* $\mathcal{A}^* \in \mathfrak{G}^{1,0}$ *and the principal symbol* $\sigma^0(\mathcal{A}^*)$ *is given by* $(\sigma^0(A)^*, \sigma^0_{\mathbb{R}^{n-1}}(\mathcal{A}^*))$ *if* $(\sigma^0(A), \sigma^0_{\mathbb{R}^{n-1}}(\mathcal{A}))$ *is the principal symbol of* $\mathcal{A}$ (*for* $\sigma^0(A)$ *the adjoint matrix is taken and for* $\sigma^0_{\mathbb{R}^{n-1}}(\mathcal{A})$ *the adjoint boundary symbol in the sense of* 2.2.5.1, *Proposition* 4).

Since it is sufficient to consider the operator components of $\mathcal{A}$ separately, the assertion follows from 1.2.3.3, Remark 2 and Proposition 10.

Now we are in the position to consider compositions of operators in $\mathrm{Op}(\mathfrak{S}^{m,d})$ without the assumption that the interior symbol is independent of x_n near $x_n = 0$. Let $A_i \in \mathrm{Op}(\mathfrak{A}_1^{m_i})$ be properly supported, $i = 1, 2$. Then we have $r^+A_1A_2 - r^+A_1r^+A_2$

$\in \mathrm{Op}\,(\mathfrak{B}^{m_1+m_2-1,\,m_2})$ and the Green symbol is given by

$$\Pi_\tau^- \sum_\alpha (1/\alpha!)\,\frac{\partial_{\xi'}^\alpha \sigma(A_1)^+(x',\xi',\nu) - \partial_{\xi'}^\alpha\,\sigma(A_1)^+(x',\xi',\tau)}{\mathrm{i}(\nu-\tau)}\,D_{x'}^\alpha \sigma(A_2)^-(x',\xi',\tau)\,, \tag{5}$$

$$\sigma(A_1)^+(x',\xi',\nu) = \Pi_\nu^+\big(\sigma(A_1)(x',\xi',\nu)\big),\ \sigma(A_2)^-(x',\xi',\nu) = \Pi_\nu^-\big(\sigma(A_2)(x',\xi',\nu)\big)\,.$$

This follows from the composition formula of boundary symbols on the line 2.1.2.3, Lemma 5. Now let A_i have symbols $\sigma(A_i)(x', x_n, \xi', \nu)|$ depending on x_n. Using the decomposition

$$\sigma(A)(x', x_n, \xi', \nu) = \sum_{j=0}^{\infty} (x_n^j/j!)\,\partial_{x_n}^j \sigma(A)(x', 0, \xi', \nu)\,\varphi(t_j x_n) + \sigma(A')(x', x_n, \xi', \nu)\,,$$

$\varphi \in C_0^\infty(\overline{\mathbb{R}}_+)$, $\varphi = 1$ near $x_n = 0$, t_j increasing sufficiently fast, where $\sigma(A')$ vanishes of infinite order at $x_n = 0$ (cf. 2.2.2.2, Lemma 3). Note that the series converges in $\mathfrak{A}^m\big((\Omega' \times \overline{\mathbb{R}}_+) \times \mathbb{R}^n\big)$. For the components of the decomposition we have $r^+A_1A_2 - r^+A_1r^+A_2 \in \mathrm{Op}\,(\mathfrak{B}^{m_1+m_2-1,\,m_2})$. In fact, take $\sigma(A_1)(x', \xi', \nu) \in \mathfrak{A}_1^{m_1}$ and

$$\sigma(A_2)(x', x_n, \xi', \nu) = x_n^j \sigma(A_2)(x', 0, \xi, \nu)\,\varphi(x_n)\,, \quad \varphi \in C_0^\infty(\overline{\mathbb{R}}_+)\,,\ \varphi = 1$$

near $x_n = 0$. We can drop φ, since the behaviour of $\sigma(A_2)$ outside a neighbourhood of $x_n = 0$ yields only smoothing operators. Taking for r^+A_2 the representation with the symbol $x_n^j\widetilde{\sigma(A_2)}(y', \xi', \nu)$ we get

$$\begin{aligned} & x_n^j\big((A_2 j^+ u)(x', x_n) - r^+A_2(j^+u)(x', x_n)\big) \\ &= (2\pi)^{-n} \int \mathrm{e}^{\mathrm{i}(x'-y')\xi' + \mathrm{i}x_n\nu}\, D_\nu^j\big(\widetilde{\sigma(A_2)}(y', \xi', \nu)\, \widehat{j^+u}(y', \nu) \\ &\quad - \Pi^+\widetilde{\sigma(A_2)}(y', \xi', \nu)\, \widehat{j^+u}(y', \nu)\big)\,\mathrm{d}y'\mathrm{d}\xi'\,\mathrm{d}\nu\,. \end{aligned}$$

Differentiation with respect to ν yields operators with x_n independent symbol applied to some ν derivative of $\widehat{j^+u}(y', \nu)$. Since the composition of the multiplication by x_n^j and $r'B \in \mathrm{Op}\,(\mathfrak{B}^{m,\,d})$ is a Green operator with the symbol $D_\tau^j b(x', \xi', \nu, \tau) \in \mathfrak{B}^{m-j,\,d-j}$, where b is the symbol of $r'B$, we obtain $r^+A_1A_2 - r^+A_1r^+A_2 \in \mathrm{Op}\,(\mathfrak{B}^{m_1+m_2-1,\,m_2})$. For general $r^+A_1 \in \mathrm{Op}\,(\mathfrak{A}^{m_1})$ and $r^+A_2 \in \mathrm{Op}\,(\mathfrak{A}^{m_2})$ the same follows, since the Green symbols depend continuously on the symbols of A_1, A_2. The symbol $\sigma(A')$ yields smoothing operators.

Note that for classical PDOs $r^+A_1 \in \mathrm{Op}\,(\mathfrak{A}^{m_1})$, $r^+A_2 \in \mathrm{Op}\,(\mathfrak{A}^{m_2})$ properly supported with principal symbols σ_{A_1}, σ_{A_2}, the principal symbol of the Green operator $r^+A_1A_2 - r^+A_1\,r^+A_2 \in \mathrm{Op}\,(\mathfrak{B}^{m_1+m_2-1,\,m_2})$ is given by

$$\Pi_\nu^+\Pi_\tau^-\left(\frac{\sigma_{A_1}^+(x', 0, \xi', \nu) - \sigma_{A_1}^+(x', 0, \xi', \tau)}{\mathrm{i}(\nu-\tau)}\,\sigma_{A_2}^-(x', 0, \xi', \tau)\right)$$

where $\sigma_{A_1}^+(x', 0, \xi', \nu) = \Pi_\nu^+\big(\sigma_{A_1}(x', 0, \xi', \nu)\big)$, $\sigma_{A_2}^-(x', 0, \xi', \tau) = \Pi_\tau^-\big(\sigma_{A_2}(x', 0, \xi', \nu)\big)$.

For $r^+A \in \mathrm{Op}\,(\mathfrak{A}^{m_1})$, $K \in \mathrm{Op}\,(\mathfrak{K}^{m_2})$, again using the composition and continuity we obtain $r^+AK \in \mathrm{Op}\,(\mathfrak{K}^{m_1+m_2})$. Passing to adjoints we get for $r'T \in \mathrm{Op}\,(\mathfrak{T}^{m_2,\,0})$ that the composition $r'T\,r^+A \in \mathrm{Op}\,(\mathfrak{T}^{m_1+m_2,\,m_1+1})$. For $r'T = Qr'D_{x_n}^j$, $Q \in L^{m_2-j}(\Omega')$ it is obvious that $r'T\,r^+A \in \mathrm{Op}\,(\mathfrak{T}^{m_1+m_2,\,m_1+j+2})$. Using Lemma 7 we get for any Green operator $r'B \in \mathrm{Op}\,(\mathfrak{B}^{m_2,\,d})$ that the operators $r^+A\,r'B$ and $r'B\,r^+A$ belong to $\mathrm{Op}\,(\mathfrak{B}^{m_1+m_2})$.

We have the following asymptotic expansions for the symbols of the composition

$$\sigma(r^+AK) \sim \sum_{\beta\in\mathbb{Z}_+^{n+1}} \Pi_\nu^+(1/\beta!)\, \partial_{\xi'}^{\beta'}\partial_\nu^{\beta_n}(-D_{x_n})^{\beta_n+\beta_{n+1}}\, \sigma(A)(x',0,\xi',\nu)$$
$$\times\, \partial_\nu^{\beta_{n+1}} D_{x'}^{\beta'}\sigma(K)(x',\xi',\nu),$$

$$\sigma(r'Tr^+A) \sim \sum_{\alpha\in\mathbb{Z}_+^n} \Pi_\nu'(1/\alpha!)\, \partial_{\xi'}^{\alpha'}\partial_\nu^{\alpha_n}\, \sigma(T)(x',\xi',\nu)\, D_{x'}^{\alpha'}D_{x_n}^{\alpha_n}\sigma(A)(x',0,\xi',\nu),$$

$$\sigma(r^+Ar'B) \sim \sum_{\beta\in\mathbb{Z}_+^{n+1}} \Pi_\nu^+\Pi_\tau^-(1/\beta!)\, \partial_{\xi'}^{\beta'}\partial_\nu^{\beta_n}\, (-D_{x_n})^{\beta_n+\beta_{n+1}}\, \sigma(A)(x',0,\xi',\nu)$$
$$\times\, \partial_\nu^{\beta_{n+1}} D_{x'}^{\beta'}\sigma(B)(x',\xi',\nu,\tau),$$

$$\sigma(r'Br^+A) \sim \sum_{\alpha\in\mathbb{Z}_+^n} \Pi_\nu^+\Pi_\tau^-(1/\alpha!)\, \partial_{\xi'}^{\alpha'}\partial_\tau^{\alpha_n}\, \sigma(B)(x',\xi',\nu,\tau)\, D_{x'}^{\alpha'}D_{x_n}^{\alpha_n}\sigma(A)(x',0,\xi',\nu).$$

The simple proof based on the above considerations is left to the reader.

Note that for classical operators in Op $(\mathfrak{K}^{m_2})$, Op $(\mathfrak{T}^{m_2,d})$, Op $(\mathfrak{B}^{m_2,d})$ and Op $(\mathfrak{A}^{m_1})$ the compositions are in the corresponding classical spaces of operators and the principal symbol of the composition is equal to the composition of the principal symbols in the sense of boundary symbols on the axis.

Therefore we proved the following

Theorem 12. *Let $\mathcal{A}_i \in$ Op $(\mathfrak{S}^{m_i,d_i})$, $i = 1, 2$, and let the composition $\mathcal{A}_1\mathcal{A}_2$ be defined. Then $\mathcal{A}_1\mathcal{A}_2 \in$ Op $(\mathfrak{S}^{m_1+m_2,d})$, $d = \max(d_1 + m_2, d_2)$. For $\mathcal{A}_i \in \mathfrak{S}^{m_i,d_i}$ it follows that $\mathcal{A}_1\mathcal{A}_2 \in \mathfrak{S}^{m_1+m_2,d}$ and the principal symbol $\sigma^0(\mathcal{A}_1\mathcal{A}_2)$ is given by*

$$\sigma^0(\mathcal{A}_1\mathcal{A}_2) = \sigma^0(\mathcal{A}_1)\,\sigma^0(\mathcal{A}_2),$$

where $\sigma^0(\mathcal{A}_i)$ are the principal symbols of $\mathcal{A}_i$ and the composition is taken in $\mathfrak{S}$, i.e. it is the matrix multiplication for the interior symbols and composition in the boundary symbols.

Corollary 11 remains true if the scalar products are defined with smooth functions $\psi\colon \Omega' \to \mathrm{GL}(k, \mathbb{C})$ with values in the positive definite Hermitean matrices as

$$(u, v)_\psi = \int_{\Omega'\times\mathbb{R}_+} \big(\psi(x')\, u(x', x_n), v(x', x_n)\big)_{\mathbb{C}^k}\, dx'\, dx_n$$

(similar for $L^2(\Omega', \mathbb{C}^j)$). This follows from the composition of principal boundary symbols.

Remark 13. Note the following useful property of the asymptotic expansions of the symbolic calculus. Let $A \in L^m$, $B \in L^{m'}$ be properly supported and $\sigma(A)$, $\sigma(B)$ their symbols defined by 1.2.3.1. (14). Then AB is properly supported with the symbol $\sigma(AB)$ and for any $N \in \mathbb{Z}_+$ the mapping $S^m \times S^{m'} \to S^{m+m'-N}$ defined by

$$\big(\sigma(A), \sigma(B)\big) \mapsto \sigma(AB) - \sigma(A)\circ\sigma(B)|_N \overset{\text{def}}{=} \sigma(AB) - \sum_{|\alpha|<N} (1/\alpha!)\, \partial_\xi^\alpha\, \sigma(A)\, D_x^\alpha\sigma(B)$$

is continuous with respect to $\sigma(A)$ and $\sigma(B)$. Similar assertions hold for all asymptotic expansions in the symbolic calculus (cf. Hörmander [7]).

Now it is not hard but rather lengthy to check this property for the symbolic calculus for operators in $\mathfrak{S}^{m,d}$.

2.3.2.4. H^s Continuity

In this section we describe the continuity of potential, trace and Green operators defined in 2.3.2.1 in the Sobolev spaces $H^s = H^{s,2}$. Moreover we consider continuous extensions of PDOs with transmission property in the half space.

Proposition 1. *Define PDOs $r^+\Lambda'^m_\pm$ in the half space with symbols $(\langle\xi'\rangle \pm i\nu)^m$, $m \in \mathbb{R}$. Then*

$$r^+ \Lambda'^m_-: H^t(\mathbb{R}^n_+) \to H^{t-s}(\mathbb{R}^n_+)$$

and

$$r^+\Lambda'^m_+: H^t_0(\overline{\mathbb{R}}^n_+) \to H^{t-s}_0(\overline{\mathbb{R}}^n_+)$$

are isomorphisms for all $t \in \mathbb{R}$.

Proof: We have $r^+\Lambda'^m_- r^+\Lambda'^{-m}_- = 1$. In fact, let $q \leqq t$ be a real number and consider a continuous operator

$$l: H^t(\mathbb{R}^n_+) \to H^q(\mathbb{R}^n)$$

with the property $lu|_{\mathbb{R}^n_+} = u$, $u \in H^t(\mathbb{R}^n_+)$. Then, for $q = m$, we have per def. $r^+\Lambda'^m_- u = r^+\Lambda'^m_- j^+ u$ and the last expression is equal to $r^+\Lambda'^m_- lu$ (cf. 1.2.1.2, Proposition 8). Now $\Lambda'^m_- H^t(\mathbb{R}^n) \to H^{t-m}(\mathbb{R}^n)$ is continuous and $r^+: H^{t-m}(\mathbb{R}^n) \to H^{t-m}(\mathbb{R}^n_+)$, too. Thus $r^+\Lambda'^m_-$ has the asserted continuity property. Since $C_0^\infty(\overline{\mathbb{R}}^n_+)$ is dense in $H^t(\mathbb{R}^n_+)$, the first isomorphism is proved.

The assertion about $r^+\Lambda'^m_+$ follows by duality, since we have $\left(H^m(\mathbb{R}^n_+)\right)' \cong H_0^{-m}(\overline{\mathbb{R}}^n_+)$ defined by the Hermitean L^2 pairing and $r^+\Lambda'^m_+$ is the adjoint of $r^+\Lambda'^m_-$. □

In 2.3.2.3 we considered adjoints with respect to the L^2 scalar products. For the class $\mathfrak{G}^{m,d}$ it is more convenient to work in $H^s(\mathbb{R}^n_+) \oplus H^{s+1/2}(\mathbb{R}^{n-1})$, $s \in \mathbb{Z}_+$, with the scalar products

$$(u \oplus v, u' \oplus v')_s = (r^+\Lambda^s_- u \oplus \Lambda'^{s+1/2} v, r^+\Lambda^s_- u' \oplus \Lambda'^{s+1/2} v')_{L^2},$$

$u, u' \in H^s(\mathbb{R}^n_+)$, $v, v' \in H^{s+1/2}(\mathbb{R}^{n-1})$ and Λ' the PDO with symbol $\langle\xi'\rangle$. Unfortunately the operators in Proposition 1 are not PDOs in $\mathbb{R}^n$. In fact, the derivatives $\partial^\alpha_{\xi'}(\langle\xi'\rangle \pm i\nu)^m$ can be estimated by $c(\langle\xi'\rangle + |\nu|)^m \langle\xi'\rangle^{-|\alpha|}$. This is weaker than the desired S^m estimate. One can consider the operators $r^+\Lambda'^m_-$, $r^+\Lambda'^m_+$ as PDOs with continuous (but non-smooth) principal symbol $(|\xi'| \pm i\nu)^m$. A theory of those PDOs in the half space will be developed in 2.4. In order to have the complete symbolic calculus we shall replace $r^+\Lambda'^m_\pm$ by a suitable operator in the class $\mathfrak{A}^m$ (for $m \in \mathbb{Z}$).

Let χ_ε be a real-valued function in $C^\infty(\overline{\mathbb{R}}_+)$, $0 \leqq \chi_\varepsilon(t) \leqq 1$ for $t \in \overline{\mathbb{R}}_+$, $\chi_\varepsilon(t) = 0$ when $t \leqq \varepsilon$, $\chi_\varepsilon(t) = 1$ when $t \geqq 2\varepsilon$. The constant $\varepsilon > 0$ will be chosen later on in a suitable way. Set $\delta_\varepsilon(\xi', \nu) = \chi_\varepsilon\left(\dfrac{|\xi'|^2}{|\xi'|^2 + \nu^2}\right)$. Then $\delta_\varepsilon(\xi', \nu)\,|\xi'| \pm i\nu$ is a smooth function on the sphere $|\xi| = 1$. For $m \in \mathbb{Z}$ the functions $(\delta_\varepsilon(\xi', \nu)\,|\xi'| \pm i\nu)^m$ are positively homogeneous and smooth for $(\xi', \nu) \neq 0$. Hence $\psi(\xi', \nu)\,\left(\delta_\varepsilon(\xi', \nu)\,|\xi'| \pm i\nu\right)^m$, ψ an excision function, is in S^m and even in $\mathfrak{A}^m$, since for any fixed $\xi' \neq 0$ $\delta_\varepsilon(\xi, \nu)$ vanishes for large ν. Denote the corresponding PDO by $\Lambda^m_\pm(\varepsilon)$ and $r^+\Lambda^m_\pm(\varepsilon)$ if they are considered in the half space.

Proposition 2. *The operators $r^+\Lambda^m_\pm(\varepsilon) \in \mathrm{Op}\,(\mathfrak{A}^m)$, $m \in \mathbb{Z}$ define isomorphisms*

$$r^+\Lambda^m_-(\varepsilon): H^t(\mathbb{R}^n_+) \to H^{t-m}(\mathbb{R}^n_+),$$

and

$$r^+\Lambda_+^m(\varepsilon): H_0^t(\overline{\mathbb{R}}_+^n) \to H_0^{t-m}(\overline{\mathbb{R}}_+^n), \quad t > -1/2,$$

for $0 < \varepsilon < \varepsilon_0(t)$, $\varepsilon_0(t)$ *sufficiently small.*

Proof: In view of Proposition 1 for the first assertion it is sufficient to show that

$$r^+\Lambda_-'^{t-m} r^+\Lambda_-^m(\varepsilon)\, r^+\Lambda_-'^{-t}: L^2(\mathbb{R}_+^n) \to L^2(\mathbb{R}_+^n)$$

is an isomorphism. Since we have operators with constant coefficients the composition rules of boundary symbols on the line can be applied (cf. 2.1.2.3). Then we have

$$r^+\Lambda_-'^{t-m} r^+\Lambda_-^m(\varepsilon)\, r^+\Lambda_-'^{-t} = r^+\Lambda_-'^{t-m}\Lambda_-^m(\varepsilon)\, r^+\Lambda_-'^{-t} = r^+\Lambda_-'^{t-m}\Lambda_-^m(\varepsilon)\,\Lambda_-'^{-t} + r'B$$

where $r'B$ is a Green operator. Now it is easily seen that the operator is different from the identity by an operator with small norm. The assertion about $r^+\Lambda_-^m$ follows if we show that $r'B$ has a small norm in L^2 for ε near 0. The symbol of the Green operator is given by

$$\sigma(B)(\xi', \nu, \tau) = \Pi_\nu^+\Pi_\tau^-\big((\langle\xi'\rangle - i\nu)^{t-m}\,(\delta_\varepsilon(\xi', \nu) - i\nu)^m - (\langle\xi'\rangle - i\tau)^{t-m}$$
$$\times\,(\delta_\varepsilon(\xi', \tau) - i\tau))^m\,(i\nu - i\tau)^{-1}\,(\langle\xi'\rangle - i\tau)^{-t}\big)$$

which belongs to $L^2(\mathbb{R} \times \mathbb{R})$ for any fixed $\xi' \neq 0$. The norm of $r'B$ in $L^2(\mathbb{R}_+^n)$ is bounded by $\sup_{\xi'} \{\int |\sigma(B)(\xi', \nu, \tau)|^2\, d\nu\, d\tau\}^{1/2}$. For $\nu^2 < (1 - 2\varepsilon)/2\varepsilon$, $\tau^2 < (1 - 2\varepsilon)/2\varepsilon$ we have

$$\big((\langle\xi'\rangle - i\nu)^{t-m}\,(\delta_\varepsilon(\xi', \nu) - i\tau)^m - (\langle\xi'\rangle - i\tau)^{t-m}\,(\delta_\varepsilon(\xi', \tau) - i\tau)^m\big)\,(i\nu - i\tau)^{-1}$$
$$\times\,(\langle\xi'\rangle - i\tau)^{-t} = \left((\langle\xi'\rangle - i\nu)^t\left(\frac{|\xi'| - i\nu}{\langle\xi'\rangle - i\nu}\right)^m - (\langle\xi'\rangle - i\tau)^t\left(\frac{|\xi'| - i\tau}{\langle\xi'\rangle - i\tau}\right)^m\right)$$
$$\times\,(i\nu - i\tau)^{-1}\,(\langle\xi'\rangle - i\nu)^{-t} \in H_\nu^- \otimes H_\tau^*\,.$$

Since $\Pi_\nu^+\Pi_\tau^-|_{H_\nu^+ \otimes H_\tau^-} = 0$ and Π^+, Π^- are continuous in L^2, it follows that

$$\|\Pi_\nu^+\Pi_\tau^-\big((\langle\xi'\rangle - i\nu)^{t-m}\,(\delta_\varepsilon(\xi', \nu) - i\nu)^m - (\langle\xi'\rangle - i\tau)^{t-m}\,(\delta_\varepsilon(\xi', \nu) - i\tau)^m\big)$$
$$\times\,(i\nu - i\tau)^{-1}\,(\langle\xi'\rangle - i\tau)^{-t}\|_{L^2}$$

is small for sufficiently small ε uniformly with respect to ξ'. Hence it is proved that $r^+\Lambda_-'^{t-m} r^+\Lambda_-^m(\varepsilon)\, r^+\Lambda_-'^{-t}$ differs from the identity by an operator of small norm and thus it is an isomorphism. The assertion about $r^+\Lambda_+^m(\varepsilon)$ is easily proved by duality. □

In the following we shall write for simplicity $r^+\Lambda_-^m$, $r^+\Lambda_+^m$ instead of $r^+\Lambda_-^m(\varepsilon)$, $r^+\Lambda_+^m(\varepsilon)$, ε fixed and sufficiently small. Let $\mathcal{A} \subset \mathfrak{G}^{0,0}$, i.e.

$$\mathcal{A} = \begin{pmatrix} r^+A + r'B & K \\ r'T & Q \end{pmatrix}$$

with orders (of growth of the symbols)

$$\text{ord}\, r^+A = 0,\ \text{ord}\, r'B = -1,\ \text{ord}\, K = 0,\ \text{ord}\, r'T = -1,\ \text{ord}\, Q = 0\,.$$

Then

$$(\mathcal{A}(u \oplus v), u' \oplus v')_0 = \big((1 \oplus \Lambda'^{1/2})\,\mathcal{A}(u \oplus v), (1 \oplus \Lambda'^{1/2})\,(u' \oplus v')\big)_{L^2}$$
$$= \big((1 \oplus \Lambda'^{1/2})\,(u \oplus v), (1 \oplus \Lambda'^{1/2})\,\{(1 \oplus \Lambda'^{-1})\,\mathcal{A}^*(1 \oplus \Lambda')\}\,(u' \oplus v')\big)_{L^2},$$

since $(1 \oplus \Lambda'^{1/2})^2 = (1 \oplus \Lambda')$. Thus $(1 \oplus \Lambda'^{-1})\,\mathcal{A}^*(1 \oplus \Lambda')$, $\mathcal{A}^*$ the L^2 adjoint, is the adjoint with respect to the scalar product in $H^0(\mathbb{R}_+^n) \oplus H^{1/2}(\mathbb{R}^{n-1})$ and the orders are such that $(1 \oplus \Lambda'^{-1})\,\mathcal{A}^*(1 \oplus \Lambda') \in \mathfrak{G}^{0,0}$.

Later it will become clear that the formal adjoints discussed in 2.3.2.3 and above in fact are Hilbert space adjoints.

Remark 3. Consider a PDO $A \in L^m(\mathbb{R}^{n-1})$ defined by

$$Av(x') = (2\pi)^{-n+1} \int e^{i(x'-y')\xi'} a(x', \xi')\, v(y')\, dy'\, d\xi' ,$$

$a \in S^m(\mathbb{R}^{n-1} \times \mathbb{R}^{n-1})$. Assume that $a(x', \xi')$ has compact support with respect to x'. According to 1.2.3.5, Theorem 1 there is a continuous extension $A: H^s(\mathbb{R}^{n-1}) \to H^{s-m}(\mathbb{R}^{n-1})$ for any $s \in \mathbb{R}$. For an arbitrary $\varepsilon > 0$ there is a properly supported PDO A_1 such that $A - A_1 \in L^{-\infty}$ and $\|A - A_1\|_s < \varepsilon$ ($\|\cdot\|_s$ denotes the norm in $\mathcal{L}(H^s(\mathbb{R}^{n-1}), H^{s-m}(\mathbb{R}^{n-1}))$). In fact, for $\varphi \in C_0^\infty(\mathbb{R}^{n-1})$, $\varphi = 1$ on $\operatorname{supp}_{x'} a(x', \xi')$ we set $A_1 = A\varphi$. Obviously A_1 is properly supported and $B = A - A_1 = A(1 - \varphi)$ is a smoothing operator with the kernel

$$K(x', y') = (2\pi)^{-n+1} \int e^{i(x'-y')\xi'} a(x', \xi')\, \bigl(1 - \varphi(y')\bigr)\, d\xi' .$$

We can choose φ in such a way that the norm $\|B\|_s$ becomes arbitrarily small. Since we have PDOs defining isomorphisms $H^s(\mathbb{R}^{n-1}) \to L^2(\mathbb{R}^{n-1})$, we can assume that $s = m = 0$. The norm $\|B\|_0$ is bounded by the Hilbert-Schmidt norm

$$\|K(x', y')\|_{L^2(\mathbb{R}^{n-1} \times \mathbb{R}^{n-1})}$$

of B. Then it is sufficient to take the set $\{y' \in \mathbb{R}^{n-1} : 1 - \varphi(y') = 0\}$ sufficiently large.

More generally, let $\varphi, \psi \in C_0^\infty(\mathbb{R}^{n-1})$ such that $\varphi\psi = \varphi$. Then $\varphi A \psi$ is properly supported and $\varphi A(1 - \psi)$ is smoothing. The norm $\|\varphi A(1 - \psi)\|_{s,s'}$ ($\|\cdot\|_{s,s'}$ denotes the norm in $\mathcal{L}(H^s(\mathbb{R}^{n-1}), H^{s'}(\mathbb{R}^{n-1}))$) becomes arbitrarily small for suitable chosen ψ. Similarly, for $\varphi \in C_0^\infty(\overline{\mathbb{R}}^n_+)$, $\psi \in C_0^\infty(\mathbb{R}^{n-1})$, $\psi = 1$ at $\operatorname{supp}_{x'} \varphi(x', x_n)$ and any $K \in \operatorname{Op}(\mathfrak{K}^m)$ the operator $\varphi K \psi$ is properly supported and $\varphi K(1 - \psi)$ smoothing. If K is defined by 2.3.2.1.(1), the norm of $\varphi K(1 - \psi)$ in $\mathcal{L}\bigl(H^s(\mathbb{R}^{n-1}), H^{s'}(\overline{\mathbb{R}}^n_+)\bigr)$, $s, s' \in \mathbb{R}$ arbitrary, becomes arbitrarily small for suitably chosen ψ. An analogous assertion about trace operators is valid by duality.

Theorem 4. *Every potential operator $K \in \operatorname{Op}(\mathfrak{K}^m)(\Omega')$, $\Omega' \subseteq \mathbb{R}^{n-1}$ open, has a continuous extension*

$$K: H^s_{\mathrm{comp}}(\Omega') \to H^{s-m-1/2}_{\mathrm{loc}}(\Omega' \times \mathbb{R}_+)$$

for any $s \in \mathbb{R}$. If K is defined by the symbol $\sigma(K)$ and $\varphi \in C_0^\infty(\mathbb{R}^{n-1})$, there is a continuous extension

$$\varphi K: H^s(\mathbb{R}^{n-1}) \to H^{s-m-1/2}(\mathbb{R}^n_+) \tag{1}$$

for any $s \in \mathbb{R}$ and the mapping

$$\mathfrak{K}^m \ni \varphi\sigma(K) \mapsto \varphi K \in \mathcal{L}\bigl(H^s(\mathbb{R}^{n-1}), H^{s-m-1/2}(\mathbb{R}^n_+)\bigr)$$

is continuous. For $s - m - 1/2 < 1/2$ we can replace $H^{s-m-1/2}(\Omega' \times \mathbb{R}_+)$ and $H^{s-m-1/2}(\mathbb{R}^n_+)$ by $H^{s-m-1/2}_{0,\mathrm{loc}}(\Omega' \times \overline{\mathbb{R}}_+)$ and $H_0^{s-m-1/2}(\overline{\mathbb{R}}^n_+)$.

Obviously Theorem 4 implies that a properly supported potential operator $K \in \operatorname{Op}(\mathfrak{K}^m)$ defines continuous mappings $K: H^s_{\mathrm{loc}}(\mathbb{R}^{n-1}) \to H^{s-m-1/2}_{\mathrm{loc}}(\mathbb{R}^n_+)$ and $K: H^s_{\mathrm{comp}}(\mathbb{R}^{n-1}) \to H^{s-m-1/2}_{\mathrm{comp}}(\mathbb{R}^+_n)$ and for $s - m - 1/2 < 1/2$ we have $K: H^s_{\mathrm{loc}}(\mathbb{R}^{n-1}) \to H^{s-m-1/2}_{0,\mathrm{loc}}(\overline{\mathbb{R}}^n_+)$, $K: H^s_{\mathrm{comp}}(\mathbb{R}^{n-1}) \to H^{s-m-1/2}_{0,\mathrm{comp}}(\overline{\mathbb{R}}^n_+)$.

Continuity of the mapping $\varphi\sigma(K) \mapsto \varphi K$ means that the norm

$$||\varphi K||_{\mathscr{L}(H^s(\mathbb{R}^{n-1}), H^{s-m-1/2})(\mathbb{R}^n_+)}$$

of φK is bounded by a linear combination $\sum_\alpha c_\alpha p_\alpha(\varphi\sigma(K))$ of semi-norms of $\varphi\sigma(K)$ with constants $c_\alpha \in \mathbb{R}_+$ and semi-norms p_α depending only on s and $\operatorname{supp}\varphi$ but not on K.

Proof: The first assertion follows from the second one. Per def. $\sigma(K)(x', \xi', \nu)$ admits an expansion

$$\sigma(K)(x', \xi', \nu) = \sum_{j=0}^{\infty} k_j(x', \xi')\, e_j(\langle\xi'\rangle^{-1}\nu)$$

where $e_j(\nu) = (1 - i\nu)^j (1 + i\nu)^{-j-1} \in H^+$, k_j in $S^m(\mathbb{R}^{n-1} \times \mathbb{R}^{n-1})$ rapidly decreasing (cf. 2.2.3.1). The function φ is included in $\sigma(K)$. Denote by K_j the potential operator defined by the symbol $k_j(x', \xi')\, e_j(\langle\xi'\rangle^{-1}\nu)$ and let $||\cdot||_s$ be the norm in $\mathscr{L}(H^s(\mathbb{R}^{n-1}), H^{s-m-1/2}(\mathbb{R}^n_+))$. It is sufficient to show

$$||K_j||_s \leqq c \sum_{l=0}^{N} c_l p_l(k_j)\, j^{N'} \tag{2}$$

for $N, N' \in \mathbb{Z}_+$, $c_l \in \mathbb{R}_+$ independent of j (p_l denotes semi-norms on $S^m(\mathbb{R}^{n-1} \times \mathbb{R}^{n-1})$). In fact, (2) implies

$$||K||_s \leqq \sum_{j=0}^{\infty} ||K_j||_s \leqq c \sum_{l=0}^{N} c_l \sum_{j=0}^{\infty} p_l(k_j)\, j^{N'}$$

and $\sum_{j=0}^{\infty} p_l(k_j)\, j^{N'}$ is bounded by a semi-norm of $\sigma(K) \in \mathfrak{R}^m(\mathbb{R}^{n-1} \times \mathbb{R}^{n-1})$.

For the proof of (2) observe that in view of Remark 3 up to an operator with arbitrarily small norm we can replace K_j by $K_j\varphi$ for a suitable $\varphi \in C_0^\infty(\mathbb{R}^{n-1})$. The adjoint $(K_j\varphi)^*$ is a trace operator with symbol $\overline{k_j(y', \xi')}\, \overline{e_j(\langle\xi'\rangle^{-1}\nu)}\, \overline{\varphi(x')} \in \mathfrak{T}^{m,0}$ the kernel of which has by assumption compact support with respect to x' and y'. Hence it can be applied to C^∞ functions on $\mathbb{R}^{n-1}$ as PDO with operator-valued symbol (cf. 2.3.2.1). According to 2.3.3.1, Proposition 4 ist can be given by a symbol $\tilde{t}_j(x', \xi', \nu) \in \mathfrak{T}^{m,0}$ depending continuously on $\overline{k_j(y', \xi')}\, \overline{e_j(\langle\xi'\rangle^{-1}\nu)}\, \overline{\varphi(x')} \in \mathfrak{T}^{m,0}$. Going back to the potential operator with the symbol $\overline{\tilde{t}_j(y', \xi', \nu)} \in \mathfrak{R}^m$ and using the decomposition as above

$$\overline{\tilde{t}_j(y', \xi', \nu)} = \sum_{l=0}^{\infty} t_{jl}(y', \xi')\, e_l(\langle\xi'\rangle^{-1}\nu)\,,$$

t_{jl} in $S^m(\mathbb{R}^{n-1} \times \mathbb{R}^{n-1})$ is rapidly decreasing with respect to l. The potential operator K_{jl} with symbol $t_{jl}(y', \xi')\, e_l(\langle\xi'\rangle^{-1}\nu)$ is the composition of the PDO T_{jl} on $\mathbb{R}^{n-1}$ with the symbol $t_{jl}(y', \xi')$ and the potential operator $K_{(l)}$ with the symbol $e_l(\langle\xi'\rangle^{-1}\nu)$, $K_{jl} = K_{(l)}T_{jl}$. The PDOs T_{jl} have continuous extensions

$$T_{jl}\colon H^s(\mathbb{R}^{n-1}) \to H^{s-m}(\mathbb{R}^{n-1})$$

with an estimate of the norm by a sum $\sum_i c_i p_i(t_{jl})$ of semi-norms of $t_{jl} \in S^m$ (cf. 1.2.4.1, Theorem 1 and 2.3.2.3, Remark 13). Consider $K_{(l)}$ and assume that $s-m-1/2 \in \mathbb{Z}$. For $K^0_{(l)} = r^+\Lambda_-^{s-m-1/2} K_l \Lambda'^{-s+m}$ (Λ' the PDO on $\mathbb{R}^{n-1}$ with the symbol $\langle\xi'\rangle$) we have

$$||K^0_{(l)}||_{\mathscr{L}(H^q(\mathbb{R}^{n-1}), H^q(\mathbb{R}^n_+))} \leqq ||K_{(l)}||_{\mathscr{L}(H^s(\mathbb{R}^{n-1}), H^{s-m-1/2}(\mathbb{R}^n_+))}\,.$$

The symbol of $K^0_{(l)}$ is given by $\Pi^+\big((\delta_s(\xi',\nu)\,|\xi'| - i\nu)^{s-m-1/2}\, e_l(\langle\xi'\rangle^{-1}\nu)\,\langle\xi'\rangle^{-s+m}\big)$ and

$$\begin{aligned}||K^0_{(l)}v||^2_{L^2(\mathbb{R}^n_+)} &\leqq c\int \big|\Pi^+\big((\delta_s(\xi',\nu)\,|\xi'| - i\nu)^{s-m-1/2}\, e_l(\langle\xi'\rangle^{-1}\nu)\big)\,\langle\xi'\rangle^{-s+m}\big|^2\\ &\quad\times |\hat v(\xi')|^2\, d\xi'\, d\nu\\ &\leqq c\int \big|\Pi^+\big((1-i\nu)^{s-m-1/2}\, e_l(\nu)\big)\big|^2\, d\nu\ ||v||^2_{L^2(\mathbb{R}^{n-1})}\,.\end{aligned}$$

As noted in the proof of 2.1.1.1, Proposition 9 we have the estimate

$$\int \big|\Pi^+\big((1-i\nu)^{s-m-1/2}\, e_l(\nu)\big)\big|^2\, d\nu \leqq c l^N$$

with constants c, $N \in \mathbb{Z}_+$ independent of l. For $s - m - {}^1/_2 \notin \mathbb{Z}$ we obtain the same estimate by interpolation. Hence we get the norm of K_j

$$||K_j||_s \leqq \sum_{l=0}^{\infty} ||K_{(l)}T_{jl}||_s \leqq c\sum_i c_i \sum_{l=0}^{\infty} p_i(t_{jl})\, l^N\,.$$

Since $\tilde t_j(y',\xi',\nu) \in \mathfrak{R}^m$ depends continuously on $k_j(x',\xi')\, e_j(\langle\xi'\rangle^{-1}\nu)\,\varphi(y') \in \mathfrak{R}^m$, the semi-norm $\sum\limits_{l=0}^{\infty} p_i(t_{jl})\, l^N$ of $\tilde t_j$ is bounded by a semi-norm of $k_j e_j\varphi$, i.e.

$$\sum_{l=0}^{\infty} p_i(t_{jl})\, l^N \leqq c\sum c_{i'} p_{i'}(k_j\varphi)\, j^{N'}$$

with suitable constants c, $c_{i'}$, $N' \in \mathbb{Z}_+$ and semi-norms $p_{i'}$ on S^m. This implies (2) and thus the assertion for $H^{s-m-1/2}(\mathbb{R}^n_+)$.

For $s - m - {}^1/_2 < {}^1/_2$ integer and $H_0^{s-m-1/2}(\overline{\mathbb{R}^n_+})$ we conclude in the same way, but replace $r^+\Lambda^{s-m-1/2}$ by $r^+\Lambda_+^{s-m-1/2}$ when reducing the order. Then the symbol of $K^0_{(l)}$ is equal to

$$\begin{aligned}&\Pi^+\big((\delta_s(\xi',\nu)\,|\xi'| + i\nu)^{s-m-1/2}\, e_l(\langle\xi'\rangle^{-1}\nu)\,\langle\xi'\rangle^{-s+m}\big)\\ &= (\delta_s(\xi',\nu)\,|\xi'| + i\nu)^{s-m-1/2}\, e_l(\langle\xi'\rangle^{-1}\nu)\,\langle\xi'\rangle^{-s+m}\end{aligned}$$

and the norm $K_{(l)} \in \mathcal{L}\big(H^s(\mathbb{R}^{n-1}), H_0^{s-m-1/2}(\overline{\mathbb{R}^n_+})\big)$ is equal to the norm $K^0_{(l)} \in \mathcal{L}\big(H^0(\mathbb{R}^{n-1}), H^0(\mathbb{R}^n_+)\big)$. For $s - m - {}^1/_2 < {}^1/_2$ arbitrary the assertion follows by interpolation. □

Theorem 5. *Every trace operator $r'T \in \operatorname{Op}(\mathfrak{T}^{m,d})(\Omega')$, $\Omega' \subseteq \mathbb{R}^{n-1}$ open, has a continuous extension*

$$r'T\colon H^s_{\text{comp}}(\Omega' \times \mathbb{R}_+) \to H^{s-m-1/2}_{\text{loc}}(\Omega')$$

for any $s > d - {}^1/_2$. If $r'T$ is defined by the symbol $\sigma(T)$ and $\varphi \in C_0^\infty(\mathbb{R}^{n-1})$, there is a continuous extension

$$\varphi r'T\colon H^s(\mathbb{R}^n_+) \to H^{s-m-1/2}(\mathbb{R}^{n-1}) \tag{3}$$

for any $s > d - {}^1/_2$. The mapping $\mathfrak{T}^{m,d} \ni \varphi\sigma(T) \mapsto \varphi r'T \in \mathcal{L}\big(H^s(\mathbb{R}^n_+), H^{s-m-1/2}(\mathbb{R}^{n-1})\big)$ is continuous. If $d = 0$, we can replace $H^s_{\text{comp}}(\Omega' \times \mathbb{R}_+)$ and $H_0^s(\overline{\mathbb{R}^n_+})$ by $H^s_{0,\text{comp}}(\Omega' \times \overline{\mathbb{R}}_+)$ and $H_0^s(\overline{\mathbb{R}^n_+})$ for any $s \in \mathbb{R}$.

Note that the restriction on s for $d > 0$ is necessary, since the restriction operator $r'\colon C_0^\infty(\overline{\mathbb{R}^n_+}) \to C_0^\infty(\mathbb{R}^{n-1})$ has a continuous extension $H^s(\mathbb{R}^n_+) \to H^{s-1/2}(\mathbb{R}^{n-1})$ iff $s > {}^1/_2$.

Proof: It is sufficient to show the second assertion. First consider trace operators

$$r'Tu(x') = \sum_{k=0}^{d-1} T_k(x', D')\, r' D_{x_n}^k u(x')$$

where $T_k \in L^{m-k}$. The assertion follows immediately from the properties of PDOs on $\mathbb{R}^n$ (cf. 1.2.3.5, Theorem 1) and the continuity of

$$r' D_{x_n}^k : H^s(\mathbb{R}_+^n) \to H^{s-k-1/2}(\mathbb{R}^{n-1})$$

for any $s - k - {}^1/_2 > 0$. Thus we can assume $r'T \in \mathrm{Op}\,(\mathfrak{T}^{m,0})$. As observed in 2.3.2.3, Proposition 10 this can be considered as the adjoint of a potential operator K in $\mathrm{Op}\,(\mathfrak{K}^m)$. The estimate of the norm follows by analogous arguments as in the proof of Theorem 4 and is left to the reader. □

Theorem 6. *Every Green operator* $r'B \in \mathrm{Op}\,(\mathfrak{B}^{m,d})\,(\Omega')$, $\Omega' \subseteq \mathbb{R}^{n-1}$ *open, has a continuous extension*

$$r'B : H^s_{\mathrm{comp}}(\Omega' \times \mathbb{R}_+) \to H^{s-m-1}_{\mathrm{loc}}(\Omega' \times \mathbb{R}_+)$$

for any $s > d - {}^1/_2$. *If* $r'B$ *is defined by the symbol* $\sigma(B)$ *and* $\varphi \in C_0^\infty(\mathbb{R}^{n-1})$, *there is a continuous extension*

$$\varphi r'B : H^s(\mathbb{R}_+^n) \to H^{s-m-1}(\mathbb{R}_+^n) \tag{4}$$

for any $s > d - {}^1/_2$ *if* $d > 0$ *and the mapping* $\mathfrak{B}^{m,d} \ni \varphi\sigma(B) \mapsto \varphi r'B \in \mathcal{L}\big(H^s(\mathbb{R}_+^n),$ $H^{s-m-1}(\mathbb{R}_+^n)\big)$ *is continuous. If* $d = 0$ *for all* $s \in \mathbb{R}$, *we can replace* $H^s_{\mathrm{comp}}(\Omega' \times \mathbb{R}_+)$ *and* $H^s(\mathbb{R}_+^n)$ *by* $H^s_{0,\mathrm{comp}}(\Omega' \times \overline{\mathbb{R}}_+)$ *and* $H^s_0(\overline{\mathbb{R}}_+^n)$ *and* $H^{s-m-1}_{\mathrm{loc}}(\Omega' \times \mathbb{R}_+)$, $H^{s-m-1}(\mathbb{R}_+^n)$ *by* $H^{s-m-1}_{0,\mathrm{loc}}(\Omega' \times \overline{\mathbb{R}}_+)$, $H_0^{s-m-1}(\overline{\mathbb{R}}_+^n)$ *for* $s - m - 1 < {}^1/_2$.

Similarly to the case of trace operators the restriction on s for $d > 0$ is related to the properties of the restriction operator r'.

Proof: It is sufficient to prove the second assertion. Any Green symbol $\sigma(B) \in \mathfrak{B}^{m,d}$ has a decomposition

$$\sigma(B)\,(x', \xi', \nu, \tau) = \sum_{j=0}^{\infty} \lambda_j \sigma(K_j)\,(x', \xi', \nu)\,\sigma(T_j)\,(\xi', \nu)$$

where $\sum_j |\lambda_j| \leqq 1$, $\sigma(K_j)$ and $\sigma(T_j)$ are symbols of potential and trace operators tending to zero in $\mathfrak{K}^0$ and $\mathfrak{T}^{m,d}$, respectively. Let K_j and $r'T_j$ be the corresponding potential and trace operators. Since the potential and trace operators continuously depend on the symbols, we have

$$\|K_j\|_{\mathcal{L}(H^{s-m-1/2}(\mathbb{R}^{n-1}),\, H^{s-m-1}(\mathbb{R}_+^n))} \leqq \text{const}, \quad \|r'T_j\|_{\mathcal{L}(H^s(\mathbb{R}_+^n),\, H^{s-m-1/2}(\mathbb{R}^{n-1}))} \leqq \text{const}.$$

for all $j \in \mathbb{Z}_+$. Thus the desired extension is given by $r'B = \sum_j K_j r'T_j$. The continuous dependence on the Green symbol is obvious. □

Theorem 7. *Every PDO with the transmission property* $r^+A \in \mathrm{Op}\,(\mathfrak{A}^m)\,(\Omega' \times \overline{\mathbb{R}}_+)$ *has a continuous extension*

$$r^+A : H^s_{\mathrm{comp}}(\Omega' \times \mathbb{R}_+) \to H^{s-m}_{\mathrm{loc}}(\Omega' \times \mathbb{R}_+)$$

for any $s > -{}^1/_2$. *If* r^+A *is defined by the symbol* $\sigma(A)$ *and* $\varphi \in C_0^\infty(\overline{\mathbb{R}}_+^n)$, *there is a continuous extension*

$$\varphi r^+A : H^s(\mathbb{R}_+^n) \to H^{s-m}(\mathbb{R}_+^n) \tag{5}$$

for any $s > -1/2$ and the mapping

$$\mathfrak{A}^m \ni \varphi\sigma(A) \mapsto \varphi r^+A \in \mathcal{L}\left(H^s(\mathbb{R}^n_+), H^{s-m}(\mathbb{R}^n_+)\right)$$

is continuous. We can replace $H^s_{\text{comp}}(\Omega' \times \mathbb{R}_+)$ and $H^s(\mathbb{R}^n_+)$ by $H^s_{0,\text{comp}}(\Omega' \times \overline{\mathbb{R}}_+)$ and $H^s_0(\overline{\mathbb{R}}^+_n)$, s arbitrary.

Proof: The continuity $r^+A\colon H^s_0(\overline{\mathbb{R}}^n_+) \to H^{s-m}(\mathbb{R}^n_+)$ follows from 1.2.3.5, Theorem 1 and the continuity $r^+\colon H^{s-m}(\mathbb{R}^n) \to H^{s-m}(\mathbb{R}^n_+)$. The norm estimate is the same as for PDOs in $\mathbb{R}^n$.

For $|s| < 1/2$ the continuity of $\varphi r^+A\colon H^s(\mathbb{R}^n_+) \to H^{s-m}(\mathbb{R}^n_+)$ follows immediately from the standard fact that the extension operator $j^+\colon H^s(\mathbb{R}^n_+) \to H^s(\mathbb{R}^n)$ is continuous for $|s| < 1/2$ and the continuity of PDO in $\mathbb{R}^n$ (cf. 1.2.3.5, Theorem 1). The norm is bounded by a sum of semi-norms of $\varphi\sigma(A)$ in $S^m(\overline{\mathbb{R}}^n_+ \times \mathbb{R}^n)$. For $s > 1/2$, $s - 1/2 \notin \mathbb{Z}_+$ we take $k \in \mathbb{Z}_+$ such that $|s - k| < 1/2$. Set $\sigma(A_1)(x, \xi', \nu) = \sigma(A)(x, \xi' \nu)$ $(\delta_\varepsilon(\xi', \nu)\,|\xi'| - i\nu)^{-k}$ and denote the corresponding PDO by $r^+A_1 \in \operatorname{Op}(\mathfrak{A}^{m-k})$. From the calculus in 2.3.2.2 we get

$$r^+A = r^+A_1\, r^+\Lambda^k_- - r'B$$

with a Green operator $r'B \in \operatorname{Op}(\mathfrak{B}^{m-1,0})$ the symbol $\sigma(B)$ of which is given by

$$\sigma(B)(x', \xi', \nu, \tau)$$
$$= \Pi^+_\nu \left(\sigma(A_1)(x', 0, \xi', \nu)\, \frac{(\delta_\varepsilon(\xi', \nu)\,|\xi'| - i\nu)^{-k} - (\delta_\varepsilon(\xi', \nu)\,|\xi'| - i\tau)^{-k})}{i(\nu - \tau)}\right).$$

$\sigma(B)$ depends continuously on $\sigma(A)$. Then the assertion follows from Proposition 1 and Theorem 6. For $s - 1/2 \in \mathbb{Z}_+$ the result follows by interpolation. □

Corollary 8. *Every operator $\mathcal{A} \in \mathfrak{G}^{m,d}$ in the half space has a continuous extension*

$$\mathcal{A}\colon \begin{matrix} H^s_{\text{comp}}(\mathbb{R}^n_+, \mathbb{C}^k) \\ \oplus \\ H^{s+1/2}_{\text{comp}}(\mathbb{R}^{n-1}, \mathbb{C}^j) \end{matrix} \to \begin{matrix} H^{s-m}_{\text{loc}}(\mathbb{R}^n_+, \mathbb{C}^{k'}) \\ \oplus \\ H^{s-m+1/2}_{\text{loc}}(\mathbb{R}^{n-1}, \mathbb{C}^{j'}) \end{matrix} \tag{6}$$

for any $s > d - 1/2$. For $(a(x, \xi), \mathfrak{a}(x', \xi'))$ with compact support with respect to x and x', respectively, the mapping

$$(a(x, \xi), \mathfrak{a}(x', \xi')) \mapsto \mathcal{A} = \operatorname{Op}(a, \mathfrak{a}) \in \mathfrak{G}^{m,d}$$

is continuous if we take in $\mathfrak{G}^{m,d}$ the topology induced from

$$\mathcal{L}\left(H^s(\mathbb{R}^n_+, \mathbb{C}^k) \oplus H^{s+1/2}(\mathbb{R}^{n-1}, \mathbb{C}^j), H^{s-m}(\mathbb{R}^n_+, \mathbb{C}^{k'}) \oplus H^{s-m+1/2}(\mathbb{R}^{n-1}, \mathbb{C}^{j'})\right),$$
$$s > d - 1/2\,.$$

If $d = 0$, there is a continuous extension

$$\mathcal{A}\colon \begin{matrix} H^s_{0,\text{comp}}(\overline{\mathbb{R}}^n_+, \mathbb{C}^k) \\ \oplus \\ H^{s+1/2}_{\text{comp}}(\mathbb{R}^{n-1}, \mathbb{C}^j) \end{matrix} \to \begin{matrix} H^{s-m}_{\text{loc}}(\mathbb{R}^n_+, \mathbb{C}^{k'}) \\ \oplus \\ H^{s-m+1/2}_{\text{loc}}(\mathbb{R}^{n-1}, \mathbb{C}^{j'}) \end{matrix}$$

for arbitrary $s \in \mathbb{R}$ and for $s - m < 1/2$ there is an extension

$$\mathcal{A}\colon \begin{matrix} H^s_{0,\text{comp}}(\overline{\mathbb{R}}^n_+, \mathbb{C}^k) \\ \oplus \\ H^{s+1/2}_{\text{comp}}(\mathbb{R}^{n-1}, \mathbb{C}^j) \end{matrix} \to \begin{matrix} H^{s-m}_{0,\text{loc}}(\overline{\mathbb{R}}^n_+, \mathbb{C}^{k'}) \\ \oplus \\ H^{s-m+1/2}_{\text{loc}}(\mathbb{R}^{n-1}, \mathbb{C}^{j'}) \end{matrix}.$$

Assume that $\mathcal{A} \in \mathfrak{G}^{m,d}$ has a continuous extension (6) for $s = 0$. Then $\mathcal{A} \in \mathfrak{G}^{m,0}$, i.e. the Green and the trace operator in $\mathcal{A}$ have the type zero. In fact, for a type > 0 such an extension cannot exist.

Remark 9. Note that for an operator $\mathcal{A} \in \mathfrak{G}^{m,d}$ with constant coefficients, i.e. the symbols are independent of x, there is a continuous extension

$$\mathcal{A}: \begin{matrix} H^s(\mathbb{R}^n_+, \mathbb{C}^k) \\ \oplus \\ H^{s+1/2}(\mathbb{R}^{n-1}, \mathbb{C}^j) \end{matrix} \to \begin{matrix} H^{s-m}(\mathbb{R}^n_+, \mathbb{C}^{k'}) \\ \oplus \\ H^{s-m+1/2}(\mathbb{R}^{n-1}, \mathbb{C}^{j'}) . \end{matrix} \tag{7}$$

The proof is a combination of the estimates for boundary symbols on the half axis and the usual estimates for PDO with constant coefficients. Then it is easy to generalize Corollary 8 to the case when the symbols stabilize near infinity, i.e. they are sums of symbols with constant coefficients and symbols having compact support with respect to x and x', respectively. For such operators there are extensions in Sobolev spaces as in (7). The formulation of the analogous results concerning $H_0^s(\overline{\mathbb{R}}^n_+)$ is left to the reader.

2.3.2.5. $H^{s,p}$ Continuity

For PDOs in $\mathbb{R}^n$ the following result is valid

Theorem 1. *Every PDO $A \in L^m(\Omega)$, $\Omega \subseteq \mathbb{R}^n$ open, has a continuous extension*

$$A: H^{s,p}_{\text{comp}}(\Omega) \to H^{s-m,p}_{\text{loc}}(\Omega)$$

for any $s \in \mathbb{R}$, $1 < p < \infty$. If A is defined by the symbol $\sigma(A)$ and $\varphi \in C_0^\infty(\mathbb{R}^n)$, there is a continuous extension

$$\varphi A: H^{s,p}(\mathbb{R}^n) \to H^{s-m,p}(\mathbb{R}^n) \tag{1}$$

for any $s \in \mathbb{R}$, and the mapping

$$S^m \ni \varphi\sigma(A) \mapsto \varphi A \in \mathcal{L}\big(H^{s,p}(\mathbb{R}^n), H^{s-m,p}(\mathbb{R}^n)\big)$$

is continuous.

For a proof cf. Hörmander [4].

Note that the norm of the extension in Theorem 1 is bounded by sums of semi-norms $\sup |D_x^\alpha \varphi\sigma(A)(x, \xi)| \langle\xi\rangle^{-m}$ for some multi-indices α depending only on s and $\operatorname{supp}\varphi$ (in other words, there is a linear combination $\sum_{\alpha\in I} c_\alpha p_\alpha$, $c_\alpha \in \mathbb{R}_+$, p_α semi-norms with index set I and c_α, $\alpha \in I$ depending only on s and $\operatorname{supp}\varphi$ such that $\|\varphi A\| \leqq \sum_{\alpha\in I} c_\alpha p_\alpha(\varphi\sigma(A))$).

Theorem 2. (i) *Every potential operator $K \in \mathrm{Op}(\mathfrak{K}^m)(\Omega')$, $\Omega' \subseteq \mathbb{R}^{n-1}$ open, has a continuous extension*

$$K: H^{s,p}_{\text{comp}}(\Omega') \to H^{s-m-1/p',p}_{\text{loc}}(\Omega' \times \mathbb{R}_+)$$

for any $s \in \mathbb{R}$, $1 < p < \infty$. If K is defined by the symbol $\sigma(K)$ and $\varphi \in C_0^\infty(\mathbb{R}^{n-1})$, there is a continuous extension

$$\varphi K: H^{s,p}(\mathbb{R}^{n-1}) \to H^{s-m-1/p',p}(\mathbb{R}^n_+) \tag{2}$$

for any $s \in \mathbb{R}$ *and the mapping*

$$\mathfrak{K}^m \ni \varphi\sigma(K) \mapsto \varphi K \in \mathcal{L}(H^{s,p}(\mathbb{R}^{n-1}), H^{s-m-1/p',p}(\mathbb{R}^n_+))$$

is continuous. For $s - m - 1/p' < 1/p$ *we can replace* $H^{s-m-1/p',p}_{\mathrm{loc}}(\Omega' \times \mathbb{R}_+)$ *and* $H^{s-m-1/p',p}(\mathbb{R}^n_+)$ *by* $H^{s-m-1/p',p}_{0,\mathrm{loc}}(\Omega' \times \overline{\mathbb{R}}_+)$ *and* $H_0^{s-m-1/p',p}(\overline{\mathbb{R}}^n_+)$.

(ii) *Every* $r'T \in \mathrm{Op}(\mathfrak{T}^{m,d})(\Omega')$ *has a continuous extension*

$$r'T\colon H^{s,p}_{\mathrm{comp}}(\Omega' \times \overline{\mathbb{R}}_+) \to H^{s-m-1/p,p}_{\mathrm{loc}}(\Omega')$$

for any $s \in \mathbb{R}$, $s - 1/p \notin \mathbb{Z}$, $s > d - 1/p'$, $1 < p < \infty$. *If* $r'T$ *is defined by the symbol* $\sigma(T)$ *and* $\varphi \in C_0^\infty(\mathbb{R}^{n-1})$, *there is a continuous extension*

$$\varphi r'T\colon H^{s,p}(\mathbb{R}^n_+) \to H^{s-m-1/p,p}(\mathbb{R}^{n-1}) \tag{3}$$

for any $s > d - 1/p'$ *and the mapping*

$$\mathfrak{T}^{m,d} \ni \varphi\sigma(T) \mapsto \varphi r'T \subset \mathcal{L}(H^{s,p}(\mathbb{R}^n_+), H^{s-m-1/p,p}(\mathbb{R}^{n-1}))$$

is continuous. If $d = 0$, *we can replace* $H^{s,p}_{\mathrm{comp}}(\Omega' \times \overline{\mathbb{R}}_+)$ *and* $H^{s,p}(\mathbb{R}^n_+)$ *by* $H^{s,p}_{0,\mathrm{comp}}(\Omega' \times \overline{\mathbb{R}}_+)$ *and* $H_0^{s,p}(\overline{\mathbb{R}}^n_+)$ *for any* $s \in \mathbb{R}$.

(iii) *Every* $r'B \in \mathrm{Op}(\mathfrak{B}^{m,d})(\Omega')$ *has a continuous extension*

$$r'B\colon H^{s,p}_{\mathrm{comp}}(\Omega' \times \overline{\mathbb{R}}_+) \to H^{s-m-1,p}_{\mathrm{loc}}(\Omega' \times \mathbb{R}_+)$$

for $s - 1/p \notin \mathbb{Z}$, $s > d - 1/p'$, $1 < p < \infty$. *If* $r'B$ *is defined by the symbol* $\sigma(B)$ *and* $\varphi \in C_0^\infty(\mathbb{R}^{n-1})$, *there is a continuous extension*

$$\varphi r'B\colon H^{s,p}(\mathbb{R}^n_+) \to H^{s-m-1,p}(\mathbb{R}^n_+) \tag{4}$$

for any $s \in \Omega$, $s > d - 1/p'$, $1 < p < \infty$, *and the mapping*

$$\mathfrak{B}^{m,d} \ni \varphi\sigma(B) \mapsto \varphi r'B \in \mathcal{L}(H^{s,p}(\mathbb{R}^n_+), H^{s-m-1,p}(\mathbb{R}^n_+))$$

is continuous. If $d = 0$, *we can replace* $H^{s,p}_{\mathrm{comp}}(\Omega' \times \overline{\mathbb{R}}_+)$, $H^{s,p}(\mathbb{R}^n_+)$ *by* $H^{s,p}_{0,\mathrm{comp}}(\Omega' \times \overline{\mathbb{R}}_+)$, $H_0^{s,p}(\overline{\mathbb{R}}^n_+)$ *for any* $s \in \mathbb{R}$ *and* $H^{s-m-1,p}_{\mathrm{loc}}(\Omega' \times \mathbb{R}_+)$, $H^{s-m-1,p}(\mathbb{R}^n_+)$ *by* $H_0^{s-m-1,p}(\Omega' \times \overline{\mathbb{R}}_+)$, $H_0^{s-m-1,p}(\overline{\mathbb{R}}^n_+)$ *for* $s - m - 1 < 1/p$.

(iv) *Every* $r^+A \in \mathrm{Op}(\mathfrak{A}^m)(\Omega' \times \mathbb{R}_+)$ *has a continuous extension*

$$r^+A\colon H^{s,p}_{\mathrm{comp}}(\Omega' \times \overline{\mathbb{R}}_+) \to H^{s-m,p}_{\mathrm{loc}}(\Omega' \times \mathbb{R}_+)$$

for any $s \in \mathbb{R}$, $s > -1/p'$, $1 < p < \infty$. *If* r^+A *is defined by the symbol* $\sigma(A)$ *and* $\varphi \in C_0^\infty(\overline{\mathbb{R}}^n_+)$, *there is a continuous extension*

$$\varphi r^+A\colon H^{s,p}(\mathbb{R}^n_+) \to H^{s-m,p}(\mathbb{R}^n_+) \tag{5}$$

for any $s > -1/p'$ *and the mapping*

$$\mathfrak{A}^m \ni \varphi\sigma(A) \mapsto \varphi r^+A \in \mathcal{L}(H^{s,p}(\mathbb{R}^n_+), H^{s-m,p}(\mathbb{R}^n_+))$$

is continuous. We can replace $H^{s,p}_{\mathrm{comp}}(\Omega' \times \overline{\mathbb{R}}_+)$, $H^{s,p}(\mathbb{R}^n_+)$ *by* $H^{s,p}_{0,\mathrm{comp}}(\Omega' \times \overline{\mathbb{R}}_+)$, $H_0^{s,p}(\overline{\mathbb{R}}^n_+)$ *for arbitrary* s *and* $H^{s-m,p}_{\mathrm{loc}}(\Omega' \times \mathbb{R}_+)$, $H^{s-m,p}(\mathbb{R}^n_+)$ *by* $H^{s-m,p}_{0,\mathrm{loc}}(\Omega' \times \overline{\mathbb{R}}_+)$, $H_0^{s-m,p}(\overline{\mathbb{R}}^n_+)$ *for* $s - m < 1/p$.

The proof closely follows that of the case $p = 2$ (cf. 2.3.2.4).

Proof: (i) As in the proof of 2.3.2.4, Theorem 3 it is sufficient to consider a potential operator K with the symbol $e_j(\langle\xi'\rangle^{-1}\nu)$ and to show that it defines a continuous mapping $H^{s,p}(\mathbb{R}^{n-1}) \to H^{s-1/p',p}(\mathbb{R}^n_+)$ with norm bounded by $cj^N, N \in \mathbb{Z}_+$. Then the assertion follows from the decomposition of the potential symbol as in 2.3.2.4, Theorem 3 using Theorem 1. Assume $s - 1/p' \in \mathbb{Z}$. As in the proof of 2.3.2.4, Theorem 3 consider $K_0 = r^+\Lambda^{s-1/p'} K \Lambda'^{-s} \in \operatorname{Op}(\mathfrak{K}^{-1/p'})$ having the symbol $\Pi^+((\delta_\varepsilon(\xi', \nu)\,|\xi'| - i\nu)^{s-1/p'} \times e(\langle\xi'\rangle^{-1}\nu)\,\langle\xi'\rangle^{-2}) = k_0(\xi', \tau)$ which depends continuously on $e_j(\langle\xi'\rangle^{-1}\nu) \in \mathfrak{K}^0$. In view of the isomorphisms $\Lambda'^s: H^{s,p}(\mathbb{R}^{n-1}) \to L^p(\mathbb{R}^{n-1})$ and $r^+\Lambda^s_-: H^{s,p}(\mathbb{R}^n_+) \to L^2(\mathbb{R}^n_+)$ it is sufficient to show the continuity

$$K_0: L^p(\mathbb{R}^{n-1}) \to L^p(\mathbb{R}^n_+)$$

and the norm estimate. Set

$$\check{k}^0(\xi', x_n) = (2\pi)^{-1} \int e^{ix_n\nu}\, k^0(\xi', \nu)\, d\nu\,.$$

Then $x_n \to \check{k}^0(\xi', x_n)$ defines an L^p function on $\mathbb{R}_+$ with values in $S^0(\mathbb{R}^{n-1} \times \mathbb{R}^{n-1})$. In fact, the decomposition

$$D^\alpha_{\xi'} \check{k}^0(\xi', x_n) = \sum_{l=0}^N g_l(\langle\xi'\rangle\, x_n)\, b_l(\xi')\,, \quad g_l \in \mathscr{S}(\overline{\mathbb{R}}_+)\,, \quad b_l \in S^{-|\alpha|-1/p'+1}(\mathbb{R}^{n-1})$$

$N \in \mathbb{Z}_+$, implies the estimate

$$\int |D^\alpha_{\xi'} \check{k}^0(\xi', x_n)|^p\, dx_n = \int \left| \sum_{l=0}^N g_l(\langle\xi'\rangle\, x_n)\, b_l(\xi') \right|^p dx_n$$

$$\leqq \int \left| \sum_{l=0}^N g_l(t)\, b_l(\xi')\, \langle\xi'\rangle^{-1/p} \right|^p dt \leqq c\, \langle\xi'\rangle^{-|\alpha|\,p}\,.$$

Considering K_0 as PDO of order 0 depending on the parameter x_n we get

$$\|K_0 v(\cdot, x_n)\|_{L^p(\mathbb{R}^{n-1})} \leqq c q(\check{k}^0(\cdot, x_n))\, \|v\|_{L^p(\mathbb{R}^{n-1})}$$

for some semi-norm q on $S^0(\mathbb{R}^{n-1})$ and hence applying Theorem 1

$$\|K_0 v\|_{L^p(\mathbb{R}^n_+)} \leqq c\, \{\int q(\check{k}^0(\cdot, x_n))^p\, dx_n\}^{1/p}\, \|v\|_{L^p(\mathbb{R}^{n-1})}\,.$$

For arbitrary $s - 1/p' \in \mathbb{R}$ the assertion follows by interpolation. For the assertion about $H^{s,p}_0(\overline{\mathbb{R}}^n_+)$ cf. the proof of 2.3.2.4, Theorem 3.

(ii) For a trace symbol $\sigma(T)(x', \xi', \nu) = a(x', \xi')\,\nu^k$, $a \in S^{m-k}(\mathbb{R}^{n-1} \times \mathbb{R}^{n-1})$, the corresponding operator is the composition of $r' D^k_{x_n}$ and a PDO on $\mathbb{R}^{n-1}$ with symbol $a(x', \xi')$. Then the assertion follows from Theorem 1 and the well-known continuity

$$r' D^k_{x_n}: H^{s,p}(\mathbb{R}^n_+) \to H^{s-k-1/p,\,p}(\mathbb{R}^{n-1})$$

for $s - 1/p \notin \mathbb{Z}$ (cf. 1.2.1.2). For trace operators of type 0 we conclude as in the proof of 2.3.2.4, Theorem 3 using that the adjoints of trace operators of type zero are potential operators of the same order and duality of $H^{s,p}$ (cf. 1.2.1.2).

(iii) As in the proof of 2.3.2.3, Theorem 4 continuity and the norm estimates immediately follow from the decomposition of Green operators in compositions of trace and potential operators and the results from (i) and (ii).

(iv) The extension by zero j^+ is a continuous operator $H^{s,p}(\mathbb{R}^n_+) \to H^{s,p}(\mathbb{R}^n)$ for $-1/p' < s < 1/p$. Then continuity and the norm estimate for $r^+A: H^{s,p}(\mathbb{R}^n_+) \to H^{s-m,p}(\mathbb{R}^n_+)$ follow from Theorem 1. By Theorem 1 we also have the continuity $r^+A: H^{s,p}_0(\overline{\mathbb{R}}^n_+) \to H^{s-m,p}(\mathbb{R}^n_+)$ for all s. For $s > 1/p$ one concludes exactly as in the proof of 2.3.2.4, Theorem 2. Then the continuous extension $r^+A: H_0^{s,p}(\overline{\mathbb{R}}_+^n) \to H_0^{s-m,p}(\overline{\mathbb{R}}_+^n)$ for $s - m < 1/p$ follows by duality. □

From the interpolation property of the Sobolev spaces $H^{s,p}$ and the Besov spaces $B^{s,p}$ mentioned in 1.2.1.2 we get the

Corollary 3. *The assertions of Theorem 2 remain valid for $B^{s,p}$ instead of $H^{s,p}$ even if the restriction $s - 1/p \notin \mathbb{Z}$ in* (ii) *and* (iii) *is dropped.*

Finally discuss conditions on the orders of PDO, potential, trace and Green operators implying continuity from $H^{s,p}$ to $H^{s,q}$ (in general $p \neq q$).

Theorem 4. *Assume $1 < p \leqq 2 \leqq q < \infty$. Let $\sigma(A)(x, \xi) \in S^m(\mathbb{R}^n \times \mathbb{R}^n)$ have compact support with respect to x. The corresponding PDO A admits a continuous extension*

$$A: H^{s,p}(\mathbb{R}^n) \to H^{s',q}(\mathbb{R}^n),$$

when $m \leqq -n\,(1/p - 1/q) + s - s'$ and the norm is bounded by a sum of semi-norms of $\sigma(A)$. Let the condition on the order m be satisfied. Then any $A \in L^m(\mathbb{R}^n)$ has a continuous extension

$$A: H^{s,p}_{\text{comp}}(\mathbb{R}^n) \to H^{s',q}_{\text{loc}}(\mathbb{R}^n).$$

For a proof cf. Hörmander [4]. The norm estimate is left to the reader.
By similar arguments as in the proof of Theorem 2 one can show

Theorem 5. *Let p, q be as in Theorem 4.* (i) *For $s - s' - m - 1/p' \geqq n(1/p - 1/q)$ any potential operator $K \in \text{Op}\,(\mathfrak{K}^m)$ has a continuous extension*

$$K: H^{s,p}_{\text{comp}}(\mathbb{R}^{n-1}) \to H^{s',q}_{\text{loc}}(\mathbb{R}^n_+).$$

(ii) *For $s - s' - m - 1/p \geqq n\,(1/p - 1/q)$ any trace operator $r'T \in \text{Op}\,(\mathfrak{T}^{m,d})$ has a continuous extension*

$$r'T: H^{s,p}_{\text{comp}}(\overline{\mathbb{R}}^n_+) \to H^{s',q}_{\text{loc}}(\mathbb{R}^{n-1}),$$

$s > d - 1/p'$, $s - 1/p \notin \mathbb{Z}$.

(iii) *For $s - s' - m - 1 \geqq n\,(1/p - 1/q)$ any Green operator $r'B \in \text{Op}\,(\mathfrak{B}^{m,d})$ has a continuous extension*

$$r'B: H^{s,p}_{\text{comp}}(\overline{\mathbb{R}}^n_+) \to H^{s',q}_{\text{loc}}(\mathbb{R}^n_+),$$

$s > d - 1/p'$, $s - 1/p \notin \mathbb{Z}$.

(iv) *For $s - s' - m \geqq n\,(1/p - 1/q)$, any PDO with the transmission property $r^+A \in \text{Op}\,(\mathfrak{A}^m)$ has a continuous extension*

$$r^+A: H^{s,p}_{\text{comp}}(\overline{\mathbb{R}}^n_+) \to H^{s',q}_{\text{loc}}(\mathbb{R}^n_+),$$

$s > -1/p'$.

Moreover if the symbols have compact support with respect to x and x', respectively, one can drop the subscripts comp *and* loc *and the norm is bounded by sums of semi-norms of the symbols.*

There are continuity results in H_0^s formulated in the following

Theorem 6. *Let p, q be as in Theorem 4.*

(i) *For* $s - s' - m - 1/p' \geqq n(1/p - 1/q)$ *any potential operator* $K \in \mathrm{Op}(\mathfrak{K}^m)$ *has a continuous extension*

$$K: H^{s,p}_{\mathrm{comp}}(\mathbb{R}^{n-1}) \to H^{s',q}_{0,\mathrm{loc}}(\overline{\mathbb{R}}^n_+) \quad \text{for } s' < 1/q .$$

(ii) *For* $s - s' - m - 1/p \leqq n(1/p - 1/q)$ *any trace operator* $r'T \in \mathrm{Op}(\mathfrak{T}^{m,0})$ *has a continuous extension*

$$r'T: H^{s,p}_{0,\mathrm{comp}}(\overline{\mathbb{R}}^n_+) \to H^{s',q}_{\mathrm{loc}}(\mathbb{R}^{n-1})$$

for any $s \in \mathbb{R}$, $s - 1/p \notin \mathbb{Z}$.

(iii) *For* $s - s' - m - 1 \geqq n(1/p - 1/q)$ *any Green operator* $r'B \in \mathrm{Op}(\mathfrak{B}^{m,0})$ *has a continuous extension*

$$r'B: H^{s,p}_{0,\mathrm{comp}}(\mathbb{R}^n_+) \to H^{s',q}_{\mathrm{loc}}(\mathbb{R}^n_+) ,$$

$s \in \mathbb{R}$, $s - 1/p \notin \mathbb{Z}$, *and for* $s' < 1/q$

$$r'B: H^{s,p}_{0,\mathrm{comp}}(\overline{\mathbb{R}}^n_+) \to H^{s',q}_{0,\mathrm{loc}}(\overline{\mathbb{R}}^n_+) .$$

(iv) *For* $s - s' - m \geqq n(1/p - 1/q)$ *any PDO* $r^+A \in \mathrm{Op}(\mathfrak{A}^m)$ *has a continuous extension*

$$r^+A: H^{s,p}_{0,\mathrm{comp}}(\overline{\mathbb{R}}^+_n) \to H^{s',q}_{\mathrm{loc}}(\mathbb{R}^n_+)$$

for any $s \in \mathbb{R}$ *and*

$$r^+A: H^{s,p}_{0,\mathrm{comp}}(\overline{\mathbb{R}}^n_+) \to H^{s',q}_{0,\mathrm{loc}}(\overline{\mathbb{R}}^n_+)$$

if $s' < 1/q$.

Moreover if the symbols have compact support with respect to x and x', respectively, one can drop the subscripts comp *and* loc *and the norm of the extensions is bounded by a sum of semi-norms of the symbols.*

Note that any PDO $A \in L^m$ has an extension as in the first assertion in (iv), i.e. the transmission property is not necessary for this extension. By interpolation a similar result is easily obtained for the Besov spaces $B^{s,p}$ and $B_0^{s,p}$. There is an obvious modification of 2.3.2.4, Remark 9 applicable to the spaces $H^{s,p}$ and $H_0^{s,p}$.

2.3.2.6. Continuity in Hölder spaces

Denote by $C^m(\mathbb{R}^n)$, $m \in \mathbb{Z}_+$ the space of all functions u on $\mathbb{R}^n$ having continuous bounded derivatives up to the order m with the norm

$$||u||_{C^m} = \sum_{|\alpha| \leqq m} \sup_x |D^\alpha u(x)| .$$

For $0 < t < 1$ let $C^t(\mathbb{R}^n)$ be the space of all continuous functions u on $\mathbb{R}^n$ with

$$||u||_{C^t} = \sup_{x \neq y} \frac{|u(x) - u(y)|}{|x - y|^t} < \infty .$$

Define $C^{m+t}(\mathbb{R}^n)$, $m \in \mathbb{Z}_+$, $0 < t < 1$ as the space of all functions $u \in C^m(\mathbb{R}^n)$ such that

$$\|u\|_{C^{m+t}} = \|u\|_{C^m} + \sum_{|\alpha|=m} \sup_{x \neq y} \frac{|D^\alpha u(x) - D^\alpha u(y)|}{|x-y|^t} < \infty .$$

These spaces are Banach spaces.

Note that $C^{m+t}(\mathbb{R}^n)$ and $C^{m'+t'}(\mathbb{R}^n)$, $m \neq m'$ or $t \neq t'$, is an interpolation pair. There is an interpolation method (cf. TRIEBEL [1]) with the property

$$\left(C^{m+t}(\mathbb{R}^n), C^{m'+t'}(\mathbb{R}^n)\right)_\theta = C^s(\mathbb{R}^n)\,, \quad 0 < \theta < 1$$

$s = (1-\theta)(m+t) + \theta(m'+t')$ for $s \notin \mathbb{Z}_+$. For $s \in \mathbb{Z}_+$ define $\mathscr{C}^s(\mathbb{R}^n) = \left(C^{m+t}(\mathbb{R}^n), C^{m'+t'}(\mathbb{R}^n)\right)$. Then $\mathscr{C}^s(\mathbb{R}^n)$ is different from $C^s(\mathbb{R}^n)$. The same definitions make sense for domains and for manifolds (maybe with boundary).

Proposition 1. *For X a compact manifold the embedding*

$$\mathscr{C}^t(X) \subset \mathscr{C}^{t'}(X) \quad \textit{for} \quad t > t'$$

is compact.

Proposition 2. *For any $t \in \mathbb{R}_+$ the restriction operator r' defines a continuous mapping*

$$r' : \mathscr{C}^t(\overline{\mathbb{R}}^n_+) \to \mathscr{C}^t(\mathbb{R}^{n-1})\,.$$

Denote by $\mathscr{C}^t_{\text{loc}}(\Omega)$ the space of all continuous u in Ω with the property that $\varphi u \in \mathscr{C}^t(\Omega)$ follows for any $\varphi \in C_0^\infty(\Omega)$. After reduction of the order of a PDO to zero and since classical PDOs of order zero can be represented as singular integrals, classical results yield (cf. MEYER [1])

Theorem 3. *Any PDO $A \in L^m_{\text{cl}}(\Omega)$, $\Omega \subseteq \mathbb{R}^n$ open, defines a continuous operator*

$$A : \mathscr{C}^t_{\text{comp}}(\Omega) \to \mathscr{C}^{t-m}_{\text{loc}}(\Omega) \tag{1}$$

for any $t \in \mathbb{R}_+$, $t > m$. If A is defined by $\sigma(A)(x, \xi)$ having compact support with respect to x, the norm of the extension (1) *is bounded by a sum of semi-norms of $\sigma(A) \in S^m_{\text{cl}}$.*

Theorem 4. *Any potential operator $K \in$* Op $(\mathfrak{K}^m)$ *defines a continuous mapping*

$$K : \mathscr{C}^t_{\text{comp}}(\mathbb{R}^{n-1}) \to \mathscr{C}^{t-m-1}_{\text{loc}}(\overline{\mathbb{R}}^n_+) \tag{2}$$

for any $t > m+1$. If K is defined by $\sigma(K)(x', \xi', v) \in \mathfrak{K}^m$ having compact support with respect to x', then the norm of the extension (2) *(without* comp *and* loc) *is bounded by a sum of semi-norms of $\sigma(K) \in \mathfrak{K}^m$.*

For the proof we need the following Lemma from AGMON/DOUGLIS/NIRENBERG [1, I].

Lemma 5. *Let $f(x', x_n) \in C^\infty(\overline{\mathbb{R}}^n_+ \setminus \{0\})$ satisfy the conditions*

(i) $Kv(x', x_n) \overset{\text{def}}{=} \int f(x'-y', x_n)\, v(y')\, dy'$, $v \in C_0^\infty(\mathbb{R}^{n-1})$, *is in* $C^\infty(\overline{\mathbb{R}}^n_+)$ *for any* $v \in C_0^\infty(\mathbb{R}^{n-1})$,

(ii) *for any multi-index $\alpha \in \mathbb{Z}^n_+$ and $x_n > 0$ there are estimates*

$$|D^\alpha_x f(x', x_n)| \leq c(|x'|^2 + x_n^2)^{-(n+m+|\alpha|)/2}$$

with some $m \in \mathbb{Z}_+$.

Then the operator K from (i) *defines a continuous mapping*

$$K \in C^t(\mathbb{R}^{n-1}) \to C^{t-m}(\overline{\mathbb{R}}^n_+)$$

for any $t > m$, $t \in \mathbb{Z}_+$, and the norm is bounded by a sum of constants from the estimate in (ii).

Proof of Theorem 4. We check the conditions of Lemma 5 for the function

$$f(x', x_n) = (2\pi)^{-n} \int e^{ix'\xi' + ix_n\nu} \sigma(K)(\xi', \nu) \, d\xi' \, d\nu \,. \tag{3}$$

This implies the assertion for potential operators with constant coefficients. For general $\sigma(K)$ use the arguments from the proof of 2.3.2.4, Theorem 3.

As shown in 2.3.2.1 the condition (i) is satisfied. After multiplication with $\langle\xi'\rangle^{-m-1}$ we can assume that $m = -1$. We have $f \in C^\infty(\overline{\mathbb{R}}^n_+ \setminus \{0\})$. Since

$$D^\alpha f(x', x_n) = (2\pi)^{-n} \int e^{ix'\xi' + ix_n\nu} \xi'^{\alpha'} \nu^{\alpha_n} k(\xi', \nu) \, d\xi' \, d\nu$$

and $\xi'^{\alpha'} \nu^{\alpha_n} k(\xi', \nu) \in \mathfrak{K}^{-1+|\alpha|}$, it is sufficient to prove (ii) for $\alpha = 0$. Substitution of $\nu = \langle\xi'\rangle \nu_1$ in (3) yields

$$f(x', x_n) = (2\pi)^{-n} \int e^{ix'\xi' + ix_n\langle\xi'\rangle\nu_1} \langle\xi'\rangle \, k_{[0]}(\xi', \nu_1) \, d\xi' \, d\nu_1 \,,$$

where $k_{[0]}(\xi', \nu_1) = k(\xi', \langle\xi'\rangle \nu_1) \in S^{-1}(\mathbb{R}^{n-1} \times \mathbb{R}^{n-1})$ (cf. the definition of $\mathfrak{K}^{-1}$ in 2.2.3.1) Then

$$g(\xi', x_n) = (2\pi)^{-1} \int e^{ix_n\nu} \langle\xi'\rangle \, k_{[0]}(\xi', \nu) \, d\nu \in S^0(\mathbb{R}^{n-1}) \otimes \mathcal{S}(\overline{\mathbb{R}}_+) \,.$$

The estimates

$$|g(\xi', x_n\langle\xi'\rangle)| \leqq c(1 + x_n\langle\xi'\rangle)^{-N}$$

for an arbitrary $N \in \mathbb{Z}_+$ and a suitable constant $c = c_N$ which is bounded by a sum of semi-norms of $k \in \mathfrak{K}^{-1}$ implies

$$\sup_{x'} |f(x', x_n)| \leqq c \int (1 + x_n\langle\xi'\rangle)^{-N} \, d\xi' \leqq c \int_0^\infty \varrho^{n-2}(1 + x_n\langle\varrho\rangle)^{-N} \, d\varrho \leqq c x_n^{-n+1} \,.$$

Using the estimate

$$|D^\alpha_{\xi'} g_1(x_n\langle\xi'\rangle)| \leqq c\langle\xi'\rangle^{-|\alpha|} \,, \quad g_1 \in \mathcal{S}(\overline{\mathbb{R}}_+) \,,$$

we get from

$$|x'^{\alpha'} f(x', x_n)| = |(2\pi)^{-n+1} \int e^{ix'\xi'} (-D_{\xi'})^{\alpha'} g(\xi', x_n\langle\xi'\rangle) \, d\xi'|$$

in the same way as above

$$|x'^{\alpha'} f(x', x_n)| \leqq c x_n^{-n+1} (x_n/|x'|)^l$$

for an arbitrary $l \in \mathbb{Z}_+$ and the constant c is bounded by a sum of semi-norms of k. Now the desired estimate in (ii) follows and thus the Theorem. □

Theorem 6 *Any trace operator $r'T \in \mathrm{Op}(\mathfrak{T}^{m,d})$ defines a continuous operator*

$$r'T \colon \mathcal{C}^t_{\mathrm{comp}}(\overline{\mathbb{R}}^n_+) \to \mathcal{C}^{t-m}_{\mathrm{loc}}(\mathbb{R}^{n-1}) \tag{4}$$

for any $t > \max(m, d)$. If $r'T$ is defined by $\sigma(T)(x', y', \nu)$ having compact support with respect to x', then the norm of the extension (4) *(without* comp *and* loc*) is bounded by a sum of semi-norms of $\sigma(T)$.*

Proof: For trace operators $r'T$ of the form $Qr'D^k_{x_n}$, $Q \in L^{m-k}_{\mathrm{cl}}(\mathbb{R}^{n-1})$ the assertion follows from Proposition 2 and Theorem 3. For general $\sigma(T)$ use the decomposition

$$\sigma(T)(x', \xi', \nu) = \sum_j t_j(x', \xi') \, e_j(\langle\xi'\rangle^{-1} \nu)$$

converging in $\mathfrak{T}^{m,d}$, $t_j \in S^m_{\text{cl}}$, $e_j \in H^-_0$. In view of Theorem 3 and the calculus in 2.3.2.2 it is sufficient to consider a trace operator with the symbol $e_j(\langle\xi'\rangle^{-1}\nu)$. We can assume that $\check{e}_j = F^{-1}e \in C_0^\infty(\overline{R}_+)$. Then the kernel

$$f(x'-y', x_n) = (2\pi)^{-n+1} \int e^{i(x'-y')\xi'} \langle\xi'\rangle^{-1} \check{e}(\langle\xi'\rangle x_n)\, d\xi'$$

admits near $(x'-y', x_n) = 0$ the estimate

$$|f(x'-y', x_n)| \leqq c\, |x'-y'|^{-n+1}$$

with a constant c bounded by a sum of semi-norms of $e \in H^+$ and for $(x'-y', x_n) \geqq \text{const.} > 0$ an estimate $|f(x'-y', x_n)| \leqq c(1+|x'-y'|)^{-N}$ for an arbitrary $N \in Z_+$ with suitable constant $c = c_N$. Thus we have

$$\begin{aligned} |r'Tu(x') - r'Tu(x_1')| &= \Big| \int f(x'-y', y_n)\, u(y', y_n)\, dy'\, dy_n \\ &\qquad - \int f(x_1'-y', y_n)\, u(y', y_n)\, dy'\, dy_n \Big| \\ &\leqq \int |f(y', y_n)|\, |u(x'-y', y_n) - u(x_1'-y', y_n)|\, dy'\, dy_n \\ &\leqq c\, \|u\|_{C^t}\, \|x'-x_1'\|^t, \end{aligned}$$

$0 < t < 1$, where the constant c is bounded by a sum of semi-norms of $e \in H^+$. This shows the assertion for $0 < t < 1$. For $s < t < s+1$, $s \in Z_+$, we have $D^\alpha u \in C^{t-s}$, $\alpha \in Z^n_+$, $|\alpha| \leqq s$ and $D^\alpha_{x'} r'Tu = \sum_\beta r'T_\beta D^\beta_{x'} u$ with $r'T_\beta \in \text{Op}(\mathfrak{T}^{m-|\beta|,d})$, $|\beta| \leqq |\alpha|$. Then the assertion follows for $t \notin Z_+$ and for arbitrary t by interpolation. □

Theorem 7. *Any Green operator $r'B \in \text{Op}(\mathfrak{B}^{m,d})$ defines a continuous operator*

$$r'B\colon \mathscr{C}^t_{\text{comp}}(\overline{R}^n_+) \to \mathscr{C}^{t-m-1}_{\text{loc}}(\overline{R}^n_+) \tag{5}$$

for any $t > \max(m, d)$. If $r'B$ is defined by $\sigma(B)(x', \xi', \nu, \tau)$ having compact support with respect to x', then the norm of the extension (5) (*without* comp *and* loc) *is bounded by a sum of semi-norms of $\sigma(B) \in \mathfrak{B}^{m,d}$.*

Proof: The assertions immediately follow from Theorem 4 and Theorem 6, since there is a decomposition $\sigma(B) = \sum_j k_j t_j$ converging in $\mathfrak{B}^{m,d}$ (cf. 2.2.4.1). □

Theorem 8. *Any PDO with the transmission property $r^+A \in \text{Op}(\mathfrak{A}^m)$ defines a continuous operator*

$$r^+A\colon \mathscr{C}^t_{\text{comp}}(\overline{R}^n_+) \to \mathscr{C}^{t-m}_{\text{loc}}(\overline{R}^n_+) \tag{6}$$

for any $t > m$. If A is defined by $\sigma(A)(x, \xi)$ having compact support with respect to x the norm of the extension (6) (*without* comp *and* loc) *is bounded by a sum of semi-norms of $\sigma(A) \in \mathfrak{A}^m$.*

Proof: We can consider r^+A as a sum of the PDO A applied to an extension $lu \in \mathscr{C}^t(R^n)$ of u into R^n_- and restriction to R^n_+ and a Green operator $r'B_A \in \text{Op}(\mathfrak{B}^{m-1,m})$ the symbol of which is given in 2.3.2.3. Hence it depends continuously on the symbol $\sigma(A)$. The assertion now follows from Theorem 3 and Theorem 7. □

Summarize the above results in the following

Corollary 9. *Any $\mathcal{A} \in \mathfrak{S}^{m,d}(\overline{R}^n_+, R^{n-1}; \mathbb{C}^k, \mathbb{C}^{k'}, \mathbb{C}^j, \mathbb{C}^{j'})$ defines a continuous operator*

$$\mathcal{A}\colon \begin{matrix} \mathscr{C}^t_{\text{comp}}(\overline{R}^n_+, \mathbb{C}^k) \\ \oplus \\ \mathscr{C}^{t+1}_{\text{comp}}(R^{n-1}, \mathbb{C}^j) \end{matrix} \to \begin{matrix} \mathscr{C}^{t-m}_{\text{loc}}(\overline{R}^n_+, \mathbb{C}^{k'}) \\ \oplus \\ \mathscr{C}^{t-m+1}_{\text{loc}}(R^{n-1}, \mathbb{C}^{j'}) \end{matrix}$$

for any $t > \max(m, d)$. *For* $(a(x, \xi), \mathfrak{a}(x', \xi'))$ *with compact support with respect to* x *and* x', *respectively, the mapping*

$$\mathfrak{A}^m \times \mathfrak{R}^{m,d} \ni (a(x, \xi), \mathfrak{a}(x', \xi')) \mapsto \mathcal{A} = \mathrm{Op}\,(a, \mathfrak{a})$$
$$\in \mathcal{L}(\mathcal{E}^t(\overline{R}^n_+, C^k) \oplus \mathcal{E}^{t+1}(R^{n-1}, C^j), \mathcal{E}^{t-m}(\overline{R}^n_+, C^{k'}) \oplus \mathcal{E}^{t-m+1}(R^{n-1}, C^{j'}))$$

is continuous. For operators with constant coefficients or with stabilizing coefficients there is an analogon to 2.3.2.4, *Remark* 9.

2.3.3. Operators on Manifolds

2.3.3.1. The Class $\mathfrak{G}$

In this section we study PDOs, potential, trace and Green operators on manifolds with boundary. The local definitions are given in 2.3.2.1. Most of the results immediately follow from the local theory and we only sketch the proofs. An essential point is the interpretation of the principal boundary symbol as a bundle morphism.

In the following let X be a compact C^∞ manifold with boundary $Y = \partial X$. The definition of a PDO A with the transmission property on X was given in 2.3.1.2. It defines a continuous mapping

$$r^+A: C^\infty(X, E) \to C^\infty(X, F)$$

$(E, F \in \mathrm{Vect}\,(X))$. Point out that in an arbitrary coordinate neighbourhood away from the boundary this is a usual PDO. Near Y the PDO r^+A is applied to functions extended by zero into R^n_-.

If X^0 is a neighbouring manifold of X and j^+ denotes the extension of sections in bundles over X by zero, for an arbitrary extension A^0 of the operator A to X^0 we have

$$r^+Au = r^+A^0j^+u\,, \quad u \in C^\infty(X, E)\,,$$

where A^0 is applied to j^+ in L^2 sense.

In the following we shall restrict ourselves to the subspace $\mathrm{Op}\,(\mathfrak{A}^m) \subset \mathrm{Op}\,(\boldsymbol{\mathfrak{A}}^m)$ of classical PDOs with transmission property, i.e. in arbitrary local coordinates and local trivializations of the bundles the symbol has an asymptotic expansion in homogeneous functions.

Denote by Ω the interior of the manifold X and set $E_\Omega = E|_\Omega$ for $E \in \mathrm{Vect}\,(X)$. As noted in 1.2.4.1 any PDO $A \in L^m_{\mathrm{cl}}(\Omega; E_\Omega, F_\Omega)$ possesses a uniquely defined principal symbol σ_A. In local coordinates over U (for fixed trivializations of the bundles) $\sigma_A(x, \xi)$ is the principal part of the local symbol $a_U(x, \xi)$ $((x, \xi)$ coordinates in $T^*U)$, i.e. the matrix consisting of the principal parts of the homogeneity m of the elements of the matrix $a_U(x, \xi)$. The behaviour under coordinate transformations and change of the trivializations gives rise to the interpretation as a bundle morphism

$$\sigma_A: \pi^*_\Omega E_\Omega \to \pi^*_\Omega F_\Omega$$

$\pi_\Omega: T^*\Omega \setminus 0 \to \Omega$ being homogeneous with respect to the R_+ action in the fibres of $T^*\Omega$ of order m, i.e.

$$\sigma_A(x, \lambda\xi) = \lambda^m \sigma_A(x, \xi)\,, \quad \lambda > 0\,,$$

$(x, \xi) \in T^*\Omega \setminus 0\,.$

For PDOs possessing an extension to X^0 the principal symbol yields a homogeneous bundle morphism

$$\sigma_A: \pi^*E \to \pi^*F\,,$$

$\pi: T^*X \setminus 0 \to X$.

Denote by $\mathfrak{A}^{(m)}(X; E, F)$ the space of all bundle morphisms $\pi^* E \to \pi^*F$ being positively homogeneous of order m and having the transmission property with respect to Y (cf. 2.2.2.3).

According to 2.2.2.3, Proposition 1, among all homogeneous bundle morphisms $\pi^*E \to \pi^*F$ of order m, the elements of $\mathfrak{A}^{(m)}(X; E, F)$ are characterized by the symmetry condition

$$D_{x_n}^k D_{\xi'}^\alpha \sigma_A(x, \xi)\big|_{x_n=0,\, \xi'=0,\, \xi_n=1} = e^{i\pi(m-|\alpha|)} D_{x_n}^k D_{\xi'}^\alpha \sigma_A(x, \xi)\big|_{x_n=0,\, \xi'=0,\, \xi_n=-1}$$

for arbitrary $k \in \mathbb{Z}_+, \alpha \in \mathbb{Z}_+^{n-1}$ in local coordinates near the boundary.

As a consequence of 1.2.4.2, Theorem 1 we get

Theorem 1. *The sequence*

$$0 \to \operatorname{Op}\left(\mathfrak{A}^{m-1}(X; E, F)\right) \xrightarrow{i} \operatorname{Op}\left(\mathfrak{A}^m(X; E, F)\right) \xrightarrow{\sigma} \mathfrak{A}^{(m)}(X; E, F) \to 0$$

(i denotes the natural inclusion and σ the principal symbol map) is exact.

For a proof it is sufficient to remark that the construction of an operator with given principal symbol in the proof of 1.2.4.1, Theorem 1 yields an operator with the transmission property, since the lower order parts of the symbol in local coordinates are obtained from the principal symbol by derivatives and multiplication with smooth functions.

A fixed right inverse of σ is denoted by Op. Since for fixed m there is a bijective correspondence between a positively homogeneous bundle morphism $\pi^*E \to \pi^*F$ and their restrictions to the sphere bundle $\pi_1^*E \to \pi_1^*F$ ($\pi_1: S^*X \to X$, S^*X defined with respect to some fixed Riemannian metric on X), we can apply Op to bundle morphisms $\pi_1^*E \to \pi_1^*F$.

Similarly as in the half space situation consider matrices of operators

$$\mathcal{A} = \begin{pmatrix} r^+A + r'B & K \\ r'T & Q \end{pmatrix} : \begin{matrix} C^\infty(X, E) \\ \oplus \\ C^\infty(Y, J) \end{matrix} \to \begin{matrix} C^\infty(X, F) \\ \oplus \\ C^\infty(Y, Q) \end{matrix} \qquad (1)$$

$Y = \partial X$, $E, F \in \operatorname{Vect}(X)$, $J, G \in \operatorname{Vect}(Y)$, where $r^+A \in \operatorname{Op}(\mathfrak{A}^m)(X; E, F)$, $Q \in L^{m'}(Y; J, G)$ for some m, m'. The classes of Green operators $r'B: C^\infty(X, E) \to C^\infty(X, F)$, of potential operators $K: C^\infty(Y, J) \to C^\infty(X, F)$ and trace operators $r'T: C^\infty(X, E) \to C^\infty(Y, G)$ are to be defined.

Let $U \subset X$ be open, $U' \overset{\text{def}}{=} U \cap Y$ (U' may be empty) such that $E|_U$, $F|_U$ and $J|_{U'}$, $G|_{U'}$ are trivial. Denote by $\chi_E: E|_U \to \Omega \times \mathbb{C}^k$, $\chi_F: F|_U \to \Omega \times \mathbb{C}^{k'}$, $\chi_J: J|_{U'} \to \Omega' \times \mathbb{C}^j$, $\chi_G: G|_{U'} \to \Omega' \times \mathbb{C}^{j'}$ trivializations (k, k', j, j' are the fibre dimensions of E, F, J and G, respectively). The commutative diagram

$$\begin{array}{ccc} C^\infty(X, E) \oplus C^\infty(Y, J) & \xrightarrow{\mathcal{A}} & C^\infty(X, F) \oplus C^\infty(Y, G) \\ \uparrow \chi_E^* \oplus \chi_J^* & & \downarrow (\chi_F)_* \oplus (\chi_G)_* \\ C_0^\infty(\Omega, \mathbb{C}^k) \oplus C_0^\infty(\Omega', \mathbb{C}^j) & \xrightarrow{\mathcal{A}_U} & C^\infty(\Omega, \mathbb{C}^{k'}) \oplus C^\infty(\Omega', \mathbb{C}^{j'}) \end{array}$$

defines an operator $\mathcal{A}_U$ called a *local representation of* $\mathcal{A}$ over U (with respect to given trivializations).

Definition 2. An operator $\mathcal{A}$ of the form (1) belongs per def. to Op $(\mathfrak{S}^{m,d})$ $(X, Y; E, F, J, G)$ or shorter Op $(\mathfrak{S}^{m,d})$ (X, Y) if for any open set $U \subset X$, $U' = U \cap Y$ such that $E|_U$, $F|_U$, $J|_{U'}$, and $G|_{U'}$ are trivial the induced operator $\mathcal{A}_U$ belongs to Op $(\mathfrak{S}^{m,d})$ (Ω, Ω') in the half space. Any complete symbol $\sigma(\mathcal{A}_U) = (a_U, \mathfrak{a}_{U'}) \in \mathfrak{S}^{m,d}$ of $\mathcal{A}_U$ is called local symbol of $\mathcal{A}$ (with respect to given trivializations).

The subspace Op $(\mathfrak{S}^{m,d}) \subset$ Op $(\mathfrak{S}^{m,d})$ is characterized by the property that any local symbol has an asymptotic expansion in homogeneous ones. We shall mainly deal with Op $(\mathfrak{S}^{m,d})$ (X, Y) and denote this space for abbreviation by $\mathfrak{G}^{m,d}(X, Y)$ $\left(\mathfrak{G}^m = \bigcup_{d \geq 0} \mathfrak{G}^{m,d}\right)$.

Note that in the above definition we can take $-\infty \leqq m \leqq m$. Operators in $\mathfrak{G}^{-\infty,d} =$ Op $(\mathfrak{S}^{-\infty,d})$ are called *smoothing operators of type d*.

Any smoothing operator $\mathcal{A} \in \mathfrak{G}^{-\infty,d}(X, Y)$

$$\mathcal{A}\colon \begin{matrix} C^\infty(X, E) \\ \oplus \\ C^\infty(Y, J) \end{matrix} \to \begin{matrix} C^\infty(X, F) \\ \oplus \\ C^\infty(Y, G) \end{matrix}$$

$E, F \in$ Vect (X), $J: G \in$ Vect (Y), defines a compact mapping.

For $\mathcal{A} \in$ Op $(\mathfrak{S}^{m,d})$ (X, Y) of the form (1) the component $r'B$ is called *Green operator*, K is called *potential operator* and $r'T$ is called *trace operator*. The space of all such Green operators is denoted by Op $(\mathfrak{B}^{m-1,d})$, the space of all such potential operators by Op $(\mathfrak{K}^m)$ and the space of all such trace operators by Op $(\mathfrak{T}^{m-1,d})$.

From the results about Op $(\mathfrak{S}^{m,d})$ in the half space (cf. 2.3.1 and 2.3.2) it is clear that we could start from operators

$$\mathcal{A} = \begin{pmatrix} r^+A + r'B & K \\ r'T & Q \end{pmatrix} : \begin{matrix} C^\infty(X, E) \\ \oplus \\ C^\infty(Y, J) \end{matrix} \to \begin{matrix} \mathcal{D}'(X, F) \\ \oplus \\ \mathcal{D}'(Y, G) . \end{matrix}$$

Then the condition that this operator is locally in Op $(\mathfrak{S}^{m,d})$ of the half space implies that the image of $\mathcal{A}$ in fact belongs to C^∞ (cf. 2.3.1.2, Theorem 1 and 2.3.2.2, Corollary 3).

Let $\mathcal{A} \in \mathfrak{G}^{m,d}(X, Y)$, $U \subset X$, be an open set for which the bundles E, F, J, G admit trivializations. If U is an interior coordinate neighbourhood, that means $U \cap Y = \emptyset$, then $\mathcal{A}_U$ is modulo $\mathfrak{G}^{-\infty,d}$ equal to the PDO A_U, the local representation of r^+A over U. In fact, as mentioned in 2.3.2.2, trace and Green operators restricted to functions with support in a fixed compact set in $\mathbb{R}^n_+$ are smoothing ones (of given type d). If U is a boundary coordinate neighbourhood, i.e. $U' = U \cap Y \neq \emptyset$, the operator $\mathcal{A}_U$ actually belongs to Op $(\mathfrak{S}^{m,d})$ and has a complete symbol

$$\sigma(\mathcal{A}_U) = (a_U, \mathfrak{a}_{U'}) \in \mathfrak{S}^{m,d}.$$

In order to give a global interpretation of the principal boundary symbol we study the behaviour of $\sigma(\mathcal{A}_U)$ under the change of trivializations.

Theorem 3. *Denote by* $\mathfrak{a}^0_U(x', \xi') \in \mathfrak{R}^{(m),d}(\Omega' \times \mathbb{R}^{n-1})$ *the principal part being positively homogeneous in* ξ' *of order* m *of the boundary symbol*

$$\mathfrak{a}_U(x', \xi') \in \mathfrak{R}^{m,d}(\Omega' \times \mathbb{R}^{n-1})$$

with respect to a trivialization over U. Then $\mathfrak{a}^0_U(x', \xi')$ is a local representation of a bundle morphism

$$
\sigma_Y(\mathcal{A}):\quad \begin{array}{c} p^*E' \otimes H^+ \\ \oplus \\ p^*J \end{array} \quad \to \quad \begin{array}{c} p^*F' \otimes H^+ \\ \oplus \\ p^*G \end{array} \tag{2}
$$

$p\colon S^*Y \to Y,\ E' = E|_Y,\ F' = F|_Y$.

The bundle morphism $\sigma_Y(\mathcal{A})$ is called *global principal boundary symbol* of $\mathcal{A}$.

Proof: Let U, V be coordinate neighbourhoods near the boundary and $U \cap V \neq \emptyset$. Let

$$
\begin{gathered}
\chi^E_U\colon E|_U \to \Omega \times \mathbb{C}^k, \quad \chi^E_V\colon E|_V \to \Omega' \times \mathbb{C}^k, \quad \chi^F_U\colon F|_U \to \Omega \times \mathbb{C}^{k'}, \\
\chi^F_V\colon F|_V \to \Omega_1 \times \mathbb{C}^{k'}, \quad \chi^J_{U'}\colon J|_{U'} \to \Omega' \times \mathbb{C}^j, \quad \chi^J_{V'}\colon J|_{V'} \to \Omega_1' \times \mathbb{C}^j, \\
\chi^G_{U'}\colon G|_{U'} \to \Omega' \times \mathbb{C}^{j'}, \quad \chi^G_{V'}\colon G|_{V'} \to \Omega' \times \mathbb{C}^{j'}
\end{gathered}
$$

be trivializations. Over $U \cap V$ and $U' \cap V'$, respectively, we have

$$
\begin{gathered}
\chi^E_V \circ (\chi^E_U)^{-1} = g^E_{VU}\varkappa^*_{UV}, \qquad \chi^F_V \circ (\chi^F_U)^{-1} = g^F_{VU}\varkappa^*_{UV}, \\
\chi^J_{V'} \circ (\chi^J_{U'})^{-1} = g^J_{V'U'}\varkappa^*_{U'V'}, \quad \chi^G_{V'} \circ (\chi^G_{U'})^{-1} = g^G_{V'U'}\varkappa^*_{U'V'}
\end{gathered}
$$

where $\varkappa^*_{\dots}$ denotes the change of coordinates and g^E_{VU} ... the cocycle in the bundle E and so on. Since there is a neighbourhood of Y in X diffeomorphic to $Y \times [0, 1)$, the coordinate transformations $\varkappa^*$ can be assumed to be independent of the normal variable x_n. Moreover the retraction of $Y \times [0, 1)$ to Y shows that we can assume that the cocycle of the bundles E, F is independent of x_n.

Then we get the following commutative diagram

$$
\begin{array}{c|ccc|c}
 & C_0^\infty(\Omega_1, \mathbb{C}^k) \oplus C_0^\infty(\Omega_1', \mathbb{C}^j) & \xrightarrow{\mathcal{A}_V} & C^\infty(\Omega_1, \mathbb{C}^{k'}) \oplus C^\infty(\Omega_1', \mathbb{C}^{j'}) & \\
g^E_{VU}\varkappa^*_{UV} & \big\downarrow (\chi^E_V)^* \oplus (\chi^J_{V'})^* & & \big\uparrow (\chi^F_V)_* \oplus (\chi^G_{V'})_* & g^F_{VU}\varkappa^*_{UV} \\
\oplus & C^\infty(X, E) \oplus C^\infty(Y, J) & \xrightarrow{\mathcal{A}} & C^\infty(X, F) \oplus C^\infty(Y, G) & \oplus \\
g^J_{V'U'}\varkappa^*_{U'V'} & \big\uparrow (\chi^E_U)^* \oplus (\chi^J_{U'})^* & & \big\downarrow (\chi^F_U)_* \oplus (\chi^G_{U'})_* & g^G_{V'U'}\varkappa^*_{U'V'} \\
 & C_0^\infty(\Omega, \mathbb{C}^k) \oplus C_0^\infty(\Omega', \mathbb{C}^j) & \xrightarrow{\mathcal{A}_U} & C^\infty(\Omega, \mathbb{C}^{k'}) \oplus C^\infty(\Omega', \mathbb{C}^{j'}) &
\end{array}
$$

It follows that

$$
\begin{aligned}
\mathcal{A}_V &= (g^F_{VU}\varkappa^*_{UV} \oplus g^G_{V'U'}\varkappa^*_{U'V'})\, \mathcal{A}_U (g^E_{VU}\varkappa^*_{UV} \oplus g^J_{V'U'}\varkappa^*_{U'V'})^{-1} \\
&= (g^{F'}_{V'U'} \oplus g^G_{V'U'})\, \varkappa^*_{U'V'} \mathcal{A}_U \varkappa^*_{V'U'} (g^{E'}_{V'U'} \oplus g^J_{V'U'})^{-1},
\end{aligned}
$$

where the special properties of $\varkappa^*_{U'V'}$ and g_{UV} have been used. Under the assumption that near the boundary the interior symbol is independent of x_n we get from 2.3.2.2, Proposition 4 and 5 that

$$
\mathfrak{a}_{V'}(y', \eta') \sim (g^{F'}_{V'U'} \oplus g^G_{V'U'}) \circ \big(\mathfrak{a}_{U'}(x', \xi')\big)_{\varkappa U', V'} \circ (g^{E'}_{V'U'} \oplus g^J_{V'U'})^{-1} \tag{3}
$$

where $\circ$ denotes the composition of complete boundary symbols according to 2.3.2.2. Proposition 4 and $(\dots)_\varkappa$ denotes the boundary symbol of $\varkappa_*\mathcal{A}$ (cf. 2.3.2.2, Proposition 5).

For the principal symbols we get

$$
\begin{aligned}
&\mathfrak{a}^0_V(y', \eta') \\
&= \big(g^{F'}_{V'U'}(x') \oplus g^G_{V'U'}(x')\big)\, \mathfrak{a}^0_U(x', \xi') \big(g^{E'}_{V'U'}(x') \oplus g^J_{V'U'}(x')\big)^{-1}\Big|_{\substack{x' = \varkappa^{-1}_{U'V'}(y') \\ \xi' = {}^t d\varkappa(x')\eta'}}
\end{aligned} \tag{4}
$$

We finish the proof by the remark that this remains true, if a depends on x_n near the boundary (cf. 2.3.2.3, Theorem 12). □

Let

$$\sigma_Y(\mathcal{A}) = \begin{pmatrix} \Pi^+\sigma_A + \Pi'\sigma_B & \sigma_K \\ \Pi'\sigma_T & \sigma_Q \end{pmatrix} \in \mathfrak{R}^{(m),d}(Y).$$

In view of Theorem 3 the components $\sigma_B, \sigma_K, \sigma_T, \sigma_Q$ of a principal boundary symbol are sections in the following bundles

$$\begin{aligned}
\sigma_B &\in C^\infty(S^*Y, \hom(p^*E', p^*F') \otimes H^+ \otimes H_d^-) \\
&\cong C^\infty(S^*Y, (p^*E')^* \otimes p^*F' \otimes H^+ \otimes H_d^-), \\
\sigma_K &\in C^\infty(S^*Y, \hom(p^*J, p^*F') \otimes H^+) \cong C^\infty(S^*Y, (p^*J)^* \otimes p^*F' \otimes H^+), \\
\sigma_T &\in C^\infty(S^*Y, \hom(p^*E', p^*G) \otimes H_d^-) \cong C^\infty(S^*Y, (p^*E')^* \otimes p^+G \otimes H_d^-), \\
\sigma_Q &\in C^\infty(S^*Y, \hom(p^*J, p^*G)) \cong C^\infty(S^*Y, (p^*J)^* \otimes p^*G).
\end{aligned}$$

We shall consider Fréchet bundles with the fibre H^+ embedded into the associated Hilbert bundle with the fibre $V^+ = (H^+)_0$ (cf. 2.1.1.1). Similarly the embedding $H_d^- \subset H_0^- \oplus \mathbb{C}^d$ induces an embedding of any bundle with the fibre H_d^- into a Hilbert bundle.

Next formulate a global version of 2.1.2.2, Lemma 4.

Proposition 4. *Let $\tilde{V} \in p^*(H^+ \otimes E')$, $\tilde{W} \subset p^*(H^+ \otimes F')$ be finite-dimensional subbundles which are pull backs under $p: S^*Y \to Y$ of bundles $V, W \in \mathrm{Vect}(Y)$.*

(i) *Let $\beta: \tilde{V} \to \tilde{W}$ be an isomorphism. Then there exists a Green boundary symbol $\Pi'\sigma_B: p^*(H^+ \otimes E') \to p^*(H^+ \otimes F')$ (of type 0) with $\beta = \Pi'\sigma_B|_{\tilde{V}}$. Moreover, a projection $\lambda: p^*(H^+ \otimes E') \to \tilde{V}$ can be represented by the Green boundary symbol $\Pi'\sigma_L$ (of type 0).*

(ii) *Any isomorphism $\varkappa: p^*W \to \tilde{W}$ can be represented as potential boundary symbol σ_K, i.e. $\sigma_K = j \circ \varkappa$, $j: \tilde{W} \to p^*(H^+ \otimes F')$ the canonical embedding.*

(iii) *Any isomorphism $\tau: \tilde{V} \to p^*V$ is induced by some trace boundary symbol $\Pi'\sigma_T$, i.e. $\tau = \Pi'\sigma_T|_{\tilde{V}}$ (σ_T of type 0).*

Proof: (ii) follows if one represents $\varkappa$ as non-vanishing section in $(p^*W)^* \otimes \tilde{W}$. It is obvious that this section in any local coordinates (for which the bundles are trivial) is a matrix of potential symbols in the usual definition possessing the invariance characterizing global potential symbols. Similarly, (iii) follows by representing τ as a non-vanishing section in $(\tilde{V})^* \otimes p^*V$. Note that the vectors in the fibres of $(V)^*$ are component-wise functions in H_0^-. Now it is of course obvious that β in (i) corresponds to a non-vanishing section in $(\tilde{V})^* \otimes W$ which is locally a matrix of usual Green symbols with the invariance characterizing global Green boundary symbols. If we apply this to the identity in $\tilde{V}^*$, the resulting Green symbol gives rise to a projection $p^*(H^+ \otimes E') \to \tilde{V}$. □

Denote by $\mathfrak{R}^{(m),d}(Y)$ the space of all boundary symbols $a(x', \xi')$ being positively homogeneous for all $\xi' \neq 0$ of orders

$$\begin{pmatrix} m, m-1 & m \\ m-1 & m \end{pmatrix} \tag{5}$$

(the integer in the matrix corresponds to the homogeneity of the functions $\sigma_A, \sigma_B \dots$). Note that multiplication with an excision function $\chi(\xi')$ vanishing near $\xi' = 0$ yields a boundary symbol in $\mathfrak{R}^{m,d}(Y)$.

Denote by $\mathfrak{S}^{(m),d}(X, Y)$ the space of all homogeneous principal symbols of operators in $\mathfrak{G}^{m,d}(X, Y)$, i.e. $\sigma(\mathcal{A}) \in \mathfrak{S}^{(m),d}(X, Y)$ consists of an interior principal symbol $\sigma_A \in \mathfrak{A}^{(m)}(X)$ and a boundary principal symbol $\sigma_Y(\mathcal{A}) \in \mathfrak{R}^{(m),d}(Y)$, which are compatible in the sense of the definition in 2.2.5.3.

Theorem 5. *The sequence*

$$0 \to \mathfrak{G}^{m-1,d}(X, Y) \xrightarrow{i} \mathfrak{G}^{m,d}(X, Y) \xrightarrow{\sigma} \mathfrak{S}^{(m),d}(X, Y) \to 0$$

is exact (i the natural inclusion and σ the principal symbol mapping).

Proof: In view of Theorem 1 and 2.1.2.3, Proposition 11 we only have to consider potential, trace and Green operators. Construct operators with given principal symbols. In the following we assume about all boundary symbols that the symbol σ_A vanishes. Extend a given principal boundary symbol from $|\xi'| = 1$ to all $\xi' \neq 0$ with the homogeneity of the components indicated in (5) and multiply with an excision function $\chi(\xi')$. In a small neighbourhood of the boundary construct potential, trace and Green operators as in the proof of 1.2.4.1, Theorem 1 using a partition of unity on Y and multiplication with a function $\varphi(x_n) \in C_0^\infty(\overline{\mathbb{R}}_+)$, $\varphi \equiv 1$ near 0. It is easily seen that we obtain operators of the desired order with given boundary symbols. □

Theorem 4 states in particular that to any global principal symbol $\sigma = (\sigma_A, \sigma_Y(\mathcal{A})) \in \mathfrak{S}^{(m),d}$ there exists modulo lower order operators a uniquely determined operator $\mathcal{A} \in \mathfrak{G}^{m,d}$ having σ as the principal symbol. For a fixed atlas of X, a fixed partition of unity and fixed local trivializations of the bundles the construction in the proof of Theorem 5 yields a right inverse of the principal symbol mapping denoted by Op.

Theorem 6. *Let $\mathcal{A}_j \in \mathfrak{G}^{m_j,d}(X, Y)$, $j \in \mathbb{Z}_+$, $m_j \to -\infty$ for $j \to \infty$. Then there exists a unique modulo $\mathfrak{G}^{-\infty,d}(X, Y)$ operator $\mathcal{A} \in \mathfrak{G}^{m,d}(X, Y)$, $m = \max_j m_j$, such that for any $N \in \mathbb{Z}_+$ there is an $N' \in \mathbb{Z}_+$ with*

$$\mathcal{A} - \sum_{j=0}^{N} \mathcal{A}_j \in \mathfrak{G}^{-N',d}(X, Y).$$

Then we write $\mathcal{A} \sim \sum_j \mathcal{A}_j$.

Proof: Consider a covering $\{U_k\}$ of X by coordinate neighbourhoods as in the proof of 1.2.4.1, Theorem 1 and a subordinated partition of unity $\{\varphi_k\}$ of functions being x_n independent near Y. Then we have $\mathcal{A}_j = \sum_{k,l} \varphi_k \mathcal{A}_j \varphi_l$, where φ denotes the direct sum of multiplication operators with φ and $\varphi|_Y$. Thus, in coordinates of $U_{kl} \supset U_k \cup U_l$, we are reduced to the local situation and take the uniquely modulo $\mathfrak{G}^{-\infty,d}$ defined operator with symbol $\sim \sum_j \sigma(\mathcal{A}_j)$ for $\mathcal{A}_{U_{kl}}$. □

Until now, in order to get short notations, we have assumed that the orders of the PDOs, the Green, potential and trace operators are related as indicated in (5). Assuming that all operators except one vanish in the matrix $\mathcal{A}$, we get the corresponding results for the various operator classes. The formulation of the corresponding assertions is left to the reader. In connection with elliptic boundary value problems we prefer to admit arbitrary orders (cf. Chapter 3).

2.3.3.2. Compositions and Adjoints

In this section we collect further results on operators in $\mathfrak{G}^{m,d}$ on manifolds. They are simple consequences of the local results. We restrict ourselves to $\mathfrak{G}^{m,d}$. The case Op $(\mathfrak{S}^{m,d})$ is similar and left to the reader.

Theorem 1. *Let*

$$\mathcal{A}_1\colon \begin{matrix} C^\infty(X,E) \\ \oplus \\ C^\infty(Y,J)\end{matrix} \to \begin{matrix} C^\infty(X,F) \\ \oplus \\ C^\infty(Y,G)\end{matrix}, \quad \mathcal{A}_2\colon \begin{matrix} C^\infty(X,F) \\ \oplus \\ C^\infty(Y,G)\end{matrix} \to \begin{matrix} C^\infty(X,F') \\ \oplus \\ C^\infty(Y,G')\end{matrix},$$

$\mathcal{A}_i \in \mathfrak{G}^{m_i,d_i}(X,Y)$. *Then* $\mathcal{A} = \mathcal{A}_2\mathcal{A}_1 \in \mathfrak{G}^{m,d}(X,Y)$, $m = m_1 + m_2$, $d = \max(m_1 + d_2, d_1)$ *and the principal symbol* $\sigma(\mathcal{A}) = (\sigma_A, \sigma_Y(\mathcal{A}))$ *is given by*

$$\sigma_A(x,\xi) = \sigma_{A_2}(x,\xi)\,\sigma_{A_1}(x,\xi),$$
$$\sigma_Y(\mathcal{A})(x',\xi') = \sigma_Y(\mathcal{A}_2)(x',\xi')\,\sigma_Y(\mathcal{A}_1)(x',\xi')$$

$\big(\sigma(\mathcal{A}_i) = (\sigma_{A_i}, \sigma_Y(\mathcal{A}_i))$ the principal symbol of $\mathcal{A}_i$ and the compositions in the sense of bundle morphisms and in the sense of the mapping 2.3.3.1. (2)$\big)$.

In particular, $\mathfrak{G}^{0,0}(X, Y; E, E, J, J)$ is an algebra.

The proof follows by localization from 2.3.2.3, Theorem 12.

In the following it will be convenient to consider operators in $\mathfrak{G}(X, Y)$ given by 2.3.3.1.(1) with leading orders of homogeneity

$$\operatorname{ord}\sigma_A = 0, \quad \operatorname{ord}\sigma_B = -1, \quad \operatorname{ord}\sigma_K = -1/2, \quad \operatorname{ord}\sigma_T = -1/2, \quad \operatorname{ord}\sigma_Q = 0. \tag{1}$$

Fix strictly positive smooth densities dx and dy on the manifolds X and Y. Then we have scalar products in $L^2(X, \mathbb{C})$ and $L^2(Y, \mathbb{C})$ given by

$$(f_1, f_2) = \int_X f_1(x)\,\overline{f_2(x)}\,dx, \quad f_1, f_2 \in L^2(X, \mathbb{C}),$$

$$(g_1, g_2) = \int_Y g_1(y)\,\overline{g_2(y)}\,dy, \quad g_1, g_2 \in L^2(Y, \mathbb{C}).$$

Let $E \in \operatorname{Vect}(X)$ be a Hermitean vector bundle, i.e. in the fibres E_x, $x \in X$, a positive definite Hermitean form $(\cdot,\cdot)_{E_x}$ smoothly depending on x is fixed. Then

$$(f_1, f_2)_E = \int_X \big(f_1(x), f_2(x)\big)_{E_x}\,dx, \quad f_1, f_2 \in L^2(X, E),$$

defines a scalar product in $L^2(X, E)$.

Let $E \in \operatorname{Vect}(X)$ be a Hermitean vector bundle and $f\colon X' \to X$ a smooth mapping. In the pull back $f^*E \in \operatorname{Vect}(X')$ a Hermitean metric induced from E is defined, i.e. with the isomorphism of the fibres $j(x)\colon (f^*E)_{f(x)} \to E_x$, $x \in X$, we set

$$\big(f_1(x_1), f_2(x)\big)_{(f^*E)_{f(x)}} = \big(j(x)\,f_1(f(x)), j(x)\,f_2(f(x))\big)_x, \quad f_1, f_2 \in L^2(X_1, f^*E).$$

For Hermitean vector bundles $E_1, E_2 \in \operatorname{Vect}(X)$ a Hermitean metric in $E_1 \otimes E_2$ is defined by

$$(f_1 \otimes g_1, f_2 \otimes g_2)_{(E_1\otimes E_2)_x} = (f_1, f_2)_{(E_1)_x}\,(g_1, g_2)_{(E_2)_x}, \quad x \in X.$$

It is called tensor product of the metrics in E_1 and E_2.

Theorem 2. *Let $\mathcal{A} \in \mathfrak{G}(X, Y; E, F, J, G)$ with order given by* (1) *where $E, F \in \operatorname{Vect}(X)$, $J, G \in \operatorname{Vect}(Y)$ are Hermitean vector bundles. Then there is a uniquely defined operator*

$\mathcal{A}^*$ in $\mathfrak{G}(X, Y; F, E, G, J)$ such that

$$(\mathcal{A}u_1, u_2)_{F\oplus G} = (u_1, \mathcal{A}^*u_2)_{E\oplus J}, \quad u_1 \in C^\infty(X, E) \oplus C^\infty(Y, J), \quad u_2 \in C^\infty(X, F) \oplus C^\infty(Y, G).$$

The principal symbol $\sigma(\mathcal{A}^*)$ of $\mathcal{A}^*$ is given by the adjoints of the principal symbols $(\sigma_A, \sigma_Y(\mathcal{A}))$ of $\mathcal{A}$, i.e.

$$\big(\sigma_{A^*}(x, \xi) f_1, f_2\big)_{p^*E(x,\xi)} = \big(f_1, \sigma_A(x, \xi) f_2\big)_{p^*F(x,\xi)}, \quad f_1 \in F_x,\ f_2 \in E_x,$$

$$p\colon S^*X \to X,$$

$$\big(\sigma_Y(\mathcal{A}^*)(x', \xi') h_1, h_2\big)_{(p_1^*E\otimes H^+\oplus p_1^*J)(x,\xi)} = \big(h, \sigma_Y(\mathcal{A})(x', \xi') h_2\big)_{(p_1^*F\otimes H^+\oplus p_1^*G)(x',\xi')},$$

$$h_1 \in F_{x'} \otimes H^+ \oplus G_{x'}, \quad h_2 \in E_{x'} \otimes H^+ \oplus J_{x'}, \quad p_1\colon S^*Y \to Y,$$

where on $p^*E, p^*F, p_1^*E \otimes H^+ \oplus p_1^*J, p_1^*F \otimes H^+ \oplus p_1^*G$ we take the induced metrics from E, F, J, G and the L^2 scalar product on H^+. Hence $\mathcal{A}^*$ has orders as in (1).

The proof easily follows by localization from 2.3.2.3, Corollary 11.

2.3.3.3. $H^{s,p}$ and Hölder Continuity on Manifolds

In this section we formulate continuity results for operators in $\mathfrak{G}(X, Y)$. The local situation was discussed in 2.3.2.4, 2.3.2.5 and 2.3.2.6. We always assume the manifolds to be compact. The corresponding results for paracompact manifolds are left to the reader.

Assume the orders of homogeneity of the components of $\mathcal{A} \in \mathfrak{G}(X, Y)$ to be given by

$$\left.\begin{aligned} &\operatorname{ord}\sigma_A = \alpha, \quad \operatorname{ord}\sigma_B = \alpha - 1, \quad \operatorname{ord}\sigma_K = \lambda, \\ &\operatorname{ord}\sigma_T = \gamma, \quad \operatorname{ord}\sigma_Q = 1 - \alpha + \lambda + \gamma. \end{aligned}\right\} \tag{1}$$

For $\alpha = \lambda = \gamma + 1 = m$ we get $\mathfrak{G}^m(X, Y)$.

Theorem 1. *Every $\mathcal{A} \in \mathfrak{G}(X, Y)$ with orders* (1) *and type $d \in \mathbb{Z}_+$ defines a continuous mapping*

$$\mathcal{A}\colon \begin{matrix} H^s(X, E) \\ \oplus \\ H^{t+\lambda+1/2}(Y, J) \end{matrix} \to \begin{matrix} H^t(X, F) \\ \oplus \\ H^{s-\gamma-1/2}(Y, G) \end{matrix}, \quad t = s - \alpha, \tag{2}$$

for any $s > d - 1/2$. For $1 < p < \infty$, $s - 1/p \notin \mathbb{Z}$, $s > d - 1/p$ there is a continuous extension in Sobolev spaces

$$\mathcal{A}\colon \begin{matrix} H^{s,p}(X, E) \\ \oplus \\ H^{t+\lambda+1/p',p}(Y, J) \end{matrix} \to \begin{matrix} H^{t,p}(X, F) \\ \oplus \\ H^{s-\gamma-1/p,p}(Y, G) \end{matrix}, \quad t = s - \alpha. \tag{3}$$

For $1 < p < \infty$, $s - 1/p \notin \mathbb{Z}$ there is a continuous extension

$$\mathcal{A}\colon \begin{matrix} H_0^{s,p}(X, E) \\ \oplus \\ H^{t+\lambda+1/p',p}(Y, J) \end{matrix} \to \begin{matrix} H^{t,p}(X, F) \\ \oplus \\ H^{s-\gamma-1/p,p}(Y, G) \end{matrix}$$

and for $t < 1/p$

$$\mathcal{A}\colon \begin{matrix} H_0^{s,p}(X, E) \\ \oplus \\ H^{t+\lambda+1/p',p}(Y, J) \end{matrix} \to \begin{matrix} H_0^{t,p}(X, F) \\ \oplus \\ H^{s-\gamma-1/p,p}(Y, G) \end{matrix}.$$

For $s > d - 1/p$ arbitrary there is a continuous extension in Besov spaces

$$\mathcal{A}: \begin{matrix} B^{s,p}(X, E) \\ \oplus \\ B^{t+\lambda+1/p',p}(Y, J) \end{matrix} \to \begin{matrix} B^{t,p}(X, F) \\ \oplus \\ B^{s-\gamma-1/p,p}(Y, G) \,. \end{matrix} \tag{4}$$

For $s > \max(\alpha, \alpha - \lambda - 1, \gamma, d)$ there is a continuous extension in Hölder spaces

$$\mathcal{A}: \begin{matrix} \mathcal{C}^s(X, E) \\ \oplus \\ \mathcal{C}^{t+\lambda+1}(Y, J) \end{matrix} \to \begin{matrix} \mathcal{C}^t(X, F) \\ \oplus \\ \mathcal{C}^{s-\gamma}(Y, G) \,, \quad t = s - \alpha \,. \end{matrix} \tag{5}$$

Any fixed right inverse Op *of the principal symbol mapping (cf. 2.3.3.1) defines a continuous mapping from the space of principal symbols (considered as C^∞ sections over the corresponding sphere bundle) into the Banach space of bounded linear operators between the given function spaces.*

Recall that $\mathcal{C}^t(X, E)$ is the space of all Hölder continuous sections of the bundle E for $t \notin \mathbb{Z}_+$ (defined by using local coordinates and local trivializations of the bundle) and that it is defined by interpolation of spaces with integer t.

In view of the embedding theorem 1.2.1.2, Proposition 12 and 2.3.2.6, Proposition 1 we get

Corollary 2. *If all orders of homogeneity of the components of $\mathcal{A} \in \mathfrak{G}(X, Y)$ are strongly less than those in (1), or in other words, if the principal symbols of given orders of homogeneity vanish, the closures (2), (3), (4), and (5) are compact operators.*

2.3.4. The Norms of Operators Modulo Compact Operators

2.3.4.1. Pseudo-Differential Operators in the Half Space

In this section we generalize the estimate of the norm of a PDO modulo compact operators by its principal symbol (cf. 1.2.4.1, Theorem 7) to PDO with the transmission property in the half space.

Theorem 1. *Let $A \in L^0_{\text{cl}}(\overline{\mathbb{R}}^n_+)$ be compactly supported and denote its principal symbol by $\sigma_A(x, \xi)$. Let $(x_0, \xi_0) \in \mathbb{R}^n_+ \times \mathbb{R}^n$, $|\xi_0| = 1$, be fixed and set*

$$R_\lambda u(x) = \lambda^{n/4} e^{i\lambda x\xi_0} u(\lambda^{1/2}(x - x_0)) \,, \quad u \in C_0^\infty(\mathbb{R}^n) \,.$$

Then we have for sufficiently large $\lambda \in \mathbb{R}_+$

(i) $||R_\lambda u||_{L^2(\mathbb{R}^n_+)} = 1$;

(ii) $R_\lambda u \to 0$ *weakly in* $L^2(\mathbb{R}^n_+)$ *for* $\lambda \to \infty$;

(iii) $\lim\limits_{\lambda \to \infty} ||R_\lambda^{-1} r^+ A R_\lambda u - \sigma_A(x_0, \xi_0)\, u||_{L^2(\mathbb{R}^n_+)} = 0$.

The proof is based on the same considerations as for the case of a classical PDO in $\mathbb{R}^n$, since, for λ sufficiently large (depending on x_0 and supp u), we have supp $R_\lambda u \subset \mathbb{R}^n_+$.

Proof: The assertion (i) is obvious, since for sufficiently large λ supp $R_\lambda u \subset \mathbb{R}^n_+$. For $u, v \in C_0^\infty(\mathbb{R}^n_+)$ and $\lambda \to \infty$

$$\begin{aligned} |(R_\lambda u(x), v(x))| &\le \lambda^{n/4} \int |u(\lambda^{1/2}(x - x_0))\, \overline{v(x)}\,|\, dx \\ &\le \lambda^{-n/4} \int |u(x)\, \overline{v(\lambda^{-1/2}\, x + x_0)}|\, dx \end{aligned}$$

tends to zero. Then

$$|(R_\lambda u(x), v(x))| \leqq ||R_\lambda u||_{L^2(\mathbb{R}^n_+)} ||v||_{L^2(\mathbb{R}^n_+)} = ||u||_{L^2(\mathbb{R}^n_+)} ||v||_{L^2(\mathbb{R}^n_+)}$$

implies for arbitrary $u, v \in L^2(\mathbb{R}^n_+)$

$$(R_\lambda u(x), v(x)) \to 0 \quad \text{for} \quad \lambda \to \infty ,$$

i.e. $R_\lambda u \to 0$ weakly in $L^2(\mathbb{R}^n_+)$. It remains to prove (iii). In view of (ii) we can assume that r^+A is defined by the symbol $\chi(\xi)\, \sigma_A(x, \xi)$ (χ an excision function), since compact operators map $R_\lambda u$ to a sequence converging to zero in the norm of $L^2(\mathbb{R}^n_+)$. From

$$\widehat{R_\lambda u}(\xi) = \lambda^{-n/4}\, e^{-ix_0(\xi-\lambda\xi_0)}\, \hat{u}((\xi - \lambda\xi_0)\, \lambda^{-1/2})$$

we get

$$R_\lambda^{-1} r^+ A R_\lambda u(x) = (2\pi)^{-n} \int e^{ix\xi}\, \chi(\lambda^{1/2}\xi + \lambda\xi_0)\, \sigma_A(x_0 + \lambda^{-1/2}x, \lambda\xi_0 + \lambda^{1/2}\xi)\, \hat{u}(\xi)\, d\xi$$
$$= (2\pi)^{-n} \int e^{ix\xi}\, \chi(\lambda^{1/2}\xi + \lambda\xi_0)\, \sigma_A(x_0 + \lambda^{-1/2}x, \xi_0 + \lambda^{-1/2}\xi)\, \hat{u}(\xi)\, d\xi .$$

For any fixed x and ξ for $\lambda \to \infty$ $\chi(\lambda^{1/2}\xi + \lambda\xi_0)\, \sigma_A(x_0 + \lambda^{-1/2}x, \xi_0 + \lambda^{-1/2}\xi)\, \hat{u}(\xi)$ tends to $\sigma_A(x_0, \xi_0)\, \hat{u}(\xi)$ and

$$|\chi(\lambda^{1/2}\xi + \lambda\xi_0)\, \sigma_A(x_0 + \lambda^{-1/2}x, \xi_0 + \lambda^{-1/2}\xi) - \sigma_A(x_0, \xi_0)|\, |\hat{u}(\xi)|$$
$$\leqq 2 \sup |\sigma_A(x, \xi)|\, |\hat{u}(\xi)| .$$

Thus $R_\lambda^{-1} r^+ A R_\lambda u(x)$ tends to $(2\pi)^{-n}\, \sigma_A(x_0, \xi_0) \int e^{ix\xi}\, \hat{u}(\xi)\, d\xi = \sigma_A(x_0, \xi_0)\, u(x)$ for any fixed x. In order to show convergence in $L^2(\mathbb{R}^n)$ let $\alpha \in \mathbb{Z}^n_+$ be an arbitrary multi-index and consider $x^\alpha R_\lambda^{-1} r^+ A R_\lambda u(x)$. Partial integration yields

$$x^\alpha R_\lambda^{-1} r^+ A R_\lambda u(x)$$
$$= (2\pi)^{-n} \int e^{ix\xi}\, D_\xi^\alpha (\chi(\lambda^{1/2}\xi + \lambda\xi_0)\, \sigma_A(x_0 + \lambda^{-1/2}x, \lambda\xi_0 + \lambda^{1/2}\xi)\, \hat{u}(\xi))\, d\xi \tag{1}$$

and $\chi(\xi)\, \sigma_A(x, \xi) \in S^0$ implies for arbitrary $\beta \in \mathbb{Z}^n_+$ $|D_\xi^\beta(\chi(\xi)\, \sigma_A(x, \xi))| \leqq c\langle\xi\rangle^{-|\beta|}$ for all $x \in \mathbb{R}^+_n$, $\xi \in \mathbb{R}$ (the x support of σ_A is compact) with a suitable constant c and thus

$$|D_\xi^\beta(\chi(\lambda^{1/2}\xi + \lambda\xi_0)\, \sigma_A(x_0 + \lambda^{-1/2}x, \lambda\xi_0 + \lambda^{1/2}\xi))| \leqq c\lambda^{|\beta|/2}\, \langle\lambda\xi_0 + \lambda^{1/2}\xi\rangle^{-|\beta|} .$$

For c one can take the semi-norm $\sup\limits_{x,\xi} |D_\xi^\beta(\chi(\xi)\, \sigma_A(x, \xi))|\, \langle\xi\rangle^{|\beta|}$ of $\chi\sigma_A \in S^0$.

Using the inequality $(1 + |a - b|^2)^{-1} = c\dfrac{1 + |a|^2}{1 + |b|^2}$, $a, b \in \mathbb{R}^n$ with a suitable constant c and taking $a = \lambda^{1/2}\xi$, $b = -\lambda\xi_0$ we get

$$\lambda^{|\beta|/2}(1 + |\lambda\xi_0 + \lambda^{1/2}\xi|^2)^{-|\beta|/2} \leqq c\, \frac{\lambda^{|\beta|/2}(1 + \lambda\, |\xi|^2)^{|\beta|/2}}{(1 + \lambda^2)^{|\beta|/2}} \leqq c(1 + |\xi|^2)^{|\beta|/2} . \tag{2}$$

In view of the estimate $|\hat{u}(\xi)| \leqq c\langle\xi\rangle^{-N}$ for arbitrary $N \in \mathbb{Z}_+$ with $c = c_N$ from (2) and the Leibniz formula we get that (1) is bounded by

$$c \int \langle\xi\rangle^{|\alpha| - N}\, d\xi < \infty$$

for $N > |\alpha| + n$. Note that this bound is independent of λ and bounded by a sum of semi-norms of the symbol $\chi(\xi)\, \sigma_A(x, \xi) \in S^0$. Together with pointwise convergence to $\sigma_A(x_0, \xi_0)\, u$ this implies convergence in $L^2(\mathbb{R}^n)$ according to Lebesgue's theorem on bounded convergence. □

Let $r^+A \in \mathrm{Op}\,(\mathfrak{A}^m) \otimes \hom\,(\mathbb{C}^k, \mathbb{C}^{k'})$, i.e. a $k' \times k$ matrix of classical PDOs with the transmission property in $\overline{\mathbb{R}}^n_+$ of order m. Assume that r^+A is compactly supported and denote the matrix of principal symbols by $\sigma_A(x, \xi)$. For fixed $(x, \xi) \in \overline{\mathbb{R}}^n_+ \times \mathbb{R}^n$ let $||\sigma_A(x, \xi)||$ be the norm of the matrix $\sigma_A(x, \xi)$ considered as continuous linear operator $\mathbb{C}^k \to \mathbb{C}^{k'}$ (the norm on $\mathbb{C}^k$ is given by the standard Hermitean scalar product). According to 2.3.2.4, Theorem 6 r^+A defines an operator in $\mathcal{L}(H^s(\mathbb{R}^n_+, \mathbb{C}^k), H^{s-m}(\mathbb{R}^n_+, \mathbb{C}^{k'}))$ for any $s > {}^1/_2$. Let $r'B \in \mathrm{Op}\,(\mathfrak{B}^{m-1,d}) \otimes \mathrm{hom}\,(\mathbb{C}^k, \mathbb{C}^{k'})$ be compactly supported. It defines an operator in $\mathcal{L}(H^s(\mathbb{R}^n_+, \mathbb{C}^k), H^{s-m}(\mathbb{R}^n_+, \mathbb{C}^{k'}))$ for any $s > -d - {}^1/_2$.

The isomorphism

$$r^+\Lambda^s_- : H^s(\mathbb{R}^n_+, \mathbb{C}^k) \to L^2(\mathbb{R}^n_+ ; \mathbb{C}^k)$$

(cf. 2.3.2.4, Proposition 2) defines a scalar product on $H^s(\mathbb{R}^n_+, \mathbb{C}^k)$ by

$$(u, v)_{H^s(\mathbb{R}^n_+, \mathbb{C}^k)} = (r^+\Lambda^s_- u, r^+\Lambda^s_- v)_{L^2(\mathbb{R}^n_+, \mathbb{C}^k)}$$

and a norm

$$||u||_{H^s(\mathbb{R}^n_+, \mathbb{C}^k)} = ||r^+\Lambda^s_- u||_{L^2(\mathbb{R}^n_+, \mathbb{C}^k)}, \quad u, v \in H^s(\mathbb{R}^n_+, \mathbb{C}^k)\,.$$

We shall consider $H^s(\mathbb{R}^n_+, \mathbb{C}^k)$ with this norm. As pointed out in 2.3.2.4 the operators with symbols $(|\xi'| \pm i\nu)^s$ in the half space are not pseudo-differential operators, since the symbols are not smooth on the sphere $|\xi| = 1$ but only continuous. Therefore the operators $\Lambda^s_\pm$ lie in the norm closure of PDOs. The first derivatives of $(|\xi'| \pm i\nu)^s$ define integrable functions on the sphere $|\xi| = 1$. If we take these functions as symbols of "PDOs" we get continuous operators in $\mathcal{L}(H^t, H^{t-s})$. The usual arguments show that $\Lambda^s_\pm$ is well-defined on manifolds modulo $\mathcal{L}(H^t, H^{t-s+1})$. Moreover the calculus of principal symbols remains valid.

Corollary 2. *Let $r^+A \in \mathrm{Op}\,(\mathfrak{A}^m) \otimes \mathrm{hom}\,(\mathbb{C}^k, \mathbb{C}^{k'})$ with principal symbol $\sigma_A(x, \xi)$ and $r'B \in \mathrm{Op}\,(\mathfrak{B}^{m-1,d}) \otimes \mathrm{hom}\,(\mathbb{C}^k, \mathbb{C}^{k'})$, both compactly supported. Then for any integer $s > d - {}^1/_2$, we have*

$$\inf_C ||r^+ A + r'B + C||_{\mathcal{L}(H^s(\mathbb{R}^n_+, \mathbb{C}^k), H^{s-m}(\mathbb{R}^n_+, \mathbb{C}^{k'}))} \geqq \sup_{x \in \mathbb{R}^n_+, |\xi| = 1} ||\sigma_A(x, \xi)|| \tag{3}$$

where the infimum is taken over all compact operators in

$$\mathcal{L}(H^s(\mathbb{R}^n_+, \mathbb{C}^k), H^{s-m}(\mathbb{R}^n_+, \mathbb{C}^{k'}))\,.$$

Proof. Using the operators $r^+\Lambda^s_-$ and $r^+\Lambda^{s-m}_-$ reduce the order to zero. The diagram

$$\begin{array}{ccc}
H^s(\mathbb{R}^n_+, \mathbb{C}^k) & \xrightarrow{r^+A + r'B + C} & H^{s-m}(\mathbb{R}^n_+, \mathbb{C}^{k'}) \\
{\scriptstyle r^+\Lambda^s_-}\downarrow & & \downarrow{\scriptstyle r^+\Lambda^{s-m}_-} \\
L^2(\mathbb{R}^n_+, \mathbb{C}^k) & \xrightarrow{r^+A_0 + r'B_0 + C_0} & L^2(\mathbb{R}^n_+, \mathbb{C}^{k'})
\end{array}$$

defines operators $r^+A_0 \in \mathrm{Op}\,(\mathfrak{A}^0) \otimes \mathrm{hom}\,(\mathbb{C}^k, \mathbb{C}^{k'})$, $r'B_0 \in \mathrm{Op}\,(\mathfrak{B}^{-1,0}) \otimes \mathrm{hom}\,(\mathbb{C}^k, \mathbb{C}^{k'})$ and $C_0 \in \mathcal{K}(L^2(\mathbb{R}^n_+, \mathbb{C}^k), L^2(\mathbb{R}^n_+, \mathbb{C}^{k'}))$. The principal symbol $\sigma_{A_0}(x, \xi)$ of A_0 is given by

$$\sigma_{A_0}(x, \xi) = \left(\delta_\varepsilon(\xi', \nu)\, |\xi'| - i\nu\right)^{s-m} \sigma_A(x, \xi)\, \left(\delta_\varepsilon(\xi', \nu)\, |\xi'| - i\nu\right)^{-s}$$

and hence $||\sigma_{A_0}(x, \xi)|| = ||\sigma_A(x, \xi)||$ for $|\xi| = 1$. Since $r^+\Lambda^s_-$ and $r^+\Lambda^{s-m}_-$ are isometries, the assertion is reduced to the case $s = m = 0$.

For $s = m = 0$ the assertion follows from Theorem 1. In fact, we shall prove that for any arbitrary Green operator $r'B \in \mathrm{Op}\,(\mathfrak{B}^{-1,0}) \otimes \mathrm{hom}\,(\mathbb{C}^k, \mathbb{C}^{k'})$ and arbitrary

$C \in \mathcal{K}(L^2(\mathbb{R}^n_+, \mathbb{C}^k), L^2(\mathbb{R}^n_+, \mathbb{C}^{k'}))$

$$\|r^+A + r'B + C\| \geqq \sup_{x \in \mathbb{R}^n_+, |\xi|=1} \|\sigma_A(x, \xi)\| . \tag{4}$$

Set $M = \sup_{x \in \mathbb{R}^n_+, |\xi|=1} \|\sigma_A(x, \xi)\|$ and let $\varepsilon > 0$ be arbitrary. Choose $(x_0, \xi_0) \in \mathbb{R}^n_+ \times \mathbb{R}^n$ and $v \in \mathbb{C}^k$, $\|v\|_{\mathbb{C}^k} = 1$ such that $\|\sigma_A(x_0, \xi_0) v\|_{\mathbb{C}^{k'}} > M - \varepsilon$. For an arbitrary scalar function $u \in C_0^\infty(\mathbb{R}^n)$, $\|u\|_{L^2(\mathbb{R}^n)} = 1$ we have $\|u \otimes v\|_{L^2(\mathbb{R}^n, \mathbb{C}^k)} = 1$. According to Theorem 1 the sequence of functions $R_\lambda(u \otimes v) = R_\lambda u \otimes v$ weakly tends to zero in $L^2(\mathbb{R}^n, \mathbb{C}^k)$ for $\lambda \to \infty$ and $\|R_\lambda u \otimes v\|_{L^2(\mathbb{R}^n, \mathbb{C}^k)} = 1$ for all λ sufficiently large. Then $r'B(R_\lambda u \otimes v)$ and $C(R_\lambda u \otimes v)$ tend to zero in the norm of $L^2(\mathbb{R}^n, \mathbb{C}^{k'})$, since C is compact, and $r'B$ restricted to functions with support in a compact set in $\mathbb{R}^n_+$ is compact, too. Moreover, by Theorem 1 (iii), we have

$$\|r^+A(R_\lambda u \otimes v)\|_{L^2(\mathbb{R}^n, \mathbb{C}^k)} \to \|\sigma_A(x_0, \xi_0) v\|_{\mathbb{C}^{k'}} .$$

This implies (4). □

Consider now classical PDOs in the half space without transmission property.

Lemma 3. *Let $r^+A \in L^0_{\mathrm{cl}}(\overline{\mathbb{R}}^n_+) \otimes \mathrm{hom}\,(\mathbb{C}^k, \mathbb{C}^{k'})$ be compactly supported with the principal symbol $\sigma_A(x, \xi)$. Then we have*

$$\inf_C \|r^+A + C\|_{\mathcal{L}(L^2(\mathbb{R}^n_+, \mathbb{C}^k), L^2(\mathbb{R}^n_+, \mathbb{C}^{k'}))} = \sup_{x \in \mathbb{R}^n_+, |\xi|=1} \|\sigma_A(x, \xi)\| , \tag{5}$$

where the infimum is taken over all operators $C \in \mathcal{K}(L^2(\mathbb{R}^n_+, \mathbb{C}^k), L^2(\mathbb{R}^n_+, \mathbb{C}^{k'}))$.

Lemma 3 corresponds to the analogous result about PDOs on $\mathbb{R}^n$ (cf. 1.2.4.1, Theorem 7). The proof of Lemma 3 will be reduced to this case.

Proof: Let $C \in \mathcal{L}(L^2(\mathbb{R}^n_+, \mathbb{C}^k), L^2(\mathbb{R}^n_+, \mathbb{C}^{k'}))$ and define $C' = j^+Cr^+$ (r^+ the restriction to $\mathbb{R}^n_+$ and j^+ the extension from $\mathbb{R}^n_+$ to $\mathbb{R}^n$ by zero), hence $C = r^+C'j^+$. For $u \in L^2(\mathbb{R}^n_+, \mathbb{C}^k)$ we have

$$\begin{aligned} \|Cu\|_{L^2(\mathbb{R}^n_+, \mathbb{C}^{k'})} &= \|r^+C'j^+u\|_{L^2(\mathbb{R}^n_+, \mathbb{C}^{k'})} \leqq \|C'j^+u\|_{L^2(\mathbb{R}^n, \mathbb{C}^{k'})} \\ &\leqq \|C'\|_{\mathcal{L}(L^2(\mathbb{R}^n, \mathbb{C}^k), L^2(\mathbb{R}^n, \mathbb{C}^k))} \|j^+u\|_{L^2(\mathbb{R}^n, \mathbb{C}^k)} \\ &\leqq \|C'\|_{\mathcal{L}(L^2(\mathbb{R}^n, \mathbb{C}^k), L^2(\mathbb{R}^n, \mathbb{C}^{k'}))} \|u\|_{L^2(\mathbb{R}^n_+, \mathbb{C}^k)} \end{aligned}$$

and therefore

$$\|C\|_{\mathcal{L}(L^2(\mathbb{R}^n_+, \mathbb{C}^k), L^2(\mathbb{R}^n_+, \mathbb{C}^{k'}))} \leqq \|C'\|_{\mathcal{L}(L^2(\mathbb{R}^n, \mathbb{C}^k), L^2(\mathbb{R}^n, \mathbb{C}^{k'}))} .$$

Note that C' is compact for a compact operator C. Then for $A \in L^0_{\mathrm{cl}}(\overline{\mathbb{R}}^n_+)$

$$\inf_C \|r^+Aj^+ + C\|_{\mathcal{L}(L^2(\mathbb{R}^n_+, \mathbb{C}^k), L^2(\mathbb{R}^n_+, \mathbb{C}^{k'}))} \leqq \inf_{C'} \|A + C'\|_{\mathcal{L}(L^2(\mathbb{R}^n, \mathbb{C}^k), L^2(\mathbb{R}^n, \mathbb{C}^{k'}))}$$

(here an arbitrary extension of the complete symbol of A into $\mathbb{R}^n_-$ is taken). By 1.2.4.1, Theorem 7 the right hand side is equal to $\sup_{x \in \mathbb{R}^n_+, |\xi|=1} \|\sigma_A(x, \xi)\|$, but the left hand side is independent of the choice of the extension of the symbol $\sigma_A(x, \xi)$. Thus we get

$$\inf_C \|r^+Aj^+ + C\|_{\mathcal{L}(L^2(\mathbb{R}^n_+, \mathbb{C}^k), L^2(\mathbb{R}^n_+, \mathbb{C}^{k'}))} \leqq \sup_{x \in \mathbb{R}^n_+, |\xi|=1} \|\sigma_A(x, \xi)\|$$

and the assertion follows from Corollary 2. □

Consider now operators in L^m_{cl}. Using the isometries

$$r^+\Lambda^s_+ : H^s_0(\overline{\mathbb{R}}^n_+) \to L^2(\mathbb{R}^n_+)$$

and

$$r^+\Lambda_-^{s-m}: H^{s-m}(\mathbb{R}_+^n) \to L^2(\mathbb{R}_+^n)$$

(cf. 2.3.2.4, Proposition 1) Lemma 3 implies

Corollary 4. *Let $r^+A \in L_{\mathrm{cl}}^m(\overline{\mathbb{R}}_+^n) \otimes \mathrm{hom}\,(\mathbb{C}^k, \mathbb{C}^{k'})$ be compactly supported with the principal symbol $\sigma_A(x, \xi)$. Then we have*

$$\inf_C \|r^+A + C\|_{\mathcal{L}(H_0^s(\overline{\mathbb{R}}_+^n, \mathbb{C}^k), H^{s-m}(\mathbb{R}_+^n, \mathbb{C}^{k'}))} = \sup_{x \in \mathbb{R}_+^n, |\xi|=1} \|\sigma_A(x, \xi)\| \tag{6}$$

where the infimum is taken over all operators $C \in \mathcal{K}(H_0^s(\overline{\mathbb{R}}_+^n, \mathbb{C}^k), H^{s-m}(\mathbb{R}_+^n, \mathbb{C}^{k'}))$.

The simple proof is left to the reader.

Remark 5. For compactly supported $A \in \mathrm{Op}\,(\mathfrak{A}^m) \otimes \mathrm{hom}\,(\mathbb{C}^k, \mathbb{C}^{k'})$ with principal symbol $\sigma_A(x, \xi)$, in general, one does not have

$$\inf_C \|r^+A + C\|_{\mathcal{L}(H^s(\mathbb{R}_+^n, \mathbb{C}^k), H^{s-m}(\mathbb{R}_+^n, \mathbb{C}^{k'}))} \leqq \sup_{x \in \mathbb{R}_+^n, |\xi|=1} \|\sigma_A(x, \xi)\|, \tag{7}$$

$C \in \mathcal{K}(H^s(\mathbb{R}_+^n, \mathbb{C}^k), H^{s-m}(\mathbb{R}_+^n, \mathbb{C}^{k'}))$. Consider, for instance, $r^+\Lambda_+^{-1} \in \mathrm{Op}\,(\mathfrak{A}^{-1})$, $r^+\Lambda_+^{-1}: H^1(\mathbb{R}_+^n) \to H^2(\mathbb{R}_+^n)$. Then $r^+\Lambda_-^2 r^+\Lambda_+^{-1} r^+\Lambda_-^{-1}: L^2(\mathbb{R}_+^n) \to L^2(\mathbb{R}_+^n)$ has the same norm and $r^+\Lambda_-^2 r^+\Lambda_+^{-1} r^+\Lambda_-^{-1} = r^+\Lambda_+^{-1}\Lambda_- + r'B$, $r'B \in \mathrm{Op}\,(\mathfrak{B}^{-1,0})$, with non-vanishing principal symbol. In 2.3.4.3 we shall prove that the norm of $r^+\Lambda_+^{-1}\Lambda_- + r'B$ is bounded from below by $\left(\sup_{|\xi|=1} |\sigma_{\Lambda_+^{-1}\Lambda_-}(\xi)|^2 + \sup_{|\xi'|=1} \|\Pi'\sigma_B(\xi')\|^2\right)^{1/2}$ and (7) is impossible.

A sufficient condition for (7) is given in the following

Lemma 6. *Let $A \in \mathrm{Op}\,(\mathfrak{A}^m) \otimes \mathrm{hom}\,(\mathbb{C}^k, \mathbb{C}^{k'})$ be compactly supported with the principal symbol $\sigma_A(x, \xi)$. Assume that r^+Alu, $u \in C_0^\infty(\overline{\mathbb{R}}_+^n, \mathbb{C}^k)$ is independent of the extension operator $l: C_0^\infty(\overline{\mathbb{R}}_+^n) \to H^t(\mathbb{R}^n)$ for some t depending on l. Then we have $r^+Aj^+ \in \mathcal{L}(H^s(\mathbb{R}_+^n, \mathbb{C}^k), H^{s-m}(\mathbb{R}_+^n, \mathbb{C}^{k'}))$ for any $s \in \mathbb{Z}$ and*

$$\inf_C \|r^+Aj^+ + C\|_{\mathcal{L}(H^s(\mathbb{R}_+^n, \mathbb{C}^k), H^{s-m}(\mathbb{R}_+^n, \mathbb{C}^{k'}))} = \sup_{x \in \mathbb{R}_+^n, |\xi|=1} \|\sigma_A(x, \xi)\| \tag{8}$$

where $C \in \mathcal{K}(H^s(\mathbb{R}_+^n, \mathbb{C}^k), H^{s-m}(\mathbb{R}_+^n, \mathbb{C}^{k'}))$.

The conditions of Lemma 6 are satisfied, for instance, for all differential operators.

Proof: Let l be a continuous extension operator $H^s(\mathbb{R}_+^n) \to H^s(\mathbb{R}^n)$. Define $C' = lCr^+$ for $C \in \mathcal{L}(H^s(\mathbb{R}_+^n, \mathbb{C}^k), H^{s-m}(\mathbb{R}_+^n, \mathbb{C}^{k'}))$. Then $C = r^+C'l$ and

$$\|C\|_{\mathcal{L}(H^s(\mathbb{R}_+^n, \mathbb{C}^k), H^{s-m}(\mathbb{R}_+^n, \mathbb{C}^{k'}))} \leqq \|C'\|_{\mathcal{L}(H^s(\mathbb{R}^n, \mathbb{C}^k), H^{s-m}(\mathbb{R}^n, \mathbb{C}^{k'}))}$$

as in the proof of Lemma 3 and

$$\inf_C \|r^+A + C\|_{\mathcal{L}(H^s(\mathbb{R}_+^n, \mathbb{C}^k), H^{s-m}(\mathbb{R}_+^n, \mathbb{C}^{k'}))} \leqq \sup_{x \in \mathbb{R}_+^n, |\xi|=1} \|\sigma_A(x, \xi)\|.$$

Then the assertion follows from Corollary 2. □

For simplicity, the results of this section have been formulated for classical PDOs only. Similar results are valid for PDOs defined with symbols $a(x, \xi) \in S^m$ for which $\lim_{\lambda \to \infty} a(x, \lambda\xi)\, \lambda^{-m}$ exists for all x, ξ.

2.3.4.2. Estimates for Boundary Symbols

The main result of this section is a theorem for Green, trace and potential operators which is analogous to 2.3.4.1, Theorem 1. This implies an estimate of the norm of Green, trace and potential operators from below by the norm of the principal boundary symbol.

Let $(x_0', \xi_0') \in \mathbb{R}^{n-1} \times \mathbb{R}^{n-1}$ be fixed, $|\xi'| = 1$. Define operators

$$R_\lambda : L^2(\mathbb{R}^{n-1}) \to L^2(\mathbb{R}^{n-1})$$

by $R_\lambda v(x') = \lambda^{(n-1)/4}\, e^{i\lambda x'\xi_0'}\, v((x' - x_0')\,\lambda^{1/2})$, $\lambda \in \mathbb{R}_+$ (cf. 2.3.4.1, Theorem 1). R_λ is unitary and for $\lambda \to \infty$ the image $R_\lambda v$ weakly tends to zero. Define a family of operators

$$S_\lambda : L^2(\mathbb{R}_+) \to L^2(\mathbb{R}_+)\,, \quad \lambda \in \mathbb{R}_+\,,$$

by $S_\lambda w(x_n) = \lambda^{1/2} w(\lambda x_n)$. Obviously, S_λ is unitary for every $\lambda \in \mathbb{R}_+$. It is easily checked that $S_\lambda w$ weakly tends to zero for $\lambda \to \infty$. In fact, for $w, w' \in C_0^\infty(\overline{\mathbb{R}}_+)$ we have

$$\begin{aligned}(S_\lambda w, w') &= \lambda^{1/2} \int w(\lambda x_n)\, \overline{w'(x_n)}\, dx_n \\ &= \lambda^{-1/2} \int w(x_n)\, \overline{w'(\lambda^{-1} x_n)}\, dx_n \to 0 \quad \text{for} \quad \lambda \to +\infty\,,\end{aligned}$$

and we have

$$|(S_\lambda w, w')| \leqq \|S_\lambda w\|_{L^2(\mathbb{R}_+)}\, \|w'\|_{L^2(\mathbb{R}_+)} = \|w\|_{L^2(\mathbb{R}_+)}\, \|w'\|_{L^2(\mathbb{R}_+)}\,.$$

Theorem 1. *Let $r'T \in \mathrm{Op}\,(\mathfrak{T}^{-1/2,0})\,(\mathbb{R}^{n-1})$ be compactly supported and denote by $\sigma_T(x', \xi', \nu)$ its principal symbol. Let $(x_0', \xi_0') \in \mathbb{R}^{n-1} \times \mathbb{R}^{n-1}$, $|\xi_0'| = 1$, be fixed. Then we have*

$$\lim_{\lambda \to \infty} \|R_\lambda^{-1} r'T(R_\lambda u \otimes S_\lambda v) - u \otimes \Pi'(\sigma_T(x_0', \xi_0', \nu)\, \hat{v}(\nu))\|_{L^2(\mathbb{R}^{n-1})} = 0$$

for arbitrary $u \in C_0^\infty(\mathbb{R}^{n-1})$, $v \in C_0^\infty(\overline{\mathbb{R}}_+)$.

Proof: Set $u_\lambda(x') = R_\lambda u(x')$, $v_\lambda(x_n) = S_\lambda v(x_n)$, $\lambda \in \mathbb{R}_+$, $u \in C_0^\infty(\mathbb{R}^{n-1})$, $v \in C_0^\infty(\overline{\mathbb{R}}_+)$. Then

$$\begin{aligned}\hat{u}_\lambda(\xi') &= F(R_\lambda u)\,(\xi') = \lambda^{n-1/4}\, e^{-ix_0'(\xi' - \lambda\xi')}\, \hat{u}((\xi' - \lambda\xi_0')\,\lambda^{-1/2})\,, \\ \hat{v}_\lambda(\nu) &= F(S_\lambda v)\,(\nu) = \lambda^{-1/2}\hat{v}(\lambda^{-1}\nu)\,.\end{aligned}$$

Since $u_\lambda \otimes v_\lambda$ weakly tends to zero in $L^2(\mathbb{R}^n_+)$, we can assume that $r'T$ is defined by the trace symbol $\chi(\xi')\,\sigma_T(x', \xi', \nu)$ (χ an arbitrary excision function). Then

$$\begin{aligned}&R_\lambda^{-1} r'T(u_\lambda \otimes v_\lambda)\,(x') \\ &= (2\pi)^{-n+1} \int e^{ix'\xi'}\, \lambda^{-1/2} \chi(\xi')\, \Pi'_\nu(\sigma_T(x_0' + \lambda^{-1/2}x', \lambda\xi_0' + \lambda^{1/2}\xi', \nu)\, \hat{u}(\xi')\, \hat{v}(\lambda^{-1}\nu))\, d\xi' \\ &= (2\pi)^{-n+1} \int e^{ix'\xi'}\, \chi(\lambda\xi_0' + \lambda^{1/2}\xi')\, \lambda^{1/2}\, \Pi'_\nu(\sigma_T(x_0' + \lambda^{-1/2}x', \lambda\xi_0' + \lambda^{1/2}\xi', \lambda\nu)\, \hat{v}(\nu))\, \hat{u}(\xi')\, d\xi'\,.\end{aligned}$$

By Lebesgue's theorem on bounded convergence for any fixed $x' \in \mathbb{R}^{n-1}$ this converges to

$$\Pi'_\nu(\sigma_T(x_0', \xi_0', \nu)\, \hat{v}(\nu))\, (2\pi)^{-n+1} \int e^{ix'\xi'}\, \hat{u}(\xi')\, d\xi' = u(x')\, \Pi'_\nu(\sigma_T(x_0', \xi_0', \nu)\, \hat{v}(\nu))\,.$$

We show now convergence in $L^2(\mathbb{R}^{n-1})$. Consider for $\alpha \in \mathbb{Z}_+^{n-1}$ arbitrary

$$x'^\alpha R_\lambda^{-1} r' T(u_\lambda \otimes v_\lambda)(x') = (2\pi)^{-n+1} \lambda^{1/2} \int e^{ix'\xi'} (-D_{\xi'})^\alpha \left[\chi(\lambda\xi_0' + \lambda^{1/2}\xi') \Pi_\nu' (\sigma_T(x_0' + \lambda^{-1/2}x', \lambda\xi_0' + \lambda^{1/2}\xi', \lambda\nu)\, \hat{v}(\nu))\, \hat{u}(\xi')\right] d\xi' .$$

According to the Leibniz formula this is bounded by

$$\sum_{|\beta| \leq |\alpha|} c_\beta \lambda^{|\beta|+1/2} \int \left|\Pi_\nu' (D_{\xi'}^\beta(\chi\sigma_T)(x_0' + \lambda^{-1/2}x', \lambda\xi_0' + \lambda^{1/2}\xi', \lambda\nu)\, \hat{v}(\nu))\, \hat{u}_\beta(\xi')\right| d\xi' \quad (1)$$

with suitable constants c_β and $\hat{u}_\beta \in \mathcal{S}(\mathbb{R}^{n-1})$. From $D_{\xi'}^\beta(\chi\sigma_T) \in \mathfrak{T}^{-|\beta|,0}$ we get an expansion

$$D_{\xi'}^\beta(\chi\sigma_T)(x', \xi', \nu) = \sum_{j=0}^{\infty} t_j(x', \xi') \frac{(1 + i\nu\langle\xi'\rangle^{-1})^j}{(1 - i\nu\langle\xi'\rangle^{-1})^{j+1}}$$

with $t_j \in S^{-|\beta|}$ rapidly decreasing, thus

$$|t_j(x_0' + \lambda^{-1/2}x', \lambda\xi_0' + \lambda^{1/2}\xi')| \leq c\langle\lambda\xi_0' + \lambda^{1/2}\xi'\rangle^{-|\beta|}$$

and for c we can take a semi-norm of t_j in $S^{-|\beta|}$. From

$$\left|\Pi_\nu'\left(\frac{(1 + i\nu\lambda\langle\lambda\xi_0' + \lambda^{1/2}\xi'\rangle^{-1})^j}{(1 - i\nu\lambda\langle\lambda\xi_0' + \lambda^{1/2}\xi'\rangle^{-1})^{j+1}}\hat{v}(\nu)\right)\right| \leq (\lambda\langle\lambda\xi_0' + \lambda^{1/2}\xi'\rangle^{-1})^{-1/2} \|\hat{v}\|_{L^2(\mathbb{R})}$$

we obtain

$$\begin{aligned} |\Pi_\nu'(D_{\xi'}^\beta(\chi\sigma_T)(x_0' + \lambda^{-1/2}x', \lambda\xi_0' + \lambda^{1/2}\xi', \lambda\nu)\, \hat{v}(\nu))| &\leq c\lambda^{-1/2}\langle\lambda\xi_0' + \lambda^{1/2}\xi'\rangle^{-|\beta|+1/2} \\ &\leq c\lambda^{-1/2}\langle\lambda^{1/2}\xi'\rangle^{|\beta|-1/2}\langle\lambda\rangle^{-|\beta|+1/2} \end{aligned}$$

and hence (1) is bounded by

$$c\int \langle\xi'\rangle^{|\alpha|-1/2-N} d\xi' < \infty$$

$(N > |\alpha| - 1/2 + n - 1)$ since $\hat{u}_\beta \in \mathcal{S}(\mathbb{R}^{n-1})$.

The pointwise convergence of $R_\lambda^{-1} r' T(u_\lambda \otimes v_\lambda)(x')$ to $u(x') \otimes \Pi'(\sigma_T(x_0', \xi_0', \nu)\, \hat{v}(\nu))$ and the estimate

$$|R_\lambda^{-1} r' T(u_\lambda \otimes v_\lambda)(x')| \leq c(1 + |x'|^2)^{-n} \in L^2(\mathbb{R}^{n-1})$$

yield the assertion using Lebesgue's theorem on bounded convergence. □

Similarly as in Theorem 1 we have for potential operators

Theorem 2. *Let $K \in \mathrm{Op}(\mathfrak{K}^{-1/2})(\mathbb{R}^{n-1})$ be compactly supported and denote by $\sigma_K(x', \xi', \nu)$ its principal symbol. Let $(x_0', \xi_0') \in \mathbb{R}^{n-1} \times \mathbb{R}^{n-1}$, $|\xi_0'| = 1$, be fixed. Then*

$$\lim_{\lambda\to\infty} \|R_\lambda^{-1} \otimes S_\lambda^{-1} K(R_\lambda v) - v \otimes F^{-1}\sigma_K(x_0', \xi_0', \nu)\|_{L^2(\mathbb{R}_+^n)} = 0$$

for arbitrary $v \in C_0^\infty(\mathbb{R}^{n-1})$ (F denotes the one-dimensional Fourier transform).

Theorem 3. *Let $r'B \in \mathrm{Op}(\mathfrak{B}^{-1,0})(\mathbb{R}^{n-1})$, be compactly supported and denote by $\sigma_B(x', \xi', \nu, \tau)$ its principal symbol. Let $(x_0', \xi_0') \in \mathbb{R}^{n-1} \times \mathbb{R}^{n-1}$, $|\xi_0'| = 1$, be fixed. Then*

$$\lim_{\lambda\to\infty} \|R_\lambda^{-1} \otimes S_\lambda^{-1} r'B(R_\lambda u \otimes S_\lambda v) - u \otimes F^{-1}\Pi_\nu'\sigma_B(x_0', \xi_0', \nu, \tau)\, Fv\|_{L^2(\mathbb{R}_+^n)} = 0$$

for arbitrary $u \in C_0^\infty(\mathbb{R}^{n-1})$, $v \in C_0^\infty(\overline{\mathbb{R}}_+)$.

Theorem 4. *Let $r^+A \in \mathrm{Op}(\mathfrak{A}^0)$ be compactly supported with the principal symbol $\sigma_A(x, \xi)$ not depending on x_n. Let $(x_0', \xi_0') \in \mathbb{R}^{n-1} \times \mathbb{R}^{n-1}$, $|\xi_0'| = 1$, be fixed. Then*

$$\lim_{\lambda\to\infty} \|R_\lambda^{-1} \otimes S_\lambda^{-1} r^+A(R_\lambda u \otimes S_\lambda v) - u \otimes F^{-1}\Pi^+\sigma_A(x_0', \xi_0', \nu)\, Fv\|_{L^2(\mathbb{R}^n_+)} = 0$$

for arbitrary $u \in C_0^\infty(\mathbb{R}^{n-1})$, $v \in C_0^\infty(\overline{\mathbb{R}}_+)$.

The proofs of Theorem 2, 3 and 4 are essentially the same as the proof of Theorem 1 with obvious modifications. They are left to the reader.

Consider now $\mathcal{A} \in \mathrm{Op}(\mathfrak{R}^{m,d})(\mathbb{R}^n_+, \mathbb{R}^{n-1}; \mathbb{C}^k, \mathbb{C}^{k'}; \mathbb{C}^j, \mathbb{C}^{j'})$,

$$\mathcal{A} = \begin{pmatrix} r^+A + r'B & K \\ r'T & Q \end{pmatrix} : \begin{matrix} C_0^\infty(\overline{\mathbb{R}}^n_+, \mathbb{C}^k) \\ \oplus \\ C_0^\infty(\mathbb{R}^{n-1}, \mathbb{C}^j) \end{matrix} \to \begin{matrix} C^\infty(\overline{\mathbb{R}}^n_+, \mathbb{C}^{k'}) \\ \oplus \\ C^\infty(\mathbb{R}^{n-1}, \mathbb{C}^{j'}) \end{matrix},$$

$r^+A \in \mathrm{Op}(\mathfrak{A}_1^m)$, $r'B \in \mathrm{Op}(\mathfrak{B}^{m-1,d})$, $K \in \mathrm{Op}(\mathfrak{K}^m)$, $r'T \in \mathrm{Op}(\mathfrak{T}^{m-1,d})$ and $Q \in L^m_{\mathrm{cl}}$ all compactly supported. Assume that the principal symbols are σ_A, σ_B, σ_K, σ_T, σ_Q and σ_A does not depend on x_n.

Set $\mathcal{H}^s(k, j) = H^s(\mathbb{R}^n_+, \mathbb{C}^k) \oplus H^{s+1/2}(\mathbb{R}^{n-1}, \mathbb{C}^j)$. In view of 2.3.2.4, Corollary 7 there is an extension of $\mathcal{A}$ as an operator in $\mathcal{L}(\mathcal{H}^s(k, j), \mathcal{H}^{s-m}(k', j'))$ for $s > d - 1/2$ arbitrary. The norm of $\mathcal{H}^s(k, j)$ let be induced from $L^2(\mathbb{R}^n_+, \mathbb{C}^k) \oplus L^2(\mathbb{R}^{n-1}, \mathbb{C}^j)$ by the isomorphism $r^+\Lambda_-^s \oplus \Lambda'^{s+1/2}$.

The principal boundary symbol $\sigma_{\mathbb{R}^{n-1}}(\mathcal{A})$ of $\mathcal{A}$

$$\sigma_{\mathbb{R}^{n-1}}(\mathcal{A})(x', \xi') = \begin{pmatrix} \Pi^+\sigma_A(x', \xi', \nu) + \Pi'\sigma_B(x', \xi', \nu, \tau) & \sigma_K(x', \xi', \nu) \\ \Pi'\sigma_T(x', \xi', \nu) & \sigma_Q(x', \xi') \end{pmatrix} : \begin{matrix} H^+ \otimes \mathbb{C}^k \\ \oplus \\ \mathbb{C}^j \end{matrix} \to \begin{matrix} H^+ \otimes \mathbb{C}^{k'} \\ \oplus \\ \mathbb{C}^{j'} \end{matrix}$$

has, for arbitrary fixed (x', ξ'), a continuous extension $\sigma_{\mathbb{R}^{n-1}}(\mathcal{A})(x', \xi') \in \mathcal{L}((H^+)_s \otimes \mathbb{C}^k \oplus \mathbb{C}^j, (H^+)_{s-m} \otimes \mathbb{C}^{k'} \oplus \mathbb{C}^{j'})$ for $s > d - 1/2$ (cf. 2.1.2.3, Corollary 10). Denote the norm of the extension by $\|\sigma_{\mathbb{R}^{n-1}}(\mathcal{A})(x', \xi')\|_s$. The norm in $(H^+)_s$ let be induced from $(H^+)_0$ by the isomorphism

$$\Pi^+(1 - i\nu)^s : (H^+)_s \to (H^+)_0 .$$

Corollary 5. *Let $\mathcal{A} \in \mathrm{Op}(\mathfrak{R}^{m,d})$ be compactly supported with the principal boundary symbol $\sigma_{\mathbb{R}^{n-1}}(\mathcal{A})(x', \xi')$. Then for any integer $s > d - 1/2$*

$$\inf_{\mathcal{C}} \|\mathcal{A} + \mathcal{C}\|_{\mathcal{L}(\mathcal{H}^s(k,j), \mathcal{H}^{s-m}(k',j'))} \geq \sup_{(x',\xi') \in S^*\mathbb{R}^{n-1}} \|\sigma_{\mathbb{R}^{n-1}}(\mathcal{A})(x', \xi')\|_s \tag{2}$$

where the infimum is taken over all $\mathcal{C} \in \mathcal{K}(\mathcal{H}^s(k, j), \mathcal{H}^{s-m}(k', j'))$.

Proof: First reduce the order and type of $\mathcal{A}$ using the isometric isomorphisms

$$\mathcal{L}^s = r^+\Lambda_-^s \oplus \Lambda'^{s+1/2} : \mathcal{H}^s(k, j) \to L^2(\mathbb{R}^n_+, \mathbb{C}^k) \otimes L^2(\mathbb{R}^{n-1}, \mathbb{C}^j)$$

to zero. The diagram

$$\begin{array}{ccc} \mathcal{H}^s(k, j) & \xrightarrow{\mathcal{A}+\mathcal{C}} & \mathcal{H}^{s-m}(k', j') \\ \downarrow{\scriptstyle \mathcal{L}^s} & & \downarrow{\scriptstyle \mathcal{L}^{s-m}} \\ L^2(\mathbb{R}^n_+, \mathbb{C}^k) \oplus L^2(\mathbb{R}^{n-1}, \mathbb{C}^j) & \xrightarrow{\mathcal{A}_0+\mathcal{C}_0} & L^2(\mathbb{R}^n_+, \mathbb{C}^{k'}) \oplus L^2(\mathbb{R}^{n-1}, \mathbb{C}^{j'}) \end{array}$$

defines the operator $\mathcal{A}_0 + \mathcal{C}_0$, $\mathcal{C}_0$ is compact. The principal boundary symbol $\sigma_{R^{n-1}}(\mathcal{A}_0)$ of $\mathcal{A}_0$ is obtained from $\sigma_{R^{n-1}}(\mathcal{A})$ by the diagram 2.1.2.3. (7) and thus

$$||\sigma_{R^{n-1}}(\mathcal{A})(x', \xi')||_s = ||\sigma_{R^{n-1}}(\mathcal{A}_0)(x', \xi')||_0 .$$

Now the orders of the components of $\mathcal{A}_0$ are such that the considerations of the proofs of Theorems 1, 2, 3, 4 can be applied. The assertion follows as in 2.3.4.1, Corollary 2. □

For the class Op $(\mathfrak{R}^{m,d})$ there is an analogous result, where the right hand side in (2) is replaced by $\overline{\lim\limits_{\xi' \to \infty}} \sup\limits_{x'} ||\mathfrak{a}(x', \xi')||$ ($\mathfrak{a}$ the complete boundary symbol). Consider now operators in $\mathfrak{G}^{m,d}(\overline{R}^n_+, R^{n-1})$, i.e. we have to take into account that the PDO symbol $\sigma_A(x, \xi)$ may depend on x_n, too. Let $\mathcal{A} \in \mathfrak{G}^{m,d}(\overline{R}^n_+, R^{n-1})$ be compactly supported with the principal symbol $\sigma(\mathcal{A}) = (\sigma_A, \sigma_{R^{n-1}}(\mathcal{A}))$. According to 2.3.2.4 there is a continuous extension

$$\mathcal{A}: \mathcal{H}^s(k, j) \to \mathcal{H}^{s-m}(k', j')$$

for any $s > d - 1/2$.

Define

$$||\sigma(\mathcal{A})||_s = \max \left\{ \sup_{x \in \overline{R}^n_+,\, |\xi|=1} ||\sigma_A(x, \xi)||, \sup_{x' \in R^{n-1},\, |\xi'|=1} ||\sigma_{R^{n-1}}(\mathcal{A})(x', \xi')||_s \right\} .$$

Corollary 6. *Let $\mathcal{A} \in \mathfrak{G}^{m,d}(\overline{R}^n_+, R^{n-1})$ be compactly supported and $\sigma(\mathcal{A})$ its principal symbol. Then, for any integer $s > d - 1/2$,*

$$\inf_{\mathcal{C}} ||\mathcal{A} + \mathcal{C}||_{\mathcal{L}(\mathcal{H}^s(k, j), \mathcal{H}^{s-m}(k', j'))} \geqq ||\sigma(\mathcal{A})||_s$$

where the infimum is taken over all $\mathcal{C} \in \mathcal{K}(\mathcal{H}^s(k, j), \mathcal{H}^{s-m}(k', j'))$.

Proof: For $\sup\limits_{x \in \overline{R}^n_+,\, |\xi|=1} ||\sigma_A(x, \xi)|| = ||\sigma(\mathcal{A})||_s$ we get for $u \in C_0^\infty(R^n_+, \mathbb{C}^k)$ using the construction of the proof of 2.3.4.1, Corollary 2 after reduction of the order to zero

$$||(R_\lambda^{-1} \oplus 1)\, \mathcal{A}(R_\lambda u \oplus 0)|| \to ||\sigma_A(x_0, \xi_0)\, u||_{\mathbb{C}^k} \quad \text{for } \lambda \to \infty ,$$

since Green and trace operators restricted to functions with support in a fixed compact set in R^n_+ are compact.

For $\sup\limits_{x' \in R^{n-1},\, |\xi'|=1} ||\sigma_{R^{n-1}}(\mathcal{A})(x', \xi')||_s = ||\sigma(\mathcal{A})||_s$ the construction of the proof of Corollary 5 implies the desired estimate. □

Denote by $\overline{\mathfrak{G}}^{m,d}_s(\overline{R}^n_+, R^{n-1}; \mathbb{C}^k, \mathbb{C}^{k'}, \mathbb{C}^j, \mathbb{C}^{j'})$ the closure of all compactly supported operators in $\mathfrak{G}^{m,d}(\overline{R}^n_+, R^{n-1}; \mathbb{C}^k, \mathbb{C}^{k'}, \mathbb{C}^j, \mathbb{C}^{j'})$ in

$$\mathcal{L}(\mathcal{H}^s(k, j), \mathcal{H}^{s-m}(k', j')) , \quad s > d - 1/2 .$$

Denote by $\overline{\mathfrak{R}}^{m,d}_s(k, k', j, j')$ the closure of $\mathfrak{R}^{m,d}(k, k', j, j')$ in

$$\mathcal{L}((H^+)_s \otimes \mathbb{C}^k \oplus \mathbb{C}^j, (H^+)_{s-m} \otimes \mathbb{C}^{k'} \oplus \mathbb{C}^{j'}) , \quad s > d - 1/2 .$$

For an arbitrary locally compact topological space X and an arbitrary Banach space B denote by $C_{(0)}(X, B)$ the completion of $C_0(X, B)$, the space of all continuous mappings with compact support, with respect to uniform convergence.

Denote by $\overline{\mathfrak{R}}^{m,d}_s(\overline{R}^n_+, R^{n-1}; \mathbb{C}^k, \mathbb{C}^{k'}; \mathbb{C}^j, \mathbb{C}^{j'})$ the closure of the space of all boundary symbols $\sigma_{R^{n-1}}(\mathcal{A})(x', \xi') \in \mathfrak{R}^{m,d}(\overline{R}^n_+, R^{n-1}; \mathbb{C}^k, \mathbb{C}^{k'}; \mathbb{C}^j, \mathbb{C}^{j'})$ with compact support with respect to x' in $C\big(S^*\mathbb{R}^{n-1}, \mathcal{L}((H^+)_s \otimes \mathbb{C}^k \oplus \mathbb{C}^j, (H^+)_{s-m} \otimes \mathbb{C}^{k'} \oplus \mathbb{C}^{j'})\big)$. Obviously, $\mathfrak{R}_s^{m,d}(\overline{\mathbb{R}}_n^+, \mathbb{R}^{n-1}; \mathbb{C}^k, \mathbb{C}^{k'}; \mathbb{C}^j, \mathbb{C}^{j'}) \subset C_{(0)}\big(S^*\mathbb{R}^{n-1}, \mathcal{L}((H^+)_s \otimes \mathbb{C}^k \oplus \mathbb{C}^j, (H^+)_{s-m} \otimes \mathbb{C}^{k'} \oplus \mathbb{C}^{j'})\big)$.

Corollary 7. *Let $\mathcal{A}_j \in \mathfrak{G}^{m,d}(\overline{\mathbb{R}}_+^n, \mathbb{R}^{n-1})$ be a sequence of operators converging in $\mathcal{L}(\mathcal{H}^s, \mathcal{H}^{s-m})$, $s > d - 1/2$. Then a uniquely defined $\sigma(\mathcal{A}) = \big(\sigma_A(x, \xi), \sigma_{\mathbb{R}^{n-1}}(\mathcal{A})(x', \xi')\big) \in C_{(0)}(S^+\mathbb{R}_+^n) \otimes \hom(\mathbb{C}^k, \mathbb{C}^{k'}) \oplus \overline{\mathfrak{R}}_s^{m,d}(\overline{\mathbb{R}}_+^n, \mathbb{R}^{n-1}; \mathbb{C}^k, \mathbb{C}^{k'}; \mathbb{C}^j, \mathbb{C}^{j'})$ corresponds to the limit $\mathcal{A} \in \overline{\mathfrak{G}}_s^{m,d}(\overline{\mathbb{R}}_+^n, \mathbb{R}^{n-1})$.*

Denote $C_{(0)}(S^*\mathbb{R}_+^n) \otimes \hom(\mathbb{C}^k, \mathbb{C}^{k'}) \oplus \overline{\mathfrak{R}}_s^{m,d}(\overline{\mathbb{R}}_+^n, \mathbb{R}^{n-1})$ by $\overline{\mathfrak{S}}_s^{(m),d}$. Then Corollary 7 means that the principal symbol mapping

$$\sigma\colon \mathfrak{G}^{m,d}(\overline{\mathbb{R}}_+^n, \mathbb{R}^{n-1}) \to \mathfrak{S}^{(m),d}$$

has a continuous extension

$$\overline{\mathfrak{G}}_s^{m,d}(\overline{\mathbb{R}}_+^n, \mathbb{R}^{n-1}) \to \overline{\mathfrak{S}_s^{(m),d}}. \tag{3}$$

In the next section we shall show that the mapping (3) is surjective.

2.3.4.3. Boundary Operators with Continuous Principal Symbols

The main result of this section is that in the estimate in 2.3.4.2, Corollary 6 in fact equality holds.

Consider the Sobolev spaces $H^s(\mathbb{R}_+^n, \mathbb{C}^k)$, $H^s(\mathbb{R}^{n-1}, \mathbb{C}^j)$ and $(H^+)_s$ with the norms as in the previous section.

Lemma 1. *Let $\sigma_A(x', x_n, \xi', \nu) \in \mathfrak{A}^{(m)}(\overline{\mathbb{R}}_+^n)$ and*

$$\mathfrak{a}(x', \xi') = \begin{pmatrix} \Pi^+\sigma_A(x', 0, \xi', \nu) + \Pi'\sigma_B(x', \xi', \nu, \tau) & \sigma_K(x', \xi', \nu) \\ \Pi'\sigma_T(x', \xi', \nu) & \sigma_Q(x', \xi') \end{pmatrix} \in \mathfrak{R}^{(m),d}(\overline{\mathbb{R}}_+^n, \mathbb{R}^{n-1}).$$

Then, for any $s > d - 1/2$ and x', ξ' fixed, we have

$$\|\mathfrak{a}(x', \xi')\|_{\mathcal{L}((H^+)_s \otimes \mathbb{C}^k \oplus \mathbb{C}^j,\, (H^+)_{s-m} \otimes \mathbb{C}^{k'} \oplus \mathbb{C}^{j'})} \geq \|\Pi^+\sigma_A(x', 0, \xi', \nu)\|_{\mathcal{L}((H^+)_s \otimes \mathbb{C}^k,\, (H^+)_{s-m} \otimes \mathbb{C}^{k'})} \tag{1}$$

and hence

$$\sup_{(x', \xi') \in S^*\mathbb{R}^{n-1}} \|\mathfrak{a}(x', \xi')\|_s \geq \sup_{\substack{(x', \xi') \in S^*\mathbb{R}^{n-1} \\ \nu \in \mathbb{R}}} \|\sigma_A(x', 0, \xi', \nu)\,(1 + \nu^2)^{-m/2}\|. \tag{2}$$

Proof: After reduction of s, m, d to zero (cf. the proof of 2.3.4.2, Corollary 5) we get from 2.1.2.3, Lemma 12

$$\|\Pi^+\sigma_A(x', 0, \xi', \nu)\| \geq \sup_{\nu \in \mathbb{R}} \|\sigma_A(x', 0, \xi', \nu)\|$$

and hence (1) implies (2). Moreover $\|\Pi^+\sigma_A(x', 0, \xi', \nu)\| = \sup_{\nu \in \mathbb{R}} \|\sigma_A(x', 0, \xi', \nu)\|$, since we have for $h \in V^+$

$$\|\Pi^+\big(\sigma_A(x', 0, \xi', \nu)\, h(\nu)\big)\|_{L^2(\mathbb{R})} \leq \|\sigma_A(x', 0, \xi', \nu)\, h(\nu)\|_{L^2(\mathbb{R})}.$$

Then the assertion follows from 2.1.2.3, Lemma 12 since the Green, potential and trace symbols are compact. □

Lemma 2. *The following conditions are equivalent:*

(i) $\mathfrak{a} = \begin{pmatrix} \Pi^+\sigma_A + \Pi'\sigma_B & \sigma_K \\ \Pi'\sigma_T & \sigma_Q \end{pmatrix} \in \overline{\mathfrak{R}}_s^{m,d}(k, k', j, j')$;

(ii) $(1 + i\nu)^{-m} \sigma_A(\nu) \in C(\overline{R}) \otimes \hom(\mathbb{C}^k, \mathbb{C}^{k'})$,

$\Pi'\sigma_B(\nu, \tau) \in \mathcal{K}((H^+)_s \otimes \mathbb{C}^k, (H^+)_{s-m} \otimes \mathbb{C}^{k'})$,

$\sigma_K \in (H^+)_{s-m} \otimes \hom(\mathbb{C}^j, \mathbb{C}^{k'})$, $\sigma_T \in (H_0^-)_{-s} \otimes \hom(\mathbb{C}^k, \mathbb{C}^{j'})$,

$\sigma_Q \in \hom(\mathbb{C}^j, \mathbb{C}^{j'})$ ($C(\overline{R})$ denotes the spaces of all continuous functions on R having limits for $\nu = \pm\infty$ which are equal).

Proof: For $s = m = d = 0$ in 2.1.2.3 it has been proved that if (ii) is satisfied, then $\mathfrak{a} \in \mathcal{L}((H^+)_0 \otimes \mathbb{C}^k \oplus \mathbb{C}^j, (H^+)_0 \otimes \mathbb{C}^{k'} \oplus \mathbb{C}^{j'})$. Moreover any

$$\Pi^+\sigma_A(\nu) + \Pi'\sigma_B, \quad \sigma_A \in C(\overline{R}) \otimes \hom(\mathbb{C}^k, \mathbb{C}^{k'}),$$

$$\Pi'\sigma_K \in \mathcal{K}((H^+)_0 \otimes \mathbb{C}^k, (H^+)_0 \otimes \mathbb{C}^{k'})$$

can be approximated by

$$\Pi^+\sigma_A'(\nu) + \Pi'\sigma_B', \quad \sigma_A' \in H_1 \otimes \hom(\mathbb{C}^k, \mathbb{C}^{k'}),$$

$$\sigma_B' \in H^+ \otimes H_0^- \otimes \hom(\mathbb{C}^k, \mathbb{C}^{k'}).$$

In fact, since

$$\|\Pi^+\sigma_A\|_{\mathcal{L}((H^+)_0 \otimes \mathbb{C}^k, (H^+)_0 \otimes \mathbb{C}^{k'})} = \sup_{\nu \in R} \|\sigma_A(\nu)\| \tag{3}$$

after the transformation $\varkappa: \overline{R} \to S^1$, take the convolution of $(\varkappa^{-1})^* \sigma_A$ with a smooth function φ which can be chosen such that $\varkappa^*(\varphi * (\varkappa^{-1})^* \sigma_A) - \sigma_A$ becomes small. For $\Pi'\sigma_B \in \mathcal{K}((H^+)_0 \otimes \mathbb{C}^k, (H^+)_0 \otimes \mathbb{C}^{k'})$ consider $\Pi'\sigma^N\Pi'\sigma_B\Pi'\sigma^N$ where

$$\sigma^N \in H^+ \otimes H_0^-, \quad \sigma^N(\nu, \tau) = \sum_{j=0}^{N} \frac{(1 - i\nu)^j}{(1 + i\nu)^{j+1}} \frac{(1 + i\tau)^j}{(1 - i\tau)^{j+1}}.$$

Since $\Pi'\sigma^N \to 1$ for $N \to \infty$ weakly in $(H^+)_0$, it follows that

$$\Pi'\sigma_B - \Pi'\sigma^N\Pi'\sigma_B\Pi'\sigma^N = \Pi'\sigma^N\Pi'\sigma_B(1 - \Pi'\sigma^N) + (1 - \Pi'\sigma_N)\Pi'\sigma_B$$

tends to zero in the norm of $\mathcal{L}((H^+)_0 \otimes \mathbb{C}^k, (H^+)_0 \otimes \mathbb{C}^{k'})$ for $N \to \infty$ and $\Pi'\sigma^N\Pi'\sigma_B\Pi'\sigma^N = \Pi'\sigma_B^N$ is a Green boundary symbol

$$\sigma_B^N(\nu, \tau) = \sum_{j,k=1}^{N} \frac{(1 - i\nu)^k}{(1 + i\nu)^{k+1}} \Pi'_{\tau'} \left(\frac{(1 + i\tau_1)^k}{(1 - i\tau_1)^{k+1}}\right) \left(\Pi'\sigma_B \left(\frac{(1 - i\nu')^j}{(1 + i\nu')^{j+1}}\right)(\tau_1)\right) \frac{(1 + i\tau)^j}{(1 - i\tau)^{j+1}}.$$

From

$$\|\sigma_K\|_{\mathcal{L}(\mathbb{C}^j, (H^+)_0 \otimes \mathbb{C}^{k'})} = \|\sigma_K\|_{(H^+)_0 \otimes \hom(\mathbb{C}^j, \mathbb{C}^{k'})}, \tag{4}$$

$$\|\Pi'\sigma_T\|_{\mathcal{L}((H^+)_0 \otimes \mathbb{C}^k, \mathbb{C}^{j'})} = \|\sigma_T\|_{(H_0^-)_0 \otimes \hom(\mathbb{C}^k, \mathbb{C}^{j'})} \tag{5}$$

follows (ii) $\Rightarrow$ (i).

Let $\mathfrak{a} \in \overline{\mathfrak{R}}_0^{0,0}(k, k'; j, j')$ and $\{\mathfrak{a}_j\}_{j \in \mathbb{Z}_+}$ be a sequence in $\mathfrak{R}^{0,0}(k, k', j, j')$ such that $\mathfrak{a}_j \to \mathfrak{a}$ for $j \to \infty$ in $\mathcal{L}((H^+)_0 \otimes \mathbb{C}^k \oplus \mathbb{C}^j, (H^+)_0 \otimes \mathbb{C}^{k'} \otimes \mathbb{C}^{j'})$.

If we put

$$\mathfrak{a}_j = \begin{pmatrix} \Pi^+\sigma_A^j + \Pi'\sigma_B^j & \sigma_K^j \\ \Pi'\sigma_T^j & \sigma_Q^j \end{pmatrix},$$

this implies that $\sigma_K^j \to \sigma_K$ for $j \to \infty$ in $\mathcal{L}(\mathbb{C}^j, (H^+)_0 \otimes \mathbb{C}^{k'})$, $\Pi'\sigma_T^j \to \Pi'\sigma_T$ in $\mathcal{L}((H^+)_0 \otimes \mathbb{C}^k, \mathbb{C}^{j'})$ and $\Pi^+\sigma_A^j + \Pi'\sigma_B^j \to \Pi^+\sigma_A + \Pi'\sigma_B$ in $\mathcal{L}((H^+)_0 \otimes \mathbb{C}^k, (H^+)_0 \otimes \mathbb{C}^{k'})$. According to (4) and (5) we get $\sigma_K \in (H^+)_0 \otimes \hom(\mathbb{C}^j, \mathbb{C}^{k'})$, $\sigma_T \in (H_0^-)_0 \otimes \hom(\mathbb{C}^k, \mathbb{C}^{j'})$. In view of Lemma 1 $\Pi^+\sigma_A^j \to \Pi^+\sigma_A$ in $\mathcal{L}((H^+)_0 \otimes \mathbb{C}^k, (H^+)_0 \otimes \mathbb{C}^{k'})$ and (3) implies $\sigma_A \in C(\overline{\mathbb{R}}) \otimes \hom(\mathbb{C}^k, \mathbb{C}^{k'})$. Then $\Pi'\sigma_K^j \to \Pi'\sigma_B$ in $\mathcal{L}((H^+)_0 \otimes \mathbb{C}^k, (H^+)_0 \otimes \mathbb{C}^{k'})$ and $\Pi'\sigma_B$ is compact since $\Pi'\sigma_B^j$ is compact for any j and the space $\mathcal{K}((H^+)_0 \otimes \mathbb{C}^k, (H^+)_0 \otimes \mathbb{C}^{k'})$ is closed in $\mathcal{L}((H^+)_0 \otimes \mathbb{C}^k, (H^+)_0 \otimes \mathbb{C}^{k'})$. This shows (i) $\Rightarrow$ (ii). □

From Lemma 2 we obtain that

$$\sigma_{\mathbb{R}^{n-1}}(\mathcal{A})(x', \xi') = \begin{pmatrix} \Pi^+\sigma_A + \Pi'\sigma_B & \sigma_K \\ \Pi'\sigma_T & \sigma_Q \end{pmatrix}$$

$s > d - 1/2$, with

$$\begin{aligned}
&(1 + i\nu)^{-m} \sigma_A(x', \xi', \nu) \in C_{(0)}(S^*\mathbb{R}^{n-1}, C(\overline{\mathbb{R}}) \otimes \hom(\mathbb{C}^k, \mathbb{C}^{k'})), \\
&\Pi'\sigma_B(x', \xi', \nu) \in C_{(0)}(S^*\mathbb{R}^{n-1}, \mathcal{K}((H^+)_0 \otimes \mathbb{C}^k, (H^+)_0 \otimes \mathbb{C}^{k'})), \qquad (6)\\
&\sigma_K(x', \xi', \nu) \in C_{(0)}(S^*\mathbb{R}^{n-1}, (H^+)_{s-m} \otimes \hom(\mathbb{C}^j, \mathbb{C}^{k'})), \\
&\sigma_T(x', \xi', \nu) \in C_{(0)}(S^*\mathbb{R}^{n-1}, (H_0^-)_{-s} \otimes \hom(\mathbb{C}^k, \mathbb{C}^{j'})), \\
&\sigma_Q(x', \xi') \in C_{(0)}(S^*\mathbb{R}^{n-1}, \hom(\mathbb{C}^j, \mathbb{C}^{j'}))
\end{aligned}$$

belongs to $C_{(0)}(S^*\mathbb{R}^{n-1}, \mathcal{L}((H^+)_s \otimes \mathbb{C}^k \oplus \mathbb{C}^j, (H^+)_{s-m} \otimes \mathbb{C}^{k'} \oplus \mathbb{C}^{j'}))$. Moreover the construction of the proof of Lemma 2 together with the expansions of Green, potential and trace symbols in 2.2.3 and 2.2.4 yields

Lemma 3. $\sigma_{\mathbb{R}^{n-1}}(\mathcal{A})(x', \xi') \in \overline{\mathfrak{R}}_s^{m,d}(\overline{\mathbb{R}}_+^n, \mathbb{R}^{n-1}; \mathbb{C}^k, \mathbb{C}^{k'}, \mathbb{C}^j, \mathbb{C}^{j'})$ *iff the conditions* (6) *are satisfied.*

The simple proof is left to the reader.

As in the previous section denote $\mathcal{H}^s = H^s(\mathbb{R}_+^n) \oplus H^{s+1/2}(\mathbb{R}^{n-1})$.

Lemma 4. *For any $\varepsilon > 0$ there exists a properly supported operator $R_\varepsilon \in \mathfrak{G}^{-\infty,0}$ with $\|u - \mathcal{R}_\varepsilon u\|_{\mathcal{H}^s} \leqq \|u\|_{\mathcal{H}^s}$, $\|u - \mathcal{R}_\varepsilon u\|_{\mathcal{H}^{-1}} \leqq c(\varepsilon) \|u\|_{\mathcal{H}^s}$, $u \in C_0^\infty(\overline{\mathbb{R}}_+^n) \oplus C_0^\infty(\mathbb{R}^{n-1})$, where $C(\varepsilon) \to 0$ for $\varepsilon \to 0$.*

Proof: Construct $\mathcal{R}_\varepsilon$ in the form

$$\mathcal{R}_\varepsilon = \begin{pmatrix} r^+R_\varepsilon & 0 \\ 0 & R_\varepsilon' \end{pmatrix}$$

and R_ε, R_ε' are found as in the case without boundary. First consider a function $\psi \in C_0^\infty(\mathbb{R}^n)$ such that $\psi \geqq 0$, $\int \psi(x)\, dx = 1$ and $0 \leqq \hat\psi(\xi) \leqq 1$. Take $\psi_0 \in C_0^\infty(\mathbb{R}^n)$, $\psi_0 \geqq 0$, $\int \psi_0(x)\, dx = 1$ and thus $|\hat\psi_0(\xi)| \leqq 1$. Set $\psi(x) = \int \psi_0(x - y)\, \psi_0(y)\, dy$. Then, in view of $\hat\psi(\xi) = \hat\psi_0(\xi)\, \overline{\hat\psi_0(\xi)} = |\hat\psi_0(\xi)|^2$, the function ψ has the desired properties. Now set $\psi_\varepsilon(x) = \varepsilon^{-n}\psi(\varepsilon^{-1}x)$ and $R_\varepsilon u(x) = \psi_\varepsilon * u(x)$. Similarly define $R_\varepsilon' v(x') = \Lambda'^{-1/2}(\psi_\varepsilon' * \Lambda'^{1/2} v)(x')$, $v \in C_0^\infty(\mathbb{R}^{n-1})$, $\psi_\varepsilon'(x') = \varepsilon^{-n+1}\psi(\varepsilon^{-1}x')$, and $r^+R_\varepsilon u(x) = r^+(\psi_\varepsilon * j^+u)(x)$, $u \in C_0^\infty(\overline{\mathbb{R}}_+^n)$. Obviously, r^+R_ε and R_ε' are smoothing and properly supported. The properties of ψ imply

$$||u - r^+ R_\varepsilon u||_{L^2(\mathbb{R}^n_+)} \leqq ||j^+ u - R_\varepsilon j^+ u||_{L^2(\mathbb{R}^n)} \leqq ||j^+ u||_{L^2(\mathbb{R}^n)} = ||u||_{L^2(\mathbb{R}^n_+)}$$

and

$$\begin{aligned} ||v - R'_\varepsilon v||_{H^{1/2}(\mathbb{R}^{n-1})} &= ||\Lambda'^{1/2} v - \Lambda'^{1/2}(\Lambda'^{-1/2}(\psi'_\varepsilon * \Lambda'^{1/2} v))||_{L^2(\mathbb{R}^{n-1})} \\ &= ||\Lambda'^{1/2} v - \psi'_\varepsilon * \Lambda^{1/2'} v||_{L^2(\mathbb{R}^{n-1})} \\ &= ||(1 - \hat{\psi}_\varepsilon(\xi'))\, F(\Lambda'^{1/2} v)\,(\xi')||_{L^2(\mathbb{R}^{n-1})} \\ &\leqq ||\Lambda'^{1/2} v||_{L^2(\mathbb{R}^{n-1})} = ||v||_{H^{1/2}(\mathbb{R}^{n-1})}\,. \end{aligned}$$

Moreover $u - r^+ R_\varepsilon u$ weakly tends to zero in $L^2(\mathbb{R}^n_+)$ for $\varepsilon \to 0$ and $v - R'_\varepsilon v$ weakly tends to zero in $H^{1/2}(\mathbb{R}^{n-1})$. Since the embeddings $L^2(\mathbb{R}^n_+) \subset H^{-1}(\mathbb{R}^n_+)$ and $H^{1/2}(\mathbb{R}^{n-1}) \subset H^{-1/2}(\mathbb{R}^{n-1})$ for functions with support in a fixed compact set are compact mappings it follows that

$$\begin{aligned} ||u - r^+ R_\varepsilon u||_{H^{-1}(\mathbb{R}^n_+)} &\leqq c(\varepsilon)\, ||u||_{L^2(\mathbb{R}^n_+)}\,, \\ ||v - R'_\varepsilon v||_{H^{-1/2}(\mathbb{R}^{n-1})} &\leqq c(\varepsilon)\, ||v||_{H^{1/2}(\mathbb{R}^{n-1})} \end{aligned}$$

with $c(\varepsilon) \to 0$ for $\varepsilon \to 0$. □

Theorem 5. *For any compactly supported $\mathcal{A} \in \mathfrak{S}^{m,d}(\overline{\mathbb{R}}^n_+, \mathbb{R}^{n-1})$ we have*

$$\inf_{\mathcal{C}} ||\mathcal{A} + \mathcal{C}||_{\mathcal{L}(\mathcal{H}^s, \mathcal{H}^{s-m})} \leqq ||\sigma(\mathcal{A})||_s\,, \quad s > d - {}^1\!/_2\,, \tag{7}$$

where the infimum is taken over all $\mathcal{C} \in \mathcal{K}(\mathcal{H}^s, \mathcal{H}^{s-m})$.

Proof: The usual reduction of the order and type to zero can be applied, i.e. the diagram

$$\begin{array}{ccc} \mathcal{H}^s & \xrightarrow{\mathcal{A}+\mathcal{C}} & \mathcal{H}^{s-m} \\ {\scriptstyle r^+\Lambda^s_- \oplus \Lambda'^s}\big\downarrow & & \big\downarrow{\scriptstyle r^+\Lambda^{s-m}_- \oplus \Lambda'^{s-m}} \\ \mathcal{H}^0 & \xrightarrow{\mathcal{A}_0+\mathcal{C}_0} & \mathcal{H}^0 \end{array}$$

defines $\mathcal{A}_0 \in \mathfrak{S}^{0,0}$ and $\mathcal{C}_0 \in \mathcal{K}(\mathcal{H}^0, \mathcal{H}^0)$ for $\mathcal{C} \in \mathcal{K}(\mathcal{H}^s, \mathcal{H}^{s-m})$ with $||\mathcal{A} + \mathcal{C}||_{\mathcal{L}(\mathcal{H}^s, \mathcal{H}^{s-m})} = ||\mathcal{A}_0 + \mathcal{C}_0||_{\mathcal{L}(\mathcal{H}^0, \mathcal{H}^0)}$ and $||\sigma(\mathcal{A})||_s = ||\sigma(\mathcal{A}_0)||_0$. Thus it can be assumed that $s = m = d = 0$.

Let $M > ||\sigma(\mathcal{A})||_0$. Then $M^2 - \sigma^*_{\mathcal{A}}(x, \xi)\, \sigma_{\mathcal{A}}(x, \xi)$ $(\sigma^*_{\mathcal{A}}(x, \xi)$ is the adjoint matrix with respect to the Hermitean scalar products in $\mathbb{C}^k$ and $\mathbb{C}^{k'}$, $(\sigma_{\mathcal{A}}(x, \xi)\, u, u')_{\mathbb{C}^{k'}} = (u, \sigma^*_{\mathcal{A}}(x, \xi)\, u')_{\mathbb{C}^{k'}}$, $u \in \mathbb{C}^k$, $u' \in \mathbb{C}^{k'})$ is positive definite. Denote by $\sigma_{\mathcal{A}_1}(x, \xi)$ its positive square root, $\sigma_{\mathcal{A}_1}(x, \xi)^2 = M^2 - \sigma^*_{\mathcal{A}}(x, \xi)\, \sigma_{\mathcal{A}}(x, \xi)$. Consider now the boundary symbols. $M^2 - \sigma_{\mathbb{R}^{n-1}}(\mathcal{A})^*(x', \xi')\, \sigma_{\mathbb{R}^{n-1}}(\mathcal{A})(x', \xi')$ is a family of positive definite operators in $\mathcal{L}((H^+)_0 \otimes \mathbb{C}^k \oplus \mathbb{C}^j, (H^+)_0 \otimes \mathbb{C}^k \oplus \mathbb{C}^j)$. Thus, for arbitrary (x', ξ'), the positive square root $\sigma_{\mathbb{R}^{n-1}}(\mathcal{A}_1)(x', \xi')$ is defined by

$$\sigma_{\mathbb{R}^{n-1}}(\mathcal{A}_1)(x', \xi')^2 = M - \sigma_{\mathbb{R}^{n-1}}(\mathcal{A})^*(x', \xi')\, \sigma_{\mathbb{R}^{n-1}}(\mathcal{A})(x', \xi')$$

smoothly depending on (x', ξ'). $\sigma_{\mathbb{R}^{n-1}}(\mathcal{A}_1)(x', \xi')$ is given by

$$\begin{aligned} &\sigma_{\mathbb{R}^{n-1}}(\mathcal{A}_1)(x', \xi') \\ &= (2\pi i)^{-1} \int_\Gamma \lambda^{1/2} (M^2 - \sigma_{\mathbb{R}^{n-1}}(\mathcal{A})^*(x', \xi')\, \sigma_{\mathbb{R}^{n-1}}(\mathcal{A})(x', \xi') - \lambda)^{-1}\, d\lambda \end{aligned}$$

where Γ is a contour in $\mathbb{C}$, $\Gamma \in \{\operatorname{Re} \lambda > 0\}$ surrounding $[\varepsilon, M^2]$, $||\sigma_{\mathbb{R}^{n-1}}(\mathcal{A})|| + \varepsilon < M$, with positive orientation. $(M^2 - \sigma_{\mathbb{R}^{n-1}}(\mathcal{A})^*(x', \xi')\, \sigma_{\mathbb{R}^{n-1}}(\mathcal{A})(x', \xi') - \lambda)^{-1}$ for fixed

(x', ξ') belongs to $\mathfrak{R}^{0,0}(k, k; j, j)$ (cf. 2.1.2.4, Proposition 6) and hence has the form

$$\begin{pmatrix} \Pi^+\sigma_{A_2} + \Pi'\sigma_{B_2} & \sigma_{K_2} \\ \Pi'\sigma_{T_2} & \sigma_{Q_2} \end{pmatrix}.$$

Therefore

$$\sigma_{R^{n-1}}(\mathcal{A}_1)(x', \xi') = \begin{pmatrix} \Pi^+\sigma_{A_1} + \Pi'\sigma_{B_1} & \sigma_{K_1} \\ \Pi'\sigma_{T_1} & \sigma_{Q_1} \end{pmatrix} \in \overline{\mathfrak{R}}_0^{0,0}(k, k; j, j)$$

smoothly depends on (x', ξ'). Form the calculus of boundary symbols on the line in 2.1.2.3, which is valid for $\overline{\mathfrak{R}}_0^{0,0}$, too, we get $\sigma'_{A_1}(x', \xi', \nu) = \sigma_{A_1}(x', 0, \xi', \nu)$. Approximating $\sigma_{R^{n-1}}(\mathcal{A}_1)(x', \xi')$ by boundary symbols in $\mathfrak{R}^{0,0}(\overline{R}^n_+, R^{n-1})$ we obtain a

$$\sigma_{R^{n-1}}(\mathcal{A}_2)(x', \xi') = \begin{pmatrix} \Pi^+\sigma_{A_1} + \Pi'\sigma_{B_2} & \sigma_{K_2} \\ \Pi'\sigma_{T_2} & \sigma_{Q_2} \end{pmatrix} \in \mathfrak{R}^{0,0}(\overline{R}^n_+, R^{n-1})$$

such that

$$M^2 - \sigma_{R^{n-1}}(\mathcal{A})^* \sigma_{R^{n-1}}(\mathcal{A}) - \sigma_{R^{n-1}}(\mathcal{A}_2)^2$$

is arbitrarily small in $C^N(S^* R^{n-1}, \mathfrak{R}^{0,0}(k, k; j, j))$ for any fixed $N \in Z_+$ (cf. the expansions of Green, potential and trace symbols in 2.2.3 and 2.2.4).

For the corresponding operators we have $M^2 - \mathcal{A}^*\mathcal{A} = \mathcal{A}_2^* + \mathcal{A}_3 + \mathcal{A}_4$, where $\mathcal{A}_4 \in \mathfrak{G}^{-1,0}(\overline{R}^n_+, R^{n-1})$ and the PDO in $\mathcal{A}_3$ vanishes and $\|\mathcal{A}_3\|_{\mathcal{L}(\mathcal{H}^0, \mathcal{H}^0)} < \delta$ is small (cf. 2.3.4.2, Corollary 7).

Therefore

$$\|\mathcal{A}u\|^2_{\mathcal{H}^0} \leq M^2\|u\|^2_{\mathcal{H}^0} + c\,\|u\|_{\mathcal{H}^{-1}}\,\|u\|_{\mathcal{H}^0} + \delta\|u\|_{\mathcal{H}^0}.$$

Using Lemma 4 we get

$$\|\mathcal{A}(u - \mathcal{R}_\varepsilon u)\|^2_{\mathcal{H}^0} \leq M^2\,\|u\|^2_{\mathcal{H}^0} + c(\varepsilon)\,\|u\|^2_{\mathcal{H}^0} + \delta\|u\|_{\mathcal{H}^0}$$

and $\mathcal{A}\mathcal{R}_\varepsilon \in \mathfrak{G}^{-\infty,0}$ is compact in $\mathcal{H}^0$ (even smoothing). Since ε and δ are arbitrary and $c(\varepsilon) \to 0$ for $\varepsilon \to 0$, the assertion follows. □

The proof of Theorem 5 shows that it is sufficient to take the infimum in (7) over all compactly supported operators $\mathcal{C} \in \mathfrak{G}^{-\infty,d}(\overline{R}^n_+, R^{n-1})$. In fact, the role of $\mathcal{C}$ in the proof is played by the composition of $\mathcal{A}$ and the properly supported smoothing operator $\mathcal{R}_\varepsilon$.

Corollary 6. *For any compactly supported* $\mathcal{A} \in \mathfrak{G}^{m,d}(\overline{R}^n_+, R^{n-1})$ *there is a compactly supported* $\mathcal{A}_1 \in \mathfrak{G}^{m,d}(\overline{R}^n_+, R^{n-1})$ *such that*

(i) $\mathcal{A} - \mathcal{A}_1 \in \mathfrak{G}^{-\infty,d}(\overline{R}^n_+, R^{n-1})$,

(ii) $\|\mathcal{A}_1\|_{\mathcal{L}(\mathcal{H}^s, \mathcal{H}^{s-m})} \leq \|\sigma(\mathcal{A})\|_s$, $s > d - 1/2$.

Theorem 5 and 2.3.4.2, Corollary 6 together imply

Corollary 7. *For any compactly supported* $\mathcal{A} \in \mathfrak{G}^{m,d}(\overline{R}^n_+, R^{n-1})$ *we have*

$$\inf_{\mathcal{C}} \|\mathcal{A} + \mathcal{C}\|_{\mathcal{L}(\mathcal{H}^s, \mathcal{H}^{s-m})} = \|\sigma(\mathcal{A})\|_s, \quad s > d - 1/2 \tag{8}$$

where the infimum is taken over all $\mathcal{C} \in \mathcal{K}(\mathcal{H}^s, \mathcal{H}^{s-m})$.

Note the special cases of (8) if all operators except one in the matrix vanish. So we have for any compactly supported $r'B \in \mathrm{Op}\,(\mathfrak{B}^{m-1,d})\,(\mathbb{R}^{n-1})$

$$\inf_{C \in \mathcal{K}(H^s(\mathbb{R}^n_+, \mathbb{C}^k),\, H^{s-m}(\mathbb{R}^n_+, \mathbb{C}^{k'}))} \|r'B + C\|_s = \|\Pi'\sigma_B\|_{\mathcal{L}((H^+)_s \otimes \mathbb{C}^k,\, (H^+)_{s-m} \otimes \mathbb{C}^{k'})},$$

$s > -{}^1/_2$, for compactly supported $K \in \mathrm{Op}\,(\mathfrak{K}^{m-1/2})\,(\mathbb{R}^{n-1})$

$$\inf_{C \in \mathcal{K}(H^s(\mathbb{R}^{n-1}, \mathbb{C}^j),\, H^{s-m}(\mathbb{R}^n_+, \mathbb{C}^{k'}))} \|K + C\|_s = \|\sigma_K\|_{\mathcal{L}(\mathbb{C}^j,\, (H^+)_{s-m} \otimes \mathbb{C}^{k'})},$$

$s \in \mathbb{R}$, for compactly supported $r'T \in \mathrm{Op}\,(\mathfrak{T})_{m-1/2,\,d}\,(\mathbb{R}^{n-1})$

$$\inf_{C \in \mathcal{K}(H^s(\mathbb{R}^n_+, \mathbb{C}^k),\, H^{s-m}(\mathbb{R}^{n-1}, \mathbb{C}^{j'}))} \|r'T + C\|_s = \|\Pi'\sigma_T\|_{\mathcal{L}((H^+)_s \otimes \mathbb{C}^k,\, \mathbb{C}^{j'})}, \quad s > d - {}^1/_2.$$

By Lemma 3 and Corollary 7 we get

Corollary 8. *The principal symbol mapping* 2.3.4.2. (3)

$$\sigma\colon \overline{\mathfrak{G}}^{m,d}_s(\overline{\mathbb{R}}^n_+, \mathbb{R}^{n-1}) \to \overline{\mathfrak{S}}^{(m),d}_s, \quad s > d - {}^1/_2,$$

is surjective.

Corollary 8 implies a corresponding result for Green, potential and trace operators. It is sufficient to take operators only consisting of operators of the considered type.

Corollary 9. *A compactly supported operator* $\mathcal{A} \in \overline{\mathfrak{G}}^{m,d}_s(\overline{\mathbb{R}}^n_+, \mathbb{R}^{n-1})$ *belongs to* $\mathcal{K}(\mathcal{H}^s, \mathcal{H}^{s-m})$, $s > d - {}^1/_2$, *iff its principal symbol (i. e. both the interior principal symbol and the principal boundary symbol) vanishes.*

This is an immediate consequence of (8).

2.3.4.4. Operators on Manifolds

The results of the previous sections carry over to operators acting between sections of vector bundles over a compact manifold with boundary (the definitions are given in 2.3.3).

Let X be a compact C^∞ manifold with boundary $Y = \partial X$. Consider $\mathcal{A} \in \mathfrak{G}^{m,d}(X, Y)$, i.e. there are vector bundles $E, F \in \mathrm{Vect}\,(X)$, $J, G \in \mathrm{Vect}\,(Y)$ such that $\mathcal{A}$ defines a mapping

$$\mathcal{A}\colon \begin{matrix} C^\infty(X, E) \\ \oplus \\ C^\infty(Y, J) \end{matrix} \to \begin{matrix} C^\infty(X, F) \\ \oplus \\ C^\infty(Y, G)\,. \end{matrix}$$

As above we set $\mathcal{H}^s(E, J) = H^s(X, E) \oplus H^{s+1/2}(Y, J)$. According to 2.3.3.3, Theorem 1, $\mathcal{A}$ has a continuous extension

$$\mathcal{A}\colon \mathcal{H}^s(E, J) \to \mathcal{H}^{s-m}(F, G)$$

for any $s > d - {}^1/_2$. Thus we have $\mathcal{A} \in \mathcal{L}\big(\mathcal{H}^s(E, J), \mathcal{H}^{s-m}(F, G)\big)$. Let $\sigma(\mathcal{A}) = \big(\sigma_A, \sigma_Y(\mathcal{A})\big)$ be the principal symbol of $\mathcal{A}$

$$\sigma_A\colon p_1^*E \to p_1^*F\,, \quad p_1\colon S^*X \to X\,,$$

and

$$\sigma_Y(\mathcal{A})\colon p^*E' \otimes H^+ \oplus p^*J \to p^+F' \otimes H^+ \oplus p^*G\,,$$

$p: S^*Y \to Y$, are bundle morphisms (cf. 2.3.3.1). Let the scalar products and the norm of $H^s(X, E)$ and $H^s(Y, J)$, $s \in \mathbb{Z}$, be defined by operators $r^+\Lambda^s_-, \Lambda'^s \in L^s(Y)$

$$r^+\Lambda^s_-: H^s(X, E) \to L^2(X, E)\,,$$
$$\Lambda'^s: H^s(Y, J) \to L^2(Y, J)$$

being isomorphisms (cf. 3.1.2.1)

$$\|u\|_{\mathcal{H}^s(X,E)} = \|r^+\Lambda^s_- u\|_{L^2(X,E)}\,, \quad \|v\|_{H^s(Y,J)} = \|\Lambda'^s v\|_{L^2(Y,J)}\,.$$

Set

$$\|\sigma_A\| = \sup_{(x,\xi)\in S^*X} \|\sigma_A(x,\xi)\|_{\hom(p_1^*E(x,\xi)),\, p_1^*F(x,\xi))}$$

(σ_A considered as a section in the bundle $\hom(p_1^*E, p_1^*E)$ over X). This is the norm of the Banach space $C(S^*X, \hom(p_1^*E, p_1^*F))$. For the boundary symbol set

$$\|\sigma_Y(\mathcal{A})\|_s = \sup_{(x',\xi')\in S^*Y} \|\sigma_Y(\mathcal{A})(x',\xi')\|_{\mathcal{L}\left(\begin{smallmatrix}(H^+)_s\otimes p^*E'_{(x',\xi')}\\ \oplus \\ p^*J_{(x',\xi')}\end{smallmatrix},\ \begin{smallmatrix}(H^+)_{s-m}\otimes p^*F'_{(x',\xi')}\\ \oplus\\ p^*G_{(x',\xi')}\end{smallmatrix}\right)}$$

and $\|\sigma(\mathcal{A})\|_s = \max(\|\sigma_A\|, \|\sigma_Y(\mathcal{A})\|_s)$.

Theorem 1. *For any* $\mathcal{A} \in \mathfrak{G}^{m,d}(X, Y)$ *we have*

$$\inf_{\mathcal{C}} \|\mathcal{A} + \mathcal{C}\|_{\mathcal{L}(\mathcal{H}^s, \mathcal{H}^{s-m})} = \|\sigma(\mathcal{A})\|_s\,, \quad s > d - 1/2\,, \tag{1}$$

where the infimum is taken over all $\mathcal{C} \in \mathcal{K}(\mathcal{H}^s, \mathcal{H}^{s-m})$.

Denote by $\overline{\mathfrak{G}}{}^{m,d}_s(X, Y)$ the closure of $\mathfrak{G}^{m,d}(X, Y)$ in $\mathcal{L}(\mathcal{H}^s, \mathcal{H}^{s-m})$, $s > -1/2$.

Corollary 2. *The principal symbol mapping*

$$\sigma: \mathfrak{G}^{m,d}(X, Y; E, F, J, G) \to \begin{matrix} C^\infty(S^*X, \hom(p_1^*E, p_1^*F)) \\ \oplus \\ C^\infty(S^*Y, \mathcal{L}(p^*E' \otimes H^+ \oplus p^*J, p^*F' \otimes H^+ \oplus p^*G)) \end{matrix}$$

has a continuous extension

$$\sigma: \overline{\mathfrak{G}}{}^{m,d}_s(X, Y; E, F, J, G) \to \begin{matrix} C(S^*X, \hom(p_1^*E, p_1^*F)) \\ \oplus \\ C(S^*Y, \mathcal{L}(p^*E' \otimes (H^+)_s \oplus p^*J, p^*F' \otimes (H^+)_{s-m} \oplus p^*G)) \end{matrix}$$

which is surjective onto the continuous operator-valued functions on S^*Y *with* $(H^+)_s \otimes p^*E' \to (H^+)_{s-m} \otimes p^*F'$ *of the form* $\Pi^+\sigma_A + \Pi'\sigma_B$, $(1 + i\nu)^{-m}\,\sigma_A(x', \xi', \nu) \in C(\overline{\mathbb{R}}) \otimes \hom(p^*E'_{(x',\xi')}, p^*F'_{(x',\xi')})$, $\Pi'\sigma_B(x', \xi') \in \mathcal{K}((H^+)_s \otimes p^*E'_{(x',\xi')}, (H^+)_{s-m} \otimes p^*F'_{(x',\xi')})$.

Denote the subspace in $C(S^*Y, (p^*E' \otimes (H^+)_s \oplus p^*J, p^*F \otimes (H^+)_{s-m} \oplus p^*G))$ described in Corollary 2 by $C(S^*Y, \overline{\mathfrak{R}}{}^{m,d}_s(p^*E', p^*F'; p^*J, p^*G))$. Note that $\mathfrak{G}^{0,0}(X, Y; E, F, J, J)$ is an algebra with an involution given by the adjoint with respect to the scalar product in $\mathcal{H}^0(E, J) = L^2(X, E) \oplus H^{1/2}(Y, J)$. Then $\overline{\mathfrak{G}}{}^{0,0}_0(X, Y; E, E, J, J)$ is a C^* algebra in $\mathcal{L}(\mathcal{H}^0(E, J), \mathcal{H}^0(E, J))$. The principal symbol mapping

$$\sigma: \overline{\mathfrak{G}}{}^{0,0}_0 \to \begin{matrix} C(S^*X, \hom(p_1^*E, p_1^*F)) \\ \oplus \\ C(S^*Y, \overline{\mathfrak{R}}{}^{0,0}_0(p^*E', p^*F'; p^*J, p^*G)) \end{matrix}$$

takes values in a C^* algebra ($*$ is taking the adjoint symbol) and the induced mapping

$$\overline{\mathfrak{G}}_0^{0,0}\mathcal{K}(\mathcal{H}^0, \mathcal{H}^0) \to C(S^*X, \hom(p_1^*, E, p_1^*E)) \\ \oplus C(S^*Y, \overline{\mathfrak{R}}_0^{0,0}(p^*E', p^*F', p^*J, p^*G))$$

defines a C^* isomorphism.

Theorem 3. *The calculus of principal symbols from* $\mathfrak{G}^{m,d}$ *carries over to* $\overline{\mathfrak{G}}_s^{m,d}$, *i.e.*

(i) *Let* $\mathcal{A}_i \in \overline{\mathfrak{G}}_s$ *with the principal symbols* $\sigma(\mathcal{A}_i)$, $i = 1, 2$, *and* $\mathcal{A}_1\mathcal{A}_2$ *let be defined. Then* $\sigma(\mathcal{A}_1\mathcal{A}_2) = \sigma(\mathcal{A}_1)\,\sigma(\mathcal{A}_2)$.

(ii) *Let* $\mathcal{A}^*$ *be the adjoint of* $\mathcal{A} \in \overline{\mathfrak{G}}_0^{0,0}$ *with respect to a fixed scalar product in* $\mathcal{H}^0$. *Then* $\mathcal{A}^* \in \overline{\mathfrak{G}}_0^{0,0}$ *and the principal symbol* $\sigma(\mathcal{A}^*)$ *is the adjoint of the principal symbol* $\sigma(\mathcal{A})$.

Corollary 4. $\mathcal{A} \in \overline{\mathfrak{G}}_s^{m,d}$ *is compact in* $\mathcal{L}(\mathcal{H}^s, \mathcal{H}^{s-m})$ *iff its principal symbol* $\sigma(\mathcal{A}) \in \mathfrak{S}^{(m),d}$ *vanishes.*

In the half space situation from the expansion of a Green symbol (cf. 2.2.4.1, Proposition 2 (iv)) and the estimate of Sobolev space norms by the complete symbol (cf. 2.3.2.4, Theorem 5) it is obvious that any compactly supported Green operator $r'B \in \mathrm{Op}\,(\mathfrak{B}^{m,d})\,(\mathbb{R}^{n-1})$ can be approximated in $\mathcal{L}(H^s(\mathbb{R}^n_+), H^{s-m}(\mathbb{R}^n_+))$ by finite sums

$$\sum_j K_j r' T_j\,, \quad r'T_j \in \mathrm{Op}\,(\mathfrak{T}^{0,d})\,(\mathbb{R}^{n-1})\,, \quad K_j \in \mathrm{Op}\,(\mathfrak{K}^m)\,(\mathbb{R}^{n-1})$$

compactly supported. In the global situation we have

Proposition 5. *Any* $r'B \in \mathrm{Op}\,(\mathfrak{B}^{m,d})\,(X, Y)$ *can be approximated in* $\mathcal{L}(H^s(X, E), H^{s-m}(X, F))$, $s > d - {}^1/_2$, *by finite sums* $\sum_j K_j r' T_j$, $r'T_j \in \mathrm{Op}\,(\mathfrak{T}^{0,d})\,(X, Y)$, $K_j \in \mathrm{Op}\,(\mathfrak{K}^m)\,(X, Y)$.

Proof. The principal Green symbol $\Pi'\sigma_B \in C^\infty(S^*Y, \mathcal{L}(p^*E' \otimes H^+, p^*F' \otimes H^+))$ can be approximated by a finite sum $\sum_j \sigma_K^j \Pi' \sigma_T^j$, $\sigma_T^j \in C^\infty(S^*Y, \mathcal{L}(p^*E' \otimes H^+, \mathbb{C}))$, $\sigma_K^j \in C^\infty(S^*Y, \mathcal{L}(\mathbb{C}, p^*F' \otimes H^+))$ with respect to the norm $\|\cdot\|_s$ of the boundary symbol, $\left\|\Pi'\sigma_B - \sum_{j=1}^N \sigma_K^j \Pi' \sigma_T^j\right\|_s < \varepsilon$, for given $\varepsilon > 0$ and N sufficiently large. According to Corollary 2 we can find potential and trace operators K_j and $r'T_j$ with the given principal symbols such that

$$\left\| r'B - \sum_{j=1}^N K_j r' T_j \right\|_{\mathcal{L}(H^s(X,E),\, H^{s-m}(X,F))} < \varepsilon$$

changing arbitrary K_j and $r'T_j$ by negligible operators. □

3. Elliptic Operators on Manifolds with Boundary

3.1. Elliptic Boundary Value Problems

3.1.1. Ellipticity and Fredholm Property

3.1.1.1. Ellipticity of Boundary Problems

Consider a compact C^∞ manifold X with boundary Y and complex vector bundles E, F and J, G over X and Y, respectively.

Let $\mathcal{A} \in \mathfrak{G}$ be an operator

$$\mathcal{A} = \begin{pmatrix} r^+A + r'B & K \\ r'T & Q \end{pmatrix} : \begin{matrix} C^\infty(X, E) \\ \oplus \\ C^\infty(Y, J) \end{matrix} \to \begin{matrix} C^\infty(X, F) \\ \oplus \\ C^\infty(Y, G) \end{matrix} \tag{1}$$

with the following orders of homogeneity of the principal symbols

$$\begin{aligned} &\alpha = \operatorname{ord} \sigma_A \in \boldsymbol{Z}\,, \quad \alpha - 1 = \operatorname{ord} \sigma_B\,, \quad \lambda = \operatorname{ord} \sigma_K\,, \\ &\gamma = \operatorname{ord} \sigma_T\,, \quad 1 - \alpha + \lambda + \gamma = \operatorname{ord} \sigma_Q,\ \lambda, \gamma \in \boldsymbol{R}\,. \end{aligned} \tag{2}$$

Definition 1. The operator (1) is called *elliptic* if the interior symbol

$$\sigma_\Omega(\mathcal{A}) = \sigma_A : \pi^*E \to \pi^*F \tag{3}$$

$(\pi\colon T^*X \setminus 0 \to X)$ and the boundary symbol

$$\sigma_Y(\mathcal{A}) = \begin{pmatrix} \Pi^+\sigma_A + \Pi'\sigma_B & \sigma_K \\ \Pi'\sigma_T & \sigma_Q \end{pmatrix} : \begin{matrix} p^*E^+ \\ \oplus \\ p^*J \end{matrix} \to \begin{matrix} p^*F^+ \\ \oplus \\ p^*G \end{matrix}\,, \tag{4}$$

$$E^+ = E|_Y \otimes H^+\,, \quad F^+ = F|_Y \otimes H^+\,,$$

$(p\colon S^*Y \to Y)$ are isomorphisms. $\mathcal{A}$ is then called an *elliptic boundary value problem* (or simply elliptic).

If $\mathcal{A}$ is elliptic, the operators $r'B$, $r'T$, K, Q are called the *elliptic boundary conditions* belonging to the operator r^+A. Sometimes $r'T$ and K are called *trace* and *potential conditions*, respectively.

Note that the conditions to (3), (4) are related to the homogeneous principal symbols of the given orders (2). We do not point out this convention everywhere. Nevertheless, the principal symbols of these orders could vanish and lower order homogeneous symbols could satisfy ellipticity. Then the assertions about properties in Sobolev spaces are to be modified in an obvious way.

The condition that (4) is an isomorphism is a generalization of the classical Shapiro-Lopatinski condition. Therefore, this notation shall be used in our case, too. In the classical form the Shapiro-Lopatinski condition is usually formulated as the condition on unique solvability of an initial value problem for a system of ordinary differential equations on the half axis in spaces of bounded functions. Since we want to use the symbolic calculus systematically, we prefer to take the Fourier image in our constructions. Set $\mathcal{E}_1 = C^\infty(X, E) \oplus C^\infty(Y, J)$, $\mathcal{E}_2 = C^\infty(X, F) \oplus C^\infty(Y, G)$.

Theorem 2. *Let $\mathcal{A} \in \mathfrak{G}$ be elliptic. Then there exists an elliptic operator $\mathcal{P} \in \mathfrak{G}$ so that the operators $\mathcal{J}_1 - \mathcal{P}\mathcal{A}$, $\mathcal{J}_2 - \mathcal{A}\mathcal{P}$ belong to $\mathfrak{G}^{-\infty}$.*

Here $\mathcal{J}_j$ denotes the identity operator in $\mathcal{E}_j$ $(j = 1, 2)$. $\mathcal{P}$ is called a C^∞ parametrix of $\mathcal{A}$. A consequence of Theorem 2 is that $\mathcal{A}: \mathcal{E}_1 \to \mathcal{E}_2$ is a Fredholm operator if $\mathcal{A}$ is elliptic. Besides $\mathcal{A}$ as operator between spaces of C^∞ sections, $\mathcal{A}$ is also considered as an operator in Sobolev spaces

$$\mathcal{A}: \mathcal{H}_1 \to \mathcal{H}_2 \tag{5}$$

with $\mathcal{H}_1 = H^s(X, E) \oplus H^{t+\lambda+1/2}(Y, J)$, $\mathcal{H}_2 = H^t(X, F) \oplus H^{s-\gamma-1/2}(Y, G)$, $s \in \mathbb{R}$, sufficiently large, $t = s - \alpha$ and α, γ, λ fixed. The existence of a continuous closure (5) is a consequence of 2.3.3.3. Theorem 1. The closure (5) is again denoted by $\mathcal{A}$. If necessary use $\mathcal{A}_{(s)}$ instead of $\mathcal{A}$.

Theorem 2'. *Let $\mathcal{A} \in \mathfrak{G}$ be elliptic. Then $\mathcal{A}_{(s)}: \mathcal{H}_1 \to \mathcal{H}_2$ is a Fredholm operator. A parametrix $\mathcal{P}_0: \mathcal{H}_2 \to \mathcal{H}_1$ of $\mathcal{A}_{(s)}$ can be obtained by closure of the C^∞ parametrix $\mathcal{P}$ mentioned in Theorem 2.*

If, more general, $\mathcal{R} \in \mathfrak{G}$ is an operator with continuous extension $\mathcal{R}: \mathcal{H}_2 \to \mathcal{H}_1$ so that $\mathcal{J} - \mathcal{R}\mathcal{A}$ and $\mathcal{J}_2 - \mathcal{A}\mathcal{R}$ are compact in $\mathcal{H}_1$ and $\mathcal{H}_2$, respectively (i.e. of course the Sobolev space closures), then $\mathcal{R}$ is called a *Sobolev space parametrix* of $\mathcal{A}$.

Theorem 2 and 2' shall be proved in 3.1.1.2. A simple consequence is the following fact. If $\mathcal{P}$, $\mathcal{P}'$ are C^∞ parametrices of an elliptic operator $\mathcal{A}$, then $\mathcal{P} - \mathcal{P}'$ belongs to $\mathfrak{G}^{-\infty}$. Conversely, if $\mathcal{P}$ is a C^∞ parametrix of $\mathcal{A}$ and $\mathcal{K} \in \mathfrak{G}^{-\infty}$, $\mathcal{P}' = \mathcal{P} + \mathcal{K}$ is a C^∞ parametrix of $\mathcal{A}$, too. Since according to Theorem 2' a Sobolev space parametrix always exists in the form of a closure of a C^∞ parametrix, we will simply speak about a parametrix in the future.

If $\mathcal{A}: \mathcal{E}_1 \to \mathcal{E}_2$ is elliptic and

$$\mathcal{P} = \begin{pmatrix} r^+P + r'N & L \\ r'S & R \end{pmatrix}: \mathcal{E}_2 \to \mathcal{E}_1 \tag{6}$$

a parametrix of $\mathcal{A}$, the compatibility between compositions of operators and the corresponding interior and boundary symbols implies $\sigma_P = \sigma_A^{-1}$, $\sigma_Y(\mathcal{P}) = \big(\sigma_Y(\mathcal{A})\big)^{-1}$. Moreover

$$\begin{aligned} &\operatorname{ord} \sigma_P = -\alpha\,, \quad \operatorname{ord} \sigma_N = -\alpha - 1\,, \quad \operatorname{ord} \sigma_L = -\gamma - 1\,, \\ &\operatorname{ord} \sigma_S = -\lambda - 1\,, \quad \operatorname{ord} \sigma_R = \alpha - \lambda - \gamma - 1\,. \end{aligned} \tag{7}$$

Note that an arbitrary operator $\mathcal{R} \in \mathfrak{G}$ with $\sigma_\Omega(\mathcal{R}) = \sigma_A^{-1}$, $\sigma_Y(\mathcal{R}) = \big(\sigma_Y(\mathcal{A})\big)^{-1}$ always defines a Sobolev space parametrix of $\mathcal{A}$ which is no C^∞ parametrix in general. We shall see that the main part of the proof of the existence of a Sobolev space parametrix as well as of a C^∞ parametrix is to show the existence of $\big(\sigma_Y(\mathcal{A})\big)^{-1}$ in $\mathfrak{R}^{(-d)}(Y)$. Note that invertibility of $\sigma_Y(\mathcal{A})$ is not necessary for the Fredholm property of $\mathcal{A}: \mathcal{E}_1 \to \mathcal{E}_2$ (contrarily to $\mathcal{A}_{(s)}: \mathcal{H}_1 \to \mathcal{H}_2$, cf. Theorem 7).

Corollary 3. *Let $p, q \in \mathbb{R}$, $p < s$, $q < t + \lambda + 1/2$ and $\mathcal{H}_0 = H^p(X, E) \oplus H^q(Y, J)$. If $\mathcal{A} \in \mathfrak{G}$, $\mathcal{A}: \mathcal{H}_1 \to \mathcal{H}_2$ is elliptic, there exists a constant $c > 0$ so that*

$$||u||_1 \leqq c\,(||\mathcal{A}u||_2 + ||u||_0) \quad \textit{for all} \quad u \in \mathcal{H}_1 \tag{8}$$

($||\cdot||_j$ denotes the norm in $\mathcal{H}_j$, $j = 0, 1, 2$).

Proof: According to Theorem 2′ choose a parametrix $\mathcal{P}$ of $\mathcal{A}$. Then $\mathcal{P}\mathcal{A} = \mathcal{J}_1 + \mathcal{K}$ with $\mathcal{K} \in \mathfrak{G}^{-\infty}$ and $||\mathcal{P}f||_1 \leqq c\,||f||_2$ with a constant $c > 0$ for all $f \in \mathcal{H}_2$. With $f = \mathcal{A}u$ it follows that

$$||u||_1 = ||\mathcal{P}\mathcal{A}u - \mathcal{K}u||_1 \leqq c\,||\mathcal{A}u||_2 + ||\mathcal{K}u||_1\,.$$

Since $\mathcal{K} \in \mathfrak{G}^{-\infty}$, there is a continuous closure $\mathcal{K}: \mathcal{H}_0 \to \mathcal{H}_1$, i.e. $||\mathcal{K}u||_1 \leqq c\,||u||_0$ for some constant $c > 0$. Thus the assertion is proved. □

Note that in the proof of Corollary 3 only a left parametrix was used.

Lemma 4. *Let $M_j \subset \mathcal{E}_j$ be finite-dimensional subspaces with $m = \dim M_j$ ($j = 1, 2$). Then for each isomorphism $\chi: M_1 \to M_2$, there exists an operator $\mathcal{K} \in \mathfrak{G}^{-\infty}$, $\mathcal{K}: \mathcal{E}_1 \to \mathcal{E}_2$ inducing the isomorphism μ and vanishing on a prescribed complement of M_1 in $\mathcal{E}_1$. An analogous assertion holds for the closure of $\mathcal{K}$ as operator $\mathcal{H}_1 \to \mathcal{E}_2$.*

Proof: Consider in $\mathcal{E}_1$ the Hermitean scalar product $(\cdot,\cdot)$ which is induced by the L^2 scalar product. Let $\mathcal{N}$ be the prescribed complement of M_1 mentioned above and let M be the orthogonal complement of $\mathcal{N}$ in $\mathcal{E}_1$. Choose an orthonormal base $u^1, \dots, u^m$ in M. Then by

$$\mathcal{K}_0 u = \sum_{k=1}^{m} (u, u^k)\, u^k$$

is defined the orthogonal projection to M along $\mathcal{N}$. The operator $\mathcal{K}_0$ induces an isomorphism $\lambda: M_1 \to M$, $\mathcal{K}_0\lambda^{-1} = 1$ on M and $\lambda^{-1}\mathcal{K}_0$ is obviously a projection to M_1 along $\mathcal{N}$. Now define $v^k = \mu\lambda^{-1}u^k$, $k = 1, \dots, m$, which is a base in M_2. Then $\mathcal{K}$ has the form

$$\mathcal{K}u = \sum_{k=1}^{m} (u, u^k)\, v^k\,,$$

i.e. $\mathcal{K} \in \mathfrak{G}^{-\infty}$. □

Theorem 5. *Let $\mathcal{A} \in \mathfrak{G}$ be an elliptic boundary problem (of the form (1)). Then there exist finite-dimensional subspaces $N_+ \subset \mathcal{E}_1$, $N_- \subset \mathcal{E}_2$ with $N_+ = \ker \mathcal{A} = \ker \mathcal{A}_{(s)}$, $N_- \oplus \operatorname{im} \mathcal{A} = \mathcal{E}_2$, $N_- \oplus \operatorname{im} \mathcal{A}_{(s)} = \mathcal{H}_2$, i.e.*

$$\operatorname{ind} \mathcal{A} = \operatorname{ind} \mathcal{A}_{(s)} \tag{9}$$

independently of s ($s \in \mathbb{R}$ sufficiently large).

Proof: From (8) follows $\mathcal{A}_{(s)}u = f$, $f \in \mathcal{E}_2 \Rightarrow u \in \mathcal{E}_1$. Thus $N_+ = \ker \mathcal{A}_{(s)} \subset \mathcal{E}_1$ independently of s and $\dim N_+ < \infty$. Let $N_s \subset \mathcal{H}_2$ be a finite-dimensional complement of $\operatorname{im} \mathcal{A}_{(s)}$. Approximating an arbitrary base of N_s by elements of $\mathcal{E}_2$ we obtain a complement $N_- \subset \mathcal{E}_2$ of $\operatorname{im} \mathcal{A}_{(s)}$ in $\mathcal{H}_2$, $\operatorname{im} \mathcal{A}_{(s)} \oplus N_- = \mathcal{H}_2$. The embedding $\operatorname{im} \mathcal{A} \subset \operatorname{im} \mathcal{A}_{(s)}$ implies the inclusion $\operatorname{im} \mathcal{A} \subset \operatorname{im} \mathcal{A}_{(s)} \cap \mathcal{E}_2$. The a-priori estimate yields that $\operatorname{im} \mathcal{A} = \operatorname{im} \mathcal{A}_{(s)} \cap \mathcal{E}_2$. An arbitrary $u \in \mathcal{E}_2 \subset \mathcal{H}_2$ has a unique decomposition $u = u_{(s)} + u_-$, $u_{(s)} \in \operatorname{im} \mathcal{A}_{(s)}$, $u_- \in N_- \subset \mathcal{E}_2$. It follows $u_{(s)} = u - u_- \in \operatorname{im} \mathcal{A}_{(s)} \cap \mathcal{E}_2 = \operatorname{im} \mathcal{A}$. Hence we have $\mathcal{E}_2 = \operatorname{im} \mathcal{A} \oplus N_-$. □

Corollary 6. *Let $\mathcal{A} \in \mathfrak{G}$ be elliptic and $M \subset \ker \mathcal{A}$, $M' \subset \mathcal{E}_2$, $M' \cap \operatorname{im} \mathcal{A} = \{0\}$ subspaces of the same dimension (necessarily finite). Then there exists a $\mathcal{K} \in \mathfrak{G}^{-\infty}$ so that $\ker(\mathcal{A} + \mathcal{K}) \oplus M = \ker \mathcal{A}$, $\operatorname{im}(\mathcal{A} + \mathcal{K}) = \operatorname{im} \mathcal{A} \oplus M'$. Especially, there exists some $\mathcal{K} \in \mathfrak{G}^{-\infty}$ so that $\mathcal{A} + \mathcal{K}$ is surjective for $\operatorname{ind} \mathcal{A} \geqq 0$ ($\mathcal{A} + \mathcal{K}$ injective for $\operatorname{ind} \mathcal{A} \leqq 0$ and $\mathcal{A} + \mathcal{K}$ an isomorphism for $\operatorname{ind} \mathcal{A} = 0$, respectively). This holds for $\mathcal{A}: \mathcal{E}_1 \to \mathcal{E}_2$ as well as for the closure $\mathcal{H}_1 \to \mathcal{H}_2$.*

Corollary 6 is a simple consequence of Lemma 4.

Let $\mathcal{A}$ be an elliptic operator (of the form (1)). Then $\operatorname{ind} \mathcal{A}$ only depends on the homogeneous principal symbols $\sigma_A, \sigma_B, \sigma_K, \sigma_T, \sigma_Q$. Namely, if $\mathcal{A}'$ is a second operator in $\mathfrak{G}$ with $\sigma_\Omega(\mathcal{A}) = \sigma_\Omega(\mathcal{A}')$, $\sigma_Y(\mathcal{A}) = \sigma_Y(\mathcal{A}')$, the difference $\mathcal{A} - \mathcal{A}': \mathcal{H}_1 \to \mathcal{H}_2$ is compact, i.e. $\operatorname{ind} \mathcal{A} = \operatorname{ind} \mathcal{A}'$.

Next we prove the necessity of the Shapiro-Lopatinski condition for the Fredholm property of operators $\mathcal{A} \in \mathfrak{G}$ in Sobolev spaces.

Theorem 7. *Let $\mathcal{A} \in \mathfrak{G}$ be of the form* (1) *with orders* (2). *Suppose that*

$$\mathcal{A}: H^s(X, E) \oplus H^{t+\lambda+1/2}(Y, J) \to H^t(X, F) \oplus H^{s-\gamma-1/2}(Y, G) \tag{10}$$

($s \in \mathbb{R}$ sufficiently large, $t = s - \alpha$) is a Fredholm operator. Then $\mathcal{A}$ is elliptic.

Proof: Restrict the consideration to the case $s = t + \lambda + 1/2 = t = s - \gamma - 1/2 = 0$. The assertion in the general case is then a simple consequence of the reduction of orders discussed in 3.1.2.1. The notations for $\mathcal{H}_1$, $\mathcal{H}_2$ have now a corresponding meaning. Denote by $\mathcal{R}$ a parametrix of $\mathcal{A}: \mathcal{H}_1 \to \mathcal{H}_2$ and put $\mathcal{K} = \mathcal{I}_1 - \mathcal{R}\mathcal{A}$. Then, similarly as in Corollary 3, we have

$$\|u\|_1 \leqq c_2 (\|\mathcal{A}u\|_2 + \|\mathcal{K}u\|_1) \tag{11}$$

with a constant $c_2 > 0$ for all $u \in \mathcal{H}_1$. Since $\mathcal{A}^*: \mathcal{H}_2 \to \mathcal{H}_1$ is Fredholm, too, we have a similar estimate for $\mathcal{A}^*$. It will be proved that (11) implies the injectivity of $\sigma_\Omega(\mathcal{A})$ and $\sigma_Y(\mathcal{A})$. This consideration applied to $\mathcal{A}^*$ yields surjectivity in view of $\sigma_\Omega(\mathcal{A}^*) = (\sigma_\Omega(\mathcal{A}))^*$, $\sigma_Y(\mathcal{A}^*) = (\sigma_Y(\mathcal{A}))^*$ (cf. 2.3.3.2). Consider first the interior symbol σ_A. Fix some point $(x_0, \xi_0) \in S^*X$, $x_0 \in Y$. In local coordinates near x_0 the symbol σ_A is a matrix-valued function and we have to prove the injectivity of this matrix. Let w be an arbitrary smooth section in E supported by a small neighbourhood of x_0 with empty intersection with Y. The family of operators $R_\lambda w(x) = \lambda^{-n/2} e^{i\lambda\langle x, \xi_0\rangle} \times w((x - x_0) \lambda^{-1/2})$, $\lambda > 0$, has the well-known property that $\|R_\lambda w\|_{L^2(X, E)} = \|w\|_{L^2(X, E)}$ and $R_\lambda w$ weakly converges to 0 for $\lambda \to \infty$. Then the estimate (11) implies for $u = \begin{pmatrix} R_\lambda w \\ 0 \end{pmatrix}$

$$\|w\|_{L^2(X, E)} \leqq c_2 \|\sigma_A(x_0, \xi_0) w(x_0)\|$$

because $r'B$ and $r'T$ act like compact operators on functions with support in $\Omega = X \setminus Y$ and so transform $R_\lambda w$ into a norm convergent sequence for $\lambda \to \infty$. Since the constant c_2 does not depend on (x_0, ξ_0), we obtain injectivity of σ_A up to the boundary. The calculation for the boundary symbols is quite similar (cf. 2.3.4.2). The details are left to the reader. □

A special and well-known elliptic boundary value problem is the Dirichlet problem for the Laplace operator in a bounded smooth domain $\Omega \subset \mathbb{R}^n$. In the notation (1) the Dirichlet problem corresponds to the map

$$\mathcal{D} = \begin{pmatrix} r^+ \Delta \\ r' \end{pmatrix} : C^\infty(X) \to \begin{matrix} C^\infty(X) \\ \oplus \\ C^\infty(Y) \end{matrix} \tag{12}$$

$(X = \overline{\Omega},\ Y = \partial\Omega)$. E, F, G are in this case the corresponding one-dimensional trivial bundles over X and Y, respectively, and the fibre dimension of J is zero. The map (12) defines an isomorphism and according to Theorem 2 (and the remarks in this connection) $\mathcal{D}^{-1}$ belongs to $\mathfrak{G}$ and has the form

$$\mathcal{D}^{-1} = (r^+P + r'N, L): \begin{array}{c} C^\infty(X) \\ \oplus \\ C^\infty(Y) \end{array} \to C^\infty(X) \tag{13}$$

with a pseudo-differential operator r^+P, a Green operator $r'N$ and a potential operator L. Denote by $G(x, z)$ the Green function of Ω and by $L(x, y)$ the Poisson kernel. Then, for arbitrary $f \in C^\infty(X)$, $g \in C^\infty(Y)$, the uniquely determined solution of the Dirichlet problem $r^+\Delta u = f$, $r'u = g$ has the form

$$u(x) = \int_X G(x, z)\, f(z)\, dz + \int_Y L(x, y)\, g(y)\, dy\,.$$

Thus

$$(r^+P + r'N)\, f = \int G(x, z)\, f(z)\, dz\,, \quad Lg = \int L(x, y)\, g(y)\, dy\,.$$

Since $\mathcal{D}^{-1}$ is the two-sided inverse of $\mathcal{D}$, it follows that

$$r^+\Delta \circ L = 0\,. \tag{14}$$

The operator (13), which is an elliptic element in $\mathfrak{G}$, justifies the point of view that it is reasonable to take into account pseudo-differential operators, Green operators and potential operators in the boundary conditions. For calculating the index of operators it is useful to consider compositions. For instance, let

$$\mathcal{A} = \begin{pmatrix} r^+\Delta \\ r'T \end{pmatrix} : C^\infty X) \to \begin{array}{c} C^\infty(X) \\ \oplus \\ C^\infty(Y) \end{array} \tag{15}$$

be another elliptic boundary problem for $r^+\Delta$ (e.g. the Neumann problem or the third boundary problem). In view of (14) and $(r^+\Delta) \circ (r^+P + r'N) = 0$ we get

$$\mathcal{E} = \mathcal{A}\mathcal{D}^{-1} = \begin{pmatrix} r^+\Delta \\ r'T \end{pmatrix} (r^+P + r'N, L) = \begin{pmatrix} 1 & 0 \\ r'T(r^+P + r'N) & r'T \circ L \end{pmatrix}$$

and $\mathcal{E}$ is Fredholm (as composition of elliptic boundary problems). Thus the operator $R = r'T \circ L\colon C^\infty(Y) \to C^\infty(Y)$ is Fredholm, too, and obviously

$$\operatorname{ind} R = \operatorname{ind} \mathcal{E} = \operatorname{ind} \mathcal{A} - \operatorname{ind} \mathcal{D} = \operatorname{ind} \mathcal{A}\,.$$

Applying 1.2.4.2, Example (3) we obtain

Remark 8. Let $n = \dim \Omega \geqq 3$. Then every elliptic boundary value problem for the Laplace operator of the form (15) has the index zero.

The Fredholm property of operators of the form (1) admits the following interpretation. For $k = \operatorname{codim}\big(\operatorname{im}(r^+A + r'B)\big) = \infty$ the potential operator K "fills" the cokernel by potentials of densities on Y so that

$$\operatorname{codim}\big(\operatorname{im}(r^+A + r'B) + \operatorname{im} K\big) < \infty\,,$$

$$\dim\big(\operatorname{im}(r^+A + r'B) \cap \operatorname{im} K\big) < \infty\,.$$

Of course this holds for $k < \infty$, too, but then K is superfluous for achieving the Fredholm property. In the case $l = \dim\ker(r^+A + r'B, K) = \infty$ the operators

$r'T$, Q define a Fredholm operator

$$(r'T, Q) : \ker (r^+A + r'B, K) \to C^\infty(Y, G) .$$

For $l < \infty$ this row in $\mathcal{A}$ is superfluous for the Fredholm property.

For the Dirichlet problem $\mathfrak{D}$ we have in particular that $r^+\Delta: C^\infty(X) \to C^\infty(X)$ is surjective and $r': \ker (r^+\Delta) \to C^\infty(Y)$ is one-to-one. For $\mathfrak{D}^{-1}$ the situation is inverse: $(r^+P + r'N): C^\infty(X) \to C^\infty(X)$ is injective and $L: C^\infty(Y) \to C^\infty(X)$ is an isomorphism onto a complement of im $(r^+P + r'N)$.

In 3.1.1.2 elliptic boundary problems will be constructed using corresponding considerations on boundary symbol level. For this reason we discuss now some properties of boundary symbols.

According to 2.1.2.1, Proposition 3 we can make the following

Remark 9. Let $\sigma_A \in \mathfrak{A}^{(\alpha)}$, $\sigma_A: \pi^*E \to \pi^*F$ be an elliptic symbol. Then

$$\Pi^+\sigma_A(x', 0, \xi', \nu): H^+ \otimes \mathbb{C}^k \to H^+ \otimes \mathbb{C}^k \tag{16}$$

is a Fredholm operator for every fixed $(x', \xi') \in S^*Y$ (k denotes the fibre dimension of E, F and in (16) there are used local coordinates and trivializations of the bundles). From the invariance of (16) formulated in 3.3.3.1 $\Pi^+\sigma_A$ can be interpreted as a Fredholm family over S^*Y

$$\Pi^+\sigma_A: p^*E^+ \to p^*F^+ . \tag{17}$$

The notation Fredholm family is justified because of the existence of closures in Hilbert spaces for each fibre (cf. 2.1.2.1, Proposition 5, 2.1.2.3. (7)) preserving the index, and because of the arising Hilbert bundles are trivial (cf. 1.1.3.4, Corollary 5). Thus we can carry over all notations and constructions about Fredholm families for Hilbert spaces to families of the form (17) for elliptic $\sigma_A \in \mathfrak{A}^{(\alpha)}$. Thus e.g.

$$\dim \ker \Pi^+\sigma_A \leqq \text{const}\,, \quad \dim \operatorname{coker} \Pi^+\sigma_A \leqq \text{const}$$

where the constants are independent of the parameter point on S^*Y and there exists a finite-dimensional trivial subbundle W in p^*F^+ so that

$$(\operatorname{im} \sigma_A)_{(x', \xi')} + W_{(x', \xi')} = (p^*F^+)_{(x', \xi')}$$

for all $(x', \xi') \in S^*Y$. We get an index element (cf. 1.1.3.4)

$$\operatorname{ind}_{S^*Y} \Pi^+\sigma_A \in K(S^*Y) . \tag{18}$$

Since (18) depends only on the homotopy class of the Fredholm family, we obtain

Proposition 10. *Let* $(\sigma^{(t)})_{0 \leqq t \leqq 1}$ *be a homotopy through elliptic symbols in* $\mathfrak{A}^{(\alpha)}$. *Then*

$$\operatorname{ind}_{S^*Y} \Pi^+\sigma^{(0)} = \operatorname{ind}_{S^*Y} \Pi^+\sigma^{(1)} . \tag{19}$$

Proposition 11. *Let* $\mathcal{A} \in \mathfrak{G}$ *be an elliptic operator of the form* (1), $\sigma_A = \sigma_\Omega(\mathcal{A})$. *Then*

$$\operatorname{ind}_{S^*Y} \Pi^+\sigma_A = [p^*G] - [p^*J] . \tag{20}$$

Thus

$$\operatorname{ind}_{S^*Y} \Pi^+\sigma_A \in p^*K(Y) \tag{21}$$

($p^*K(Y)$ *denotes the image of* $K(Y)$ *under pull back with respect to* $p: S^*Y \to Y$).

Proof: Set for abbreviation

$$\sigma_Y(\mathcal{A}) = \begin{pmatrix} \alpha & \varkappa \\ \tau & \varrho \end{pmatrix} : \begin{matrix} p^*E^+ \\ \oplus \\ p^*J \end{matrix} \to \begin{matrix} p^*F^+ \\ \oplus \\ p^*G . \end{matrix}$$

Using standard properties of the composition of boundary symbols (cf. 2.1.2.3) we get that $\alpha^{-1} = \Pi^+\sigma_A^{-1}$ is a family of parametrices of $\Pi^+\sigma_A$. Using constructions as in 3.1.1.5 we get a homotopy through Fredholm families over S^*Y

$$\sigma_Y(\mathcal{A}) \simeq \begin{pmatrix} \alpha & 0 \\ 0 & \varrho - \tau\alpha^{-1}\varkappa \end{pmatrix}.$$

Since the homotopy preserves the index element and $\sigma_Y(\mathcal{A})$ is an isomorphism, we get

$$\mathrm{ind}_{S^*Y}\,\Pi^+\sigma_A = -\,\mathrm{ind}_{S^*Y}\,(\varrho - \tau\alpha^{-1}\varkappa)\,. \tag{22}$$

$\varrho - \tau\alpha^{-1}\varkappa$ as a morphism of finite dimensional bundles $p^*J \to p^*G$ is obviously a Fredholm family with the index $[p^*J] - [p^*G]$. Thus the assertion follows from (22). □

Theorem 12. *Let* $r^+A\colon C^\infty(X, E) \to C^\infty(X, F)$ *be an elliptic pseudo-differential operator with the transmission property and let* $\sigma_A\colon \pi^*E \to \pi^*F$ *be the homogeneous principal symbol. Then the following conditions are equivalent:*

(i) σ_A *satisfies* (21);

(ii) *there exist vector bundles* J, G *over* Y *and an elliptic operator* $\mathcal{A} \in \mathfrak{G}$ *of the form* (1) *with the given operator* r^+A *in the left upper corner.*

(ii) ⇒ (i) is a consequence of Proposition 11. In 3.1.1.2 (i) ⇒ (ii) will be proved.

In 3.1.2.1 a reduction of orders shall be discussed so that the assumption $\sigma_A \in \mathfrak{A}^{(0)}$ is no restriction of generality for questions concerning the index. If $\sigma_A \in \mathfrak{A}^{(0)}$, $\sigma_A\colon \pi^*E \to \pi^*F$, $\sigma = \sigma_A|_Y$, defines an isomorphism $\sigma(\cdot, \nu)\colon p^*E' \to p^*F'$ for each fixed $\nu \in \mathbb{R}$ in local coordinates we have

$$\sigma_0(x') = \lim_{\nu\to+\infty} \sigma(x', \xi', \nu) = \lim_{\nu\to-\infty} \sigma(x', \xi', \nu) = \sigma(x', 0, \pm 1) \tag{23}$$

and (23) globally defines an isomorphism $\sigma_0\colon E' \to F'$, $E' = E|_Y$, $F' = F|_Y$ (cf. 3.2.1.2). Identifying $\mathbb{R}_\nu \cup \{\pm\infty\}$ with S^1, σ can be interpreted as an isomorphism $\sigma\colon s^*p^*E' \to s^*p^*F'$ $\big(s\colon S^1 \times S^*Y \to S^*Y,\ s\colon (\nu, x', \xi') \to (x', \xi')\big)$. Another interpretation is that σ represents a function on S^1 with values in the set of isomorphisms $E' \to F'$. Note that the distinction of E' and F' for the consideration of $\mathrm{ind}_{S^*Y}\,\Pi^+\sigma(\sigma \in \mathfrak{A}^{(0)}|_Y)$ is not necessary. Namely, by $\sigma_0\colon E' \xrightarrow{\cong} F'$ an isomorphism $\gamma\colon p^*(H^+ \otimes E') \to p^*(H^+ \otimes F')$ is induced and obviously

$$\mathrm{ind}_{S^*Y}\,\Pi^+\sigma = \mathrm{ind}_{S^*Y}\,(\gamma^{-1} \circ \Pi^+\sigma)\,.$$

Since H^+ can be replaced by the L^2 closure $V^+ = (H^+)_0$ without changing the index, $\Pi^+\sigma$ can be replaced by the Fredholm family

$$\gamma^{-1} \circ \Pi^+\sigma\colon p^*\big((H^+)_0 \otimes E'\big) \to p^*\big((H^+)_0 \otimes E'\big)\,.$$

Finally, since Hilbert space bundles are trivial (cf. 1.1.3.4, Corollary 5), there is a fixed Hilbert space $\mathcal{H}$ with $S^*Y \times \mathcal{H} \cong p^*\big((H^+)_0 \otimes E'\big)$. Thus $\Pi^+\sigma$ can be regarded, except bundle isomorphisms, as a Fredholm family in $\mathcal{H}$ with parameter space S^*Y. Particularly the bundle structures of E', F' play no role.

3.1.1.2. Construction of Elliptic Boundary Problems

First we finish the proof of 3.1.1.1, Theorem 12 $\big($(i) → (ii)$\big)$. Let

$$r^+A\colon C^\infty(X, E) \to C^\infty(X, F) \tag{1}$$

be an elliptic operator with the transmission property and principal symbol σ_A: $\pi^*E \to \pi^*F$, $\alpha = \operatorname{ord} \sigma_A \in Z$. Suppose the condition 3.1.1.1. (21) is satisfied and construct vector bundles J, G over Y and an isomorphism of the form 3.1.1.1.(4). Then the symbols in 3.1.1.4. represent sections

$$\left.\begin{aligned} &\sigma_B \in C^\infty(S^*Y, (p^*E')^* \otimes p^*F' \otimes H_\nu^+ \otimes H_\tau^-)\,, \\ &\sigma_T \in C^\infty(S^*Y, (p^*E')^* \otimes p^*G \otimes H^-)\,, \\ &\sigma_K \in C^\infty(S^*Y, (p^*J)^* \otimes p^*F' \otimes H^+)\,, \\ &\sigma_Q \in C^\infty(S^*Y, (p^*J)^* \otimes p^*G)\,. \end{aligned}\right\} \tag{2}$$

Choosing arbitrary real numbers γ, λ we can invariantly define the symbols σ_B, σ_K, σ_T, σ_Q for all $\xi' \neq 0$ by the formulas

$$\left.\begin{aligned} \sigma_B(x', \xi', \nu, \tau) &= |\xi'|^{\alpha-1}\, \sigma_B\left(x', \frac{\xi'}{|\xi'|}, \frac{\nu}{|\xi'|}, \frac{\tau}{|\xi'|}\right), \\ \sigma_T(x', \xi', \tau) &= |\xi'|^{\gamma}\, \sigma_T\left(x', \frac{\xi'}{|\xi'|}, \frac{\tau}{|\xi'|}\right), \\ \sigma_K(x', \xi', \nu) &= |\xi'|^{\lambda}\, \sigma_K\left(x', \frac{\xi'}{|\xi'|}, \frac{\nu}{|\xi'|}\right), \\ \sigma_Q(x', \xi') &= |\xi'|^{1-\alpha+\lambda+\gamma}\, \sigma_Q\left(x', \frac{\xi'}{|\xi'|}\right). \end{aligned}\right\} \tag{3}$$

Then the operator 3.1.1.1.(1) with the elements r^+A, $\operatorname{Op}(\sigma_B) = r'B$, $\operatorname{Op}(\sigma_K) = K$, $\operatorname{Op}(\sigma_Q) = Q$ is elliptic. Thus we are reduced to the construction of an isomorphism 3.1.1.1.(4).

Lemma 1. *Let* $G_1 \in \operatorname{Vect}(S^*Y)$ *and* $[G_1] - [\mathbb{C}^N] \in p^*K(Y)$, *i.e. there exist bundles* G, $J \in \operatorname{Vect}(Y)$ *with*

$$[G_1] - [\mathbb{C}^N] = [p^*G] - [p^*J]\,. \tag{4}$$

Then there exist bundles J_1, $G_0 \in \operatorname{Vect}(Y)$ *with* $G_1 \oplus p^*J_1 \cong p^*G_0$.

Proof: (4) means per def. that there exist bundles H', $H'' \in S^*Y$ with

$$G_1 \oplus H' \cong p^*G \oplus H'', \quad \mathbb{C}^N \oplus H' \cong p^*J \oplus H''\,.$$

Let $(H')^\perp$ be a bundle over S^*Y so that $H' \oplus (H')^\perp \cong \mathbb{C}^k$ for suitable k. Put $H = H'' \oplus (H')^\perp$. Then $G_1 \oplus \mathbb{C}^k \cong p^*G \oplus H$, $\mathbb{C}^{N+k} \cong p^*J \oplus H$, i.e.

$$G_1 \oplus \mathbb{C}^k \oplus p^*J \cong p^*G \oplus H \oplus p^*J \cong p^*G \oplus \mathbb{C}^{N+k}\,.$$

Thus $J_1 = J \oplus \mathbb{C}^k$, $G_0 = G \oplus \mathbb{C}^{N+k}$ satisfy the assertion. □

Note that an analogous statement holds for $\mathbb{C}^N$ replaced by p^*L, $L \in \operatorname{Vect}(Y)$ arbitrary.

Lemma 2. *Let* $\sigma_A \in \mathfrak{A}^{(\alpha)}$ *be elliptic and suppose that* 3.1.1.1.(21) *is satisfied. Then there exist a Green symbol* $\sigma_{B_0} \in \mathfrak{B}$ *and bundles* G_0, $J_0 \in \operatorname{Vect}(Y)$ *so that*

$$\ker_{S^*Y}(\Pi^+\sigma_A + \Pi'\sigma_{B_0}) \cong p^*G_0\,, \quad \operatorname{coker}_{S^*Y}(\Pi^+\sigma_A + \Pi'\sigma_{B_0}) \cong p^*J_0\,.$$

Proof: In the definition of 3.1.1.1.(18) a trivial subbundle $W \subset p^*F^+$, $W \cong \mathbb{C}^N$ with $W + \operatorname{im} \Pi^+\sigma_A = p^*F^+$ was used (this means the vector sum in each fibre). According to 2.1.2.2., Lemma 4 there exists a Green symbol σ_{B_1} so that $\Pi'\sigma_{B_1}$:

$p^*F^+ \to W$ is a projection. Then W is the kernel of $1 - \Pi'\sigma_{B_1}: p^*F^+ \to p^*F^+$. Now $(1 - \Pi'\sigma_{B_1})\Pi^+\sigma_A = \Pi^+\sigma_A + \Pi'\sigma_{B_2}$ is a Fredholm family with coker $(\Pi^+\sigma_A + \Pi'\sigma_{B_2}) \cong \mathbb{C}^N$. Thus $\ker(\Pi^+\sigma_A + \Pi'\sigma_{B_2})$ has a locally constant dimension so that there exists a vector bundle G_1 over S^*Y with $G_1 \cong \ker(\Pi^+\sigma_A + \Pi'\sigma_{B_2})$, i. e. $\ker_{S^*Y}\Pi^+\sigma_A = [G_1] - [\mathbb{C}^N]$. Using Lemma 1 we find a subbundle $\tilde{J}_1$ in p^*E^+ with $\tilde{J}_1 \cong p^*J_1$ for some $J_1 \in \mathrm{Vect}(Y)$, so that $G_1 \oplus p^*J_1 \cong p^*G_0$ for some $G_0 \in \mathrm{Vect}(Y)$. Let $\Pi'\sigma_{B_3}$: $p^*E^+ \to p^*E^+$ be a Green operator projecting onto p^*J_1. Then $(\Pi^+\sigma_A + \Pi'\sigma_{B_2}) \times (1 - \Pi'\sigma_{B_3})$ is a Fredholm family with a kernel isomorphic to p^*G_0 and with a cokernel isomorphic to p^*J_0 for $J_0 = J_1 \oplus \mathbb{C}^N$. Thus σ_{B_0} with $\Pi'\sigma_{B_0} = -\Pi^+\sigma_A \circ \Pi'\sigma_{B_3} + \Pi'\sigma_{B_2} - \Pi'\sigma_{B_2} \circ \Pi'\sigma_{B_3}$ has the desired property. □

Let J_0, G_0 and σ_{B_0} be as in Lemma 2. Then the isomorphism $p^*J_0 \to \mathrm{coker}\,(\Pi^+\sigma_A + \Pi'\sigma_B)$ can be interpreted as potential symbol $\sigma_{K_0} \in \hom(p^*J_0, p^*F^+) \otimes H^+$. Then

$$(\Pi\sigma_A + \Pi'\sigma_{B_0}, \sigma_{K_0}):\ \begin{matrix} p^*E^+ \\ \oplus \\ p^*J_0 \end{matrix} \to p^*F^+ \tag{5}$$

is surjective and $\mathrm{im}\,(\Pi^+\sigma_A + \Pi'\sigma_{B_0}) \cap \mathrm{im}\,\sigma_{K_0} = \{0\}$. By construction $\tilde{G}_0 = \ker(\Pi^+\sigma_A + \Pi'\sigma_{B_0})$ is isomorphic to p^*G_0. The choice of an isomorphism $\tilde{G}_0 \to p^*G_0$ corresponds to the choice of a never-vanishing section in $\tilde{G}_0^* \otimes p^*G_0$. This corresponds to a section in $(p^*E^+)^* \otimes p^*G_0 \oplus (p^*J_0)^* \otimes p^*G_0$. Note that $(p^*E^+)^* = \big(p^*(E' \otimes H^+)\big)^* = (p^*E')^* \otimes H_0^-$. Here the pairing $H_0^- \times H^+ \to \mathbb{C}$ is given by $(t, k) \to \Pi'(tk)$, $t \in H_0^-$, $k \in H^+$. Thus the isomorphism $\tilde{G}_0 \to p^*G_0$ can be represented in the form $(\Pi'\sigma_{T_0}, \sigma_{Q_0})$ for a trace symbol $\sigma_{T_0} \in C^\infty\big(S^*Y, (p^*E')^* \otimes p^*G_0 \otimes H_0^-\big)$ and a pseudo-differential symbol $\sigma_{Q_0} \in C^\infty\big(S^*Y, (p^*J_0)^* \otimes p^*G_0\big)$ on Y. The map

$$(\Pi'\sigma_{T_0}, \sigma_{Q_0}):\ p^*\begin{pmatrix} E^+ \\ \oplus \\ J_0 \end{pmatrix} \to p^*G_0$$

inducing the given isomorphism over $\tilde{G}_0$ vanishes on a complement of $\tilde{G}_0$. Thus we get an isomorphism

$$\begin{pmatrix} \Pi^+\sigma_A + \Pi'\sigma_{B_0} & \sigma_{K_0} \\ \Pi'\sigma_{T_0} & \sigma_{Q_0} \end{pmatrix}:\ \begin{matrix} p^*E^+ \\ \oplus \\ p^*J_0 \end{matrix} \to \begin{matrix} p^*F^+ \\ \oplus \\ p^*G_0 \end{matrix}, \tag{6}$$

i.e. an elliptic boundary symbol and 3.1.1.1, Theorem 12 is proved. □

In the situation discussed above we have automatically $\sigma_{Q_0} = 0$. For other J_0 and σ_{B_0}, σ_{K_0}, for which (5) is surjective and $\tilde{G}_0$ (the kernel of (5)) isomorphic to some p^*G_0, $G_0 \in \mathrm{Vect}(Y)$, the described construction gives an isomorphism of the form $(\Pi'\sigma_{T_0}, \sigma_{Q_0})$ where in general $\sigma_{Q_0} \neq 0$.

The construction of elliptic boundary problems in the proof of 3.1.1.1, Theorem 12 shows that a reasonable subclass of problems contains potential operators K for which $\sigma_K: p^*J_0 \to p^*F^+$ is injective. For a better understanding of constructions with boundary symbols it is advisable to consider this case first.

Remark 3. Let $\sigma_A \in \mathfrak{A}^{(\alpha)}$ be elliptic and satisfy the condition 3.1.1.1.(21). A small modification of the constructions in the proof of 3.1.1.1, Theorem 12, (i) $\Rightarrow$ (ii), shows that one can find elliptic boundary symbols with $\Pi^+\sigma_A$ in the left upper corner without Green symbol. Namely, let W be as in that proof and let $\sigma_{K_1}: \mathbb{C}^N \to W \subset p^*F^+$ be an injective map onto W in the form of a potential symbol. Then we have $G_1 = \ker_{S^*Y}(\Pi^+\sigma_A, \sigma_{K_1})$ again. Next represent an isomorphism $p^*J_1 \to \tilde{J}_1 \subset p^*F^+$ by a second potential σ_{K_2}. Then $(\Pi^+\sigma_A, \sigma_{K_1}, \sigma_{K_2}): p^*E^+ \oplus \mathbb{C}^N \oplus p^*J_1 \to p^*F^+$ is surjective with kernel $\tilde{G}_0 \cong G_1 \oplus p^*J_1 \cong p^*G_0$. Representation of an isomorphism, $\tilde{G}_0 \to G_0$ again by a pair $(\Pi'\sigma_{T_0}, \sigma_{Q_0})$ gives an elliptic boundary symbol $\sigma_{K_0} = (\sigma_{K_1}, \sigma_{K_2})$, $p^*J_0 = \mathbb{C}^N \oplus p^*J_1$.

The construction of an elliptic boundary symbol (6) gave a trace symbol of type 0. Let now (6) be an arbitrary elliptic boundary symbol and $\tilde{G}_0$ the kernel of the first row. Then the second row induces an isomorphism $\tilde{G}_0 \to p^*G_0$. On the other hand, this isomorphism can always be represented in the form $(\Pi'\sigma_{T_1}, \sigma_{Q_1})$ for σ_{T_1} of type 0 (this is a consequence of the considerations above). Since $(\Pi'\sigma_{T_0}, \sigma_{Q_0})$ and $(\Pi'\sigma_{T_1}, \sigma_{Q_1})$ define the same isomorphisms $\tilde{G}_0 \to p^*G_0$, $(1-t)(\Pi'\sigma_{T_0}, \sigma_{Q_0}) + t(\Pi'\sigma_{T_1}, \sigma_{Q_1})$, $0 \leq t \leq 1$, defines this isomorphism, too. Thus one can generate a corresponding homotopy through elliptic boundary symbols without change of the first row into one with zero type trace symbol and it follows

Remark 4. Let $\mathcal{A}_0 \in \mathfrak{G}$ be elliptic. Then there exists an elliptic $\mathcal{A}_1 \in \mathfrak{G}$ homotopic to $\mathcal{A}_0$ through elliptic operators in $\mathfrak{G}$, where the trace operator in $\mathcal{A}_1$ has type zero. The above proof of 3.1.1.1, Theorem 12, (i) $\Rightarrow$ (ii), is in some sense constructive and gives rise to a lot of concrete elliptic boundary value problems. There are important special cases where one has additional informations. For instance, if the bundles E, F have fibre dimension 1 and $\dim X \geq 3$, then $\Pi^+\sigma_A: p^*E^* \to p^*F^+$ is either injective or surjective. This follows from a corresponding property for scalar Wiener-Hopf-operators and because S^*Y is connected. Then if, for instance, $p^*G \cong \ker_{S^*Y} \Pi^+\sigma_A$, there is an isomorphism of the form

$$\begin{pmatrix} \Pi^+\sigma_A \\ \\ \Pi'\sigma_T \end{pmatrix}: p^*E^+ \to \begin{matrix} p^*F^+ \\ \oplus \\ p^*G \end{matrix},$$

i.e. the corresponding boundary problem contains only trace conditions. If $\Pi^+\sigma_A$ is injective and $p^*J \oplus \operatorname{im} \Pi^+\sigma_A \cong p^*F^+$, the corresponding boundary problem contains only potential conditions.

Lemma 5. *Suppose that we are given bundles G_0, G, $J \in \operatorname{Vect}(Y)$ and $N \in \mathbb{Z}_+$ with*

$$[p^*G_0] - [\mathbb{C}^N] = [p^*G] - [p^*J]$$

in $K(S^*Y)$. Then there exists an integer M and a $V \in \operatorname{Vect}(Y)$ with

$$p^*(G_0 \oplus \mathbb{C}^M) \cong p^*(G \oplus V), \quad \mathbb{C}^{N+M} \cong p^*(J \oplus V).$$

Proof: Let $J^\perp$ be a bundle over Y with $J \oplus J^\perp = \mathbb{C}^{N_1}$ and $N_1 \geq N$. Then

$$[p^*(G_0 \oplus \mathbb{C}^{N_2})] - [\mathbb{C}^{N_1}] = [p^*(G \oplus J^\perp)] - [\mathbb{C}^{N_1}]$$

for $N_2 = N_1 - N$. Because of the definition of equality of elements in $K(S^*Y)$ there exist bundles H', $H'' \in \operatorname{Vect}(S^*Y)$ with

$$p^*(G_0 \oplus \mathbb{C}^{N_2}) \oplus H' \cong p^*(G \oplus J^\perp) \oplus H'', \quad \mathbb{C}^{N_1} \oplus H' \cong \mathbb{C}^{N_1} \oplus H''.$$

Then $p^*(G_0 \oplus \mathbb{C}^{N_2}) \oplus H' \oplus \mathbb{C}^{N_1} \cong p^*(G \oplus J^\perp) \oplus H'' \oplus \mathbb{C}^{N_1}$. Let $(\mathbb{C}^{N_1} \oplus H')^\perp$ be a bundle over S^*Y with $(\mathbb{C}^{N_1} \oplus H') \oplus (\mathbb{C}^{N_1} \oplus H')^\perp = \mathbb{C}^{N_3}$ for a certain N_3. Then

$$p^*(G_0 \oplus \mathbb{C}^{N_2}) \oplus \mathbb{C}^{N_3} \cong p^*(G \oplus J^\perp) \oplus \mathbb{C}^{N_3},$$

$$\mathbb{C}^N \oplus \mathbb{C}^{N_2} \oplus \mathbb{C}^{N_3} = \mathbb{C}^{N_1+N_3} \cong p^*(J \oplus J^\perp) \oplus \mathbb{C}^{N_3}.$$

Thus we can set $M = N_2 + N_3$, $V = J^\perp \oplus \mathbb{C}^{N_3}$. □

Next we pass to the proof of 3.1.1.1, Theorem 2. Let $\mathcal{A} \in \mathfrak{G}$ be elliptic and $\sigma_A = \sigma_\Omega(\mathcal{A}) \in \mathfrak{A}^{(\alpha)}$. Then $\sigma_P = \sigma_A^{-1} \in \mathfrak{A}^{(-\alpha)}$ is again elliptic. Suppose that $\mathcal{A}$ has the form 3.1.1.1.(1) with given bundles J, G. Then we have

Proposition 6. *There exists an elliptic boundary symbol of the form*

$$\sigma_Y(\mathcal{P}) = \begin{pmatrix} \Pi^+\sigma_P + \Pi'\sigma_N & \sigma_L \\ \Pi'\sigma_S & \sigma_R \end{pmatrix} : \begin{matrix} p^*F^+ \\ \oplus \\ p^*G \end{matrix} \to \begin{matrix} p^*E^+ \\ \oplus \\ p^*J \end{matrix} \tag{7}$$

with $\sigma_Y(\mathcal{P}) = (\sigma_Y(\mathcal{A}))^{-1}$.

Proof: The construction of (7) consists in a parameter depending version of 2.1.2.4, Proposition 6. The parameter is the point on S^*Y. First prove the existence of an elliptic boundary symbol

$$\mathfrak{b} = \begin{pmatrix} \Pi^+\sigma_P + \Pi'\sigma_{N_1} & \sigma_{L_1} \\ \Pi'\sigma_{S_1} & \sigma_{R_1} \end{pmatrix} : \begin{matrix} p^*F^+ \\ \oplus \\ p^*G \end{matrix} \to \begin{matrix} p^*E^+ \\ \oplus \\ p^*J \end{matrix} \tag{8}$$

with the given bundles J, G. Since $\Pi^+\sigma_P$ is a parametrix of $\Pi^+\sigma_A$ in the sense of Fredholm families, we have $\operatorname{ind}_{S^*Y}(\Pi^+\sigma_P) = [p^*J] - [p^*G]$. In order to find an isomorphism of the form (8) it is obviously sufficient to find a Green symbol σ_{N_1} with

$$\ker_{S^*Y}(\Pi^+\sigma_P + \Pi'\sigma_{N_1}) \cong p^*J, \quad \operatorname{coker}_{S^*Y}(\Pi^+\sigma_P + \Pi'\sigma_{N_1}) \cong p^*G.$$

The constructions at the beginning of this section show that there exists a Green symbol σ_{N_0} so that

$$\tilde{J}_0 = \ker_{S^*Y}(\Pi^+\sigma_P + \Pi'\sigma_{N_0}) \cong p^*J_0, \quad \tilde{G}_0 = \operatorname{coker}_{S^*Y}(\Pi^+\sigma_P + \Pi'\sigma_{N_0}) \cong \mathbb{C}^N$$

for suitable $J_0 \in \operatorname{Vect}(Y)$ and N, i.e. $[p^*J_0] - [\mathbb{C}^N] = [p^*J] - [p^*G]$. Lemma 5 shows that for N sufficiently large there are decompositions $\tilde{J}_0 = \tilde{V} \oplus \tilde{J}$, $\tilde{G}_0 = \tilde{W} \oplus \tilde{G}$ and $J_0 \cong V \oplus J$, $\mathbb{C}^N \cong W \oplus G$ $(V, W \in \operatorname{Vect}(Y))$ with isomorphisms $\tilde{V} \cong p^*V$, $\tilde{J} \cong p^*J$, $\tilde{W} \cong p^*W$, $\tilde{G} \cong p^*G$, where $\Pi^+\sigma_P + \Pi'\sigma_{N_0}$ induces an isomorphism $\beta \colon \tilde{W} \to \tilde{V}$. There exists a Green symbol σ_{N_2} so that the isomorphism β is induced by $-\Pi'\sigma_{N_2}$ and that $\Pi'\sigma_{N_2}$ vanishes on a complement of $\tilde{W}$ in p^*E^+. Then $\sigma_{N_1} = \sigma_{N_0} + \sigma_{N_2}$ is a Green symbol with the desired property. The main point in the following calculations is to ensure that the inverse of $\mathfrak{a} = \sigma_Y(\mathcal{A})$, which can be separately calculated for each $\varrho \in S^*Y$ (cf. 2.1.2.4, Proposition 6), smoothly depends on the parameter $\varrho \in S^*Y$. For $\mathfrak{b}$ given by (8) smoothness is obviously satisfied. Moreover, it is easily seen that compositions of boundary symbols smoothly depend on $\varrho \in S^*Y$ if the factors have this property. Thus

$$\mathfrak{c} = \mathfrak{a}\mathfrak{b} = \begin{pmatrix} 1 + \Pi'\sigma_{G_0} & \sigma_{L_0} \\ \Pi'\sigma_{S_0} & \sigma_{R_0} \end{pmatrix} : \begin{matrix} p^*F^+ \\ \oplus \\ p^*G \end{matrix} \to \begin{matrix} p^*F^+ \\ \oplus \\ p^*G \end{matrix} \tag{9}$$

is a smooth function of ϱ. If we show that $\mathfrak{c}^{-1}$ is smooth, too, $\mathfrak{a}^{-1} = \mathfrak{b}\mathfrak{c}^{-1}$ is smooth. Therefore, it is sufficient to consider (9). Decompose (9) in the form

$$\mathfrak{c} = \begin{pmatrix} 1 - \sigma_{K_1} \circ \Pi'\sigma_{T_1} & \sigma_{L_0} \\ \Pi'\sigma_{S_0} & \sigma_{R_0} \end{pmatrix} + \begin{pmatrix} \Pi'\sigma_{G_1} & 0 \\ 0 & 0 \end{pmatrix} \tag{10}$$

for suitable trace, potential and Green symbols

$$\left.\begin{aligned} &\Pi'\sigma_{T_1}\colon p^*F^+ \to p^*V\,, \qquad \sigma_{K_1}\colon p^*V \to p^*F^+\,, \\ &\Pi'\sigma_{G_1}\colon p^*F^+ \to p^*F^+ \end{aligned}\right\} \tag{11}$$

and a certain $V \in \mathrm{Vect}\,(Y)$. Denote by $\mathfrak{d}$ the first summand on the right of (10). It will be shown that the decomposition (10) can be chosen in such a way that $\mathfrak{d}$ has an inverse of the form

$$\mathfrak{d}^{-1} = \begin{pmatrix} 1 + \Pi'\sigma_{G_2} & \sigma_{L_1} \\ \Pi'\sigma_{S_1} & \sigma_{R_1} \end{pmatrix}, \tag{12}$$

which is smooth. Then

$$\mathfrak{d}^{-1}\,\mathfrak{c} = \begin{pmatrix} 1 + \Pi'\sigma_{G_4} & 0 \\ \Pi'\sigma_{S_4} & 1 \end{pmatrix} \tag{13}$$

is smooth, too. Here σ_{S_4} is a certain trace symbol and

$$\Pi'\sigma_{G_4} = (1 + \Pi'\sigma_{G_2})\,\Pi'\sigma_{G_1}\,. \tag{14}$$

For $\Pi'\sigma_{G_1}$ being small in a sense to be described we shall show the existence of a smooth inverse $(1 + \Pi'\sigma_{G_4})^{-1}$. Then

$$(\mathfrak{d}^{-1}\,\mathfrak{c})^{-1} = \begin{pmatrix} (1 + \Pi'\sigma_{G_4})^{-1} & 0 \\ -\Pi'\sigma_{S_4}(1 + \Pi'\sigma_{G_4})^{-1} & 1 \end{pmatrix}$$

is smooth and thus $\mathfrak{c}^{-1} = (\mathfrak{d}^{-1}\mathfrak{c})^{-1}\,\mathfrak{d}$, too. Therefore, it remains to show that the decomposition (10) is possible in such a way that (12) exists as well as $(1 + \Pi'\sigma_{G_4})$. First prove that if

$$\mathfrak{d} = \begin{pmatrix} 1 + \sigma_{K_1} \circ \Pi'\sigma_{T_1} & \sigma_{L_0} \\ \Pi'\sigma_{S_0} & \sigma_{R_0} \end{pmatrix} \tag{15}$$

is smooth and invertible, the inverse $\mathfrak{d}^{-1}$ is smooth. Introduce the abbreviation

$$\mathfrak{d} = \begin{pmatrix} 1 - \varkappa_2 \circ \tau_2 & \varkappa_0 \\ \tau_0 & \delta_0 \end{pmatrix}$$

with obvious notation for the matrix elements. Without loss of generality we can assume that $\operatorname{ord}\delta_0 = 0$. Otherwise multiplication by $1 \oplus \delta_1$ with some elliptic symbol δ_1 on Y and $\operatorname{ord}\delta_1 = -\operatorname{ord}\delta_0$ gives this situation. Denote by 1 also the identity in p^*V. Then we have

$$\begin{pmatrix} 1 - \varkappa_2\tau_2 & \varkappa_0 & 0 \\ \tau_0 & \delta_0 & 0 \\ 0 & 0 & 1 \end{pmatrix} = \begin{pmatrix} 1 & 0 & -\varkappa_2 \\ 0 & 1 & 0 \\ 0 & 0 & 1 \end{pmatrix} \cdot \begin{pmatrix} 1 & \varkappa_0 & \varkappa_2 \\ \tau_0 & \delta_0 & 0 \\ \tau_2 & 0 & 1 \end{pmatrix} \begin{pmatrix} 1 & 0 & 0 \\ 0 & 1 & 0 \\ -\tau_2 & 0 & 1 \end{pmatrix}. \tag{16}$$

Since

$$\begin{pmatrix} 1 & 0 & -\varkappa_2 \\ 0 & 1 & 0 \\ 0 & 0 & 1 \end{pmatrix}^{-1} = \begin{pmatrix} 1 & 0 & \varkappa_2 \\ 0 & 1 & 0 \\ 0 & 0 & 1 \end{pmatrix}, \quad \begin{pmatrix} 1 & 0 & 0 \\ 0 & 1 & 0 \\ -\tau_2 & 0 & 1 \end{pmatrix}^{-1} = \begin{pmatrix} 1 & 0 & 0 \\ 0 & 1 & 0 \\ \tau_2 & 0 & 1 \end{pmatrix},$$

the first and the third factor on the right in (16) are smoothly invertible. Thus the second factor is invertible. Set

$$\alpha = 1\,,\quad \varkappa = (\varkappa_0, \varkappa_2)\,,\quad \tau = \begin{pmatrix}\tau_0\\ \tau_2\end{pmatrix},\quad \delta = \begin{pmatrix}\delta_0 & 0\\ 0 & 1\end{pmatrix}.$$

Then the second factor in (16) has the form $\mathfrak{m} = \begin{pmatrix}\alpha & \tau\\ \varkappa & \delta\end{pmatrix}$ and

$$\omega = \delta - \tau\alpha^{-1}\varkappa = \begin{pmatrix}\delta_0 & 0\\ 0 & 1\end{pmatrix} - \begin{pmatrix}\tau_0\\ \tau_2\end{pmatrix}(\varkappa_0, \varkappa_2) = \begin{pmatrix}\delta_0 - \tau_0\varkappa_0 & -\tau_0\varkappa_2\\ -\tau_2\varkappa_0 & 1 - \tau_2\varkappa_2\end{pmatrix}$$

is a smooth elliptic symbol over G. Therefore, since we have now the finite dimensional situation, we have a smooth inverse ω^{-1}. The inverse of $\mathfrak{m}$ calculated in the form 3.1.1.5.(5) is smooth, since $\mathfrak{m}^{-1}$ contains only compositions of smooth elements. Thus we also get a smooth inverse of the left hand side of (16), which is of the form $\mathfrak{d}^{-1} \oplus 1$. Hence $\mathfrak{d}^{-1}$ is smooth.

Now consider the situation where the type of the trace and Green parts contained in (9) is zero. Consider the closure

$$\mathfrak{c}\colon p^*(V^+ \otimes F') \oplus p^*G \to p^*(V^+ \otimes F') \oplus p^*G\,, \tag{17}$$

$V^+ = (H^+)_0$. Then the decomposition $\Pi'\sigma_{G_0} = -\sigma_K \circ \Pi'\sigma_{T_1} + \Pi'\sigma_{G_1}$ is possible in such a way that the operator norm of $\Pi'\sigma_{G_1}$ is as small as we want, i.e. also the operator norm of (14). Thus $\mathfrak{d}$ is invertible in $p^*(V^+ \otimes F') \oplus p^*G$ and restriction to $p^*F^+ \otimes p^*G$ gives the inverse obtained by the above construction. Moreover, there exists $(1 + \Pi'\sigma_{G_1})^{-1}$ in $\left(p^*(V^+ \otimes F^+) \oplus p^*G\right)_{(y,\eta)}$ for each $(y, \eta) \in S^*Y$ and this inverse can be obtained by the Neumann series (cf. 2.1.2.4, Lemma 4). If we fix a number N and choose the norm of $\Pi'\sigma_{G_1}$ small enough, the series can be differentiated N times with respect to the parameters. Thus, for each fixed N, a suitable decomposition (10) ensures the corresponding differentiability. Thus the whole inverse is smooth.

In the case of non-zero types of the symbols contained in $\mathfrak{c}$ we can pass from $\mathfrak{a}$ to $\mathfrak{t}_s\mathfrak{a}\mathfrak{r}_s^{-1} = \mathfrak{a}_0$ with $s \in Z_+$ sufficiently large, $m \geq$ type of all symbols contained in $\mathfrak{a}$ and

$$\mathfrak{r}_s\colon p^*\left((H^+)_s \otimes E'\right) \oplus p^*J \to p^*(V^+ \otimes E') \oplus p^*J\,,$$
$$\mathfrak{t}_s\colon p^*\left((H^+)_s \otimes F'\right) \oplus p^*G \to p^*(V^+ \otimes F') \oplus p^*G$$

isomorphisms with smooth inverses, similarly defined as in 2.1.2.3. For instance $\mathfrak{r}_s = (\Pi^+ l_s^-) \cdot 1_{p^*E'} \oplus 1_{p^*J}$, $\mathfrak{r}_s^{-1} = (\Pi^+ l_{-s}^-) \cdot 1_{p^*E} \oplus 1_{p^*J}$. Then the procedure described above for $\mathfrak{a}_0$ immediately gives the result for $\mathfrak{a}$. □

P r o o f o f 3.1.1.1, T h e o r e m 2: Given an elliptic $\mathcal{A} \in \mathfrak{G}$ of the form 3.1.1.1.(1) consider the corresponding pair of symbols $\left(\sigma_{\mathcal{A}}, \sigma_Y(\mathcal{A})\right)$ and the pair of inverses $\left(\sigma_P, \sigma_Y(\mathcal{P})\right)$ (cf. Proposition 6). Uniquely determined symbols $\sigma_N, \sigma_L, \sigma_S, \sigma_R$ are defined by $\sigma_Y(P)$ on S^*Y (cf. 2.3.3.1). Extension with the homogeneities according to 3.1.1.1. (7) (cf. (3)) gives symbols for all $\xi' \neq 0$. Thus we find some $\mathcal{P}_0 \in \mathfrak{G}$ for which the operators in the matrix have the given orders and $\sigma_\Omega(\mathcal{P}_0) = \sigma_P$, $\sigma_Y(\mathcal{P}_0) = \sigma_Y(\mathcal{P})$. Since composition of operators is compatible with composition of interior and boundary symbols, we get $\sigma_\Omega(\mathcal{P}_0\mathcal{A}) = \sigma_\Omega(\mathcal{J}_1) = 1_{\pi^*E}$, $\sigma_Y(\mathcal{P}_0\mathcal{A}) = \sigma_Y(\mathcal{J}_1) = 1_{p^*E} \oplus 1_{p^*J}$ and analogous identities with respect to $\mathcal{A}\mathcal{P}_0$. For $\mathcal{K}_1 = \mathcal{P}_0\mathcal{A} - \mathcal{J}_1$, $\mathcal{K}_2 = \mathcal{A}\mathcal{P}_0 - \mathcal{J}_2$, we have

$$\operatorname{ord}(\mathcal{K}_j)^k \to -\infty \quad \text{for} \quad k \to \infty \tag{18}$$

in the sense that the orders of the entries in $(\mathcal{K}_j)^k$ tend to $-\infty$. Thus we can define the asymptotic sums $\mathcal{L}_j \overset{\text{def}}{=} \sum_{k=0}^{\infty} (-1)^k (\mathcal{K}_j)^k \in \mathfrak{G}$, $j = 1, 2$. With $\mathcal{P}_1 = \mathcal{L}_1 \mathcal{P}_0$, $\mathcal{P}_2 = \mathcal{P}_0 \mathcal{L}_2$ we obviously get $\mathcal{P}_1 \mathcal{A} \equiv \mathcal{J}_1 \bmod \mathfrak{G}^{-\infty}$, $\mathcal{A}\mathcal{P}_2 \equiv \mathcal{J}_2 \bmod \mathfrak{G}^{\infty}$. Thus $\mathcal{P}_1$ is a left, $\mathcal{P}_2$ a right C^∞ parametrix of $\mathcal{A}$, hence $\mathcal{P}_1 \equiv \mathcal{P}_2 \bmod \mathfrak{G}^{-\infty}$. This means $\mathcal{P} \overset{\text{def}}{=} \mathcal{P}_1$ is a two sided C^∞ parametrix of $\mathcal{A}$. □

Note that $\mathcal{P}_0$ mentioned in the proof is a Sobolev space parametrix of $\mathcal{A}$. The construction of $\mathcal{P}_0$ is possible with the help of local inverses of $\sigma_Y(\mathcal{A})$, too (i.e. local in the sense of neighbourhoods in Y which form a covering) and a partition of unity. For this a corresponding local version of Proposition 6 is needed ignoring the global structure of J and G, respectively.

The construction of $\sigma_Y(\mathcal{A})^{-1}$ in the proof of Proposition 6 is also possible on operator level. This gives a Sobolev space parametrix. The method looks like a variant of the reduction to the boundary to be discussed in 3.2.1.3. Here, in the parametrix construction a Neumann series is contained besides taking a parametrix of some elliptic operator on the boundary.

Finally, let us make some remarks on the parametrix construction in the special case of elliptic operators of the form

$$\mathcal{A} = \begin{pmatrix} r^+A + r'B \\ \\ r'T \end{pmatrix} : C^\infty(X, E) \to \begin{matrix} C^\infty(X, F) \\ \oplus \\ C^\infty(Y, G) \end{matrix} \tag{19}$$

Let

$$\mathcal{P} = (r^+P + r'NL) : \begin{matrix} C^\infty(X, F) \\ \oplus \\ C^\infty(Y, G) \end{matrix} \to C^\infty(X, E)$$

be a C^∞ parametrix of $\mathcal{A}$. Then we have modulo smoothing operators $(r^+A + r'B) \times (r^+P + r'N) = 1$ (identity in $C^\infty(X, E)$) and $(r^+A + r'B)\, L = 0$. Let now

$$\mathcal{A}_1 = \begin{pmatrix} r^+A + r'B \\ \\ r'T_1 \end{pmatrix} : C^\infty(X, E) \to \begin{matrix} C^\infty(X, F) \\ \oplus \\ C^\infty(Y, G) \end{matrix} \tag{20}$$

be another elliptic boundary problem for r^+A (with the same $r'B$). Then

$$\mathcal{A}_1\mathcal{P} = \begin{pmatrix} 1 & 0 \\ r'T(r^+P + r'N) & (r'T_1)\, L \end{pmatrix} \tag{21}$$

mod $\mathfrak{G}^{-\infty}$. The right lower corner

$$Q = (r'T_1)\, L : C^\infty(Y, G) \to C^\infty(Y, G) \tag{22}$$

is a pseudo-differential operator on Y. Ellipticity of $\mathcal{A}_1$ and $\mathcal{P}$ implies that (22) is a Fredholm operator not only in spaces of C^∞ sections but in Sobolev spaces. Thus the continuous extension of (22) in corresponding Sobolev spaces is Fredholm, too. This means that Q is elliptic (cf. 1.2.4.2, Theorem 3). Let R be a C^∞ parametrix of Q (cf. 1.2.4.2). For $R_0 = \operatorname{Op}(\sigma_Q^{-1})$ and $C = 1 - R_0Q_0$ we can set $R = \left(\sum_j (-1)^j C^j\right) R_0$ in the sense of asymptotic sums of PDOs. Then obviously

$$\mathcal{E}_1 = \begin{pmatrix} 1 & 0 \\ -R(r'T_1)\,(r^+P + r'N) & R \end{pmatrix}$$

is a C^∞ parametrix of (12). Thus $\mathcal{P}_1 = \mathcal{A}\mathcal{C}_1 \in \mathfrak{G}$ is a C^∞ parametrix of $\mathcal{A}_1$. Therefore, starting with a fixed elliptic $\mathcal{A} \in \mathfrak{G}$ of the form (19) and a given parametrix $\mathcal{P}$ we can find the parametrix of another elliptic $\mathcal{A}_1 \in \mathfrak{G}$ of the form (20) by calculating of the parametrix of an elliptic pseudo-differential operator on the boundary.

In 3.2.1.3 we return to this question under another point of view. The reduction to the boundary discussed there can be similarly used for parametrix constructions. If $\mathcal{A}$, $\mathcal{A}_1 \in \mathfrak{G}$ are elliptic boundary problems with the same left upper corners and if a parametrix $\mathcal{P} \in \mathfrak{G}$ of $\mathcal{A}$ is known, we find an elliptic PDO S on Y so that a parametrix $\mathcal{P}_1$ of $\mathcal{A}_1$ can be expressed in terms of $\mathcal{P}$ and S^{-1}, a parametrix of S.

The reader can derive this expression as an exercise (cf. the 3.1.1.5, Proposition 5 and constructions in 4.2.1).

3.1.1.3. Classical Elliptic Boundary Problems

An elliptic operator $\mathcal{A} \in \mathfrak{G}$ is called *classical* if it has the form

$$\mathcal{A} = \begin{pmatrix} r^+A \\ \\ r'T \end{pmatrix} : C^\infty(X, E) \to \begin{matrix} C^\infty(X, F) \\ \oplus \\ C^\infty(Y, G) \end{matrix} \tag{1}$$

where A is an elliptic differential operator on X and for which $\sigma_T(x', \xi', \nu)$ polynomially depends on ν for every $(x', \xi') \in S^*Y$. In this section we consider a slightly modified class of boundary problems, namely those for which there is a direct decomposition $G = \bigoplus_{j=1}^{l} G_j$ and $r'T$ corresponds to a vector $r'T_j$ $(j = 1, \dots, l)$ of trace operators with $m_j = \operatorname{ord} \sigma_{T_j} \in \mathbb{Z}_+$. By a simple reduction of orders on the boundary one can transform such problems into $\mathcal{A} \in \mathfrak{G}$ of the form (1) with fixed order $\operatorname{ord} \sigma_T$ (cf. 3.1.2.1). Thus the notion of ellipticity is generalized in an obvious way to the case of different orders of the trace operators. The theory of boundary problems of this special type is treated in a plenty of papers and books (cf. Lions/Magenes [1], Agmon/Douglis/Nirenberg [1], Schulze/Wildenhain [1], Solonnikov [1], Schechter [1, 3]).

The assertions in this section and the discussion of examples are a supplement and an illustration to the general theory of elliptic operators in $\mathfrak{G}$ and will be given without proofs. A very simple example is the Dirichlet problem for the polyharmonic operator Δ^k in a bounded smooth domain Ω in $\mathbb{R}^n$. It can be formulated in the form

$$\Delta^k u = f \in C^\infty(X)\,, \quad \left(\frac{\partial}{\partial n}\right)^{j-1} u|_Y = g_j \in C^\infty(Y)\,, \quad j = 1, \dots, k \tag{2}$$

($\partial/\partial n$ denotes the derivative in direction of the interior normal direction; the sign r^+ is usually omitted in connection with differential operators). It is well-known that the Dirichlet problem (2) is elliptic and has index zero. Other elliptic boundary conditions are defined by the trace operators $(\partial/\partial n)^{m_j} u|_Y = g_j$, $j = 1, \dots, k$. Here the m_j are integers, $m_j \neq m_k$ for $j \neq k$ and $0 \leq m_j \leq 2k - 1$. Using an explicit solution of the Dirichlet problem one can explicitly construct a parametrix of the problem with the more general boundary conditions using the constructions at the end of 3.1.1.2. For the ball an explicit formula for the solution of the Dirichlet problem is given in Schulze/Wildenhain [1].

Ellipticity of boundary conditions can be formulated as usual in local coordinates in $\overline{\mathbb{R}}^n_+$. Then the Dirichlet conditions are the restriction to $x_n = 0$ of the derivatives of u with respect to x_n.

Let $A(x, D)$ be a scalar elliptic differential operator in $\overline{\mathbb{R}}^n_+$ of order m satisfying the following root condition: If $\nu_1(x', \xi'), \dots, \nu_m(x', \xi')$ denote the roots of the equation $\sigma_A(x', 0, \xi', \nu) = 0$ $\big((x', \xi') \in S^*Y\big)$, then the upper (and lower) complex ν half plane contains precisely $m/2$ roots $\nu_{j_1}, \dots, \nu_{j_{m/2}}$. The root condition implies that $m = \operatorname{ord} \sigma_A$ is even. Then it can be proved that the boundary conditions $(\partial/\partial x_n)^{j-1} u|_{x_n=0}$ $(j = 1, \dots, m/2)$ are elliptic with respect to the operator $A(x, D)$.

Note that the root condition is always satisfied for $n - 1 \geqq 2$. For elliptic "systems" $A: C^\infty(X, E) \to C^\infty(X, F)$ the root condition is formulated with respect to $\det \sigma_A$ in each local coordinate system for which E and F are trivial.

In case $E = F = X \times \mathbb{C}$, $G = Y \times \mathbb{C}^{m/2}$, $m = \operatorname{ord} \sigma_A$, consider a boundary problem

$$A(x, D)\, u = f \in C^\infty(X)\,, \quad T_j(x, D)\, u|_Y = g_j \in C^\infty(Y) \quad (j = 1, \dots, m/2)\,. \tag{3}$$

Suppose once for all in this section that the root condition for A is satisfied. Let $m_j = \operatorname{ord} \sigma_{T_j}$, $0 \leqq m_j \leqq m - 1$ and $m_j \neq m_k$ for $j \neq k$. Introduce local coordinates near a boundary point in the usual way. By freezing the coefficients of A and of the T_j at the given point on the boundary we get a boundary problem in $\overline{\mathbb{R}}^n_+$ with constant coefficients

$$A(D)\, u = f\,, \quad T_j(D)\, u|_{x_n=0} = g_j \quad (j = 1, \dots, m/2)\,. \tag{4}$$

For simplicity the same letters were used. Next we give an equivalent formulation of the Shapiro-Lopatinski condition and a solution formula for (4). There is a standard relation between (3) and the family of problems (4). Let $\eta = \xi' \neq 0$ and

$$M^\pm(\eta, \nu) = \prod_{j=1}^{m/2} \big(\nu - \nu_j^\pm(\eta)\big) = \sum_{k=0}^{m/2} c_k^\pm(\eta)\, \nu^{(m/2)-k}\,. \tag{5}$$

Here $\nu_j^\pm(\eta)$ denote the roots of $\sigma_A(\eta, \nu) = 0$ with positive and negative imaginary parts, respectively. Then $\sigma_A(\eta, \nu) = M^+(\eta, \nu)\, M^-(\eta, \nu)$. Moreover define

$$M_j^+(\eta, \nu) = \sum_{k=0}^{j} c_k^+(\eta)\, \nu^{j-k} \quad \left(j = 0, \dots, \frac{m}{2} - 1\right). \tag{6}$$

The coefficients $c_k^+(\eta)$ are analytic functions with respect to $\eta \in \mathbb{R}^{n-1} \setminus \{0\}$ and homogeneous of degree k. Then

$$\frac{1}{2\pi i} \int_\gamma \frac{M^+_{(m/2)-j-1}(\eta, \nu)}{M^+(\eta, \nu)}\, \nu^k \, d\nu = \delta_{jk} \quad \left(0 \leqq j \leqq \frac{m}{2} - 1, 0 \leqq k \leqq \frac{m}{2} - 1\right) \tag{7}$$

for each closed smooth curve γ in the upper complex half plane around all roots $\nu_j^+(\eta)$, $j = 1, \dots, m/2$, $|\eta| = 1$. By dividing polynomials in ν with remainder we get expressions $T_j(\eta, \nu) = T_j^0(\eta, \nu)\, M^+(\eta, \nu) + T_j'(\eta, \nu)$ $(\eta \neq 0$ fixed) with polynomials $T_j'(\eta, \nu) = \sum_{k=1}^{m/2} t_{jk}(\eta)\, \nu^{k-1}$ $(j = 1, \dots, m/2)$. The following condition is called *complementing condition* $\big($of $T = (T_1, \dots, T_{m/2})$ with respect to $A\big)$: for each fixed $\eta \in \mathbb{R}^{n-1} \setminus \{0\}$ the polynomials $T_j(\eta, \nu)$ are linearly independent mod $M^+(\eta, \nu)$. This is equivalent to the condition

$$\operatorname{Det}\big(t_{jk}(\eta)\big) \neq 0\,, \quad \eta \neq 0\,. \tag{8}$$

Now we have the following proposition (a proof is given in LIONS/MAGENES [1]). The boundary problem (3) at the given boundary point satisfies the Shapiro-Lopatinski condition iff the corresponding problem (4) satisfies the complementing condition.

Suppose that (4) satisfies the complementing condition. Then the solution u for $f = 0$ in the half space can be expressed in the form

$$u(x', x_n) = \sum_{j=1}^{m/2} \int K_j(x' - z', x_n)\, g_j(z')\, dz' \tag{9}$$

with so called *Poisson kernels* $K_j(x' - z', x_n)$ $(j = 1, \dots, m/2)$. The Poisson kernels can be explicitly constructed in the following way. Denote by (t^{jk}) the inverse matrix of (t_{jk}) and consider the polynomials $N_k(\eta, \nu) = \sum_{j=1}^{m} t^{jk}(\eta)\, M^+_{(m/2)-j}(\eta, \nu)$ $(k = 1, \dots, m/2)$. A consequence of (7) is then the property

$$\frac{1}{2\pi i} \int_\gamma \frac{N_k(\eta, \nu)\, T_j(\eta, \nu)}{M^+(\eta, \nu)}\, d\nu = \delta_{jk}\,. \tag{10}$$

Define a function $F(s, l)$ for $s \in \mathbb{C}$, $l \in \mathbb{Z}$ to be equal to

$$\frac{-1}{(2\pi i)^{n-1}\, l!}\, s^l \left(\log \frac{s}{i} - \sum_{j=1}^{l} \frac{1}{j}\right) \quad \text{for} \quad l > 0\,,$$

$$\frac{-1}{(2\pi i)^{n-1}} \log \frac{s}{i} \quad \text{for} \quad l = 0\,, \qquad \frac{(-1)^l\,(-l-1)!}{(2\pi i)^{n-1}}\, s^l \quad \text{for} \quad l < 0$$

(log means the principal branch of the logarithm). Then $(d/ds)^q F(s, l) = F(s, l - q)$. Moreover one has the so called *John's identity*

$$g(x') = \int_{\mathbb{R}^{n-1}} \Delta_{z'}^{(n-1+q)/2} g(z')\, dz' \int_{|\eta|=1} F\big((x' - z')\, \eta, q\big)\, d\omega_\eta \tag{11}$$

for each $g \in C_0^\infty(\mathbb{R}^n)$ (cf. JOHN [1], SCHULZE/WILDENHAIN [1]). Here q is an integer sufficiently large with $n - 1 + q$ even. $d\omega_\eta$ denotes the surface element on the unit sphere $|\eta| = 1$. Put

$$K_j(x', x_n) = \int_{|\eta|=1} d\omega_\eta \frac{1}{2\pi i} \int_\gamma F(x'\eta + x_n\nu, m_j - n)\, \frac{N_j(\eta, \nu)}{M^+(\eta, \nu)}\, d\nu \tag{12}$$

$(m_j = \operatorname{ord} \sigma_{T_j})$. Denote by $K_{jq}(x', x_n)$ that function defined by an integral of the type (12) with $m_j + q$ substituted for $m_j - n$. Then with q as above

$$\Delta_{x'}^{(n-1+q)/2} K_{jq}(x', x_n) = K_j(x', x_n)\,.$$

It can be easily proved that $K_j(x' - z', x_n)$ are the Poisson kernels of the boundary problem (4) (for details, cf. AGMON/DOUGLIS/NIRENBERG [1], LIONS/MAGENES [1], SCHULZE/WILDENHAIN [1]). Using other methods one obtains other expressions for the Poisson kernels.

A simple example is the Dirichlet problem for the Laplace operator

$$\Delta u = 0 \text{ in } \mathbb{R}^n_+\,, \qquad u|_{x_n=0} = g \in C_0^\infty(\mathbb{R}^{n-1})\,.$$

The unique bounded solution in the half space is

$$u(x) = (2\pi)^{-n+1} \int_{\mathbb{R}^{n-1}} e^{ix'\xi'}\, e^{-x_n|\xi'|}\, \hat{g}(\xi')\, d\xi'\,.$$

Solutions in the half space of general elliptic boundary problems for elliptic pseudo-differential operators are explicitly constructed in ÈSKIN [3]. The results in ÈSKIN [3] show that the root condition means a certain special behaviour of the symbol with respect to factorization in plus and minus functions determining the number of the boundary conditions.

Next we shall discuss Green formulas for elliptic boundary problems of the form (3). Let $\sigma_{T_j}(x, \xi) = \sum_{|\alpha|=m_j} t_{j\alpha}\xi^\alpha$ be the homogeneous principal symbol of T_j. Suppose that for each $x \in Y$ and $\xi = (0, \nu) \neq 0$ follows $\sigma_{T_j}(x, \xi) \neq 0$ $(j = 1, \dots, m/2)$. In the half space this just means that the coefficient of ν^{m_j} does not vanish. Operators T_j satisfying this condition are called *normal to* Y. A tupel of differential operators

$$T_1(x, D), \dots, T_{m/2}(x, D)\,, \quad S_1(x, D), \dots, S_{m/2}(x, D) \tag{13}$$

is called *Dirichlet system of order* $m/2$, if all operators T_j, S_j are normal to Y and if the numbers $m_j = \operatorname{ord} \sigma_{T_j}$, $\overline{m}_j = \operatorname{ord} \sigma_{S_j}$ $(j = 1, \dots, m/2)$ are all integers $0, \dots, m-1$ in a suitable ordering.

Let $A^*\colon C^\infty(X) \to C^\infty(X)$ be the formal adjoint to A with respect to the Hermitean scalar product $(\cdot,\cdot)$ in L^2, i.e. $(Au, v) = (u, A^*v)$ for all $u, v \in C_0^\infty(\Omega)$. Define the vector spaces

$$V = \{u \in C^\infty(X)\colon Tu|_Y = 0\}\,,$$
$$V^* = \{v \in C^\infty(X)\colon (Au, v) = (u, A^*v) \text{ for all } u \in V\}\,.$$

Then $C_0^\infty(\Omega) \subseteq V \subseteq C^\infty(X)$ and a similar inclusion for V^*. Under the conditions about the boundary problem (3) there exist differential operators $T_j^*(x, D)$ $(j = 1, \dots, m/2)$ (normal to Y) so that $\operatorname{ord} \sigma_{T_j^*} = m - 1 - \overline{m}_j$, $V^* = \{v \in C^\infty(X)\colon T^*v|_Y = 0\}$ and

$$V = \{u \in C^\infty(X)\colon (Au, v) = (u, A^*v) \quad \text{for all } v \in V^*\}$$

and that the boundary problem

$$A^*(x, D)\, v = h \in C^\infty(X)\,, \quad T_j^*(x, D)\, v|_Y = k_j \in C^\infty(Y) \quad (j = 1, \dots, m/2) \tag{14}$$

is elliptic. Denote by $(\cdot,\cdot)'$ the Hermitean scalar product in $L^2(Y) \times \dots \times L^2(Y)$ ($m/2$ factors) with respect to the surface measure on Y. Then one can find Dirichlet systems (13) and

$$T_1^*(x, D), \dots, T_{m/2}^*(x, D)\,, \quad R_1(x, D), \dots, R_{m/2}(x, D) \tag{15}$$

of operators which are normal to Y (the T_j, T_j^* are as in (3) and (14), respectively) with $2m - 1 - m_j = \operatorname{ord} \sigma_{R_j}$ $(j = 1, \dots, m/2)$ so that

$$(Au, v) + (r'Tu, r'Rv)' = (u, A^*v) + (r'Su, r'T^*v)' \tag{16}$$

for all $u, v \in C^\infty(X)$. Here e.g. $r'R$ denotes the vector $(R_1v|_Y, \dots, R_{m/2}v|_Y)$. The Dirichlet systems for which (16) is valid are not uniquely determined. These assertions are proved in SCHECHTER [1], LIONS/MAGENES [1]. (16) is called a *Green formula* for the elliptic boundary problem (3) and (14) is called *adjoint* to (3). A consequence of (16) is that the elliptic boundary problem (3) has a solution $u \in C^\infty(X)$ iff $(f, v) + (g, r'Rv)' = 0$ for all $v \in C^\infty(X)$ satisfying (14) with $h = 0$, $k = (k_1, \dots, k_{m/2}) = 0$. This corresponds to a finite number of conditions. Green formulas are used for many concrete questions on elliptic boundary value problems, e.g. detailed regularity investigations (cf. SCHULZE/WILDENHAIN [1]), not only for scalar operators, but for systems, too.

Besides the standard L^2 estimates for elliptic boundary value problems (cf. 3.1.1.1, Corollary 3), L^p and Hölder estimates are of interest (cf. 3.1.1.4). There are other interesting estimates in certain uniform norms. Suppose as usual that the operators $T_j(x, D)$ $(j = 1, \dots, m/2)$ are given on $\bar{\Omega}$. Consider a tubular neighbourhood U of Y in X, $U \cong Y \times [0, 1)$. Then, in U, derivatives in directions "tangent to Y" can be invariantly defined. Set $\|u\|_{C^k_{\mathrm{tang}}(U)} = \sup\limits_{\{\omega\}} \sup\limits_{y \in \omega} \sup\limits_{|\gamma| \leq k} |D^\gamma_y u(y)|$ for $u \in C^\infty(\bar{U})$. Here $\{\omega\}$ denotes an open finite covering of U by sets $\omega \cong V \times [0, 1)$, V a coordinate neighbourhood in Y with local coordinates y. Let $l \geq 1$, r integers with $m_j \leq r \leq m + l$ $(j = 1, \dots, m/2)$. Then there exists a constant $c > 0$ so that for all solutions $u \in C^{m+1}(X)$ of (3) with $f = 0$ the following estimate holds

$$\sum_{j=1}^{m/2} \|T_j u\|_{C^{r-m_j}_{\mathrm{tang}}(U)} \leq c \left\{ \sum_{j=1}^{m/2} \|g_j\|_{C^{r-m_j}(Y)} + \|u\|_{L^1(X)} \right\}. \tag{17}$$

Especially for the Dirichlet problem (i.e. for $T_j = (\partial/\partial n)^{j-1}$ $(j = 1, \dots, m/2)$ near Y) it follows that

$$\|u\|_{C^{(m/2)-1}(X)} \leq c \left\{ \sum_{j=1}^{m/2} \|g_j\|_{C^{(m/2)-j}(Y)} + \|u\|_{L^1(X)} \right\}. \tag{18}$$

The estimate (17) is proved in Krasovskiĭ [1] and generalizes the maximum estimate (18) given in Agmon [1] (cf. Schulze/Wildenhain [1], Schulze [1]). An analogy of the estimates (17) for general elliptic boundary problems in the class $\mathfrak{G}$ is not yet known.

Let $A(D_{x'}, D_{x_n}) = \sum\limits_{k=0}^{m} A_k(D_{x'})\, D^k_{x_n}$ be an elliptic differential operator in the half space with constant coefficients and $m = \operatorname{ord} \sigma_A$. Let $\gamma\colon C^\infty(\bar{\mathbb{R}}^n_+) \to C^\infty(\mathbb{R}^{n-1}, \mathbb{C}^{2m})$ be the operator defined by $\gamma(u) = (\gamma_0(u), \dots, \gamma_{m-1}(u))$, $\gamma_j(u) = \mathrm{i}^{-1}(\partial/\partial x_n)^j\, u|_{x_n=0}$. Then γ is a trace operator in the sense of our usual definition. Denote by u^0 that locally integrable function in $\bar{\mathbb{R}}^n_+$, equal to a given $u \in C^\infty(\bar{\mathbb{R}}^n_+)$ and zero for $x_n < 0$. Then $A(u^0) \in \mathcal{D}'(\mathbb{R}^n)$, $\operatorname{sing\,supp} A(u^0) \subset \mathbb{R}^{n-1}_{x'}$ and $w = A(D)\,(u^0) - (r^+ A(D)\, u)^0 \in \mathcal{D}'(\mathbb{R}^n)$, $\operatorname{supp} w \subset \mathbb{R}^{n-1}_{x'}$ $(u \in C^\infty(\bar{\mathbb{R}}^+_n))$. By induction with respect to j one proves

$$D^j_{x_n}(u^0) = (r^+ D^j_{x_n} u)^0 + \frac{1}{\mathrm{i}} \sum_{k=0}^{j-1} D^k_{x_n}(\delta(x_n) \otimes \gamma_{j-1-k}(u)) .$$

Then, with the Cauchy data $v_l = \gamma_l(u)$, we get

$$w = \frac{1}{\mathrm{i}} \sum_{j=0}^{m-1} \sum_{l=0}^{m-1-j} A_{j+l+1}(D_{x'})\, D^j_{x_n}(\delta(x_n) \otimes v_l(x')) . \tag{19}$$

Let P be a fundamental solution of A represented as integral operator with kernel $E(x - z)$ and $E(x)$ defined by $h(x)\,|x|^{m-n}$ for n odd or $n > m$, $q(x) \log |x| + h(x)\,|x|^{m-n}$ for n even, $n \leq m$. In the last expression h is a function positively homogeneous of degree 0 and q a homogeneous polynomial in the x variables of degree $m - n$.

Since the left hand side of (19) depends only on the Cauchy data $v = \gamma u$, the operator

$$K_0 v = r^+ (P(A(D)\,(u^0)) - (r^+ A(D)\, u)^0) \tag{20}$$

is a potential operator ($u \in C^\infty(\bar{\mathbb{R}}^n_+)$ is supposed to be decreasing at infinity sufficiently fast). From $(r^+ A)\,(r^+ P) = 1$ it follows that $K_0 v \in C^\infty(\bar{\mathbb{R}}^n_+)$ is a solution of

$$A(D)\, u = 0 \quad \text{in} \quad \mathbb{R}^n_+ . \tag{21}$$

Thus

$$K_0\gamma(K_0v) = r^+[P(A(D))\,(K_0v)^0]$$
$$= r^+[P(A(D)\,\{r^+[P(A(D)\,(u^0)) - (A(D)\,u)^0]\})]$$
$$= r^+[P(A(D)\,(u^0) - (A(D)\,u)^0)] = K_0v\,,$$

i.e. the operator $K_0\gamma$ is a projection. Since $K_0\gamma$ reproduces the solution K_0v with the Cauchy data γK_0v, it follows that

$$\gamma(K_0\gamma K_0v) = \gamma K_0v\,, \tag{22}$$

i.e. γK_0 is a projection, too (γK_0 is a pseudo-differential operator on the boundary). γK_0 is called *Calderon-Seeley projector* belonging to A. It projects onto the space of Cauchy data of solutions of (21) which are smooth up to the boundary.

Similar constructions are valid on symbolic level, particularly for boundary symbols. Then, instead of a fundamental solution P, one can take an arbitrary parametrix P of $A(D)$ ($\sigma_P = \sigma_A^{-1}$). The boundary symbols $\sigma_Y(K_0\gamma)$ and $\sigma_Y(\gamma K_0)$ are consequently projections and we have

$$\sigma_Y(r^+Ar^+P) = 1\,, \quad \sigma_Y(r^+Pr^+A) = 1 - \sigma_Y(K_0\gamma)\,. \tag{23}$$

The Shapiro-Lopatinski condition (or ellipticity) for the boundary problem (4) can be formulated as follows. (4) is elliptic iff $\sigma_Y(TK_0)$ defines an isomorphism from im $\sigma_Y(\gamma K_0)$ onto $S^*Y \times \mathbb{C}^{m/2}$. We will return to this result in 3.1.2.4 in connection with general operators in $\mathfrak{G}$ (cf. 3.1.2.4, Theorem 3).

Let M be a closed compact manifold and $M = X_+ \cup X_-$ with smooth manifolds $X_\pm$ with common boundary $Y = X_+ \cap X_-$. Denote restrictions of bundles over M to $X_\pm$ by the same letter. Identify X_+ with X and set $\Omega = \operatorname{int} X$. Let $A\colon C^\infty(M, E) \to C^\infty(M, F)$ be an elliptic differential operator satisfying the root condition with respect to Y. Denote by $r^+A\colon C^\infty(X, E) \to C^\infty(X, F)$ the induced operator over $X = X_+$. For given $u \in C^\infty(X, E)$ define the Cauchy data $\gamma u \in \bigoplus_{j=1}^{m-1} C^\infty(Y, E')$ ($E' = E|_Y$) by applying $(1/\mathrm{i}\;\partial/\partial x_n)$ $(j = 0, \ldots, m-1)$ to u in a tubular neighbourhood U of Y and taking the limit $x_n \to 0$. Here x_n denotes a global normal variable in $U \cong Y \times [0, 1)$. Define the spaces

$$L^{s,m}(Y, E') = \bigoplus_{j=0}^{m-1} H^{s-j-(1/2)}(Y, E')\,,$$
$$\ker^s(r^+A) = \{u \in H^s(X, E)\colon Au = 0 \quad \text{in } \Omega\}$$

($s \in \mathbb{R}$). It can be proved that γ has for each $s \in \mathbb{R}$ a continuous extension

$$\gamma\colon \ker^s(r^+A) \to L^{s,m}(Y, E')\,.$$

Set

$$R^{s,m}(Y, E') = \{g \in L^{s,m}(Y, E')\colon g = \gamma u \quad \text{for some } u \in \ker^s(r^+A)\}$$

and $\ker_0(r^+A) = \{u \in C^\infty(M, E)\colon Au = 0,\ \operatorname{supp} u \subseteq X\}$. Then there exists a linear operator

$$K_0\colon \bigoplus_{j=0}^{m-1} C^\infty(Y, E') \to C^\infty(X, E)$$

with continuous extension $K_0\colon L^{s,m}(Y, E') \to \ker^s(r^+A)$, $s \in \mathbb{R}$. Then it follows that the operator

$$\gamma K_0\colon L^{s,m}(Y, E') \to L^{s,m}(Y, E')$$

is continuous for each $s \in \mathbb{R}$. Moreover, γK_0 is a projection onto $R^{s,m}(Y, E')$. Finally, there exists a direct decomposition

$$\ker^s (r^+A) = \ker_0 (r^+A) \oplus \ker_1^s (r^+A) ,$$

so that $K_0\gamma\colon \ker^s (r^+A) \to \ker_1^s (r^+A)$ is a projection along $\ker_0 (r^+A)$. These assertions are proved in SEELEY [3] (cf. TAIRA [1]).

Let us change slightly our notations and set $E_\pm = E|_{X_\pm}$. Consider the operators

$$r^\pm A\colon C^\infty(X_\pm, E_\pm) \to C^\infty(X_\pm, E_\pm)$$

induced by the elliptic differential operator $A\colon C^\infty(M, E) \to C^\infty(M, E)$ of order m. By taking the Cauchy data we get mappings $\gamma^\pm\colon \ker (r^\pm A) \to \bigoplus_{j=0}^{m-1} C^\infty(Y, E')$ possessing the corresponding continuous extensions to $\ker^s (r^\pm A)$ mentioned above. Then it is a problem (suggested by B. BOJARSKI, Warszaw) when the spaces $\gamma^+ \ker^s (r^+A)$, $\gamma^- \ker^s (r^-A)$ form a Fredholm pair in $L^{s,m}(Y, E')$ (cf. 1.1.1.2) and when $\operatorname{ind} \big(\gamma^+ \ker^s (r^+A), \gamma^- \ker^s (r^-A)\big)$ is equal to the index of the elliptic operator A on M (for a special case cf. GILKEY [2]).

For a given elliptic differential operator on M

$$A\colon C^\infty(M, E) \to C^\infty(M, F) \tag{24}$$

(satisfying the root condition with respect to Y) consider the operators

$$r^\pm A\colon C^\infty(X_\pm, E_\pm) \to C^\infty(X_\pm, F_\pm) .$$

Suppose that

$$\mathcal{A}^\pm = \begin{pmatrix} r^\pm A \\ \\ r'_{(\pm)} T_\pm \end{pmatrix} \colon C^\infty(X_\pm, E_\pm) \to \begin{matrix} C^\infty(X_\pm, F_\pm) \\ \oplus \\ C^\infty(Y, G_\pm) \end{matrix} \tag{25}$$

are elliptic boundary value problems over X_+ and X_-, respectively ($r'_\pm$ denotes the r' operator with respect to X_+ and X_-, respectively. $G_\pm$ are given vector bundles over Y). The aim of the following considerations is the construction of an elliptic pseudo-differential operator R over Y with

$$\operatorname{ind} \mathcal{A}^+ + \operatorname{ind} \mathcal{A}^- = \operatorname{ind} A - \operatorname{ind} R . \tag{26}$$

Suppose without loss of generality that $\operatorname{ord} \sigma_{T_\pm} = m - \frac{1}{2}$ so that there are continuous extensions $r'_{(\pm)} T_\pm\colon H^m(X_\pm, E_\pm) \to H^0(Y, G_\pm)$ (if this order condition is not satisfied one can modify the orders by multiplying by suitable elliptic pseudo-differential operators on the boundary). In view of the usual Sobolev space continuity of operators in $\mathfrak{G}$ we have continuous extensions

$$\mathcal{A}^\pm\colon H^m(X_\pm, E_\pm) \to H^0(X_\pm, F_\pm) \oplus H^0(Y, G_\pm) . \tag{27}$$

The operators (27) are Fredholm. Denote by $\widetilde{M}$ the disjoint union of X_+ and X_- and let $\widetilde{E}$ be the bundle on M arising as disjoint union of E_+ and E_-. Set

$$\mathcal{A} = \mathcal{A}^+ \oplus \mathcal{A}^-\colon H^m(\widetilde{M}, \widetilde{E}) \to H^0(M, F) \oplus H^0(Y, G_+ \oplus G_-) .$$

Here $H^m(\widetilde{M}, \widetilde{E}) = H^m(X_+, E_+) \times H^m(X_-, E_-)$, $H^0(\widetilde{M}, \widetilde{F}) = H^0(M, F) = H^0(X_+, F_+) \oplus H^0(X_-, F_-)$. Denote by $J\colon H^m(M, E) \to H^m(\widetilde{M}, \widetilde{E})$ the natural embedding and by $b\colon H^0(M, F) \oplus H^0(Y, G_+ \oplus G_-) \to H^0(M, F)$ the canonical projection. Then obviously

$$A = b \circ \mathcal{A} \circ J . \tag{28}$$

Moreover, let $\mathcal{P}$ be a parametrix of $\mathcal{A}$ ($\mathcal{P}$ is the direct sum of parametrices of $\mathcal{A}^+$ and $\mathcal{A}^-$). Let $a: H^0(Y, G_+ \oplus G_-) \to H^0(M, F) \oplus H^0(Y, G_+ \oplus G_-)$ be the canonical embedding. Define an operator

$$L: H^m(\widetilde{M}, \widetilde{E}) \to \bigoplus_0^{m-1} H^0(Y, E')$$

by

$$L(u_+, u_-) = \bigoplus_{j=0}^{m-1} (\Delta_{E'})^{-m+j+(1/2)} (r'_{(+)} D_m^j u_+ - r'_{(-)} D_n^j u_-) .$$

Here $\Delta_{E'}$ denotes a pseudo-differential operator on Y with the homogeneous principal symbol $-|\xi'| \cdot 1_{\pi_Y^* E'}$, $\pi_Y: T^*Y \setminus 0 \to Y$, $D_n = \mathrm{i}^{-1}(\partial/\partial x_n)$, x_n a global normal direction to Y. Then the following sequence is exact

$$0 \to H^m(M, E) \xrightarrow{J} H^m(\widetilde{M}, \widetilde{E}) \xrightarrow{L} \bigoplus_0^{m-1} H^0(Y, E') .$$

Note that there exists an isomorphism $\bigoplus_0^{m-1} p^*E' \cong p^*(G_+ \oplus G_-)$, $p: S^*Y \to Y$. This is a simple consequence of the considerations above. Now

$$R = L \circ \mathcal{P} \circ a: H^0(Y, G_+ \oplus G_-) \to H^0\left(Y, \bigoplus_0^{m-1} E'\right) \tag{29}$$

is an elliptic pseudo-differential operator on Y. In order to see this, note first that a is injective and $\operatorname{im} a = H^0(Y, G_+ \oplus G_-)$. The bundle $p^*(G_+ \oplus G_-)$ is mapped by $\sigma_Y(\mathcal{P})$ isomorphically onto $\ker \Pi^+\sigma_A \oplus \ker \Pi^-\sigma_A$. The last space is isomorphically mapped onto $p^*\left(\bigoplus_0^{m-1} E'\right)$ by $F^{-1}\sigma_Y(L)$ (F the Fourier transform in x_n direction), because the Cauchy data of bounded solutions of $\sigma_A(x', 0, \xi', D_n)\, u(x_n) = 0$ ($\xi' \neq 0$) on $\{x_n \geqq 0\}$ and $\{x_n \leqq 0\}$, respectively, are complementary subbundles in $p^*\left(\bigoplus_0^{m-1} E'\right)$ (cf. HÖRMANDER [3]).

Thus we get the following diagram

$$\begin{array}{ccccccc}
0 \to H^m(M, E) & \xrightarrow{J} & H^m(\widetilde{M}, \widetilde{E}) & \xrightarrow{L} & \bigoplus_0^{m-1} H(Y, E') \to 0 \\
\downarrow A & & \mathcal{A}\downarrow \ \uparrow \mathcal{P} & & \uparrow R \\
0 \leftarrow H^0(M, F) & \xleftarrow{b} & \begin{matrix} H^0(M, F) \\ \oplus \\ H^0(Y, G_+ \oplus G_-) \end{matrix} & \xleftarrow{a} & H^0(Y, G_+ \oplus G_-) \leftarrow 0 .
\end{array} \tag{30}$$

The rows in (30) are exact and we have the identities (28), (29). Thus the formula (26) is a consequence of the following proposition.

Let H, V, W, M, N be Hilbert spaces, and

$$\begin{array}{ccccccc}
0 \to H & \xrightarrow{J} & V & \xrightarrow{L} & W \to 0 \\
\downarrow A & & \mathcal{A}\downarrow \ \uparrow \mathcal{P} & & \uparrow R \\
0 \leftarrow M & \xleftarrow{b} & M \oplus N & \xleftarrow{a} & N \leftarrow 0
\end{array} \tag{31}$$

be a diagram with exact rows. Let a be the canonical embedding and b the canonical projection. Let A, R, $\mathcal{A}$, $\mathcal{P}$ be Fredholm operators, and

$$A = b \circ \mathcal{A} \circ J , \quad R = L \circ \mathcal{P} \circ a . \tag{32}$$

Suppose that $\mathcal{A}$ is constructed similarly as above as a direct sum of Fredholm operators acting from $V^\pm \to W^\pm$, $V = V^+ \oplus V^-$, $M \oplus N = W^+ \oplus W^-$ and let $\mathcal{P}$ be a parametrix of $\mathcal{A}$. Then

$$\operatorname{ind} A = \operatorname{ind} \mathcal{A} + \operatorname{ind} R\,. \tag{33}$$

In order to prove this replace the diagram (31) by another one, namely

$$\begin{array}{ccccccccc} 0 & \to & H & \to & V \oplus \mathbb{C}^k & \to & W \oplus \mathbb{C}^k & \to & 0 \\ & & \downarrow & & \downarrow\uparrow & & \uparrow & & \\ 0 & \leftarrow & M & \leftarrow & M \oplus N \oplus \mathbb{C}^l & \leftarrow & N \oplus \mathbb{C}^l & \leftarrow & 0 \end{array}$$

with an obvious modification of the operators. For instance, R is replaced by $N \oplus \mathbb{C}^l \xrightarrow{j} N \xrightarrow{R} W \xrightarrow{i} W \oplus \mathbb{C}^k$ with j and i the canonical embedding and projection, respectively. The modified operators are again denoted by the same letters and the identities (32) are preserved. The number $\operatorname{ind} \mathcal{A} + \operatorname{ind} R$ remained obviously unchanged, but by suitable choice of k, l one can suppose $\operatorname{ind} \mathcal{A} = 0$. Thus it is sufficient to prove (33) for this case. Consider the canonical projections $b\colon M \oplus N \to M$, $c\colon M \oplus N \to N$ and denote by M_0 and N_0, respectively, the finite dimensional images of $\operatorname{coker} \mathcal{A}$ with respect to b and c, respectively (remember that $\operatorname{coker} \mathcal{A}$ is identified with a complementary subspace of $\operatorname{im} \mathcal{A}$). Then $M_0 \oplus N_0 = \operatorname{coker} \mathcal{A} \cong \ker \mathcal{A}$ and we have direct decompositions $M = M_0 \oplus M_1$, $N = N_0 \oplus N_1$ with complementary spaces M_1 and N_1 to M_0 and N_0, respectively. From $\operatorname{ind} \mathcal{A} = 0$ we get

$$\dim \ker \mathcal{A} = \dim M_0 + \dim N_0\,. \tag{34}$$

Set $V = V_1 \oplus \ker \mathcal{A}$, then $\mathcal{A}\colon V_1 \to M_1 \oplus N_1$ is invertible. By the inverse $\mathcal{P}_0$: $M_1 \oplus N_1 \to V_1$ a regularizer of $\mathcal{A}$ is determined. Let $V_1' = \mathcal{P}_0 M_1$, $V_1'' = \mathcal{P}_0 N_1$, i.e $V_1 = V_1' \oplus V_1''$, $V = V_1' \oplus V_1'' \oplus \ker \mathcal{A}$. The last decomposition gives a direct decomposition $\operatorname{im} J = U \oplus U_1 \oplus U_0$ with $U \subseteq V_1'$, $U_1 \subseteq V_1''$, $U_0 \subseteq \ker \mathcal{A}$. The first condition in (32) and the Fredholm property of A gives a decomposition $V_1' = V_0 \oplus U$, $\dim V_0 < \infty$. Since J is injective, we obtain $\dim \ker A = \dim U_0 + \dim U_1$, $\dim \operatorname{coker} A = \dim V_0 + \dim M_0$. From $L\colon V_0 \oplus U_1^\perp \oplus U_0^\perp \cong W$ for $R_0 = L\mathcal{P}_0 a$ follows $\dim \ker R_0 = \dim N_0 + \dim U_1$, $\dim \operatorname{coker} R_0 = \dim V_0 + (\dim \ker \mathcal{A} - \dim U_0) = \dim V_0 - \dim U_0 + \dim M_0 + \dim N_0$. In the last equation (34) was used. Thus $\operatorname{ind} A = \operatorname{ind} R_0$. Now $\mathcal{P} - \mathcal{P}_0$ is compact, i.e. $R - R_0$ is compact, too. Thus $\operatorname{ind} A = \operatorname{ind} R_0$. $\square$

The formula (26) describes a connection between the indices of elliptic boundary value problems and indices of elliptic pseudo-differential operators on closed compact manifolds. In 3.2.1.3 further connections of this kind shall be studied. They are useful for calculations of indices of concrete elliptic boundary problems.

The formula (26) can be easily generalized. Let M be a compact manifold with boundary and different boundary components Y_1, Y_2 ($\partial M = Y_1 \cup Y_2$). Let X_j be compact manifolds with boundary $\partial X_j = Y_j \cup Y$ (disjoint union), $j = 1, 2$. Suppose that M is obtained from X_1, X_2 by identifying along Y. Let $\mathcal{A}_j$ be elliptic boundary problems on X_j for the corresponding restrictions of an elliptic differential operator on M and let $\mathcal{A}$ be that elliptic problem for A on M containing on Y_j the conditions belonging to $\mathcal{A}_j$ ($j = 1, 2$). Then there exists an elliptic pseudo-differential operator R on Y so that $\operatorname{ind} \mathcal{A}_1 + \operatorname{ind} \mathcal{A}_2 = \operatorname{ind} \mathcal{A} - \operatorname{ind} R$. The correspondence $(\mathcal{A}_1, \mathcal{A}_2) \to (\mathcal{A}, \mathcal{R})$ belongs to the clutching of X_1, X_2 along Y. Therefore we speak of a clutching construction for boundary value problems. Formulas of the type (26) will be proved in 3.2.1.3 for general elliptic boundary problems in $\mathfrak{G}$.

Consider again the situation in formula (26) for $A = \Delta_E$ (the Laplacian in E with principal symbol $-|\xi|^2 \cdot 1_{\pi^*E}$). Then, for $r'_{(+)}T_+ = r'$ (cf. (25)), we simply get the Dirichlet problem on X_+ for $\mathcal{A}^+$. Similarly, the Dirichlet problem $\mathcal{A}^-$ can be considered on X_-. Suppose that X_- is diffeomorphic to X_+ so that M is obtained from $X = X_+$ by clutching along Y a second exemplar of X. Then obviously ind $\mathcal{A}^+ =$ ind $\mathcal{A}^-$. Moreover, ind $\Delta_E = 0$. Thus we have 2 ind $\mathcal{A}^+ =$ ind R in this case. Since σ_R is homotopic to the identity on S^*Y, we obtain ind $R = 0$. Thus we proved the following proposition. Let X be a manifold with boundary and $E \in$ Vect (X). Let $\mathcal{D}_E$ be the Dirichlet problem for Δ_E on X. Then

$$\text{ind}\, \mathcal{D}_E = 0\,. \tag{35}$$

Note that (35) can be proved in another way, namely by constructing suitable Green formulas for which $\mathcal{D}_E$ is self-adjoint.

3.1.1.4. A Priori Estimates in $H^{s,p}$ and Hölder Spaces

In this section we briefly discuss the standard Fredholm properties of elliptic boundary value problems in Sobolev spaces with $1 < p < \infty$ and Hölder spaces. Let $\mathcal{A} \in \mathfrak{G}$ be of the from 3.1.1.1.(1) and suppose in this section that the orders of the homogeneous principal symbols $\sigma_A, \sigma_B, \sigma_T, \sigma_K, \sigma_Q$ are given by 3.1.1.1.(2). Introduce the notations

$$\mathcal{H}_1^p = B^{s,p}(X,E) \oplus B^{t+\lambda+(1/p'),p}(Y,J)\,,\quad \mathcal{H}_2^p = B^{t,p}(X,F) \oplus B^{s-\gamma-(1/p),p}(Y,G)\,,$$

$s \in \mathbb{R}$, $t = s - \alpha$, $1 < p < \infty$ and

$$\mathcal{C}_1 = \mathcal{C}^s(X,E) \oplus \mathcal{C}^{t+\lambda+1}(Y,J)\,,\quad \mathcal{C}_2 = \mathcal{C}^t(X,F) \oplus \mathcal{C}^{s-\gamma}(Y,G)\,,$$

$s \in \mathbb{R}$, $t = s - \alpha$. The operator $\mathcal{A}\colon \mathcal{C}_1 \to \mathcal{C}_2$ has then continuous extensions

$$\mathcal{A}_{(s,p)}\colon \mathcal{H}_1^p \to \mathcal{H}_2^p \quad \text{if } s > d - \frac{1}{p}\,, \tag{1}$$

(d the type of $\mathcal{A}$) and

$$\mathcal{A}_{\langle s\rangle}\colon \mathcal{C}_1 \to \mathcal{C}_2 \quad \text{if } s > \max\,(\alpha, \alpha - \lambda - 1, \gamma, d) \tag{2}$$

(cf. 2.3.3.3).

Theorem 1. *Let $\mathcal{A} \in \mathfrak{G}$ be elliptic and $\mathcal{P}$ a C^∞ parametrix of $\mathcal{A}$ (cf.* 3.1.1.1, *Theorem* 2). *Then the closure of $\mathcal{P}\colon \mathcal{C}_1 \to \mathcal{C}_1$ as an operator $\mathcal{P}_{(t,p)}\colon \mathcal{H}_2^p \to \mathcal{H}_1^p$ is a parametrix of $\mathcal{A}_{(s,p)}$. The closure $\mathcal{P}_{\langle t\rangle}\colon \mathcal{C}_2 \to \mathcal{C}_1$ is a parametrix of $\mathcal{A}_{\langle s\rangle}$.*

The proof of Theorem 1 immediately follows from 2.3.3.3, Corollary 2 if we use $\sigma_\Omega(\mathcal{P}) = \big(\sigma_\Omega(\mathcal{A})\big)^{-1}$, $\sigma_Y(\mathcal{P}) = \big(\sigma_Y(\mathcal{A})\big)^{-1}$ so that $\sigma_\Omega(\mathcal{P}\mathcal{A})$, $\sigma_Y(\mathcal{P}\mathcal{A})$, $\sigma_\Omega(\mathcal{A}\mathcal{P})$, $\sigma_Y(\mathcal{A}\mathcal{P})$ are the corresponding identities.

Note that an arbitrary operator $\mathcal{P}_0 \in \mathfrak{G}$ with $\sigma_\Omega(\mathcal{P}_0) = \sigma_\Omega(\mathcal{A})^{-1}$, $\sigma_Y(\mathcal{P}_0) = \big(\sigma_Y(\mathcal{A})\big)^{-1}$ gives rise to parametrices of $\mathcal{A}_{(s,p)}$, $\mathcal{A}_{\langle s\rangle}$, too. But for a-priori estimates it is useful to work with a C^∞ parametrix.

Corollary 2. *Define the spaces*

$$\mathcal{H}_0^p = B^{s_0,p}(X,E) \oplus B^{t_0,p}(Y,J)$$

for arbitrary fixed s_0, $t_0 \in \mathbb{R}$, $s_0 < s$, $t_0 < t + \lambda + 1/p'$ and

$$\mathcal{C}_0 = \mathcal{C}^{s_0}(X,E) \oplus \mathcal{C}^{t_0}(Y,J)$$

with $s_0, t_0 \in \mathbb{R}$, $s_0 < s$, $t_0 < t + \lambda + 1$. *Then, under the conditions of Theorem* 1, *there exists a constant* $c > 0$ *so that*

$$\|u\|_{\mathcal{H}_1^p} \leq c\,(\|\mathcal{A}_{(s,p)}u\|_{\mathcal{H}_2^p} + \|u\|_{\mathcal{H}_0^p}) \tag{3}$$

for all $u \in \mathcal{H}_1^p$ *and*

$$\|u\|_{\mathcal{E}_1} \leq c\,(\|\mathcal{A}_{\langle s\rangle}u\|_{\mathcal{E}_2} + \|u\|_{\mathcal{E}_0}) \tag{4}$$

for all $u \in \mathcal{E}_1$.

Corollary 2 can be proved in the same way as 3.1.1.1, Corollary 3. The estimate (3) is a generalization of the well-known L^p estimates for classical elliptic boundary value problems. (4) is a generalization of the Schauder estimates.

Theorem 3. *Let* $N_+ \subset \mathcal{E}_1$, $N_- \subset \mathcal{E}_2$ *be the finite-dimensional subspaces mentioned in* 3.1.1.1, *Theorem* 5. *Then* $N_+ = \ker \mathcal{A}_{(s,p)}$, $N_- \oplus \operatorname{im} \mathcal{A}_{(s,p)} = \mathcal{H}_2^p$, $N_+ = \ker \mathcal{A}_{\langle s\rangle}$, $N_- \oplus \operatorname{im} \mathcal{A}_{\langle s\rangle} = \mathcal{E}_2$, *i.e. in particular*

$$\operatorname{ind} \mathcal{A} = \operatorname{ind} \mathcal{A}_{(s,p)} = \operatorname{ind} \mathcal{A}_{\langle s\rangle}\,. \tag{5}$$

The proof is a simple generalization of the proof of 3.1.1.1, Theorem 5 and left to the reader.

If we are given an isomorphism $\mathcal{A}\colon \mathcal{E}_1 \to \mathcal{E}_2$ of the form 3.1.1.1.(1), (2), the corresponding closures (1), (2) are isomorphisms, too. This follows from $N_- = \{0\}$, $N_+ = \{0\}$.

L^p and Schauder estimates are of considerable interest in the analytical theory of elliptic boundary value problems. The results here are a simple consequence of the general theory of operators in $\mathfrak{G}$, mainly the symbolic calculus, and of 2.3.3.3. Under this point of view the proof of a-priori estimates via a parametrix explicitly expressed by inverse interior and boundary symbols is a simplification and works in very general situations. The same idea can be used for proofs of L^p and Schauder estimates for overdetermined elliptic boundary problems treated in 3.1.2.3. Since the method is obvious, we do not explicitly formulate the corresponding results in this case.

A similar method works in the case of Douglis-Nirenberg elliptic and overdetermined elliptic systems. Using reduction of orders by means of the isomorphisms constructed in 3.1.2.1 we obtain the case of usual ellipticity and overdetermined ellipticity. The L^2 estimates in the case of Douglis-Nirenberg elliptic systems are explicitly given in 3.1.2.1.

One may expect that an analogue of 3.1.1.1, Theorem 7 is valid for the spaces $\mathcal{H}_1^p$, $\mathcal{H}_2^p$ and $\mathcal{E}_1$, $\mathcal{E}_2$, respectively, too. The generalization of the methods in the proof should only require technical modifications. For closed compact manifolds the L^p case was treated by SEELEY [2] and KRUPNIK [1].

3.1.1.5. Fredholm Operators in a Special Matrix Form

The remarks in this section are thought as an appendix where we collect some simple properties of matrices of operators. Let H_i, L_i ($i = 1, 2$) be Fréchet spaces and

$$\mathfrak{a} = \begin{pmatrix} \alpha & \varkappa \\ \tau & \delta \end{pmatrix} : \begin{matrix} H_1 \\ \oplus \\ L_1 \end{matrix} \to \begin{matrix} H_2 \\ \oplus \\ L_2 \end{matrix} \tag{1}$$

be a linear continuous operator. Identical operators in various spaces are denoted by 1. If $\alpha\colon H_1 \to H_2$ is a Fredholm operator, α^{-1} denotes a parametrix of α and the inverse if α is an isomorphism.

Operators of the form

$$\mathfrak{k} = \begin{pmatrix} 1 & \varkappa \\ 0 & 1 \end{pmatrix} : \begin{matrix} H \\ \oplus \\ L \end{matrix} \to \begin{matrix} H \\ \oplus \\ L \end{matrix}; \qquad \mathfrak{t} = \begin{pmatrix} 1 & 0 \\ \tau & 1 \end{pmatrix} : \begin{matrix} H \\ \oplus \\ L \end{matrix} \to \begin{matrix} H \\ \oplus \\ L \end{matrix} \tag{2}$$

are isomorphisms with

$$\mathfrak{k}^{-1} = \begin{pmatrix} 1 & -\varkappa \\ 0 & 1 \end{pmatrix}, \qquad \mathfrak{t}^{-1} = \begin{pmatrix} 1 & 0 \\ -\tau & 1 \end{pmatrix}. \tag{3}$$

If we replace $\varkappa$ and τ by $s\varkappa$ and $s\tau$, respectively, $0 \leq s \leq 1$, we get homotopies through isomorphisms.

Proposition 1. *Let $\alpha: H_1 \to H_2$ be a Fredholm operator. Then* (1) *is a Fredholm operator iff*

$$\omega = \delta - \tau\alpha^{-1}\varkappa : L_1 \to L_2 \tag{4}$$

is Fredholm. Then a parametrix of (1) *is given by*

$$\mathfrak{a}^{-1} = \begin{pmatrix} \alpha^{-1} + \alpha^{-1}\varkappa\omega^{-1}\tau\alpha^{-1} & -\alpha^{-1}\varkappa\omega^{-1} \\ -\omega^{-1}\tau\alpha^{-1} & \omega^{-1} \end{pmatrix}. \tag{5}$$

Proof. Modulo compact operators we have

$$\begin{pmatrix} \alpha & 0 \\ 0 & \omega \end{pmatrix} = \begin{pmatrix} 1 & 0 \\ -\tau\alpha^{-1} & 1 \end{pmatrix} \begin{pmatrix} \alpha & \varkappa \\ \tau & \delta \end{pmatrix} \begin{pmatrix} 1 & -\alpha^{-1}\varkappa \\ 0 & 1 \end{pmatrix}. \tag{6}$$

Since the first and the third factor in (6) on the right are isomorphisms, the Fredholm property of ω is equivalent to that of $\mathfrak{a}$. Now (5) immediately follows from $(\alpha \oplus \omega)^{-1} = \alpha^{-1} \oplus \omega^{-1}$ and (3). □

Corollary 2. *Let $\alpha: H_1 \to H_2$ and* (1) *be Fredholm operators. Then*

$$\operatorname{ind} \mathfrak{a} = \operatorname{ind} \alpha + \operatorname{ind} \omega\,. \tag{7}$$

There is a homotopy through Fredholm operators $\mathfrak{a} \simeq \alpha \oplus \omega$.

The homotopy follows from (6) if in the first and the third factor on the right $-\tau\alpha^{-1}$ and $-\alpha^{-1}\varkappa$ are replaced by $-s\tau\alpha^{-1}$ and $-s\alpha^{-1}\varkappa$, $0 \leq s \leq 1$, respectively.

Remark 3. Let $\alpha: H_1 \to H_2$ be an isomorphism. (1) is an isomorphism iff (4) is an isomorphism. (5) is then inverse to (1) and the homotopy $\mathfrak{a} \simeq \alpha \oplus \omega$ resulting from (6) goes through isomorphisms.

Corollary 4. *Let $\alpha: H_1 \to H_2$ be Fredholm and*

$$\tau_0: H_1 \to V\,, \quad \varkappa_0: V \to H_2 \tag{8}$$

be linear continuous operators, V a certain Fréchet space. Then

$$\mathfrak{g} = \begin{pmatrix} \alpha - \varkappa_0\tau_0 & \varkappa \\ \tau & \delta \end{pmatrix} : \begin{matrix} H_1 \\ \oplus \\ L_1 \end{matrix} \to \begin{matrix} H_2 \\ \oplus \\ L_2 \end{matrix} \tag{9}$$

is Fredholm iff

$$\mathfrak{g}_1 = \begin{pmatrix} \alpha & \varkappa & \varkappa_0 \\ \tau & \delta & 0 \\ \tau_0 & 0 & 1 \end{pmatrix} : \begin{matrix} H_1 \\ \oplus \\ L_1 \\ \oplus \\ V \end{matrix} \to \begin{matrix} H_2 \\ \oplus \\ L_2 \\ \oplus \\ W \end{matrix} \tag{10}$$

is Fredholm. Then there is a homotopy through Fredholm operators

$$\mathfrak{g} \oplus 1_V \simeq \mathfrak{g}_1 \,. \tag{11}$$

A similar assertion holds for isomorphisms.

The assertions in Proposition 1 and Corollary 4 remain true if we speak about parametrices modulo an operator ideal, e.g. operators with C^∞ kernels in spaces of C^∞ functions on open sets. For instance, if we are given a matrix $\mathfrak{a}$ of PDOs on a paracompact manifold, a parametrix $\mathfrak{a}^{-1}$ is defined by $\mathfrak{a}\mathfrak{a}^{-1} =$ identity + matrix of operators with C^∞ kernels and $\mathfrak{a}^{-1}\mathfrak{a} = \ldots$ In general, operators with C^∞ kernels are not compact in C^∞, but the formal constructions remain valid.

Let $\mathfrak{a}$ be a Fredholm operator of the form (1) (α not necessarily Fredholm) and consider a second Fredholm operator

$$\mathfrak{a}_0 = \begin{pmatrix} \alpha & \varkappa_0 \\ \tau_0 & \delta_0 \end{pmatrix} : \begin{matrix} H_1 \\ \oplus \\ M_1 \end{matrix} \to \begin{matrix} H_2 \\ \oplus \\ M_2 \end{matrix} \tag{12}$$

where we suppose that the left upper corners in $\mathfrak{a}$ and $\mathfrak{a}_0$ are the same. Let

$$\mathfrak{p}_0 = \begin{pmatrix} \pi_0 & \lambda_0 \\ \sigma_0 & \varrho_0 \end{pmatrix} : \begin{matrix} H_2 \\ \oplus \\ M_2 \end{matrix} \to \begin{matrix} H_1 \\ \oplus \\ M_1 \end{matrix} \tag{13}$$

be a parametrix of $\mathfrak{a}_0$.

Proposition 5. *Under the assumptions above the operator*

$$\zeta = \begin{pmatrix} -\tau\pi_0\varkappa + \delta & \tau\lambda_0 \\ -\sigma_0\varkappa & \varrho_0 \end{pmatrix} : \begin{matrix} L_1 \\ \oplus \\ M_2 \end{matrix} \to \begin{matrix} M_2 \\ \oplus \\ L_1 \end{matrix} \tag{14}$$

is Fredholm. If $\varepsilon = \begin{pmatrix} \varepsilon_{11} & \varepsilon_{12} \\ \varepsilon_{21} & \varepsilon_{22} \end{pmatrix}$ *denotes a parametrix of* (14), *a parametrix of* (1) *is given by*

$$\mathfrak{a}^{-1} = \begin{pmatrix} \pi_0 + \pi_0\varkappa(\varepsilon\mu)_1 - \lambda_0(\varepsilon\mu)_2 & -\pi_0\varkappa\varepsilon_{11} + \lambda_0\varepsilon_{21} \\ -(\varepsilon\mu)_1 & \varepsilon_{11} \end{pmatrix}. \tag{15}$$

Here, $\mu = \begin{pmatrix} \mu_1 \\ \mu_2 \end{pmatrix} = \begin{pmatrix} \tau\pi_0 \\ \sigma_0 \end{pmatrix}$ *and* $(\varepsilon\mu)_k = (\varepsilon_{k1}\mu_1 + \varepsilon_{k2}\mu_2)$, $k = 1, 2$.

The proof of Proposition 5 can be given by a straightforward calculation. The meaning of the expression (15) is that a parametrix of $\mathfrak{a}$ is expressed by a parametrix of some given fixed $\mathfrak{a}_0$ and a parametrix of a Fredholm operator ζ acting between the second components of the spaces. In connection with boundary problems these components play the role of spaces of C^∞ sections of bundles over the boundary.

For completeness formulate further properties of operators of the form (1).

Remark 6. (1) is Fredholm iff

$$(\alpha, \varkappa) : \begin{matrix} H_1 \\ \oplus \\ L_1 \end{matrix} \to H_2 \tag{16}$$

has finite-dimensional cokernel and

$$(\tau, \delta) : \ker (\alpha, \varkappa) \to L_2 \tag{17}$$

is Fredholm. (1) is an isomorphism iff (16) is surjective and (17) an isomorphism.

Remark 7. Let (1) be an isomorphism. The maps

$$(\alpha, \varkappa) : \ker (\tau, \delta) \to H_2, \quad (\tau, \delta) : \ker (\alpha, \varkappa) \to L_2$$

are isomorphisms and hence

$$H_1 \oplus L_1 \cong H_2 \oplus L_2 \cong \ker (\alpha, \varkappa) \oplus \ker (\tau, \delta) .$$

Proposition 8. *Let* $\mathfrak{a}$ *be an operator of the form* (1). *Then*

$$\ker \mathfrak{a} = \ker (\alpha, \varkappa) \cap \ker (\tau, \delta). \tag{18}$$

If $\mathfrak{a}$ *is surjective, there exist complements* $\ker (\alpha, \varkappa)^\perp$ *and* $\ker (\tau, \delta)^\perp$ *in* $H_1 \oplus L_1$ *with*

$$\ker (\alpha, \varkappa)^\perp \cap \ker (\tau, \delta)^\perp = \{0\} \tag{19}$$

and

$$(\ker \mathfrak{a})^\perp = \ker (\alpha, \varkappa)^\perp \oplus \ker (\tau, \delta)^\perp \tag{20}$$

for some complement $(\ker \mathfrak{a})^\perp$ *of* $\ker \mathfrak{a}$.

Proof: (18) is trivial. Then

$$(\ker \mathfrak{a})^\perp = \big(\ker (\alpha, \varkappa)\big)^\perp + \big(\ker (\tau, \delta)\big)^\perp \tag{21}$$

follows for complements of the spaces in the brackets. We have to show that the complements on the right in (21) can be chosen in such a way that (19) holds. Set

$$(\alpha, \varkappa)_1 = (\alpha, \varkappa)|_{(\ker \mathfrak{a})^\perp}, \quad (\alpha, \varkappa)_0 = (\alpha, \varkappa)|_{\ker \mathfrak{a}},$$

$$(\tau, \delta)_1 = (\tau, \delta)|_{(\ker \mathfrak{a})^\perp}, \quad (\tau, \delta)_0 = (\tau, \delta)|_{\ker \mathfrak{a}}$$

(the operators with index 0 are, of course, 0).

With $H_1 \oplus L_1 = (\ker \mathfrak{a})^\perp \oplus \ker \mathfrak{a}$ we consequently have a surjective map

$$\mathfrak{a} = \begin{pmatrix} (\alpha, \varkappa)_1 & (\alpha, \varkappa)_0 \\ (\tau, \delta)_1 & (\tau, \delta)_0 \end{pmatrix} : \begin{matrix} (\ker \mathfrak{a})^\perp \\ \oplus \\ \ker \mathfrak{a} \end{matrix} \to \begin{matrix} H_2 \\ \oplus \\ L_2 \end{matrix},$$

i.e. an isomorphism

$$\begin{pmatrix} (\alpha, \varkappa)_1 \\ (\tau, \delta)_1 \end{pmatrix} : (\ker \mathfrak{a})^\perp \to \begin{matrix} H_2 \\ \oplus \\ L_2 \end{matrix}.$$

Thus $(\ker \mathfrak{a})^\perp = \big(\ker (\alpha, \varkappa)_1\big)^\perp \oplus \big(\ker (\tau, \delta)_1\big)^\perp$. Since $\ker (\alpha, \varkappa) = \ker (\alpha, \varkappa)_1 \oplus \ker \mathfrak{a}$, $\ker (\tau, \delta) = \ker (\tau, \delta)_1 \oplus \ker \mathfrak{a}$, $\big(\ker (\alpha, \varkappa)_1\big)^\perp$, and $\big(\ker (\tau, \delta)_1\big)^\perp$ can be considered as complements of $\ker (\alpha, \varkappa)$ and $\ker (\tau, \delta)$, respectively, in $H_1 \oplus L_1$. For these complements (19) is obvious. □

Remark 9. Let

$$A\colon \bigoplus_{j=1}^{N} H_j \to \bigoplus_{k=1}^{M} L_k \tag{22}$$

be a Fredholm operator where A is a matrix of continuous operators $A_{kj}\colon H_j \to L_k$ between corresponding Fréchet spaces. Then the spaces $\ker A_{kj}$ in H_j and $\operatorname{im} A_{kj}$ in L_k are closed and have topological complements.

The simple proof is left to the reader.

3.1.2. Examples and Remarks

3.1.2.1. The Operators $\Lambda^{\pm}$ and Reduction of the Orders

In this section we will discuss certain special operators to be used for the investigation of general elliptic boundary problems in $\mathfrak{G}$. One of the results is the following

Theorem 1. *For arbitrary* $E \in \operatorname{Vect}(X)$ *there exists an operator* $\mathcal{L}_E^- \in \mathfrak{G}$ *of the form*

$$r^+\Lambda_E^- : C^\infty(X, E) \to C^\infty(X, E) \tag{1}$$

with elliptic symbol $\sigma_{\Lambda_E^-} \in \mathfrak{A}^{(1)}$ *so that* (1) *is an isomorphism.*

Then the closure

$$\mathcal{L}_E^- : H^s(X, E) \to H^{s-1}(X, E) \tag{2}$$

$(s > -\frac{1}{2})$ *is an isomorphism, too.*

In view of 2.1.1.1, Theorem 7 we find that $\sigma_Y(r^+\Lambda_E^-) = \Pi^+\sigma_{\Lambda_E^-}\colon p^*E^+ \to p^*E^+$ has to be an isomorphism. The meaning of (1) is that it is an elliptic boundary problem for an elliptic pseudo-differential operator with index zero without trace and potential conditions.

Recall that isomorphisms in the form (2) have already been used in the half space (cf. 2.3.2.4). Unfortunately, in case of compact manifolds with boundary it is not so trivial to show that the corresponding operators define isomorphisms.

Theorem 1 shall be proved as a consequence of further theorems. Because of 3.1.1.1, Corollary 6 it is sufficient to show $\operatorname{ind}(r^+\Lambda_E^-) = 0$.

Corollary 2. *For given* $E \in \operatorname{Vect}(X)$, $J \in \operatorname{Vect}(Y)$, $s \in \mathbb{Z}_+$, $t \in \mathbb{R}$ *there exists an elliptic operator* $\mathcal{L}_{E,J}^{s,t} \in \mathfrak{G}$ *defining an isomorphism*

$$\mathcal{L}_{E,J}^{s,t}\colon H^s(X, E) \oplus H^t(Y, J) \to H^0(X, E) \oplus H^0(Y, J)\,. \tag{3}$$

It is sufficient to set $\mathcal{L}_{E,J}^{s,t} = (\mathcal{L}_E^-)^s \oplus R$ with an elliptic pseudo-differential operator R on Y of order t inducing an isomorphism $R\colon H^t(Y, J) \to H^0(Y, J)$. One can take for R e.g. an operator with the homogeneous principal symbol $|\xi'|^t \cdot 1_{\pi_Y^* J}$, $\pi_Y\colon T^*Y \setminus 0 \to Y$ (cf. 1.2.4.2, Theorem 5 in the case of $M = Y$). Similarly as 1.2.4.2. (7), a scalar product in $H^s(X, E)$, $s \in \mathbb{Z}_+$, can be defined by

$$(u, v)_{H^s(X, E)} = \left((r^+\Lambda_E^-)^s u, (r^+\Lambda_E^-)^s v\right)_{L^2(X, E)}\,.$$

Assume that we are given an operator $\mathcal{A} \in \mathfrak{G}$ defining a continuous mapping

$$\mathcal{A}\colon H^s(X, E) \oplus H^{t+\lambda+1/2}(Y, J) \to H^t(X, F) \oplus H^{s-\gamma-1/2}(Y, G) \tag{4}$$

(cf. 2.3.3.3, Theorem 1), $s \in \mathbb{Z}_+$, $t = s - \alpha \leqq 0$. Then $t \in \mathbb{Z}_+$ because of $\alpha \in \mathbb{Z}$ and we can define the operator

$$\mathcal{B} = \mathcal{L}^{t,s-\gamma-1/2}_{F,G} \mathcal{A} (\mathcal{L}^{s,t+\lambda+1/2}_{E,J})^{-1} : H^0(X, E) \oplus H^0(Y, J) \to H^0(X, F) \oplus H^0(Y, G), \tag{5}$$

$\mathcal{B} \in \mathfrak{G}$. Passing from (4) to (5) we speak about reduction of the order and type to zero. $\mathcal{B}$ is obviously elliptic iff $\mathcal{A}$ is elliptic.

In this section denote by U a fixed neighbourhood of Y in X for which a diffeomorphism $U \to Y \times [0, 1)$ is fixed. We shall identify U with $Y \times [0, 1)$, where $[0, 1)$ corresponds to the interval $0 \leqq x_n < 1$. Let $\chi_\varepsilon(t)$ be a real function in $C^\infty(\overline{\mathbb{R}}_+)$, $0 \leqq \chi_\varepsilon(t) \leqq 1$ $(t \in \mathbb{R}_+)$ and $\chi_\varepsilon(t) = 0$ if $t \leqq \varepsilon$, $\chi_\varepsilon(t) = 1$ when $t \geqq 2\varepsilon$. Here ε is given, $0 < \varepsilon < {}^1/_2$. Moreover, set

$$\delta_\varepsilon(\xi) = \chi_\varepsilon\left(\frac{|\xi'|^2}{|\xi'|^2 + \nu^2}\right) |\xi'| \,. \tag{6}$$

Then the symbols $\delta_\varepsilon(\xi) \pm i\nu$ $((x, \xi) \in T^*U \setminus 0)$ are elliptic and belong to $\mathfrak{A}^{(1)}$ (since $\delta_\varepsilon(\xi)$ has compact support in ν for each fixed $\xi' \neq 0$). If $\delta_\varepsilon(\xi) \pm i\nu$ is considered as a function on $S^*X|_U$ (i.e. for $|\xi| = 1$), then

$$\delta_\varepsilon(\xi) \pm i\nu \to |\xi'| \pm i\nu \quad \text{for} \quad \varepsilon \to 0 \tag{7}$$

with respect to uniform convergence.

Choose a real function $a \in C^\infty(\overline{\mathbb{R}}_+)$, $0 \leqq a \leqq 1$ with $a(x_n) = 1$ when $0 \leqq x_n < \eta_1$, $a(x_n) = 0$ if $x_n > \eta_0$, $0 < \eta_1 < \eta_0 < 1$. Set

$$l^\pm(x, \xi) = |\xi|^{1-a(x_n)} \left(\delta_\varepsilon(\xi) \pm i\nu\right)^{a(x_n)}, \tag{8}$$

where the power is defined with that branch of the logarithm which is for real positive arguments. Note that $\delta_\varepsilon(\xi) \geqq 0$, i.e. $-\pi/2 \leqq \arg\left(\delta_\varepsilon(\xi) \pm i\nu\right) \leqq \pi/2$, $\nu \in \mathbb{R}$. Thus (8) is a smooth function over $T^*U \setminus 0$ and equal to $|\xi|$ on $T^*(Y \times [\eta_0, 1]) \setminus 0$. Extending by $|\xi|$ to $T^*(X \setminus U) \setminus 0$ gives an elliptic symbol $\in \mathfrak{A}^{(1)}$. For abbreviation this shall again be denoted by $l^\pm$.

For $E \in \operatorname{Vect}(X)$ define the elliptic symbols

$$\lambda_E^\pm = l^\pm \cdot 1_{\pi^* E} \in \mathfrak{A}^{(1)}, \qquad \omega_E = \frac{l^+}{l^-} \cdot 1_{\pi^* E} \in \mathfrak{A}^{(0)} \,. \tag{9}$$

Near Y we have

$$l^+(x, \xi)/l^-(x, \xi) \to (|\xi'| + i\nu)\,(|\xi'| - i\nu)^{-1} \tag{10}$$

for $\varepsilon \to 0$ in the uniform convergence on S^*X. Suppose in the following that the number $\varepsilon > 0$ contained in $l^\pm$ is fixed and sufficiently small, $\varepsilon < \varepsilon_0$. How to fix ε_0 becomes clear in the following considerations.

Lemma 3. $\operatorname{ind}_{S^*Y} \Pi^+ \lambda_E^+ = [p^*E']$, $\operatorname{ind}_{S^*Y} \Pi^+ \lambda_E^- = 0$, $\operatorname{ind}_{S^*Y} \Pi^+ \omega_E = [p^*E']$.

Proof: In view of the invariance of boundary symbols of the type $\Pi^+\sigma$ (cf. 2.3.3.1) it is sufficient to prove $\operatorname{ind}_{S^*Y} \Pi^+ l^+ = [\mathbb{C}]$, $\operatorname{ind}_{S^*Y} \Pi^+ l^- = 0$, $\operatorname{ind}_{S^*Y} \Pi^+ (l^+/l^-) = [\mathbb{C}]$. First remark that there are homotopies

$$\Pi^+ l^\pm \simeq \Pi^+(|\xi'| \pm i\nu)\,, \qquad \Pi^+(l^+/l^-) \simeq \Pi^+\left(\frac{|\xi'| + i\nu}{|\xi'| - i\nu}\right) \tag{11}$$

$(0 \leqq \varepsilon \leqq \varepsilon_0)$ through families of Fredholm operators over S^*Y (consideration of homotopies is admissible here because of corresponding extensions in Hilbert spaces).

Thus it is sufficient to consider the operator families on the right hand sides in (11), $|\xi'| = 1$. In 2.1.2.3, Lemma 8 was proved that $\Pi^+(1 - i\nu)^k \colon H^+ \to H^+$ defines an isomorphism for each $k \in \mathbb{Z}$. The inverse is given by $\Pi^+(1 - i\nu)^{-k}$. Thus the assertion follows for l^-. Moreover, because of the composition rule (cf. 2.1.2.3, Lemma 5)

$$\Pi^+(\sigma_2\sigma_1 h) = \Pi^+\sigma_2\{\Pi^+\sigma_1 h\} + L(\Pi^+\sigma_2, \Pi^-\sigma_1)\, h \tag{12}$$

($h \in H^+$) for $\sigma_2 = (1 + i\nu)$, $\sigma_1 = (1 - i\nu)^{-1}$ it is sufficient to consider the second operator family in (11). Its symbol is $z^{-1} = (1 + i\nu)\,(1 - i\nu)^{-1}$. Interpreted as function on S^1 the index of $\Pi^+ z^{-1}$ is equal to 1, the negative of the winding number (cf. 2.1.2.1, Proposition 2).

Since kernel and cokernel are independent of the direction of ξ', the assertion is completely proved. □

Corollary 4. $\operatorname{ind}_{S^*Y} \Pi^+\sigma_{\Delta_E} = [p^*E'] \quad (\sigma_{\Delta_E} = -\,|\xi|^2 \cdot 1_{\pi^*E})$.

This follows similarly as Lemma 3 from the local scalar situation and $|\xi|^2 = (|\xi| + i\nu)(|\xi'| - i\nu)$ in view of the composition rule (12) with $\sigma_2 = 1 + i\nu$, $\sigma_1 = 1 - i\nu$ and $\operatorname{ind}_{S^*Y} \Pi^+(1 + i\nu) = [\mathbb{C}]$.

Remark 5. Since the proof of Lemma 3 and Corollary 4 is reduced to scalar operators, we have $\ker_{S^*Y} \Pi^+\lambda_E^+ = \ker_{S^*Y} \Pi^+\omega_E = \ker_{S^*Y} \Pi^+\sigma_{E\Delta} \cong p^*E'$ and the cokernels are zero. Moreover, $\Pi^+\lambda_E^- \colon p^*E^+ \to p^*E^*$ is an isomorphism.

Lemma 6. *The boundary symbols*

$$\begin{pmatrix} \Pi^+\lambda_E^+ \\ \Pi' \end{pmatrix}, \quad \begin{pmatrix} \Pi^*\omega_E \\ \Pi' \end{pmatrix}, \quad \begin{pmatrix} \Pi^+\sigma_{\Delta_E} \\ \Pi' \end{pmatrix}, \tag{13}$$

considered as mappings $p^*E^+ \to p^*E^+ \oplus p^*E'$, are isomorphisms.

Proof: It is again sufficient to consider the scalar case. Begin with the second operator in (13). Since $\Pi^+(1 + i\nu)\,(1 - i\nu)^{-1} \colon H^+ \to H^+$ is surjective, we have to prove that $\Pi' \colon \ker \Pi^+(1 + i\nu)\,(1 - i\nu)^{-1} \to \mathbb{C}$ is an isomorphism. A simple calculation shows that

$$\ker \Pi^+ \left(\frac{1 + i\nu}{1 - i\nu}\right) = \left\{\frac{b}{1 + i\nu} : b \in \mathbb{C}\right\}$$

and $\Pi'(1 + i\nu)^{-1} = 1$. Thus the assertion is proved for the second operator in (13). Moreover, $\Pi^+(1 - i\nu) \colon H^+ \to H^+$ is an isomorphism. Then

$$\begin{pmatrix} \Pi^+(1 - i\nu) & 0 \\ 0 & 1 \end{pmatrix} \begin{pmatrix} \Pi^+\left(\dfrac{1 + i\nu}{1 - i\nu}\right) \\ \Pi' \end{pmatrix} = \begin{pmatrix} \Pi^+(1 + i\nu) \\ \Pi' \end{pmatrix}$$

is an isomorphism, too (use (12) with $\sigma_2 = (1 - i\nu)$, $\sigma_1 = (1 - i\nu)\,(1 - i\nu)^{-1}$). A similar consideration which is left to the reader yields the proof for the third operator in (13). □

Denote by $r^+\Lambda_E^\pm$, $r^+\Omega_E$ and $r^+\Delta_E$ pseudo-differential operators $C^\infty(X, E) \to C^\infty(X, E)$ with the symbols $\lambda_E^\pm$, ω_E and σ_{Δ_E}, respectively. In view of Remark 5 and Lemma 6 the operators $\in \mathfrak{G}$

$$\mathcal{L}_E^- = r^+\Lambda_E^-, \quad \mathcal{L}_E^+ = \begin{pmatrix} r^+\Lambda_E^+ \\ r' \end{pmatrix}, \quad \mathcal{W}_E = \begin{pmatrix} r^+\Omega_E \\ r' \end{pmatrix}, \quad \mathcal{D}_E = \begin{pmatrix} r^+\Delta_E \\ r' \end{pmatrix} \tag{14}$$

are elliptic.

In 2.3.4.4 there was defined a class $\overline{\mathfrak{G}}$ obtained from $\mathfrak{G}$ as a certain closure. The interior symbols of operators in $\overline{\mathfrak{G}}$ are continuous for $\xi \neq 0$ and in general not differentiable. The assertions connected with the calculus of principal symbols are carried over to $\overline{\mathfrak{G}}$, particularly the correspondence between the composition of operators and symbols, and the parametrix and inverse of symbols.

Instead of (8) consider now the continuous symbols

$$l_0^{\pm}(x, \xi) = |\xi|^{1-a(x_n)} (|\xi'| \pm i\nu)^{a(x_n)}$$

and instead of (9)

$$\lambda_{0,E}^{\pm} = l_0^{\pm} \cdot 1_{\pi^*E}, \quad \omega_{0,E} = \frac{l_0^+}{l_0^-} \cdot 1_{\pi^*E}.$$

Denote the corresponding elliptic operators in $\overline{\mathfrak{G}}$ which similarly defined as in (14) by $\mathscr{L}_{0,E}^{\pm}$ and $\mathscr{W}_{0,E}$, respectively. Then there are homotopies $\mathscr{L}_{0,E}^{\pm} \simeq \mathscr{L}_E^{\pm}$, $\mathscr{W}_{0,E}^{\pm} \simeq \mathscr{W}_E$ through elliptic operators in $\overline{\mathfrak{G}}$ (the homotopies are realized in the sense of Fredholm operators in fixed Sobolev spaces; a homotopy on operator level corresponds to a homotopy on symbolic level and an additional arrangement for lower order terms arising as compact operators). Application of symbolic calculus gives $(\mathscr{L}_{0,E}^- \oplus 1) \times \mathscr{L}_{0,E}^- \simeq \mathscr{D}_E$ and $(\mathscr{L}_{0,E}^- \oplus 1)^{-1} \mathscr{L}_{0,E}^+ \simeq \mathscr{W}_{0,E}$ through elliptic elements in $\overline{\mathfrak{G}}$. Hence

$$(\mathscr{L}_E^- \oplus 1)\, \mathscr{L}_E^+ \simeq \mathscr{D}_E, \quad (\mathscr{L}_E^- \oplus 1)^{-1} \mathscr{L}_E^+ \simeq \mathscr{W}_E \tag{15}$$

through elliptic operators in $\overline{\mathfrak{G}}$ (as usual $(\ldots)^{-1}$ means a parametrix of the operator in the brackets). Note that

$$\operatorname{ind} \mathscr{D}_E = 0 \tag{16}$$

(cf. 3.1.1.3. (35)).

Theorem 7. *The operators $\mathscr{L}_E^{\pm}$ and $\mathscr{W}_E$ have index zero.*

The proof of Theorem 7 shall be reduced to the consideration of another elliptic boundary problem to be defined in the following.

Consider a bundle $G \in \operatorname{Vect}(Y)$ and a $G^{\perp} \in \operatorname{Vect}(Y)$ with $G \oplus G^{\perp} \cong Y \times \mathbb{C}^l$ for some $l \in \mathbb{Z}_+$. For the neighbourhood $U = Y \times [0, 1)$ of Y in X mentioned at the beginning we have a projection $q\colon U \to Y$. Set $L = q^*G$, $L^{\perp} = q^*G^{\perp}$, then $L \oplus L^{\perp} \cong U \times \mathbb{C}^l$. Consider the bundles $\pi_U^*L, \pi_U^*L^{\perp}, \pi_U^*(U \times \mathbb{C}^l)$ over $T^*U \setminus 0$ ($\pi_U\colon T^*U \setminus 0 \to U$). Then the projections $p_G\colon G \oplus G^{\perp} \to G$, $p_{G^{\perp}}\colon G \oplus G^{\perp} \to G^{\perp}$ induce projections

$$s_L\colon \pi_U^*(U \times \mathbb{C}^l) \to \pi_U^*L, \quad s_{L^{\perp}}\colon \pi_U^*(U \times \mathbb{C}^l) \to \pi_U^*L^{\perp}.$$

By

$$\gamma_{G,U} = \frac{l^+(x, \xi)}{l^-(x, \xi)} s_L + s_{L^{\perp}} : \pi_U^*(U \times \mathbb{C}^l) \to \pi_U^*(U \times \mathbb{C}^l) \tag{17}$$

an elliptic symbol over U is given. It is the identity over $Y \times (1 - \eta, 1)$ for $\eta > 0$ sufficiently small. Thus (17) can be extended to X as an elliptic symbol over $X \setminus U$ defined as identity. Denote the resulting elliptic symbol over X by γ_G. Then $\gamma_G \in \mathfrak{A}^{(0)}$. A simple consequence of Lemma 3 is $\operatorname{ind}_{S^*Y} \Pi^+ \gamma_G = [p^*G]$. Put $r^+\Gamma_G = \operatorname{Op}(\gamma_G)$. Denote by $r'_G\colon C^\infty(X, \mathbb{C}^l) \to C^\infty(Y, G)$ the map obtained by restriction to Y and projection corresponding to p_G. Then

$$\mathscr{B}_G = \begin{pmatrix} r^+\Gamma_G \\ \\ r'_G \end{pmatrix} : C^\infty(X, \mathbb{C}^l) \to \begin{matrix} C^\infty(X, \mathbb{C}^l) \\ \oplus \\ C^\infty(Y, G) \end{matrix} \tag{18}$$

is an elliptic operator in $\mathfrak{G}$.

Theorem 8. ind $\mathscr{B}_G = 0$.

A proof of Theorem 8 shall be given at the end of this section.

Now we give the proof of Theorem 7 using (16) and Theorem 8. Let $E^\perp$ be a vector bundle over X with $E \oplus E^\perp = X \times \mathbb{C}^l$ for some $l \in \mathbb{Z}_+$. Define the elliptic boundary problem

$$\mathscr{K}^+ = \begin{pmatrix} r^+\Lambda_E^+ & 0 \\ 0 & 1_{E^\perp} \\ \hdashline r'_E & 0 \end{pmatrix} : C^\infty(X, \mathbb{C}^l) \to \begin{matrix} C^\infty(X, \mathbb{C}^l) \\ \oplus \\ C^\infty(Y, E') \end{matrix} .$$

Here r'_E denotes the map obtained by the projection $C^\infty(X, \mathbb{C}^l) \to C^\infty(X, E)$ and then restriction to Y. Moreover, set

$$\mathscr{K}^- = \begin{pmatrix} r^+\Lambda_E^- & 0 \\ 0 & 1_{E^\perp} \end{pmatrix} : C^\infty(X, \mathbb{C}^l) \to C^\infty(X, \mathbb{C}^l) .$$

That the PDOs contained in $\mathscr{K}^\pm$ have different orders is not essential for the consideration. It is clear that ind $\mathscr{K}^\pm =$ ind $\mathscr{L}_E^\pm$. Similarly as (15) one can prove $\mathscr{B}_{E'} \simeq (\mathscr{K}^- \oplus 1_{E'})^{-1} \mathscr{K}^+$ so that

$$0 = \text{ind}\, \mathscr{B}_{E'} = \text{ind}\, \mathscr{K}^+ - \text{ind}\, \mathscr{K}^- = \text{ind}\, \mathscr{L}_E^+ - \text{ind}\, \mathscr{L}_E^- .$$

Furthermore, (16) and the first homotopy in (15) show that $0 = \text{ind}\, \mathscr{D}_E = \text{ind}\, \mathscr{L}_E^+ + \text{ind}\, \mathscr{L}_E^- = 0$ and therefore ind $\mathscr{L}_E^\pm = 0$. The second homotopy in (15) shows that ind $\mathscr{W}_E = 0$. This proves Theorem 7. □

For the proof of Theorem 8 some further steps are needed. With (18) we can connect in a canonical way an elliptic operator $\mathscr{B}'_G$ on $Y \times [0, 1]$ by restricting $r^+\Gamma_G$ to $Y \times [0, 1]$ (remember that some boundary neighbourhood U is identified with $Y \times [0, 1)$) and by taking over $Y \times \{0\}$ the trace conditions as in (18) over Y and no trace conditions over the boundary component $Y \times \{1\}$. Then obviously ind $\mathscr{B}_G$ = ind $\mathscr{B}'_G$.

Now it is clear that $r^+\Gamma_G$ can be extended by 1 to $X \stackrel{\text{def}}{=} Y \times [0, \infty)$ and so we get a boundary problem $\mathscr{B}''_G$ over the new X by obvious extension of $\mathscr{B}'_G$ to X. Since we did not change anything near Y the ellipticity of the trace condition is still satisfied. $\mathscr{B}''_G$ shall be considered as Fredholm operator

$$\mathscr{B}''_G \colon H^1(X, \mathbb{C}^l) \to H^1(X, \mathbb{C}^l) \oplus H^{1/2}(Y, G) . \tag{19}$$

Let $q\colon Y \times [0, \infty) \to Y$ be the canonical projection and $L = q^*G$, $L^\perp = q^*G^\perp$. The pseudo-differential operator in $\mathscr{B}''_G$ acts as the identity on sections in $L^\perp$. Therefore, the index of $\mathscr{B}''_G$ is equal to the index of

$$\mathscr{W}_L = \begin{pmatrix} r^+\Omega_L \\ \\ r' \end{pmatrix} : H^1(X, L) \to \begin{matrix} H^1(X, L) \\ \oplus \\ H^{1/2}(Y, G) . \end{matrix} \tag{20}$$

Because of ind $\mathscr{B}_G = $ ind $\mathscr{W}_L$ it is sufficient to prove

$$\text{ind } \mathscr{W}_L = 0\,. \tag{21}$$

Besides $r^+\Omega_L$ we also consider the operator $r^+\Omega_L^1$ defined in the following way. Let Δ_G' be a self-adjoint negative definite differential operator on Y with the homogeneous principal symbol $-|\xi'|^2 \cdot 1_{\pi_Y^* G}$ $(\pi_Y\colon T^*Y \setminus 0 \to Y)$. Then we have a system of eigenfunctions $v_k(y)$ with eigenvalues $-\lambda_k^2$, $\lambda_k > 0$ (and this system is complete in $H^0(Y, G)$ as well as in $H^1(Y, G)$).

Put for abbreviation

$$\chi(\lambda, \nu) = \chi_\varepsilon(\lambda, \nu) = \chi_\varepsilon\left(\frac{\lambda^2}{\lambda^2 + \nu^2}\right)$$

(cf. (6)) and set

$$\Omega_L^1 = \frac{\chi\left(\sqrt{-\Delta_G'}, D_n\right)\sqrt{-\Delta_G'} + \mathrm{i}D_n}{\chi\left(\sqrt{-\Delta_G'}, D_n\right)\sqrt{-\Delta_G'} - \mathrm{i}D_n}\,. \tag{22}$$

Then Ω_L^1 is an elliptic PDO on $Y \times \mathbb{R}$ with homogeneous principal symbol

$$\frac{\chi(|\xi'|, \nu)\,|\xi'| + \mathrm{i}\nu}{\chi(|\xi'|, \nu)\,|\xi'| - \mathrm{i}\nu}\,.$$

On functions of the form $v_k(y)\,w(x_n)$, $w \in \mathscr{S}(\mathbb{R})$, we have

$$\Omega_L^1(v_k w) = F_n^{-1}\left\{\frac{\chi(\lambda_k, \nu)\,\lambda_k + \mathrm{i}\nu}{\chi(\lambda_k, \nu)\,\lambda_k - \mathrm{i}\nu}\,v_k(y)\,\hat{w}(\nu)\right\} \tag{23}$$

(F_n denotes the Fourier transform in x_n direction). Similarly the operator $r^+\Omega_L^1$ can be defined by taking $w \in \mathscr{S}(\overline{\mathbb{R}}_+)$ and applying Π_ν^+ under the Fourier transform. Furthermore, we can consider the operator

$$\Omega_{0,L}^1 = \frac{\sqrt{-\Delta_G'} + \mathrm{i}D_n}{\sqrt{-\Delta_G'} - \mathrm{i}D_n} \tag{24}$$

on $Y \times \mathbb{R}$, defined by

$$\Omega_{0,L}^1(v_k w) = F_n^{-1}\left\{\frac{\lambda_k + \mathrm{i}\nu}{\lambda_k - \mathrm{i}\nu}\,v_k(y)\,\hat{w}(\nu)\right\} \tag{25}$$

and similarly $r^+\Omega_{0,L}^1$ on $Y \times \overline{\mathbb{R}}_+$ (L as a bundle over $Y \times \mathbb{R}$ denotes the pull back of C under the projection $Y \times \mathbb{R} \to Y$).

We also have to use operators of the form

$$r^+\frac{\sqrt{-\Delta_G'} - \mathrm{i}D_n}{\sqrt{-\Delta_G'} + \mathrm{i}D_n}\,, \qquad r'\frac{1}{\sqrt{-\Delta_G'} - \mathrm{i}D_n}\,, \ldots$$

defined in a similar way, and we need the following

Lemma 9. *The operator* (24) *is a continuous map* $H^1(Y \times \mathbb{R}, L) \to H^1(Y \times \mathbb{R}, L)$. *The operator* $r^+\Omega_0^1\colon H^1(X, L) \to H^1(X, L)$ *is continuous. Moreover, the operators*

$$r^+ \frac{\sqrt{-\Delta_G'} - \mathrm{i}D_n}{\sqrt{-\Delta_G'} + \mathrm{i}D_n} : H^1(X, L) \to H^1(X, L)\,, \tag{26}$$

$$r' \frac{\sqrt{-\Delta_G'}}{\sqrt{-\Delta_G'} - \mathrm{i}D_n} : H^1(X, L) \to H^{1/2}(Y, G)\,, \tag{27}$$

$$\frac{1}{\sqrt{-\Delta_G'} + \mathrm{i}D_n} : H^{1/2}(Y, G) \to H^1(X, L) \tag{28}$$

are continuous.

Proof: Consider e.g. (26). Put for abbreviation $A = \sqrt{-\Delta_G'}$. Then

$$\sigma(\nu) = \frac{A - \mathrm{i}\nu}{A + \mathrm{i}\nu}, \quad \nu \in \mathbb{R}\,,$$

can be considered as operator-valued symbol in $S^0(\mathbb{R}_{x_n} \times \mathbb{R}_\nu, \mathcal{L}(H^t(Y, G), H^t(Y, G)))$ $t \in \mathbb{R}$. Thus, in view of 2.3.2.1, we get continuous mappings

$$r^+\sigma(D_n)\colon H^0(\mathbb{R}_+, H^t(Y, G)) \to H^0(\mathbb{R}_+, H^t(Y, G))\,.$$

Using now the identification

$$H^1(Y \times \overline{\mathbb{R}}_+, L) = \left\{u \in H^0(\mathbb{R}_+, H^1(Y, G)) : \frac{\partial u}{\partial x_n} \in H^0(\mathbb{R}_+, H^0(Y, G))\right\}$$

(cf. Lions/Magenes [1]) we just obtain (26). The assertions about the other operators can be proved in a similar way. □

Lemma 10. *The operator*

$$\begin{pmatrix} r^+\Omega_{0,L}^1 \\ \\ r' \end{pmatrix} : H^1(X, L) \to \begin{matrix} H^1(X, L) \\ \oplus \\ H^{1/2}(Y, G) \end{matrix} \tag{29}$$

is an isomporhism.

It is sufficient to show that

$$\begin{pmatrix} r^+\Omega_{0,L}^1 \\ \\ Qr' \end{pmatrix} : H^1(X, L) \to \begin{matrix} H^1(X, L) \\ \oplus \\ H^1(Y, G) \end{matrix} \tag{30}$$

is an isomorphism, where Q is an elliptic PDO on Y inducing an isomorphism Q: $H^{1/2}(Y, G) \to H^1(Y, G)$.

In order to prove the latter assertion consider the operator

$$\begin{pmatrix} r^+\Omega_0^1 \\ \\ r'T_0 \end{pmatrix} : H^1(X, L) \to \begin{matrix} H^1(X, L) \\ \oplus \\ H^1(Y, G) \end{matrix} \tag{31}$$

with

$$T_0 = Q_0\,(\sqrt{-\Delta_G'} - \mathrm{i}D_n)^{-1}\,, \tag{32}$$

$Q_0 = \sqrt[4]{-\Delta'_G}$ and prove that (31) is an isomorphism. Then we pass to (30) by comparing the different boundary conditions.

Lemma 11. *The operator* (31) *is an isomorphism.*

Proof: Let $v(y)$ be an eigenfunction of Δ'_G belonging to the eigenvalue $-\lambda^2, \lambda > 0$. Then (31) acts on products $v(y)\, w(x_n)$, $w \in \mathcal{S}(\overline{\mathbb{R}}_+)$, in the form

$$v(y)\, w(x_n) \to \begin{pmatrix} F_n^{-1}\Pi^+ \dfrac{\lambda + i\nu}{\lambda - i\nu} v(y)\, \hat{w}(\nu) \\ \Pi'(\lambda - i\nu)^{-1} Q^{-1}\, v(y)\, \hat{w}(\nu) \end{pmatrix}. \tag{33}$$

Thus, an operator

$$\mathfrak{a} = \begin{pmatrix} \Pi^+ \dfrac{\lambda + i\nu}{\lambda - i\nu} \\ \Pi'(\lambda - i\nu)^{-1} \end{pmatrix} : H^+ \to \begin{matrix} H^+ \\ \oplus \\ \mathbb{C} \end{matrix} \tag{34}$$

belonging to the eigenvalue λ is connected with (31). (34) looks like a boundary symbol. We shall show that the inverse of (31) can be obtained by calculating the inverse of (34).

The adjoint of $\mathfrak{a}$ is

$$\mathfrak{a}^* = \left(\Pi^+ \frac{\lambda - i\nu}{\lambda + i\nu},\ (\lambda + i\nu)^{-1}\right).$$

Thus, with $z = (\lambda - i\nu)\,(\lambda + i\nu)^{-1}$,

$$\mathfrak{a}\mathfrak{a}^* = \begin{pmatrix} 1 & 0 \\ 0 & c\lambda \end{pmatrix}. \tag{35}$$

Here we have used $\Pi^+ z^{-1} \Pi^+ z = 1$, $\Pi^+ z^{-1} (\lambda + i\nu)^{-1} = 0$, $\Pi'(\lambda - i\nu)^{-1}(\lambda + i\nu)^{-1} = c\lambda$ with a constant $c \neq 0$ and $\Pi'(\lambda - i\nu)^{-1}\Pi^+ z = \Pi'\Pi^-(\lambda + i\nu)^{-1} = 0$. Since $\lambda > 0$, (35) is an isomorphism and

$$(\mathfrak{a}\mathfrak{a}^*)^{-1} = \begin{pmatrix} 1 & 0 \\ 0 & (c\lambda)^{-1} \end{pmatrix}.$$

By $\mathfrak{a}^{-1} = \mathfrak{a}^*(\mathfrak{a}\mathfrak{a}^*)^{-1}$ we get

$$\mathfrak{a}^{-1} = \left(\Pi^+ z\,,\ (\lambda + i\nu)^{-1}\,(c\lambda)^{-1}\right).$$

Now we check that the operator

$$\left(r^+ \frac{\sqrt{-\Delta'_G} - iD_n}{\sqrt{-\Delta'_G} + iD_n},\ (\sqrt{-\Delta'_G} + iD_n)^{-1}\,(c\sqrt{-\Delta'_G})^{-1}\, Q_0^{-1}\right) \tag{36}$$

is in fact the inverse of (31). First remark that

$$r^+ \Omega^1_{0,L}(vw) = va(w)$$

for each $w \in \mathcal{S}(\overline{\mathbb{R}}_+)$ and a certain other $a(w) \in \mathcal{S}(\overline{\mathbb{R}}_+)$, so that the set of finite linear combinations $\sum_k v_k w_k$ is transformed into itself (v_k eigenfunction of Δ'_G belonging to $-\lambda_k^2$, $\lambda_k > 0$, $w_k \in \mathcal{S}(\overline{\mathbb{R}}_+)$). Moreover, we have

$$r^+ \frac{\sqrt{-\Delta'_G} - iD_n}{\sqrt{-\Delta'_G} + iD_n}(vw) = vb(w)$$

with $b(w) \in \mathcal{S}(\overline{\mathbb{R}}_+)$ and $b(a(w)) = w$. This is an immediate consequence of the rules of boundary symbolic calculus. Since the set of finite linear combinations $\sum_k v_k w_k$ is dense in $H^1(X, L)$ and because of the continuity of

$$r^+ \frac{\sqrt{-\Delta'_G} - \mathrm{i}D_n}{\sqrt{-\Delta'_G} + \mathrm{i}D_n} : H^1(X, L) \to H^1(X, L) , \tag{37}$$

(37) defines a right inverse of $r^+ \Omega^1_{0,L}$. Therefore, $r^+ \Omega^1_{0,L}$ is surjective.

Now let again v be an eigenfunction belonging to λ. Then the rules analogous to the boundary symbolic calculus show that the function

$$u = (\sqrt{-\Delta'_G} + \mathrm{i}D_n)^{-1} (c \sqrt{-\Delta'_G})^{-1} Q_0^{-1} v$$

belongs to ker $r^+ \Omega^1_{0,L}$ and that $r' T_0 u = v$. Therefore, since the finite linear combinations of eigenfunctions are dense in $H^1(Y, G)$ and because of the continuity of all operators under consideration we get surjectivity of $r' T_0$: ker $r^+ \Omega^1_{0,L} \to H^1(Y, G)$. Thus, (31) is surjective. Now apply (36) from the left to a vector $\big(r^+ \Omega^1_{0,L}(vw), r' T_0(vw)\big)$. Then we can again use the rules of boundary symbolic calculus and we find that the result is just vw, so that in view of the continuity of the composition and of the finite linear combinations $\sum_k v_k w_k$ being dense in $H^1(X, L)$, we get that (36) is the left inverse of (31). Thus (31) is injective and we have proved that (36) is the inverse of (31). □

Proof of Lemma 10: Multiply (36) from the left by (30). Then we get the operator

$$\begin{pmatrix} 1 & 0 \\ Qr' \circ r^+ \dfrac{\sqrt{-\Delta'_G} - \mathrm{i}D_n}{\sqrt{-\Delta'_G} + \mathrm{i}D_n} & Qr' (\sqrt{-\Delta'_G} + \mathrm{i}D_n)^{-1} (c \sqrt{-\Delta'_G})^{-1} Q_0^{-1} \end{pmatrix}. \tag{38}$$

The composition in the right upper corner is (except a constant factor) QQ_0^{-1}, which is an isomorphism. This follows again by the usual arguments of boundary symbols calculus. Thus (38) is an isomorphism. Since (36) is an isomorphism, the same is proved for (30). □

Remark 12.

$$\mathcal{W}^1_L = \begin{pmatrix} r^+ \Omega^1_L \\ r' \end{pmatrix} : H^1(X, L) \to \begin{matrix} H^1(X, L) \\ \oplus \\ H^{1/2}(Y, G) \end{matrix} \tag{39}$$

is an isomorphism for $\varepsilon > 0$ sufficiently small in the definition of (22).

The proof of Remark 12 is similar to that of Lemma 10. In order to sketch the idea first consider the boundary problem

$$\begin{pmatrix} r^+ \Omega^1_L \\ r' T_0 \end{pmatrix} : H^1(X, L) \to \begin{matrix} H^1(X, L) \\ \oplus \\ H^1(Y, G) \end{matrix} \tag{40}$$

with T_0 given by (32). Then we can define the inverse of the "boundary symbol"

$$\begin{pmatrix} \Pi^+ \dfrac{\chi(\lambda, \nu)\, \lambda + \mathrm{i}\nu}{\chi(\lambda, \nu)\, \lambda - \mathrm{i}\nu} \\ \Pi' (\lambda - \mathrm{i}\nu)^{-1} \end{pmatrix}. \tag{41}$$

This is possible for all $\lambda \geqq \lambda_0$, $\lambda_0 > 0$ fixed, since, for given $\delta > 0$, we find an $\varepsilon > 0$ with

$$\left|\frac{\chi(\lambda, \nu)\, \lambda + i\nu}{\chi(\lambda, \nu)\, \lambda - i\nu} - \frac{\lambda + i\nu}{\lambda - i\nu}\right| < \delta \tag{42}$$

for all $\lambda \geqq \lambda_0$. Denoting the inverse of (41) by

$$\left(\Pi^+\sigma(\lambda, \nu) + \Pi'\beta(\lambda, \nu),\ \varkappa(\lambda, \nu)\right) \tag{43}$$

with corresponding operator, Green and potential symbols σ, β and $\varkappa$, respectively, the operator

$$\left(r^+\sigma(A, D_n) + r'\beta(A, D_n),\ \varkappa(A, D_n)\right) \tag{44}$$

$\left(A = \sqrt{-\Delta'_G}\right)$ is inverse to (40). The proof of the latter fact is based on a calculation on products vw as in the proof of Lemma 11 and a continuity result in Sobolev spaces for the operators occurring in (44) which can be proved in the same way as Lemma 9. Then one can pass to (39) using analogous arguments as in the proof of Lemma 10.

We do not go into further details here, because below we will prove this result once more by approximating (29) by operators of type (39) for $\varepsilon \to 0$. The proof of Theorem 8 is finished if we show the following

Lemma 13. *There is a homotopy*

$$\mathscr{W}_L \simeq \mathscr{W}^1_{0,L} \tag{45}$$

through Fredholm operators in $H^1(X, L) \to H^1(X, L) \oplus H^{1/2}(Y, G)$.

Proof: In order to prove Lemma 13 we need further constructions and Lemmata. Choose functions $b, c \in C^\infty(\mathbb{R})$, $0 \leqq b \leqq 1$, $0 \leqq c \leqq 1$, $b^2 + c^2 = 1$, $ab = a$ (a is the function occurring in (8)), $b = 0$ for large x_n. We can suppose that Ω_L is given by

$$\Omega_L = b(\Omega^1_L)^{a(x_n)}\, b + c^2\,,$$

where $(\Omega^1_L)^{a(x_n)}$ is defined by means of

$$(\delta_\varepsilon(\xi) + i\nu)^a / (\delta_\varepsilon(\xi) - i\nu)^a \cdot 1_{\pi^*L}\,.$$

Define a first homotopy

$$\Omega^{(t)}_L = b(\Omega^1_L)^{a(x_n) + t(1 - a(x_n))}\, b + c(\Omega^1_L)^t\, c\,, \tag{46}$$

$0 \leqq t \leqq 1$. Then $\Omega^{(0)}_L = \Omega_L$, $\Omega^{(1)}_L - \Omega^1_L$ is a PDO of order -1. The second homotopy consists of an x_n-translation to the left

$$\Omega^{(t)}_L = \Omega^{(1)}_L(x_n - t + 1)\,, \tag{47}$$

$1 \leqq t \leqq t_2$, t_2 large enough, so that $\Omega^{(t_2)}_L$ is equal to Ω^1_L in an open neighbourhood of the half axis $x_n \geqq 0$. The third homotopy $\Omega^{(t)}_L$, $t_2 \leqq t \leqq t_2 + 1$, is defined by connecting the function χ in the definition of Ω^1_L with 1,

$$\chi^{(t)} = \chi_{\varepsilon(1 + t_2 - t)}\,. \tag{48}$$

Consider the elliptic boundary problems

$$\mathscr{W}^{(t)}_L = \begin{pmatrix} r^+\Omega^{(t)}_L \\ \\ r' \end{pmatrix} : H^1(X, L) \to \begin{matrix} H^1(X, L) \\ \oplus \\ H^{1/2}(Y, G) \end{matrix}\,, \tag{49}$$

$0 \leqq t \leqq t_2 + 1$. Then $\mathscr{W}^{(0)}_L = \mathscr{W}_L$, $\mathscr{W}^{(t_2+1)}_L = \mathscr{W}^1_{0,L}$.

The proof of Lemma 13 is finished if we show that $\mathcal{W}_L^{(t)}$ continuously depends on t in the operator norm topology and that all $\mathcal{W}_L^{(t)}$, $0 \leqq t \leqq t_2 + 1$, are Fredholm operators. The continuous dependence on t for $0 \leqq t \leqq t_2$ is clear, since we have a smooth deformation of symbols. The Fredholm property along the second homotopy is clear because the principal symbols remain invariant so that a parametrix of $\mathcal{W}^{(t_1)}$ is also a parametrix of $\mathcal{W}^{(t)}$, $1 \leqq t \leqq t_2$.

Next consider the first homotopy. Note that if we are given an x_n-translation invariant pseudo-differential operator P of degree 0 on $Y \times \mathbb{R}$ and if $c(x_n)$ is a C^∞ function on $\mathbb{R}$ which is constant for large x_n, the operator $cP - Pc$ induces a compact operator on $H^1(Y \times \mathbb{R}, L)$. The PDO (22) is x_n-translation invariant. Furthermore, Ω_L^1 induces an isomorphism

$$\Omega_L^1 \colon H^1(Y \times \mathbb{R}, L) \to H^1(Y \times \mathbb{R}, L) .$$

This can be proved by regarding Ω_L^1 as PDO on $\mathbb{R}$ with the operator-valued symbol $\big(\chi(A, \nu) A + i\nu\big) \big(\chi(A, \nu) A - i\nu\big)^{-1}$, where the inverse operator is just defined by the operator connected with the inverse symbol.

In order to show the Fredholm property of $\mathcal{W}_L^{(t)}$, $0 \leqq t \leqq 1$, we have to find a parametrix $H^1(Y \times \mathbb{R}_+, L) \oplus H^{1/2}(Y, G) \to H^1(Y \times \mathbb{R}_+, L)$ for $0 \leqq t \leqq 1$. The existence of parametrices $\mathcal{P}_L^{(t)}$ in C^∞ is clear because of the ellipticity of interior and boundary symbols of $\mathcal{W}_L^{(t)}$. Let b_1, c_1 be C^∞ functions on $\mathbb{R}$ with $b_1^2 + c_1^2 = 1$, $b_1 b = b$, $b_1 = 0$ for large x_n. Set

$$\mathcal{P}_0^{(t)} = b_1 \mathcal{P}_1^{(t)}(b_1 \oplus 1) + c_1(\Omega_L^{-t}, 0)\, (c_1 \oplus 1)$$

where $b_1 \oplus 1 = \begin{pmatrix} b_1 & 0 \\ 0 & 1 \end{pmatrix}$ and similarly $c_1 \oplus 1$. Then

$$\begin{aligned} \mathcal{J}_2 - \mathcal{A}^{(t)}\mathcal{P}_0^{(t)} &= (b_1^2 + c_1^2) \oplus 1 - \mathcal{A}^{(t)} b_1 \mathcal{P}_1^{(t)}(b_1 \oplus 1) - \mathcal{A}^{(t)} c_1(\Omega_L^{-t}, 0)\, (c_1 \oplus 1) \\ &= \big(b_1^2 \oplus 1 - \mathcal{A}^{(t)} b_1 \mathcal{P}_1^{(t)}(b_1 \oplus 1)\big) \\ &\quad + \big(c_1^2 \oplus 1 - \mathcal{A}^{(t)} c_1(\Omega_L^{-t}, 0)\, (c_1 \oplus 1)\big) \end{aligned}$$

is compact in $H^1(Y \times \mathbb{R}_+, L) \oplus H^{1/2}(Y, G)$ ($\mathcal{J}_2$ denotes the identity in this space) and

$$\begin{aligned} \mathcal{J}_1 - \mathcal{P}_0^{(t)}\mathcal{A}^{(t)} &= b_1^2 + c_1^2 - \mathcal{P}_0^{(t)}\mathcal{A}^{(t)} \\ &= \big(b_1^2 - b_1 \mathcal{P}_1^{(t)}(b_1 \oplus 1)\, \mathcal{A}^{(t)}\big) + \big(c_1^2 - c_1(\Omega_L^{-t}, 0)\, (c_1 \oplus 1)\, \mathcal{A}^{(t)}\big) \end{aligned}$$

is compact in $H^1(Y \times \mathbb{R}_+, L)$ ($\mathcal{J}_1$ the identity in this space). Therefore, $\mathcal{P}_0^{(t)}$ is a family of parametrices for $\mathcal{W}_L^{(t)}$ in the considered Sobolev spaces and the Fredholm property of the operators $\mathcal{W}_L^{(t)}$ in the first homotopy is proved. It remains to consider the third homotopy.

Note that if we use Remark 12, the third homotopy is not necessary for the proof of Theorem 8 because we have already proved $\mathcal{W}_L^1 \simeq \mathcal{W}_L = \mathcal{W}_L^{(t_3)}$ through Fredholm operators and ind $\mathcal{W}_L^1 = 0$ shows Theorem 8. But it is useful to give another proof.

Let A be a self-adjoint positively definite elliptic PDO on Y of order 1. Consider A as a densely defined unbounded operator $L^2(Y) \to L^2(Y)$ (with domain equal to $H^1(Y)$). It is easily checked that this is a closed operator with compact resolvent. Hence, by a well-known result about unbounded operators in a Hilbert space (cf. Dunford/Schwartz [1]) the spectrum consists of a countable number of points $\lambda_j \in \mathbb{C}$, which must be real and positive in view of the assumptions. Moreover, there is a complete orthonormal set in $L^2(Y)$ of smooth eigenfunctions $e_j \in C^\infty(Y)$. For any

$u \in L^2(Y)$ we have

$$||u||^2_{L^2(Y)} = \sum_j |(u, e_j)|^2$$

if $u = \sum_j (u, e_j)\, e_j$ is the Fourier decomposition. Since A is positive and self-adjoint, we get $\ker A = \{0\}$ and $\operatorname{coker} A = \ker A^* = \ker A = \{0\}$. Thus A defines an isomorphism $H^1(Y) \to L^2(Y)$ and, for any $s \in \mathbb{Z}_+$,

$$A^s\colon H^s(Y) \to L^2(Y)$$

is a continuous isomorphism.

Then a norm of $H^s(Y)$ can be defined by

$$||u||^2_{H^s(Y)} = ||A^s u||^2_{L^2(Y)} = \Big\| \sum_j (u, e_j)\, \lambda_j^s e_j \Big\|^2_{L^2(Y)} = \sum_j |(u, e_j)|^2\, \lambda_j^{2s}.$$

Thus we have proved

Lemma 14. *Let A be a self-adjoint positive definite elliptic PDO on Y of order 1. Then $H^s(Y)$ can be characterized as the completion of $C^\infty(Y)$ with respect to the norm*

$$||u||_s = \Big(\sum_j |(u, e_j)|^2\, \lambda_j^{2s}\Big)^{1/2}.$$

Lemma 15. *Let A be as in Lemma 14. Then*

$$B_\varepsilon \overset{\text{def}}{=} \frac{\chi_\varepsilon(A, D_n)\, A + \mathrm{i}D_n}{\chi_\varepsilon(A, D_n)\, A - \mathrm{i}D_n} - \frac{A + \mathrm{i}D_n}{A - \mathrm{i}D_n} \to 0 \quad \textit{if} \quad \varepsilon \to 0 \tag{50}$$

with respect to norm convergence of operators in $H^s(\mathbb{R}, H^t(Y))$, $s, t \in \mathbb{R}$.

Proof: Let $v_k(y)$ be the complete sequence of eigenfunctions of A belonging to the eigenvalues $\lambda_k > 0$, $k = 1, 2, \ldots$. Then for $w \in C_0^\infty(\mathbb{R})$, we get

$$F_n\{B_\varepsilon(v_k w)\,(y, \nu)\} = v_k(y) \left(\frac{\chi_\varepsilon(\lambda_k, \nu)\, \lambda_k + \mathrm{i}\nu}{\chi_\varepsilon(\lambda_k, \nu)\, \lambda_k - \mathrm{i}\nu} - \frac{\lambda_k + \mathrm{i}\nu}{\lambda_k - \mathrm{i}\nu}\right) \hat{w}(\nu). \tag{51}$$

Multiplication by $(1 + \nu^2)^{s/2}$ on both sides and integration gives

$$\begin{aligned} ||B_\varepsilon(v_k w)||^2_{H^s(\mathbb{R}, H^t(Y))} &= (2\pi)^{-1} \int (1 + \nu^2)^s\, ||F_n(B_\varepsilon(v_k w))\,(y, \nu)||^2_{H^t(Y)}\, \mathrm{d}\nu \\ &\leqq \delta ||v_k||^2_{H^t(Y)}\, ||w||^2_{H^s(\mathbb{R})} = \delta ||v_k w||^2_{H^s(\mathbb{R}, H^t(Y))} \end{aligned} \tag{52}$$

with δ independent of k as in (42). Now the v_k form an orthonormal system in $H^0(Y)$ and therefore

$$(v_j, v_k)_{H^t(Y)} = (A^t v_j, A^t v_k)_{H^0(Y)} = \lambda_j^t \lambda_k^t (v_j, v_k)_{H^0(Y)} = \lambda_j^{2t} \delta_{jk}.$$

Choosing an orthonormal system w_j in $H^s(\mathbb{R})$, the functions $v_k w_j$ form an orthogonal base in $H^s(\mathbb{R}, H^t(Y))$ and

$$\begin{aligned} \Big\| B_\varepsilon \Big(\sum_{k,j} v_k w_j\Big) \Big\|^2_{H^s((\mathbb{R}, H^t(Y))} &\leqq \sum_{k,j} ||B_\varepsilon(v_k w_j)||^2_{H^s(\mathbb{R}, H^t(Y))} \\ &\leqq \delta \sum_{k,j} ||v_k||^2_{H^t(Y)}\, ||w_j||^2_{H^s(\mathbb{R})} = \delta \sum_{k,j} ||v_k w_j||^2_{H^s(\mathbb{R}, H^t(Y))}. \end{aligned}$$

Since the finite sums of the form $\sum_{k,j} v_k w_j$ are dense in $H^s(\mathbb{R}, H^t(Y))$, the assertion is proved. □

Corollary 16. *Consider the operator*

$$r^+ B_\varepsilon j^+ : H_0^s(\overline{\mathbb{R}}_+, H^t(Y)) \to H^s(\overline{\mathbb{R}}_+, H^t(Y)). \tag{53}$$

Then $r^+ B_\varepsilon j^+ \to 0$ in the norm topology if $\varepsilon \to 0$.

Corollary 17. *The operator*

$$r^+ \frac{\chi_\varepsilon(A, D_n)\, A + \mathrm{i}D_n}{\chi_\varepsilon(A, D_n)\, A - \mathrm{i}D_n} : H_0^1(Y \times \overline{\mathbb{R}}_+) \to H^1(Y \times \mathbb{R}_+) \tag{54}$$

is an isomorphism if ε is sufficiently small.

Proof: Because of the proof of Lemma 10 we know that

$$r^+ \frac{A + \mathrm{i}D_n}{A - \mathrm{i}D_n} : H_0^1(Y \times \overline{\mathbb{R}}_+) \to H^1(Y \times \mathbb{R}_+)$$

is an isomorphism. Here $\ker r' = H_0^1(Y \times \overline{\mathbb{R}}_+)$ is used. Corollary 17 shows that then (54) is an isomorphism, too, since the isomorphisms form an open set in the operator norm topology. □

Lemma 18. *The operator*

$$r^+ B_\varepsilon j^+ : H^1(Y \times \mathbb{R}_+) \to H^1(Y \times \mathbb{R}_+)$$

tends to zero in the norm topology if $\varepsilon \to 0$.

Proof: The assertion follows in a similar way as Lemma 15. The only change is that one has to apply Π^+ in (51) and to use

$$\int \left| \Pi_\nu^+ \left(\frac{\chi_\varepsilon(\lambda, \nu)\, \lambda + \mathrm{i}\nu}{\chi_\varepsilon(\lambda, \nu)\, \lambda - \mathrm{i}\nu} - \frac{\lambda + \mathrm{i}\nu}{\lambda - \mathrm{i}\nu} \right) \right|^2 \mathrm{d}\nu \to 0$$

instead of (42) if $\varepsilon \to 0$ uniformly for $\lambda \geqq \lambda_0 > 0$. This convergence is easily checked and left to the reader. □

The last consideration applied to $A = \sqrt{-\Delta_G'}$ gives another proof of Remark 12. Thus the third homotopy in the proof of Lemma 13 is a homotopy through isomorphisms if ε is sufficiently small. Lemma 13 is proved. □

Thus the proof of Theorem 8 is finished, too.

The isomorphisms of the form (3) can be used for the investigation of elliptic boundary problems of the Douglis-Nirenberg type (cf. Agmon/Douglis/Nirenberg [1], Solonnikov [1]) by reducing the orders. This shows that Douglis-Nirenberg elliptic systems are of the usual type except factors being isomorphisms.

Let $p, q \geqq 0$ be integers and

$$E = \bigoplus_{j=1}^{p} E_j\,, \quad F = \bigoplus_{k=1}^{q} F_k\,, \quad J = \bigoplus_{k=1}^{q} J_k\,, \quad G = \bigoplus_{j=1}^{p} G_j$$

decompositions of the bundles $E, F \in \operatorname{Vect}(X)$ and $J, G \in \operatorname{Vect}(Y)$, respectively. Moreover, let $s_j, t_k \in \mathbb{Z}$, $\lambda_k, \gamma_j \in \mathbb{R}$ ($j = 1, \ldots, p$; $k = 1, \ldots, q$). Suppose that we are given operators $\mathcal{A}_{jk} \in \mathfrak{G}$,

$$\mathcal{A}_{jk}: H^{s+s_j}(X, E_j) \oplus H^{s+\lambda_k+1/2}(Y, J_k) \to H^{s+t_k}(X, F_k) \oplus H^{s-\gamma_j-1/2}(Y, G_j)$$

($s \in \mathbb{Z}_+$ sufficiently large, $1 \leqq j \leqq p$; $1 \leqq k \leqq q$).

Definition 19. The operator matrix

$$\mathcal{A} = (\mathcal{A}_{jk}): \begin{matrix} \bigoplus_{j=1}^{p} H^{s+s_j}(X, E_j) \\ \oplus \\ \bigoplus_{k=1}^{q} H^{s+\lambda_k+1/2}(Y, J_k) \end{matrix} \to \begin{matrix} \bigoplus_{k=1}^{q} H^{s+t_k}(X, F_k) \\ \oplus \\ \bigoplus_{j=1}^{p} H^{s-\gamma_j-1/2}(Y, G_j) \end{matrix} \tag{55}$$

is *elliptic of Douglis-Nirenberg type* if the operator matrix

$$\mathcal{A}^0 = (\mathcal{A}^0_{jk}): H^0(X, E) \oplus H^0(Y, J) \to H^0(X, F) \oplus H^0(Y, G)$$

with

$$\mathcal{A}^0_{jk} = (\mathcal{L}^{s+t_k, s-\gamma_j-1/2}_{F_k, G_j}) \mathcal{A}_{jk} (\mathcal{L}^{s+s_j, s+\lambda_k+1/2}_{E_j, J_k})^{-1}$$

is elliptic in the sense of 3.1.1.1, Definition 1.

Obviously all properties concerning the Fredholm property and the index of elliptic boundary problems are transformed to Douglis-Nirenberg elliptic systems. Therefore, we omit the details. Remark only the form of a-priori estimates. For each l, $m \in \mathbb{R}_+$ there exists a constant $c > 0$ so that for all $(u, v) \in \left(\bigoplus_j H^{s+s_j}(X, E_j)\right) \oplus \left(\bigoplus_k H^{s+\lambda_k+1/2}(Y, J_k)\right)$ and $(f, g) = \mathcal{A}(u, v)$ the following estimate holds

$$\sum_{j=1}^{p} \|u_j\|_{H^{s+s_j}(X, E_j)} + \sum_{k=1}^{q} \|v_k\|_{H^{s+\lambda_k+1/2}(Y, J_k)}$$

$$\leqq c \left\{ \sum_{k=1}^{q} \|f_k\|_{H^{s+t_k}(X, F_k)} + \sum_{j=1}^{p} \|g_j\|_{H^{s-\gamma_j-1/2}(Y, G_j)} \right.$$

$$\left. + \sum_{j=1}^{p} \|u_j\|_{H^{s+s_j-l}(X, E_j)} + \sum_{k=1}^{q} \|v_k\|_{H^{s+\lambda_k+1/2-m}(Y, J_k)} \right\}.$$

The u_j, v_k, g_j and f_k denote the components of u, v, g and f, respectively.

For general Sobolev spaces $H^{s,p}$ and Hölder spaces there are corresponding a-priori estimates for elliptic boundary value problems of Douglis-Nirenberg type (cf. 3.1.1.4).

3.1.2.2. Adjoints with Respect to Green Formulas

Let $\sigma_A: \pi^*E \to \pi^*F$ be an elliptic symbol over X and $\sigma_A^*: \pi^*F \to \pi^*E$ adjoint to σ_A with respect to fixed Hermitean metrics in E and F, respectively. Then $\sigma_A \in \mathfrak{A}^{(\alpha)}$ implies $\sigma_A^* \in \mathfrak{A}^{(\alpha)}$. Consider an elliptic $\mathcal{A} \in \mathfrak{G}$ with $\sigma_\Omega(\mathcal{A}) = \sigma_A$,

$$\mathcal{A}: H^{s_0}(X, E_0) \oplus H^{t_0}(Y, J_0) \to H^{s_1}(X, E_1) \oplus H^{t_1}(Y, J_1) \tag{1}$$

continuous, s_0, $s_1 \in \mathbb{Z}_+$. In this section we discuss the question how to construct an elliptic operator $\mathcal{A}^* \in \mathfrak{G}$ with $\sigma_\Omega(\mathcal{A}^*) = \sigma_A^*$,

$$\mathcal{A}^*: H^{s_1'}(X, E_1) \oplus H^{t_1'}(Y, G_1) \to H^{s_0'}(X, E_0) \oplus H^{t_0'}(Y, G_0) \tag{2}$$

continuous for suitable $s_i' \in \mathbb{Z}_+$, $t_i' \in \mathbb{R}$ and certain bundles $G_i \in \operatorname{Vect}(Y)$, $i = 0, 1$. The operator $\mathcal{A}^*$ will be adjoint of $\mathcal{A}$ with respect to a suitable non-degenerate bilinear form so that the Fredholm alternative is obtained. Set $G_i = J_i \oplus \left(\bigoplus_1^{s_i} E_i'\right)$ $(i = 0, 1)$. Choose some fixed $r \in \mathbb{Z}$ with $r - s_j \geqq 0$ $(j = 0, 1)$.

Lemma 1. *There exists an elliptic operator* $\mathcal{N}_i \in \mathfrak{G}$ *with* $\sigma_\Omega(\mathcal{N}_i) = \left(l^+(x,\xi)\right)^{-s_i} \cdot 1_{\pi^* E_i}$ *so that*

$$\mathcal{N}_i\colon H^{r-s_i}(X, E_i) \oplus H^{r-t_i}(Y, G_i) \to H^r(X, E_i) \oplus H^r(Y, J_i) \tag{3}$$

is an isomorphism $(i = 0, 1)$.

Proof: Consider the operator $\mathcal{L}^+_{E_i}\colon H^s(X, E_i) \to H^{s-1}(X, E_i) \oplus H^{s-1/2}(Y, E_i')$, $s > -1/2$ (cf. 3.1.2.1.(14)). Since $\operatorname{ind} \mathcal{L}^+_{E_i} = 0$, we can assume (after having added a smoothing operator) that $\mathcal{L}^+_{E_i}$ is an isomorphism. Then composition s_i times gives an isomorphism

$$\mathcal{M}_i\colon H^r(X, E_i) \to H^{r-s_i}(X, E_i) \oplus \left(\bigoplus_{k=1}^{s_i} H^{r-k+1/2}(Y, E_i')\right)$$

$(s - s_i + 1 > 1/2)$. Applying isomorphisms $H^{s-k+1/2}(Y, E_i') \to H^{r-t_i}(Y, E_i')$ in the form of elliptic pseudo-differential operators on Y of suitable orders to $\mathcal{M}_i$ we get an isomorphism $\mathcal{M}_i'\colon H^r(X, E_i) \to H^{r-s_i}(X, E_i) \oplus H^{r-t_i}\left(Y, \bigoplus_1^{s_i} E_i'\right)$. Choosing now an isomorphism $R\colon H^r(Y, J_i) \to H^{s-t_i}(Y, J_i)$ in the form of an elliptic pseudo-differential operator on Y, the inverse of $\mathcal{M}_i' \oplus R$ denoted by $\mathcal{N}_i$ has the asserted property. □

Denote Hermitean scalar products in the spaces $H^0(X, E_0) \oplus H^0(Y, J_0)$ and $H^0(X, E_1) \oplus H^0(Y, J_1)$ by φ and ψ, respectively. Introduce the operator

$$\mathcal{B} = \mathcal{L}^{s_1, t_1}_{E_1, J_1} \mathcal{A} (\mathcal{L}^{s_0, t_0}_{E_0, J_0})^{-1}\colon H^0(E, E_0) \oplus H^0(Y, J_0) \to H^0(X, E_1) \oplus H^0(Y, J_1) \tag{4}$$

$(\mathcal{A} \in \mathfrak{G}$ given by (1)$)$. Then we can define an operator $\mathcal{B}^* \in \mathfrak{G}$ by

$$\psi(\mathcal{B}u, v) = \varphi(u, \mathcal{B}^* v) \tag{5}$$

for all $u \in H^0(X, E_0) \oplus H^0(Y, J_0)$, $v \in H^0(X, E_1) \oplus H^0(Y, J_1)$ and $\sigma_\Omega(\mathcal{B}^*) = \left(\sigma_\Omega(\mathcal{B})\right)^*$, $\psi'\left(\sigma_Y(\mathcal{B})\, h, l\right) = \varphi'\left(h, \sigma_Y(\mathcal{B}^*)\, l\right)$ for all $h \in p^*(E_0^+ \oplus J_0)_{(x', \xi')}$, $l \in p^*(E_1^+ \oplus J_1)_{(x', \xi')}$, where ψ' and φ' denote the scalar products in the fibres of $p^*(E_1^+ \oplus J_1)$ and $p^*(E_0^+ \oplus J_0)$ induced by ψ and φ, respectively (cf. 2.3.3.2). The ellipticity of $\mathcal{A}$ implies that of $\mathcal{B}$ and $\mathcal{B}^*$ and obviously $\operatorname{ind} \mathcal{B}^* = -\operatorname{ind} \mathcal{B}$.

From $\sigma_\Omega(\mathcal{B}) = \left(l^-(x,\xi)\right)^{s_1} \cdot 1_{\pi^* E_1} \sigma_\Omega(\mathcal{A}) \left(l^-(x,\xi)\right)^{-s_0} \cdot 1_{\pi^* E_0}$ it follows that

$$\sigma_\Omega(\mathcal{B}^*) = \left(l^+(x,\xi)\right)^{-s_0} \cdot 1_{\pi^* E_0} \sigma_\Omega^*(\mathcal{A}) \left(l^+(x,\xi)\right)^{s_1} \cdot 1_{\pi^* E_1}. \tag{6}$$

Define the "adjoint" of $\mathcal{A}$ by $\mathcal{A}^* = \mathcal{N}_0^{-1} \mathcal{B}^* \mathcal{N}_1$. It is continuous as an operator

$$\mathcal{A}^*\colon H^{r-s_1}(X, E_1) \oplus H^{r-t_1}(Y, G_1) \to H^{r-s_0}(X, E_0) \oplus H^{r-t_0}(Y, G_0) \tag{7}$$

and Fredholm, since the ellipticity of $\mathcal{A}$ implies that of $\mathcal{A}^*$. Note that $\sigma_\Omega(\mathcal{A}^*) = \left(\sigma_\Omega(\mathcal{A})\right)^*$ and $\operatorname{ind} \mathcal{A}^* = -\operatorname{ind} \mathcal{A}$ (because $\operatorname{ind} \mathcal{B} = \operatorname{ind} \mathcal{A}$ and $\operatorname{ind} \mathcal{B}^* = \operatorname{ind} \mathcal{A}^*$). Moreover, it is easily seen that the definition of $\mathcal{A}^*$ is compatible with compositions, i.e. $(\mathcal{A}_0 \mathcal{A}_1)^* = \mathcal{A}_0^* \mathcal{A}_1^*$. For operators of the form $\mathcal{A} = 1 \oplus L\colon H^s(X, E) \oplus H^{t_0}(Y, J_0) \to H^s(X, E) \oplus H^{t_1}(Y, J_1)$ it follows that $\mathcal{A}^* = 1 \oplus M$ and M is, except for lower order terms, the usual adjoint of L with respect to Hermitean scalar products in $H^0(Y, J_i)$ $(i = 0, 1)$.

Proposition 2. *Let* $\sigma \in \mathfrak{A}^{(\alpha)}$ $(\alpha \in Z)$ *be elliptic,* $\sigma\colon \pi^* E \to \pi^* F$ *and* $\operatorname{ind}_{S^* Y}(\Pi^+ \sigma) \in p^* K(Y)$. *Then*

$$\operatorname{ind}_{S^* Y}(\Pi^+ \sigma^*) = -\operatorname{ind}_{S^* Y}(\Pi^+ \sigma) + \alpha [p^* E']. \tag{8}$$

Proof: First choose an elliptic operator $\mathcal{A} \in \mathfrak{G}$ with $\sigma_\Omega(\mathcal{A}) = \sigma$ (cf. 3.1.1.1, Theorem 12). With the notation $J_0 = J$, $J_1 = G$ from 3.1.1.1.(20) we get $\mathrm{ind}_{S^*Y} \Pi^+\sigma = [p^*J_1] - [p^*J_0]$ and because of the definition of $\mathcal{A}^*$, $\mathrm{ind}_{S^*Y} \Pi^+\sigma^* = [p^*G_0] - [p^*G_1] = [p^*J_0] - [p^*J_1] + s_0[p^*E_0'] - s_1[p^*E_1'] = -\mathrm{ind}_{S^*Y} \Pi^+\sigma + \alpha[p^*E']$. In the latter equation $\alpha = s_0 - s_1$ and $E' = E_0' \cong E_1'$ were used (cf. the remarks at the end of 3.1.1.1). □

Remark 3. For abbreviation let $\mathcal{L}_j = \mathcal{L}^{s_j, t_j}_{E_j', J_j}$ $(j = 0, 1)$. Then

$$\psi(\mathcal{L}_1\mathcal{A}f, \mathcal{N}_1 g) = \varphi(\mathcal{L}_0 f, \mathcal{N}_0\mathcal{A}^* g) \tag{9}$$

for all $f \in C^\infty(X, E_0) \oplus C^\infty(Y, J_0)$, $g \in C^\infty(X, E_1) \oplus C^\infty(Y, G_1)$.

(9) obviously follows from $\psi(\mathcal{B}u, v) = (u, \mathcal{B}^*v)$ for $u = \mathcal{L}_0 f$, $v = \mathcal{N}_1 g$. Since the operators contained in $\mathcal{N}_j$ $(j = 0, 1)$ have non-positive types, the adjoint can be defined. Then (9) yields

$$\chi(\mathcal{N}_1^*\mathcal{L}_1\mathcal{A}f, g) = \omega(\mathcal{N}_0^*\mathcal{L}_0 f, \mathcal{A}^* g) \tag{10}$$

with the scalar products χ and ω in the spaces $H^0(X, E_1) \oplus H^0(Y, G_1)$ and $H^0(X, E_0) \oplus H^0(Y, G_0)$, respectively. The formula (10) can be considered as a certain analogue of the classical Green formula (cf. 3.1.1.3.). The bilinear forms

$$\chi_0(h, g) = \chi(\mathcal{N}_1^*\mathcal{L}_1 h, g),\ \omega_0(f, m) = \omega(\mathcal{N}_0^*\mathcal{L}_0 f, m)\,,$$

$g \in C^\infty(X, E_1) \oplus C^\infty(Y, G_1)$, $h \in C^\infty(X, E_1) \oplus C^\infty(Y, J_1)$, $f \in C^\infty(X, E_0) \oplus C^\infty(Y, J_0)$, $m \in C^\infty(X, E_0) \oplus C^\infty(Y, G_0)$ are non-degenerate and (10) means

$$\chi_0(\mathcal{A}f, g) = \omega_0(f, \mathcal{A}^* g) \tag{11}$$

for all smooth f and g. Thus we obtain a Fredholm alternative, namely

$$\ker \mathcal{A} \perp_{(\omega_0)} \mathrm{im}\, \mathcal{A}^*,\ \mathrm{im}\, \mathcal{A} \perp_{(\chi_0)} \ker \mathcal{A}^*\,. \tag{12}$$

Here $\perp_{(\ldots)}$ denotes orthogonality with respect to the bilinear form in the brackets. The operators are taken in spaces of smooth sections.

Note that of course $\psi(\mathcal{L}_1 h, \mathcal{N}_1 g) = \chi_0(h, g)$, $\varphi(\mathcal{L}_0 f, \mathcal{N}_0 m) = \omega_0(f, m)$ so that the adjoints of $\mathcal{N}_j$ in the definition of the bilinear forms χ_0 and ω_0 are not needed.

Finally remark that the definition of the operator $\mathcal{A}^*$ depending on $\mathcal{A} \in \mathfrak{G}$ is not restricted to elliptic $\mathcal{A}$. The operator $\mathcal{A} \in \mathfrak{G}$ can be quite arbitrary.

3.1.2.3. Over- and Underdetermined Elliptic Boundary Problems

The concept of ellipticity of boundary value problems can be generalized if one speaks about injectivity or surjectivity of interior and boundary symbols. The latter condition can be called the corresponding one-sided (left and right, respectively) Shapiro-Lopatinski condition. Operators of this type occur in theoretical physics and in the consideration of elliptic complexes on manifolds with boundary (cf. 3.2.3).

Definition 1. An operator

$$\mathcal{A}\colon C^\infty(X, E) \oplus C^\infty(Y, J) \to C^\infty(X, F) \oplus C^\infty(Y, G) \tag{1}$$

is called an *overdetermined* (*underdetermined*) *elliptic boundary value problem* if both $\sigma_A = \sigma_\Omega(\mathcal{A})\colon \pi^*E \to \pi^*F$ $(\pi\colon T^*X \setminus 0 \to X)$ and $\sigma_Y(\mathcal{A})\colon p^*(E^+ \oplus J) \to p^*(F^+ \oplus G)$ are injective (surjective). For abbreviation, we speak about *over-* and *underdetermined elliptic systems.*

For instance, if $\mathcal{A} \in \mathfrak{G}$ is an elliptic operator, the first row of operators $(r^+A + r'B, K)$ is an underdetermined and the first column of operators consisting of $(r^+A + r'B,$

$r'T$) is an overdetermined elliptic system in the sense of Definition 1. A further example of an overdetermined system in the sense of Definition 1 is the following problem

$$\operatorname{rot} v = f, \quad \operatorname{div} v = \varrho , \tag{2}$$

$$(n_1 v_1 + n_2 v_2 + n_3 v_3)|_{\partial\Omega} = g , \tag{3}$$

considered in a smoothly bounded domain $\Omega \subset \mathbb{R}^3$, $v = (v_1, v_2, v_3)$, $f = (f_1, f_2, f_3) \in C^\infty(\bar{\Omega}, \mathbb{C}^3)$, $g \in C^\infty(Y)$, (n_1, n_2, n_3) the vector of the interior normal to $Y = \partial\Omega$. This and further systems being of interest in theoretical physics of Douglis-Nirenberg type, too, are considered in SOLONNIKOV [2]. Note that the analogue of Douglis-Nirenberg ellipticity in the over- and underdetermined case can be treated in a similar way as for the usual ellipticity by the reducing of orders (cf. 3.1.2.1).

In this section we show among other things the existence of corresponding one-sided parametrices of over- and underdetermined elliptic systems. The necessity of the one-sided Shapiro-Lopatinski condition was proved in HÖPPNER [1]. By the way the necessity is an immediate consequence of the corresponding result for ordinary elliptic $\mathcal{A} \in \mathfrak{G}$.

Similarly as in 3.1.1.1 we use the notations $\mathcal{H}_1 = H^s(X, E) \oplus H^{t+\lambda+1/2}(Y, J)$, $\mathcal{H}_2 = H^t(X, F) \oplus H^{s-\nu-1/2}(Y, G)$, $t = s - \alpha$, and $\mathcal{H}_0 = H^p(X, E) + H^q(Y, J)$ with arbitrary fixed $p < s$, $q < t + \lambda + 1/2$. The norms in the corresponding spaces are denoted by $\|\cdot\|_j$ $(j = 0, 1, 2)$. As usual the orders in $\mathcal{A}$ are fixed in such a way that $\mathcal{A}: \mathcal{H}_1 \to \mathcal{H}_2$ is continuous.

Theorem 2. *Let $\mathcal{A} \in \mathfrak{G}$ be an overdetermined elliptic system. Then there exists an $\mathcal{L} \in \mathfrak{G}$ which is a left parametrix of $\mathcal{A}$, i.e. $\mathcal{L}\mathcal{A} - \mathcal{I} \in \mathfrak{G}^{-\infty}$. Moreover $\mathcal{L}$ is an underdetermined elliptic system and each left parametrix $\in \mathfrak{G}$ of $\mathcal{A}$ has this property. If $\mathcal{A}$ is injective, there exists a left inverse $\mathcal{L} \in \mathfrak{G}$.*

Proof: From the symbolic calculus of operators in $\mathfrak{G}$ it follows that

$$\mathcal{B} = \mathcal{L}_{F,G}^{t,s-\nu-1/2} \mathcal{A} (\mathcal{L}_{E,J}^{s,t+\lambda+1/2})^{-1} \tag{4}$$

is an overdetermined elliptic system (cf. 3.1.2.1.(3)). Then $\mathcal{B}^* \in \mathfrak{G}$ exists, and $\mathcal{B}^*$ is obviously an underdetermined elliptic system. Symbolic calculus shows ellipticity of the operator $\mathcal{B}^*\mathcal{B} \in \mathfrak{G}$. Let $(\mathcal{B}^*\mathcal{B})^{-1}$ be a parametrix of $\mathcal{B}^*\mathcal{B}$. Then $\mathcal{M} = (\mathcal{B}^*\mathcal{B})^{-1}\mathcal{B}^*$ is underdetermined elliptic and a left parametrix of $\mathcal{B}$. Thus

$$\mathcal{L} = (\mathcal{L}_{E,J}^{s,t+\lambda+1/2})^{-1} \mathcal{M} \mathcal{L}_{F,G}^{t,s-\nu-1/2} \tag{5}$$

is a left parametrix of $\mathcal{A}$ which is of course an underdetermined elliptic system, too.

Suppose now that $\mathcal{A}: \mathcal{H}_1 \to \mathcal{H}_2$ is injective and $\mathcal{P}: \mathcal{H}_2 \to \mathcal{H}_1$ a left parametrix of $\mathcal{A}$ in $\mathfrak{G}$, i.e. $\mathcal{P}\mathcal{A} = \mathcal{I} + \mathcal{K}_1$ for some $\mathcal{K}_1 \in \mathfrak{G}^{-\infty}$. Then there exists an operator $\mathcal{K}_2 \in \mathfrak{G}^{-\infty}$ so that $(\mathcal{I} + \mathcal{K}_2)\mathcal{P}\mathcal{A} = \mathcal{I} + \mathcal{K}_0$ with some $\mathcal{K}_0 \in \mathfrak{G}^{-\infty}$ of finite rank. $\mathcal{P}_0 = (\mathcal{I} + \mathcal{K}_2)\mathcal{P}$ is a left parametrix of $\mathcal{A}$, too. Define a decomposition $\mathcal{H}_1 = \ker \mathcal{K}_0 \oplus \mathcal{V}$, $\mathcal{V} \subset C^\infty(X, E) \oplus C^\infty(Y, J)$ some finite dimensional subspace. $\mathcal{A}|_{\ker \mathcal{K}_0}: \ker \mathcal{K}_0 \to \mathcal{A}(\ker \mathcal{K}_0)$ is injective and $\mathcal{P}' = \mathcal{P}_0|_{\mathcal{A}(\ker \mathcal{K}_0)}$ is a left inverse to $\mathcal{A}|_{\ker \mathcal{K}_0}$. The operator $\mathcal{A}|_{\mathcal{V}}: \mathcal{V} \to \mathcal{H}_2$ defines an isomorphism $\mathcal{V} \to \mathcal{W}$ for some finite dimensional subspace $\mathcal{W} \subset C^\infty(X, F) \oplus C^\infty(Y, G)$. The inverse of this isomorphism can be generated by some $\mathcal{K} \in \mathfrak{G}^{-\infty}$ vanishing on a suitable complement of $\mathcal{W}$. Then $\mathcal{L} = \mathcal{P}' \oplus \mathcal{K}$ is a left inverse of $\mathcal{A}$. □

Corollary 3. *Let $\mathcal{A} \in \mathfrak{G}$, $\mathcal{A}: \mathcal{H}_1 \to \mathcal{H}_2$ be an overdetermined elliptic system. Then there exists a constant $c > 0$ so that*

$$||u||_1 \leq c(||\mathcal{A}u||_2 + ||u||_0) \tag{6}$$

for all $u \in \mathcal{H}_1$, moreover $\dim \ker \mathcal{A} < \infty$, $\ker \mathcal{A} \subset C^\infty(X, E) \oplus C^\infty(Y, J)$ *and* $\dim \ker \mathcal{A}$ *is independent of s in the definition of the space $\mathcal{H}_1$.*

Proof: (6) follows similarly as 3.1.1.1, (8). The assertion about $\ker \mathcal{A}$ is a consequence of the ellipticity of $\mathcal{L}\mathcal{A}$ ($\mathcal{L}$ left parametrix of $\mathcal{A}$) and $\ker \mathcal{A} \subseteq \ker \mathcal{L}\mathcal{A}$. □

Let $\mathcal{L}$ be a left parametrix of the overdetermined elliptic operator $\mathcal{A}$ and $\mathcal{N} \in \mathfrak{G}$ an operator with $\ker \mathcal{N} \supseteq \operatorname{im} \mathcal{A}$. Then $\mathcal{L} + \mathcal{N}$ is a left parametrix of $\mathcal{A}$, too.

On symbolic level we have the following remark. For each left inverse of $(\sigma_\Omega(\mathcal{A}), \sigma_Y(\mathcal{A}))$ denoted by $(\sigma_\Omega(\mathcal{L}), \sigma_Y(\mathcal{L}))$ we find a corresponding left parametrix $\mathcal{L}$. Namely, if $\mathcal{L}_0$ is an operator in $\mathfrak{G}$ with $\sigma_\Omega(\mathcal{L}_0) = \sigma_\Omega(\mathcal{L})$, $\sigma_Y(\mathcal{L}_0) = \sigma_Y(\mathcal{L})$, we have negative orders of all operators contained in $\mathcal{L}_0\mathcal{A} - \mathcal{J}$. Thus $\mathcal{L} \overset{\text{def}}{=} \sum_{k=0}^{\infty} (-1)^k (\mathcal{L}_0\mathcal{A} - \mathcal{J})^k \mathcal{L}_0$ (asymptotic sum) is a left parametrix. Note that $\mathcal{L}_0$ is a left Sobolev space parametrix of $\mathcal{A}$. The construction of $\mathcal{L}$ by starting with left inverses of the symbols is another construction than given in Theorem 2. But two pairs of left inverses $(\sigma_\Omega(\mathcal{L}_i), \sigma_Y(\mathcal{L}_i))$ $(i = 1, 2)$ of $(\sigma_\Omega(\mathcal{A}), \sigma_Y(\mathcal{A}))$ have obviously the property $(\sigma_\Omega(\mathcal{L}_1) - \sigma_\Omega(\mathcal{L}_2))\, \sigma_\Omega(\mathcal{A}) = 0$, $(\sigma_Y(\mathcal{L}_1) - \sigma_Y(\mathcal{L}_2))\, \sigma_Y(\mathcal{A}) = 0$. In general, the difference $\mathcal{L}_1 - \mathcal{L}_2$ of two left parametrices of $\mathcal{A}$ is not compact.

Similarly as Theorem 2 we can prove

Theorem 4. *Let $\mathcal{A} \in \mathfrak{G}$ be an underdetermined elliptic system. Then there exists an $\mathcal{R} \in \mathfrak{G}$ which is a right parametrix of $\mathcal{A}$, i.e. $\mathcal{A}\mathcal{R} - \mathcal{J} \in \mathfrak{G}^{-\infty}$. Moreover $\mathcal{R}$ is an overdetermined elliptic system and each right parametrix $\in \mathfrak{G}$ of $\mathcal{A}$ has this property. If $\mathcal{A}$ is surjective, there exists a right inverse $\mathcal{R} \in \mathfrak{G}$.*

Corollary 5. *Let $\mathcal{A} \in \mathfrak{G}$, $\mathcal{A}: \mathcal{H}_1 \to \mathcal{H}_2$ be an underdetermined elliptic system. Then there exists a subspace $V \subset C^\infty(X, F) \oplus C^\infty(Y, G)$ with* $\dim V < \infty$, $V \oplus \operatorname{im} \mathcal{A} = \mathcal{H}_2$.

This follows from the ellipticity of $\mathcal{A}\mathcal{R}$ for each right parametrix $\mathcal{R}$ of $\mathcal{A}$ and $\operatorname{im} \mathcal{A} \supseteq \operatorname{im} \mathcal{A}\mathcal{R}$ (cf. 3.1.1.1, Theorem 5).

If $\mathcal{R}$ is a right parametrix of an underdetermined elliptic $\mathcal{A}$, it is clear that $(\sigma_\Omega(\mathcal{R}), \sigma_Y(\mathcal{R}))$ is a right inverse to $(\sigma_\Omega(\mathcal{A}), \sigma_Y(\mathcal{A}))$. Similarly as in the overdetermined case one can construct a right parametrix starting with right inverses and taking an asymptotic sum.

The properties of overdetermined and underdetermined elliptic systems are dual to each other in some sense. For instance, taking adjoints in the sense of 3.1.2.2 the adjoint of an overdetermined elliptic operator is underdetermined elliptic and conversely.

The proof of the following simple remark is left to the reader.

Remark 6. *Let $\mathcal{A}$ be an underdetermined elliptic system. Then there exists a right parametrix $\mathcal{R}$ of $\mathcal{A}$ with* $\operatorname{im} \mathcal{R} = (\ker \mathcal{A})^\perp$. *Let $\mathcal{A}$ be an overdetermined elliptic system. Then there exists a left parametrix $\mathcal{L}$ of $\mathcal{A}$ with* $\operatorname{im} \mathcal{A} = (\ker \mathcal{L})^\perp$.

Note that if $\mathcal{A}: \mathcal{H}_1 \to \mathcal{H}_2$ is surjective and $\mathcal{R}: \mathcal{H}_2 \to \mathcal{H}_1$ a right inverse of $\mathcal{A}$, the operator $\mathcal{J} - \mathcal{R}\mathcal{A}$ is a projection to $\ker \mathcal{A}$. Similarly, if $\mathcal{A}: \mathcal{H}_1 \to \mathcal{H}_2$ is injective and $\mathcal{L}$ a left inverse of $\mathcal{A}$, the operator $\mathcal{J} - \mathcal{A}\mathcal{L}$ is a projection to a complement of $\operatorname{im} \mathcal{A}$.

For the construction of overdetermined and underdetermined elliptic systems the following assertion is useful.

Lemma 7. *Let $\sigma \in \mathfrak{A}^{(\alpha)}$, $\sigma: \pi^*E \to \pi^*F$ be injective. Then for each Green symbol β*

$$\dim \ker_{(x',\xi')} (\Pi^+\sigma + \Pi'\beta) < \infty \quad \textit{for all } (x',\xi') \in S^*Y \,. \tag{7}$$

Similarly, if $\sigma \in \mathfrak{A}^{(\alpha)}$, $\sigma: \pi^*E \to \pi^*F$ is surjective,

$$\dim \operatorname{coker}_{(x',\xi')} (\Pi^+\sigma + \Pi'\beta) < \infty \quad \textit{for all } (x',\xi') \in S^*Y \tag{8}$$

for each Green symbol β. The dimensions in (7), (8) *are uniformly bounded for all $(x',\xi') \in S^*Y$.*

Proof: Let us show (7). The proof of (8) is analogous. If σ is injective, there exists a left inverse $\sigma^{-1} \in \mathfrak{A}^{(-\alpha)}$. One can set $\sigma^{-1} = (\sigma^*\sigma)^{-1}\sigma^*$, where σ^* is adjoint to σ with respect to Hermitean metrics in E and F. It is sufficient to show that each operator family of the form $\Pi^+\sigma^{-1} + \Pi'\gamma$ for an arbitrary Green symbol γ is a left parametrix of $\Pi^+\sigma + \Pi'\beta$. This is clear by the results in 2.1.2. We have $\Pi^+\sigma^{-1}\Pi^+\sigma = 1 + \Pi'\beta_0$, $\Pi'\gamma\Pi^+\sigma = \Pi'\beta_1$, $\Pi^+\sigma^{-1}\Pi'\beta = \Pi'\beta_2$, $\Pi'\gamma\Pi'\beta = \Pi'\beta_3$ with Green symbols β_j $(j = 0, \ldots, 3)$. The operators $\Pi'\beta_j$ are compact. $\mathfrak{a} = (\Pi^+\sigma^{-1} + \Pi'\gamma)(\Pi^+\sigma + \Pi'\beta)$ is a continuous Fredholm family over S^*Y and the dimension of $\ker \mathfrak{a}$ is uniformly bounded. Since $\ker (\Pi^+\sigma + \Pi'\beta) \subseteq \ker \mathfrak{a}$, the assertions are proved. □

Proposition 8. *Let $\sigma \in \mathfrak{A}^{(\alpha)}$, $\sigma: \pi^*E \to \pi^*F$, be injective. Then there exists an overdetermined elliptic system $\mathcal{A} \in \mathfrak{G}$ with $\sigma = \sigma_\Omega(\mathcal{A})$. Similarly, if σ is surjective, there exists an underdetermined elliptic system $\mathcal{A} \in \mathfrak{G}$ with $\sigma = \sigma_\Omega(\mathcal{A})$.*

Proof: It is sufficient to consider the case in which σ is surjective. Namely, if σ is injective, the adjoint σ^* is surjective. Taking some underdetermined elliptic $\mathcal{A}^* \in \mathfrak{G}$ with $\sigma^* = \sigma_\Omega(\mathcal{A}^*)$ we get by the adjoint (defined as in 3.1.2.2) some overdetermined $\mathcal{A} \in \mathfrak{G}$ with $\sigma = \sigma_\Omega(\mathcal{A})$. Thus let σ be surjective. Then, similarly as in 3.1.1.2 we find a subbundle $\tilde{J} \subset p^*F^+$ with $\tilde{J}_{(x',\xi')} + \operatorname{im} \Pi^+\sigma_{(x',\xi')} = (p^*F^+)_{(x',\xi')}$ for all $(x',\xi') \in S^*Y$ and a bundle $J \in \operatorname{Vect}(Y)$ so that $p^*J \cong \tilde{J}$. The choice of an isomorphism gives rise to a potential symbol σ_K and $(\Pi^+\sigma, \sigma_K): (p^*E^+) \oplus (p^*J) \to p^*F^+$ is surjective. Now we find an $\mathcal{A} \in \mathfrak{G}$ with $\sigma_\Omega(\mathcal{A}) = \sigma$ and $\sigma_Y(\mathcal{A}) = (\Pi^+\sigma, \sigma_K)$. □

The proof of Proposition 8 shows that for every elliptic symbol σ there always exist over- and underdetermined elliptic boundary problems $\mathcal{A} \in \mathfrak{G}$ with $\sigma_\Omega(\mathcal{A}) = \sigma$ even if no elliptic problem for σ exists (cf. 3.1.1.1, Theorem 12). This is the case e.g. for the Cauchy-Riemann operator in the unit disc in the complex plane.

Remark 9. *Let $r^+A: C^\infty(X, E) \to C^\infty(X, F)$ be an elliptic differential operator. Then $\mathcal{A} = (r^+A)$ is an underdetermined elliptic system, i.e.* $\dim \operatorname{coker} (r^+A) < \infty$.

For the proof it is sufficient to use $\Pi^+\sigma_A \circ \Pi^+\sigma_A^{-1} = 1$ since σ_A is a polynomial in ν. Thus, if r^+A is an elliptic differential operator for which the condition 3.1.1.1. (21) is satisfied, there always exists an elliptic $\mathcal{A} \in \mathfrak{G}$ with $\sigma_A = \sigma_\Omega(\mathcal{A})$, where $\mathcal{A}$ contains no potential conditions K and a pseudo-differential operator Q on Y.

Note that $\Pi^+\sigma_A$ always has a right inverse if σ_A is elliptic and belongs to H_ν^- for each fixed $(x',\xi') \in S^*Y$.

If $\mathcal{A}$ is an overdetermined elliptic system, one can ask for integrability conditions for the right hand side of $\mathcal{A}u = f$. This question is connected with the question when $\mathcal{A}$ can be considered as a first operator in a Fredholm complex (cf. 3.2.3 and SCHULZE [6]).

3.1.2.4. Generalization of the Calderon-Seeley Projectors

The Calderon-Seeley projectors mentioned in 3.1.1.3 are useful for an interpretation of the Shapiro-Lopatinski condition. They have an analogy for general operators in $\mathfrak{G}$ and will be discussed in this section.

Consider an operator $\mathcal{A}_0 \in \mathfrak{G}$ of the form

$$\mathcal{A}_0 = (r^+A + r'B, K): \begin{array}{c} C^\infty(X, E) \\ \oplus \\ C^\infty(Y, J) \end{array} \to C^\infty(X, F) \tag{1}$$

and operators $\mathcal{K}_0, \mathcal{T}_1 \in \mathfrak{G}$ of the form

$$\mathcal{K}_0 = \begin{pmatrix} K_0 \\ \\ Q_0 \end{pmatrix}: C^\infty(Y, \mathbb{C}^N) \to \begin{array}{c} C^\infty(X, E) \\ \oplus \\ C^\infty(Y, J) \end{array}, \tag{2}$$

$$\mathcal{T}_1 = (r'T_1, Q_1): \begin{array}{c} C^\infty(X, E) \\ \oplus \\ C^\infty(Y, J) \end{array} \to C^\infty(Y, \mathbb{C}^N). \tag{3}$$

If $\mathcal{A}_0$ is surjective and $\mathcal{R}$ a right inverse of $\mathcal{A}_0$, the operator $\mathcal{I} - \mathcal{R}\mathcal{A}_0$ is a projector to $\ker \mathcal{A}_0$. Note that for elliptic $\sigma_A: \pi^*E \to \pi^*F$ in $\mathfrak{A}^{(\alpha)}$ there always exists a surjective operator $\mathcal{A}_0$ of the form (1) with $\sigma_A = \sigma_\Omega(\mathcal{A}_0)$.

Theorem 1. *Let $\sigma_A: \pi^*E \to \pi^*F$ be an elliptic symbol in $\mathfrak{A}^{(\alpha)}$ and $\mathcal{A}_0 \in \mathfrak{G}$ a surjective operator of the form (1) with $\sigma_A = \sigma_\Omega(\mathcal{A}_0)$. Then there exist some $N \in \mathbb{Z}_+$ and operators $\mathcal{K}_0$ and $\mathcal{T}_1$ in $\mathfrak{G}$ of the type (2) and (3), respectively, so that $\mathcal{T}_1\mathcal{K}_0$ and $\mathcal{K}_0\mathcal{T}_1$ are projections where $\mathcal{K}_0\mathcal{T}_1 = \mathcal{I} - \mathcal{R}\mathcal{A}_0$ for some right inverse $\mathcal{R}$ of $\mathcal{A}_0$. Moreover $\mathcal{A}_0\mathcal{K}_0 = 0$.*

Proof: Let $\mathcal{R}' \in \mathfrak{G}$ be a right inverse of $\mathcal{A}_0$ (cf. 3.1.2.3, Theorem 4) and $\mathcal{P}' = \mathcal{I} - \mathcal{R}'\mathcal{A}_0$. Since $\mathcal{P}'$ contains only a Green operator in the left upper corner, $\mathcal{P}'$ can be written in the form

$$\mathcal{P}' = \mathcal{K}'\mathcal{T}_1 + \mathcal{C} \tag{4}$$

with operators

$$\mathcal{T}_1 = (r'T_1, Q_1): \begin{array}{c} C^\infty(X, E_1) \\ \oplus \\ C^\infty(Y, J) \end{array} \to C^\infty(Y, \mathbb{C}^N),$$

$$\mathcal{K}' = \begin{pmatrix} M \\ \\ Q \end{pmatrix}: C^\infty(Y, \mathbb{C}^N) \to \begin{array}{c} C^\infty(X, E) \\ \oplus \\ C^\infty(Y, J) \end{array}$$

and an operator $\mathcal{C} \in \mathfrak{G}$ with small norm so that $(\mathcal{I} - \mathcal{C})^{-1} \in \mathfrak{G}$ exists (the norm is defined corresponding to closures of the operators in Sobolev spaces). In the case $\mathcal{C} = 0$ we can set $\mathcal{R} = \mathcal{R}'$, $(\mathcal{I} - \mathcal{R}\mathcal{A}_0) = \mathcal{K}_0$ because $\mathcal{A}_0(\mathcal{I} - \mathcal{R}\mathcal{A}_0) = 0$ implies $\mathcal{A}_0\mathcal{K}_0 = \mathcal{A}_0(\mathcal{I} - \mathcal{R}\mathcal{A}_0)\mathcal{K}' = 0$ and $\mathcal{K}_0\mathcal{T}_1\mathcal{K}_0 = (\mathcal{I} - \mathcal{R}\mathcal{A}_0)\mathcal{K}_0 = \mathcal{K}_0$, i.e. $\mathcal{T}_1\mathcal{K}_0$ is a projection. In order to prove the assertion for $\mathcal{C} \neq 0$ it is sufficient to find a right inverse $\mathcal{R}$ and an operator $\mathcal{K}''$ so that $\mathcal{J} - \mathcal{R}\mathcal{A}_0 = \mathcal{K}''\mathcal{T}_1$ and one can define $\mathcal{K}_0 = (\mathcal{J} - \mathcal{R}\mathcal{A}_0)\,\mathcal{K}''$. Set $\mathcal{P}' = \mathcal{J} - \mathcal{R}'\mathcal{A}_0$, $\mathcal{M} = \mathcal{P}'(\mathcal{J} - \mathcal{C})^{-1}\mathcal{K}'\mathcal{T}_1$ and show that $\mathcal{R} = (\mathcal{J} - \mathcal{M})\,\mathcal{R}'$ is a right inverse of $\mathcal{A}_0$, too, and that because of $\mathcal{P} = \mathcal{J} - \mathcal{R}\mathcal{A}_0 = \mathcal{P}'(\mathcal{J} - \mathcal{C})^{-1}\mathcal{K}'\mathcal{T}_1$ we can define $\mathcal{K}'' = \mathcal{P}'(\mathcal{J} - \mathcal{C})^{-1}\mathcal{K}'$. From $\mathcal{A}_0\mathcal{P}' = 0$ it follows that $\mathcal{A}_0\mathcal{M} = 0$, i.e. $\mathcal{A}_0\mathcal{R} = \mathcal{A}_0(\mathcal{J} - \mathcal{M})\,\mathcal{R}' = \mathcal{J}$. Set $\mathcal{S} = (\mathcal{J} - \mathcal{C})^{-1}\mathcal{K}'\mathcal{T}_1$, i.e. $\mathcal{M} = \mathcal{P}'\mathcal{S}$. Then

$$\begin{aligned}\mathcal{P} &= \mathcal{J} - \mathcal{R}\mathcal{A}_0 = \mathcal{J} - (\mathcal{J} - \mathcal{M})\,\mathcal{R}'\mathcal{A}_0 = \mathcal{J} - \mathcal{R}'\mathcal{A}_0 + \mathcal{M}\mathcal{R}'\mathcal{A}_0\\ &= \mathcal{J} - \mathcal{R}'\mathcal{A}_0 - \mathcal{M}(\mathcal{J} - \mathcal{R}'\mathcal{A}_0) + \mathcal{M} = (\mathcal{J} - \mathcal{M})(\mathcal{J} - \mathcal{R}'\mathcal{A}_0) + \mathcal{M}\\ &= (\mathcal{J} - \mathcal{M})\,\mathcal{P}' + \mathcal{M} = (\mathcal{J} - \mathcal{P}'\mathcal{S})\,\mathcal{P}' + \mathcal{M}.\end{aligned}$$

Using (4) we get $\mathcal{S} = (\mathcal{J} - \mathcal{C})^{-1}\mathcal{P}' - (\mathcal{J} - \mathcal{C})^{-1}\mathcal{C}$, i.e.

$$\begin{aligned}\mathcal{P}'\mathcal{S}\mathcal{P}' &= \mathcal{P}'(\mathcal{J} - \mathcal{C})^{-1}\mathcal{P}' - \mathcal{P}'(\mathcal{J} - \mathcal{C})^{-1}\mathcal{C}\mathcal{P}'\\ &= \mathcal{P}'(\mathcal{J} - \mathcal{C})^{-1}(\mathcal{P}' - \mathcal{C}\mathcal{P}') = \mathcal{P}'(\mathcal{J} - \mathcal{C})^{-1}(\mathcal{J} - \mathcal{C})\,\mathcal{P}' = \mathcal{P}',\end{aligned}$$

hence $\mathcal{J} - \mathcal{R}\mathcal{A}_0 = \mathcal{M} = \mathcal{P}'(\mathcal{J} - \mathcal{C})^{-1}\mathcal{K}'\mathcal{T}_1$. □

The operator

$$L = \mathcal{T}_1\mathcal{K}_0\colon C^\infty(Y, \mathbb{C}^N) \to C^\infty(Y, \mathbb{C}^N) \tag{5}$$

which is a pseudo-differential operator over Y is a generalization of the Calderon-Seeley projector defined in 3.1.1.3 ($\mathcal{K}_0$ plays the role of K_0 and $\mathcal{T}_1$ the role of γ). The second formula in 3.1.1.3. (23) is a consequence of $\mathcal{J} - \mathcal{R}\mathcal{A} = \mathcal{K}_0\mathcal{T}_1$ on boundary symbol level. Since (5) is a projection, $\sigma_L\colon p^*(Y \times \mathbb{C}^N) \to p^*(Y \times \mathbb{C}^N)$ (the homogeneous principal symbol of L) is a projection, too, i.e. $G^{(+)} = \operatorname{im}\sigma_L$ is a subbundle of $p^*(Y \times \mathbb{C}^N)$.

Lemma 2. *Under the conditions of Theorem* 1 *we have*

$$\tilde{G} \overset{\text{def}}{=} \ker\sigma_Y(\mathcal{A}_0) = \operatorname{im}\sigma_Y(\mathcal{K}_0) \tag{6}$$

and there is an isomorphism

$$\sigma_Y(\mathcal{K}_0)\colon G^{(+)} \to \tilde{G}. \tag{7}$$

Proof: Using $\mathcal{A}_0\mathcal{K}_0 = 0$ we get $\operatorname{im}\sigma_Y(\mathcal{K}_0) \subseteq \tilde{G}$. Since $\mathcal{K}_0\mathcal{T}_1$ projects to $\ker\mathcal{A}_0$, $\sigma_Y(\mathcal{K}_0\mathcal{T}_1)$ is a projection to $\tilde{G}$. Now $\operatorname{im}\sigma_Y(\mathcal{K}_0\mathcal{T}_1) \supseteq \operatorname{im}\sigma_Y(\mathcal{K}_0)$ implies $\operatorname{im}\sigma_Y(\mathcal{K}_0) \subseteq \tilde{G}$ and thus (6). From the definition of $G^{(+)}$ it follows that $\operatorname{im}\left(\sigma_Y(\mathcal{K}_0)|_{G^{(+)}}\right) = \operatorname{im}\sigma_Y(\mathcal{K}_0\mathcal{T}_1\mathcal{K}_0)$. Moreover, $\operatorname{im}\sigma_Y(\mathcal{K}_0\mathcal{T}_1\mathcal{K}_0) \supseteq \operatorname{im}\sigma_Y(\mathcal{K}_0\mathcal{T}_1\mathcal{K}_0\mathcal{T}_1) = \tilde{G}$. Thus

$$\sigma_Y(\mathcal{K}_0)|_{G^{(+)}}\colon G^{(+)} \to \tilde{G} \tag{8}$$

is surjective. $\sigma_Y(\mathcal{T}_1\mathcal{K}_0)$ projects to $G^{(+)}$, i.e. it is the identity over $G^{(+)}$. Because of $\sigma_Y(\mathcal{T}_1\mathcal{K}_0) = \sigma_Y(\mathcal{T}_1)\,\sigma_Y(\mathcal{K}_0\mathcal{T}_1\mathcal{K}_0) = \sigma_Y(\mathcal{T}_1)\left(\sigma_Y(\mathcal{K}_0)|_{G^{(+)}}\right)$, (8) is injective, hence an isomorphism. □

Theorem 3. *Let $\mathcal{A} \in \mathfrak{G}$ be a boundary problem of the form* 3.1.1.1.(1) *with σ_A elliptic. Suppose that the first row* (1) *is surjective. Define the pseudo-differential operator $S = \mathcal{T}\mathcal{K}_0\colon C^\infty(Y, \mathbb{C}^N) \to C^\infty(Y, \mathbb{C}^N)$ ($\mathcal{T}$ denotes the second row of $\mathcal{A}$). The operator $\mathcal{A}$ is elliptic iff σ_S induces an isomorphism $\sigma_S\colon G^{(+)} \to p^*G$.*

Proof: $\sigma_Y(\mathcal{A})$ is an isomorphism iff $\sigma_Y(\mathcal{A}_0)$ is surjective and $\sigma_Y(\mathcal{T})\colon \tilde{G} \to p^*G$ is an isomorphism $\left(\tilde{G} = \ker\sigma_Y(\mathcal{A}_0)\right)$. The first condition follows from the surjectivity of $\mathcal{A}_0$. The second follows from the isomorphism (8). □

3.2. Index of Elliptic Boundary Value Problems

3.2.1. The Group Ell(X, Y)

3.2.1.1. Stable Equivalence and Homotopies of Operators

The index of elliptic boundary problems has (as well as the index of elliptic pseudo-differential operators on closed compact manifolds) some formal properties concerning direct sums, compositions, passing to a parametrix, and homotopies. This gives rise to a certain equivalence relation of elliptic boundary problems and it will be shown that the corresponding classes form an Abelian group.

Denote by $\mathfrak{E}_0$ the set of all elliptic operators $\mathcal{A} \in \mathfrak{G}$ of the form 3.1.1.1. (1) with arbitrary $E, F \in \operatorname{Vect}(X)$, $J, G \in \operatorname{Vect}(Y)$ and $\operatorname{ord} \sigma_A = 0$, $\operatorname{ord} \sigma_T + {}^1/_2 = 0$, $\operatorname{ord} \sigma_K + {}^1/_2 = 0$ so that $\mathcal{A}$ is Fredholm as operator

$$\mathcal{A}: H^0(X, E) \oplus H^0(Y, J) \to H^0(X, F) \oplus H^0(Y, G)\,. \tag{1}$$

Let for abbreviation $\mathcal{E} = C^\infty(X, E) \oplus C^\infty(Y, J)$, $\mathcal{F} = C^\infty(X, F) \oplus C^\infty(Y, G)$. If there are indices at the bundles, write corresponding indices at the spaces of sections.

For given $\mathcal{A}_j \in \mathfrak{E}_0$, $\mathcal{A}_j: \mathcal{E}_j \to \mathcal{F}_j$ $(j = 1, 2)$ a direct sum $\mathcal{A}_1 \oplus \mathcal{A}_2: \mathcal{E}_1 \oplus \mathcal{E}_2 \to \mathcal{F}_1 \oplus \mathcal{F}_2$ can be defined in an obvious way, $\mathcal{A}_1 \oplus \mathcal{A}_2 \in \mathfrak{E}_0$. Then

$$\operatorname{ind}(\mathcal{A}_1 \oplus \mathcal{A}_2) = \operatorname{ind} \mathcal{A}_1 + \operatorname{ind} \mathcal{A}_2\,. \tag{2}$$

Moreover, $\mathcal{A}, \mathcal{B} \in \mathfrak{E}_0$, $\mathcal{A}: \mathcal{E}_1 \to \mathcal{E}_2$, $\mathcal{B}: \mathcal{E}_0 \to \mathcal{E}_1$ implies $\mathcal{A}\mathcal{B} \in \mathfrak{E}_0$ and

$$\operatorname{ind} \mathcal{A}\mathcal{B} = \operatorname{ind} \mathcal{A} + \operatorname{ind} \mathcal{B}\,. \tag{3}$$

Let $\mathcal{A} \in \mathfrak{E}_0$ and $\mathcal{A}^{-1} \in \mathfrak{E}_0$ be a parametrix of $\mathcal{A}$. Then

$$\operatorname{ind} \mathcal{A}^{-1} = -\operatorname{ind} \mathcal{A}\,. \tag{4}$$

Finally, for $\mathcal{A}_0, \mathcal{A}_1 \in \mathfrak{E}_0$

$$\mathcal{A}_0 \simeq \mathcal{A}_1 \Rightarrow \operatorname{ind} \mathcal{A}_0 = \operatorname{ind} \mathcal{A}_1\,. \tag{5}$$

(The sign $\simeq$ means homotopy through elements in $\mathfrak{E}_0$; for the continuous dependence on t, $0 \leqq t \leqq 1$, in the homotopy one can take the norm of operators in fixed Sobolev spaces).

The identical mapping $1_{\mathcal{E}}$ in $\mathcal{E}$ belongs to $\mathfrak{E}_0$ and has the index zero. If $h: L \to M$ is a bundle isomorphism, denote by h^* the induced isomorphism of the space of C^∞ sections in M to the space of C^∞ sections in L.

Definition 1. Two operators $\mathcal{A}, \mathcal{A}' \in \mathfrak{E}_0$, $\mathcal{A}: \mathcal{E} \to \mathcal{F}$, $\mathcal{A}': \mathcal{E}' \to \mathcal{F}'$ are called *equivalent* if there exist bundles $E_1, E_1' \in \operatorname{Vect}(X)$, $J_1, J_1' \in \operatorname{Vect}(Y)$ and bundle isomorphisms

$$e: E \oplus E_1 \to E' \oplus E_1', \quad f: F \oplus E_1 \to F' \oplus E_1',$$
$$j: J \oplus J_1 \to J' \oplus J_1', \quad g: G \oplus J_1 \to G' \oplus J_1'$$

with

$$\mathcal{A} \oplus 1_{\mathcal{E}_1} \simeq \begin{pmatrix} f^* & 0 \\ 0 & g^* \end{pmatrix} (\mathcal{A}' \oplus 1_{\mathcal{E}_1'}) \begin{pmatrix} e^* & 0 \\ 0 & j^* \end{pmatrix}^{-1}. \tag{6}$$

Obviously, by Definition 1 an equivalence relation in $\mathfrak{E}_0$ is correctly defined. Denote by $[\mathcal{A}]$ the equivalence class represented by $\mathcal{A} \in \mathfrak{E}_0$. Sometimes we speak about stable homotopy of elliptic boundary problems instead of equivalence. Denote by Ell (X, Y) the set of equivalence classes of elements in $\mathfrak{E}_0$.

Proposition 2. Ell (X, Y) *is an additive Abelian group with respect to* $[\mathcal{A}_1] + [\mathcal{A}_2] \overset{\text{def}}{=} [\mathcal{A}_1 \oplus \mathcal{A}_2]$ *and by* ind: $\mathfrak{E}_0 \to Z$ *there is induced a group homomorphism*

$$\text{ind}\colon \text{Ell}\,(X, Y) \to Z\,. \tag{7}$$

Moreover $[\mathcal{A}\mathcal{B}] = [\mathcal{A}] + [\mathcal{B}]$, $[\mathcal{A}^{-1}] = -[\mathcal{A}]$.

Proof: Let $\mathcal{A}_j \in \mathfrak{E}_0$, $\mathcal{A}_j\colon \mathcal{E}_j \to \mathcal{F}_j$ $(j = 1, 2)$. In order to show $[\mathcal{A}_1 \oplus \mathcal{A}_2] = [\mathcal{A}_2 \oplus \mathcal{A}_1]$ it is sufficient to remark that a change of components in the underlying bundles defines isomorphisms $\mathcal{E}_1 \oplus \mathcal{E}_2 \to \mathcal{E}_2 \oplus \mathcal{E}_1$, $\mathcal{F}_1 \oplus \mathcal{F}_2 \to \mathcal{F}_2 \oplus \mathcal{F}_1$. Now let $\mathcal{A}, \mathcal{B} \in \mathfrak{E}_0$, $\mathcal{A}\colon \mathcal{E} \to \mathcal{F}$, $\mathcal{B}\colon \mathcal{H} \to \mathcal{E}$. Consider a continuous family of complex unitary 2×2 matrices $a(t) = a_{ij}(t))$ $(0 \leqq t \leqq 1)$ connecting the identity $a(0)$ with the matrix $a(1)$ for which $a_{12}(1) = a_{21}(1) = 1$, $a_{11} = a_{22} = 0$. Then

$$\begin{pmatrix} 1_{\mathcal{E}} & 0 \\ 0 & \mathcal{A} \end{pmatrix} \begin{pmatrix} a_{11}(t)\, 1_{\mathcal{E}} & a_{12}(t)\, 1_{\mathcal{E}} \\ a_{21}(t)\, 1_{\mathcal{E}} & a_{22}(t)\, 1_{\mathcal{E}} \end{pmatrix} \begin{pmatrix} \mathcal{B} & 0 \\ 0 & 1_{\mathcal{E}} \end{pmatrix}, \quad 0 \leqq t \leqq 1\,,$$

obviously defines a homotopy between $\begin{pmatrix} \mathcal{B} & 0 \\ 0 & \mathcal{A} \end{pmatrix}$ and $\begin{pmatrix} 0 & 1_{\mathcal{E}} \\ \mathcal{A}\mathcal{B} & 0 \end{pmatrix}$. This proves $[\mathcal{A}\mathcal{B}] = [\mathcal{A}] + [\mathcal{B}]$. From this $[\mathcal{A}^{-1}] = -[\mathcal{A}]$ easily follows. □

The elements $\mathcal{A}$ in $\mathfrak{E}_0$ of the form $1 \oplus Q$, where Q is an elliptic (classical) pseudo-differential operator of order zero on Y, generate a subgroup of Ell (X, Y).

Proposition 3. *Let* $(\sigma_{A^t})_{0 \leqq t \leqq 1}$, $(\sigma_{B^t})_{0 \leqq t \leqq 1}$ *be continuous families in* $\mathfrak{A}^{(0)}$ *and* $\mathfrak{B}^{(-1)}$, *respectively,* σ_{A^t} *elliptic for* $0 \leqq t \leqq 1$, *and let*

$$\mathcal{A}^0 = \begin{pmatrix} r^+A^0 + r'B^0 & K^0 \\ & \\ r'T^0 & Q^0 \end{pmatrix} \colon \begin{matrix} C^\infty(X, E) \\ \oplus \\ C^\infty(Y, J) \end{matrix} \to \begin{matrix} C^\infty(X, F) \\ \oplus \\ C^\infty(Y, G) \end{matrix} \tag{8}$$

be elliptic with $\sigma_\Omega(\mathcal{A}^0) = \sigma_{A^0}$ *and* σ_{B^0} *the homogeneous principal symbol of* $r'B^0$. *Then there exists a bundle* $L \in$ Vect (Y) *and a homotopy in* $\mathfrak{E}_0$ *of the form*

$$\mathcal{A}^{(t)} = \begin{pmatrix} r^+A^{(t)} + r'B^{(t)} & K^{(t)} \\ & \\ r'T^{(t)} & Q^{(t)} \end{pmatrix} \colon \begin{matrix} C^\infty(X, E) \\ \oplus \\ C^\infty(Y, J \oplus L) \end{matrix} \to \begin{matrix} C^\infty(X, F) \\ \oplus \\ C^\infty(Y, G \oplus L) \end{matrix} \tag{9}$$

$(0 \leqq t \leqq 1)$, *with* $\mathcal{A}^{(0)} = \mathcal{A}^0 \oplus 1_L$, $\sigma_\Omega(\mathcal{A}^{(t)}) = \sigma_{A^t}$ *and* σ_{B^t} *the homogeneous principal symbol of* $r'B^{(t)}$.

Proof: It is sufficient to construct $\sigma_Y(\mathcal{A}^{(t)})$. Consider the family of boundary symbols $\Pi^+\sigma_{A^t} + \Pi'\sigma_{B^t}\colon p^*E^+ \to p^*F^+$. Then there exists a subbundle $\tilde{L} \subset p^*F^+$, some $L \in$ Vect (Y) and an isomorphism $\sigma_K\colon p^*L \to \tilde{L}$ so that im $(\Pi^+\sigma_{A^t} + \Pi'\sigma_{B^t})$ + im $\sigma_K = p^*F^+$ independent of t. Thus

$$(\Pi^+\sigma_{A^t} + \Pi'\sigma_{B^t}, \sigma_{K^0}, t\sigma_K)\colon \begin{matrix} p^*E^+ \\ \oplus \\ p^*J \\ \oplus \\ p^*L \end{matrix} \to p^*F^+ \tag{10}$$

is surjective for all $0 \leq t \leq 1$. The kernel of (10) for $t = 0$ is mapped by $\begin{pmatrix} \Pi'\sigma_{T^0} & \sigma_{Q^0} & 0 \\ 0 & 0 & 1_L \end{pmatrix}$ isomorphically to $p^*(G \oplus L)$. Because of the contractibility of $[0, 1]$ this isomorphism can be extended to an isomorphism of the kernel of (10) to $\tau^*(p^*(G \oplus L))$ ($\tau: [0, 1] \to 0$ the contraction). This extension can be described by an operator family of the form $(\Pi'\sigma_{T^{(t)}}, \sigma_{Q^{(t)}})$. Thus $\sigma_Y(\mathcal{A}^{(t)})$ is constructed. □

Corollary 4. *Let $\mathcal{A} \in \mathfrak{E}_0$ be of the form* 3.1.1.1. (1). *Then there exists an $L \in$ Vect (Y) and a homotopy $\mathcal{A} \oplus 1_L \cong \mathcal{A}^{(1)}$ in $\mathfrak{E}_0$ for which r^+A remains fixed for all t and so that $\mathcal{A}^{(1)}$ contains no Green operator.*

This immediately follows from Proposition 3 for $\sigma_{A^t} = \sigma_A$, $\sigma_{B^t} = (1 - t)\,\sigma_B$, $0 \leq t \leq 1$.

The Proposition 3 can be modified as follows. Let $(\sigma_{A^t})_{0 \leq t \leq 1}$ be a continuous family in $\mathfrak{A}^{(0)}$ and let (8) be elliptic with $\sigma_\Omega(\mathcal{A}^0) = \sigma_{A^0}$. Then there exists a homotopy $(\sigma_{B^t})_{0 \leq t \leq 1}$ of Green symbols in $\mathfrak{B}^{(-1)}$ and a homotopy in $\mathfrak{E}_0$

$$\mathcal{A}^t = \begin{pmatrix} r^+A^t + r'B^t & K^t \\ r'T^t & Q^t \end{pmatrix} : \begin{matrix} C^\infty(X, E) \\ \oplus \\ C^\infty(Y, J) \end{matrix} \to \begin{matrix} C^\infty(X, F) \\ \oplus \\ C^\infty(Y, G) \end{matrix},$$

$0 \leq t \leq 1$, which is the given $\mathcal{A}^0$ at $t = 0$ and σ_{B^t} the homogeneous principal symbol of $r'B^t$. The simple proof of the last remark is left to the reader (cf. the constructions in the proof of 3.1.1.2, Lemma 2).

Finally, note that the assertions on homotopies can be generalized in an obvious way to symbols and operators of arbitrary orders.

3.2.1.2. Homotopies of Symbols

In connection with 3.2.1.1, Proposition 3 it is interesting that near the boundary every elliptic symbol $\sigma \in \mathfrak{A}^{(0)}$ can be deformed into a certain normal form if some trivial symbol is added before. In 3.2.1.3 the homotopies of elliptic symbols give rise to a lot of assertions on the index of elliptic boundary value problems. The homotopies of symbols to be discussed in this section are also closely related to a proof of the Bott periodicity theorem to which we return in 3.2.2.2 under another point of view.

Let $(x, \xi) = (x', x_n, \xi', \nu)$ be local coordinates in the cotangent bundle of a tubular neighbourhood U of Y. Identify U with $Y \times [0, 1)$ and denote by (y, x_n) points in U. Then we have a projection $q: U \to Y$, $q(y, x_n) = y$. Denote by $\pi_U: T^*U \setminus 0 \to U$ the canonical projection. Consider in U the scalar elliptic symbol

$$z = \big(\delta_\varepsilon(\xi) - i\nu\big)\,\big(\delta_\varepsilon(\xi) + i\nu\big)^{-1} \tag{1}$$

(cf. 3.1.2.1.(6), $\varepsilon > 0$ sufficiently small).

By

$$\sigma(y, x_n, \xi', \nu) = z^k \cdot 1_{\pi_U^*(q^*G)}$$

($k \in Z$, $G \in$ Vect (Y)) an elliptic symbol over U homogeneous of order zero is defined. For abbreviation denote it by $z^k \cdot 1_{p^*G}$. This notation is justified, because in the following the constructions for $x_n = 0$ and $|\xi'| = 1$ are valid for $T^*U \setminus 0$ by pull back.

Theorem 1. *Let $\sigma \in \mathfrak{A}^{(0)}$ be elliptic, $\sigma: \pi^*E \to \pi^*F$ $(E, F \in \mathrm{Vect}\,(X))$. Then the following assertions are equivalent:*

(i) *there exist bundles $J, G \in \mathrm{Vect}\,(Y)$ with*

$$\mathrm{ind}_{S^*Y}\, \Pi^+\sigma = [p^*G] - [p^*J]\,; \tag{2}$$

(ii) *there exist bundles $J_0, G_0, L, M \in \mathrm{Vect}\,(Y)$ so that over U there is a homotopy through elliptic symbols in $\mathfrak{A}^{(0)}$*

$$\begin{pmatrix} \sigma|_U(x, \xi', \nu) & 0 \\ 0 & 1_{p^*L} \end{pmatrix} \simeq a \begin{pmatrix} z^{-1}\cdot 1_{p^*G_0} & 0 & 0 \\ 0 & z\cdot 1_{p^*J_0} & 0 \\ 0 & 0 & 1_{p^*M} \end{pmatrix} \tag{3}$$

with

$$[p^*G_0] - [p^*J_0] = [p^*G] - [p^*J]^\circ \tag{4}$$

and $a: q^*(E' \oplus L) \to q^*(F' \oplus M)$ a certain bundle isomorphism $\big($thus $[p^*(E' \oplus L)] = [p^*(G_0 \oplus J_0 \oplus M)]$ in $K(S^*Y)\big)$.

Proof: 3.1.2.1, Lemma 3 shows

$$\mathrm{ind}_{S^*Y}\, \Pi^+(z^{-1}\cdot 1_{p^*G_0}) = [p^*G_0]\,, \quad \mathrm{ind}_{S^*Y}\, \Pi^+(z\cdot 1_{p^*J_0}) = [p^*J_0]\,.$$

Therefore, (ii) $\Rightarrow$ (i) is clear because of the homotopy invariance of the index element. Thus (i) $\Rightarrow$ (ii) has to be proved. For the neighbourhood U of Y use local coordinates in such a way that the coordinate diffeomorphisms are independent of x_n and that the Jacobian is equal to ± 1. This corresponds to a choice of a Riemannian metric over U and $|\xi'|$ is independent of the local coordinate system. Set $z_1 = (|\xi'| - i\nu)\,(|\xi'| + i\nu)^{-1}$. It is clear that $z_1 = z_1(\xi', \nu)$ is continuous but not C^∞ on the sphere $|\xi'|^2 + \nu^2 = 1$ because even the first order derivatives of z_1 at the points $p^\pm$ with coordinates $\xi' = 0$, $\nu = \pm 1$ do not exist. We have $z_1(p^\pm) = \lim\limits_{p \to p^\pm} z_1(p) = -1\,(p = (\xi', \nu), |\xi'|^2 + \nu^2 = 1)$ and $z(p^\pm) = -1$. The function z is different from z_1 in those points where $\delta_\varepsilon(\xi)\,|\xi'|^{-1} < 1$ and the set of these points is contracted to one point if $\varepsilon \to 0$.

Since z and z_1 are both continuous for any $\delta_0 > 0$, there exists an $\varepsilon_0 > 0$ so that $|z - z_1| = \sup\limits_p |z(p) - z_1(p)| < \delta$ for $\varepsilon < \varepsilon_0$.

By $\tilde{\mathfrak{A}}^{(0)}$ denote the set of all continuous morphisms $\sigma: \pi_U^*(q^*E') \to \pi_U^*(q^*F')$ $\big($for arbitrary $E', F' \in \mathrm{Vect}\,(Y)\big)$, positively homogeneous of degree 0, having an expansion of the form

$$\sigma(x', \xi', \nu) = \sum_{k=-\infty}^{\infty} b_k(x', \xi')\, z_1^k \tag{5}$$

with a rapidly decreasing sequence of C^∞ bundle morphisms $b_k: \pi_Y^*E' \to \pi_Y^*F'$, $k \in \mathbb{Z}$. For $\sigma \in \tilde{\mathfrak{A}}^{(0)}$ we obviously have

$$\sigma(x', 0, +1) = \sigma(x', 0, -1)\,. \tag{6}$$

Any $\sigma \in \mathfrak{A}^{(0)}$ over U can be uniformly approximated by a sequence of symbols in $\tilde{\mathfrak{A}}^{(0)}$, namely by $\sum\limits_{k=-\infty}^{\infty} b_k(x', \xi')\, z^k$ if σ is written in the form (5) and ε is chosen sufficiently small in the definition of z. If we have a homotopy over Y of the type as asserted in Theorem 1 through elliptic symbols in $\tilde{\mathfrak{A}}^{(0)}$, we obviously get a homotopy over U through elliptic symbols in $\mathfrak{A}^{(0)}$, too. Hence it is sufficient to consider homotopies over Y in $\tilde{\mathfrak{A}}^{(0)}$. Note that an element σ in $\tilde{\mathfrak{A}}^{(0)}$ is uniquely determined by $\sigma|_{S^*Y}$ because of the homogeneity and independence of x_n. Since from now on the variable z in the form (1) does not occur in the proof, set for abbreviation $z = (|\xi'| - i\nu)(|\xi'| + i\nu)^{-1}$. Despite of the fact that it is sufficient to consider $(y, \xi') \in S^*Y$, take over $|\xi'|$ in the formulas as parameter in order to make it clear at which places a number 1 plays the role of the module of ξ'.

Set in local coordinates $a_{\pm}(\xi) = \nu \mp i|\xi'|$ and

$$w = -z^{-1} = a_+(\xi)\, a_-^{-1}(\xi)\,. \tag{7}$$

Then $\nu = -i|\xi'|\,(w + 1)(w - 1)^{-1}$, i.e.

$$\sigma(x', \xi', \nu) = \sigma\left(x', \xi', -i|\xi'|\frac{w+1}{w-1}\right).$$

Here $\sigma(x', \xi', \nu) \overset{\text{def}}{=} \sigma(x', 0, \xi', \nu)$, $\sigma \in \mathfrak{A}^{(0)}$ the given elliptic symbol. Thus, σ can be considered as isomorphism

$$\sigma\colon p_1^*E' \to p_1^*F' \tag{8}$$

($p_1\colon (S^*Y) \times S \to Y$ the canonical projection, $S^1 = \{|w| = 1\}$, $E' = E|_Y$, $F' = F|_Y$.)

We obtain a uniformly converging series

$$\sigma(x', \xi', \nu) = \sum_{j=-\infty}^{\infty} a_j(\eta)\, w^j \tag{9}$$

(with uniform convergence also in $\eta = (x', \xi') \in S^*Y$). The coefficients in (9) represent bundle morphisms $p^*E' \to p^*F'$. Let $M \in Z_+$ be sufficiently large. Then, because of the ellipticity of σ and the uniform approximation,

$$\sum_{j=-M}^{M} a_j(\eta)\colon p^*E' \to p^*F'$$

is an isomorphism and

$$\sigma^1(x', \xi', \nu) = \sum_{j=-M}^{M} a_j(\eta)\, w^j \left(\sum_{k=-M}^{M} a_k(\eta)\right)^{-1} \sigma(x', 0, \pm 1)$$

elliptic. Moreover $\sigma^1 \simeq \sigma$, $\sigma^1(x', 0, \pm 1) = \sigma(x', 0, \pm 1)$ ($\simeq$ denotes homotopies through elliptic symbols in $\widetilde{\mathfrak{A}}^{(0)}$ over Y). Thus we have

$$\sigma^1(\eta, \nu) = \sum_{j=-M}^{M} b_j(\eta)\, w^j$$

with

$$b_j(\eta) = a_j(\eta)\left(\sum_{k=-M}^{M} a_k(\eta)\right)^{-1} \sigma(x', 0, \pm 1)\colon p^*E' \to p^*F'\,.$$

Set $c_{k+M} = b_k$ and

$$\sigma^2(\eta, \nu) = \sigma^1(\eta, \nu)\, w^M = \sum_{j=0}^{2M} c_j(\eta)\, w^j\,. \tag{10}$$

Define the bundle $W = \bigoplus_{1}^{2M} E'$ and use the following homotopy (cf. 1.1.3.3, Lemma 3)

$$\begin{pmatrix} \sigma^2(\eta, \nu) & 0 \\ 0 & 1_{p^*W} \end{pmatrix} \simeq \begin{pmatrix} c_0(\eta) & c_1(\eta) & c_2(\eta) & \dots & c_{2M}(\eta) \\ -w \cdot 1_{p^*E'} & 1_{p^*E'} & 0 & \dots & 0 \\ 0 & -w \cdot 1_{p^*E'} & 1_{p^*E'} & \dots & 0 \\ \vdots & \vdots & \vdots & & \vdots \\ 0 & 0 & 0 & \dots & 1_{p^*E'} \end{pmatrix}. \tag{11}$$

Denote the right hand side in (11) by $\sigma^3(\eta, \nu) = \alpha w + \beta(\eta)$. Using the homotopy defined above we get

$$\begin{pmatrix} \sigma^1(\eta, \nu) & 0 \\ 0 & 1_{p^*W} \end{pmatrix} = \begin{pmatrix} \sigma^2(\eta, \nu) & 0 \\ 0 & 1_{p^*W} \end{pmatrix} \begin{pmatrix} w^{-M} \cdot 1_{p^*E'} & 0 \\ 0 & 1_{p^*W} \end{pmatrix}$$

$$\simeq \sigma^3(\eta, \nu) \begin{pmatrix} w^{-M} \cdot 1_{p^*E'} & 0 \\ 0 & 1_{p^*W} \end{pmatrix}. \tag{12}$$

Since $\sigma^2(\eta, \pm\infty) = \sigma^2(x', 0, \pm 1)$ $(\eta = (x', \xi'))$, the isomorphism

$$\alpha + \beta(\eta): p^*(E' \oplus W) \to p^*(F' \oplus W) \tag{13}$$

depends only on the base point x' and thus defines an isomorphism

$$a_0: E' \oplus W \to F' \oplus W. \tag{14}$$

Define

$$\gamma(\eta) = \mathrm{i}(\alpha + \beta(\eta))^{-1} (-\alpha + \beta(\eta)): p^*(E' \oplus W) \to p^*(E' \oplus W).$$

Then

$$\sigma^3(\eta, \nu) = \alpha w + \beta(\eta) = a_-^{-1}(\xi)\, a_0(\nu \cdot 1_{p^*(E' \oplus W)} + |\xi'|\, \gamma(\eta)). \tag{15}$$

Since σ^3 and a_0 are elliptic, $\nu \cdot 1_{p^*(E' \oplus W)} + |\xi'|\, \gamma(\eta): p^*(E' \oplus W) \to p^*(E' \oplus W)$ is an isomorphism for all $\nu \in \mathbb{R}$. Especially for $\nu = 0$ it follows that $\gamma: p^*(E' \oplus W) \to p^*(E' \oplus W)$ is an isomorphism. Consider γ in local coordinates for fixed $\eta \in S^*Y$. Then the corresponding matrix has no real eigenvalues λ. There exist smooth curves $\Gamma_+(\Gamma_-)$ in the lower (upper) complex λ-half plane, which are the boundary of bounded domains containing the eigenvalues of $\gamma(\eta)$ with negative (positive) imaginary part. $\Gamma_\pm$ can be chosen independently of $\eta \in S^*Y$ because of the compactness of S^*Y. Let $M_\pm(\eta)$ be the direct sum of the eigenspaces for $\gamma(\eta)$ belonging to the eigenvalues with $\operatorname{Im} \lambda \lessgtr 0$. Because of the invariant meaning, $M_+(\eta)$ and $M_-(\eta)$ can be regarded as fibres of bundles M_+ and M_-, respectively, over S^*Y for which

$$M_+ \oplus M_- = p^*(E' \oplus W). \tag{16}$$

Then we have projections $\pi_\pm: p^*(E' \oplus W) \to M_\pm$ with $\pi_+ + \pi_- = 1_{p^*(E' \oplus W)}$. One can set

$$\pi_\pm(\eta) = -\frac{1}{2\pi \mathrm{i}} \int_{\Gamma_\pm} (\gamma(\eta) - \lambda)^{-1}\, \mathrm{d}\lambda.$$

We obtain

$$a_-^{-1}(\xi)\, (\nu \cdot 1_{p^*(E' \oplus W)} + |\xi'|\, \gamma(\eta))$$
$$= a_-^{-1}(\xi)\, \{[\nu\pi_+(\eta) + |\xi'|\, \gamma(\eta)\, \pi_+(\eta)] + [\nu\pi_-(\eta) + |\xi'| \gamma(\eta)\, \pi_-(\eta)]\}.$$

Since $\pi_+(\eta)$ acts on $M_+(\eta)$ as identity and $\gamma(\eta)\, m_+ = \lambda m_+$ for each eigenvector m_+ with $\operatorname{Im} \lambda < 0$, we obtain an isomorphism

$$(\nu + \mathrm{i}\mu)\, \pi_+ + |\xi'|\, \gamma\pi_+: M_+ \to M_+$$

for each $\mu \leqq 0$. Similarly

$$(\nu + \mathrm{i}\mu)\, \pi_- + |\xi'|\, \gamma\pi_-: M_- \to M_-$$

is an isomorphism for each $\mu \geqq 0$. Therefore

$$(\nu + \mathrm{i}t_1|\xi'|)^{-1}\, \{[(\nu - \mathrm{i}t_2|\xi'|)\, \pi_+ + |\xi'|\, \gamma\pi_+] + [(\nu + \mathrm{i}t_3|\xi'|)\, \pi_- + |\xi'|\, \gamma\pi_-]\}:$$
$$p^*(E' \oplus W) \to p^*(E' \oplus W)$$

is an isomorphism for all $\nu \in \mathbb{R}$, $t_1 > 0$, $t_2 \geqq 0$, $t_3 \geqq 0$. Assume $t = t_j > 0$ to be sufficiently large ($j = 1, 2, 3$). Then

$$\frac{\nu - it|\xi'|}{\nu + it|\xi'|}\pi_+ + \pi_-: p^*(E' \oplus W) \to p^*(E' \oplus W)$$

is an isomorphism for all $\nu \in \mathbb{R}$ and t sufficiently large independently of ν. Then of course

$$\frac{\nu - i|\xi'|}{\nu + i|\xi'|}\pi_+ + \pi_- = \frac{a_+(\xi)}{a_-(\xi)}\pi_+ + \pi_-: p^*(E' \oplus W) \to p^*(E' \oplus W)$$

is an isomorphism, too, and we have a homotopy

$$\sigma^3 \simeq a_0(w\pi_+ + \pi_-) . \tag{17}$$

In view of (12) we get a homotopy

$$\begin{pmatrix}\sigma^1 & 0\\ 0 & 1_{p^*W}\end{pmatrix} \simeq a_0(w\pi_+ + \pi_-)\begin{pmatrix}w^{-M}\cdot 1_{p^*E'} & 0\\ 0 & 1_{p^*W}\end{pmatrix}. \tag{18}$$

From (2) it follows that

$$\mathrm{ind}_{S^*Y}\,\Pi^+\sigma^1 = [p^*G] - [p^*J] , \tag{19}$$

and from 3.1.2.1, Lemma 3

$$\mathrm{ind}_{S^*Y}\,\Pi^+(w^{-M}\cdot 1_{p^*E'}) = -\left[p^*\bigoplus_1^M E'\right],$$

$$\mathrm{ind}_{S^*Y}\,\Pi^+(a_0\, w\pi_+ + \pi_-) = [M_+] .$$

Hence, in view of the homotopy invariance of the index element, (18) shows

$$\mathrm{ind}_{S^*Y}\,\Pi^+\sigma^1 = [M_+] - \left[p^*\bigoplus_1^M E'\right].$$

Compairing with (19) gives

$$[p^*G] - [p^*J] = [M_+] - \left[p^*\bigoplus_1^M E'\right]. \tag{20}$$

Applying 3.1.1.2, Lemma 1 we get bundles $W_\pm, L_\pm \in \mathrm{Vect}\,(Y)$ with

$$M_\pm \oplus p^*W_\pm \cong p^*L_\pm . \tag{21}$$

Next use the obvious homotopy

$$(w\pi_+ + \pi_-)\begin{pmatrix}w^{-M}\cdot 1_{p^*E'} & 0\\ 0 & 1_{p^*W}\end{pmatrix} \oplus 1_{p^*W}$$

$$\simeq \begin{pmatrix} w\pi_+ + \pi_- & 0 & 0\\ 0 & w^{-1}\cdot 1_{p^*\bigoplus_1^M E'} & 0\\ 0 & 0 & 1_{p^*\bigoplus_1^M E'}\end{pmatrix}$$

(cf. (18)). Then the proof of Theorem 1 is reduced to the symbol

$$\sigma^4(\eta, \nu) = w\pi_+(\eta) + \pi_-(\eta) . \tag{22}$$

Because of (16), σ^4 can be written as isomorphism

$$w\cdot 1_{M_+} \oplus 1_{M_-}: M_+ \oplus M_- \to M_+ \oplus M_-$$

for each $w \in S^1$. Thus, using (21) and

$$1_{p^*W_+} \oplus 1_{p^*W_+} \simeq w \cdot 1_{p^*W_+} \oplus w^{-1} \cdot 1_{p^*W_+},$$

we get

$$\begin{aligned} & w \cdot 1_{M_+} \oplus 1_{p^*W_+} \oplus 1_{p^*W_+} \oplus 1_{M_-} \oplus 1_{p^*W_-} \\ & \cong w \cdot 1_{M_+} \oplus w \cdot 1_{p^*W_+} \oplus w^{-1} \cdot 1_{p^*W_+} \oplus 1_{M_-} \oplus 1_{p^*W_-} \\ & = w \cdot 1_{p^*L_+} \oplus w^{-1} \cdot 1_{p^*W_+} \oplus 1_{p^*L_-}: \\ & p^*L_+ \oplus p^*W_+ \oplus p^*L_- \to p^*L_+ \oplus p^*W_+ \oplus p^*L_-. \end{aligned}$$

This is just the desired homotopy for σ^4 and Theorem 1 is proved if the order of diagonal elements is changed (if necessary) by homotopies. □

Note that with the notations in (3) we have

$$L \cong \left(\bigoplus_1^{2M} E'\right) \oplus \left(\bigoplus_1^{2M} E'\right) \oplus W_+ \oplus W_+ \oplus W_-, \quad M \cong L_- \oplus \left(\bigoplus_1^{M} E'\right),$$

$$G_0 \cong L_+, \quad J_0 \cong \left(\bigoplus_1^{M} E'\right) \oplus W_+$$

with the bundles $W_\pm$, $L_\pm$ introduced in (21). The isomorphism a is a direct sum of a_0 in (14) and identities in bundles which occurred as direct summands during the stabilization procedure.

Denote by $\mathfrak{A}_\Omega^{(0)}$ the set of all elliptic symbols σ on X belonging to $\mathfrak{A}^{(0)}$ for which there exists a tubular neighbourhood $U = U(\sigma) \cong Y \times [0, 1)$ of Y so that $\sigma|_U$ depends only on the base point $x \in U$, i.e. extension of $\sigma|_U$ to T^*U by homogeneity including the zero section gives a smooth function on U.

Note that symbols $\sigma: \pi^*E \to \pi^*F$ in $\mathfrak{A}_\Omega^{(0)}$ generate elements in the K group $K(T^*\Omega)$, $\Omega = X \setminus Y$.

Remark 2. Note that a symbol σ over $U \cong Y \times [0, 1)$ of the type $z^m \cdot 1_{p^*J} \oplus 1_{p^*J^\perp}$ ($J, J^\perp \in \operatorname{Vect}(Y)$, $J \oplus J^\perp \cong \mathbb{C}^j$ for some $j \in \mathbb{Z}_+$), $m \in \mathbb{Z}$, can be considered as the restriction to $U_{1/2} \cong Y \times [0, 1/2)$ of an elliptic symbol on X belonging to $\mathfrak{A}^{(0)}$, because z^m is homotopic through elliptic symbols (not in $\mathfrak{A}^{(0)}$) to the identiy (cf. 3.1.2.1). This homotopy can be applied in $U \setminus U_{1/2}$ and then one can extend the symbol as identity to $X \setminus U$.

The size of the strip U is not essential. After deformation one can always assume $U = Y \times [0, 1)$. We will not point out this in future if we use neighbourhoods of Y.

Remark 3. Let $\sigma \in \mathfrak{A}^{(0)}$, $\sigma: \pi^*E \to \pi^*F$ be an elliptic symbol with (2). Then there exists an $N \in \mathbb{Z}_+$ and an elliptic symbol $\sigma_1 \in \mathfrak{A}_\Omega^{(0)}$ with

$$\sigma \oplus 1_{\mathbb{C}^N}|_{X \setminus U} \cong \sigma_1|_{X \setminus U}. \tag{23}$$

If $r^+A_1 = \operatorname{Op}(\sigma_1)$,

$$r^+A_1: C^\infty(X, E \oplus \mathbb{C}^N) \to C^\infty(X, F \oplus \mathbb{C}^N) \tag{24}$$

is Fredholm without additional conditions (boundary and potential conditions). For the construction of σ_1 it is sufficient to add trivial bundles and to apply, besides the homotopy (3), a homotopy near Y corresponding to $z^m \simeq 1$ (through elliptic symbols without the transmission property).

Corollary 4. *Let $\sigma \in \mathfrak{A}^{(0)}$ be elliptic. Then the following conditions are equivalent:*

(i) *there exists an $N \in \mathbb{Z}^+$ so that there exists a homotopy $\sigma \oplus 1_{\mathbb{C}^N} \simeq \sigma_1$ through elliptic symbols in $\mathfrak{A}^{(0)}$ with $\sigma_1 \in \mathfrak{A}^{(0)}_{\Omega}$;*

(ii) $\operatorname{ind}_{S^*Y} \Pi^+\sigma = 0.$ (25)

Proof: Suppose $[p^*G_0] - [p^*J_0] = 0$ for $G_0, J_0 \in \operatorname{Vect}(Y)$. Then there is a $k \in \mathbb{Z}_+$ with $p^*(G_0 \oplus \mathbb{C}^k) \cong p^*(J_0 \oplus \mathbb{C}^k)$.

(ii) $\Rightarrow$ (i): From (ii) and (3) $[p^*G_0] - [p^*J_0] = 0$ follows because of the homotopy invariance of the index (cf. 3.1.2.1, Lemma 3). Now, passing from the right hand side of (3) (excepting a) to

$$(z^{-1} \cdot 1_{p^*G_0}) \oplus (z \cdot 1_{p^*J_0}) \oplus 1_{p^*\mathbb{C}^k} \oplus 1_{p^*\mathbb{C}^k} \oplus 1_{p^*M}, \tag{26}$$

we get a homotopy in $\mathfrak{A}^{(0)}$ near Y through elliptic symbols

$$\begin{aligned}(26) &\simeq (z^{-1} \cdot 1_{p^*(G_0 \oplus \mathbb{C}^k)}) \oplus (z \cdot 1_{p^*(J_0 \oplus \mathbb{C}^k)}) \oplus 1_{p^*M} \\ &\simeq 1_{p^*(G_0 \oplus \mathbb{C}^k)} \oplus 1_{p^*(J_0 \oplus \mathbb{C}^k)} \oplus 1_{p^*M}.\end{aligned}$$

Let $L^\perp \in \operatorname{Vect}(Y)$ so that $L \oplus L^\perp = \mathbb{C}^l$ for some $l \in \mathbb{Z}_+$. Then $\sigma \oplus 1_{\mathbb{C}^l}$ admits the desired homotopy. An argument as in Remark 2 gives the homotopy over the whole of X. The other direction (i) $\Rightarrow$ (ii) is trivial because of the homotopy invariance of the index and $\Pi^+\sigma_1 = 0$. □

Let $\sigma \in \mathfrak{A}^{(0)}$ be elliptic and consider a neighbourhood V of a point in Y, $V \cong \overline{\mathbb{R}}^n_+$, so that in local coordinates σ corresponds to a map $\sigma\colon (T^*\overline{\mathbb{R}}^+_n) \times \mathbb{C}^k \to (T^*\overline{\mathbb{R}}^n_+) \times \mathbb{C}^k$. Then $\sigma|_{Y \cap V}$ can be interpreted as smooth map

$$\sigma\colon S^*(\mathbb{R}^n)|_{x_n=0} \to \operatorname{GL}(k, \mathbb{C}).$$

Because of (3) and $z^m \simeq 1$ the condition (2) ($\Leftrightarrow \operatorname{ind} \Pi^+\sigma \in p^*K(Y)$) means that $\sigma \oplus 1_{\mathbb{C}^N}$ has an extension to a continuous map

$$\sigma \oplus 1_{\mathbb{C}^N}\colon \{(x', \xi)\colon x' \in \mathbb{R}^{n-1},\ \xi \in \mathbb{R}^n,\ |\xi| \leq 1\} \to \operatorname{GL}(N + k, \mathbb{C}).$$

This is always satisfied for $n - 1$ even (cf. 1.1.3.3, Theorem 4). Thus we have the following

Remark 5. Let $\sigma \in \mathfrak{A}^{(\alpha)}$ be elliptic and $n = 2l + 1$, $l \in \mathbb{Z}_+$. Then $\operatorname{Op}(\sigma)$ possesses elliptic boundary conditions over each neighbourhood V of Y (sufficiently small), i.e. there exists a bijective boundary symbol over $S^*(Y \cap V)$

$$\sigma_Y = \begin{pmatrix} \Pi^+\sigma + \Pi'\sigma_B & \sigma_K \\ \Pi'\sigma_T & \sigma_Q \end{pmatrix} \colon \begin{matrix} p^*(E \oplus \mathbb{C}^N)^+ \\ \oplus \\ p^*J \end{matrix} \to \begin{matrix} p^*(F \oplus \mathbb{C}^N)^+ \\ \oplus \\ p^*G \end{matrix} \tag{27}$$

for suitable $J, G \in \operatorname{Vect}(X)$.

For the construction of σ_Y it is sufficient to reduce the order of σ by means of an elliptic symbol λ_E^- described in 3.1.2.1 (i.e. pass from σ to $\sigma(\lambda_E^-)^{-\alpha} \in \mathfrak{A}^{(0)}$) and to use the fact that the condition $\operatorname{ind} \Pi^+\sigma \in p^*K(Y \cap V)$ ensures locally the constructions in the proof of 3.1.1.1, Theorem 12.

Globally, from Theorem 1 and $z^m \simeq 1$ ($m \in \mathbb{Z}$) we get the following

Remark 6. An elliptic symbol $\sigma \in \mathfrak{A}^{(0)}$ satisfies the condition

$$\operatorname{ind}_{S^*Y} \Pi^+\sigma \in p^*K(Y) \tag{28}$$

iff $\sigma \oplus 1_{C^N}$ for $N \in Z_+$ sufficiently large has an extension to an isomorphism

$$\sigma \oplus 1_{C^N}: \pi_1^*(E' \oplus C^N) \to \pi_1^*(F' \oplus C^N), \tag{29}$$

$\pi_1: B^*X|_Y \to Y$, B^*X denotes the unit ball bundle induced by S^*X.

In 3.2.2.1 we shall return to this remark and give another proof. The extension of $\sigma \oplus 1_{C^N}$ is not necessarily homogeneous, but after applying a homotopy near Y one can achieve homogeneity. Note that in view of 3.1.1.1, Theorem 12 the Remark 6 gives an equivalent criterion for the existence of an elliptic boundary problem $\mathcal{A} \in \mathfrak{G}$ with $\sigma = \sigma_\Omega(\mathcal{A})$.

This observation shows, from another point of view, that the condition (28) is a topological obstruction for the existence of an elliptic $\mathcal{A} \in \mathfrak{G}$ for $r^+A = \mathrm{Op}(\sigma)$. The difference construction for elliptic boundary problems to be discussed in 3.2.2.4 will show that the choice of some concrete extension of (29) to an isomorphism over $B^*X|_Y$ (which is not uniquely determined) corresponds to a choice of an elliptic $\mathcal{A} \in \mathfrak{G}$ with $\sigma = \sigma_\Omega(\mathcal{A})$ up to stable equivalence, cf. in 3.2.1.1.

Remark 7. Let $\sigma: \pi^*E \to \pi^*F$ be an elliptic symbol in $\mathfrak{A}^{(\alpha)}$ over X, $n = \dim X > 2$, and fibre dimension of E and F equal to 1. Then there exists a neighbourhood V in X for each $y \in Y$ so that there exists a bijective boundary symbol over $S^*(Y \cap V)$ of the form (27).

For the proof one can consider a local expression in $\overline{R}^n_+ (\cong V)$. In order to show that $\sigma: R^{n-1} \times \{|\xi| = 1\} \to C \setminus \{0\}$ has an extension to a continuous mapping $R^{n-1} \times \{|\xi| \leq 1\} \to C \setminus \{0\}$ passing to $\sigma/|\sigma|: R^{n-1} \times S^{n-1} \to S^1$ it is sufficient to remark that the homotopy groups $\pi_j(S^1)$ are trivial for $j \geq 2$ and that R^{n-1} is contractible.

The Remarks 5 and 7 show that the topological obstruction mentioned in Remark 6 locally vanishes in certain cases. On the other hand there are examples that the condition in Remark 6 is not empty. The following example also shows that the dimension condition in Remark 7 is necessary. Consider the Cauchy-Riemann operator

$$\frac{1}{2}\left(\frac{\partial}{\partial x} - \frac{1}{i}\frac{\partial}{\partial y}\right): C^\infty(X) \to C^\infty(X) \tag{30}$$

in $\Omega = \{z \in C: |z| < 1\}$, $X = \overline{\Omega}$, $z = x + iy$. The symbol of (30) is $(i/2)(\xi + i\eta)$ if ξ and η denote the dual variables to x and y, respectively. Identify $SX|_Y$ with $S^1 \times S^1$. Then the symbol defines a map $S^1 \to S^1$ (except for the factor $i/2$) for a fixed point $y \in S^1$ by multiplication by the complex number $\zeta = \xi + i\eta$. It is well-known thas there exists no stabilization $S^1 \to U(k)$ ($U(k)$ unitary group in C^k) with a continuous extension $\{|\zeta| \leq 1\} \to U(k)$ (cf. 1.1.3.3, Theorem 4). Thus the Cauchy-Riemann operator does not admit an elliptic boundary problem in $\mathfrak{G}$. Nevertheless there exist Fredholm boundary problems for (30) possessing another type (where conditions are separately posed to real or imaginary parts of the function).

3.2.1.3. Reduction to the Boundary and Theorems of Agranovič-Dynin Type

An essential tool for the investigation of boundary problems is the so-called *reduction to the boundary*. By this method assertions on boundary problems can be reduced to assertions on pseudo-differential operators on the boundary.

Theorem 1. *Let*

$$\mathcal{A}_i = \begin{pmatrix} r^+A + r'B_i & K_i \\ r'T_i & Q_i \end{pmatrix} : \begin{matrix} C^\infty(X, E) \\ \oplus \\ C^\infty(Y, J_i) \end{matrix} \to \begin{matrix} C^\infty(X, F) \\ \oplus \\ C^\infty(Y, G_i) \end{matrix} \tag{1}$$

($i = 1, 2$) be elliptic boundary problems in $\mathfrak{G}$. Then there exists an elliptic pseudo-differential operator S over Y the symbol of which can be expressed in terms of σ_A, σ_{B_i}, σ_{K_i}, σ_{T_i}, σ_{Q_i} ($i = 1, 2$) with

$$\operatorname{ind} \mathcal{A}_2 - \operatorname{ind} \mathcal{A}_1 = \operatorname{ind} S\,. \tag{2}$$

Proof: In view of 3.2.1.1, Corollary 4 one can assume without loss of generality that $r'B_i = 0$ for $i = 1, 2$. Moreover, after the reduction of orders (cf. 3.1.2.1) assume $\mathcal{A}_1, \mathcal{A}_2 \in \mathfrak{E}_0$ (cf. 3.2.1.1). We pass to the operators

$$\tilde{\mathcal{A}}_1 = \begin{pmatrix} r^+A & K_2 & K_1 \\ r'T_1 & 0 & Q_1 \\ 0 & 1 & 0 \end{pmatrix} : \begin{matrix} C^\infty(X, E) \\ \oplus \\ C^\infty(Y, J_2) \\ \oplus \\ C^\infty(Y, J_1) \end{matrix} \to \begin{matrix} C^\infty(X, F) \\ \oplus \\ C^\infty(Y, G_1) \\ \oplus \\ C^\infty(Y, J_2) \end{matrix}, \tag{3}$$

and

$$\tilde{\mathcal{A}}_2 = \begin{pmatrix} r^+A & K_2 & K_1 \\ r'T_2 & Q_2 & 0 \\ 0 & 0 & 1 \end{pmatrix} : \begin{matrix} C^\infty(X, E) \\ \oplus \\ C^\infty(Y, J_2) \\ \oplus \\ C^\infty(Y, J_1) \end{matrix} \to \begin{matrix} C^\infty(X, F) \\ \oplus \\ C^\infty(Y, G_2) \\ \oplus \\ C^\infty(Y, J_1) \end{matrix}. \tag{4}$$

Obviously $\tilde{\mathcal{A}}_i \in \mathfrak{E}_0$ ($i = 1, 2$) and

$$\operatorname{ind} \mathcal{A}_i = \operatorname{ind} \tilde{\mathcal{A}}_i\,, \quad i = 1, 2. \tag{5}$$

Denote by

$$\mathcal{P}_1 = \begin{pmatrix} r^+P + r'N_1 & L_1 \\ r'S_1 & R_1 \end{pmatrix}$$

a parametrix of $\mathcal{A}_1$. Then

$$\tilde{\mathcal{P}}_1 = \begin{pmatrix} r^+P + r'N_1 & L_1 & -(r^+P + r'N_1)\,K_2 \\ 0 & 0 & 1 \\ r'S_1 & R_1 & -(r'S_1)\,K_2 \end{pmatrix}$$

is a parametrix of $\tilde{\mathcal{A}}_1$ and we have mod $\mathfrak{G}^{-\infty}$

$$\tilde{\mathcal{A}}_2\tilde{\mathcal{P}}_1 = \begin{pmatrix} 1 & 0 & 0 \\ (r'T_2)\,(r^+P + r'N_1) & (r'T_2)\,L_1 & -(r'T_2)\,(r^+P + r'N_1)\,K_2 + Q_2 \\ r'S_1 & R_1 & -(r'S_1)\,K_2 \end{pmatrix}.$$

Here $\mathcal{A}_1\mathcal{P}_1 \equiv \mathcal{J}$ mod $\mathfrak{G}^{-\infty}$ is used. Set

$$S = \begin{pmatrix} (r'T_2)\,L_1 & -(r'T_2)\,(r^+P + r'N_1)\,K_2 + Q_2 \\ R_1 & -(r'S_1)\,K_2 \end{pmatrix} : \begin{matrix} C^\infty(Y, G_1) \\ \oplus \\ C^\infty(Y, J_2) \end{matrix} \to \begin{matrix} C^\infty(Y, G_2) \\ \oplus \\ C^\infty(Y, J_1) \end{matrix}.$$

Since $\tilde{\mathcal{A}}_2\tilde{\mathcal{P}}_1 \in \mathfrak{E}_0$, the boundary symbol $\sigma_Y(\tilde{\mathcal{A}}_2\tilde{\mathcal{P}}_1)$ is bijective. $\sigma_Y(\tilde{\mathcal{A}}_2\tilde{\mathcal{P}}_1)$ contains σ_S as the right lower corner. The first row of $\sigma_Y(\tilde{\mathcal{A}}_2\tilde{\mathcal{P}}_1)$ is equal to $(1, 0)$, i.e. σ_S is bijective, too, and hence S is an elliptic pseudo-differential operator on the boundary. Now (2) follows from (5) and $\operatorname{ind}(\tilde{\mathcal{A}}_2\mathcal{P}_1) = \operatorname{ind}\tilde{\mathcal{A}}_2 - \operatorname{ind}\tilde{\mathcal{A}}_1 = \operatorname{ind} S$. □

The formula (2) is called *Agranovič-Dynin formula.*

The reduction to the boundary gives a slightly weaker version of 3.1.1.1, Theorem 7, namely the following

Remark 2. *Let $\sigma_A\colon \pi^*E \to \pi^*F$ be elliptic, $\sigma_A \in \mathfrak{A}^{(\alpha)}$, $\operatorname{ind}_{S^*Y} \Pi^+\sigma_A \in p^*K(Y)$ and let $\mathcal{A}_2 \in \mathfrak{G}$ be an operator of the form (1) with $\sigma_A = \sigma_\Omega(\mathcal{A}_2)$. Assume that a Sobolev space closure of $\mathcal{A}_2$ is Fredholm. Then $\mathcal{A}_2$ is elliptic.*

Proof: According to 3.1.1.1, Theorem 12 there exists an elliptic operator $\mathcal{A}_1 \in \mathfrak{G}$ with $\sigma_\Omega(\mathcal{A}_1) = \sigma_A$. Without loss of generality consider operators with reduced orders (cf. 3.1.2.1) so that

$$\mathcal{A}_j\colon H^0(X, E) \oplus H^0(Y, J_j) \to H^0(X, F) \oplus H^0(Y, G_j)$$

$(i = 1, 2)$ are Fredholm. We find $\mathcal{A}_1$ in such a way that the left upper corners $r^+A + r'B_j$ in $\mathcal{A}_j$ are equal for $j = 1, 2$. Now pass to operators $\tilde{\mathcal{A}}_j$ similarly defined as (3), (4) and denote by $\tilde{\mathcal{P}}_1 \in \mathfrak{G}$ a parametrix of $\tilde{\mathcal{A}}_1$ which exists because of the ellipticity of $\tilde{\mathcal{A}}_1$. Then we get analogous formulas as in the proof of Theorem 1. Since $\tilde{\mathcal{A}}_2\tilde{\mathcal{P}}_1$ is Fredholm, the operator S has to be Fredholm. Since S is an elliptic PDO on Y we can apply 1.2.4.2, Theorem 3. Therefore, S is elliptic. This means that $\sigma_Y(\tilde{\mathcal{A}}_2\tilde{\mathcal{P}}_1)$ is bijective. Since $\sigma_Y(\tilde{\mathcal{A}}_1)$ is bijective, the same follows for the composition $\sigma_Y(\tilde{\mathcal{A}}_2) = \sigma_Y(\tilde{\mathcal{A}}_2\tilde{\mathcal{P}}_1\tilde{\mathcal{A}}_1)$. Then $\sigma_Y(\tilde{\mathcal{A}})$ is bijective, too. □

Lemma 3. *Let $\mathcal{A} \in \mathfrak{E}_0$,*

$$\mathcal{A} = \begin{pmatrix} r^+A + r'B & K \\ r'T & Q \end{pmatrix} : \begin{matrix} C^\infty(X, E) \\ \oplus \\ C^\infty(Y, J) \end{matrix} \to \begin{matrix} C^\infty(X, F) \\ \oplus \\ C^\infty(Y, G) \end{matrix}. \tag{6}$$

Then there exists a neighbourhood U of Y and some $N \in \mathbb{Z}_+$ and an operator

$$\mathcal{A}_0 = \begin{pmatrix} r^+A_0 + r'B_0 & K_0 \\ r'T_0 & Q_0 \end{pmatrix} : \begin{matrix} C^\infty(X, E \oplus \mathbb{C}^N) \\ \oplus \\ C^\infty(Y, J) \end{matrix} \to \begin{matrix} C^\infty(X, E \oplus \mathbb{C}^N) \\ \oplus \\ C^\infty(Y, G) \end{matrix}, \tag{7}$$

$\mathcal{A}_0 \in \mathfrak{E}_0$, with

$$\sigma_\Omega(\mathcal{A}_0)|_{X\setminus U} = 1_{\pi^*(E\oplus\mathbb{C}^N)}|_{X\setminus U} \tag{8}$$

and there exists an elliptic pseudo-differential operator

$$r^+A_1\colon C^\infty(X, E \oplus \mathbb{C}^N) \to C^\infty(X, F \oplus \mathbb{C}^N) \tag{9}$$

which in Fredholm (without boundary conditions) so that

$$\operatorname{ind}\mathcal{A}_0 = \operatorname{ind}\mathcal{A} - \operatorname{ind}(r^+A_1). \tag{10}$$

One can choose $\mathcal{A}_0$ in such a way that $B_0 = 0$.

Proof: Choose $N \in \mathbb{Z}_+$ sufficiently large and take the operator r^+A_1 as in 3.2.1.2, Remark 3. Then

$$\mathcal{A}_1 = \begin{pmatrix} r^+A_1 & 0 \\ 0 & 1_J \end{pmatrix} : \begin{matrix} C^\infty(X, E \oplus \mathbb{C}^N) \\ \oplus \\ C^\infty(Y, J) \end{matrix} \to \begin{matrix} C^\infty(X, F \oplus \mathbb{C}^N) \\ \oplus \\ C^\infty(Y, J) \end{matrix}$$

belongs to $\mathfrak{E}_0$ and ind $(r^+A_1) =$ ind $\mathcal{A}_1$. Let $\mathcal{A}_1^{-1}$ be a parametrix of $\mathcal{A}_1$. Then the operator $\mathcal{A}_0 = \mathcal{A}\mathcal{A}_1^{-2}$ obviously has the property (10) and (8) follows from 3.2.1.2. (23). In order to get $B_0 = 0$ one can apply 3.2.1.1, Corollary 4. □

Lemma 3 shows that for each $\mathcal{A} \in \mathfrak{E}_0$ of the form (6) one can find an operator $\mathcal{A}_0 \in \mathfrak{E}_0$ for which $\sigma_\Omega(\mathcal{A}_0): \pi^*E \to \pi^*E$ is the identity outside a neighbourhood U of Y where $\mathcal{A}_0$ essentially contains the same boundary conditions as $\mathcal{A}$. Putting $X_0 = Y \times [0, 1]$, from $\mathcal{A}_0$ we get an elliptic operator $\mathcal{A}_0'$ on X_0 with same boundary conditions on the boundary component $Y \times \{0\}$ as in $\mathcal{A}_0$ and no boundary conditions on $Y \times \{1\}$. Obviously, ind $\mathcal{A}_0 =$ ind $\mathcal{A}_0'$.

The construction of $\mathcal{A}_0$ and r^+A_1 in Lemma 3 can be modified in the following way. First apply the homotopy 3.2.1.2.(3) to $\sigma_\Omega(\mathcal{A})$ near Y. Then, applying 3.2.1.2, Remark 2 and 3.2.1.1, Corollary 4, we get a homotopy in $\mathfrak{E}_0$ from $\mathcal{A} \oplus 1_{\mathbb{C}^N} \oplus 1_{\mathbb{C}^{N'}}$ ($\mathbb{C}^N$, $\mathbb{C}^{N'}$ denote trivial bundles over X and Y, respectively, with the corresponding fibre dimensions N, N' sufficiently large) to some $\mathcal{A}'$ with

$$\sigma_\Omega(\mathcal{A}')|_Y = a(z^{-1} \cdot 1_{p^*G_0} \oplus z \cdot 1_{p^*J_0} \oplus 1_{p^*M}), \tag{11}$$

$M \in$ Vect (Y) a suitable bundle. Let $U = Y \times [0, 1)$ be a neighbourhood of Y in X and $q: \overline{U} \to Y$ the canonical projection to Y, $G_1 = q^*G_0$, $J_1 = q^*J_0$. In 3.1.2.1 has been constructed a homotopy $z \simeq 1$ through elliptic symbols over U (of course without transmission property). Moreover, there were considered elliptic symbols of the type μ_G^+ over $\overline{U} = Y \times [0, 1]$ with

$$\mu_{G_1}^+|_{Y\times\{0\}} = z^{-1} \cdot 1_{p^*G_0}, \quad \mu_{G_1}^+|_{Y\times\{1\}} = 1_{p^*G_0}.$$

We defined an elliptic boundary problem $\mathcal{M}_{G_1}^+$ over $\overline{U}$ with $\sigma_\Omega(\mathcal{M}_{G_1}^+) = \mu_{G_1}^+$ Dirichlet conditions over $Y \times \{0\}$, no conditions over $Y \times \{1\}$ and ind $\mathcal{M}_{G_1}^+ = 0$. A parametrix $(\mathcal{M}_{J_1}^+)^{-1}$ of $\mathcal{M}_{J_1}^+$ can be considered similarly. Then

$$\sigma_\Omega((\mathcal{M}_{J_1}^+)^{-1})|_{Y\times\{0\}} = z^{-1} \cdot 1_{p^*J_0}, \quad \sigma_\Omega((\mathcal{M}_{J_1}^+)^{-1})|_{Y\times\{1\}} = 1_{p^*J_0}$$

(compare the convention about the notations at the beginning of 3.2.1.2).

It is clear that the homotopy 3.2.1.2. (3) near the boundary in a strip $U_2 \cong Y \times [0,2)$ can be arranged in such a way that

$$\sigma_Y(\mathcal{A}')|_U = a(\mu_{G_1}^+ \oplus (\mu_{J_1}^+)^{-1} \oplus 1_{p^*M}). \tag{12}$$

Apply the construction in the proof of Lemma 3 to the operator A'. Then one obtains operators A_0', r^+A_1' with

$$\text{ind } \mathcal{A} = \text{ind } \mathcal{A}' = \text{ind } \mathcal{A}_0' + \text{ind } r^+A_1'. \tag{13}$$

Here the symbol of r^+A_1' belongs to $\mathfrak{A}_\Omega^{(0)}$ near Y and it is stable homotopic to σ_A outside a neighbourhood.

Lemma 4. *Let A_0' be the operator mentioned above. Then there exists an elliptic pseudo-differential operator R_0 over Y where σ_{R_0} can be expressed in terms of the principal symbols of the operators contained in $\mathcal{A}$ so that*

$$\text{ind } \mathcal{A}_0' = \text{ind } R_0. \tag{14}$$

Proof: Restriction of $\mathcal{A}_0'$ to $\overline{U}$ gives an elliptic boundary problem $\mathcal{A}_0''$ over $\overline{U}$ with the same conditions as in $\mathcal{A}_0'$ on $Y \times \{0\}$ and no conditions on $Y \times \{1\}$. Moreover, from (12) we obtain the existence of some elliptic boundary problem $\mathcal{A}_1''$ over $\overline{U}$ with ind $\mathcal{A}_1'' = 0$ and $\sigma_\Omega(\mathcal{A}_1'') = \sigma_\Omega(\mathcal{A}_0'')$. It is sufficient to use the boundary conditions from $\mathcal{M}_{G_1}^+$ and $(\mathcal{M}_{J_1}^+)^{-1}$. Then the assertion follows from Theorem 1 applied to $\mathcal{A}_0''$, $\mathcal{A}_1''$. □

Next we consider a generalization of the clutching construction for boundary problems discussed in 3.1.1.3. Let M be a closed compact C^∞ manifold and $M = X_+ \cup X_-$ with smooth compact submanifolds $X_\pm$ and the common boundary $Y = X_+ \cap X_-$. Set $E_\pm = E_{X_\pm}$, $E \in \mathrm{Vect}\,(M)$.

Theorem 5. *Let*

$$A: C^\infty(M, E) \to C^\infty(M, E)$$

be an elliptic PDO on M of order zero with homogeneous principal symbol σ_A. Assume that $\sigma_A^\pm = \sigma_A|_{X_\pm} \in \mathfrak{A}^{(0)}$ over $X_\pm$ and let

$$\mathcal{A}^\pm: \begin{matrix} C^\infty(X_\pm, E_\pm) \\ \oplus \\ C^\infty(Y, J_\pm) \end{matrix} \to \begin{matrix} C^\infty(X_\pm, F_\pm) \\ \oplus \\ C^\infty(Y, G_\pm) \end{matrix} \tag{15}$$

be elliptic boundary problems belonging to the class $\mathfrak{G}$ over X_+ and X_-, respectively, for the PDOs $r^\pm A^\pm = \mathrm{Op}\,(\sigma_A^\pm)$ over $X_\pm$. Then there exists an elliptic PDO R over Y where σ_R can be expressed in terms of the homogenous principal symbols of the operators occurring in $\mathcal{A}^\pm$ so that

$$\mathrm{ind}\,\mathcal{A}^+ + \mathrm{ind}\,\mathcal{A}^- = \mathrm{ind}\,A - \mathrm{ind}\,R\,. \tag{16}$$

Proof: Apply the formula (13) for $\mathcal{A}^\pm$. Then with operators $r^\pm(A_1')^\pm$ (with obvious notation) we have

$$\mathrm{ind}\,\big(r^+(A_1')\big)^+ + \mathrm{ind}\,\big(r^-(A_1')\big)^- = \mathrm{ind}\,A \tag{17}$$

because of the homotopy invariance of the index. Applying Lemma 4 to the operators $(\mathcal{A}_0')^\pm$ we get elliptic PDOs $R_0^\pm$ for which ind $(\mathcal{A}_0')^\pm = \mathrm{ind}\,R_0^\pm$. From (13), (14) it follows that ind $\mathcal{A}^\pm = R_0^\pm + \mathrm{ind}\,r^\pm(A_1')^\pm$. Addition gives (16) in view of (17) with R defined as parametrix of the direct sum $R_0^+ \oplus R_0^-$. □

Remark 6. *Let $\mathcal{A} \in \mathfrak{E}$ be an operator of the form* (6). *Then for sufficiently large M, N and Fredholm operators*

$$\left.\begin{aligned} r^+A_0 + r'B_0&: C^\infty(X, E \oplus \mathbb{C}^M) \to C^\infty(X, E \oplus \mathbb{C}^M)\,, \\ r^+A_1 &: C^\infty(X, E \oplus \mathbb{C}^N) \to C^\infty(X, F \oplus \mathbb{C}^N) \end{aligned}\right\} \tag{18}$$

we have

$$\mathrm{ind}\,\mathcal{A} = \mathrm{ind}\,(r^+A_0 + r'B_0) + \mathrm{ind}\,r^+A_1\,. \tag{19}$$

Here r^+A_1 can be chosen as in formula (10) *and $r'B_0$ is a suitable Green operator.*

Proof: In order to show the formula (19) it is sufficient to define $r^+A_0 + r'B_0$ in such a way that ind $\mathcal{A}_0 = \mathrm{ind}\,(r^+A_0 + r'B_0)$ for the operator $\mathcal{A}_0$ in (10). $\mathcal{A}_0$ is considered as an elliptic boundary problem over $\overline{U} = Y \times [0, 1]$ without conditions over $Y \times \{1\}$,

$$\mathcal{A}_0: \begin{matrix} C^\infty(\overline{U}, E_1) \\ \oplus \\ C^\infty(Y, J) \end{matrix} \to \begin{matrix} C^\infty(\overline{U}, E_1) \\ \oplus \\ C^\infty(Y, G) \end{matrix}$$

($Y \times \{0\}$ is identified with Y, $E_1 = E|_{\overline{U}}$). Denote by J_1 and G_1 the pull backs of J and G, respectively, under $q: \overline{U} \to Y$. Then we can pass to the elliptic boundary problem

$$\mathscr{B} = \mathscr{A}_0 \oplus 1_{J_1} \oplus 1_{G_1}: \begin{matrix} C^\infty(\overline{U}, E_1 \oplus J_1 \oplus G_1) \\ \oplus \\ C^\infty(Y, J) \end{matrix} \to \begin{matrix} C^\infty(\overline{U}, E_1 \oplus J_1 \oplus G_1) \\ \oplus \\ C^\infty(Y, G) . \end{matrix}$$

Obviously ind $\mathscr{B}$ = ind $\mathscr{A}_0$. Consider the elliptic boundary problems

$$\mathscr{M}_{G_1}^+: \begin{matrix} C^\infty(\overline{U}, G_1) \\ \oplus \\ C^\infty(Y, G) \end{matrix} \to C^\infty(\overline{U}, G_1), \ (\mathscr{M}_{J_1}^+)^{-1}: C^\infty(\overline{U}, J_1) \to \begin{matrix} C^\infty(\overline{U}, J_1) \\ \oplus \\ C^\infty(Y, J) \end{matrix}$$

with index zero. Set $L_1 = E_1 \oplus J_1 \oplus G_1$ and

$$\mathscr{C} = (1_{E_1} \oplus 1_{J_1} \oplus \mathscr{M}_{G_1}^+)\, \mathscr{B} \big(1_{E_1} \oplus (\mathscr{M}_{J_1}^+)^{-1} \oplus 1_{G_1}\big): C^\infty(\overline{U}, L_1) \to C^\infty(\overline{U}, L_1) .$$

Let L be a vector bundle on $\overline{U}$ with $J_1 \oplus G_1 \oplus L = \mathbb{C}^M$. Then $\mathscr{C} \oplus 1_{\mathbb{C}^M}$ has a canonical extension to an elliptic boundary problem over X of the form $r^+A_0 + r'B_0$: $C^\infty(X, E \oplus \mathbb{C}^M) \to C^\infty(X, E \oplus \mathbb{C}^M)$ and ind $(r^+A_0 + r'B_0)$ = ind $\mathscr{A}_0$. □

In 3.2.2 we shall return to expressions of the type (19) and get a sharper version, namely the existence of a Fredholm operator $r^+A_\Omega: C^\infty(X, E \oplus \mathbb{C}^l) \to C^\infty(X, F \oplus \mathbb{C}^l)$ ($\mathbb{Z}_+$ sufficiently large) with $\sigma_{A\Omega} \in \mathfrak{A}_\Omega^{(0)}$ and ind $\mathscr{A}$ = ind r^+A_Ω.

3.2.2. K-Theoretic Aspects

3.2.2.1. A Connection between the Index Element and the Difference Element of the Symbol

In this section we give a little more insight into the K-theoretic behaviour of elliptic boundary problems. First let us introduce some abbreviations. The choice of a Riemannian métric on T generates an isomorphism $T^*X \to TX$. For this reason the star at T^*X will be omitted here. The closed unit ball bundles induced by TX and TY shall be denoted by BX and BY, respectively. Then SX, SY are the corresponding unit sphere bundles. Moreover set

$$B'X = BX|_Y, \ S'X = SX|_Y .$$

We have a direct decomposition $TX|_Y = TY \oplus N$ with the conormal bundle $N = Y \times \mathbb{R}$ of Y with respect to the identical embedding $Y \to X$. Set $N_1 = BX \cap N$.

Consider an elliptic symbol $\sigma_A: \pi^*E \to \pi^*F$ over X, $\sigma_A \in \mathfrak{A}^{(0)}$. Set for abbreviation $\sigma = \sigma_A|_Y$. Because of the homogeneity we then have

$$\lim_{|\nu| \to \infty} \sigma(x', \xi', \nu) = \lim_{|\nu| \to \infty} \sigma\left(x', \frac{\xi'}{|\nu|}, \frac{\nu}{|\nu|}\right) = \sigma(x', 0, \pm 1) \tag{1}$$

and the transmission property implies

$$\sigma(x', 0, -1) = \sigma(x', 0, +1) \tag{2}$$

for all $x' \in Y$. From (2) we get a canonical extension of the isomorphism $\sigma|_{S'X}$ to an isomorphism over $S'X \cup N_1$, i.e. an isomorphism between the pull backs of E', F' with respect to the projection $S'X \cup N_1 \to Y$. An isomorphism $\sigma_0: E' \to F'$ is induced by (2). The symbol σ represents a difference element $\mathrm{d}(\sigma) \in K(B'X, S'X)$ (cf. 1.1.3.1

and 1.2.4.2). The extension of the isomorphism to $S'X \cup N_1$ gives rise to an element $d_1(\sigma) \in K(B'X, S'X \cup N_1)$. Denote the restriction homomorphism of relative K-groups by

$$\varrho\colon K(B'X, S'X \cup N_1) \to K(B'X, S'X) \tag{3}$$

(this is induced by the pull back with respect to the inclusion of pairs $(B'X, S'X) \to (B'X, S'X \cup N_1)$). Obviously $\varrho(d_1(\sigma)) = d(\sigma)$.

Note that in the definition of the difference element $d(\sigma)$ (as well as $d_1(\sigma)$) represented by $0 \to \pi^*E|_Y \xrightarrow{\sigma} \pi^*F|_Y \to 0$ one can assume that $E = F$ because of the isomorphism $E' \cong F'$.

Lemma 1. *There is a canonical isomorphism*

$$\tau\colon K(B'X, S'X \cup N_1) \to K(\mathbb{R}^2 \times SY) \tag{4}$$

Proof: Denote by Z' the bundle $BY \times N_1|_{\mathrm{diag}\, Y}$ over Y, i.e. the bundle of elements (y, η, ν) with $y \in Y$, $|\eta| \leqq 1$, $|\nu| \leqq 1$. Set $B'' = S'X \cup N_1$, $Z'' = \partial Z' \cup N_1$. Then we have a continuous map $f\colon (Z', Z'') \to (B', B'')$ defined by $f(y, \eta, \nu) = (y, \eta, \nu)$ for $(y, \eta, \nu) \in B'$ and $f(y, \eta, \nu) = (y, \eta', \nu)$ with $\eta' = |c|\, \eta$, $c^2|\eta|^2 + |\nu|^2 = 1$ for $(y, \eta, \nu) \notin B'$. Obviously, f induces an isomorphism $f^*\colon K(B', B'') \to K(Z', Z'')$. There is a natural identification $Z' \setminus Z'' = (0, 1) \times (-1, 1) \times SY$ $((-1, 1) = \operatorname{int} N_1)$ and an isomorphism $l\colon K(Z', Z'') \to K(Z' \setminus Z'')$. By substituting homeomorphisms $(0, 1) \cong \mathbb{R}$, $(-1, 1) \cong \mathbb{R}$ we get an isomorphism $g\colon K(Z' \setminus Z'') \to K(\mathbb{R}^2 \times SY)$. Then $\tau = g \circ l \circ f^*$ is an isomorphism as desired. □

Set

$$d_2(\sigma) = \tau(d_1(\sigma))\,. \tag{5}$$

Applying 1.1.3.3, Theorem 1 we get an isomorphism

$$\beta\colon K(SY) \to K(\mathbb{R}^2 \times SY)\,. \tag{6}$$

Lemma 1 gives a map $\sigma \to d_2(\sigma) \in K(\mathbb{R}^2 \times SY)$. Moreover we have a map $\sigma \to \operatorname{ind}_{SY} \Pi^+\sigma \in K(SY)$ (cf. 3.1.1.1. (18)).

Theorem 2. *Let $\sigma_\Delta \in \mathfrak{A}^{(0)}$ be elliptic and $\sigma = \sigma_\Delta|_Y$. Then, with the Bott isomorphism* (6) *we have*

$$\beta(\operatorname{ind}_{SY} \Pi^+\sigma) = d_2(\sigma)\,. \tag{7}$$

A proof of Theorem 2 will be given in 3.2.2.2. Using (7) we can give another proof of 3.2.1.2, Remark 6 formulated as the following

Proposition 3. *Let $\sigma \in \mathfrak{A}^{(0)}$ be elliptic. Then*

$$d(\sigma) = 0 \Leftrightarrow \operatorname{ind}_{SY} \Pi^+\sigma \in p^*K(Y)\,. \tag{8}$$

For the proof we need two Lemmata.

Lemma 4. *There is a canonical isomorphism*

$$\lambda\colon K(\mathbb{R}^1 \times BY, \mathbb{R}^1 \times SY) \to K(B'X, S'X)\,. \tag{9}$$

Proof: In view of the definition of the K-functor for locally compact spaces (cf. 1.1.3.2) each element in $K(\mathbb{R}^1 \times BY, \mathbb{R}^1 \times SY)$ can be represented by an isomorphism $\varphi\colon E|_U \to F|_U$ with vector bundles E, F over $\mathbb{R}^1 \times BY$ and an open set $U \subset \mathbb{R}^1 \times BY$ with $\mathbb{R}^1 \times SY \subset U$ and a compact complement. The identification λ

then easily follows from the homeomorphisms $B'X \cong [-1, 1] \times BY$, $S'X \cong ([-1, 1] \times SY) \cup (\{-1\} \times \{1\} \times BY)$ and the excision axiom of the K-theory. □

Lemma 5. *The following diagram commutes*

$$\begin{array}{ccc} K(\mathbb{R}^2 \times SY) & \xrightarrow{\delta} & K(\mathbb{R}^1 \times BY, \mathbb{R}^1 \times SY) \\ \cong \uparrow \tau & & \cong \uparrow \lambda \\ K(B'X, S'X \cup N_1) & \xrightarrow{\varrho} & K(B'X, S'X) \end{array} \tag{10}$$

with the homomorphisms (3), (4), (9) *and the coboundary operator* δ *of the K-theory* (cf. 1.1.3.3). *If* $\sigma \in \mathfrak{A}^{(0)}$ *is elliptic, we have*

$$\delta\tau(d_1(\sigma)) = \lambda(d(\sigma)) \,. \tag{11}$$

Proof: Any element $a \in K(\mathbb{R}^1 \times \mathbb{R}^1 \times SY)$ can be represented by an isomorphism $\varphi: E|_V \to F|_V$ with vector bundles E, F over $\mathbb{R}^1 \times \mathbb{R}^1 \times SY$ and an open subset V with a compact complement. One can suppose that E, F are trivial over V. Then φ can be represented by a continuous map $\varphi': \mathbb{R}^1 \times SY \to \mathrm{GL}\,(k, \mathbb{C})$ if k is the fibre dimension of E and F, and $\varphi'(x', \xi', -\infty) = \varphi'(x', \xi', +\infty)$ for all $(x', \xi') \in SY$. Let $\mathbb{C}^k$ be the trivial k-dimensional bundle over $\mathbb{R}^1 \times BY$. Then φ' corresponds to an isomorphism $\mathbb{C}^k|_U \to \mathbb{C}^k|_U$ for some open set $U \subset \mathbb{R}^1 \times BY$ with a compact complement containing $\mathbb{R}^1 \times SY$. This isomorphism represents an element in $K(\mathbb{R}^1 \times BY, \mathbb{R}^1 \times SY)$ which is equal to δa by definition of the coboundary operator δ. For $a = d_2(\sigma)$ we obviously have $\delta a = \lambda^{-1}(d(\sigma))$ (note that the isomorphism $E' \cong F'$ admits the consideration of trivial bundles). The assertion then follows immediately from the definition of τ, ϱ, λ. □

Proof of Proposition 3. The following diagram commutes

$$\begin{array}{ccccc} K(\mathbb{R}^2 \times BY) & \longrightarrow & K(\mathbb{R}^2 \times SY) & \xrightarrow{\delta} & K(\mathbb{R}^1 \times BY, \mathbb{R}^1 \times SY) \\ \cong \uparrow \beta & & \cong \uparrow \beta & & \cong \uparrow \lambda \\ K(BY) & \longrightarrow & K(SY) & \xrightarrow{\gamma} & K(B'X, S'X) \\ \cong \uparrow & \nearrow p^* & & & \\ K(Y) & & & & \end{array} \tag{12}$$

if we define $\gamma = \lambda^{-1} \circ \delta \circ \beta$ (the first homomorphisms in the rows are the restrictions). The first row in (12) is a part of the exact sequence of K-theory for the pair (BY, SY) (cf. 1.1.3.3.(6)). Thus the second row in (12) is exact and hence

$$K(Y) \xrightarrow{p^*} K(SY) \xrightarrow{\gamma} K(B'X, S'X) \tag{13}$$

is exact, too. Let $a = \mathrm{ind}_{SY}(\Pi^+\sigma)$. Then (7) means that $d_2(\sigma) = \beta(a)$, i.e. $\delta(d_2(\sigma)) = \delta\beta(a)$. Since per def. $d_2(\sigma) = \tau(d_1(\sigma))$, we get from (11) $\lambda(d(\sigma)) = \delta\beta(a)$, i.e. $d(\sigma) = \gamma(a)$. Since (13) is exact, we get $a \in p^*K(Y) \Rightarrow d(\sigma) = 0$ and $a \notin p^*K(Y) \Rightarrow d(\sigma) \neq 0$. □

3.2.2.2. On the Analytical Proof of the Bott Periodicity Theorem

The proof of 3 2.2.1, Theorem 2 is essentially the same as a proof of the Bott periodicity theorem (cf. 1.1.3.3) given in Atiyah [3]. This proof is called analytical because of the use of families of Wiener-Hopf operators (cf. 2.1.2.1). Similarly as in the preceding sections an elementary knowledge on vector bundles and K-theory is supposed.

Let X be a locally compact space. Then each element $a \in K(X)$ can be represented by a triple (E, F, α), where E, F are (complex) vector bundles over X, and $\alpha: E \to F$ is a bundle morphism which is an isomorphism on a complement of a compact set $C \subset X$. If Y, X are locally compact spaces, there is defined an (external) multiplication

$$t: K(Y) \otimes K(X) \to K(Y \times X) \tag{1}$$

(cf. 1.1.3.2). Let for abbreviation $uv = t(uv)$. In the case $Y = \mathbb{R}^2$ we have an isomorphism $K(\mathbb{R}^2) \cong \mathbb{Z}$ between the additive Abelian groups. $\mathbb{R}^2$ can be identified with $\mathbb{C}$. Let

$$B_0 = \{z \in \mathbb{C}: |z| \leqq 1\}, \quad B_\infty = \{z \in \mathbb{C}: |z| \geqq 1\}$$

so that $S^1 = B_0 \cap B_\infty$. By $(z, e) \to (z, z^{-1} e)$ an isomorphism $B_\infty \times \mathbb{C} \to B_\infty \times \mathbb{C}$ is defined. It can be extended as a bundle homomorphism $l: \mathbb{C} \times \mathbb{C} \to \mathbb{C} \times \mathbb{C}$ to the whole of $\mathbb{C}$. The element in $K(\mathbb{C}) = K(\mathbb{R}^2)$ represented by the triple $(\mathbb{C} \times \mathbb{C}, \mathbb{C} \times \mathbb{C}, l)$ is denoted by b (the Bott element). It generates the additive Abelian group $K(\mathbb{R}^2)$. Then multiplication by b in the sense of (1) gives a homomorphism

$$\beta: K(X) \to K(\mathbb{R}^2 \times X). \tag{2}$$

(2) is an isomorphism (cf. 1.1.3.3, Theorem 1). For the proof recall some well known facts about vector bundles and K-theory.

(a) Let X be a paracompact space. Then there is a canonical isomorphism

$$K(S^n \times X) \cong K(\mathbb{R}^n \times X) \oplus K(X) \tag{3}$$

$(n = 1, 2, \ldots)$. Especially for $X = \{p\}$ (a single point) we have $K(S^n) \cong K(\mathbb{R}^n) \oplus \mathbb{Z}$.

(b) Identify S^2 with the Riemannian number sphere. Let $S^2 = B_+ \cup B_-$, where $B_\pm$ denotes the upper and lower closed hemisphere of S^2. Let $\mathbb{C}^+$ be the one-point-compactification of $\mathbb{C}$ and $r: \mathbb{C}^+ \to S^2$ a map corresponding to the stereographic projection where B_0 transforms into B_+, B_∞ to B_- and S^1 to the equator.

Let X be a compact space and $s: S^1 \times X \to X$ the canonical projection. Suppose that we are given a bundle $E \in \operatorname{Vect}(X)$ and an isomorphism

$$\sigma: s^*E \to s^*E. \tag{4}$$

Using the clutching construction for the bundles $s_\pm^* E$ over $B_\pm \times X$ ($s_\pm: B_\pm \times X \to X$ the canonical projection, $S^1 \times X = B_+ \times X \cap B_- \times X$) (cf. 1.1.2.1) and the clutching isomorphism (4) we get a vector bundle $F = v(E, \sigma) \in \operatorname{Vect}(S^2 \times X)$. Conversely it is easily seen that for any $F \in \operatorname{Vect}(S^2 \times X)$ there exists some $E \in \operatorname{Vect}(X)$ and an isomorphism (4) with $F \cong v(E, \sigma)$. Apply the clutching construction for the trivial bundle $E = \mathbb{C}^k$ over X with the clutching isomorphism

$$\sigma: S^1 \times \mathbb{C}^k \to S^1 \times \mathbb{C}^k. \tag{5}$$

For $X = \{p\}$ we get some $F \in \operatorname{Vect}(S^2)$ then.

(c) Any isomorphism (5) corresponds to a Fredholm operator $A: L_+^2(S^1) \otimes \mathbb{C}^k \to L_+^2(S^1) \otimes \mathbb{C}^k$ (cf. 2.1.2.1, Proposition 3 and 2.1.1.1) with

$$\operatorname{ind} A = -\deg(\det \sigma). \tag{6}$$

Let $\mathcal{H}$ be a Hilbert space and $L_+^2(S^1) \otimes \mathbb{C}^k \to \mathcal{H}$ a fixed isomorphism. Consider the space $\mathcal{F}(\mathcal{H})$ of Fredholm operators in $\mathcal{H}$ in the norm topology. Then each isomorphism (4) corresponds to a continuous family of operators in $\mathcal{F}(\mathcal{H})$ depending on $x \in X$

(X compact), and we get a continuous map

$$\operatorname{op}(E, \sigma): X \to \mathcal{F}(\mathcal{H}). \tag{7}$$

In the definition of (7) the fact that Hilbert space bundles are trivial (cf. 1.1.3.4, Corollary 5) is exploited. Thus, applying the constructions of 1.1.3.4 we get an index element

$$\operatorname{ind}_X \operatorname{op}(E, \sigma) \in K(X). \tag{8}$$

Denote by $[X, \mathcal{F}(\mathcal{H})]$ the set of homotopy classes of continuous maps $X \to \mathcal{F}(\mathcal{H})$. Then, from (7) one can pass to a map

$$[\mathrm{op}]: (E, \sigma) \to [\operatorname{op}(E, \sigma)]$$

($[\cdot]$ denotes the homotopy class of the operator in the brackets). Note that [op] only depends on the homotopy class of σ in the set of isomorphisms. Denote by $\mathcal{V}(S^1 \times X)$ the set of pairs (E, σ) introduced above and consider the maps

$$v: \mathcal{V}(S^1 \times X) \to \operatorname{Vect}(S^2 \times X), \tag{9}$$

$$[\mathrm{op}]: \mathcal{V}(S^1 \times X) \to [X, \mathcal{F}(\mathcal{H})], \tag{10}$$

$$\operatorname{ind}: [X, \mathcal{F}(\mathcal{H})] \to K(X). \tag{11}$$

(9) is bijective (up to an obvious equivalence between paris (E, σ)), cf. (b). Thus we get a map

$$\operatorname{ind} \circ [\mathrm{op}] \circ v^{-1}: \operatorname{Vect}(S^2 \times X) \to K(X). \tag{12}$$

Since (12) is obviously additive, there is induced a map

$$\alpha_1: K(S^2 \times X) \to K(X). \tag{13}$$

Using (3) and restricting (13) to $K(\mathbb{R}^2 \times X)$ gives us a homomorphism

$$\alpha: K(\mathbb{R}^2 \times X) \to K(X). \tag{14}$$

(d) The definition of (11) is functorial in the following sense. If Y is another compact space and $f: Y \to X$ a continuous map, the following diagram commutes

$$\begin{array}{ccc} [X, \mathcal{F}(\mathcal{H})] & \xrightarrow{\operatorname{ind}} & K(X) \\ \downarrow f^* & & \downarrow f^* \\ [Y, \mathcal{F}(\mathcal{H})] & \xrightarrow{\operatorname{ind}} & K(Y) \end{array} \tag{15}$$

(the definition of $f^*: [X, \mathcal{F}(\mathcal{H})] \to [Y, \mathcal{F}(\mathcal{H})]$ is obvious).

(e) The clutching construction mentioned in (b) can be applied to $X = \{p\}$ and $E = \mathbb{C}$, $\sigma = z^{-1}$. In this case (5) is equal to $(z, e) \to (z, z^{-1} e)$, $e \in \mathbb{C}$. Denote by H_{-1} the resulting vector bundle over S^2. H_{-1} is called *Hopf bundle*. Set

$$b_0 = [H_{-1}] - [\mathbb{C}] \in \widetilde{K}(S^2). \tag{16}$$

By $r: \mathbb{C}^+ \to S^2$ an element $r^* b_0 = b \in \widetilde{K}(\mathbb{C}^+) = K(\mathbb{R}^2)$ is defined and b is just the Bott element mentioned at the beginning. For $X = \{p\}$ (14) has the form $\alpha: K(\mathbb{R}^2) \to \mathbb{Z}$ and (6) means that

$$\alpha(b) = 1. \tag{17}$$

The proof of 1.1.3.3, Theorem 1 is based on the following

Theorem 1. *For each locally compact space X there exists a homomorphism*

$$\alpha = \alpha_X : K(\mathbb{R}^2 \times X) \to K(X) \tag{18}$$

with the following properties:

(i) α *is functorial with respect to* X;

(ii) *if Y is another locally compact space, the following diagram commutes*

$$\begin{array}{ccc} K(\mathbb{R}^2 \times X) \otimes K(Y) & \xrightarrow{t_1} & K(\mathbb{R}^2 \times X \times Y) \\ \downarrow{\scriptstyle \alpha_X \otimes \mathrm{id}_Y} & & \downarrow{\scriptstyle \alpha_{X \times Y}} \\ K(X) \otimes K(Y) & \xrightarrow{t_2} & K(X \times Y) \end{array} \tag{19}$$

(*the t_j ($j = 1, 2$) denote the corresponding external multiplications*);

(iii) *for $X = \{p\}$ and the Bott class $b \in K(\mathbb{R}^2)$ we have*

$$\alpha(b) = 1 . \tag{20}$$

Proof: For compact X define α by (14). For this α, (20) and (17) are equivalent. The functoriality of a is a simple consequence of (15). In order to verify (19) for compact spaces we have to show that the difference

$$t_1(\alpha_X \otimes \mathrm{id}) (u \otimes v) - \alpha_{X \times Y}(t_2(u \otimes v)) \tag{21}$$

vanishes for arbitrary $u \in K(\mathbb{R}^2 \times X)$, $v \in K(Y)$. Since all mappings occurring here are $K(Y)$ module homomorphisms, it is sufficient to restrict the consideration to the case $v = [\mathbb{C}]$ (i.e. the class represented by the trivial bundle $Y \times \mathbb{C}$). Denote by $\pi: X \times Y \to X$ the canonical projection. Then (21) is equal to $\pi^*\alpha_X(u) - \alpha_{X \times Y}(\pi^* u)$ and this vanishes because of the functoriality of α. Thus (19) commutes.

For locally compact X define α corresponding to the diagram with exact rows

$$\begin{array}{ccccccc} 0 & \to & K(\mathbb{R}^2 \times X) & \to & K(\mathbb{R}^2 \times X^+) & \to & K(\mathbb{R}^2) \\ & & \downarrow & & \downarrow{\scriptstyle \alpha} & & \downarrow{\scriptstyle \alpha} \\ 0 & \to & K(X) & \to & K(X^+) & \to & K(+) \end{array} \tag{22}$$

(the map to be defined is on the left). As usual X^+ denotes the one-point-compactification of X and $+$ the infinity point. The right square in (22) commutes because of the functoriality of α for compact spaces. By passing from X to X^+ functoriality can be easily proved for locally compact spaces, too. □

Proof of 1.1.3.3, Theorem 1: We shall prove that for every locally compact space X the mappings $\beta: K(X) \to K(\mathbb{R}^2 \times X)$ and $\alpha: K(\mathbb{R}^2 \times X) \to K(X)$ are inverse to each other. In order to show $\alpha\beta = \mathrm{id}$ substitute $X = \{p\}$ and replace Y by X in (19). Then for $u \in K(X)$ $u = \alpha(b) \cdot u = \alpha\beta u$ follows from (20). Next prove $\beta\alpha = \mathrm{id}$. Let $u \in K(\mathbb{R}^2 \times X)$ and consider the map $\varphi: X \times \mathbb{R}^2 \to \mathbb{R}^2 \times X$ defined by $\varphi((x, y)) = (y, x)$. Then the pull back

$$\varphi^*: K(\mathbb{R}^2 \times X) \to K(X \times \mathbb{R}^2) \tag{23}$$

is an isomorphism. Now $\beta\alpha(u) = b\alpha(u) = u$ is equivalent to $(\alpha(u))\, b = \varphi^*(u)$. From (19) it follows that $(\alpha(u))\, b = \alpha(ub)$. Here $ub \in K(\mathbb{R}^2 \times X \times \mathbb{R}^2)$. By $(x, y, z) \to (z, y, x)$ a map $\tau: \mathbb{R}^2 \times X\ \mathbb{R}^2 \to \mathbb{R}^2 \times X \times \mathbb{R}^2$ is defined. It is easily seen that τ is homotopic through diffeomorphisms to the identity. Thus the induced map τ^* for K groups is equal to the identity. From $\tau^*(ub) = b\varphi^*(u)$ it follows that

$$\alpha(ub) = \alpha\big(\tau^*(ub)\big) = \alpha\big(b\varphi^*(u)\big) = \varphi^*(u)$$

(in the latter equation $\alpha\beta = \mathrm{id}$ was used). Since (23) is an isomorphism, β has to be an isomorphism. □

Proof of 3.2.2.1, Theorem 2: Let $\sigma \in \mathfrak{A}^{(0)}$, $\sigma: \pi^*E \to \pi^*F$, be elliptic. By σ there is induced an isomorphism $\sigma_0: E' \to F'$. Pull back of σ_0 to SY gives an isomorphism $p^*\sigma_0: p^*E' \to p^*F'$. Denote by $s: S^1 \times SY \to SY$ the canonical projection. Then we get an isomorphism $s^*p^*\sigma_0: s^*p^*E' \to s^*p^*F'$. Applying the remarks made at the end of 3.1.1.1 another isomorphism $s^*p^*E' \to s^*p^*F'$ is connected with σ. Composing with $(s^*p^*\sigma_0)^{-1}$ we get an isomorphism

$$\sigma^1: s^*p^*E' \to s^*p^*E' . \tag{24}$$

Following (c) we get a family of Fredholm operators

$$\pi^+\sigma': p^*\big(L^2_+(S^1) \otimes E'\big) \to p^*\big(L^2_+(S^1) \otimes E'\big)$$

(cf. 2.1.2.1) and the following diagram commutes

$$\begin{array}{ccc} p^*\big(L^2_+(S^1) \otimes E'\big) & \xrightarrow{\pi^+\sigma^1} & p^*\big(L^2_+(S^1) \otimes E'\big) \\ \big\downarrow & & \big\downarrow \\ p^*(V^+ \otimes E') & \xrightarrow{\Pi^+\sigma^2} & p^*(V^+ \otimes E') \end{array} \tag{25}$$

with $\sigma^2 = \sigma_0^{-1}(\sigma|_Y)$. The vertical isomorphisms are induced by $A: L^2_+(S^1) \to V^+$, $A = \varkappa^* \circ (1 + z)$, $\varkappa$: $\mathbb{R}_\nu \to S^1_z$, $\varkappa(\nu) = z = (1 - i\nu)\,(1 + i\nu)^{-1}$ (cf. 2.1.1.1). The operator family

$$\Pi^+\sigma: p^*(V^+ \otimes E') \to p^*(V^+ \otimes F') \tag{26}$$

satisfies $\Pi^+\sigma = \gamma \circ \Pi^+\sigma^2$. Here $\gamma: p^*(V^+ \otimes E') \to p^*(V^+ \otimes F')$ is an isomorphism induced by $\sigma_0: E' \to F'$. Thus $\mathrm{ind}_{SY}\Pi^+\sigma = \mathrm{ind}_{SY}\Pi^+\sigma^2$. From (25) it follows that $\mathrm{ind}_{SY}\Pi^+\sigma^1 = \mathrm{ind}_{SY}\Pi^+\sigma^2$ and hence

$$\mathrm{ind}_{SY}\Pi^+\sigma = \mathrm{ind}_{SY}\Pi^+\sigma^1 . \tag{27}$$

Because of 2.1.2.3, Proposition 7 it is not essential whether $\Pi^+\sigma$ is considered as a family of operators (26) or as

$$\Pi^+\sigma: p^*(H^+ \otimes E') \to p^*(H^+ \otimes F') .$$

The element in $K(\mathbb{R}^2 \times SY)$ defined by the clutching isomorphism (24) (cf. (b)) is equal to the element $d_2(\sigma)$ in 3.2.2.1.(5). Now Theorem 1 and the proof of 1.1.3.3, Theorem 1 show that $\beta(\mathrm{ind}_{SY}\Pi^+\sigma) = d_2(\sigma)$. □

Conclude this section with some remarks about the connection of the previous consideration with 3.2.1.2. Let X be a compact C^∞ manifold with boundary Y. Compare the variable $z = (1 - i\nu)\,(1 + i\nu)^{-1}$ introduced in connection with the map $\varkappa$: $\mathbb{R}_\nu \to S^1$ with the symbol

$$\zeta = \big(\delta_\varepsilon(\xi) - i\nu\big)\,\big(\delta_\varepsilon(\xi) + i\nu\big)^{-1} \in \mathfrak{A}^{(0)} \tag{28}$$

over U ($U \cong Y \times [0, 1)$ a neighbourhood of Y in X). Since $\zeta \simeq \zeta_1 = (|\xi'| - i\nu)$ $(|\xi'| + i\nu)^{-1}$ through elliptic symbols in $\tilde{\mathfrak{A}}^{(0)}$ (cf. the proof of 3.2.1.2, Theorem 1), we have $\mathrm{ind}_{SY}\Pi^+\zeta = \mathrm{ind}_{SY}\Pi^+\zeta_1$. Moreover,

$$\zeta_1|_{SY} = z . \tag{29}$$

Take in (b) SY as the parameter space and for E the trivial one-dimensional bundle $\mathbb{C}$ over SY. Then $v(\mathbb{C}, z^{-1})$ corresponds to the bundle on $S^2 \times SY$ obtained by pull back of H_{-1} with respect to the canonical projection $S^2 \times SY \to S^2$. The corresponding difference element $d_2(z^{-1}) \in K(\mathbb{R}^2 \times SY)$ (cf. 3.2.2.1.(5)) is identical with the pull back of the Bott element to $\mathbb{R}^2 \times SY$ with respect to $\mathbb{R}^2 \times SY \to \mathbb{R}^2$. In view of 3.1.2.1, Lemma 3 we have $\operatorname{ind}_{SY} \Pi^+ z^{-1} = [\mathbb{C}]$. In this sence the identification (29) is compatible with the constructions in the proof of Theorem 1 and 1.1.3.3, Theorem 1.

More generally, we have the following situation. Let v be the clutching construction mentioned in (b). For $G_0 \in \operatorname{Vect}(Y)$ set

$$m_+ = v(p^*G_0, z^{-1} \cdot 1_{p^*G_0}) - [q^*G_0] \in K(S \times SY)$$

($q: S^2 \times SY \to Y$ denotes the canonical projection). Then m_+ corresponds to some element in $K(\mathbb{R}^2 \times SY)$. With the isomorphism $\alpha: K(\mathbb{R}^2 \times SY) \to K(SY)$ considered in Theorem 1 we have $\alpha(m^+) = [p^*G_0] \in K(SY)$. Similarly, for $J_0 \in \operatorname{Vect}(Y)$ we have $\alpha(m_-) = -[p^*J_0]$ with $m_- = v(p^*J_0, z \cdot 1_{p^*J^0}) - [q^*J_0] \in K(S^2 \times SY)$. Now consider the symbol σ in 3.2.1.2, Theorem 1. Then

$$\alpha\big(d_2(\sigma)\big) = \alpha(m_+) + \alpha(m_-)\,.$$

This gives a certain connection between 3.2.1.2, Theorem 1 and 3.2.2.1, Theorem 2.

3.2.2.3. The Index Homomorphism

Following Atiyah/Singer [2, 1] we introduce the index homomorphism (called *topological index*)

$$\operatorname{ind}_t: K(T\Omega) \to \mathbb{Z}\,, \tag{1}$$

Ω being a manifold. Here we only give the basic definitions and results without proofs. First consider a compact manifold X. Let U be another manifold and $i: X \to U$ an embedding. Define a homomorphism

$$i_!: K(TX) \to K(TU')\,. \tag{2}$$

For this let N be an open tubular neighbourhood of X in U. Then N may be identified with the normal bundle of X in U and TN is an open neighbourhood of TX in TU. There is a natural identification between TN and $\pi^*(N \otimes_{\mathbb{R}} \mathbb{C})$, $\pi: TX \to X$. So we have the Thom homomorphism

$$\lambda: K(TX) \to K(TN) \tag{3}$$

(cf. 1.1.3.3.(7)). Since TN is open in TU, we have the pull back

$$k_*: K(TN) \to K(TU) \tag{4}$$

with respect to $(TU)^+ \to (TN)^+$ (cf. 1.1.3.2.(3)). Now define $i_! = k_* \circ \lambda$. In particular, if $X = \{p\}$ is a single point and $j: \{p\} \to \mathbb{R}^n$ the embedding as origin, we have $j_!: K(T\{p\}) \cong \mathbb{Z} \to K(\mathbb{C}^n)$ and $j_!$ is just the Thom homomorphism for the vector space $\mathbb{C}^n$ regarded as a bundle over $\{p\}$. $j_!$ can be interpreted as repeated application of Bott isomorphisms $K(\mathbb{C}^l) \to K(\mathbb{C}^{l+1})$, $l = 0, \ldots, n$. Now put $U = \mathbb{R}^n$, n sufficiently large, and

$$\operatorname{ind}_t = (j_!)^{-1} \circ i_!: K(TX) \to \mathbb{Z}\,. \tag{5}$$

It can be proved that ind_t is independent of the choice of the embedding. Moreover, ind_t has the following properties. If $f\colon X \to Y$ is a diffeomorphism, the diagram

$$\begin{array}{ccc} K(TX) & \xleftarrow{f^*} & K(TY) \\ & {\scriptstyle \mathrm{ind}_t}\searrow \quad \swarrow{\scriptstyle \mathrm{ind}_t} & \\ & \mathbb{Z} & \end{array}$$

commutes (ind_t is related to the space under consideration). The index satisfies the following axioms:

(A1) If X is a point, ind_t is the identity on $\mathbb{Z}$;

(A2) if X, Y are compact manifolds and $i\colon X \to Y$ an inclusion, the diagram

$$\begin{array}{ccc} K(TX) & \xrightarrow{i_!} & K(TY) \\ & {\scriptstyle \mathrm{ind}_t}\searrow \quad \swarrow{\scriptstyle \mathrm{ind}_t} & \\ & \mathbb{Z} & \end{array} \tag{6}$$

commutes.

These axioms uniquely characterize the homomorphism (5).

In order to extend ind_t to non-compact manifolds Ω, use the following excision axiom. Let Ω be a non-compact manifold and $j\colon \Omega \to X$, $j'\colon \Omega \to X'$ two open embeddings into compact manifolds X, X'. Then the diagram

$$\begin{array}{ccccc} & & K(TX) & & \\ & {\scriptstyle j_*}\nearrow & & \searrow{\scriptstyle \mathrm{ind}_t} & \\ K(T\Omega) & & & & \mathbb{Z} \\ & {\scriptstyle j'_*}\searrow & & \nearrow{\scriptstyle \mathrm{ind}'_t} & \\ & & K(TX') & & \end{array}$$

commutes. j_* refers to the open embedding $T\Omega \to TX$, similarly does j'_*. Thus (1) can be defined by composing $\mathrm{ind}_t^X \circ j_* = \mathrm{ind}_t^\Omega$.

Finally, mention a multiplicative axiom. If X, Y are manifolds and $a \in K(TX)$, $b \in K(TY)$, then

$$\mathrm{ind}_t^{X \times Y}(ab) = \mathrm{ind}_t^X a \cdot \mathrm{ind}_t^Y b\,. \tag{7}$$

Here, we speak about the external multiplication $K(TX) \times K(TY) \to K(TX \times TY) = K\big(T(X \times Y)\big)$.

3.2.2.4. The Index Theorem for Boundary Value Problems

In this section we shall give a proof of an analogue of the Atiyah-Singer index theorem (cf. 1.2.4.2, Theorem 7) for elliptic boundary value problems in $\mathfrak{G}$. The proof essentially consists in a reduction to the case of closed compact manifolds by a suitable difference construction which assigns an element $d(\mathcal{A}) \in K(T\Omega)$ to any elliptic operator $\mathcal{A} \in \mathfrak{G}$ so that

$$\mathrm{ind}_a \mathcal{A} = \mathrm{ind}_t\, d(\mathcal{A})\,. \tag{1}$$

The right hand side is given by 3.2.2.3.(1). Consider a neighbourhood U of Y in X, $U \cong Y \times [0, 1)$ and let Y' be the submanifold of X corresponding to $Y \times \{\frac{1}{2}\}$. Then we have a diffeomorphism $Y \to Y'$ and thus an embedding $i\colon Y \to \Omega$. The normal bundle W of Y' in U is diffeomorphic to $Y' \times (0, 1)$ and can be identified with $\mathrm{int}\, U$. Since $TW \cong TY' \times \mathbb{R}^2$ and $TY' \cong TY$, we have the Bott isomorphism

$K(TY) \to K(TW)$. Moreover, TW is an open subset of $T\Omega$. Applying 1.1.3.2.(3) we get a homomorphism $K(TW) \to K(T\Omega)$. Composing with the Bott isomorphism given a homomorphism

$$i_!: K(TY) \to K(T\Omega)\,. \tag{2}$$

(2) is of course a special case of 3.2.2.3.(2).

Define the set

$$\mathfrak{E}(X, Y) = \{\mathcal{A} \in \mathfrak{G}: \mathcal{A} \text{ elliptic}\}\,.$$

Similarly, let $\mathfrak{E}(M)$ be the set of all classical elliptic PDOs on the closed compact manifold M. In 1.2.4.2 we mentioned a difference construction d. This means nothing else but to interprete an isomorphism $\sigma_A: \pi^*E \to \pi^*F$ ($E, F \in \mathrm{Vect}\,(M)$, $\pi: TM \setminus 0 \to M$) as an element in $K(TM)$. Thus if σ_A is the homogeneous principal symbol of $A \in \mathfrak{E}(M)$, there is a well defined map

$$d: \mathfrak{E}(M) \to K(TM)\,. \tag{3}$$

For simplicity denote (3) as difference construction, too.

Note that an elliptic operator r^+A_1: $C^\infty(X, E) \to C^\infty(X, F)$ with homogeneous principal symbol $\sigma_{A_1} \in \mathfrak{A}_\Omega^{(0)}$ (i.e. $r^+A_1 \in \mathfrak{E}(X, Y)$) corresponds in a natural way to an element $d(r^+A_1) \in K(T\Omega)$ and $\mathrm{ind}_a\,(r^+A_1) = \mathrm{ind}_t\, d(r^+A_1)$ (cf. 1.2.4.2, Theorem 7 and 3.2.2.3).

Theorem 1. *There exists a uniquely determined map*

$$d: \mathfrak{E}(X, Y) \to K(T\Omega) \tag{4}$$

with the following properties:

(i) $d(\mathcal{A} \oplus \mathcal{B}) = d(\mathcal{A}) + d(\mathcal{B})$, $d(\mathcal{A}_1\mathcal{A}_2) = d(\mathcal{A}_1) + d(\mathcal{A}_2)$ *(if the composition is defined)*;

(ii) *if* $\mathcal{A}^0 \simeq \mathcal{A}^1$ *is a homotopy through operators in* $\mathfrak{E}(X, Y)$, $d(\mathcal{A}^0) = d(\mathcal{A}^1)$;

(iii) *if* $\mathcal{A} = \begin{pmatrix} r^+A_1 & 0 \\ 0 & R \end{pmatrix}$ *with* $\sigma_{A_1} \in \mathfrak{A}_\Omega^{(0)}$, $d(\mathcal{A}) = d(r^+A_1) + i_! d(R)$;

(iv) $d(\mathcal{A}) = 0$ *if* $\mathcal{A}$ *is one of the operators* 3.1.2.1.(14).

Proof: First prove the uniqueness of (4). Put $\mathfrak{E}_0(X, Y) = \mathfrak{E}_0$ (cf. 3.2.1.1). The reduction of orders yields a map

$$\mathcal{L}: \mathfrak{E}(X, Y) \to \mathfrak{E}_0(X, Y)\,, \tag{5}$$

where $\mathcal{B} = \mathcal{L}(\mathcal{A})$ is defined by 3.1.2.1.(5). From (i), (iii), (iv), $d(\mathcal{L}^{s,t}_{E,J}) = 0$ follows for arbitrary $s \in \mathbb{Z}$, $t \in \mathbb{R}$, $E \in \mathrm{Vect}\,(X)$, $J \in \mathrm{Vect}\,(Y)$. Thus $(d\mathcal{L}(\mathcal{A})) = d(\mathcal{A})$ for arbitrary $\mathcal{A} \in \mathfrak{E}(X, Y)$. Now let $\mathcal{A} \in \mathfrak{E}_0(X, Y)$ and r^+A the elliptic PDO in the left upper corner. If $\mathcal{A}$ is written in the form of 3.1.1.1.(1), $\mathrm{ind}_{SY}\Pi^+\sigma_A = [p^*G] - [p^*J]$ (cf. 3.1.1.1, Proposition 11). Consider the operator $\mathcal{B}_G$ given by 3.1.2.1. (18) and let $\mathcal{B}_G^{-1}$ be a parametrix of $\mathcal{B}_G$. Define

$$\mathcal{A}_0 = \mathcal{A} \oplus \mathcal{L}(\mathcal{B}_G^{-1}) \oplus \mathcal{L}(\mathcal{B}_J)\,.$$

Using (ii), (iv) we get $d\big(\mathcal{L}(\mathcal{B}_G^{-1})\big) = 0$, $d\big(\mathcal{L}(\mathcal{B}_J)\big) = 0$. Denote by r^+A_0 the PDO in the left upper corner of $\mathcal{A}_0$. In view of 3.1.1.1, Proposition 11 it follows that $\mathrm{ind}_{SY}\Pi^+\sigma_{A_0}' = 0$. Using 3.2.1.2, Corollary 4, 3.2.1.1, Proposition 3 and Corollary 4 we find an operator $\mathcal{C}_1 \in \mathfrak{E}_0(X, Y)$ with $d(\mathcal{C}_1) = 0$ and $\mathcal{A}_0 \oplus \mathcal{C}_1 \simeq \mathcal{A}_1$ through operators in $\mathfrak{E}_0(X, Y)$, where

$\mathcal{A}_1$ has the form

$$\mathcal{A}_1 = \begin{pmatrix} r^+A_1 & K_1 \\ r'T_1 & Q_1 \end{pmatrix} \tag{6}$$

with $\sigma_{A_1} \in \mathfrak{A}_\Omega^{(0)}$. Since r^+A_1 is Fredholm, we have a homotopy in $\mathfrak{E}_0(X, Y)$

$$\mathcal{A}_1 \simeq \begin{pmatrix} r^+A_1 & 0 \\ 0 & R_1 \end{pmatrix}, \quad R_1 = Q_1 - (r'T_1)(r^+A_1)^{-1} K_1 \tag{7}$$

with elliptic R_1 (cf. 3.1.1.5). Then, by (i), (ii), (iii), we get a necessary representation

$$d(\mathcal{A}) = d(r^+A_1) + i_! d(R_1) . \tag{8}$$

In order to prove the existence we have to check that the element in $K(T\Omega)$ on the right in (8) does not depend on the choice of the homotopies. Let

$$\mathcal{A}_0 \oplus \mathcal{E}_i \simeq r^+A_i \oplus R_i ,$$

$i = 1, 2$, be two homotopies of the type described above. Passing to $\mathcal{A}_0 \oplus \mathcal{E}_i \oplus \mathcal{J}_i$, where $\mathcal{J}_i$ are the identical operators in $C^\infty(X, E_i) \oplus C^\infty(Y, J_i)$ for suitable $E_i \in \operatorname{Vect}(X)$, $J_i \in \operatorname{Vect}(Y)$, $i = 1, 2$, and denoting $\mathcal{E}_i \oplus \mathcal{J}_i$ again by $\mathcal{E}_i$, we can suppose without loss of generality that there is a homotopy in $\mathfrak{E}_0(X, Y)$

$$r^+A_1 \oplus R_1 \simeq r^+A_2 \oplus R_2 . \tag{9}$$

Multiplication by a parametrix of $r^+A_2 \oplus R_2$ gives a homotopy

$$(r^+A_1)(r^+A_2)^{-1} \oplus R_1R_2^{-1} \simeq 1 \oplus 1 \tag{10}$$

with obvious notation on the right. Now the proof of Theorem 1 is finished if we show the following

Lemma 2. *From* (9) *it follows that*

$$d(r^+A_1) + i_! d(R_1) = d(r^+A_2) + i_! d(R_2) . \tag{11}$$

Proof: (10) shows that we have to prove the following assertion. If we are given an elliptic operator $r^+A \oplus Q \in \mathfrak{E}_0(X, Y)$ with $\sigma_A \in \mathfrak{A}_\Omega^{(0)}$ and $Q \in \mathfrak{E}_0(Y)$ with

$$r^+A \oplus Q \simeq 1 \oplus 1 \tag{11}$$

(homotopy in $\mathfrak{E}_0(X, Y)$), we have

$$d(r^+A) + i_! d(Q) = 0 . \tag{12}$$

Denote the homotopy (11) by

$$\mathcal{A}^t = \begin{pmatrix} r^+A^t + r'B^t & K^t \\ r'T^t & Q^t \end{pmatrix} \tag{13}$$

with $\mathcal{A}^0 = (r^+A) \oplus Q$, A^1 equal to the identiy. It is useful to take the whole t half axis $\mathbb{R}_+$ as the parameter space and to define $\mathcal{A}^t$ as the identity for $t \geqq 1$. Let U be a tubular neighbourhood of Y in X and $U \to Y \times [0, \infty)$ $(x \to (x', x_n))$ a diffeomorphism. Define a new symbol

$$\sigma_{A_1^t} = \begin{cases} \sigma_{A^{t\delta}} & \text{if } t \leqq 1 , \\ \sigma_{A^{(t-1)+(2-t)\delta}} & \text{if } 1 \leqq t \leqq 2 , \\ 1 & \text{if } t \geqq 2 \end{cases}$$

with $b \in C^\infty(\overline{\mathbb{R}}_+)$, $0 \leqq b \leqq 1$, $b = 0$ near Y, $b = 1$, if $x_n \geqq 1$. Then obviously $d(r^+A^0) = d(r^+A_1^0) = d(r^+A_1^1)$. Define elliptic operators by

$$\mathcal{A}_1^t = \begin{cases} \begin{pmatrix} r^+A_1^t + r'B^0 & K^0 \\ r'T^0 & Q^0 \end{pmatrix} & \text{if } 0 \leqq t \leqq 1\,, \\ \begin{pmatrix} r^+A_1^t + r'B^{t-1} & K^{t-1} \\ r'T^{t-1} & Q^{t-1} \end{pmatrix} & \text{if } t \geqq 1\,, \end{cases}$$

$r^+A_1^t = \mathrm{Op}\,(\sigma_{A_1^t})$. Then $\sigma_\Omega(\mathcal{A}_1^t)$ is constantly equal to 1 if $t \geqq 1$ and $x_n \geqq 1$. Replacing $\mathcal{A}^t$ by $\mathcal{A}_1^{t-1}$ for $t \geqq 1$ we see that we can assume the interior symbol of the considered homotopy to be equal to 1 if $x_n \geqq 1$ or $t \geqq 1$. Thus we shall suppose that $X = Y \times \overline{\mathbb{R}}_+$.

Let $\sigma^t(x, \xi)$ be a continuous family of symbols, $t \in \overline{\mathbb{R}}_+$, $\sigma^t \in \mathfrak{A}^{(0)}$ on $Y \times [0'\ \infty)$ and $\sigma^t(x, \xi) = 1$ if $x_n \geqq 1$ or $t \geqq 1$. Then σ^t defines a difference element

$$d_1(\sigma^t) \in K\big(\mathbb{R}_t^+ \times B'X,\ \mathbb{R}_t^+ \times (S'X \cup N_1)\big)$$

(cf. 3.2.2.1) and an index element

$$\mathrm{ind}_{\mathbb{R}_t^+ \times SY}\, \Pi^+\sigma^t \in K(\mathbb{R}_t^+ \times SY)$$

($\mathbb{R}_t^+ \overset{\text{def}}{=} \mathbb{R}_{+,t}$). The following Lemmata can be similarly proved as the Lemmata in 3.2.2.1 with the corresponding numbers. The slight modifications are left to the reader.

Lemma 1′. *There is a canonical isomorphism*

$$\tau\colon K\big(\mathbb{R}_t^+ \times B'X,\ \mathbb{R}_t^+ \times (S'X \cup N_1)\big) \to K(\mathbb{R}_{|\xi'|}^+ \times \mathbb{R}_\nu \times \mathbb{R}_t^+ \times SY)\,.$$

Set $d_2(\sigma^t) = \tau d_1(\sigma^t)$.

Lemma 5′. *The following diagram commutes*

$$\begin{array}{ccc} K\big(\mathbb{R}_t^+ \times B'X,\ \mathbb{R}_t^+ \times (S'X \cup N_1)\big) & \xrightarrow[\cong]{\ \tau\ } & K(\mathbb{R}_{|\xi'|}^+ \times \mathbb{R}_\nu \times \mathbb{R}_t^+ \times SY) \\ \Big\downarrow \varrho & & \Big\downarrow \delta \\ K(\mathbb{R}_t^+ \times B'X,\ \mathbb{R}_t^+ \times S'X) & \xrightarrow[\cong]{\ \lambda\ } & K(\mathbb{R}_\nu \times \mathbb{R}_t^+ \times BY,\ \mathbb{R}_\nu \times \mathbb{R}_t^+ \times SY)\,. \end{array}$$

Here ϱ is the natural restriction homomorphism and λ is similarly defined as in 3.2.2.1, Lemma 4 with additional change of the order of spaces.

The following diagram commutes

$$\begin{array}{ccc} K(\mathbb{R}_{|\xi'|}^+ \times \mathbb{R}_\nu \times \mathbb{R}_t^+ \times SY) & \xleftarrow{\ \beta\ } & K(\mathbb{R}_t^+ \times SY) \\ \Big\downarrow \delta & & \Big\downarrow \delta \\ K(\mathbb{R}_\nu \times \mathbb{R}_t^+ \times BY,\ \mathbb{R}_\nu \times \mathbb{R}_t^+ \times SY) & \xleftarrow{\ \tilde\beta\ } & K(BY, SY)\,. \end{array} \tag{14}$$

Here $\tilde\beta$ denotes the Bott isomorphism with respect to $\mathbb{R}_\nu \times \mathbb{R}_t^+$. The commutativity of the corresponding diagram is clear if β is the Bott isomorphism with respect to the same variables as in the second rows of (14) (cf. 1.1.3.3.(6)). Since the map $\mathbb{R}_{|\xi'|}^+ \times \mathbb{R}_\nu \times \mathbb{R}_t^+ \to \mathbb{R}_\nu \times \mathbb{R}_t^+ \times \mathbb{R}_{|\xi'|}^+$ of interchanging variables is homotopic to the identity, (14) commutes, too.

The following assertion is again a parameter depending variant of the corresponding theorem in 3.2.2.1.

Theorem 2′. *Under the conditions mentioned above we have*

$$\beta^{-1}\left(d_2(\sigma^t)\right) = \operatorname{ind}_{\mathbb{R}_t^+ \times SY}(\Pi^+\sigma^t)\,. \tag{15}$$

Remark 3. By $t \to x_n$ there is induced an isomorphism

$$\varrho_1\colon K(\mathbb{R}_t^+ \times B'X,\, \mathbb{R}_t^+ \times S'X) \to K(BX, SX \cup B'X) \cong K(T\Omega)$$

$(\Omega = Y \times \mathbb{R}^+)$ and the composition

$$\varrho_1 \circ \lambda^{-1} \circ \tilde{\beta}^{-1}\colon \quad K(BY, SY) \to K(T\Omega)$$

corresponds to the map $i_!\colon K(TY) \to K(T\Omega)$ (note that $K(BY, SY) \cong K(TY)$).

Remark 4. The family σ^t, $t \in \overline{\mathbb{R}}_t^+$, defines an element

$$d(\sigma^t) \in K\left(\overline{\mathbb{R}}_t^+ \times BX, (\{0\} \times (SX \cup B'X)) \cup (\overline{\mathbb{R}}_t^+ \times N_1)\right).$$

The inclusion

$$\left(\overline{\mathbb{R}}_t^+ \times BX, \{0\} \times (SX \cup B'X)\right) \hookrightarrow \left(\overline{\mathbb{R}}_t^+ \times BX, (\{0\} \times (SX \cup B'X)) \cup (\overline{\mathbb{R}}_t^+ \times N_1)\right)$$

is a homotopy equivalence of pairs. Thus the restriction to Y

$$\varrho_0\colon K\left(\overline{\mathbb{R}}_t^+ \times BX, (\{0\} \times (SX \cup B'X)) \cup (\overline{\mathbb{R}}_t^+ \times N_1)\right) \to K\left(\mathbb{R}_t^+ \times B'X, \mathbb{R}_t^+ \times (S'X \cup N_1)\right)$$

is an isomorphism.

The composition

$$\varrho_1 \circ \varrho \circ \varrho_0\colon K\left(\overline{\mathbb{R}}_t^+ \times BX, (\{0\} \times (SX \cup B'X)) \cup (\overline{\mathbb{R}}_t^+ \times N_1)\right) \to K(BX, SX \cup B'X)$$

is equal to the homomorphism corresponding to the restriction to $t = 0$, since the mappings $\overline{\mathbb{R}}^+ \ni t \to (0, t) \in (\overline{\mathbb{R}}^+)^2$ and $\overline{\mathbb{R}}^+ \ni t \to (t, 0) \in (\overline{\mathbb{R}}^+)^2$ are homotopic.

Collecting all information we get the following commutative diagram

$$\begin{array}{ccccc}
K\left(\overline{\mathbb{R}}_t^+ \times BX, (\{0\} \times (SX \cup B'X)) \cup (\overline{\mathbb{R}}_t^+ \times N_1)\right) & & & & \\
\cong \downarrow \varrho_0 & & & & \\
K\left(\mathbb{R}_t^+ \times B'X, \mathbb{R}_t^+ \times (S'X \cup N_1)\right) & \xrightarrow[\cong]{\tau} & K(\mathbb{R}_{|\xi'|}^+ \times \mathbb{R}_\nu \times \mathbb{R}_t^+ \times SY) & \xrightarrow[\cong]{\beta^{-1}} & K(\mathbb{R}_t^+ \times SY) \\
\downarrow \varrho & & \downarrow \delta & & \downarrow \delta \\
K(\mathbb{R}_t^+ \times B'X, \mathbb{R}_t^+ \times S'X) & \xrightarrow[\cong]{\lambda} & K(\mathbb{R}_\nu \times \mathbb{R}_t^+ \times BY, \mathbb{R}_\nu \times \mathbb{R}_t^+ \times SY) & \xrightarrow[\cong]{\tilde{\beta}^{-1}} & K(BY, SY) \\
\cong \downarrow \varrho_1 & & & & \\
K(T\Omega) & & & &
\end{array} \tag{16}$$

Now we can finish the proof of Lemma 2. With the given homotopy $\mathcal{A}^t$, $0 \leqq t < \infty$, a continuous family σ_{A^t} with $\sigma_{A^t} = 1$ for $t \geqq 1$ or $x_n \geqq 1$ is connected. Thus we have an element

$$d(\sigma_{A^t}) \in K\left(\overline{\mathbb{R}}_t^+ \times BX, (\{0\} \times (SX \cup B'X)) \cup (\overline{\mathbb{R}}_t^+ \times N_1)\right).$$

For $t = 0$ we get a difference element $d(\sigma_{A^0}) \in K(T\Omega)$ (note that $\sigma_{A^0} \in \mathfrak{A}_\Omega^{(0)}$) and in view of Remark 4

$$\varrho_1 \varrho \varrho_0 \left(d(\sigma_{A^t})\right) = d(\sigma_{A^0}) \tag{17}$$

The element $d_2(\sigma_{A^t}) = \tau\varrho_0 d(\sigma_{A^t})$ corresponds to an index element

$$\mathrm{ind}_{\mathbb{R}^+\times SY}\,\Pi^+\sigma_{A^t}\in K(\mathbb{R}^+\times SY)\,.$$

Using (17) and Remark 3 we obtain from (16)

$$i_!\delta(\mathrm{ind}_{\mathbb{R}^+\times SY}\,\Pi^+\sigma_{A^t}) = d(\sigma_{A^0})\,. \tag{18}$$

First assume that

$$\mathrm{ind}_{\mathbb{R}^+\times SY}\,\Pi^+\sigma_{A^t} = 0 \tag{19}$$

and apply a version of 3.2.1.2, Corollary 4 to families of symbols. Then we find that there is an $N\in\mathbb{Z}_+$ so that $\sigma_{A^t}\oplus 1_{\mathbb{C}^N}$ can be deformed through families of symbols in $\mathfrak{A}^{(0)}$ into a family $\sigma_{A_1^t}$ with $\sigma_{A_1^t}\in\mathfrak{A}_1^{(0)}$ for all t. In view of (18) we get

$$d(\sigma_{A_1^0}) = 0\,. \tag{20}$$

Thus, after addition of some identity operator belonging to a bundle over Y, we can suppose that the given homotopy is of the form

$$\begin{pmatrix} 1+r'B^t & K^t\\ r'T^t & Q^t\end{pmatrix},\quad t\geqq 0\,, \tag{21}$$

equal to the identity for $t\geqq 1$. Here we used a version of 3.2.1.1, Proposition 3 for families. Thus the given homotopy $\mathcal{A}^t$ can be replaced by $1\oplus Q^0\simeq 1\oplus 1$. Note that after the addition of an identity (21) can be replaced by a homotopy with $r'B^t\equiv 0$ (cf. 3.2.1.1, Corollary 4). Take into account this identity in Q^0, too. Then

$$Q^t - (r'T^t)\,K^t\,,\quad t\geqq 0$$

is a homotopy of elliptic PDOs on Y connecting 1 and Q^0 (cf. 3.1.1.5). Therefore $d(Q^0) = 0$, and (12) is proved in the case (19).

In order to prove (12) in general, i.e. for

$$\mathrm{ind}_{\mathbb{R}^+\times SY}\,\Pi^+\sigma_{A^t} \neq 0\,, \tag{22}$$

it is sufficient to consider a special homotopy of symbols with the given index element. That the special choice does not influence the result follows from direct addition of the given homotopy and a family of parametrices of the special homotopy where the direct sum satisfies (19), for which the assertion is proved.

The element $q = \mathrm{ind}_{\mathbb{R}^+\times SY}\,\Pi^+\sigma_{A^t}\in K^{-1}(SY) = K(\mathbb{R}_t^+\times SY)$ can be represented as a C^∞ matrix function

$$\tilde{q}(x',\xi')\colon SY\to \mathrm{GL}\,(N,\mathbb{C})$$

(cf. 1.1.3.1, Theorem 7; do not mix up the notations!). Denote by $c_s(x',\xi')$, $s\geqq 0$, a homotopy connecting $\tilde{q}\pm\tilde{q}^{-1}$ with the identical matrix (for $s\geqq 1$) (cf. 1.1.3.1, Lemma 8) and let $2N$ be the number of corresponding rows and columns. Similarly as in 3.1.2.1 let $\omega = \omega_{\mathbb{C}^N} = -l^+(l^-)^{-1}\cdot 1_{\pi^*\mathbb{C}^N}$ and

$$\sigma^t(x,\xi) = c_{t+x_n}\left(x',\frac{\xi'}{|\xi'|}\right)\begin{pmatrix}\omega^a & 0\\ 0 & 1\end{pmatrix} c_{t+x_n}^{-1}\left(x',\frac{\xi'}{|\xi'|}\right).$$

The function a is defined in 3.1.2.1. The symbol σ^t is defined for $\xi'\neq 0$ and can be extended as identity for $\xi' = 0$. Thus σ^t is the identity for small ξ' or $t+x_n$ large. Now $\sigma^t(\sigma^0)^{-1}$ is a continuous family and in view of 3.1.2.1, Lemma 3 (with obvious notations)

$$q = \operatorname{ind}_{\mathbb{R}^+ \times SY} \Pi^+ (\sigma^t(\sigma^0)^{-1}) = \operatorname{ind}_{\mathbb{R}^+ \times SY} \Pi^+ \sigma^t - \operatorname{ind}_{\mathbb{R}^+ \times SY} \Pi^+ \sigma^0$$
$$= (c_t \mathbb{C}^N) - (\mathbb{C}^N)$$

in $K(\mathbb{R}^+ \times SY)$. The last $\mathbb{C}^N$ corresponds to the first factor in $\mathbb{C}^N \times \mathbb{C}^N$. Similarly as in 3.1.2.1

$$\mathscr{C}^t = \begin{pmatrix} r^+ C^t \\ \pi c_t^{-1} r' \end{pmatrix}$$

with $C^t = \operatorname{Op}(\sigma^t)$, $\pi\colon \mathbb{C}^N \times \mathbb{C}^N \to \mathbb{C}^N$ projection to the first factor, is a family of elliptic boundary value problems. $\mathcal{A}^t = \mathscr{C}^t(\mathscr{C}^\infty)^{-1}$ is a family of elliptic operators with $\mathcal{A}^\infty = 1$, $\mathcal{A}^0 = (r^+ A^0) \oplus Q^0$, $\sigma_{Q_0} = \tilde{q}^{-1}$. With the corresponding element $d(\tilde{q}^{-1}) \in K(TY)$

$$d(\sigma_{\mathcal{A}^0}) + i_! d(\tilde{q}^{-1}) = i_! d(\tilde{q}) - i_! d(\tilde{q}) = 0\,.$$

follows from (18). Thus Lemma 2 and hence Theorem 1 are completely proved. □

Note that (4) induces a homomorphism

$$d\colon \operatorname{Ell}(X, Y) \to K(T\Omega)$$

with $\operatorname{ind} \circ d = d \circ \operatorname{ind}_t$ (cf. 3.2.1.1).

Note also that for the calculation of the index of $\mathcal{A}$ it is not necessary to transform elliptic symbols occurring on Y to Ω by $i_!$. In view of (7) we have a homotopy of the form

$$\mathcal{A} \oplus \mathcal{L}(\mathcal{B}_G^{-1}) \oplus \mathcal{L}(\mathcal{B}_J) \oplus \mathscr{C}_1 \simeq (r^+ A_1) \oplus R_1$$

with $\sigma_{A_1} \in \mathfrak{A}_\Omega^{(0)}$, $R_1 \in \mathscr{E}(Y)$ and therefore

$$\operatorname{ind} \mathcal{A} = \operatorname{ind}(r^+ A_1) + \operatorname{ind} R_1\,.$$

3.2.3. Elliptic Complexes

3.2.3.1. Generalities on Complexes

Various problems of geometry and analysis give rise to complexes of pseudo-differential operators on manifolds. In a natural way complexes occur in the index theory (cf. Atiyah/Bott [2]). Therefore, we discuss complexes of PDOs on manifolds with boundary and generalize the concept of elliptic boundary value problems to complexes.

First recall some definitions and assertions about complexes in general and about complexes of PDOs on closed compact manifolds. If we speak about abstract complexes of operators, for simplicity we restrict ourselves to Hilbert spaces. In connection with PDOs it is convenient to consider not only Sobolev spaces, but spaces of C^∞ sections in the corresponding vector bundles, too. The assertions have to be transformed in an obvious way to this case, because here one has an analogue to the elliptic regularity which has already been used for ordinary elliptic boundary value problems (cf. 3.1.1.1, Theorem 5). The role of compact operators is played in this case by the operators with C^∞ kernels.

Let $\mathscr{H}_0, \ldots, \mathscr{H}_N$ be Hilbert spaces. A sequence of linear continuous operators $A_k\colon \mathscr{H}_k \to \mathscr{H}_{k+1}$ $(k = 0, \ldots, N-1)$ is called a *complex* if $A_{k+1}A_k = 0$ for all k. It is convenient to set $\mathscr{H}_{-1} = \{0\}$, $\mathscr{H}_{N+1} = \{0\}$ and $A_{-1} = 0$, $A_N = 0$.

Definition 1. Let

$$\mathfrak{A}: 0 \to \mathcal{H}_0 \xrightarrow{A_0} \mathcal{H}_1 \xrightarrow{A_1} \dots \longrightarrow \mathcal{H}_{N-1} \xrightarrow{A_{N-1}} \mathcal{H}_N \to 0 \tag{1}$$

be a complex. A complex

$$\mathfrak{P}: 0 \leftarrow \mathcal{H}_0 \xleftarrow{P_0} \mathcal{H}_1 \xleftarrow{P_1} \dots \longleftarrow \mathcal{H}_{N-1} \xleftarrow{P_{N-1}} \mathcal{H}_N \leftarrow 0 \tag{2}$$

is called *parametrix* of $\mathfrak{A}$ if

$$A_{k-1}P_{k-1} + P_kA_k = 1 + C_k \tag{3}$$

for $k = 0, \dots, N$ and compact operators $C_k: \mathcal{H}_k \to \mathcal{H}_k$ (by 1 we denote the identity operator in various spaces).

If (1) is a complex, we have $\operatorname{im} A_k \subseteqq \ker A_{k+1}$. Therefore, we can speak about the factor spaces

$$H^k(\mathfrak{A}) = \ker A_{k+1}/\operatorname{im} A_k\,. \tag{4}$$

The spaces $H^k(\mathfrak{A})$ $(k = 0, \dots, N)$ are called *cohomology spaces* of the complex $\mathfrak{A}$.

Definition 2. (1) is called a *Fredholm complex* if $\dim H^k(\mathfrak{A}) < \infty$ $(k = 0, \dots, N)$. The finite number

$$\operatorname{ind} \mathfrak{A} = \sum_{k=0}^{N} (-1)^k \dim H^k(\mathfrak{A})$$

is called *index* of $\mathfrak{A}$. A complex $\mathfrak{A}$ is called *exact* if $\dim H^k(\mathfrak{A}) = 0$ for all k.

Any Fredholm operator $A: \mathcal{H}_0 \to \mathcal{H}_1$ corresponds to a Fredholm complex $\mathfrak{A}: 0 \to \mathcal{H}_0 \xrightarrow{A} \mathcal{H}_1 \to 0$ and obviously $\operatorname{ind} A = \operatorname{ind} \mathfrak{A}$. It turns out that all standard definitions and properties on Fredholm operators can be generalized to Fredholm complexes.

Theorem 3. *The following assertions are equivalent*

(i) $\mathfrak{A}$ *is a Fredholm complex,*

(ii) $\mathfrak{A}$ *has a parametrix* $\mathfrak{P}$.

Moreover if $\mathfrak{A}$ *is a Fredholm complex,* $\operatorname{im} A_k$ *is closed in* $\mathcal{H}_{k+1}$ *for all* k. *If* $\mathfrak{P}$ *is a parametrix of* $\mathfrak{A}$, *we have* $\operatorname{ind} \mathfrak{A} = -\operatorname{ind} \mathfrak{P}$.

Theorem 3 shall be proved after Theorem 4. Denote the scalar products in the Hilbert spaces $\mathcal{H}_k$ by $(\cdot,\cdot)_k$. Then, to a complex (1), an adjoint complex can be defined

$$\mathfrak{A}^*: 0 \leftarrow \mathcal{H}_0 \xleftarrow{A_0^*} \mathcal{H}_1 \xleftarrow{A_1^*} \dots \longleftarrow \mathcal{H}_{N-1} \xleftarrow{A_{N-1}^*} \mathcal{H}_N \leftarrow 0 \tag{5}$$

(obviously $(A_kA_{k-1})^* = A_{k-1}^*A_k^* = 0$). Note that the operators

$$D_k = A_{k-1}A_{k-1}^* + A_k^*A_k: \mathcal{H}_k \to \mathcal{H}_k \tag{6}$$

are self-adjoint.

Theorem 4. *Let* $\mathfrak{A}$ *be a Fredholm complex. Then* $\mathfrak{A}^*$ *is a Fredholm complex, too, the operators* D_k $(k = 0, \dots, N)$ *given by* (6) *are Fredholm operators and*

$$\operatorname{ind} \mathfrak{A} = -\operatorname{ind} \mathfrak{A}^* = \sum_{k=0}^{N} (-1)^k \dim \ker D_k\,. \tag{7}$$

Moreover for all k *there is an orthogonal decomposition*

$$\mathcal{H}_k = \operatorname{im} A_k^* \oplus \operatorname{im} A_{k-1} \oplus \ker D_k\,. \tag{8}$$

Proof: The proof of the Fredholm property of $\mathfrak{A}^*$ and of the first equation in (7) is quite simple and left to the reader. The Fredholm property of the operators D_k is obvious. First prove (8). Since D_k is self-adjoint, we have an orthogonal decomposition $\mathcal{H}_k = \ker D_k \oplus \operatorname{im} D_k$ and the Fredholm property of D_k implies $\dim \ker D_k < \infty$. Every $u \in \mathcal{H}_k$ has the form $u = v + (A_{k-1}A_{k-1}^* + A_k^*A_k)\, w$ with $v \in \ker D_k$, $w \in \mathcal{H}_k$. Obviously, $\operatorname{im} A_{k-1} + \operatorname{im} A_k^* = \operatorname{im} D_k^*$. From $(A_{k-1}u, A_k^*v) = (A_kA_{k-1}u, v) = 0$ for all $u, v \in \mathcal{H}_k$ it follows that $\operatorname{im} D_k = \operatorname{im} A_{k-1} \oplus \operatorname{im} A_k^*$. Thus (8) is proved. In order to show the second equation in (7) it is sufficient to show

$$\ker A_k = \operatorname{im} A_{k-1} \oplus \ker D_k . \tag{9}$$

Since $\mathfrak{A}$ is a complex, we have $\ker A_k \supseteq \operatorname{im} A_{k-1}$. Moreover, $\ker D_k = \ker A_{k-1}^* \cap \ker A_k$ because of $D_ku = 0 \Leftrightarrow (D_ku, u) = 0 \Leftrightarrow (A_{k-1}^*u, A_{k-1}^*u) + (A_ku, A_ku) = 0$. Thus $\ker D_k \subseteq \ker A_k$. Now let $u \in \ker A_k$. Then a decomposition

$$u = v + A_{k-1}A_{k-1}^*w + A_k^*A_kw = v + A_{k-1}w_1 + A_k^*w_2$$

follows with $v \in \ker D_k$ and hence

$$0 = A_ku = A_kv + A_kA_{k-1}w_1 + A_kA_k^*w_2 = A_kA_k^*w_2 = 0$$

so that $A_k^*w_2 = 0$. This shows $\ker A_k = \operatorname{im} A_{k-1} + \ker D_k$. From $(A_{k-1}u, v) = (u, A_{k-1}^*v) = 0$ for all $u \in \mathcal{H}_{k-1}$, $v \in \ker D_k \subseteq \ker A_{k-1}^*$ we finally obtain (9). □

Proof of Theorem 3: Let $\mathfrak{A}$ be Fredholm. We construct a parametrix $\mathfrak{P}$ for $\mathfrak{A}$. First let R_k be a parametrix of D_k inducing an isomorphism $\operatorname{im} A_k^* \to \operatorname{im} A_k^*$ and $\ker R_k = \ker D_k$. Set $P_k = R_kA_k^*$ and show that the complex of the P_k defines a parametrix of $\mathfrak{A}$. From $\operatorname{im} P_k = \operatorname{im} A_k^*$ it follows that $P_{k+1}P_k = 0$. Next remark that

$$R_{k+1}A_k \sim A_kR_k \tag{10}$$

($B_1 \sim B_2$ means $B_1 - B_2$ compact). From $A_kA_{k+1} = 0$ it follows that $A_kD_k = D_kA_k^*A_k = D_{k+1}A_k$. This yields (10) because of $R_kD_k \sim 1$, $D_kR_k \sim 1$ after multiplication from the left by R_{k+1} and the right by R_k. Using (10) we get

$$\begin{aligned} P_kA_k - A_{k-1}P_{k-1} &= R_kA_k^*A_k + A_{k-1}R_{k-1}A_{k-1}^* \\ &\sim R_kA_k^*A_k + R_kA_{k-1}A_{k-1}^* = R_kD_k \sim 1 , \end{aligned}$$

i.e. the P_k form a parametrix. Conversely, assume that $\mathfrak{A}$ is a complex possessing a parametrix so that $A_{k-1}P_{k-1} + P_kA_k = 1 + C_k$ with compact operators C_k. For $h \in \ker A_k$ we have

$$A_{k-1}P_{k-1}h = (1 + C_k)\, h , \tag{11}$$

i.e. $\operatorname{im} A_{k-1} \supseteq (1 + C_k)(\ker A_k)$. From (11) it follows that by restriction of C_k to $\ker A_k$ a map $\ker A_k \to \ker A_k$ is induced. Thus $(1 + C_k)(\ker A_k)$ is closed and has finite codimension in $\ker A_k$. Therefore, $\operatorname{im} A_{k-1}$ is closed and of finite codimension in $\ker A_k$, i.e. $\dim H^k(\mathfrak{A}) < \infty$. It is now also clear that $\operatorname{im} A_k$ in $\mathcal{H}_{k+1}$ is closed. □

Let $\mathfrak{A}$ be a complex of the form (1) and $\mathfrak{C}$ a second complex with operators Q_k and spaces $\mathcal{L}_k$. A commutative diagram

$$\begin{array}{ccccccccc} 0 \to & \mathcal{H}_0 & \xrightarrow{A_0} & \mathcal{H}_1 & \xrightarrow{A_1} & \dots & \xrightarrow{A_{N-1}} & \mathcal{H}_N & \to 0 \\ & \downarrow T_0 & & \downarrow T_1 & & & & \downarrow T_N & \\ 0 \to & \mathcal{L}_0 & \xrightarrow{Q_0} & \mathcal{L}_1 & \xrightarrow{Q_1} & \dots & \xrightarrow{Q_{N-1}} & \mathcal{L}_N & \to 0 \end{array} \tag{12}$$

with continuous operators $T_k: \mathscr{H}_k \to \mathscr{L}_k$ is called a morphism $\mathfrak{A} \to \mathfrak{C}$. With (12) there are connected further complexes in a natural way. The operators A_k induce a complex

$$\ker T: 0 \to \ker T_0 \to \ker T_1 \to \dots \to \ker T_N \to 0 \tag{13}$$

and the Q_k a complex

$$\operatorname{coker} T: 0 \to \operatorname{coker} T_0 \to \operatorname{coker} T_1 \to \dots \to \operatorname{coker} T_N \to 0 \,. \tag{14}$$

Thus one can define the cohomology spaces $H^k(\ker T)$, $H^k(\operatorname{coker} T)$ $(k = 0, \dots, N)$. To any morphism (12) corresponds a mapping cone (cf. SPANIER [1]), namely the complex

$$\mathfrak{K}: 0 \to \begin{array}{c}\mathscr{H}_0\\ \oplus\\ \{0\}\end{array} \xrightarrow{\begin{pmatrix}-A_0 & 0\\ T_0 & 0\end{pmatrix}} \begin{array}{c}\mathscr{H}_1\\ \oplus\\ \mathscr{L}_0\end{array} \xrightarrow{\begin{pmatrix}-A_1 & 0\\ T_1 & Q_0\end{pmatrix}} \begin{array}{c}\mathscr{H}_2\\ \oplus\\ \mathscr{L}_1\end{array} \xrightarrow{\begin{pmatrix}-A_2 & 0\\ T_2 & Q_1\end{pmatrix}} \dots$$

$$\dots \to \begin{array}{c}\mathscr{H}_{N-1}\\ \oplus\\ \mathscr{L}_{N-2}\end{array} \xrightarrow{\begin{pmatrix}-A_{N-1} & 0\\ T_{N-1} & Q_{N-2}\end{pmatrix}} \begin{array}{c}\mathscr{H}_N\\ \oplus\\ \mathscr{L}_{N-1}\end{array} \xrightarrow{\begin{pmatrix}0 & 0\\ T_N & Q_{N-1}\end{pmatrix}} \begin{array}{c}\{0\}\\ \oplus\\ \mathscr{L}_N\end{array} \to 0 \,.$$

Mention without proof the following.

Proposition 5. *$\mathfrak{K}$ is a Fredholm complex iff both complexes* $\ker T$ *and* $\operatorname{coker} T$ *are Fredholm. Then*

$$\operatorname{ind} \mathfrak{K} = \operatorname{ind}(\ker T) - \operatorname{ind}(\operatorname{coker} T) \,.$$

For the proof one has to show $H^0(\mathfrak{K}) \cong H^0(\ker T)$, $H^{N+1}(\mathfrak{K}) \cong H^N(\operatorname{coker} T)$. $H^k(\mathfrak{K}) \cong H^k(\ker T) \oplus H^{k-1}(\operatorname{coker} T)$, $1 \leq k \leq N$. Especially, one obtains that $\mathfrak{K}$ is exact iff both $\ker T$ and $\operatorname{coker} T$ are exact.

A morphism (12) of complexes $\mathfrak{A}$, $\mathfrak{C}$ is called isomorphism if the operators T_k $(0 \leq k \leq N)$ are isomorphisms. Then we have the following trivial

Remark 6. Let (12) be an isomorphism. $\mathfrak{A}$ is Fredholm iff $\mathfrak{C}$ is Fredholm. If the operators $S_k: \mathscr{L}_{k+1} \to \mathscr{L}_k$ form a parametrix of $\mathfrak{C}$, the operators

$$P_k = T_k^{-1} S_k T_{k+1}: \mathscr{H}_{k+1} \to \mathscr{H}_k \,, \quad 0 \leq k \leq N \,,$$

form a parametrix of $\mathfrak{A}$.

A family of complexes

$$\mathfrak{A}^t: 0 \to \mathscr{H}_0 \xrightarrow{A_0^t} \mathscr{H}_1 \longrightarrow \dots \xrightarrow{A_{N-1}^t} \mathscr{H}_N \to 0 \tag{15}$$

$(0 \leq t \leq 1)$ is called homotopy if the operators A_k^t, $0 \leq k \leq N - 1$, continuously depend on t (with respect to the operator norm). Then write $\mathfrak{A}^0 \simeq \mathfrak{A}^1$.

A direct addition of complexes is defined by adding operators and spaces with the same indices. If the length of the complexes is different, one can get the same length by putting additional zero spaces and operators $0 \to 0 \to \dots \to 0$ on both sides to the considered complex. A complex of the form $0 \to \dots \to 0 \to \mathscr{H} \to \mathscr{H} \to 0 \to \dots \to 0$ is called *elementary* if it is exact. Two complexes $\mathfrak{A}$, $\mathfrak{B}$ are called *stable homotopic* if there are elementary complexes $\mathfrak{M}_k$ $(k = 1, \dots, p)$, $\mathfrak{N}_l$ $(l = 1, \dots, q)$ and a homotopy $\mathfrak{A}' \cong \mathfrak{B}'$ of complexes with

$$\mathfrak{A}' \cong \mathfrak{A} \oplus \mathfrak{M}_1 \oplus \dots \oplus \mathfrak{M}_p \,, \quad \mathfrak{B}' \cong \mathfrak{B} \oplus \mathfrak{N}_1 \oplus \dots \oplus \mathfrak{N}_q \,. \tag{16}$$

Proposition 7. *Any Fredholm complex $\mathfrak{A}$ is stable homotopic to a Fredholm complex of the form*

$$\mathfrak{B}: 0 \to \mathcal{H} \xrightarrow{B} \mathcal{L} \to 0 \tag{17}$$

with a Fredholm operator B and

$$\operatorname{ind} \mathfrak{A} = \operatorname{ind} B . \tag{18}$$

Proof: Let $\mathfrak{P}$ be a parametrix of $\mathfrak{A}$. By adding a trivial complex we pass to a Fredholm complex

$$0 \to \mathcal{H}_0 \xrightarrow{A_0} \dots \to \mathcal{H}_{N-3} \xrightarrow{\binom{A_{N-3}}{0}} \begin{matrix}\mathcal{H}_{N-2}\\ \oplus \\ \mathcal{H}_N\end{matrix} \xrightarrow{\begin{pmatrix}A_{N-2} & 0\\ 0 & 1\end{pmatrix}} \begin{matrix}\mathcal{H}_{N-1}\\ \oplus \\ \mathcal{H}_N\end{matrix} \xrightarrow{(A_{N-1},\,0)} \mathcal{H}_N \to 0 .$$

Then for each t, $0 \leqq t \leqq 1$, the following complex is Fredholm

$$0 \to \mathcal{H}_0 \xrightarrow{A_0} \dots \to \mathcal{H}_{N-3} \xrightarrow{\binom{A_{N-3}}{0}} \begin{matrix}\mathcal{H}_{N-2}\\ \oplus \\ \mathcal{H}_N\end{matrix} \xrightarrow{\begin{pmatrix}A_{N-2} & tP_{N-1}\\ 0 & (1-t)\end{pmatrix}} \begin{matrix}\mathcal{H}_{N-1}\\ \oplus \\ \mathcal{H}_N\end{matrix}$$

$$\xrightarrow{((1-t)A_{N-1},\,-t)} H_N \to 0 .$$

For $t = 1$ we obtain except the trivial complex $0 \to \mathcal{H}_N \xrightarrow{-1} \mathcal{H}_N \to 0$ the Fredholm complex

$$0 \to \mathcal{H}_0 \to \dots \to \mathcal{H}_{N-3} \xrightarrow{\binom{A_{N-3}}{0}} \begin{matrix}\mathcal{H}_{N-2}\\ \oplus \\ \mathcal{H}_N\end{matrix} \xrightarrow{(A_{N-2},\,P_{N-1})} \mathcal{H}_{N-1} \to 0 \tag{19}$$

and it is easily seen that the index of (19) is equal to the index of $\mathfrak{A}$. Iterating this construction we get a Fredholm operator

$$B = \begin{pmatrix} A_0 & P_1 & 0 & \dots & 0\\ 0 & A_2 & P_3 & \dots & 0\\ 0 & 0 & A_4 & \dots & 0\\ \vdots & \vdots & \vdots & \ddots & \vdots\\ 0 & 0 & 0 & \dots & A_{N-1}\end{pmatrix} : \begin{matrix}\mathcal{H}_0\\ \oplus\\ \mathcal{H}_2\\ \oplus\\ \vdots\\ \oplus\\ \mathcal{H}_{N-1}\end{matrix} \to \begin{matrix}\mathcal{H}_1\\ \oplus\\ \mathcal{H}_3\\ \oplus\\ \vdots\\ \oplus\\ \mathcal{H}_N\end{matrix} .$$

Without loss of generality it is assumed here that N is odd, otherwise the length of $\mathfrak{A}$ can be modified by adding $\to 0 \to 0$. □

Proposition 8. *Let $(\mathfrak{A}^t)_{0 \leqq t \leqq 1}$ be a homotopy of Fredholm complexes. Then there exists a homotopy of Fredholm complexes $(\mathfrak{P}^t)_{0 \leqq t \leqq 1}$ where $\mathfrak{P}^t$ is a parametrix of $\mathfrak{A}^t$ for all t, $0 \leqq t \leqq 1$. Moreover,*

$$\operatorname{ind} \mathfrak{A}^0 = \operatorname{ind} \mathfrak{A}^1 . \tag{20}$$

Proof: The first assertion is a simple consequence of the parametrix construction for complexes given in the proof of Theorem 3 and a corresponding fact for homotopies of Fredholm operators. (20) follows from the constructions in the proof of Proposition 7 applied to $\mathfrak{A}^t$ and $\mathfrak{P}^t$, respectively. □

Let X be a compact topological space. Suppose that for each $x \in X$ we are given a complex

$$\mathfrak{A}(x): 0 \to \mathcal{H}_0 \xrightarrow{A_0(x)} \mathcal{H}_1 \xrightarrow{A_1(x)} \dots \xrightarrow{A_{N-1}(x)} \mathcal{H}_N \to 0, \tag{21}$$

$A_k(x)$, $k = 0, \dots, N-1$ continuously depending on $x \in X$. Then (21) is called a *family of complexes* with parameter space X. For $X = [0, 1]$ we have a homotopy of complexes.

A family of complexes over $X \times [0, 1]$ is considered as homotopy of families of complexes over X. Direct addition and stable homotopy for families of complexes can be similarly defined as for single complexes.

If $\mathfrak{A}(\cdot)$ is a family of Fredholm complexes over X, one can easily show (similarly as for $X = [0, 1]$) that there exists a family $\mathfrak{P}(\cdot)$ of Fredholm complexes over X so that $\mathfrak{P}(x)$ is a parametrix of $\mathfrak{A}(x)$ for all $x \in X$. Denote by $\mathfrak{F}(X)$ the set of stable homotopy classes of families of Fredholm complexes over X. Direct addition induces the structure of an Abelian group on $\mathfrak{F}(X)$. If $\mathfrak{A}$ represents an element in $\mathfrak{F}(X)$, the negative is represented by a parametrix of $\mathfrak{A}$.

Similarly as in the case of families of Fredholm operators (cf. 1.1.3.4) there is an isomorphism

$$\operatorname{ind}_X: \mathfrak{F}(X) \to K(X). \tag{22}$$

(22) could be defined by reducing the length following the construction in the proof of Proposition 7 applied to families of Fredholm complexes, i.e. to find a stable equivalent Fredholm family B over X for a given family $\mathfrak{A}$ of Fredholm complexes and to define $\operatorname{ind}_X \mathfrak{A} = \operatorname{ind}_X B$. Because of a connection to boundary problems we give a more direct construction of the index element (then it is of course an exercise to check the equivalence of the various definitions). First prove the following

Lemma 9. *Let X be a compact topological space, $\mathcal{H}_1$, $\mathcal{H}_2$ Hilbert spaces and $A: X \to \mathcal{L}(\mathcal{H}_1, \mathcal{H}_2)$ a continuous map (with respect to the norm topology in $\mathcal{L}(\mathcal{H}_1, \mathcal{H}_2)$). If* $\dim \operatorname{coker} A(x) < \infty$ *for all $x \in X$, there exists a finite dimensional subspace $\mathcal{J} \subset \mathcal{H}_2$ with $\mathcal{J} + \operatorname{im} A(x) = \mathcal{H}_2$ for all $x \in X$, i.e.* $\dim \operatorname{coker} A(x) < \infty$ *uniformly for all $x \in X$.*

Proof: For each $x_0 \in X$ we find a finite dimensional subspace $\mathcal{J}(x_0) \subset \mathcal{H}_2$ with $\operatorname{im} A(x_0) + \mathcal{J}(x_0) = \mathcal{H}_2$. If $K: \mathbb{C}^j \to \mathcal{J}(x_0)$ is a linear surjective map, we get a surjective operator $A(x_0) \oplus K: \mathcal{H}_1 \oplus \mathbb{C}^j \to \mathcal{H}_2$ defined by $(A(x_0) \oplus K)(u \oplus v) = A(x_0)u + Kv$. The set of surjective operators in $\mathcal{L}(\mathcal{H}_1 \oplus \mathbb{C}^j, \mathcal{H}_2)$ is open. Thus $A(x) \oplus K$ is surjective for all x in some open neighbourhood $U(x_0)$. The sets $U(x_2)$, $x_0 \in X$, form an open covering of X. Since X is compact, there is a finite subcovering $U(x_1), \dots, U(x_N)$. Then we can set $\mathcal{J} = \mathcal{J}(x_1) + \dots + \mathcal{J}(x_n)$. □

Lemma 10. *Let $A: X \to \mathcal{L}(\mathcal{H}_1, \mathcal{H}_2)$ be as in Lemma 9 and $A(x): \mathcal{H}_1 \to \mathcal{H}_2$ surjective for all $x \in X$. Then $\ker A$ is a Hilbert space bundle over X.*

Proof: Let $x_0 \in X$ be fixed and choose a continuous projection $q(x_0): \mathcal{H}_1 \to \ker A(x_0)$. Then

$$\begin{pmatrix} A(x) \\ q(x_0) \end{pmatrix}: \mathcal{H}_1 \to \begin{matrix} \mathcal{H}_2 \\ \oplus \\ \ker A(x_0) \end{matrix} \tag{23}$$

is an isomorphism at $x = x_0$. Since the set of isomorphisms is open, (23) is an isomorphism for all x in an open neighbourhood $U(x_0)$ of x_0. Thus $q(x_0)$: $\ker A(x) \to \ker A(x_0)$ has to be an isomorphism for all $x \in U(x_0)$ and $\{x\} \times \ker A(x) \to \{x\} \times \ker A(x_0)$ defines a trivialization of the family of kernels of $A(x)$. □

From the proof of Lemma 10 we get a system of trivializations of $X \times \mathcal{H}_1$ of the form

$$U(x_0) \times \mathcal{H}_1 \xrightarrow{\binom{q(x_0)}{1-q(x_0)}} U(x_0) \times \begin{matrix}\mathcal{L}_0\\ \oplus \\ \mathcal{M}_0\end{matrix}$$

with $\mathcal{L}_0 = \ker A(x_0)$, $\mathcal{M}_0 = \mathcal{L}_0^\perp$ ($\perp$ denotes an orthogonal complement) and transition functions

$$U(x_i) \cap U(x_j) \times (\mathcal{L}_j \oplus \mathcal{M}_j) \to U(x_i) \cap U(x_j) \times (\mathcal{L}_i \oplus \mathcal{M}_i)$$

of the form $l_{ij} \oplus m_{ij}$ with families of isomorphisms $l_{ij}\colon \mathcal{L}_j \to \mathcal{L}_i$, $m_{ij}\colon \mathcal{M}_j \to \mathcal{M}_i$.

Lemma 11. *Let $\mathcal{H}_1 \xrightarrow{A_1} \mathcal{H}_2 \xrightarrow{A_2} \mathcal{H}_3 \to 0$ be a family of complexes of operators over a compact space X with* $\dim (\ker A_2/\operatorname{im} A_1) < \infty$ *for all $x \in X$. Let* $\ker A_2$ *be a Hilbert space bundle. Then there exists a finite-dimensional subbundle $\mathcal{J}_2 \subset \ker A_2$ so that* $(\mathcal{J}_2)_x + \operatorname{im} A_1(x) = \ker A_2(x)$ *for all $x \in X$.*

Proof: Suppose that the operator families are given as systems of families over a finite number of open neighbourhoods $W \subset X$, $W \times \mathcal{H}_1 \xrightarrow{A_{1W}} W \times \mathcal{H}_2 \xrightarrow{A_{2W}} W \times \mathcal{H}_3$ where $W \times \mathcal{H}_2 = W \times (\mathcal{L}_W \oplus \mathcal{M}_W)$ with $\mathcal{L}_W = \ker A_{2W}(x_0)$, $x_0 \in W$, $\mathcal{M}_W = \mathcal{L}_W^\perp$. Over homeomorphic neighbourhoods U, V we have families of isomorphisms

$$(\chi_{UV}, l_{UV} \oplus m_{UV})\colon V \times (\mathcal{L}_V \oplus \mathcal{M}_V) \to U \times (\mathcal{L}_U \oplus \mathcal{M}_U).$$

According to Lemma 9 choose a finite-dimensional subspace $\mathcal{J}_W$ with $\operatorname{im} A_{1W} + \mathcal{J}_W = \mathcal{L}_W$ over each W. These $\mathcal{J}_W$ span over each $x \in X$ some finite-dimensional subspace in $\ker A_1(x)$. It is now easily seen that one can choose a finite-dimensional subbundle $\mathcal{J}_2 \subset \ker A_2$ so that the space mentioned above is contained in $(\mathcal{J}_2)_x$. □

Now let (21) be a family of Fredholm complexes. Then $\dim \operatorname{coker} A_{N-1}(x) < \infty$ for all $x \in X$. Applying Lemma 9 we find a finite-dimensional vector space J_N and an isomorphism $K_N\colon \mathcal{J}_N \to \mathcal{J}_N$ onto a subspace $\mathcal{J}_N'$ in $\mathcal{H}_N$ so that the operator family

$$(A_{N-1}, K_N)\colon \begin{matrix}\mathcal{H}_{N-1}\\ \oplus \\ \mathcal{J}_N\end{matrix} \to \mathcal{H}_N$$

is surjective. In view of Lemma 10 $\ker (A_{N-1}, K_N)$ is a Hilbert space bundle. Thus applying Lemma 11 there is a finite-dimensional subbundle $\mathcal{J}_{N-1} \subset \ker (A_{N-1}, K_N)$ with

$$\operatorname{im}\begin{pmatrix}A_{N-2}(x)\\ 0\end{pmatrix} + (\mathcal{J}_{N-1})_x = \ker \big(A_{N-1}(x), K_N\big)\,.$$

Choose a finite-dimensional bundle $J_{N-1} \in \operatorname{Vect}(x)$ and an isomorphism

$$\begin{pmatrix}K_{N-1}\\ \\ Q_{N-1}\end{pmatrix}\colon \mathcal{J}_{N-1} \to \mathcal{J}_{N-1} \subset \begin{matrix}\mathcal{H}_{N-1}\\ \oplus \\ \mathcal{J}_N\,.\end{matrix}$$

Then we get a family of Fredholm complexes of the form

$$0 \to \mathscr{H}_0 \xrightarrow{A_0} \cdots \longrightarrow \mathscr{H}_{N-3} \xrightarrow{\binom{A_{N-3}}{0}} \begin{matrix}\mathscr{H}_{N-2}\\ \oplus \\ \mathscr{J}_{N-1}\end{matrix} \xrightarrow{\begin{pmatrix}A_{N-2} & K_{N-1}\\ 0 & Q_{N-1}\end{pmatrix}} \begin{matrix}\mathscr{H}_{N-1}\\ \oplus \\ \mathscr{J}_{N}\end{matrix}$$

$$\xrightarrow{(A_{N-1},\, K_N)} \mathscr{H}_N \to 0$$

which is exact at the Nth and $(N-1)^{\text{th}}$ places. It is easily seen (cf. the proof of Lemma 10) that

$$\mathscr{H} = \ker \begin{pmatrix} A_{N-2} & K_{N-1} \\ 0 & Q_{N-1} \end{pmatrix} \tag{24}$$

is a Hilbert space bundle, too, and that there is a finite dimensional bundle $\mathscr{J}_{N-2} \in \operatorname{Vect}(X)$ and an operator family $\begin{pmatrix} K_{N-2} \\ Q_{N-2} \end{pmatrix} : \mathscr{J}_{N-2} \to \mathscr{J}_{N-2}$ defining an isomorphism, where $\mathscr{J}_{N-2}$ is a finite dimensional subbundle of (24) with $\mathscr{H} = (\mathscr{J}_{N-2})_x + \operatorname{im} \begin{pmatrix} A_{N-3}(x) \\ 0 \end{pmatrix}$. Iterating this procedure we finally get a family of complexes of the form

$$0 \to \begin{matrix}\mathscr{H}_0\\ \oplus \\ \mathscr{J}_1\end{matrix} \xrightarrow{\begin{pmatrix}A_0 & K_1\\ 0 & Q_1\end{pmatrix}} \begin{matrix}\mathscr{H}_1\\ \oplus \\ \mathscr{J}_2\end{matrix} \xrightarrow{\begin{pmatrix}A_1 & K_2\\ 0 & Q_2\end{pmatrix}} \cdots \xrightarrow{\begin{pmatrix}A_{N-2} & K_{N-1}\\ 0 & Q_{N-1}\end{pmatrix}} \begin{matrix}\mathscr{H}_{N-1}\\ \oplus \\ \mathscr{J}_N\end{matrix} \xrightarrow{(A_{N-1},\, K_N)} \mathscr{H}_N \to 0 \tag{25}$$

which is exact except at the 0th place. The kernel of $\begin{pmatrix} A_0 & K_1 \\ 0 & Q_1 \end{pmatrix}$ is isomorphic to some $J_0 \in \operatorname{Vect}(X)$ and the choice of an isomorphism

$$\begin{pmatrix} K_0 \\ Q_0 \end{pmatrix} : J_0 \to \ker \begin{pmatrix} A_0 & K_1 \\ 0 & Q_1 \end{pmatrix}$$

gives an exact family

$$0 \to J_0 \xrightarrow{\binom{K_0}{Q_0}} \begin{matrix}\mathscr{H}_0\\ \oplus \\ \mathscr{J}_1\end{matrix} \xrightarrow{\begin{pmatrix}A_1 & K_2\\ 0 & Q_2\end{pmatrix}} \cdots \tag{26}$$

where the last pieces in (25) and (26) are the same.

Define

$$\operatorname{ind}_X \mathfrak{A} = \sum_{k=0}^{N} (-1)^k [\mathscr{J}_k] \in K(X)\,. \tag{27}$$

It is a simple (but a little voluminous) exercise to check that (27) is independent of the concrete choice of the objects J_k, K_k, Q_k and that $\operatorname{ind}_X \mathfrak{A}$ only depends on the equivalence class in the sense of stable homotopy. Thus (22) is constructed.

A further exercise is the following

Remark 12. Let (21) be a family of Fredholm complexes over X. Then there exist families of compact operators $C_0, \ldots, C_{N-1}$ so that

$$0 \to \mathscr{H}_0 \xrightarrow{(A_0+C_0)(x)} \mathscr{H}_1 \xrightarrow{(A_1+C_1)(x)} \mathscr{H}_2 \to \cdots \xrightarrow{(A_{N-1}+C_{N-1})(x)} \mathscr{H}_N \to 0$$

is also a family of Frédholm complexes with the property that there exist $L_k \in \mathrm{Vect}\,(X)$ with $L_k \cong \ker\,(A_k + C_k)/\mathrm{im}\,(A_{k-1} + C_{k-1})$, $0 \leq k \leq N$. Then

$$\mathrm{ind}_X\,\mathfrak{A} = \sum_{k=0}^{N} (-1)^k\,[L_k]\,. \tag{28}$$

Next consider complexes of PDOs on manifolds. Let M be a closed compact C^∞ manifold, $n = \dim M$. Consider a complex of PDOs $A_k \in L^{m_k}_{\mathrm{cl}}(M; E_k, E_{k+1})$, $k = 0, \ldots, N-1$,

$$\mathfrak{A}: 0 \to C^\infty(M, E_0) \xrightarrow{A_0} C^\infty(M, E_1) \longrightarrow \ldots \xrightarrow{A_{N-1}} C^\infty(M, E_N) \to 0\,. \tag{29}$$

If $\sigma_k: \pi^*E_k \to \pi^*E_{k+1}$ is the homogeneous principal symbol of A_k, from $A_{k+1}A_k = 0$ it follows that $\sigma_{k+1}\sigma_k = 0$, i.e. we get a complex

$$0 \to \pi^*E_0 \xrightarrow{\sigma_0} \pi^*E_1 \longrightarrow \ldots \xrightarrow{\sigma_{N-1}} \pi^*E_N \to 0 \tag{30}$$

($\pi: T^*M \setminus \{0\} \to M$), (29) is called *elliptic* if the complex (30) is exact. The closures of the operators A_k in Sobolev spaces form a complex

$$0 \to H^{s_0}(M, E_0) \xrightarrow{A_0} H^{s_1}(M, E_1) \longrightarrow \ldots \xrightarrow{A_{N-1}} H^{s_N}(M, E_N) \to 0 \tag{31}$$

($s_k - s_{k+1} = m_k$). Because of standard regularity properties of elliptic complexes to be formulated below, the essential assertions about (31) are valid in a similar version for (29) and conversely.

Consider isomorphisms $\Lambda^{s_k}_{E_k}: C^\infty(M, E_k) \to C^\infty(M, E_k)$, $\Lambda^{s_k}_{E_k} \in L^{s_k}_{\mathrm{cl}}(M; E_k, E_k)$ as in 1.2.4.2, Theorem 5. Then the operators $B_k = \Lambda^{s_{k+1}}_{E_{k+1}} A_k (\Lambda^{s_k}_{E_k})^{-1}$, $k = 0, \ldots, N-1$, form again a complex. Since ord $B_k = 0$, closure in Sobolov spaces gives a complex

$$0 \to H^0(M, E_0) \xrightarrow{B_0} H^0(M, E_1) \longrightarrow \ldots \xrightarrow{B_{N-1}} H^0(M, E_N) \to 0\,. \tag{32}$$

The operators $\Lambda^{s_k}_{E_k}$ define an isomorphism between the complexes (31), (32). Thus for parametrix constructions and other considerations it is sufficient to consider the case $m_k = 0$ ($k = 0, \ldots, N-1$) (cf. Remark 6).

Fix on M a Riemannian metric and Hermitean metrics in the bundles E_k ($k = 0, \ldots, N$). Then we have Hermitean scalar products in the spaces $H^0(M, E_k)$. The considerations at the beginning of this section can be applied to complexes of PDOs where we speak about operators with C^∞ kernels instead of compact operators. In particular we have the notion of a parametrix.

Theorem 13. *Let* (29) *be elliptic. Then both* (29) *and* (31) *are Fredholm complexes with the same index. Conversely, if* (31) *is Fredholm, then it is elliptic.*

Proof: The first part of the theorem is nearly obvious after the considerations at the beginning of this section. Let $\mathfrak{A}$ be elliptic and without loss of generality $m_k = 0$ ($k = 0, \ldots, N-1$). First define the elliptic PDOs

$$D_k = A_{k-1}A^*_{k-1} + A^*_k A_k \tag{33}$$

and define a parametrix of (31) using C^∞ parametrices R_k of operators (33) (cf. the proof of Theorem 3). Then we immediately get the Fredholm property of (31). Since $\ker D_k \subset C^\infty(M, E_k)$ ($0 \leq k \leq N$), we get by applying (8) that the indices of (29) and (31) are equal and moreover, that the cohomology spaces can be represented by finite-dimensional subspaces of smooth sections in the corresponding bundles. Thus the first part of the theorem is proved. Now assume that (31) is Fredholm. Then the adjoint complex is Fredholm, too, and hence the operators (33) are Fredholm. In view of 1.2.4.2, Theorem 3 the D_k are elliptic PDOs. Using the constructions in the proof of Theorem 3 we get a pseudo-differential parametrix of our complex. Applying the construction in the proof of Proposition 7 to (31) we obtain a PDO B which is Fredholm and hence elliptic. Application of these constructions with stable homotopies on symbolic level in the inverse direction shows that ellipticity of complexes is preserved. Then ellipticity of (31) follows. □

Remark 14. *Let* (29) *be elliptic. Then* $A_k(C^\infty(M, E_k))$ *is closed in* $C^\infty(M, E_{k+1})$. *Moreover,* $\ker A_k$ *has a topological complement in* $C^\infty(M, E_k)$ *for all* k.

Proof: The first assertion follows similarly as in the proof of Theorem 3. In order to show the second assertion set $Z_k = \ker A_k$, $B_k = \operatorname{im} A_{k-1}$. Consider a parametrix of (29) with operators P_k, i.e. $A_{k-1}P_{k-1} + P_kA_k = 1 + C_k$ with operators C_k with C^∞ kernels. Consider the map $F_k\colon Z_k \oplus B_k \to C^\infty(M, E_k)$, $F_k(z \oplus b) = z + P_kb$. Then $\operatorname{im} F_k \supseteqq \operatorname{im}(1 + C_k)$. Thus $\operatorname{im} F_k$ is closed and has finite codimension. Moreover $(1 - C_{k+1})\, b = (A_kP_k + P_{k+1}A_{k+1})\, b = A_kP_kb = -A_kF_k(z \oplus b)$, i.e. $\dim\ker F_k \leqq \dim\ker(1 + C_{k+1}) < \infty$. Therefore, F_k is a Fredholm operator and F_k induces an isomorphism of $Z_k \oplus \left(B_{k+1}P_k^{-1}(Z_k) \cap B_{k+1}\right)$ onto a closed subspace W_k of finite codimension in $C^\infty(M, E_k)$. If $W_k^\perp$ denotes a complement of W_k, we obtain that $P_kB_{k+1}P_kB_{k+1} \cap Z_k + W_k^\perp$ is a complement of Z_k in $C^\infty(M, E_k)$. □

Similarly as (8) it can be proved that there is a topological decomposition

$$C^\infty(M, E_k) = \operatorname{im} A_k^* \oplus \operatorname{im} A_{k+1} \oplus \ker D_k\,, \tag{34}$$

where the operators are considered in the C^∞ spaces.

Propositon 15. *For a given exact symbol sequence* (30) *there exists a complex of PDOs of the form* (29) *with* $\sigma_{A_k} = \sigma_k$, $0 \leqq k \leqq N - 1$.

Proof: First choose arbitrary PDOs $A_j \in L_{\text{cl}}^{m_j}(M; E_j, E_{j+1})$ with $\sigma_{A_j} = \sigma_j$. We have to construct operators K_j with C^∞ kernels so that the operators $A_j + K_j$ form a complex. Set for abbreviation $\Gamma(E_j) = C^\infty(M, E_j)$ and consider the last piece of the sequence

$$\ldots \to \Gamma(E_{N-2}) \xrightarrow{A_{N-2}} \Gamma(E_{N-1}) \xrightarrow{A_{N-1}} \Gamma(E_N) \to 0\,. \tag{35}$$

The constructions in the proof of Theorem 3 formally applied to the exact symbol sequence (30) give an exact symbol sequence

$$0 \leftarrow \pi^*E_0 \xleftarrow{\sigma_0^{-1}} \pi^*E_1 \xleftarrow{\sigma_1^{-1}} \ldots \xleftarrow{\sigma_{N-1}^{-1}} \pi^*E_N \leftarrow 0$$

with

$$\sigma_{k+1}\sigma_{k+1}^{-1} + \sigma_k^{-1}\sigma_k = 1_{\pi^*E_k} \tag{36}$$

for all k. Let P_{N-1} and P_{N-2} PDOs with the symbols σ_{N-1}^{-1} and σ_{N-2}^{-1}, respectively. Since σ_{N-1} is surjective, we can assume without loss of generality that

$$\operatorname{im} P_{N-1} = (\ker A_{N-1})^\perp \tag{37}$$

for a certain complement of the kernel of A_{N-1}. From (36) we get $A_{N-2}P_{N-2} + P_{N-1}A_{N-1} = 1 - C$ for some PDO C with vanishing homogeneous principal symbol. Thus the composition

$$\Gamma(E_{N-1}) \xrightarrow{\binom{P_{N-2}}{A_{N-1}}} \begin{matrix}\Gamma(E_{N-2})\\ \oplus\\ \Gamma(E_N)\end{matrix} \xrightarrow{(A_{N-2},\, P_{N-1})} \Gamma(E_{N-1})$$

defines an elliptic PDO, hence $\dim(\operatorname{im} P_{N-2} \cap \ker A_{N-2})$, $\dim(\operatorname{im} A_{N-1} \cap \ker P_{N-1}) < \infty$ (because of (37) the second dimension is zero). Moreover,

$$\dim(\ker A_{N-1} \cap \ker P_{N-2}) < \infty, \operatorname{codim}(\ker A_{N-1} + \ker P_{N-2}) < \infty .$$

Thus, in view of (36), $\dim(\operatorname{im} P_{N-1} \cap \operatorname{im} A_{N-2}) < \infty$, i.e. the space $L_{N-1} = (\ker A_{N-1})^{\perp} \cap \operatorname{im} A_{N-2}$ is finite-dimensional. Denote by $K_{N-1}: \Gamma(E_{N-1}) \to L_{N-1}$ a projection. K_{N-1} is an operator with C^∞ kernel. Then $C_{N-1} = A_{N-1}(1 - K_{N-1}) - A_{N-1}$ is an operator with C^∞ kernel and the operator $A'_{N-1} = A_{N-1} + C_{N-1}$ satisfies $A'_{N-1}A_{N-2} = 0$. Now consider the piece

$$\ldots \to \Gamma(E_{N-3}) \xrightarrow{\binom{A_{N-3}}{0}} \begin{matrix}\Gamma(E_{N-2})\\ \oplus \\ \Gamma(E_N)\end{matrix} \xrightarrow{(A_{N-2},\, P_{N-1})} \Gamma(E_{N-1}) .$$

The operators $\begin{pmatrix}A_{N-3}\\ 0\end{pmatrix}$ and (A_{N-2}, P_{N-1}) satisfy similar assumptions as the operators A_{N-2}, A_{N-1} above. Thus we get again

$$\dim\left((\ker(A_{N-2}, P_{N-1}))^{\perp} \cap \operatorname{im}\begin{pmatrix}A_{N-3}\\ 0\end{pmatrix}\right) < \infty .$$

From

$$\operatorname{im}\begin{pmatrix}A_{N-3}\\ \\ 0\end{pmatrix} \subset \begin{matrix}\Gamma(E_{N-2})\\ \oplus\\ \{0\}\end{matrix}, \quad \ker(A_{N-2}, P_{N-1})^{\perp} \cap \begin{matrix}\Gamma(E_{N-2})\\ \oplus\\ \{0\}\end{matrix} = \begin{matrix}(\ker A_{N-2})\\ \oplus\\ \{0\}\end{matrix}$$

it follows that the space $L_{N-2} = (\ker A_{N-2})^{\perp} \cap \operatorname{im} A_{N-3}$ is finite-dimensional. Let $K_{N-2}: \Gamma(E_{N-2}) \to L_{N-2}$ be a projection. K_{N-2} is an operator with C^∞ kernel. Then $C_{N-2} = A_{N-2}(1 - K_{N-2}) - A_{N-2}$ has C^∞ kernel, too. The operator $A'_{N-2} = A_{N-2}(1 - K_{N-2})$ satisfies $A'_{N-2}A_{N-3} = 0$, $\operatorname{im} A'_{N-2} \subseteqq \operatorname{im} A_{N-2}$, i.e. also $A'_{N-1}A'_{N-2} = 0$. Iterating this construction gives us the desired operator complex. □

Concerning further properties and applications of elliptic complexes of PDOs cf. Atiyah/Bott [2], Palais [1], Gilkey [1], Pillat/Schulze [1]. As a classical example consider the de Rham complex on a Riemannian manifold M. Let $\Lambda^p(T^*M)$ be the pth exterior product of T^*M (cf. 1.1.2.2). Set $\Omega^p_{\mathbb{C}}(M) = C^\infty(M, \Lambda^p(T^*M) \otimes \mathbb{C})$. Then $\Omega^p_{\mathbb{C}}(M)$ is per def. the space of complex differential forms on M. Denote by $d_p: \Omega^p_{\mathbb{C}}(M) \to \Omega^{p+1}_{\mathbb{C}}M)$ the operator of exterior differentiation. This is a differential operator with the homogeneous principal symbol $\sigma_{d_p}(x, \xi) = \xi \wedge$ (exterior multiplication by ξ). Since $d_{p+1}d_p = 0$, we get a complex

$$\Re: 0 \to \Omega^0_{\mathbb{C}}(M) \xrightarrow{d_0} \Omega^1_{\mathbb{C}}(M) \longrightarrow \ldots \xrightarrow{d_{n-1}} \Omega^n_{\mathbb{C}}(M) \to 0 . \tag{38}$$

(38) is called *de Rham complex*. The operators

$$\Delta_p = d_{p-1}d^*_{p-1} + d^*_p d_p: \Omega^p_{\mathbb{C}}(M) \to \Omega^p_{\mathbb{C}}(M)$$

are called the *Laplace operators* connected with (38). The forms $u \in \ker \Delta_p$ are called *harmonic*.

Let $H^k(M, \mathbb{C})$ be the kth cohomology group of M (with coefficients in $\mathbb{C}$). Then the classical de Rham theorem says that $H^k(\Re) \cong H^k(M, \mathbb{C})$ so that

$$\chi(M) = \operatorname{ind} \Re = \sum_{k=0}^{n} (-1)^k \dim H^k(M, \mathbb{C}) \tag{39}$$

is just the Euler number of M.

A special case of the assertion (34) is the following decomposition theorem (cf. also (7)).

Theorem 16. *There is a topological decomposition*

$$\Omega^p_{\mathbb{C}}(M) = \{\text{harmonic } p\text{-forms}\} \oplus d_{p-1}\big(\Omega^{p-1}_{\mathbb{C}}(M)\big) \oplus d^*_{p+1}\big(\Omega^{p+1}_{\mathbb{C}}(M)\big)\,.$$

Each cohomology class of the de Rham cohomology (with coefficients in $\mathbb{C}$) can be represented by a uniquely determined harmonic form.

Similar assertions hold for real forms and the cohomology groups $H^k(M, \mathbb{R})$.

3.2.3.2. Complexes on Manifolds with Boundary

Let X be a compact smooth manifold with boundary Y. In this section we shall discuss complexes $\mathfrak{A}$ of operators in $\mathfrak{G}$ of the form

$$\mathcal{A}_k = \begin{pmatrix} r^+A_k + r'B_k & K_k \\ r'T_k & Q_k \end{pmatrix} : \begin{matrix} C^\infty(X, E_k) \\ \oplus \\ C^\infty(Y, G_k) \end{matrix} \to \begin{matrix} C^\infty(X, E_{k+1}) \\ \oplus \\ C^\infty(Y, G_{k+1}) \end{matrix} \tag{1}$$

$(k = 0, \dots, N-1)$ with bundles $E_k \in \operatorname{Vect}(X)$, $G_k \in \operatorname{Vect}(Y)$ $(k = 0, \dots, N)$. As usual set $\mathcal{A}_{-1} = 0$, $\mathcal{A}_N = 0$ and similarly consider the spaces with the numbers $k = -1$ and $k = N+1$ as the zero vector.

Let $\alpha_k = \operatorname{ord}\sigma_{A_k}$, $\alpha_k - 1 = \operatorname{ord}\sigma_{B_k}$, $\lambda_k = \operatorname{ord}\sigma_{K_k}$, $\gamma_k = \operatorname{ord}\sigma_{T_k}$, $1 - \alpha_k + \lambda_k + \gamma_k = \operatorname{ord}\sigma_{Q_k}$ be the orders of the homogeneous principal symbols. Choose an $s_0 \in \mathbb{R}_+$ sufficiently large and define $s_{k+1} = s_k - \alpha_k$ for $k = 0, \dots, N-1$. Assume that

$$\lambda_{k+1} = \alpha_k + \alpha_{k+1} - \gamma_k - 1 \tag{2}$$

$(k = 0, \dots, N-1)$ and set

$$t_k = s_{k+1} + \lambda_k + \tfrac{1}{2}\left(= s_{k-1} - \gamma_{k-1} - \tfrac{1}{2}\right). \tag{3}$$

Then the Sobolev space closure of the operators (1)

$$\mathcal{A}_k\colon H^{s_k}(X' E_k) \oplus H^{t_k}(Y, G_k) \to H^{s_{k+1}}(X, E_{k+1}) \oplus H^{t_{k+1}}(Y, G_k) \tag{4}$$

forms a complex of operators between Hilbert spaces. For each $r \in \mathbb{R}_+$ the numbers $s_k' = s_k + r$, $t_k' = t_k + r$ satisfy the condition (3), too.

With the given complex $\mathfrak{A}$ there are connected the complex of interior symbols

$$\sigma_\Omega(\mathfrak{A})\colon 0 \to \pi^*E_0 \xrightarrow{\sigma_{A_0}} \pi^*E_1 \xrightarrow{\sigma_{A_1}} \dots \xrightarrow{\sigma_{A_{N-1}}} \pi^*E_N \to 0 \tag{5}$$

$(\pi\colon T^*X \setminus 0 \to X)$ and the complex of boundary symbols

$$\sigma_Y(\mathfrak{A})\colon 0 \to \begin{matrix} p^*E_0^+ \\ \oplus \\ p^*G_0 \end{matrix} \xrightarrow{\sigma_Y(\mathcal{A}_0)} \begin{matrix} p^*E_1^+ \\ \oplus \\ p^*G_1 \end{matrix} \xrightarrow{\sigma_Y(\mathcal{A}_1)} \dots \xrightarrow{\sigma_Y(\mathcal{A}_{N-1})} \begin{matrix} p^*E_N^+ \\ \oplus \\ p^*G_N \end{matrix} \to 0 \tag{6}$$

$(p\colon S^*Y \to Y)$. That (5), (6) are complexes is a consequence of the symbolic calculus for operators in $\mathfrak{G}$, i.e. $\sigma_\Omega(\mathcal{A}\mathcal{B}) = \sigma_\Omega(\mathcal{A})\,\sigma_\Omega(\mathcal{B})$, $\sigma_Y(\mathcal{A}\mathcal{B}) = \sigma_Y(\mathcal{A})\,\sigma_Y(\mathcal{B})$ for $\mathcal{A}, \mathcal{B} \in \mathfrak{G}$. It is clear that the restrictions on Y of the symbols in (5) correspond to the operator symbols in the left upper corners in (6).

Denote complexes of operators in $\mathfrak{G}$ also as complexes over (X, Y).

Definition 1. A complex $\mathfrak{A}$ over (X, Y) is called *elliptic* if both $\sigma_\Omega(\mathfrak{A})$ and $\sigma_Y(\mathfrak{A})$ are exact.

Theorem 2. *Let $\mathfrak{A}$ be an elliptic complex on (X, Y). Then $\mathfrak{A}$ is a Fredholm complex both for spaces of C^∞ sections and for Hilbert spaces and the index is the same. Conversely, if the complex of operators (4) is Fredholm, it is necessarily elliptic.*

Proof: First pass by reducing of orders (cf. 3.1.2.1) to a complex $\mathfrak{B}$ of operators

$$\mathcal{B}_k = \mathcal{L}^{s_{k+1},\, t_{k+1}}_{E_{k+1},\, G_{k+1}} \mathcal{A}_k (\mathcal{L}^{s_k,\, t_k}_{E_k,\, G_k})^{-1} : H^0(X, E_k) \oplus H^0(Y, G_k)$$
$$\to H^0(X, E_{k+1}) \oplus H^0(Y, G_{k+1}) \,. \tag{7}$$

Since the $\mathcal{L}^{s_j,\, t_j}_{E_j,\, G_j}$ are elliptic operators in $\mathfrak{G}$, the symbolic calculus shows that the complex $\mathfrak{B}$ is also elliptic. Fix Riemannian metrics on X and Y in the usual sense and Hermitean metrics in the vector bundles. Then we have Hermitean scalar products in the spaces $H^0(X, E_k)$ and $H^0(Y, G_k)$. According to the considerations in 2.3.3.2 the adjoint operators $\mathcal{B}_k^*$ belong again to $\mathfrak{G}$ and form an elliptic complex $\mathfrak{B}^*$. The operators

$$\mathcal{D}_k = \mathcal{B}_{k-1}\mathcal{B}_{k-1}^* + \mathcal{B}_k^*\mathcal{B}_k \tag{8}$$

are elliptic. As in the proof of 3.2.3.1 Theorem 3 we find a parametrix $\mathfrak{S}$ of $\mathfrak{B}$ consisting of operators in $\mathfrak{G}$. Since all the steps in the construction of $\mathfrak{S}$ can be made on symbolic level, it is quite clear that $\mathfrak{S}$ is again elliptic. Applying 3.2.3.1, Remark 6 we immediately obtain a parametrix $\mathfrak{P}$ of $\mathfrak{A}$. Now the Fredholm property of $\mathfrak{A}$ realized in Sobolev spaces follows from 3.2.3.1, Theorem 3. The other assertions follow in a similar way as the corresponding assertions in 3.2.3.1, Theorem 13. The only additional remark is that one has to consider the complex $\mathfrak{B}$ first and then to apply the complex isomorphism $\mathfrak{B} \to \mathfrak{A}$. □

The last remark in the proof of Theorem 2 shows that all standard properties of elliptic complexes of PDOs on closed compact manifolds formulated in 3.2.3.2 have an analogon for elliptic complexes on (X, Y). Therefore, we can omit the details of the proofs. Let $\mathfrak{B}$ be the complex defined in the proof of Theorem 2.

Theorem 3. *Let $\mathfrak{A}$ be an elliptic complex on (X, Y). Then*

$$\operatorname{ind} \mathfrak{A} = \sum_{k=0}^{N} (-1)^k \dim \ker \mathcal{D}_k \tag{9}$$

with the elliptic operators (8). *There are isomorphisms*

$$H^k(\mathfrak{A}) \cong \ker \mathcal{D}_k \tag{10}$$

and direct decompositions

$$H^0(X, E_k) \oplus H^0(Y, G_k) = \operatorname{im} \mathcal{B}_k^* \oplus \operatorname{im} \mathcal{B}_{k-1} \oplus \ker \mathcal{D}_k \,. \tag{11}$$

$\mathcal{B}_k\big(C^\infty(X, E_k) \oplus C^\infty(Y, G_k)\big)$ *is closed in* $C^\infty(X, E_{k+1}) \oplus C^\infty(Y, G_{k+1})$ *and* $\ker \mathcal{B}_k$ *has a topological complement in* $C^\infty(X, E_k) \oplus C^\infty(Y, G_k)$. *There are direct decompositions*

$$C^\infty(X, E_k) \oplus C^\infty(Y, G_k) = \operatorname{im} \mathcal{B}_k^* \oplus \operatorname{im} \mathfrak{B}_{k-1} \oplus \ker \mathcal{D}_k \,. \tag{12}$$

The proof of (9), (10) is an immediate consequence of $\operatorname{ind} \mathcal{B} = \sum_{k=0}^{N} (-1)^k \dim \ker \mathcal{D}_k$ (cf. 3.2.3.1. (7)) and $H^k(\mathfrak{A}) \cong H^k(\mathfrak{B}) \cong \ker \mathcal{D}_k$. (11), (12) can be proved in a similar way as the corresponding assertions in 3.2.3.1, i.e. (8) and (34).

Remark 4. Let $\mathfrak{A}$ be an elliptic complex on (X, Y). Then one can choose a parametrix $\mathfrak{P}$ of $\mathfrak{A}$ in such a way that

$$H^{s_k}(X, E_k) \oplus H^{t_k}(Y, G_k) = \operatorname{im} \mathcal{P}_k \oplus \operatorname{im} \mathcal{A}_{k-1} \oplus L_k \tag{13}$$

for all $k = 0, \ldots, N$. Here $L_k \subset C^\infty(X, E_k) \oplus C^\infty(Y, G_k)$ are finite-dimensional subspaces with $H^k(\mathfrak{A}) \cong L_k (k = 0, \ldots, N)$. Restriction of the $\mathcal{P}_k$, $\mathcal{A}_{k-1}$ to spaces of C^∞ sections gives a direct decomposition of $C^\infty(X, E_k) \oplus C^\infty(Y, G_k)$ similar to (13).

In order to construct $\mathfrak{P}$ it is sufficient to choose the parametrix $\mathcal{R}_k$ of $\mathcal{D}_k$ (cf. the proof of 3.2.3.1, Theorem 3) in such a way that $\mathcal{R}_k$ induces an isomorphism $(\ker \mathcal{D}_k)^\perp \to (\ker \mathcal{D}_k)^\perp$ and $\ker \mathcal{R}_k = \ker \mathcal{D}_k$ for all k. Then $L_k = (\mathcal{L}^{s_k, t_k}_{E_k, G_k})^{-1} (\ker \mathcal{D}_k)$. If $\mathfrak{A}$ is an elliptic complex of operators $\mathcal{A}_k \in \mathfrak{G}$ on (X, Y) $(k = 0, \ldots, N-1)$, A_0 is an overdetermined, A_{N-1} an underdetermined elliptic system in the sense of 3.1.2.3. One can ask whether or when an over- or underdetermined elliptic system can be considered as a part of an elliptic complex. We will not discuss this question here. Some results are given in SCHULZE [6].

There are several possibilities to construct examples of elliptic complexes over (X, Y). One of them is to realize elliptic complexes as mapping cones $\mathfrak{K}$ of morphisms of complexes of the form

$$\begin{array}{ccccc} 0 \to \ldots \to H^{s_k}(X, E_k) & \xrightarrow{r^+A_k - r'B_k} & H^{s_k+1}(X, E_{k+1}) & \to \ldots \to 0 \\ \downarrow r'T_k & & \downarrow r'T_{k+1} & \\ 0 \to \ldots \to H^{t_k}(Y, G_k) & \xrightarrow{Q_k} & H^{t_k+1}(Y, G_{k+1}) & \to \ldots \to 0 \end{array} \tag{14}$$

(cf. 3.2.3.1. (12)). Here r^+A_k are pseudo-differential operators on X, $r'B_k$ Green operators, $r'T_k$ trace operators, and Q_k pseudo-differential operators on Y. (14) is called *elliptic* if the corresponding mapping cone $\mathfrak{K}$ is elliptic. Then one obtains complexes in the sense of DYNIN [4]. Instead of (14) one can consider morphisms with potential operators $K_k: H^{t_k}(Y, G_k) \to H^{s_k}(X, E_k)$. Since only the order of the complexes is changed in the abstract context, one has in this case, too, the notion of ellipticity. (14) could be denoted as elliptic boundary problem for the complex in the first row. For elliptic problems one has an assertion similar to 3.2.3.1, Proposition 5. Details are left to the reader. Let $N = 1$ in (14), i.e.

$$\begin{array}{ccccc} 0 \to H^{s_0}(X, E_0) & \xrightarrow{-r^+A - r'B} & H^{s_1}(X, E_1) & \to 0 \\ \downarrow r'T_0 & & \downarrow r'T_1 & \\ 0 \to H^{t_0}(Y, G_0) & \xrightarrow{Q} & H^{t_1}(Y, G_1) & \to 0 \end{array} \tag{15}$$

(the minus signs have only technical reasons). The cone $\mathfrak{K}$ corresponding to (15) is

$$0 \to H^{s_0}(X, E_0) \xrightarrow{\begin{pmatrix} r^+A + r'B \\ r'T_0 \end{pmatrix}} \begin{matrix} H^{s_1}(X, E_1) \\ \oplus \\ H^{t_0}(Y, G_0) \end{matrix} \xrightarrow{(r'T_1, Q)} H^{t_1}(Y, G_1) \to 0 \tag{16}$$

with the boundary symbol

$$\sigma_Y(\mathfrak{K}): 0 \to p^*E_0^+ \xrightarrow{\begin{pmatrix} \Pi^+\sigma_A + \Pi'\sigma_B \\ \Pi'\sigma_{T_0} \end{pmatrix}} \begin{matrix} p^*E_1^+ \\ \oplus \\ p^*G_0 \end{matrix} \xrightarrow{(\Pi'\sigma_{T_1}, \sigma_Q)} p^*G_1 \to 0 . \tag{17}$$

Ellipticity of (15) means that σ_A is elliptic and (17) is exact. Now it is quite simple to construct a right inverse

$$\begin{pmatrix} \sigma_{K_0} \\ \sigma_{Q_0} \end{pmatrix} \quad \text{of} \quad (\Pi'\sigma_{T_1}, \sigma_Q)$$

in the class of boundary symbols so that

$$\begin{pmatrix} \Pi^+\sigma_A + \Pi'\sigma_{B_0} & \sigma_{K_0} \\ \Pi'\sigma_{T_0} & \sigma_{Q_0} \end{pmatrix} : \begin{matrix} p^*E_0^+ \\ \oplus \\ p^*G_1 \end{matrix} \to \begin{matrix} p^*E_1^+ \\ \oplus \\ p^*G_0 \end{matrix} \tag{18}$$

is an isomorphism. Thus with σ_A and (18) we can connect an elliptic operator $\in \mathfrak{G}$ with an index equal to the index of (16) (cf. the constructions in the proof of 3.2.3.1, Proposition 7). The last remarks show that complexes in the sense of DYNIN [4] are a generalization of elliptic operators in $\mathfrak{G}$.

Let

$$0 \to \pi^*E_0 \xrightarrow{\sigma_{A_0}} \pi^*E_1 \xrightarrow{\sigma_{A_1}} \dots \xrightarrow{\sigma_{A_{N-1}}} \pi^*E_N \to 0 \tag{19}$$

be an exact symbol sequence over X and $\operatorname{ord} \sigma_{A_k} = \alpha_k$. Suppose that the σ_{A_k} have the transmission property. We shall study the existence of elliptic complexes $\mathfrak{A}$ on (X, Y) for which $\sigma_\Omega(\mathfrak{A})$ is equal to (19). Such an $\mathfrak{A}$ exists iff there is an exact sequence of boundary symbols

$$\begin{pmatrix} \Pi^+\sigma_{A_k} + \Pi'\sigma_{B_k} & \sigma_{K_k} \\ \Pi'\sigma_{T_k} & \sigma_{Q_k} \end{pmatrix} : \begin{matrix} p^*E_k^+ \\ \oplus \\ p^*G_k \end{matrix} \to \begin{matrix} p^*E_{k+1}^+ \\ \oplus \\ p^*G_{k+1} \end{matrix} \tag{20}$$

with suitable $G_k \in \operatorname{Vect}(Y)$. The existence of $\mathfrak{A}$ can be proved in a similar way as 3.2.3.1, Proposition 15 starting with given exact $\sigma_\Omega(\mathfrak{A})$, $\sigma_Y(\mathfrak{A})$.

Lemma 5. *Let* (19) *be an exact sequence of homogeneous symbols with the transmission property. Then there exist Green symbols* σ_{B_k} $(k = 0, \dots, N)$ *so that*

$$\mathfrak{a}: 0 \to p^*E_0^+ \xrightarrow{\Pi^+\sigma_{A_0} + \Pi'\sigma_{B_0}} p^*E_1^+ \to \dots \xrightarrow{\Pi^+\sigma_{A_{N-1}} + \Pi'\sigma_{B_{N-1}}} p^*E_N^+ \to 0 \tag{21}$$

is a family of Fredholm complexes over S^*Y. The index of (21) *as an element of $K(S^*Y)$ is independent of the choice of the σ_{B_k}.*

The proof of Lemma 5 is similar to the proof of 3.2.3.1, Proposition 15. Here the role of operators with C^∞ kernels is played by Green boundary symbols. An additional aspect is that we have families of operators over S^*Y. This gives rise to simple technical considerations. One can pass to closures of the operators in Sobolev spaces (cf. 2.1.2.1). The Green operators are then compact in the usual sense. After reduction of orders, i.e. replacing the operators $\Pi^+\sigma_{A_k} + \Pi'\sigma_{B_k}$ by $\Pi^+\sigma_{A_{\overline{s}_{k+1}}}(\Pi^+\sigma_{A_k} + \Pi'\sigma_{B_k}) \times (\Pi^+\sigma_{A_{\overline{s}_k}})^{-1}$ (cf. 3.1.2.1), one can work with adjoint operators and use all constructions in 3.2.3.1 concerning the cohomology of complexes, index, parametrices and so on for families of the form (21). In particular we get the following

Remark 6. One can choose the Green symbols in (21) in such a way that for suitable finite dimensional subbundles L_k of $p^*E_k^+$ there are decompositions

$$L_k \oplus \operatorname{im}(\Pi^+\sigma_{A_{k-1}} + \Pi'\sigma_{B_{k-1}}) = \ker(\Pi^+\sigma_{A_k} + \Pi'\sigma_{B_k})$$

for all k. The index element of (21) (cf. 3.2.3.1. (28)) belongs to $p^*K(Y)$ iff $L_k \cong p^*G_k$ for a suitable choice of the σ_{B_k} with bundels $G_k \in \operatorname{Vect}(Y)$. Moreover

$$\operatorname{ind}_{S^*Y} \mathfrak{a} = \sum_{k=0}^{N} (-1)^k [L_k]. \tag{22}$$

Now we easily get

Theorem 7. *For a given exact sequence* (19) *of symbols with the transmission property there exists an elliptic complex* $\mathfrak{A}$ *over* Y *with* $\sigma_\Omega(\mathfrak{A})$ *given by* (19) *iff the index element* (2) *belongs to* $p^*K(Y)$.

The necessity of $(22) \in p^*K(Y)$ follows from $\operatorname{ind}_{S^*Y} \mathfrak{a} = \sum_{k=0}^{N} (-1)^k [p^*G_k]$ if $\mathfrak{a}$ denotes the complex of left upper corners in (6). That this condition is sufficient follows in a similar way as the construction of 3.2.3.1. (26). This means that the construction of an exact boundary symbol sequence is compatible with (19). Then there exists a corresponding operator complex $\mathfrak{A}$.

For more details cf. the constructions in 3.1.1.2. Note that Theorem 7 is a generalization of 3.1.1.1, Theorem 12.

Remark 8. For a given exact sequence (19) of symbols with the transmission property there exists an elliptic boundary problem of the form (14) with $\sigma_\Omega(r^+A_k) = \sigma_{A_k}$ iff $\operatorname{ind}_{S^*Y} \mathfrak{a} \in p^*K(Y)$ for $\mathfrak{a}$ as in (21). A similar assertion holds for morphisms with potential operators instead of trace operators.

The definitions and assertions about stable homotopy equivalence of complexes formulated in 3.2.3.1 can be applied to elliptic complexes on (X, Y). Application of 3.2.3.1, Proposition 7 shows that every elliptic complex $\mathfrak{A}$ on (X, Y) is stable homotopic to an elliptic boundary value problem $\mathcal{A} \in \mathfrak{G}$ with $\operatorname{ind} \mathfrak{A} = \operatorname{ind} \mathcal{A}$. Then the index theorem for elliptic operators in $\mathfrak{G}$ (cf. 3.2.2.4) can be applied to elliptic complexes. Moreover, there is an analogy of the Agranovič-Dynin formula and the clutching construction, cf. 3.2.1.3. Details are left to the reader.

Finally, discuss an "external multiplication" between an elliptic boundary problem $\mathcal{A} \in \mathfrak{G}$ (on X) and an elliptic PDO

$$S: C^\infty(M, V) \to C^\infty(M, W), \tag{23}$$

where M is a closed compact C^∞ manifold and $V, W \in \operatorname{Vect}(M)$. Assume $\pi = \operatorname{ord} \sigma_S > 0$. Let $\mathcal{A}$ be given in the form 3.1.1.1. (1) and $\alpha - 1 = \operatorname{ord} \sigma_B > 0$ (with $\alpha = \operatorname{ord} \sigma_A$), $\gamma = \operatorname{ord} \sigma_T > 0$, $\lambda = \operatorname{ord} \sigma_K > 0$, $1 - \alpha + \lambda + \gamma = \operatorname{ord} \sigma_Q > 0$. For abbreviation denote the spaces of C^∞ sections by $\Gamma(\ldots)$, e.g. $\Gamma(M, V) = C^\infty(M, V)$.

$X \times M$ is a smooth compact manifold with boundary $Y \times M$. Consider the canonical projections $p_1: X \times M \to X$, $p_2: X \times M \to M$, $q_1: Y \times M \to Y$, $q_2: Y \times M \to M$. For abbreviation, external products between vector bundles are again denoted by the usual sign $\otimes$, i.e. $E \otimes V \overset{\text{def}}{=} (p_1{}^*E) \otimes (p_1{}^*V)$ for $E \in \operatorname{Vect}(X)$, $V \in \operatorname{Vect}(M)$ and $G \otimes W \overset{\text{def}}{=} (q_1^*G) \otimes (q_2^*W)$ for $G \in \operatorname{Vect}(Y)$, $W \in \operatorname{Vect}(M)$.

From $\mathcal{A}$ we pass to the operator

$$\mathcal{A} \otimes 1_V = \begin{pmatrix} (r^+A + r'B) \otimes 1_V & K \otimes 1_V \\ (r'T) \otimes 1_V & Q \otimes 1_V \end{pmatrix} : \begin{matrix} \Gamma(X \times M, E \otimes V) \\ \oplus \\ \Gamma(Y \times M, J \otimes V) \end{matrix} \to \begin{matrix} \Gamma(X \times M, F \otimes V) \\ \oplus \\ \Gamma(Y \times M, G \otimes V), \end{matrix} \tag{24}$$

$V \in \operatorname{Vect}(M)$. Since the orders of homogeneity of all operators contained in $\mathcal{A}$ are supposed to be positive, the operator $\mathcal{A} \otimes 1_V$ belongs to the class $\overline{\mathfrak{G}}$ of operators with continuous symbols. Similarly the operator

$$1_E \otimes S\colon \Gamma(X \times M, E \otimes V) \to \Gamma(X \times M, E \otimes W) \tag{25}$$

belongs to the class of PDOs on $X \times M$ with continuous symbol, $E \in \operatorname{Vect}(X)$. Note that $1_E \otimes S$ acts in tangent direction to the boundary of $X \times M$.

Let $\mathcal{A}$ be elliptic and $\mathcal{A}^{-1}$ a parametrix of $\mathcal{A}$. Then the operator

$$\mathcal{R} = \mathcal{L}^{\alpha,\delta}_{E,J}\mathcal{A}^{-1}(\mathcal{L}^{-\alpha,-\delta}_{F,G})^{-1}, \quad \delta = 1 - \alpha + \lambda + \gamma,$$

(cf. 3.1.2.1. (3)) is elliptic, $\operatorname{ind} \mathcal{R} = -\operatorname{ind} \mathcal{A}$, and the operators contained in $\mathcal{R}$ have the same orders as the corresponding operators in $\mathcal{A}$, i.e. the PDO has the order α like the Green operator, the trace operator the order $\gamma + \frac{1}{2}$ and so on.

Similarly, if S^{-1} is a parametrix of the elliptic operator S, set

$$P = \Lambda^{\mu}_{V} S^{-1}(\Lambda^{-\mu}_{W})^{-1}$$

(cf. 1.2.4.2, Theorem 5). Then obviously $\operatorname{ind} P = -\operatorname{ind} S$ and $\operatorname{ord} P = \operatorname{ord} S$. Moreover, let

$$S_\delta = \Lambda^{\delta-\mu}_{W} S, \quad P_\delta = P\Lambda^{\delta-\mu}_{W}.$$

Define the following operator $\in \overline{\mathfrak{G}}$ over $X \times M$

$$\mathcal{A} \otimes S = \left(\begin{array}{cc:cc} \multicolumn{2}{c:}{\mathcal{A} \otimes 1_V} & -1_F \otimes P & 0 \\ & & 0 & -1_G \otimes P_\delta \\ \hdashline 1_E \otimes S & 0 & \multicolumn{2}{c}{\mathcal{R} \otimes 1_W} \\ 0 & 1_J \otimes S_\delta & & \end{array}\right):$$

$$\begin{matrix} \Gamma(X \times M, E \otimes V) & & \Gamma(X \times M, V \otimes F) \\ \oplus & & \oplus \\ \Gamma(Y \times M, J \otimes V) & & \Gamma(Y \times M, G \otimes V) \\ \oplus & \to & \oplus \\ \Gamma(X \times M, F \otimes W) & & \Gamma(X \times M, E \otimes W) \\ \oplus & & \oplus \\ \Gamma(Y \times M, G \otimes W) & & \Gamma(Y \times M, J \otimes W). \end{matrix} \tag{26}$$

Note that $\mathcal{A} \otimes S$ depends on the choice of the reducing operators in 3.1.2.1.(3), 1.2.4.2.(5) and on the choice of the parametrices.

Theorem 9. *Suppose that* $\mathcal{A} \in \mathfrak{G}$ *of the form* 3.1.1.1.(1) *is elliptic,* $\alpha - 1 > 0$, $\gamma > 0$, $\lambda > 0$, $1 - \alpha + \lambda + \gamma > 0$ *and let* $S \in L^{\mu}_{\mathrm{cl}}(M; V, W)$ *be elliptic,* $\mu > 0$. *Then the operator* $\mathcal{A} \otimes S$ *defined by* (26) *is elliptic and*

$$\operatorname{ind}(\mathcal{A} \otimes S) = \operatorname{ind} \mathcal{A} \operatorname{ind} S. \tag{27}$$

The proof of Theorem 9 is not difficult but unfortunately rather voluminous. Therefore, we give only the idea. Details can be easily checked by the reader. The ellipticity of $\mathcal{A} \otimes S$ follows by trivial arguments of linear algebra, namely that external tensor products are bijective if one of the factors is bijective. In order to prove (27) one can apply the multiplicative properties of the index homomorphism ind_t (cf. Atiyah/Singer [2], Palais [1]). Let $d(\mathcal{A}) \in K(T^*\Omega)$, $d(S) \in K(T^*N)$ be the difference elements belonging to $\mathcal{A}$ and S, respectively (cf. 3.2.2.4, 1.2.4.2). We shall use the fact that the external multiplication of elements of the K groups $K(T^*\Omega) \otimes K(T^*M) \to K(T^*(\Omega \times M))$ corresponds just to

$$d(\mathcal{A}) \otimes d(S) = d(\mathcal{A} \otimes S). \tag{28}$$

(28) is only needed in an obvious special case to be described below. Since $\operatorname{ind}_t d(\mathcal{A} \otimes S) = \operatorname{ind}_t d(\mathcal{A}) \operatorname{ind}_t d(S)$ (because of a general property of the topological index, cf. ATIYAH/SINGER [2, I]), (27) follows from the index theorems in 1.2.4.2 and 3.2.2.4. The construction of the difference element $d(\mathcal{A})$ shows the existence of an elliptic operator $\mathcal{E} \in \mathfrak{G}$ over X with $\operatorname{ind} \mathcal{E} = 0$ (even an isomorphism), where the orders of homogeneity of the principal symbols of all operators contained in $\mathcal{E}$ are positive and so that $\mathcal{A} \otimes \mathcal{E}$ is homotopic through elliptic operators in $\mathfrak{G}$ to an operator of the form $r'A_1 \oplus Q_1$. Here r^+A_1 is a PDO on X with the transmission property, $\operatorname{ord} \sigma_{A_1} > 0$, r^+A_1 is Fredholm without boundary conditions and Q_1 is an elliptic PDO over Y with $\operatorname{ord} \sigma_{Q_1} > 0$. Thus

$$(\mathcal{A} \oplus \mathcal{E}) \otimes S \simeq (r^+A_1 \otimes Q_1) \otimes S = \big((r^+A_1) \otimes S\big) \oplus (Q_1 \otimes S) \tag{29}$$

and hence $\operatorname{ind} (\mathcal{A} \oplus \mathcal{E}) \otimes S = \operatorname{ind} \big((r^+A_1) \otimes S\big) + \operatorname{ind} (Q_1 \otimes S)$. Now $d\big((r^+A_1)\big) \otimes d(S = d\big((r^+A_1) \otimes S\big)$ is quite obvious so that

$$\operatorname{ind} \big((r^+A_1) \otimes S\big) = \operatorname{ind} (r^+A_1) \operatorname{ind} S\,. \tag{30}$$

Moreover,

$$\operatorname{ind} (Q_1 \otimes S) = \operatorname{ind} Q_1 \operatorname{ind} S \tag{31}$$

is well-known (cf. 1.2.4.2). Thus (29), (30), (31) yield

$$\operatorname{ind} \big((\mathcal{A} \oplus \mathcal{E}) \otimes S\big) = \operatorname{ind} (\mathcal{A} \oplus \mathcal{E}) \operatorname{ind} S = \operatorname{ind} \mathcal{A} \operatorname{ind} S\,. \tag{32}$$

Since $\mathcal{E}$ is an isomorphism, $\operatorname{ind} (\mathcal{E} \otimes S) = 0$. Thus (27) is a consequence of (32).

The external multiplication between $\mathcal{A}$ and S gives rise to an elliptic complex on $X \times M$ similarly as 1.1.3.2.(8). Using reduction of the length in the sense of 3.2.3.1, Proposition 7 it can be proved that the index of the arising complex is equal to the index of $\mathcal{A} \otimes S$.

4. Further Results on Boundary Value Problems

4.1. Analytical Index Formulas

The index theorem for boundary value problems states that the index of an arbitrary elliptic boundary problem can be expressed in terms of topological invariants of the symbol, more exactly of the corresponding difference element in the K-group of $T^*\Omega$ (Ω the interior of the considered manifold X with boundary). In the cohomological formulation (cf. Atiyah/Singer [2, III]) this means that there are well-defined cohomology classes in the de Rham cohomology $\mu \in H^{2n}(T^*X)$ and $\nu \in H^{2n-2}(T^*Y)$ with compact support (H^k denotes the k-dimensional cohomology space with values in $\mathbb{C}$) expressed in terms of characteristic classes of the manifold X and of the principal symbol of the operator $\mathcal{A} \in \mathfrak{G}(X, Y)$ such that

$$\operatorname{ind} \mathcal{A} = \mu[T^*X] + \nu[T^*Y] .$$

Similarly as for PDOs on closed compact manifolds there is the interesting problem to derive explicit (and simple) analytical expressions for differential forms representing the cohomology classes of μ and ν in terms of the symbol. For several elliptic differential operators occurring in differential geometry this attempt yields relations between local invariants (of the Riemannian structure or the curvature) and global topological invariants. For example, the index of the de Rham complex of exterior differentials is equal to the Euler characteristic of the manifold (without boundary). The Gauss-Bonnet theorem states that this index is equal to the integral over the Euler form of an arbitrary Riemannian structure on X. Further examples are discussed in Gilkey [1].

On the other hand, one looks for special cases for analytical proofs of the index theorem. For manifolds without boundary a proof is given in Atiyah/Bott/Patodi [1] for simple operators by the heat equation method. For general operators (under some assumptions about the manifold) a proof has been given by Fedosov in [3]. Following the ideas of Fedosov we shall consider the case of boundary value problems.

4.1.1. The "Coarse" Index Formula

The starting point is the classical formula

$$\operatorname{ind} \mathcal{A} = \operatorname{tr} (1 - \mathcal{R}\mathcal{A})^N - \operatorname{tr} (1 - \mathcal{A}\mathcal{R})^N$$

(cf. 1.1.1.1, Proposition 8), where $\mathcal{R}$ is a parametrix of $\mathcal{A} \in \mathfrak{G}$ with the property that $(1 - \mathcal{R}\mathcal{A})^N$ and $(1 - \mathcal{A}\mathcal{R})^N$ are operators of trace class for sufficiently large $N \in \mathbb{Z}_+$.

The boundary value problems $\mathcal{A}$ and $\mathcal{R}$ can be described in terms of its complete symbols (mod $\mathfrak{G}^{-\infty}$). Applying the calculus of complete symbols we get theorems on regularized trace. These are used in the transformation of the above formula into an expression containing only the principal symbols.

4.1.1.1. Formal Complete Symbols

In the context of analytical index formulas to be discussed below it is useful to introduce so-called *formal symbols*, i.e. formal power series in a new variable λ with usual symbols as coefficients. The λ powers fix an ordering of the components of the symbol similar to the ordering by homogeneity. This is important if the symbols are only given for small $|\xi|$ and $|\xi'|$, respectively.

Formal symbols of PDOs are introduced in Fedosov [3]. We shall give all definitions and results about formal boundary symbols. Then it is obvious which modifications yield the classes of formal symbols with the transmission property and so on.

Definition 1. Let $\Omega' \subseteq \mathbb{R}^{n-1}$ be open. Denote by $\mathfrak{R}_\lambda^{m,d}$ the space of all formal power series

$$\sigma = \sum_{j=0}^{\infty} \lambda^j \sigma_j ,$$

where $\sigma_j \in \mathfrak{R}^{m_j, d}$, $m_j \to -\infty$ monotonic, $m = m_0$.

For any $N \in \mathbb{Z}_+$ define a mapping $\mathfrak{R}_\lambda^{m,d} \to \mathfrak{R}^{m,d}$ by $\sigma \mapsto \sigma|_N \overset{\text{def}}{=} \sum_{j<N} \sigma_j$. The image of this mapping for $N = 1$ is called *principal symbol* of σ (this is exactly the coefficient of λ^0).

Point out the difference between the principal symbol defined above and the notion used in chapter 2 and 3. In the earlier definition the principal symbol was completely defined by its values on $S^*\Omega' = \{(x', \xi'): |\xi'| = 1\}$. Now, moreover, there is fixed an extension as a smooth function for all ξ' (including $\xi' = 0$). This can be done by choosing an excision function $\chi(\xi')$ vanishing near $\xi' = 0$, which will be multiplied with the extension by homogeneity of the values on the sphere.

We have $\sigma = \sigma'$ iff $\sigma|_N = \sigma'|_N$ for every $N \in \mathbb{Z}_+$. There is a mapping

$$\mathfrak{R}_\lambda^{m,d} \to \mathfrak{R}^{m,d}/\mathfrak{R}^{-\infty,d}$$

defined by

$$\sigma = \sum_{j=0}^{\infty} \lambda^j \sigma_j \mapsto \sigma' \sim \sum_{j=0}^{\infty} \sigma_j .$$

Let $\sigma = \sum_j \lambda^j \sigma_j$ and $\sigma' = \sum_j \lambda^j \sigma_j'$ be formal boundary symbols, $\sigma \in \mathfrak{R}_\lambda^{m,d}$, $\sigma' \in \mathfrak{R}_\lambda^{m',d'}$. Define the composition of formal boundary symbols by

$$\sigma \circ \sigma' = \sum_\alpha \frac{\lambda^{|\alpha|}}{\alpha!} \partial_{\xi'}^\alpha \sigma D_{x'}^\alpha \sigma' , \tag{1}$$

where $\partial^\alpha \sigma$ denotes the formal power series obtained from σ by elementwise differentiation of the coefficients, and on the right hand side the multiplication of formal power series with the composition rule of boundary symbols for the coefficients is taken.

(1) defines an element of $\mathfrak{R}_\lambda^{m+m',d''}$, $d'' = \max(m' + d, d')$. It is easily checked that the multiplication in the algebra of formal symbols is associative.

Definition 2. Let $\sigma \in \mathfrak{R}_\lambda^{m,d}$ and $\mathcal{A} \in \mathrm{Op}(\mathfrak{R}^{m,d})$. Then σ is called a *symbol* of $\mathcal{A}$ if, for any $N \in \mathbb{Z}_+$ there is an $N' \in \mathbb{Z}_+$ such that $\mathcal{A} - \mathrm{Op}(\sigma|_{N'}) \in \mathrm{Op}(\mathfrak{R}^{-N,d})$.

Observe that for a given operator $\mathcal{A} \in \mathrm{Op}(\mathfrak{R}^{m,d})$ no coefficient of the formal symbol is uniquely defined but only modulo lower order terms. Operators $\mathcal{A}_1$, $\mathcal{A}_2$ with a given formal symbol σ may differ by a smoothing operator.

2.3.2.2, Proposition 4 and the definition of the composition of formal symbols (1) imply

Proposition 3. *Let $\mathcal{A}_i \in \mathrm{Op}(\mathfrak{R}^{m_i,d_i})$, $i = 1, 2$, one of them properly supported with formal symbols $\sigma_i \in \mathfrak{R}^{m_i,d_i}$. Assume that $\mathcal{A}_1\mathcal{A}_2$ is defined. Then $\mathcal{A}_1\mathcal{A}_2 \in \mathrm{Op}(\mathfrak{R}^{m_1+m_2,d'})$, $d' = \max(d_1 + m_2, d_2)$ with $\sigma_1 \circ \sigma_2$ as formal symbol.*

Let Ω', Ω_1' be open sets in $\mathbb{R}^{n-1}$ and $\varkappa\colon \Omega' \to \Omega_1'$ a diffeomorphism. Let (x', ξ') and (y', η') be coordinates in $T^*\Omega'$ and $T^*\Omega_1'$, respectively. Then we get from 2.3.2.2, Proposition 5

Proposition 4. *Let $\mathcal{A} \in \mathrm{Op}(\mathfrak{R}^{m,d})(\Omega')$ and $\sigma(\mathcal{A})(x', \xi') \in \mathfrak{R}_\lambda^{m,d}(\Omega')$ its formal boundary symbol. Then the operator $\varkappa_*\mathcal{A} \in \mathrm{Op}(\mathfrak{R}^{m,d})(\Omega_1')$ has*

$$\sigma(\mathcal{A})_\varkappa(y', \eta') = \sum_\alpha (\lambda^{|\alpha|}/\alpha!)\, \partial_{\xi'}^\alpha\, \sigma(\mathcal{A})\left(x', {}^t(\mathrm{d}\varkappa)(x')\,\eta'\right) D_{z'}^\alpha\, \mathrm{e}^{\mathrm{i}h(z',x')\eta'}\big|_{z'=x'},$$

$$x' = \varkappa^{-1}(y'), h(z', x') = \varkappa(z') - \varkappa(x') - \mathrm{d}\varkappa(x')(z' - x'),$$

as a formal symbol.

Proposition 5. (i) *If $\sigma_i \in \mathfrak{R}_\lambda^{m_i,d_i}$, $i = 1, 2$, then $(\sigma_1)_\varkappa \circ (\sigma_2)_\varkappa = (\sigma_1 \circ \sigma_2)_\varkappa$.*

(ii) *For diffeomorphisms $\varphi\colon \Omega' \to \Omega_1'$, $\varkappa\colon \Omega_1' \to \Omega_2'$ and $\sigma \in \mathfrak{R}_\lambda^{m,d}$ we have $\sigma_{\varkappa\circ\varphi} = (\sigma_\varphi)_\varkappa$.*

For formal symbols of PDOs a proof is given in FEDOSOV [3]. It can be immediately generalized to our situation. If we replace $\mathfrak{R}^{m,d}$ by $\mathfrak{A}^m$ or $\mathfrak{S}^{m,d}$ in the above definitions we obtain the spaces of formal symbols $\mathfrak{A}_\lambda^m$ and $\mathfrak{S}_\lambda^{m,d}$. The formulation of analogous assertions to Proposition 3, 4, 5 are left to the reader.

Consider now operators on manifolds. Let $\mathcal{A} \in \mathfrak{G}^{m,d}(X, Y; E, F, J, G)$, E, $F \in \mathrm{Vect}(X)$, J, $G \in \mathrm{Vect}(Y)$. Let $\mathfrak{U} = \{U_j\}$ be a finite covering of X by open contractible sets (hence any bundle over U_j admits a trivialization). Let for any $U \in \mathfrak{U}$ a formal symbol $\sigma_U \in \mathfrak{S}_\lambda^{m,d}$ of $\mathcal{A}_U$ be given. This collection of formal symbols is called *global formal symbol* if for arbitrary U, $V \in \mathfrak{U}$ with non-empty intersection over $U \cap V$ the local expressions σ_U and σ_V are related by

$$\sigma_V = \psi_{VU} \circ (\sigma_U)_\varkappa \circ \psi'_{UV}. \tag{2}$$

Here ψ and ψ' near the boundary denote the cocycle of $F' \oplus G$ and $E' \oplus J$, respectively. In the interior ψ and ψ' denote the cocycle of F and E, respectively. These matrix functions are considered as coefficient of λ^0.

Denote by $\mathfrak{S}_\lambda^{m,d}(X, Y)$ the space of all global formal symbols. Of course, there is no preferred atlas of the manifold and no preferred local trivializations of the bundles. But from σ_U for a fixed atlas using (2) we get local representations of $\sigma \in \mathfrak{S}_\lambda^{m,d}(X, Y)$ in arbitrary other local coordinates. Note that (2) obviously implies 2.3.3.1. (3) if we replace the formal symbols by their asymptotic sums.

Proposition 6. *For any $\mathcal{A} \in \mathfrak{G}^{m,d}(X, Y)$ there is a global formal symbol $\sigma \in \mathfrak{S}_\lambda^{m,d}(X, Y)$.*

Proof: Fix the coefficient of λ^0 in one coordinate neighbourhood (of each connected component of X) in such a way that it is equal to the local symbol in $\mathfrak{S}^{m,d}$ of $\mathcal{A}$

modulo $\mathfrak{S}^{m-1,d}$. In other coordinates the coefficient of λ^0 is obtained from (2) and uniquely defined modulo $\mathfrak{S}^{m-1,d}$. In the same way we obtain the coefficients of higher λ powers. Proposition 5 (ii) shows the correctness of this procedure. □

Now, the role of a homogeneous principal symbol for classical PDOs or boundary value problems is played by the coefficient $\sigma_0 = (\sigma_0(A), \sigma_Y(\mathcal{A})_0)$. In fact, (2) implies that

$$\sigma_0(A): \pi^*E \to \pi^*F\,, \tag{3}$$

$\pi: T^*X \to X$ is a bundle morphism and

$$\sigma_Y(\mathcal{A})_0: p^*E' \otimes H^+ \oplus p^*J \to p^*F' \otimes H^+ \oplus p^*G\,, \tag{4}$$

$p: T^*Y \to Y$ is a bundle morphism. They are compatible in the sense of the definition in 2.2.5.3.

Proposition 7. *Let bundle morphisms $\sigma_0(A)$ and $\sigma_Y(\mathcal{A})_0$ be given (cf. (3), (4)). Assume that $\sigma_0(A)\,(x, \xi) \in \hom(E_x, F_x)$ is positively homogeneous of order m for $|\xi| \geqq 1$, i.e.*

$$\sigma_0(A)\,(x, \lambda\xi) = \lambda^m \sigma_0(A)\,(x, \xi),\ |\xi| \geqq 1\,,\ \lambda \geqq 1\,,$$

and

$$\sigma_Y(\mathcal{A})_0\,(x', \xi') = \begin{pmatrix} \Pi^+\sigma_0(A)\,(x', \xi', \nu) + \Pi'\sigma_0(B)\,(x', \xi', \nu, \tau) & \sigma_0(K)\,(x', \xi', \nu) \\ \Pi'\sigma_0(T)\,(x', \xi', \nu) & \sigma_0(Q)\,(x', \xi') \end{pmatrix},$$

where, for $|\xi'| \geqq 1$, the symbols $\sigma_0(B)$, $\sigma_0(T)$ are positively homogeneous with respect to (ξ', ν, τ) of order $m-1$, and $\sigma_0(K)$ and $\sigma_0(Q)$ are positively homogeneous with respect to (ξ', ν) of order m. Then there exists a global formal symbol $\sigma \in \mathfrak{S}_\lambda^{m,d}(X, Y)$ with $(\sigma_0(A), \sigma_Y(\mathcal{A})_0)$ as principal symbol.

The simple proof based on the construction in the proof of Proposition 6 is left to the reader.

4.1.1.2. Theorems on the Regularized Trace

In this section we consider the trace of operators in $\mathfrak{G}(X, Y)$ and prove theorems on the regularized trace, which shall be used in the proof of the coarse index formula. A linear operator $A: H_1 \to H_2$ (H_i Hilbert spaces) belongs to the trace class if

$$||A||_{\mathrm{tr}} \overset{\mathrm{def}}{=} \sup \sum_i |(Ae_i, f_i)| < \infty\,,$$

where the supremum is taken over all orthonormal systems $\{e_i\}$ in H_1 and $\{f_i\}$ in H_2. This number is called *trace norm* of A. Denote by $\mathcal{T}(H_1, H_2)$ the space of all operators in $\mathcal{L}(H_1, H_2)$ of trace class. This is a Banach space with the trace norm. $\mathcal{T}(H, H)$ is a two sided ideal in $\mathcal{L}(H, H)$.

Consider $\mathcal{A} \in \mathfrak{G}^{m,d}(X, Y; E, F, J, G)$. Set $\mathcal{H}^s(E, J) = H^s(X, E) \oplus H^{s+1/2}(Y, J)$, where H^s denotes the usual Sobolev space of sections in the corresponding bundle. According to 2.3.3.3, Theorem 1 the operator $\mathcal{A}$ admits a continuous extension

$$\mathcal{A}: \mathcal{H}^s(E, J) \to \mathcal{H}^{s-m}(F, G)$$

for any $s \in \mathbb{R}$, $s > d - {}^1/_2$. For $m < 0$ this leads to a continuous operator

$$\mathcal{A}: \mathcal{H}^s(E, J) \to \mathcal{H}^s(F, G)\,,\quad s > d - {}^1/_2$$

using the continuous embedding $\mathcal{H}^{s-m}(F, G) \subset \mathcal{H}^s(F, G)$.

Theorem 1. *Let $\mathcal{A} \in \mathfrak{G}^{m,d}(X, Y; E, F, J, G)$. The Sobolev space extension*

$$\mathcal{A}: \mathcal{H}^s(E, J) \to \mathcal{H}^s(F, G) \tag{1}$$

for any $s > d - {}^1/_2$ is an operator of trace class when $m < -n = -\dim X$.

The proof is a consequence of some lemmata.

Lemma 2. *Let H_i, H_i', $i = 1, 2$, be Hilbert spaces and let*

$$A = \begin{pmatrix} A_{11} & A_{12} \\ A_{21} & A_{22} \end{pmatrix} : \begin{matrix} H_1 \\ \oplus \\ H_1' \end{matrix} \to \begin{matrix} H_2 \\ \oplus \\ H_2' \end{matrix}$$

be linear and continuous. Then A is an operator of trace class iff A_{ij} is of the trace class with respect to the corresponding spaces for all $i, j = 1, 2$.

Proof: If $(\cdot,\cdot)$ and $(\cdot,\cdot)'$ denote the scalar products in Hilbert spaces H and H', a scalar product in $H \oplus H'$ is given by $((u_1, v_1), (u_2' v_2)) = (u_1, u_2) + (v_1, v_2)'$, $(u_i, v_i) \in H \oplus H'$. Orthonormal systems $\{e_j\}$ and $\{g_i\}$ in H and H' yield the orthonormal system $\{(e_i, 0), (0, g_i)\}$ in $H \oplus H'$. Then

$$\begin{aligned} \|A\|_{tr} &= \sup_{\{f_i^{(1)}\}, \{f_i^{(2)}\}} \sum_i |(Af_i^{(1)}, f_i^{(2)})| \\ &\geq \sup_{\{e_i^{(1)}\}, \{e_i^{(2)}\}} \sum_i \left|\left(A\begin{pmatrix} e_i^{(1)} \\ 0 \end{pmatrix}, \begin{pmatrix} e_i^{(2)} \\ 0 \end{pmatrix}\right)\right| + \sup_{\{g_i^{(1)}\}, \{g_i^{(2)}\}} \sum_i \left|\left(A\begin{pmatrix} 0 \\ g_i^{(1)} \end{pmatrix}, \begin{pmatrix} 0 \\ g_i^{(2)} \end{pmatrix}\right)\right| \\ &= \|A_{11}\|_{tr} + \|A_{22}\|_{tr}. \end{aligned}$$

In fact, on the left hand side of the inequality the supremum is taken over all orthonormal systems $\{f_i^{(1)}\}$ in $H_1 \oplus H_1'$ and $\{f_i^{(2)}\}$ in $H_2 \oplus H_2'$, while on the right hand side the supremum is taken only over orthonormal systems of the form $\{(e_i^{(1)}, 0), (0, g_i^{(1)})\}$ in $H_1 \oplus H_1'$ and $\{(e_i^{(2)}, 0), (0, g_i^{(2)})\}$ in $H_2 \oplus H_2'$, respectively. Similarly, changing the orders in the orthonormal systems we get

$$\|A\|_{tr} \geq \|A_{12}\|_{tr} + \|A_{21}\|_{tr}.$$

Therefore all A_{ij} are operators of trace class if A is of trace class.

In order to prove the converse implication we use the following characterization of operators of trace class. A belongs to $\mathcal{T}(H_1, H_2)$ iff there is an orthonormal base $\{e_i\}$ in H_1 such that $\sum_i \|Ae_i\| < \infty$. Let $\{e_i\}$ and $\{e_i'\}$ be orthonormal bases in H_1 and H_1' such that $\sum_i \|A_{11}e_i\| < \infty$ and $\sum_i \|A_{12}e_i'\| < \infty$. Then the criterion is satisfied for the operator $A_1 = (A_{11}, A_{12})$: $H_1 \oplus H_1' \to H_2$ with the orthonormal base $\{(e_i, 0), (0, e_i')\}$. In the same way the operator $A_2 = (A_{21}, A_{22})$: $H_1 \oplus H_1' \to H_2'$ is a trace class operator. Then it is sufficient to show that for trace class operators $A_1: H \to H_2$, $A_2: H \to H_2'$ the operator $A = (A_1, A_2): H \to H_2 \oplus H_2'$ is trace class.

Consider $A_1' \overset{\text{def}}{=} (A_1, 0): H \to H_2 \oplus H_2'$. Let $\{g_i\}$ be an orthonormal base in H such that $\sum_i \|A_1 g_i\| < \infty$. Then $\sum_i \|A_1' g_i\| = \sum_i \|A_1 g_i\| < \infty$ and A_1' is trace class. The same argument can be applied to $A_2' \overset{\text{def}}{=} (0, A_2)$. Then $A = A_1' + A_2'$ is trace class, too. □

Recall that $A \in \mathcal{L}(H_1, H_2)$ (H_1, H_2 Hilbert spaces) is called a *Hilbert-Schmidt operator* if $\sum_j \|Ae_j\|^2 < \infty$ for a complete orthonormal system $\{e_j\}$ in H_1. It can be proved that the number $\left(\sum_j ||Ae_j||^2\right)^{1/2}$ is independent of the choice of $\{e_j\}$. It is called *Hilbert-Schmidt norm* $||\cdot||_{HS}$. The space of all operators of finite rank is dense in the space of all Hilbert-Schmidt operators $\mathcal{H}(H_1, H_2)$ with respect to the Hilbert-Schmidt norm.

Lemma 3. *Let $A \in L^m(\mathbb{R}^n)$ be compactly supported. Then, for $m < -n/2$, we have $A \in \mathcal{H}\left(L^2(\mathbb{R}^n), L^2(\mathbb{R}^n)\right)$ and the Hilbert-Schmidt norm is equal to $(2\pi)^{-n} \int |\sigma(A)(x,\xi)|^2 \, dx \, d\xi$, where A is defined by $\sigma(A)(x, \xi)$.*

Proof: Denote by $K_A(x, y)$ the distributional kernel of A. Then we have $A \in \mathcal{H}\left(L^2(\mathbb{R}^n), L^2(\mathbb{R}^n)\right)$ iff $K_A(x, y) \in L^2(\mathbb{R}^n \times \mathbb{R}^n)$ and

$$||A||_{HS}^2 = \int |K_A(x, y)|^2 \, dx \, dy .$$

The kernel of the PDO A with symbol $\sigma(A)(x, \xi)$ is given by

$$K_A(x, y) = (2\pi)^{-n} \int e^{i(x-y)\xi} \sigma(A)(x, \xi) \, d\xi$$

and

$$\sigma(A)(x, \xi) = \int e^{-it\xi} K_A(x, x - t) \, dt .$$

By Parseval's equality

$$\int |K_A(x, y)|^2 \, dx \, dy = \int |K_A(x, x - t)|^2 \, dx \, dt = \int |\sigma(A)(x, \xi)|^2 \, dx \, d\xi$$

follows. This shows the assertion. □

Corollary 4. *Let $A \in L^m(\mathbb{R}^n)$ be compactly supported. Then, for $m < -n$, we have $A \in \mathcal{T}\left(L^2(\mathbb{R}^n), L^2(\mathbb{R}^n)\right)$.*

In fact, we can decompose A into a composition of two PDOs of order $< -n/2$ using the invertible PDO $\Lambda^{-n/2+\varepsilon}$, $\varepsilon > 0$ sufficiently small.

Corollary 5. *Let X be a compact manifold without boundary and $A \in L^m(X; E, F)$, $E, F \in \operatorname{Vect}(X)$. Then $A \in \mathcal{H}\left(H^s(X, E), H^s(X, F)\right)$ when $m < -n/2$, and $A \in \mathcal{T}\left(H^s(X, E), H^s(X, F)\right)$ when $m < -n$ for any $s \in \mathbb{R}$.*

Proof: In view of the isomorphisms

$$\Lambda_E^s : H^s(X, E) \to H^0(X, E)$$

we can assume that $s = 0$. Then the assertion follows from Corollary 4. □

For $m > 0$ the natural embedding

$$i_m : H^{s+m}(X, E) \to H^s(X, E)$$

is of trace class iff

$$\Lambda_E^{-m} : H^{s+m}(X, E) \to H^{s+m}(X, E)$$

is of trace class. Therefore Corollary 5 implies

Lemma 6. *The embedding*

$$i_m : H^{s+m}(X, E) \to H^s(X, E)$$

is of trace class when $m > n = \dim X$.

For non-compact X the same is true for the operator φi_m, where φ denotes the multiplication operator with the function $\varphi \in C_0^\infty(X)$.

Proof of Theorem 1: In view of Lemma 2 it is sufficient to show that all components r^+A, $r'A$, $r'B$, $\mathcal{A}$, $r'T$, Q of $\mathcal{A}$ are trace class operators. According to 2.3.3.3, Theorem 1 $\mathcal{A}$ defines a continuous operator

$$\mathcal{A}: \mathcal{H}^s(E, J) \to \mathcal{H}^{s-m}(F, G)\,, \quad s > d - {}^1/_2\,.$$

Then the mapping (1) is the composition

$$\mathcal{H}^s(E, J) \xrightarrow{\ \mathcal{A}\ } \mathcal{H}^{s-m}(F, G) = \begin{matrix} H^{s-m}(X, F) \\ \oplus \\ H^{s-m+1/2}(Y, G) \end{matrix} \xrightarrow{\ i_{-m} \oplus i_{-m}\ } \begin{matrix} H^s(X, F) \\ \oplus \\ H^{s+1/2}(Y, G) \end{matrix}\,,$$

which is of trace class by Lemma 6.

Proposition 7. *Let $K: H^0(X, E) \to H^0(X, E)$ be an operator of trace class defined by*

$$Ku(x) = \int_X K(x, y)\, u(y)\, dy\,, \quad u \in H^0(X, E)$$

(dy is a fixed density on X), where K is a continuous section of hom (p_1^*E, p_2^*E), *$p_1: X \times X \to X$ and $p_2: X \times X \to X$ the projection to the first and the second factor of the product space, respectively. Then*

$$\operatorname{tr} K = \int_X \operatorname{Tr} K(y, y)\, dy \tag{2}$$

(Tr denotes the trace of hom (E_y, E_y)).

A proof of this classical result is given in Hörmander [7].

Remark that if $K: H^s(X, E) \to H^s(X, E)$, $s \in \mathbb{R}$, is a PDO of order $< -n = -\dim X$, its distributional kernel is continuous and it defines a trace class operator. For $s = 0$ tr K is given by (2). Note that any trace class operator has a discrete countable spectrum and $\operatorname{tr} K = \sum_j \lambda_j$, where the sum is taken over all eigenvalues λ_j counted with their multiplicities. The eigenvalues of K considered as an operator on $H^s(X, E)$ are independent of s and so is the trace.

Lemma 8. *Let H, H' be Hilbert spaces with the scalar products $(\cdot, \cdot)$ and $(\cdot, \cdot)'$. Consider $H \oplus H'$ as a Hilbert space. Let*

$$A = \begin{pmatrix} A_{11} & A_{12} \\ A_{21} & A_{22} \end{pmatrix} : \begin{matrix} H \\ \oplus \\ H' \end{matrix} \to \begin{matrix} H \\ \oplus \\ H' \end{matrix}$$

be a trace class operator. Then

$$\operatorname{tr} A = \operatorname{tr} A_{11} + \operatorname{tr} A_{22}\,. \tag{3}$$

Proof: For orthonormal bases $\{e_i\}$ and $\{e_i'\}$ in H and H' an orthonormal base in $H \oplus H'$ is given by $\{(e_i, 0), (0, e_i')\}$. We have

$$\operatorname{tr} A = \sum_i \left\{ \left(A \begin{bmatrix} e_i \\ 0 \end{bmatrix}, \begin{bmatrix} e_i \\ 0 \end{bmatrix} \right) + \left(A \begin{bmatrix} 0 \\ e_i' \end{bmatrix}, \begin{bmatrix} 0 \\ e_i' \end{bmatrix} \right) \right\}$$

$$= \sum_i (A_{11} e_i, e_i) + \sum_i (A_{22} e_i', e_i') = \operatorname{tr} A_{11} + \operatorname{tr} A_{22}\,. \quad \square$$

Let $\mathcal{A} \in \mathfrak{G}^{m,d}(\overline{\mathbb{R}}^n_+, \mathbb{R}^{n-1}; \mathbb{C}^k, \mathbb{C}^k, \mathbb{C}^j, \mathbb{C}^j)$, $m < -n$, with the symbol

$$\sigma(\mathcal{A}) = \big(\sigma(A)\,(x, \xi), \sigma_{\mathbb{R}^{n-1}}(\mathcal{A})\,(x', \xi')\big)\,.$$

In $\mathcal{H}^s(\overline{\mathbb{R}}^n_+, \mathbb{R}^{n-1}; \mathbb{C}^k, \mathbb{C}^j) = H^s(\mathbb{R}^n_+, \mathbb{C}^k) \oplus H^{s+1/2}(\mathbb{R}^{n-1}, \mathbb{C}^j)$ $\mathcal{A}$ defines a trace class operator. For $d = 0$ the components of $\mathcal{A}$ have continuous kernels. In fact, the kernel of r^+A is given by

$$(2\pi)^{-n} \int e^{i(x-y)\xi} \sigma(A)(x, \xi) \, d\xi$$

and for $m < -n$ this integral converges absolutely. Similarly, Green and trace operators of type zero and potential operators have continuous kernels. Hence we can apply Proposition 7 and get from Lemma 8 that

$$\operatorname{tr} \mathcal{A} = (2\pi)^{-n} \int_{\mathbb{R}^n_+ \times \mathbb{R}^n} \operatorname{Tr} \sigma(A)(x, \xi) \, d\xi \, dx$$
$$+ (2\pi)^{-n+1} \int_{\mathbb{R}^{2n-2}} \left(\operatorname{Tr} \Pi'_\nu \sigma_B(x', \xi', \nu, \nu) + \operatorname{Tr} \sigma_Q(x', \xi')\right) d\xi' \, dx' . \tag{4}$$

(4) is valid for Green operators of trace class and type $d > 0$, too. Set $\operatorname{Tr}' \sigma_Y(x', \xi') = \Pi'_\nu \sigma_B(x', \xi', \nu, \nu) + \operatorname{Tr} \sigma_Q(x', \xi')$.

Consider $\mathcal{A}_1 \in \mathfrak{G}^{m,d}(\overline{\mathbb{R}}^n_+, \mathbb{R}^{n-1}; \mathbb{C}^{k'}, \mathbb{C}^k, \mathbb{C}^{j'}, \mathbb{C}^j)$ and $\mathcal{A}_2 \in \mathfrak{G}^{m',d'}(\overline{\mathbb{R}}^n_+, \mathbb{R}^{n-1}; \mathbb{C}^k, \mathbb{C}^{k'}, \mathbb{C}^j, \mathbb{C}^{j'})$, both compactly supported with formal symbols $\sigma_i = \left(\sigma(A_i), \sigma_{\mathbb{R}^{n-1}}(\mathcal{A}_i)\right) \in \mathfrak{S}_\lambda$. Assume that the interior symbols $\sigma(A_i)$ are of the form $\sigma(A_i)(x', x_n, \xi) = \varphi(x_n) \sigma(A_i)(x', 0, \xi)$ for some $\varphi \in C_0^\infty(\overline{\mathbb{R}}_+)$, $\varphi = 1$ near zero. Then

$$\mathcal{A}_1 \circ \mathcal{A}_2 - \operatorname{Op}(\sigma_1 \circ \sigma_2|_N)$$

for $N > m + m' + n$ is a trace class operator in the sense of Theorem 1 (for the definition of $\sigma_1 \circ \sigma_2|_N$ cf. 4.1.1.1). The expression $\operatorname{tr}\left(\mathcal{A}_1 \circ \mathcal{A}_2 - \operatorname{Op}(\sigma_1 \circ \sigma_2|_N)\right)$ is called *regularized trace* of the composition.

Theorem 9. *The regularized trace of the composition is independent of the order of the factors, i.e. we have*

$$\operatorname{tr}\left(\mathcal{A}_1 \circ \mathcal{A}_2 - \operatorname{Op}(\sigma_1 \circ \sigma_2|_N)\right) = \operatorname{tr}\left(\mathcal{A}_2 \circ \mathcal{A}_1 - \operatorname{Op}(\sigma_2 \circ \sigma_1|_N)\right) \tag{5}$$

for any $N > m + m' + n$.

For the proof we need the following

Lemma 10. *Let* $\sigma(A_i) \in \mathfrak{A}^{m_i}$ *be independent of* x_n, $\sigma(B_i) \in \mathfrak{B}^{m_i-1}$, $i = 1, 2$, $\sigma(K) \in \mathfrak{K}^{m_2}$, $\sigma(T) \in \mathfrak{T}^{m_1-1}$ *with compact support with respect to* x' *and* $m_1 + m_2 < -n$. *Then*

$$\operatorname{tr} \operatorname{Op}\left(\Pi'_\nu(\sigma(T)(x', \xi', \nu) \circ \sigma(K)(x', \xi', \nu))|_N\right)$$
$$= \operatorname{tr} \operatorname{Op}\left(\sigma(K)(x', \xi', \nu) \circ \sigma(T)(x', \xi', \tau)|_N\right),$$

$$\operatorname{tr} \operatorname{Op}\left(\Pi'_{\tau'}(\sigma(B_1)(x', \xi', \nu, \tau_1) \circ \sigma(B_2)(x', \xi', \tau_1, \tau))|_N\right)$$
$$= \operatorname{tr} \operatorname{Op}\left(\Pi'_{\tau'}(\sigma(B_2)(x', \xi', \nu, \tau_1) \circ \sigma(B_1)(x', \xi', \tau_1, \tau))|_N\right),$$

$$\operatorname{tr} \operatorname{Op}\left(\sigma(A_1)(x', \xi', \nu) \circ \sigma(B_2)(x', \xi', \nu, \tau)|_N\right)$$
$$= \operatorname{tr} \operatorname{Op}\left(\sigma(B_2)(x', \xi', \nu, \tau) \circ \sigma(A_1)(x', \xi', \tau)|_N\right),$$

$$\operatorname{tr} \operatorname{Op}\left(\varphi_1(x_n) \sigma(A_1)(x', \xi', \nu) \circ \varphi_2(x_n) \sigma(A_2)(x', \xi', \nu)|_N\right)$$
$$+ \operatorname{tr} \operatorname{Op}\left((\sigma(A_1)^+(x', \xi', \nu) - \sigma(A_1)^+(x', \xi', \tau))(i\nu - i\tau)^{-1} \circ \sigma(A_2)(x', \xi', \tau)|_N\right)$$
$$= \operatorname{tr} \operatorname{Op}\left(\varphi_2(x_n) \sigma(A_2)(x', \xi', \nu) \circ \varphi_1(x_n) \sigma(A_1)(x', \xi', \nu)|_N\right)$$
$$+ \operatorname{tr} \operatorname{Op}\left((\sigma(A_2)^+(x', \xi', \nu) - \sigma(A_2)^+(x', \xi', \tau))(i\nu - i\tau)^{-1}\right.$$
$$\left. \circ \, \sigma(A_1)(x', \xi', \tau)|_N\right), \tag{6}$$

$$\sigma(A_i)^+(x', \xi', \nu) = \Pi^+_\nu\left(\sigma(A_i)(x', \xi', \nu)\right),$$

where $\operatorname{Op} \sigma$ *denotes the operator of the corresponding class defined with the symbol* σ, $\varphi_i \in C_0^\infty(\overline{\mathbb{R}}_+)$, $\varphi_i \equiv 1$ *near* 0, $i = 1, 2$.

Proof: The first three equalities follow from (6) by partial integration. Set for abbreviation $\sigma_i = \sigma(A_i)$. First we show

$$\operatorname{tr} \operatorname{Op} (\varphi_1\sigma_1 \circ \varphi_2\sigma_2|_N) - \operatorname{tr} \operatorname{Op} (\varphi_2\sigma_2 \circ \varphi_1\sigma_1|_N)$$
$$= i \sum_{|\alpha'| \leq N} (1/\alpha'!) (2\pi)^{-n+1} \int \operatorname{Tr} (\partial_{\xi'}^{\alpha'} \partial_\nu \sigma_1(x', \xi', \nu) D_{x'}^{\alpha'} \sigma_2(x', \xi', \nu)) \, d\xi' \, dx' . \quad (7)$$

For $\alpha = (\alpha', \alpha_n) \in \mathbb{Z}_+^n$ and $\alpha_n = 0$ partial integration with respect to x', ξ' yields

$$\operatorname{tr} \operatorname{Op} (\partial_{\xi'}^{\alpha'}(\varphi_1(x_n) \sigma_1(x', \xi'', \nu)) D_{x'}^{\alpha'}(\varphi_2(x_n) \sigma_2(x', \xi', \nu)))$$
$$= (2\pi)^{-n} \int_{R^+} \varphi_1(x_n) \varphi_2(x_n) \, dx_n \int \operatorname{Tr} (\partial_{\xi'}^{\alpha'} \sigma_1(x', \xi', \nu) D_{x'}^{\alpha'} \sigma_2(x', \xi', \nu)) \, d\xi' \, d\nu \, dx'$$
$$= (2\pi)^{-n} \int_{R^+} \varphi_2(x_n) \varphi_1(x_n) \, dx_n \int \operatorname{Tr} (\partial_{\xi'}^{\alpha'} \sigma_2(x', \xi', \nu) D_{x'}^{\alpha'} \sigma_1(x', \xi', \nu)) \, d\xi' \, d\nu \, dx'$$
$$= \operatorname{tr} \operatorname{Op} (\partial_{\xi'}^{\alpha'}(\varphi_2(x_n) \sigma_2(x', \xi', \nu)) D_{x'}^{\alpha'}(\varphi_1(x_n) \sigma_2(x', \xi', \nu))) .$$

Similarly, for $\alpha_n \leq 2$ we get

$$\operatorname{tr} \operatorname{Op} (\varphi_1(x_n) \partial_{\xi'}^{\alpha'} \partial_\nu^{\alpha_n} \sigma_1(x', \xi', \nu) D_{x_n}^{\alpha_n} \varphi_2(x_n) D_{x'}^{\alpha'} \sigma_2(x', \xi', \nu))$$
$$= (2\pi)^{-n} \int_{R^+} \varphi_1(x_n) D_{x_n}^{\alpha_n} \varphi_2(x_n) \, dx_n \int \operatorname{Tr} (\partial_{\xi'}^{\alpha'} \partial_\nu^{\alpha_n} \sigma_1(x', \xi', \nu) D_{x'}^{\alpha'} \sigma_2(x', \xi', \nu)) \, d\xi' \, d\nu \, dx'$$
$$= (2\pi)^{-n} \int_{R^+} \varphi_2(x_n) D_{x_n}^{\alpha_n} \varphi_1(x_n) \, dx_n \int \operatorname{Tr} (\partial_{\xi'}^{\alpha'} \partial_\nu^{\alpha_n} \sigma_2(x', \xi', \nu) D_{x'}^{\alpha'} \sigma_1(x', \xi', \nu)) \, d\xi' \, d\nu \, dx'$$
$$= \operatorname{tr} \operatorname{Op} (\varphi_2(x_n) \partial_{\xi'}^{\alpha'} \partial_\nu^{\alpha_n} \sigma_2(x', \xi', \nu) D_{x_n}^{\alpha_n} \varphi_1(x_n) D_{x'}^{\alpha'} \sigma_1(x', \xi', \nu)) ,$$

since all derivatives of φ_i vanish at $x_n = 0$ and hence no boundary term occurs in the partial integration with respect to x_n. For $\alpha_n = 1$ we have

$$\operatorname{tr} \operatorname{Op} (\varphi_1(x_n) \partial_{\xi'}^{\alpha'} \partial_\nu \sigma_1(x', \xi', \nu) D_{x_n} \varphi_2(x_n) D_{x'}^{\alpha'} \sigma_2(x', \xi', \nu))$$
$$= (2\pi)^{-n} \int_{R^+} \varphi_1(x_n) D_{x_n} \varphi_2(x_n) \, dx_n \int \operatorname{Tr} (\partial_{\xi'}^{\alpha'} \partial_\nu \sigma_1(x', \xi', \nu) D_{x'}^{\alpha'} \sigma_2(x', \xi', \nu)) \, d\xi' \, d\nu \, dx'$$
$$= (2\pi)^{-n} i \int \operatorname{Tr} (\partial_{\xi'}^{\alpha'} \partial_\nu \sigma_1(x', \xi', \nu) D_{x'}^{\alpha'} \sigma_2(x', \xi', \nu)) \, d\xi' \, d\nu \, dx'$$
$$- (2\pi)^{-n} \int_{R^+} \varphi_2(x_n) D_{x_n} \varphi_1(x_n) \, dx_n \int \operatorname{Tr} (\partial_{\xi'}^{\alpha'} \partial_\nu \sigma_1(x', \xi', \nu) D_{x'}^{\alpha'} \sigma_2(x', \xi', \nu)) \, d\xi' \, d\nu \, dx'$$
$$= (2\pi)^{-n} i \int \operatorname{Tr} (\partial_{\xi'}^{\alpha'} \partial_\nu \sigma_1(x', \xi', \nu) D_{x'}^{\alpha'} \sigma_2(x', \xi', \nu)) \, d\xi' \, d\nu \, dx'$$
$$+ \operatorname{tr} \operatorname{Op} (\varphi_2(x_n) \partial_{\xi'}^{\alpha'} \partial_\nu \sigma_2(x', \xi', \nu) D_{x_n} \varphi_1(x_n) D_{x'}^{\alpha'} \sigma_1(x', \xi', \nu)) .$$

Summation over all α yields (7). Now the assertion follows from

$$\operatorname{tr} \operatorname{Op} ((\sigma_1^+(x', \xi', \nu) - \sigma_1^+(x', \xi', \tau)) (i\nu - i\tau)^{-1} \circ \sigma_2(x', \xi', \tau)|_N)$$
$$- \operatorname{tr} \operatorname{Op} ((\sigma_2^+(x', \xi', \nu) - \sigma_2^+(x', \xi', \tau)) (i\nu - i\tau)^{-1} \circ \sigma_1(x', \xi', \tau)|_N)$$
$$= -i \sum_{|\alpha'| < N} (1/\alpha'!) (2\pi)^{-n+1} \int \operatorname{Tr} (\Pi_\nu'(\partial_{\xi'}^{\alpha'} \partial_\nu \sigma_1(x', \xi', \nu) D_{x'}^{\alpha'} \sigma_2(x', \xi', \nu))) \, d\xi' \, dx' .$$

According to (4) we have

$$\operatorname{tr} \operatorname{Op} ((\partial_{\xi'}^{\alpha'} \sigma_1^+(x', \xi', \nu) - \partial_{\xi'}^{\alpha'} \sigma_1^+(x', \xi', \tau)) (i\nu - i\tau)^{-1} D_{x'}^{\alpha'} \sigma_2(x', \xi', \tau))$$
$$= -i(2\pi)^{-n+1} \int \operatorname{Tr} (\Pi_\nu'(\partial_{\xi'}^{\alpha'} \partial_\nu \sigma_1^+(x', \xi', \nu) D_{x'}^{\alpha'} \sigma_2(x', \xi', \nu))) \, d\xi' \, dx'$$

and

$$\operatorname{tr} \operatorname{Op} ((\partial_{\xi'}^{\alpha'} \sigma_2^+(x', \xi', \nu) - \partial_{\xi'}^{\alpha'} \sigma_2^+(x', \xi', \tau)) (i\nu - i\tau)^{-1} D_{x'}^{\alpha'} \sigma_1(x', \xi', \tau))$$
$$= -i(2\pi)^{-n+1} \int \operatorname{Tr} (\Pi_\nu'(\partial_{\xi'}^{\alpha'} \partial_\nu \sigma_2^+(x', \xi', \nu) D_{x'}^{\alpha'} \sigma_1^-(x', \xi', \nu))) \, d\xi' \, dx'$$
$$= i (2\pi)^{-n+1} \int \operatorname{Tr} (\Pi_\nu'(\partial_{\xi'}^{\alpha'} \partial_\nu \sigma_1^-(x', \xi', \nu) D_{x'}^{\alpha'} \sigma_2(x', \xi', \nu))) \, d\xi' \, dx'$$

and the assertion follows. □

Lemma 10 implies Theorem 9 in the case of $\mathcal{A}_1 \circ \mathcal{A}_2$ being a trace class operator.

In order to show Theorem 9 in the general case we need a suitable extension of the space of symbols with the transmission property. Set $\lambda_z^-(\nu) = (1 - i\nu)^z$, $z \in \mathbb{C}$, $\nu \in \mathbb{R}$. For $z \in \mathbb{Z}$ we have $\lambda_z^- \in H$, but in general λ_z^- does not belong to H.

Denote by L the algebra generated by H and the functions λ_z^-, $z \in \mathbb{C}$. The decomposition of $\sigma \in L$ as $\sigma = \sum_{j=1}^{N} \lambda_{z_j}^- \sigma_j$ is, of course, not unique and we can assume $\operatorname{Re} z_j < -c$ for any fixed c, take for instance $c = 1$. Then $\lambda_z^- \in L^2(\mathbb{R})$, and $\Pi_L^+(\lambda_z^- \sigma^-) = 0$ follows for an arbitrary $\sigma^- \in H^-$ where Π_L^+ denotes the Fourier transform of r^+ extended to $\mathcal{S}'(\mathbb{R}_-)$ by zero. Define $\Pi_L^+ \lambda_z^- \sigma^+$, $\sigma^+ \in H^+$, $\lambda_z^- \in L^2(\mathbb{R})$ as the L^2 extension of Π^+ on H_0. It follows that

$$\Pi_L^+ \lambda_z^- \sigma^+ = \Pi^+ \lambda_z^- (\sigma^+ + \sigma^-)$$

for arbitrary $\sigma^- \in H^-$. Choosing $\sigma^- \in H_0^-$ such that $\sigma^+ + \sigma^- \in \mathcal{S}(\mathbb{R})$ we obtain $\Pi_L^+ : L \to H^+$ and $(\Pi_L^+)^2 = \Pi_L^+$.

The definition of Π_L^+ yields that 2.1.1.2. (2) remains valid for Π_L^+ and for functions in $L \cap L^2(\mathbb{R})$. Set $\Pi_L^- = 1 - \Pi_L^+$, $L^- = \Pi_L^-(L)$. An extension Π_L' of Π' to L is defined by $\Pi_L'|_{L^-} = 0$, $\Pi_L'|_{L^+} = \Pi'$.

In the same way as in 2.1.2 one can define boundary symbols in the one-dimensional case. Since $\Pi^+ \lambda_z^- : H^+ \to H^+$ is an isomorphism with the inverse $\Pi^+ \lambda_{-z}^-$, the results about the Fredholm property of $\Pi^+ \sigma$, $\sigma \in L$ remain valid. Then, as in 2.1.2.2, we can define boundary symbols on the line with L instead of H, L^- instead of H^- and Π^+, Π' replaced by Π_L^+, Π_L'. The composition formulas from 2.1.2.3 remain valid.

Similarly as in 2.2.2 — 2.2.5 one defines symbol classes $\mathfrak{A}' \supset \mathfrak{A}$, $\mathfrak{R}' \supset \mathfrak{R}$ and $\mathfrak{S}' \supset \mathfrak{S}$ with H, H^- replaced by L, L^-. The corresponding operator classes are defined as in 2.3.2.1. The composition rule of 2.3.2.2 remains valid for Op $(\mathfrak{R}')$ instead of Op $(\mathfrak{R})$. The proof of 2.3.2.4, Proposition 1 shows that for $r^+ A \in \operatorname{Op}(\mathfrak{A}'^m)$ the usual Sobolev space extension exists. Hence operators in Op $(\mathfrak{A}')$ are smooth in the half-space, but may not have the transmission property.

Define $\mathcal{L}^z = r^+ \Lambda_-^z \oplus \Lambda'^z$, where $r^+ \Lambda_-^z$ is defined in 2.3.2.4. We use the same notation for $r^+ \Lambda_-^z \otimes 1_k \oplus \Lambda'^z \oplus 1_j$.

Lemma 11. *The functions*

$$\operatorname{tr}\left(\mathcal{A}_1 \circ \mathcal{L}^z \circ \mathcal{A}_2 - \operatorname{Op}\left(\sigma_1 \circ \sigma(\mathcal{L}^z) \circ \sigma_2|_N\right)\right)$$

and

$$\operatorname{tr}\left(\mathcal{A}_2 \circ \mathcal{A}_1 \circ \mathcal{L}^z - \operatorname{Op}\left(\sigma_2 \circ \sigma_1 \circ \sigma(\mathcal{L}^z)|_N\right)\right)$$

are analytic in $\operatorname{Re} z < \varepsilon$, $\varepsilon > 0$, *where* $N > m_1 + m_2 + n + \varepsilon$.

Proof: $\mathcal{L}^z$ defines an analytic family of continuous operators in $\mathcal{L}(\mathcal{H}^s, \mathcal{H}^{s-\varepsilon})$ $(\mathcal{H}^s = H^s(\mathbb{R}_+^n) \oplus H^{s+1/2}(\mathbb{R}^{n-1}))$ for $\operatorname{Re} z < \varepsilon$, $\varepsilon > 0$, $\mathcal{L}^z \in \operatorname{Op}(\mathfrak{S}'^{\varepsilon,0})$. We have

$$\mathcal{A}_2 \circ \mathcal{A}_1 \circ \mathcal{L}^z - \operatorname{Op}\left(\sigma_2 \circ \sigma_1 \circ \sigma(\mathcal{L}^z)|_N\right) = \left(\mathcal{A}_2 \circ \mathcal{A}_1 - \operatorname{Op}\left(\sigma_2 \circ \sigma_1|_N\right)\right) \mathcal{L}^z,$$

since $\sigma(\mathcal{L}^z)$ is independent of x. Hence $\mathcal{A}_2 \circ \mathcal{A}_1 \circ \mathcal{L}^z - \operatorname{Op}\left(\sigma_2 \circ \sigma_1 \circ \sigma(\mathcal{L}^z)|_N\right)$ is an analytic family in $\mathcal{L}(\mathcal{H}^s, \mathcal{H}^{s+n-\varepsilon})$ and according to Theorem 1 an analytic family of trace class operators in $\mathcal{L}(\mathcal{H}^s, \mathcal{H}^s)$. Then the function

$$\operatorname{tr}\left(\mathcal{A}_2 \circ \mathcal{A}_1 \circ \mathcal{L}^z - \operatorname{Op}\left(\sigma_2 \circ \sigma_1 \circ \sigma(\mathcal{L}^z)|_N\right)\right)$$

is analytic for $\operatorname{Re} z < \varepsilon$, since tr is a continuous linear functional on the space of all trace class operators.

It remains to prove that $\mathcal{A}_1 \circ \mathcal{L}^z \circ \mathcal{A}_2 - \mathrm{Op}\,(\sigma_1 \circ \sigma(\mathcal{L}^z) \circ \sigma_2|_N)$ is an analytic family of operators in $\mathrm{Op}\,(\mathfrak{S}'^{-n-\varepsilon})$. Denote by $\mathcal{A}_1^{(z)}$ the composition $\mathcal{A}_1 \circ \mathcal{L}^z$ and $\sigma(\mathcal{A}_1^{(z)}) = (\sigma_{\mathcal{A}_1^{(z)}}, \sigma_Y(\mathcal{A}_1^{(z)}))$. We can assume that $\mathcal{A}_2$ is properly supported. Then there is a symbol $\tilde{\sigma}^2 = (\tilde{\sigma}_{\mathcal{A}_2}(y, \xi), \tilde{\sigma}_Y^2(y', \xi'))$ such that

$$\begin{bmatrix} F & 0 \\ 0 & F' \end{bmatrix} \mathcal{A}_2 \begin{bmatrix} u \\ v \end{bmatrix} = \int e^{-iy\xi} \left(\tilde{\sigma}_A^2(y, \xi) - \tilde{\sigma}_A^2(y', 0, \xi)\right) u(y)\, dy$$

$$+ \int e^{-iy'\xi'}\, \tilde{\sigma}_Y^2(y', \xi') \begin{bmatrix} \hat{u}(y', \nu) \\ v(y') \end{bmatrix} dy',$$

where $\hat{u} = F_n j^+ u$, $(u, v) \in C_0^\infty(\mathbb{R}^{n-1}, \mathfrak{S}(\overline{\mathbb{R}}_+, \mathbb{C}^k) \oplus \mathbb{C}^l)$. Hence

$$\mathcal{A}_1^{(z)} \circ \mathcal{A}_2 - \mathrm{Op}\,(\sigma(\mathcal{A}_1^{(z)}) \circ \sigma_2|_N)$$
$$= \mathrm{Op}\,(\sigma(\mathcal{A}_1^{(z)})\,(x, \xi)\, \tilde{\sigma}(\mathcal{A}_2)\,(y, \xi) - \sigma(\mathcal{A}_1^{(z)})\,(x, y) \circ \sigma_2(x, \xi)|_N)$$

and

$$\sigma_{\mathcal{A}_1^{(z)}}(x, \xi)\, \tilde{\sigma}_{\mathcal{A}_2}(y, \xi) - \sigma_{\mathcal{A}_1^{(z)}}(x, \xi) \circ \sigma_{\mathcal{A}_2}(x, \xi)|_N$$
$$= \sum_{|\alpha| = N} (N/\alpha!) \int_0^1 \partial_\xi^\alpha \sigma_{\mathcal{A}_1^{(z)}}(x, \xi)\, D_y^\alpha \tilde{\sigma}_{\mathcal{A}_2}(x + t(y - x), \xi)\, (1 - t)^{N-1}\, dt\,,$$

$$\sigma_Y(\mathcal{A}_1^{(z)})\,(x', \xi')\, \tilde{\sigma}_Y^2(y', \xi') - \sigma_Y(\mathcal{A}_1^{(z)})\,(x', \xi') \circ \sigma_Y^2(x', \xi')|_N$$
$$= \sum_{|\alpha'| = N} (N/\alpha'!) \int_0^1 \partial_{\xi'}^{\alpha'} \sigma_Y(\mathcal{A}_1^{(z)})\,(x', \xi')\, D_{y'}^{\alpha'} \tilde{\sigma}_Y^2(x' + t(y' - x'), \xi'(\,(1 - t)^{N-1}\, dt\,.$$

This is an analytic family of symbols in the closure of $\mathfrak{S}'^{-n-\varepsilon}$ and the assertion follows. □

Proof of Theorem 9: First consider the case that $m_1 + m_2 < -n$, i.e. $\mathcal{A}_1 \circ \mathcal{A}_2$ and $\mathcal{A}_2 \circ \mathcal{A}_1$ are trace class operators. Then $\mathrm{tr}\, \mathcal{A}_1 \circ \mathcal{A}_2 = \mathrm{tr}\, \mathcal{A}_2 \circ \mathcal{A}_1$ and according to Lemma 10 $\mathrm{tr\,Op}\,(\sigma_1 \circ \sigma_2|_N) = \mathrm{tr\,Op}\,(\sigma_2 \circ \sigma_1|_N)$. Hence for $m_1 + m_2 < -n$ the assertion is proved. For m_1, m_2 arbitrary, use the analytic family $\mathcal{L}^z \in \mathrm{Op}\,(\mathfrak{S}'^z)$, $\mathrm{Re}\, z < \varepsilon$. Then

$$\mathrm{tr}\,(\mathcal{A}_1 \circ \mathcal{L}^z \circ \mathcal{A}_2 - \mathrm{Op}\,(\sigma_1 \circ \sigma(\mathcal{L}^z) \circ \sigma_2|_N))$$
$$= \mathrm{tr}\,(\mathcal{A}_2 \circ \mathcal{A}_1 \circ \mathcal{L}^z - \mathrm{Op}\,(\sigma_2 \circ \sigma_1 \circ \sigma(\mathcal{L}^z)|_N))$$

for $\mathrm{Re}\, z \leq -m_1 - m_2 - n$. By Lemma 11 both sides of this equation are analytic functions in $\mathrm{Re}\, z < \varepsilon$ and equality holds in particular for $z = 0$. □

Corollary 12. *Let* $\mathcal{A}_1 \in \mathfrak{G}^{m,d}(\overline{\mathbb{R}}^n_+, \mathbb{R}^{n-1}; \mathbb{C}^{k'}, \mathbb{C}^k, \mathbb{C}^{j'}, \mathbb{C}^j)$ *and* $\mathcal{A}_2 \in \mathfrak{G}^{m',d'}(\overline{\mathbb{R}}_n^+, \mathbb{R}^{n-1}; \mathbb{C}^k, \mathbb{C}^{k'}, \mathbb{C}^j, \mathbb{C}^{j'})$, $\varphi_i^{(1)}(x') \in C_0^\infty(\mathbb{R}^{n-1})$, $\varphi_i^{(2)}(x) \in C_0^\infty(\overline{\mathbb{R}}_+)$,

$$\varphi_i^{(2)}(x_n) = 1 \quad \textit{near } 0,\ \varphi_i(x', x_n) = \varphi_i^{(1)}(x')\, \varphi_i^{(2)}(x_n)\,, \quad i = 1, 2, 3, 4.$$

Then, for any $N > m + m' + n$, *we have*

$$\mathrm{tr}\,(\mathcal{M}_{\varphi_1} \mathcal{A}_1 \mathcal{M}_{\varphi_2} \mathcal{M}_{\varphi_3} \mathcal{A}_2 \mathcal{M}_{\varphi_4} - \mathcal{M}_{\varphi_1}\, \mathrm{Op}\,(\sigma_1 \circ \sigma(\varphi_2 \varphi_3) \circ \sigma_2|_N)\, \mathcal{M}_{\varphi_4})$$
$$= \mathrm{tr}\,(\mathcal{M}_{\varphi_3} \mathcal{A}_2 \mathcal{M}_{\varphi_4} \mathcal{M}_{\varphi_1} \mathcal{A}_1 \mathcal{M}_{\varphi_2} - \mathcal{M}_{\varphi_3}\, \mathrm{Op}\,(\sigma_2 \circ \sigma(\varphi_4 \varphi_1) \circ \sigma_1)|_N)\, \mathcal{M}_{\varphi_2})\,,$$

where $\mathcal{M}_\varphi$ *denotes the multiplication operator with* $\varphi(x) \otimes 1 \oplus \varphi^{(1)}(x') \otimes 1$, $\sigma(\varphi) = (\varphi(x) \otimes 1 \oplus \varphi^{(1)}(x') \otimes 1)\, \lambda^0$ *and* σ_i *is a formal symbol of* $\mathcal{A}_i$.

Proof: First we note that Theorem 9 can be applied to operators $\mathcal{A} = \mathcal{M}_\varphi \mathcal{A}_1$, $\mathcal{A}_1 \in \mathfrak{G}^{m,d}$. In fact, let $\psi \in C_0^\infty(\overline{\mathbb{R}}_+)$ be a function with support in the set where

$\varphi(x)\,\sigma_A(x,\xi)$ is independent of x_n. If we use the decomposition $\mathcal{M}_\varphi\mathcal{A} = \mathcal{M}_\psi\mathcal{M}_\varphi\mathcal{A} + \mathcal{M}_{1-\psi}\mathcal{M}_\varphi\mathcal{A}$, from (5) we obtain on both sides two smoothing operators of trace class and an operator as in Theorem 9. Then

$$\begin{aligned}
&\operatorname{tr}\left(\{\mathcal{M}_{\varphi_1}\mathcal{A}_1\mathcal{M}_{\varphi_2\varphi_3}\mathcal{A}_2 - \mathcal{M}_{\varphi_1}\operatorname{Op}(\sigma_1\circ\sigma(\varphi_2\varphi_3)\circ\sigma_2|_N)\}\,\mathcal{M}_{\varphi_4}\right)\\
&= \operatorname{tr}\left(\mathcal{M}_{\varphi_4}\{\mathcal{M}_{\varphi_1}\mathcal{A}_1\mathcal{M}_{\varphi_2\varphi_3}\mathcal{A}_2 - \mathcal{M}_{\varphi_1}\operatorname{Op}(\sigma_1\circ\sigma(\varphi_2\varphi_3)\circ\sigma_2|_N)\}\right)\\
&= \operatorname{tr}\left(\mathcal{M}_{\varphi_4\varphi_1}\mathcal{A}_1\mathcal{M}_{\varphi_2\varphi_3}\mathcal{A}_2 - \operatorname{Op}(\sigma(\varphi_4\varphi_1)\circ\sigma_1\circ\sigma(\varphi_2\varphi_3)\circ\sigma_2|_N)\right)\\
&= \operatorname{tr}\left(\mathcal{M}_{\varphi_2\varphi_3}\mathcal{A}_2\mathcal{M}_{\varphi_4\varphi_1}\mathcal{A}_1 - \mathcal{M}_{\varphi_3}\operatorname{Op}(\sigma(\varphi_3)\circ\sigma_2\circ\sigma(\varphi_4\varphi_1)\circ\sigma_1|_N)\right)\\
&= \operatorname{tr}\left(\mathcal{M}_{\varphi_2}\mathcal{A}_2\mathcal{M}_{\varphi_4\varphi_1}\mathcal{A}_1\mathcal{M}_{\varphi_3} - \mathcal{M}_{\varphi_2}\operatorname{Op}(\sigma_2\circ\sigma(\varphi_4\varphi_1)\circ\sigma_1|_N)\,\mathcal{M}_{\varphi_3}\right). \quad\square
\end{aligned}$$

For the global situation we need a theorem on regularized trace under coordinate transformations. Let $\varkappa\colon \mathbb{R}^{n-1}\to\mathbb{R}^{n-1}$ be a diffeomorphism and $\mathcal{A}\in\mathfrak{G}^{m,d}(\overline{\mathbb{R}}^n_+, \mathbb{R}^{n-1}; \mathbb{C}^k, \mathbb{C}^k, \mathbb{C}^j, \mathbb{C}^j)$ with the symbol $\sigma = (\sigma(A), \sigma_{\mathbb{R}^{n-1}}(\mathcal{A}))$, $\sigma(A)$ independent of x_n. Let $\varphi_i^{(1)}$, $\varphi_i^{(2)}$, $i = 1, 2$, as in Corollary 12. Then, for $N > m + n$, the operator $\varkappa_*\mathcal{A} - \operatorname{Op}(\sigma_\varkappa|_N)$ has an order $< -n$ and $\mathcal{M}_{\varphi_1}(\varkappa_*\mathcal{A} - \operatorname{Op}(\sigma_\varkappa|_N))\,\mathcal{M}_{\varphi_2}$ is a trace class operator in the sense of Theorem 1.

Theorem 13. *Let $\mathcal{A}\in\mathfrak{G}^{m,d}(\overline{\mathbb{R}}^n_+, \mathbb{R}^{n-1}; \mathbb{C}^k, \mathbb{C}^k, \mathbb{C}^j, \mathbb{C}^j)$ be defined by the symbol $\sigma\in\mathfrak{S}^{m,d}(\overline{\mathbb{R}}^n_+, \mathbb{R}^{n-1}; \mathbb{C}^k, \mathbb{C}^k, \mathbb{C}^j, \mathbb{C}^j)$ and φ_i, $i = 1, 2$, as in Corollary 12. Then, for any $N > m + n$, we have*

$$\operatorname{tr}\left(\mathcal{M}_{\varphi_1}(\varkappa_*\mathcal{A} - \operatorname{Op}(\sigma_\varkappa|_N))\,\mathcal{M}_{\varphi_2}\right) = 0\,.$$

It is easily checked that Theorem 13 is equivalent to

Theorem 13′. *Under the conditions of Theorem 13 the operators*

$$M_{\varphi_1}[\varkappa_*(r^+A) - \operatorname{Op}(\sigma(A)_\varkappa|_N)]\,M_{\varphi_2},\quad M_{\varphi_1^{(1)}}[\varkappa_*(r'B) - \operatorname{Op}(\sigma(B)_\varkappa|_N)]\,M_{\varphi_2^{(1)}}$$

and

$$M_{\varphi_1^{(1)}}[\varkappa_*(Q) - \operatorname{Op}(\sigma(Q)_\varkappa|_N)]\,M_{\varphi_2^{(1)}}$$

have trace zero.

Proof: We show the assertion for the PDO Q. With obvious modifications the conclusion works for r^+A and $r'B$, too. The definition of $\varkappa_*Q$ yields the FIO representation

$$\begin{aligned}
\varkappa_*Qv(x') &= (2\pi)^{-n+1}\int e^{i(\varkappa^{-1}(x')-y')\xi'}\,\sigma(Q)\,(\varkappa^{-1}(x'),\xi')\,v(\varkappa(y'))\,dy'\,d\xi'\\
&= (2\pi)^{-n+1}\int e^{i(\varkappa^{-1}(x')-\varkappa^{-1}(y'))\xi'}\,\sigma(Q)\,(\varkappa^{1}(x'),\xi')\,v(y')\left|\frac{D\varkappa^{-1}(y')}{Dy'}\right|dy'\,d\xi'
\end{aligned}$$

(cf. the proof of 1.2.3.4, Theorem 1). In a neighbourhood U of diag Ω' there is a smooth mapping $\psi\colon U\to \mathrm{GL}\,(n-1,\mathbb{R})$ such that

$$\begin{aligned}
&(\varkappa^{-1}(x') - \varkappa^{-1}(y'))\,\psi(x',y')\,\xi' = (x'-y')\,\xi'\,,\quad (x',y')\in U\,,\\
&\det\psi(x',x')\det d\varkappa^{-1}(x') = 1
\end{aligned}$$

(cf. 1.2.3.4, Lemma 3). In U the distributional kernel of $\varkappa_*Q$ is given by

$$(2\pi)^{-n+1}\int e^{i(x'-y')\xi'}\,\sigma(Q)\,(\varkappa^{-1}(x'),\psi(x',y')\,\xi')\,D(x',y')\,d\xi'\,,$$

where $D(x', y') = |\det d\varkappa^{-1}(x')|\,|\det \psi(x', y')|$. Consider the Taylor expansion of $\sigma(Q)\,(\varkappa^{-1}(x'), \psi(x', y')\,\xi')\,D(x', y')$ with respect to y' in $y' = x'$

$$\begin{aligned}&\sigma(Q)\,\big(\varkappa^{-1}(x'), \psi(x', y')\,\xi'\big)\,D(x', y')\\ &= \sum_{|\alpha|<N} (1/\alpha!)\,\partial_{y'}^{\alpha}\big(\sigma(Q)\,(\varkappa^{-1}(x'), \psi(x', y')\,\xi')\,D(x', y')\big)\big|_{y'=x'}\,(y'-x')^{\alpha}\\ &+ \sum_{|\alpha|=N} N/\alpha!\,(y'-x')^{\alpha}\int_0^1 \partial_{y'}^{\alpha}\big(\sigma(Q)\,(\varkappa^{-1}(x'), \psi(x', y')\,\xi')\,D(x', y')\big)\big|_{y'=x'+t(y'-x')}\\ &\times (1-t)^{N-1}\,dt\,.\end{aligned}$$

The sum over $|\alpha| < N$ defines Op $(\sigma(Q)_{\varkappa}|_N)$ and we get

$$\begin{aligned}&(2\pi)^{-n+1}\sum_{|\alpha|=N}(N/a!)\int e^{i(x'-y')\xi'}\,(y'-x')^{\alpha}\\ &\times\int_0^1 \partial_{y'}^{\alpha}(\sigma(Q))\,\big(\varkappa^{-1}(x'), \psi(x', y')\,\xi'\big)\,D(x', y')|_{y'=x'+t(y'-x')}\,(1-t)^{N-1}\,dt\,d\xi'\end{aligned}$$

for the distributional kernel of $\varkappa_* Q$ —Op $(\sigma(Q)_{\varkappa}|_N)$ on U. After having changed the order of integration we fix t. By partial integration we get

$$\begin{aligned}&(2\pi)^{-n+1}\int e^{i(x'-y')\xi'}(y'-x')^{\alpha}\,\partial_{y'}^{\alpha}\big(\sigma(Q)\,(\varkappa^{-1}(x'), \psi(x', y')\,\xi')\,D(x', y')|_{y'=x'+t(y'-x')}\big)\,d\xi'\\ &= (2\pi)^{-n+1}\int e^{i(x'-y')\xi'}\,D_{\xi'}^{\alpha}\big(\partial_{y'}^{\alpha}\,\sigma(Q)\,(\varkappa^{-1}(x'), \psi(x', y')\,\xi')\,D(x', y')|_{y'=x'+t(y'-x')}\big)\,d\xi'\,.\end{aligned}$$

At $y' = x'$ we have

$$(2\pi)^{-n+1}\int D_{\xi'}^{\alpha}\,\partial_{y'}^{\alpha}\big(\sigma(Q)\,(\varkappa^{-1}(x'), \psi(x', y')\,\xi')\,D(x', y')|_{y'=x'}\big)\,d\xi'\,.$$

The integrand is a sum of terms of the form $c(x')\,\xi'^{\gamma}\partial_{\xi'}^{\beta}\sigma(Q)\,\big(\varkappa^{-1}(x'), \psi(x', y')\,\xi'\big)$, where $|\gamma| + |\alpha| \leqq |\beta| \leqq 2\,|\alpha|$, $|\gamma| \leqq |\beta|/2$. For $N = |\alpha| > m + n\ |\gamma| + 1$ partial integrations are possible and

$$\int \xi'^{\gamma}\,\partial_{\xi'}^{\beta}\,\sigma(Q)\,\big(\varkappa^{-1}(x'), \psi(x', x')\,\xi'\big)\,d\xi' = 0\,.$$

Hence the trace vanishes. □

Let $\sigma \in \mathfrak{S}_{\lambda}^{m,d}(X, Y;\ E, E, J, J)$, $E \in$ Vect (X), $J \in$ Vect (Y) be a formal global symbol, i.e. a collection of formal interior and boundary symbols in local coordinates of a fixed covering and with respect to local trivializations of the bundles with a behaviour under changes of trivializations described by 4.1.1.1.(2). Let U, V coordinate neighbourhoods, $U \cap V \neq \emptyset$ and $\varrho_i \in C_0^{\infty}(U \cap V)$ arbitrary functions having, in local coordinates near the boundary the form $\varrho_i(x', x_n) = \varrho_i^{(1)}(x')\,\varrho_i^{(2)}(x_n)$ ($\{x_n = 0\}$ is the boundary), $\varrho_i^{(2)} \equiv 1$ near 0. Set $\mathcal{A} = \mathcal{M}_{\varrho_1}$ Op $(\sigma_U|_N)\,\mathcal{M}_{\varrho_2}$, $\mathcal{A}' = \mathcal{M}_{\varrho_1}$ Op $(\sigma_V|_N)\,\mathcal{M}_{\varrho_2}$ for $N > m + n$ (σ_U denotes the symbol in local coordinates over U).

Corollary 14. *Let $\mathcal{A} + \mathcal{B}$ be a trace class operator in the sense of Theorem* 1, $\mathcal{B} \in \mathfrak{G}^{m,d}(X, Y; E, E, J, J)$. *Then $\mathcal{A}' + \mathcal{B}$ is of trace class and*

$$\operatorname{tr}(\mathcal{A} + \mathcal{B}) = \operatorname{tr}(\mathcal{A}' + \mathcal{B})\,.$$

Proof: Consider $\mathcal{A} - \mathcal{A}'$ in coordinates of V. This is a trace class operator and according to Theorem 13 we have

$$\operatorname{tr}(\mathcal{A} - \mathcal{A}') = \operatorname{tr}\big(\mathcal{M}_{\varrho_1}(\mathrm{Op}\,(\sigma_U|_N)_V - \mathrm{Op}\,(\sigma_V|_N))\,\mathcal{M}_{\varrho_2}\big) = 0$$

where $(\cdot)_V$ denotes the push forward of an operator over U to coordinates of V. □

4.1.1.3. The "Coarse" Index Formula

In this section we derive an expression for the index of an elliptic operator in $\mathfrak{G}$ as an integral over a form given by the formal interior and boundary symbol. In the next section we shall deal with simplifications especially eliminating higher derivatives of the principal symbol.

Introduce the notion of the trace of a formal symbol.

Lemma 1. *Let* $\sigma(\mathcal{A}) \in \mathfrak{S}_\lambda^{m,d}(X, Y; E, E, J, J)$ *be a formal symbol,* $m < -n$, $\sigma(\mathcal{A}) = \sum_k \lambda^k \sigma^k$. *Then the local expressions*

$$(2\pi)^{-n} \int \mathrm{Tr}\, (\sigma(A)_U^k (x, \xi))\, d\xi\,, \quad (2\pi)^{-n+1} \int \mathrm{Tr}\, (\Pi'_\nu \sigma(B)_U^k (x', \xi', \nu, \nu))\, d\xi'$$

and

$$(2\pi)^{-n+1} \int \mathrm{Tr}\, \sigma(Q)_U^k (x', \xi')\, d\xi'$$

define local expressions of densities on X and Y.

Proof: Set $\mathcal{A}_N = \mathrm{Op}\, (\sigma(\mathcal{A})|_N)$ and let $K_{\mathcal{A}_N}$ be the distributional kernel of $\mathcal{A}_N$. Under the coordinate transform $z' = \varkappa(y')$, $z_n = y_n$, the components of $K_{\mathcal{A}_N}$ are multiplied by $|D\varkappa(y')/Dy'|$. In view of the proof of 4.1.1.2, Theorem 13′ we have, for the components $K_{r^+A_N}(x, y)$, $K_{r'B_N}(x, y)$ and $K_{Q_N}(x', y')$ of $K_{\mathcal{A}_N}$

$$\mathrm{Tr}\, K_{\varkappa_*(r^+\mathcal{A}_N)}(x, x) = \mathrm{Tr}\, K_{\mathrm{Op}\,((\sigma_{\mathcal{A}}|_N)_\varkappa|_N)}(x, x)$$

and so on. Then we get

$$(2\pi)^{-n} \sum_{k=0}^{N-1} \int \mathrm{Tr}\, \sigma(A)^k (x, \xi)\, d\xi\, |\det d\varkappa(x')|$$

$$= (2\pi)^{-n} \sum_{k=0}^{N-1} \int \mathrm{Tr}\, (\sigma(A)^k)_\varkappa (x, \xi)|_N\, d\xi\,. \quad \square$$

Set

$$\mathrm{Sp}_N \sigma(\mathcal{A}) = \sum_{k=0}^{N-1} \left\{ \int_X (2\pi)^{-n} \int \mathrm{Tr}\, \sigma(A)^k (x, \xi) \right\} d\xi\, dx$$

$$+ \int_Y (2\pi)^{-n+1} \left(\int \mathrm{Tr}\, \Pi'_\nu \sigma(B)^k (x', \xi', \nu, \nu)\, d\xi' + \int \mathrm{Tr}\, \sigma(Q)^k (x', \xi')\, d\xi' \right) dy'$$

for $\sigma(\mathcal{A}) = \sum_k \lambda^k \sigma(\mathcal{A})^k \in \mathfrak{S}_\lambda^m(X, Y; E, E, J, J)$, $m < -n$. We shall give other expressions for $\mathrm{Sp}_N \sigma(\mathcal{A})$.

Let $\{U_j\}$ be a covering of X and $\{\varphi_j\}$ a subordinated partition of unity. Define

$$\mathcal{A}'_N = \sum_{i, j,\, U_i \cap U_j \neq \emptyset} \mathcal{M}_{\varphi_i}\, \mathrm{Op}\, (\sigma(\mathcal{A})|_N)\, \mathcal{M}_{\varphi_j}$$

where $\sigma(\mathcal{A})|_N$ is taken in local coordinates of $U_{ij} \supset U_i \cup U_j$. For $m < -n$ the operator $\mathcal{A}'_N$ is of trace class.

Lemma 2. *For any* $N \in \mathbb{Z}_+$ *we have*

$$\mathrm{Sp}_N\, \sigma(\mathcal{A}) = \mathrm{tr}\, \mathcal{A}'_N\,.$$

Proof: According to 4.1.1.2, Corollary 14

$$\sum_{i, j} (2\pi)^{-n} \int_X \left\{ \int \varphi_i(x)\, (\mathrm{Tr}\, \sigma(A)\, (x, \xi)_{ij}|_N)\, \varphi_j(x)\, d\xi \right\} dx$$

$$= \sum_i (2\pi)^{-n} \int_X \left\{ \int \varphi_i(x)\, (\mathrm{Tr}\, \sigma(A)\, (x, \xi)_i|_N)\, d\xi \right\} dx$$

($\sigma(A)(x, \xi)_{ij}$ denotes the local representation of $\sigma(A)$ over $U_{ij} \supset U_i \cup U_j$, $\sigma(A)(x, \xi)_i$ the local representation of $\sigma(A)$ over U_i). Similar equalities are valid for the Green symbol $\sigma(B)$ and the symbol $\sigma(Q)$. □

Let $\mathcal{A} \in \mathfrak{G}^{m,d}(X, Y)$ be an elliptic operator with a symbol $\sigma(\mathcal{A}) \in \mathfrak{S}_\lambda^{m,d}(X, Y)$. Denote by $\sigma(\mathcal{R})$ a symbol in $\mathfrak{S}_\lambda^{-m,d'}(X, Y)$ the principal part of which is $\sigma^0(\mathcal{A})^{-1}$ outside some compact set, i.e. for the principal interior symbols we have

$$1 - \sigma(A)^0(x, \xi)\, \sigma(R)^0(x, \xi) = 0\,, \quad 1 - \sigma(R)^0(x, \xi)\, \sigma(A)^0(x, \xi) = 0$$

when $|\xi| \geqq c_1$ and for the principal boundary symbols

$$1 - \sigma_Y(\mathcal{A})^0(x', \xi')\, \sigma_Y(\mathcal{R})^0(x', \xi') = 0\,, \tag{1}$$

$$1 - \sigma_Y(\mathcal{R})^0(x', \xi')\, \sigma_Y(\mathcal{A})^0(x', \xi') = 0 \quad \text{when} \quad |\xi'| \geqq c_1\,.$$

Then the corresponding operator $\mathcal{R} \in \mathfrak{G}^{-m,d}(X, Y)$ is a (Sobolev space) parametrix of $\mathcal{A}$, and $1 - \mathcal{R}\mathcal{A}$ and $1 - \mathcal{A}\mathcal{R}$ have order -1. Then $(1 - \mathcal{R}\mathcal{A})^N$ and $(1 - \mathcal{A}\mathcal{R})^N$ are trace class operators for $N > n$ (cf. 4.1.1.2, Theorem 1) and the abstract index formula can be applied.

Theorem 3. *Let $\mathcal{A} \in \mathfrak{G}^{m,d}(X, Y)$ and $\sigma(\mathcal{A}) \in \mathfrak{S}_\lambda^{m,d}(X, Y)$ be its formal symbol with the principal part $\sigma^0(\mathcal{A}) = (\sigma(A)^0, \sigma_Y^0(\mathcal{A}))$. Let $\sigma(\mathcal{R}) \in \mathfrak{S}_\lambda^{-m,d'}(X, Y)$ be a formal symbol with the principal part $\sigma^0(\mathcal{R}) = (\sigma(R)^0, \sigma_Y^0(\mathcal{R}))$ such that*

$$\sigma(R)^0(x, \xi) = (\sigma(A)^0(x, \xi))^{-1} \quad \text{when} \quad |\xi| \geqq c_1$$

and

$$\sigma_Y^0(\mathcal{R})(x', \xi') = (\sigma_Y^0(\mathcal{A})(x', \xi'))^{-1} \quad \text{when} \quad |\xi'| \geqq c_1$$

(c_1 a suitable constant). Then, for any $N > n = \dim X$, we have

$$\operatorname{ind} \mathcal{A} = \mathrm{Sp}_N\left((1 - \sigma(\mathcal{R}) \circ \sigma(\mathcal{A}))^N\right) - \mathrm{Sp}_N\left((1 - \sigma(\mathcal{A}) \circ \sigma(\mathcal{R}))^N\right), \tag{2}$$

where the Nth power is taken in the symbol algebra $\mathfrak{S}_\lambda$.

Write (2) in a more explicit way

$$\begin{aligned}
\operatorname{ind} \mathcal{A} &= (2\pi)^{-n} \int_X \Big\{ \int_{R^n} [\mathrm{Tr}\,(1 - \sigma(R^0)(x, \xi) \circ \sigma(A)^0(x, \xi))^N|_N \\
&\quad - \mathrm{Tr}\,(1 - \sigma(A)^0(x, \xi) \circ \sigma(R)^0(x, \xi))^N|_N]\, d\xi \Big\}\, dx \\
&\quad + (2\pi)^{-n+1} \int_X \Big\{ \int_{R^{n-1}} [\mathrm{Tr}'\,(1 - \sigma_Y^0(\mathcal{R})(x', \xi') \circ \sigma_Y^0(\mathcal{A})(x', \xi'))^N|_N \\
&\quad - \mathrm{Tr}'\,(1 - \sigma_Y^0(\mathcal{A})(x', \xi') \circ \sigma_Y^0(\mathcal{R})(x', \xi'))^N|_N]\, d\xi' \Big\}\, dx' \\
&= (2\pi)^{-n} \int_X \Big\{ \int_{R^n} \sum_{k=0}^N \binom{N}{k} (-1)^k [\mathrm{Tr}\,(\sigma(R)^0(x, \xi) \circ \sigma(A)^0(x, \xi))^k|_N \\
&\quad - \mathrm{Tr}\,(\sigma(A)^0(x, \xi) \circ \sigma(R)^0(x, \xi))^k|_N]\, d\xi \Big\}\, dx \\
&\quad + (2\pi)^{-n+1} \int_Y \Big\{ \int_{R^{n-1}} \sum_{k=0}^N \binom{N}{k} (-1)^k [\mathrm{Tr}'\,(\sigma_Y^0(\mathcal{R})(x, \xi') \circ \sigma_Y^0(\mathcal{A})(x', \xi'))^k|_N \\
&\quad - \mathrm{Tr}'\,(\sigma_Y^0(\mathcal{A})(x', \xi') \circ \sigma_Y^0(\mathcal{R})(x', \xi'))^k|_N]\, d\xi' \Big\}\, dx'
\end{aligned} \tag{3}$$

where $\circ$ denotes the composition in the algebra of complete interior and boundary symbols. The symbols are taken in arbitrary local coordinates. Then the partial derivatives involved in the composition rules in the symbol algebra make sense.

Looking carefully at the integrands it turns out that the ξ and ξ' integration gives contributions over $|\xi| < c_1$ and $|\xi'| < c_1$ only (cf. (3)). In fact, define $\sigma_j(x, \xi)$ by

$$\left(1 - \sigma(R)^0 (x, \xi) \circ \sigma(A)^0 (x, \xi)\right)^N = \sum_{j=0}^{\infty} \sigma_{j'}(x, \xi) \lambda^j .$$

Observe that for $j < N$ all σ_j contain a factor of the form $\partial^\alpha_{x,\xi}\left(1 - \sigma(R)^0 (x, \xi) \sigma(A)^0 (x, \xi)\right)$, $\alpha \in Z^{2n}_+$. and vanish where $\sigma(R)^0 (x, \xi) = \left(\sigma(A)^0 (x, \xi)\right)^{-1}$. The same is true for $\left(1 - \sigma(A)^0 (x, \xi) \circ \sigma(R)^0 (x, \xi)\right)^N|_N$. Using the calculus of boundary symbols a similar conclusion is valid for the integral over ξ'.

On the other hand, it is easily seen that the right hand side in (3) is independent of the behaviour of $\sigma(A)^0 (x, \xi)$ and $\sigma(R)^0 (x, \xi)$ near $\xi = 0$, and the behaviour of $\sigma^0_Y(\mathcal{A}) (x', \xi')$ and $\sigma^0_Y(\mathcal{R}) (x', \xi')$ near $\xi' = 0$.

Remark 4. Let $\sigma(A)^0$ be an elliptic principal symbol on X, i.e. there are bundles $E, F \in \text{Vect}(X)$ such that

$$\sigma(A)^0 : p^*_X E \to p^*_X F , \quad p_X : S^*X \to X ,$$

is an isomorphism. Assume that $\sigma(A)^0$ admits elliptic boundary conditions in the class $\mathfrak{G}$. In view of 3.2.1.2, Remark 6 this is equivalent to the existence of an extension of $\sigma(A)^0 \oplus 1 : p^*_X(E \oplus L) \to p^*_X(F \oplus L)$, $p_X : S^*X \to X$ to the ball bundle $B^*X|_U$ as isomorphism (U a neighbourhood of the boundary).

Choosing this extension of $\sigma(A)^0 \oplus 1$ in the formula (3) we obtain that the form integrated over X has its support in the interior of X. Then $\Pi^+\sigma(A)^0 (x', \xi', \nu)$: $E'_{x'} \otimes H^+ \to F'_{x'} \otimes H^+$, $(x', \xi') \in B^*Y$, $B^*Y = \{(x', \xi') \in T^*Y : |\xi'| \leq c_1\}$ is a Fredholm family. Then the principal boundary symbol $\sigma^0_Y(\mathcal{A}) (x', \xi')$ defines an isomorphism for $|\xi'| \geq c_1$ and a Fredholm operator for all $(x', \xi') \in B^*Y$.

Assume that the components $\Pi'\sigma(B)^0$, $\sigma(K)^0$, $\Pi'\sigma(T)^0$, $\sigma(Q)^0$ admit an extension, to B^*Y such that the induced extension of $\sigma^0_Y(\mathcal{A}) (x', \xi')$ is an isomorphism for all $(x', \xi') \in B^*Y$ (the operator symbol $\sigma(A)^\circ$ of the extension of $\sigma^0_Y(\mathcal{A})$ to B^*U is, of course, the above extension of $\sigma(A)^0$ to $B^*X|_Y$ as isomorphism).

By the above remarks we get

Corollary 5. *Let $\mathcal{A} \in \mathfrak{G}(X, Y)$ be elliptic and the principal symbols $\sigma(A)^0 (x, \xi)$: $p^*_X E \to p^*_X F$, $p_X : S^*X \to X$ and $\sigma^0_Y(\mathcal{A}) (x', \xi')$: $p^*_Y E' \otimes H^+ \oplus p^*_Y J \to p^*_Y F' \otimes H^+ \oplus p^*_Y G$, $p_Y : S^*Y \to Y$ admit compatible extensions as isomorphisms to the ball bundle B^*X and B^*Y, respectively. Then* ind $\mathcal{A} = 0$.

In fact, under the above conditions the integrals in (3) are taken over sets of measure zero.

In particular Corollary 5 yields another proof of 3.1.2.1, Theorem 7, 8. Consider, for instance, the operator with the symbol $(|\xi'| + i\nu)/(|\xi'| - i\nu)\, 1_{p^*E}$ on $S^*X|_U$ (U a suitable neighbourhood of Y). Set $\tilde{\sigma}(\xi', \nu) = \left((1 - \nu^2)^{1/2} + i\nu\right) / \left((1 - \nu^2)^{1/2} - i\nu\right)$ when $|\xi| \leq 1$, i.e. the extension is independent of ξ' for $|\xi'|^2 \leq 1 - \nu^2$ and $|\xi| \leq 1$. For $|\xi'| \leq 1$, $\nu \in \mathbb{R}$ put $\tilde{\sigma}(y', \nu) = (|\xi'| + i\nu)/(|\xi'| - i\nu)$ as before. Obviously, σ extends σ given on $S^*X|_U$ to a non-vanishing function on $B^+X|_U$. Hence, $\Pi^+\tilde{\sigma}$ defines an extension of $\Pi^+\sigma$ from S^*Y as continuous Fredholm family over B^*Y. The boundary symbol $\begin{bmatrix} \Pi^+\tilde{\sigma} \\ \Pi' \end{bmatrix}$ is an isomorphism for all $|\xi'| \leq 1$, which is the condition in Corollary 5.

Note that Corollary 5 gives a condition for the vanishing of the index of boundary problems which can be easily checked for large classes of classical boundary value problems in domains of the $\mathbb{R}^n$ with smooth boundary.

The proof of Theorem 3 is similar to that in the case of manifolds without boundary in FEDOSOV [3].

Proof of Theorem 3: Fix a finite covering $\{U_j\}$ of X by open sets with the property that if $U_{i_1} \cap \ldots \cap U_{i_k} \neq \emptyset$, $k \in \mathbb{Z}_+$ arbitrary, there is an open set $U \supset U_{i_1} \cup \ldots \cup U_{i_k}$ which is contractible and hence all bundles are trivial over U. Let $\{\varphi_j\}$ be a partition of unity subordinated $\{U_j\}$ which is locally near the boundary of the form as in 4.1.1.2, Corollary 14. Let $\sigma(\mathcal{R}) \in \mathfrak{S}_\lambda^{-m}(X, Y)$ be a formal global symbol with the principal symbol $\sigma^0(\mathcal{R})$ and $\mathcal{R} \in \mathfrak{G}^{-m}$ the corresponding operator.

The operator

$$\mathcal{A}' = \sum_{\gamma,\delta} \mathcal{A}_{\gamma\delta}, \quad \mathcal{A}_{\gamma\delta} = \mathcal{M}_{\varphi\gamma} \operatorname{Op}(\sigma(\mathcal{A})|_N) \mathcal{M}_{\varphi\delta}$$

($\sigma(\mathcal{A})$ taken in local coordinates of $U_{\gamma\delta} \supset U_\gamma \cup U_\delta$) is different from $\mathcal{A}$ by an operator of lower order and thus has the same index. Similarly,

$$\mathcal{R}' = \sum_{\alpha,\beta} \mathcal{R}_{\alpha\beta}, \quad \mathcal{R}_{\alpha\beta} = \mathcal{M}_{\varphi\alpha} \operatorname{Op}(\sigma(\mathcal{R})|_N) \mathcal{M}_{\varphi\beta}$$

($\sigma(\mathcal{R})$ in local coordinates of $U_{\alpha\beta} \supset U_\alpha \cup U_\beta$) is a parametrix of $\mathcal{A}$ and $\mathcal{A}'$. In view of the abstract index formula 1.1.1.1, Proposition 8 we have

$$\begin{aligned}
\operatorname{ind} \mathcal{A} &= \operatorname{tr}\left((1 - \mathcal{R}'\mathcal{A}')^N\right) - \operatorname{tr}\left((1 - \mathcal{A}'\mathcal{R}')^N\right) \\
&= \operatorname{tr}\left(\sum_{k=0}^N \binom{N}{k} (-1)^k (\mathcal{R}'\mathcal{A}')^k\right) - \operatorname{tr}\left(\sum_{k=0}^N \binom{N}{k} (-1)^k (\mathcal{A}'\mathcal{R}')^k\right) \\
&= \operatorname{tr}\left(\sum_{k=0}^N \binom{N}{k} (-1)^k \sum \mathcal{R}_{\alpha_1\beta_1}\mathcal{A}_{\gamma_1\delta_1} \ldots \mathcal{R}_{\alpha_k\beta_k}\mathcal{A}_{\gamma_k\delta_k}\right) \\
&\quad - \operatorname{tr}\left(\sum_{k=0}^N \binom{N}{k} (-1)^k \sum \mathcal{A}_{\gamma_1\delta_1}\mathcal{R}_{\alpha_1\beta_1} \ldots \mathcal{A}_{\gamma_k\delta_k}\mathcal{R}_{\alpha_k\beta_k}\right),
\end{aligned} \tag{4}$$

where $\alpha_i, \beta_i, \gamma_i, \delta_i$, $i = 1, \ldots, k$, independently run over the index set of the covering. For $\alpha_1, \ldots, \delta_k$ with $U_{\alpha_1} \cap \ldots \cap U_{\delta_k} = \emptyset$ the operators $\mathcal{R}_{\alpha_1\beta_1} \ldots \mathcal{A}_{\gamma_k\delta_k}$ and $\mathcal{A}_{\gamma_1\delta_1} \ldots \mathcal{R}_{\alpha_k\beta_k}$ are smoothing and

$$\operatorname{tr}(\mathcal{R}_{\alpha_1\beta_1}\mathcal{A}_{\gamma_1\delta_1} \ldots \mathcal{A}_{\alpha_k\delta_k}) = \operatorname{tr}(\mathcal{A}_{\gamma_1\delta_1} \ldots \mathcal{A}_{\gamma_k\delta_k}\mathcal{R}_{\alpha_1\beta_1}).$$

Thus we can restrict ourselves in (4) to such $\alpha_1, \ldots, \delta_k$ that $U_{\alpha_1} \cap \ldots \cap U_{\delta_k} = \emptyset$. By the same argument it follows that we can replace $\mathcal{R}_{\alpha_i,\beta_i}$ and $\mathcal{A}_{\gamma_i\delta_i}$ by $\mathcal{R}'_{\alpha_i,\beta_i}$ and $\mathcal{A}'_{\gamma_i,\delta_i}$ obtained by taking $\sigma(\mathcal{R})$ and $\sigma(\mathcal{A})$ in local coordinates of $U_{\alpha_1} \cup \ldots \cup U_{\delta_k}$. According to 4.1.1.2, Corollary 12 we get

$$\begin{aligned}
&\sum \operatorname{tr}\left(\mathcal{R}'_{\alpha_1\beta_1} \ldots \mathcal{A}'_{\gamma_k\delta_k} - \mathcal{M}_{\varphi_{\alpha_1}} \operatorname{Op}(\sigma(\mathcal{R}) \circ \varphi_{\beta_1}\varphi_{\gamma_1} \circ \sigma(\mathcal{A}) \circ \ldots \circ \sigma(\mathcal{A})|_N) \mathcal{M}_{\varphi_{\delta_k}}\right) \\
&= \sum \operatorname{tr}\left(\mathcal{A}'_{\gamma_1\delta_1} \ldots \mathcal{R}'_{\alpha_1\beta_1} - \mathcal{M}_{\varphi_{\gamma_1}} \operatorname{Op}(\sigma(\mathcal{A}) \circ \varphi_{\delta_1}\varphi_{\alpha_1} \circ \sigma(\mathcal{R}) \circ \ldots \circ \sigma(\mathcal{R})|_N) \mathcal{M}_{\varphi_{\beta_k}}\right)
\end{aligned}$$

and hence

$$\begin{aligned}
\operatorname{ind} \mathcal{A} = {} & \operatorname{tr}\left(\sum_{k=0}^N \binom{N}{k} (-1)^k \left(\sum \mathcal{M}_{\varphi_{\alpha_1}} \operatorname{Op}(\sigma(\mathcal{R}) \circ \varphi_{\beta_1}\varphi_{\delta_1} \circ \ldots \circ \sigma(\mathcal{A})|_N) \mathcal{M}_{\varphi_{\delta_k}}\right)\right) \\
& - \operatorname{tr}\left(\sum_{k=0}^N \binom{N}{k} (-1)^k \left(\sum \mathcal{M}_{\varphi_{\gamma_1}} \operatorname{Op}(\sigma(\mathcal{A}) \circ \varphi_{\delta_1}\varphi_{\alpha_1} \circ \ldots \circ \sigma(\mathcal{R})|_N) \mathcal{M}_{\varphi_{\beta_k}}\right)\right).
\end{aligned}$$

According to 4.1.1.2, Corollary 14 the trace does not change if we turn to local coordinates of $U_{\alpha_1} \cup U_{\delta_k}$ and $U_{\gamma_1} \cup U_{\beta_k}$, respectively, i.e.

$$\operatorname{tr}\Big(\sum_{\beta_1\ldots\gamma_k} \mathcal{M}_{\varphi_{\alpha_1}} \operatorname{Op}\left(\sigma(\mathcal{R}) \circ \varphi_{\beta_1}\varphi_{\delta_1} \circ \ldots \circ \sigma(\mathcal{A})|_N\right) \mathcal{M}_{\varphi_{\delta_k}}$$
$$- \mathcal{M}_{\varphi_{\alpha_1}} \operatorname{Op}\left((\sigma(\mathcal{R}) \circ \sigma(\mathcal{A}))^k|_N\right) \mathcal{M}_{\varphi_{\delta_k}}\Big) = 0 .$$

It follows that

$$\begin{aligned}\operatorname{ind} \mathcal{A} &= \operatorname{tr}\left(\sum_{k=0}^{N} \binom{N}{k} (-1)^k \Big(\sum_{\alpha,\beta} \mathcal{M}_{\varphi_\alpha} \operatorname{Op}\left((\sigma(\mathcal{R}) \circ \sigma(\mathcal{A}))^k|_N\right) \mathcal{M}_{\varphi_\beta}\Big)\right)\\ &\quad - \operatorname{tr}\left(\sum_{k=0}^{N} \binom{N}{k} (-1)^k \Big(\sum_{\alpha,\beta} \mathcal{M}_{\varphi_\alpha} \operatorname{Op}\left((\sigma(\mathcal{A}) \circ \sigma(\mathcal{R}))^k|_N\right) \mathcal{M}_{\varphi_\beta}\Big)\right)\\ &= \operatorname{Sp}_N\left((1 - \sigma(\mathcal{R}) \circ \sigma(\mathcal{A}))^N\right) - \operatorname{Sp}_N\left((1 - \sigma(\mathcal{A}) \circ \sigma(\mathcal{R}))^N\right) . \quad \square\end{aligned}$$

4.1.2. Improved Index Formulas

4.1.2.1. Elliptic Tupels

In the following sections we shall deal with the problem to derive from the coarse index formula in 4.1.1.3, Theorem 3 simplified explicit expressions for differential forms on T^*X and T^*Y the sum of the integrals of which is equal to the index of an elliptic operator $\mathcal{A} \in \mathfrak{G}$. The improved index formulas contain a considerably smaller number of derivatives. Lower order parts of the complete symbols are eliminated. Similarly as in the case without boundary one has to pose a topological condition to the underlying manifold, namely that the Todd class of the complexified tangent bundle vanishes.

We follow the program developed for elliptic PDOs on manifolds without boundary in Fedosov [3]. As motivation for the investigation of elliptic tupels recall the relation between elliptic symbols and elliptic sets of morphisms of trivial bundles.

Let E be a (complex) vector bundle over X (X a compact manifold without boundary). Denote by $E^\perp$ a complementary bundle to E, i.e. $E \oplus E^\perp \cong X \times \mathbb{C}^N$ for some $N \in \mathbb{Z}_+$. There is induced an embedding of the fibre E_x over $x \in X$ as a k dimensional subspace L_x into $\mathbb{C}^N$ (k the fibre dimension of the bundle E). Denote by $p^E(x)$ the function on X with values in the projectors in $\mathbb{C}^N$ assigning to any $x \in X$ the projector onto L_x parallel to $L_x^\perp$, the image of the fibre of $E_x^\perp$ in $\mathbb{C}^N$. Any vector bundle E can be represented by a function $p^E(x)$, $x \in X$, with values in the projectors in $\mathbb{C}^N$. The image of $p^E(x)$ represents the fibre E_x. Sections of E are then $\mathbb{C}^N$ valued functions with the property $p^E(x)\, s(x) = s(x)$ if $s \in C^\infty(X, \mathbb{C}^N)$ denotes the section. A local trivialization over $U \subset X$ of the bundle p^E is given by an $N \times k$ matrix $f_U(x)$ with column vectors $f_U^{(i)}(x) \in \mathbb{C}^N$, $i = 1, \ldots, k$, defining a base in $\operatorname{im} p^E(x) \cong E_x$, $x \in U$. The induced mapping $f_U(x)\colon \mathbb{C}^k \to \mathbb{C}^N$ defines an isomorphism onto $E_x \subset \mathbb{C}^N$. Then $p^E(x)\, f_U(x) = f_U(x)$. There is a uniquely defined $k \times N$ matrix function $e_U(x)$, $x \in U$, satisfying

$$e_U(x)\, f_U(x) = 1\,, \quad f_U(x)\, e_U(x) = p^E(x), \quad x \in U\,.$$

For $U \cap V = \emptyset$ and trivializations over U and V the cocycle g_{VU}^E of E is given by $g_{VU}^E(x) = f_V^{-1}(x)\, f_U(x)\colon \mathbb{C}^k \to \mathbb{C}^k$, $x \in U \cap V$. Define $\tilde{g}_{VU}(x) = f_V(x)\, f_U^{-1}(x)\, p^E(x)\colon \mathbb{C}^N \to \mathbb{C}^N$, $x \in U \cap V$. Then $p^E(x)\, \tilde{g}_{VU}(x)\, p^E(x) = \tilde{g}_{VU}(x)$ and $\tilde{g}_{VU}(x)$ defines an isomorphism of the fibres im $p^E(x)$ over $U \cap V$. Then the transformation of the matrix function f is given by $f_V(x) = \tilde{g}_{VU}(x) f_U(x)$. For the matrix function e we get $e_U(x) = e_V(x) \tilde{g}_{VU}(x)$.

Now let E, F be vector bundles over X represented by projector valued functions p^E, p^F. Denote by a_U the $k' \times k$ matrix function obtained from a using local trivializations of E and F over U (k and k' are the fibre dimensions of E and F, respectively). Set

$$a'_U = f_U^F a_U e_U^E . \tag{1}$$

This defines a bundle morphism $a' : X \times \mathbb{C}^N \to X \times \mathbb{C}^N$. In fact, we have

$$a'_V = f_V^F a_V e_V^E = f_V^F g_{VU}^F a_U g_{UV}^E e_V^E = f_U^F a_U e_U^E .$$

It is easily checked that

$$a' p^E = p^F a' = a' .$$

Moreover, an arbitrary matrix function $a' : X \to \hom(\mathbb{C}^N, \mathbb{C}^N)$ satisfying $a' p^E = p^F a' = a'$ corresponds to a bundle morphism locally defined by

$$a_U = e_U^F a'_U f_U^E .$$

The program to be carried out in the following sections can be roughly described as replacing the vector bundles and isomorphisms between vector bundles by trivial bundles and elliptic tupels (cf. Definition 2). Since we shall use, besides the matrix algebra, the algebra of boundary symbols, too, we give the basic definitions in a general formulation (cf. Fedosov [5]).

Definition 1. An associative algebra A over $\mathbb{C}$ is called *algebra with trace* if there is given a two-sided ideal $I \subset A$ and a linear functional Sp defined on I satisfying the condition

$$\operatorname{Sp}(ab) = \operatorname{Sp}(ba) \quad \text{for any} \quad a \in I,\ b \in A .$$

Definition 2. A tupel $d = (p_1, p_2, a, r)$ of elements of the algebra A with the properties

(i) $p_j^2 = p_j, \quad j = 1, 2$,

(ii) $p_2 a = a p_1 = a, \ p_1 r = r p_2 = r$,

is called *elliptic* if

(iii) $p_1 - ra$ and $p_2 - ar$ belong to I.

Then $\operatorname{Sp}(p_1 - ra) - \operatorname{Sp}(p_2 - ar)$ is well-defined and is called *index* of the elliptic tupel d.

Let $d = (p_1, p_2, a, r)$ be an elliptic tupel in the algebra A with trace, $d' = (p'_1, p'_2, a', r')$ an elliptic tupel in the algebra A' with trace. Assume that there is given an isomorphism $A \xrightarrow{\cong} A'$ such that the trace functional on A is equal to the pull back of the trace functional on A'. Then d and d' are called *isomorphic* if the components of d are mapped under the isomorphism into the components of d'.

The direct sum $d \oplus d'$ is defined by $d \oplus d' = (p_1 \oplus p'_1, p_2 \oplus p'_2, a \oplus a', r \oplus r')$ with components in $A \oplus A'$.

The tensor product $d \otimes d' = (P_1, P_2, E, F)$ is defined by

$$P_1 = \begin{pmatrix} p_1 \otimes p_1' & 0 \\ 0 & p_2 \otimes p_2' \end{pmatrix}, \quad P_2 = \begin{pmatrix} p_2 \otimes p_1' & 0 \\ 0 & p_1 \otimes p_2' \end{pmatrix},$$

$$E = \begin{pmatrix} a \otimes (p_1' - ra) & p_2 \otimes r' \\ -p_1 \otimes a' & r \otimes p_2' \end{pmatrix}, \quad F = \begin{pmatrix} r \otimes p_1' & -p_1 \otimes r \\ p_2 \otimes a' & a \otimes (p_2' - a'r') \end{pmatrix}$$

taking values in the algebra $A \otimes A' \otimes \hom(\mathbb{C}^2, \mathbb{C}^2)$ with trace $\operatorname{Sp}(a \otimes a') = \operatorname{Sp} a \operatorname{Sp} a'$.

An elliptic tupel is called *trivial* if $p_1 - ra = 0$ and $p_2 - ar = 0$.

The elliptic tupels with $p_1 - p_2 \in I$ play a special role. Then $d = (p_1, p_2, p_2p_1, p_1p_2)$ is elliptic and has the index $\operatorname{ind} d = \operatorname{Sp}(p_1 - p_2)$. Such tupels are called *canonical* ones. We define a canonical elliptic tupel $d' = (P_1, P_2, P_2P_1, P_1P_2)$ corresponding to an arbitrarily given elliptic tupel $d = (p_1, p_2, a, r)$ in the following way. The projectors P_1, P_2 are given by

$$P_1 = \begin{pmatrix} p_1 - ra & r \\ a(p_1 - ra) & ar \end{pmatrix}, \quad P_2 = \begin{pmatrix} 0 & r \\ 0 & p_2 \end{pmatrix}.$$

They take values in the algebra $A \otimes \hom(\mathbb{C}^2, \mathbb{C}^2)$ with trace. Then $\operatorname{ind} d = \operatorname{ind} d'$.

Fix a topology in A so that it is reasonable to speak about smooth functions defined on a manifold X with values in A. An elliptic family of tupels $d(x)$ consists of a tupel $(p_1(x), p_2(x), a(x), r(x))$ of smooth functions with values in A where the conditions (i), (ii) and (iii) are satisfied for any fixed x. If the manifold X is non-compact, we assume that $d(x)$ is trivial outside a compact set, i.e. for x outside a compact set we have

$$p_1(x) - r(x)\,a(x) = 0\,, \quad p_2(x) - a(x)\,r(x) = 0\,.$$

The complement of the largest open set in X where the elliptic tupel is trivial is called *support* of the elliptic family. Two elliptic families $d_0(x)$ and $d_1(x)$, $x \in X$, are called *homotopic* if there exists an elliptic family $d(x, t)$, $(x, t) \in X \times (-\varepsilon, 1 + \varepsilon)$ for some $\varepsilon > 0$ such that $d(x, 0) = d_0(x)$ and $d(x, 1) = d_1(x)$. An elliptic family $d(x)$ is called *canonical* if the elliptic tupel $d(x)$ is canonical for any fixed $x \in X$. The above construction of a canonical elliptic tupel out of an arbitrary one can be generalized to elliptic families.

Definition 3. Let $p(x) \in C^\infty(X, A)$ be a projector-valued function. The differential form $\omega = -\frac{1}{2\pi i} p\, dp\, dp$ is called *curvature form* of p.

The curvature of a projector-valued function is understood as differential form with values in the algebra A. If $(x_1, \ldots, x_n)$ are local coordinates on $U \subset X$, any differential form with values in A has the form $\sum a_{i_1 \ldots i_k}(x)\, dx_{i_1} \ldots dx_{i_k}$, $x \in U$, where the functions $a_{i_1 \ldots i_k}(x)$ take values in A. If $a_{i_1 \ldots i_k}(x)$ has values in the ideal I, the trace functional can be applied to differential forms with values in A

$$\operatorname{Sp}\left(\sum a_{i_1 \ldots i_k}(x)\, dx_{i_1} \ldots dx_{i_k}\right) = \sum \operatorname{Sp}\left(a_{i_1 \ldots i_k}(x)\right) dx_{i_1} \ldots dx_{i_k}\,.$$

The results are usually (complex-valued) differential forms on X.

Definition 4. Let $d(x) = (p_1(x), p_2(x), p_2(x)\,p_1(x), p_1(x)\,p_2(x))$ be a canonical elliptic family and ω_k the curvature form of the projector p_k, $k = 1, 2$. It is easily checked that the trace functional Sp is defined on $\omega_1^j - \omega_2^j$, since $\omega_1^j - \omega_2^j$ takes values in I

and has compact support. The complex valued differential form

$$\operatorname{ch} d(x) = \operatorname{Sp}\,(e^{\omega_1} - e^{\omega_2}) = \operatorname{Sp}\,\big(p_1(x) - p_2(x)\big) + \sum_{j=1}^{\infty} (1/j!)\,\operatorname{Sp}\,(\omega_1^j - \omega_2^j)$$

is called *Chern character* of $d(x)$.

For an arbitrary elliptic family $d(x)$ define the Chern character as Chern character of the associated canonical elliptic family. Let $d(x)$ be an elliptic family. Direct addition of a trivial elliptic family (i.e. a family with empty support) is called *stabilization* of $d(x)$. Homotopies of stabilizations are called *stable homotopies.*

Theorem 5. *The form* $\operatorname{ch} d(x) \in \bigoplus_{l=0}^{n} C^\infty(X, \Lambda^l(T^*X))$ *is closed and has compact support. Its cohomology class* $[\operatorname{ch} d(x)] \in H^*(X, \mathbb{C})$ *is invariant under stable homotopies.*

A proof is given in Fedosov [5].

In the case that A is the matrix algebra with the usual trace, a canonical tupel $d(x) = \big(p_1(x), p_2(x), p_2(x)\,p_1(x), p_1(x)\,p_2(x)\big)$ defines a pair of vector bundles over X and thus an element of $d_1 \in K(X)$ (with compact support). It can be proved that $[\operatorname{ch} d(x)]$ is equal to the usual Chern character $[\operatorname{ch} d_1] \in H^*(X, \mathbb{C})$ of $d_1 \in K(X)$.

Consider two elliptic families $d(x) = \big(p_1(x), p_2(x), a(x), r(x)\big)$ and $d'(x) = \big(p_1'(x), p_2'(x), a'(x), r'(x)\big)$, $x \in X$. For $p_2'(x) = p_1(x)$ we can define a family $d''(x) = \big(p_1'(x), p_2(x), a(x)\,a'(x), r'(x)\,r(x)\big)$, called the *composition* $d(x)\,d'(x)$ of $d(x)$ and $d'(x)$. It is easily checked that the composition of elliptic families is an elliptic family. For $p_1(x) = p_2(x)$ and $a(x) = r(x) = 1$ we have $d(x)\,d'(x) = d'(x)$. Note that

$$[\operatorname{ch}\,(d_1 \oplus d_2)] = [\operatorname{ch} d_1] + [\operatorname{ch} d_2]$$

and

$$[\operatorname{ch} d_1 d_2] = [\operatorname{ch} d_1] + [\operatorname{ch} d_2]\,,$$

if the composition is defined.

The first equation is obvious. The second follows from the homotopy $d_1 \oplus d_2 = d_1 d_2 + \mathfrak{t}$, where $\mathfrak{t}$ is a family with empty support (cf. 3.2.1.1, Proposition 2).

Let $A \in L^m(X)$ be an elliptic PDO,

$$A: C^\infty(X, E) \to C^\infty(X, F)\,,$$

E, $F \in \operatorname{Vect}(X)$. Denote by

$$\sigma_A^0: \pi^*E \to \pi^*F\,, \quad \pi: T^*X \to X\,,$$

the principal symbol which is an isomorphism outside a compact subset of T^*X. Define an elliptic family $d_A(x, \xi) = \big(p^E(x), p^F(x), \sigma_A(x, \xi), \sigma_R(x, \xi)\big)$ of $N \times N$ matrix functions on T^*X. p^E, p^F are the projector valued functions associated with the bundles E and F. The matrix-valued symbol $\sigma_A(x, \xi)$ is defined from $\sigma_A^0(x, \xi)$ by (1). $\sigma_R(x, \xi)$ is an $N \times N$ matrix-valued function on T^*X defined by (1) from an inverse of $\sigma_A^0(x, \xi)$ out sidea compact set. One checks easily that $d_A(x, \xi)$ is in fact an elliptic family with values in the $N \times N$ matrices (with the usual trace functional).

Of course, neither $\sigma_A(x, \xi)$ nor $\sigma_R(x, \xi)$ are invertible as mappings $\mathbb{C}^N \to \mathbb{C}^N$, but define only an isomorphism on the subspaces given by the projectors $p^E(x)$ and $p^F(x)$.

Theorem 6. *Assume that X admits an embedding into $\mathbb{R}^N$, N sufficiently large with trivial normal bundle. Let $A \in L^m(X; E, F)$ be elliptic and $d_A(x, \xi)$ the associated elliptic* *family on T^*X. Then*

$$\operatorname{ind} A = \int_{T^*X} \operatorname{ch} d_A(x, \xi),$$

where the orientation of T^*X is given by the form $d\xi_1\, dx_1 \dots d\xi_n\, dx_n$.

Theorem 6 can be considered as an analytical variant of the Atiyah-Singer index theorem and is proved in FEDOSOV [3]. In fact, the Atiyah-Singer theorem in its cohomological formulation states

$$\operatorname{ind} A = \int_{T^*X} (\operatorname{ch} \sigma_A)\, \tau(X). \tag{2}$$

$\tau(X)$ denotes a differential form representing the Todd class of $T_{\mathbb{C}}X$ in the cohomology. ch σ_A denotes the Chern character of the difference element defined by σ_A in $K(T^*X)$. Now, the assumption on X in Theorem 6 implies that the Todd class is trivial. Theorem 6 follows from (2) if we note that the analytical Chern character ch $d_A(x, \xi)$ and ch σ_A define the same cohomology class.

The above concept of elliptic tupels turned out ot be useful to prove index formulas for elliptic operators with random coefficients (cf. FEDOSOV/ŠUBIN [1]) and to recover a special case of the index formula for elliptic families of PDOs (cf. ATIYAH/SINGER [2, IV], FEDOSOV [5]).

4.1.2.2. The Half Space Situation

In this section we consider the special case of elliptic pseudo-differential boundary problems in the half space. It is assumed that the operators have constant coefficients near infinity (i.e. they stabilize). Topologically this is equivalent to the situation of operators in $\mathfrak{G}$ on the semi-sphere. Moreover, we assume all bundles to be trivial. Then we get an analogue to the index formula for elliptic PDOs in $\mathbb{R}^n$ derived in FEDOSOV [2] and HÖRMANDER [7] for a considerably larger class of PDOs.

Consider symbols $\sigma = (\sigma(A), \sigma_{\mathbb{R}^{n-1}}(\mathcal{A})) \in \mathfrak{S}^{m,d}(\overline{\mathbb{R}}^n_+, \mathbb{R}^{n-1})$ such that

$$\sigma(A)\,(x, \xi) = \sigma(A)\,(\infty, \xi) \quad \text{when} \quad |x| \geqq c,$$
$$\sigma_{\mathbb{R}^{n-1}}(\mathcal{A})\,(x', \xi') = \sigma_{\mathbb{R}^{n-1}}(\mathcal{A})\,(\infty, \xi') \quad \text{when} \quad |x'| \geqq c,$$

for a suitable constant c. Such symbols are called ***stabilizing near infinity***.

A symbol $\sigma \in \mathfrak{S}^{m,d}(\overline{\mathbb{R}}^n_+, \mathbb{R}^{n-1})$ is called ***elliptic*** if $\sigma(A)\,(x, \xi)$ is an invertible matrix for $|x|^2 + |\xi|^2 \geqq c_0$, $\sigma_{\mathbb{R}^{n-1}}(\mathcal{A})\,(x', \xi')$ is an invertible boundary symbol on $\mathbb{R}_+$ for $|x'|^2 + |\xi'|^2 \geqq c_0$ and $|\sigma(A)^{-1}\,(x, \xi)| \leqq c_1\, \langle\xi\rangle^{-m}$ for $|x|^2 + |\xi|^2 \geqq c_0$, $\sigma_{\mathbb{R}^{n-1}}(\mathcal{A})\,(x', \xi')^{-1} \in \mathfrak{R}^{-m}$ for $|x'|^2 + |\xi'|^2 \geqq c_0$ (the latter condition means that, for any excision function $\chi(x', \xi')$ vanishing for $|x'|^2 + |\xi'|^2 \leqq c_0$, the boundary symbol $\chi(x', \xi')\, \sigma_{\mathbb{R}^{n-1}}(\mathcal{A})\,(x', \xi')^{-1}$ belongs to $\mathfrak{R}^{-m}$).

The assumption about the symbols of stabilizing near infinity implies that the corresponding operator $\mathcal{A} \in \mathfrak{G}^{m,d}(\overline{\mathbb{R}}^n_+, \mathbb{R}^{n-1})$ has a continuous extension to Sobolev spaces

$$\mathcal{A}: \mathcal{H}^s \to \mathcal{H}^{s-m}$$

for $s > d - 1/2$ ($\mathcal{H}^s = H^s(\mathbb{R}^n_+, \mathbb{C}^k) \oplus H^{s+1/2}(\mathbb{R}^{n-1}, \mathbb{C}^j)$). In fact, use the decomposition of $\mathcal{A}$ into a sum of an operator with 'constant coefficients', i.e. with a symbol independent of x, and an operator the symbol of which has compact support with respect to x and x', respectively. For both operators the asserted extension to Sobolev spaces exists (cf. 2.3.2.4).

In order to use the calculus of complete boundary symbols in its simple version (cf. 2.3.2.2), we further assume that the interior symbol $\sigma(A)(x, \xi)$ is independent of x_n near $x_n = 0$.

Theorem 1. *Let $\mathcal{A} \in \mathfrak{S}^{m,d}(\overline{R}^n_+, R^{n-1})$ be defined with the symbol $\sigma(\mathcal{A})$ stabilizing near infinity and being elliptic. Then*

$$\mathcal{A}: \mathcal{H}^s \to \mathcal{H}^{s-m},$$

$s > d - {}^1/_2$, is a Fredholm operator. If the interior symbol is independent of x_n near $x_n = 0$, the index is given by

$$\operatorname{ind} \mathcal{A} = \operatorname{Sp}_N\left((1 - \sigma(\mathcal{R}) \circ \sigma(\mathcal{A}))^N\right) - \operatorname{Sp}_N\left((1 - \sigma(\mathcal{A}) \circ \sigma(\mathcal{R}))^N\right) \tag{1}$$

$(\sigma(\mathcal{R})$ the symbol of a parametrix$)$.

Proof: Define $\sigma(R) \in \mathfrak{A}^{-m}$ such that

$$\sigma(R)(x, \xi) = \sigma(A)(x, \xi)^{-1} \quad \text{for} \quad |x|^2 + |\xi|^2 \geqq c_0$$

and $\sigma_{R^{n-1}}(\mathcal{R}) \in \mathfrak{R}^{-m}$ such that

$$\sigma_{R^{n-1}}(\mathcal{R})(x', \xi') = \sigma_{R^{n-1}}(\mathcal{A})(x', \xi')^{-1} \quad \text{for} \quad |x'|^2 + |\xi'| \geqq c_0 .$$

Then $\sigma(\mathcal{R}) = \left(\sigma(R), \sigma_{R^{n-1}}(\mathcal{R})\right)$ is the symbol of a parametrix $\mathcal{R}$ of $\mathcal{A}$. In fact, $1 - \mathcal{R}\mathcal{A}$ and $1 - \mathcal{A}\mathcal{R}$ are compact in $\mathcal{H}^s$, since $1 - \sigma(R) \circ \sigma(A)$ belongs to $\mathfrak{A}^{-1}$ and vanishes for $|x| \geqq c$, $1 - \sigma_{R^{n-1}}(\mathcal{R}) \circ \sigma_{R^{n-1}}(\mathcal{A})$ belongs to $\mathfrak{R}^{-1}$ and vanishes for $|x'| \geqq c$. Similar assertions are valid for $1 - \sigma(\mathcal{A}) \circ \sigma(\mathcal{R})$. The index formula follows exactly as in 4.1.1.3, Theorem 3 but without reference to changes of coordinates. □

As in the case of compact manifold observe that the index depends only on the values of $\left(\sigma(A)(x, \xi), \sigma_{R^{n-1}}(\mathcal{A})(x', \xi')\right)$ for (x, ξ): $|x|^2 + |\xi|^2 \leqq c'$ and (x', y'): $|x'|^2 + |\xi'|^2 \leqq c'$, where c' is a constant with the property that for some $c < c'$ the interior symbol $\sigma(A)(x, \xi)$ is invertible in $|x|^2 + |\xi|^2 \geqq c$ and the boundary symbol $\sigma_{R^{n-1}}(\mathcal{A})(x', \xi')$ is invertible in $|x'|^2 + |\xi'|^2 \leqq c$. Thus we can assume that $\sigma(A)$ and $\sigma_{R^{n-1}}(\mathcal{A})$ are given only for $|x|^2 + |\xi|^2 \geqq c'$ and $|x'|^2 + |\xi'|^2 \leqq c'$. Then ellipticity means that the interior and the boundary symbol are invertible near the boundary of the domain where they are defined. Similarly as in 4.1.2.1 consider $\left(\sigma(\mathcal{A}), \sigma(\mathcal{R})\right)$ as an elliptic tupel. Then it is reasonable to speak about homotopies through elliptic tupels (preserving the condition of stabilization near ∞).

Theorem 2. *The right hand side of* (1) *is invariant under homotopies of symbols.*

Proof: Let $\sigma^t(\mathcal{A})$ be a homotopy of elliptic symbols and $\sigma^t(\mathcal{R})$ the corresponding homotopy of symbols of a parametrix. For abbreviation set $\sigma^t(\mathcal{A}) = \sigma_1(t)$, $\sigma^t(\mathcal{R}) = \sigma_2(t)$. We shall prove that

$$\frac{\mathrm{d}}{\mathrm{d}t}\left(\operatorname{Sp}_N\left((1 - \sigma_2(t) \circ \sigma_1(t))^N\right) - \operatorname{Sp}_N\left((1 - \sigma_1(t) \circ \sigma_2(t))^N\right)\right) = 0 . \tag{2}$$

Obviously $\frac{\mathrm{d}}{\mathrm{d}t} \operatorname{Sp}_N \sigma(t) = \operatorname{Sp}_N \dot{\sigma}(t)$, where $\dot{\sigma}(t) = \frac{\mathrm{d}}{\mathrm{d}t}\sigma(t)$. Then

$$\frac{\mathrm{d}}{\mathrm{d}t}\left(\operatorname{Sp}_N\left((1 - \sigma_2(t) \circ \sigma_1(t))^N\right)\right) = \operatorname{Sp}_N\left(\sum_{j=1}^{N} (1 - \sigma_2 \circ \sigma_1) \ldots \frac{\mathrm{d}}{\mathrm{d}t}(1 - \sigma_2 \circ \sigma_1) \ldots (1 - \sigma_2 \circ \sigma_1)\right),$$

where the derivative $\frac{\mathrm{d}}{\mathrm{d}t}$ is applied to the jth factor. Since symbols commute under the trace functional Sp_N, we get

$$\frac{\mathrm{d}}{\mathrm{d}t}\left(\mathrm{Sp}_N\left(\left(1-\sigma_2(t)\circ\sigma_1(t)\right)^N\right)\right) = N\,\mathrm{Sp}_N\left(\frac{\mathrm{d}}{\mathrm{d}t}(1-\sigma_2\circ\sigma_1)\,(1-\sigma_2\circ\sigma_1)^{N-1}\right).$$

In view of

$$\frac{\mathrm{d}}{\mathrm{d}t}\left(1-\sigma_2(t)\circ\sigma_1(t)\right) = -\,\dot\sigma_2(t)\circ\sigma_1(t) - \sigma_2(t)\circ\dot\sigma_1(t)$$

we get

$$\begin{aligned}
&\mathrm{Sp}_N\left(\frac{\mathrm{d}}{\mathrm{d}t}(1-\sigma_2\circ\sigma_1)\,(1-\sigma_2\circ\sigma_1)^{N-1}\right) - \mathrm{Sp}_N\left(\frac{\mathrm{d}}{\mathrm{d}t}(1-\sigma_1\circ\sigma_2)\,(1-\sigma_1\circ\sigma_2)^{N-1}\right)\\
&= -\,\mathrm{Sp}_N\left(\dot\sigma_2\circ\sigma_1\circ(1-\sigma_2\circ\sigma_1)^{N-1}\right) - \mathrm{Sp}_N\left(\sigma_1\circ\dot\sigma_2\circ(1-\sigma_2\circ\sigma_1)^{N-1}\right)\\
&\quad + \mathrm{Sp}_N\left(\dot\sigma_1\circ\sigma_2\circ(1-\sigma_1\circ\sigma_2)^{N-1}\right) + \mathrm{Sp}_N\left(\sigma_1\circ\dot\sigma_2\circ(1-\sigma_1\circ\sigma_1)^{N-1}\right).
\end{aligned}$$

Then (2) follows from

$$\begin{aligned}
\mathrm{Sp}_N\left(\dot\sigma_2\circ\sigma_1\circ(1-\sigma_2\circ\sigma_1)^{N-1}\right) &= \mathrm{Sp}_N\left(\dot\sigma_2\circ(1-\sigma_1\circ\sigma_2)^{N-1}\circ\sigma_1\right)\\
&= \mathrm{Sp}_N\left(\sigma_1\circ\dot\sigma_2\circ(1-\sigma_1\circ\sigma_2)^{N-1}\right)
\end{aligned}$$

and the assertion is proved. □

Applying Theorem 2 we can simplify the right hand side of the index formula (1).

Theorem 3. *Assume that* $\mathcal{A}\in\mathfrak{G}^{m,d}(\overline{\mathbb{R}}^n_+, \mathbb{R}^{n-1})$ *satisfies the conditions of Theorem* 1 *and let* $\sigma(\mathcal{A}) = \left(\sigma(A), \sigma_{\mathbb{R}^{n-1}}(\mathcal{A})\right)$ *be its (principal) symbol. Then*

$$\begin{aligned}
\operatorname{ind}\mathcal{A} = &-\frac{(n-1)!}{(2n-1)!\,(2\pi\mathrm{i})^n}\int\limits_{S(\mathbb{R}^n_+)}\mathrm{Tr}\left(\left(\sigma(A)^{-1}\,\mathrm{d}\sigma(A)\right)^{2n-1}\right)\\
&-\frac{(n-1)!}{\mathrm{i}(2n-2)!\,(2\pi\mathrm{i})^{n-1}}\int\limits_{S(\mathbb{R}^{n-1})}\mathrm{Tr}\,\Pi'_\nu\left(\left(\sigma(A)^{-1}\,\mathrm{d}'\sigma(A)\right)^{2n-3}\sigma(A)^{-1}\,\partial_\nu\sigma(A)\right)\\
&-\frac{(n-2)!}{(2n-3)!\,(2\pi\mathrm{i})^{n-1}}\int\limits_{S(\mathbb{R}^{n-1})}\mathrm{Tr}'\left(\left(\sigma_{\mathbb{R}^{n-1}}(\mathcal{A})^{-1}\,\mathrm{d}'\sigma_{\mathbb{R}^{n-1}}(\mathcal{A})\right)^{2n-2}\right), \qquad (3)
\end{aligned}$$

where $S(\mathbb{R}^n_+)$ *and* $S(\mathbb{R}^{n-1})$ *denote cosphere bundles* $|\xi| = c$ *and* $|\xi'| = c$ *for* c *such that* $\sigma(A)\,(x,\xi)^{-1}$ *and* $\sigma_{\mathbb{R}^{n-1}}(\mathcal{A})\,(x',\xi')^{-1}$ *exist on* $S(\mathbb{R}^n_+)$ *and* $S(\mathbb{R}^{n-1})$, d' *denotes the differential with respect to* (x',ξ') *and the orientation of the spheres is induced by the forms* $\mathrm{d}\xi_1\,\mathrm{d}x_1\ldots\mathrm{d}\xi_n\,\mathrm{d}x_n$ *and* $\mathrm{d}\xi_1\,\mathrm{d}x_1\ldots\mathrm{d}\xi_{n-1}\,\mathrm{d}x_{n-1}$, *respectively.*

In (3) $\sigma(A)\,(x,\xi)$ is considered as matrix function, $\mathrm{d}\sigma(A)\,(x,\xi)$ is a matrix valued 1-form. In the exterior power $\left(\sigma(A)^{-1}\,\mathrm{d}\sigma(A)\right)^{2n-1}$ one must take matrix multiplication with exterior multiplication of the entries. The result is a matrix valued $2n-1$ form the trace of which is a usual $2n-1$ form. In a similar way other terms are to be understood with the difference that in the last term the exterior forms take values in the boundary symbols on the line.

Recall that for

$$\sigma_{\mathbb{R}^{n-1}}(\mathcal{A}) = \begin{pmatrix} \Pi^+\sigma(A) + \Pi'\sigma(B) & \sigma(K) \\ \Pi'\sigma(T) & \sigma(Q) \end{pmatrix}$$

the trace functional Tr' was defined in 4.1.1.2 and is given by

$$\mathrm{Tr}'\, \sigma_{\mathbb{R}^{n-1}}(\mathcal{A}) = \mathrm{Tr}\, \Pi'\sigma(B)\,(\nu, \tau) + \mathrm{Tr}\, \sigma(Q)\,.$$

The proof is similar to that in the case of elliptic PDOs in $\mathbb{R}^n$. One has to take into account the compatibility of interior and boundary symbols. Some troubles are due to the fact that the trace functional Tr′ involved in formula (3) does not vanish on commutators. It is possible to generalize the method of the proof to the case of non-trivial bundles. Then the formula becomes much more involved. Since this is a special case of the formulas given in 4.1.2.3, we shall not go into details.

The index formula (3) remains valid for elliptic operators in the class $\mathfrak{G}$ acting between trivial bundles under the assumption that the manifold X admits an embedding into $\mathbb{R}^N$, $N \in \mathbb{Z}_+$ sufficiently large, with trivial normal bundle. This can be proved in a similar way as the 'embedding proof' for the index of usual elliptic PDO.

Proof: In view of 3.2.1.2, Remark 6 near the boundary $x_n = 0$ the interior symbol $\sigma(A)\,(x, \xi)$, after stabilization, can be extended from $|\xi| = c$ to $|\xi| \leqq c$ as an isomorphism (the constant c is such that $\sigma(A)\,(x, \xi)$ is an isomorphism for $|\xi| = c$). The choice of an extension of $\sigma(A)\,(x, \xi)$ and $\sigma(R)\,(x, \xi)$ for small ξ has no influence on the value of (1). Thus the integral over $T^*\mathbb{R}^n_+$ is equal to the integral over $T^*\mathbb{R}^n_+|_{x_n > \varepsilon}$ for some $\varepsilon > 0$. Then we can ignore the presence of the boundary, the transmission property and so on. It will follow as a special case of the considerations for boundary symbols that the term integrated over $T^*\mathbb{R}^n_+$ is equal to

$$-\frac{(n-1)!}{(2n-1)!\,(2\pi i)^n} \int\limits_{S^*\mathbb{R}^n_+} \mathrm{Tr}\,\left((\sigma(A)^{-1}\, \mathrm{d}\sigma(A))^{2n-1}\right),$$

where $S^*\mathbb{R}^n_+ = \{(x, \xi) \in T^*\mathbb{R}^n_+ : |x|^2 + |\xi|^2 = c\}$ and c such that on $S^*\mathbb{R}^n_+$ the matrix $\sigma(A)\,(x, \xi)$ is invertible, d denotes the differential with respect to x, ξ, and the orientation of $S^*\mathbb{R}^n_+$ is induced by $\mathrm{d}\xi_1\, \mathrm{d}x_1 \dots \mathrm{d}\xi_n\, \mathrm{d}x_n$.

Consider now the integral over $T^*\mathbb{R}^{n-1}$. The choice of an extension of $\sigma(A) \oplus 1$ for the sphere bundle as an isomorphism to the whole ball bundle is assumed to be fixed. Homotopies of the boundary symbol must preserve the property that $\sigma(A) \oplus 1$ is an isomorphism.

Let $\varepsilon = (\varepsilon_1, \dots, \varepsilon_{n-1}) \in (\mathbb{R}'_+)^{n-1}$, $\delta = (\delta_1, \dots, \delta_{n-1}) \in (\mathbb{R}_+)^{n-1}$. Set

$$\sigma^{\varepsilon,\delta}_{\mathbb{R}^{n-1}}(\mathcal{A})\,(x', \xi') = \sigma_{\mathbb{R}^{n-1}}(\mathcal{A})\,(\varepsilon x', \delta\xi')\,,$$

$$\varepsilon x' = (\varepsilon_1 x_1, \dots, \varepsilon_{n-1} x_{n-1}),\ \delta\xi' = (\delta_1\xi_1, \dots, \delta_{n-1}\xi_{n-1})\,.$$

or more explicitly

$$\sigma^{\varepsilon,\delta}_{\mathbb{R}^{n-1}}(\mathcal{A})\,(x', \xi') = \begin{pmatrix} \Pi^+\sigma(A)\,(\varepsilon x', \delta\xi', \nu) + \Pi'\sigma(B)\,(\varepsilon x', \delta\xi', \nu, \tau) & \sigma(K)\,(\varepsilon x', \delta\xi'\, \nu) \\ \Pi'\sigma(T)\,(\varepsilon x', \delta\xi', \nu) & \sigma(Q)\,(\varepsilon x', \delta\xi') \end{pmatrix}.$$

For ε_i, δ_i near 1 this is an elliptic boundary symbol and for $|\xi'| \geqq c$ the inverse is given by $\sigma^{\varepsilon,\delta}_{\mathbb{R}^{n-1}}(\mathcal{R})\,(x', \xi') = \sigma_{\mathbb{R}^{n-1}}(\mathcal{R})\,(\varepsilon x', \delta\xi')$ if $\sigma_{\mathbb{R}^{n-1}}(\mathcal{R})\,(x', \xi')$ is the inverse of $\sigma_{\mathbb{R}^{n-1}}(\mathcal{A})\,(x', \xi')$ for $|\xi'| \geqq c$. Moreover, $(\sigma_{\mathbb{R}^{n-1}}(\mathcal{A}), \sigma_{\mathbb{R}^{n-1}}(\mathcal{R}))$ and $(\sigma^{\varepsilon,\delta}_{\mathbb{R}^{n-1}}(\mathcal{A}), \sigma^{\varepsilon,\delta}_{\mathbb{R}^{n-1}}(\mathcal{R}))$

are homotopic elliptic tupels. According to Theorem 2 we get

$$\begin{aligned}&(2\pi)^{-n+1}\int\big(\mathrm{Tr}'\,(1-\sigma_{R^{n-1}}(\mathcal{R})\circ\sigma_{R^{n-1}}(\mathcal{A}))^N|_N\\&\qquad-\mathrm{Tr}'\,(1-\sigma_{R^{n-1}}(\mathcal{A})\circ\sigma_{R^{n-1}}(\mathcal{R}))^N|_N\big)\,\mathrm{d}\xi'\,\mathrm{d}x'\\&=(2\pi)^{-n+1}\int\big(\mathrm{Tr}'\,(1-\sigma^{\varepsilon,\delta}_{R^{n-1}}(\mathcal{R})\circ\sigma^{\varepsilon,\delta}_{R^{n-1}}(\mathcal{A}))^N|_N\\&\qquad-\mathrm{Tr}'\,(1-\sigma^{\varepsilon,\delta}_{R^{n-1}}(\mathcal{A})\circ\sigma^{\varepsilon,\delta}_{R^{n-1}}(\mathcal{R}))^N|_N\big)\,\mathrm{d}\xi'\,\mathrm{d}x'\,.\end{aligned}\tag{4}$$

The calculus of complete boundary symbols yields

$$\begin{aligned}&\big(1-\sigma_{R^{n-1}}(\mathcal{R})\circ\sigma_{R^{n-1}}(\mathcal{A})\big)^N|_N\\&=\sum_{k=0}^{N}\binom{N}{k}(-1)^k\big(\sigma_{R^{n-1}}(\mathcal{R})\circ\sigma_{R^{n-1}}(\mathcal{A})\big)^k\\&=\sum_{k=0}^{N}\sum c_{\alpha_1\ldots\beta_{2k}}\,\partial_{\xi'}^{\alpha_1}\partial_{x'}^{\beta_1}\sigma_{R^{n-1}}(\mathcal{R})\,(x',\xi')\,\partial_{\xi'}^{\alpha_2}\partial_{x'}^{\beta_2}\sigma_{R^{n-1}}(\mathcal{A})\,(x',\xi')\\&\qquad\ldots\,\partial_{\xi'}^{\alpha_{2k}}\partial_{x'}^{\beta_{2k}}\sigma_{R^{n-1}}(\mathcal{A})\,(x',\xi')\end{aligned}$$

where $\sum_{i=1}^{2k}|\alpha_i|=\sum_{i=1}^{2k}|\beta_i|<N$ with suitable constants $c_{\alpha_1\ldots\beta_{2k}}$, $\alpha_i,\beta_i\in\mathbb{Z}_+^{n-1}$. Similarly

$$\begin{aligned}&\big(1-\sigma^{\varepsilon,\delta}_{R^{n-1}}(\mathcal{R})\circ\sigma^{\varepsilon,\delta}_{R^{n-1}}(\mathcal{A})\big)^N|_N\\&=\sum_{k=0}^{N}\sum c_{\alpha_1\ldots\beta_{2k}}\,\partial_{\xi'}^{\alpha_1}\partial_{x'}^{\beta_1}\big(\sigma_{R^{n-1}}(\mathcal{R})\,(\varepsilon x',\delta\xi')\big)\ldots\partial_{\xi'}^{\alpha_{2k}}\partial_{x'}^{\beta_{2k}}\big(\sigma_{R^{n-1}}(\mathcal{A})\,(\varepsilon x',\delta\varepsilon')\big)\\&=\sum_{k=0}^{N}\sum c_{\alpha_1\ldots\beta_{2k}}\varepsilon^{\beta_1+\ldots+\beta_{2k}}\delta^{\alpha_1+\ldots+\alpha_{2k}}\big(\partial_{\xi'}^{\alpha_1}\partial_{x'}^{\beta_1}\sigma_{R^{n-1}}(\mathcal{R})\big)\,(\varepsilon x',\delta\xi')\\&\qquad\ldots\,\big(\partial_{\xi'}^{\alpha_{2k}}\partial_{x'}^{\beta_{2k}}\sigma_{R^{n-1}}(\mathcal{A})\big)\,(\varepsilon x',\delta\xi')\,.\end{aligned}$$

With these expressions and changing $\mathrm{d}\xi'\,\mathrm{d}x'$ into $\mathrm{d}(\varepsilon x')\,\mathrm{d}(\delta\xi')$ from (4) we get on the right hand side a polynomial in ε and δ (defined for ε_i, δ_i near 1), while the left hand side is independent of ε and δ. Hence the value is given by the coefficient of $\varepsilon^0\delta^0$, i.e. exactly by the sum involving the products $\partial_{\xi'}^{\alpha_1}\partial_{x'}^{\beta_1}\sigma_{R^{n-1}}(\mathcal{R})\ldots\partial_{\xi'}^{\alpha_{2k}}\partial_{x'}^{\beta_{2k}}\sigma_{R^{n-1}}(\mathcal{A})$ with $\sum\alpha_i=(1,\ldots,1)$, $\sum\beta_i=(1,\ldots,1)$. In other words, the value of (4) is given by the sum over all products containing exactly once all derivatives with respect to x_i, ξ_i, $i=1,\ldots,n-1$. There may occur mixed derivatives of one factor and other factors without any derivative.

Consider now another homotopy. Let p be an arbitrary permutation of $2n-2$ elements. $(\delta_{ip(i)})$ is a $(2n-2)\times(2n-2)$ matrix obtained from the unit matrix by applying the permutation p to the columns. Set $T_p=(\delta_{ip(i)})\,(\operatorname{sgn}p\oplus 1_{2n-3})$. Then $T_p\simeq 1$ in SO $(2n-2,\mathbb{R})$. For an arbitrary elliptic boundary symbol $\sigma_{R^{n-1}}(x',\xi')$ set

$$\sigma^p_{R^{n-1}}(x',\xi')=\sigma_{R^{n-1}}\left(T_p\begin{pmatrix}x'\\\xi'\end{pmatrix}\right).$$

Then $\sigma_{R^{n-1}}\simeq\sigma^p_{R^{n-1}}$ through elliptic boundary symbols. We have

$$\begin{aligned}&(2\pi)^{-n+1}\int\big(\mathrm{Tr}'\,(1-\sigma_{R^{n-1}}(\mathcal{R})\circ\sigma_{R^{n-1}}(\mathcal{A}))^N|_N\\&\qquad-\mathrm{Tr}'\,(1-\sigma_{R^{n-1}}(\mathcal{A})\circ\sigma_{R^{n-1}}(\mathcal{R}))^N|_N\big)\,\mathrm{d}\xi'\,\mathrm{d}x'\\&=(2\pi)^{-n+1}\operatorname{sgn}p\int\big(\mathrm{Tr}'\,(1-\sigma^p_{R^{n-1}}(\mathcal{R})\circ\sigma^p_{R^{n-1}}(\mathcal{A}))^N|_N\\&\qquad-\mathrm{Tr}'\,(1-\sigma_{R^{n-1}}(\mathcal{A})\circ\sigma_{R^{n-1}}(\mathcal{R}))^N|_N\big)\,\mathrm{d}\xi'\,\mathrm{d}x'\,.\end{aligned}$$

In fact, for products of the form $\partial_{\xi'}^{\alpha_1}\partial_{x'}^{\beta_1}\sigma_{R^{n-1}}(\mathcal{R}) \dots \partial_{\xi'}^{\alpha_{2k}}\partial_{x'}^{\beta_{2k}}\sigma_{R^{n-1}}(\mathcal{A})$, where any derivative occurs exactly once,

$$\partial_{\xi'}^{\alpha_1}\partial_{x'}^{\beta_1}\sigma^p_{R^{n-1}}(\mathcal{R})\,(x',\xi') \dots \partial_{\xi'}^{\alpha_{2k}}\partial_{x'}^{\beta_{2k}}\sigma^p_{R^{n-1}}(\mathcal{A})\,(x',\xi')$$
$$= \operatorname{sgn} p\,\left(\partial_{\xi'}^{\alpha_1}\partial_{x'}^{\beta_1}\sigma_{R^{n-1}}(\mathcal{R})\right)\left(T_p\begin{pmatrix}x'\\ \xi'\end{pmatrix}\right) \dots \left(\partial_{\xi'}^{\alpha_{2k}}\partial_{x'}^{\beta_{2k}}\sigma_{R^{n-1}}(\mathcal{A})\right)\left(T_p\begin{pmatrix}x'\\ \xi'\end{pmatrix}\right).$$

Hence it follows that

$$(2\pi)^{-n+1}\int \left(\operatorname{Tr}'\,(1-\sigma_{R^{n-1}}(\mathcal{R})\circ\sigma_{R^{n-1}}(\mathcal{A}))^N|_N\right.$$
$$\left.-\operatorname{Tr}'\,(1-\sigma_{R^{n-1}}(\mathcal{A})\circ\sigma_{R^{n-1}}(\mathcal{R}))^N|_N\right)\,d\xi'\,dx'$$
$$=\frac{1}{(2n-2)!\,(2\pi)^{n-1}}\sum_p \operatorname{sgn} p\int \left(\operatorname{Tr}'\,(1-\sigma^p_{R^{n-1}}(\mathcal{R})\circ\sigma^p_{R^{n-1}}(\mathcal{A}))^N|_N\right.$$
$$\left.-\operatorname{Tr}'\,(1-\sigma^p_{R^{n-1}}(\mathcal{A})\circ\sigma^p_{R^{n-1}}(\mathcal{R}))^N|_N\right)\,d\xi'\,dx'. \tag{5}$$

Then the mixed derivatives of a factor cancel and the products with every factor differentiated exactly once are alternated.

For abbreviation set $\sigma_1(x',\xi') = \sigma_{R^{n-1}}(\mathcal{A})\,(x',\xi')$, $\sigma_2(x',\xi') = \sigma_{R^{n-1}}(\mathcal{R})\,(x',\xi')$. In order to compute the right hand side of (5) return to formula (1). Note that instead of $\sigma_2\circ\sigma_1$ and $\sigma_1\circ\sigma_2$ we can take

$$\sigma_2\sigma_1 - i\sum_{j=1}^{n-1}\partial_{\xi_j}\sigma_2\partial_{x_j}\sigma_1$$

and

$$\sigma_1\sigma_2 - i\sum_{j=1}^{n-1}\partial_{\xi_j}\sigma_1\partial_{x_j}\sigma_2\,,$$

since the other terms of the asymptotic expansion contain factors with higher than first order derivatives.

Moreover, we can introduce an explicit expression of the boundary symbol of a parametrix σ_2. Set

$$\sigma_2(x',\xi') = \psi(x;\xi')\,\sigma_1(x',\xi')^{-1}\,,$$

where ψ is an excision function vanishing for those (x',ξ') where σ_1 is not invertible. Hence, from (1) and the above remarks we get that (5) is equal to

$$(2\pi)^{-n+1}\int\left(\operatorname{Tr}'\,((1-\psi+i\partial_{x'}\sigma_2\partial_{\xi'}\sigma_1)^N|_N)\right.$$
$$\left.-\operatorname{Tr}'\,((1-\psi+i\partial_{\xi'}\sigma_1\partial_{x'}\sigma_2))^N|_N\right)\,d\xi'\,dx'. \tag{6}$$

Since in $\partial_{\xi'}\sigma_2\partial_{x'}\sigma_1 = \sum_{j=1}^{n-1}\partial_{\xi_j}\sigma_2(x',\xi')\,\partial_{x_j}\sigma_1(x',\xi')$ all derivatives with respect to x_j, ξ_j, $j=1,\dots,n-1$, occur, we can drop products where $(\partial_{\xi'}\sigma_2\,\partial_{x'}\sigma_1)$ is differentiated. Exactly $2n-2$ derivatives with respect to $x_1,\dots,x_{n-1},\xi_1,\dots,\xi_{n-1}$ occur in the terms with $n-1$ factors $\partial_{\xi'}\sigma_2\,\partial_{x'}\sigma_1$.

Since the scalar function ψ commutes with an arbitrary boundary symbol under Tr', this is given by

$$(2\pi)^{-n+1}\binom{N}{n-1}(2-\psi)^{N-n+1}\,i^{n-1}(\partial_{\xi'}\sigma_2\,\partial_{x'}\sigma_1)^{n-1}\,.$$

By alternation of the derivatives we get that this is equal to

$$-\frac{(n-1)!}{(2n-2)!\,(2\pi i)^{n-1}}\binom{N}{n-1}\int \left(\operatorname{Tr}'\,(d'\sigma_2\, d'\sigma_1)^{n-1} - \operatorname{Tr}'\,(d'\sigma_1\, d'\sigma_2)^{n-1}\right) \times (1-\psi)^{N-n+1}\,, \tag{7}$$

where the sign comes from the choice of the orientation of $T^*\mathbb{R}^{n-1}$ given by $d\xi_1\, dx_1 \ldots d\xi_{n-1}\, dx_{n-1}$. From $\sigma_2(x', \xi') = \psi(x', \xi')\, \sigma_1(x', \xi')^{-1}$ it follows that

$$d'\sigma_2 = d'\psi\sigma_1^{-1} + \psi d'(\sigma_1^{-1}) = d'\psi\, \sigma_1^{-1} - \psi\sigma_1^{-1}\, d'\sigma_1\, \sigma_1^{-1}\,,$$

since, on the support of ψ, we have

$$0 = d'(\sigma_1\sigma_1^{-1}) = d'\sigma_1\, \sigma_1^{-1} + \sigma_1 d'(\sigma_1^{-1})\,.$$

(7) implies

$$-\frac{(n-1)!}{(2n-2)!\,(2\pi i)^{n-1}}\binom{N}{n-1}\int \Big\{\operatorname{Tr}'\,\big(((d'\psi\sigma_1^{-1} - \psi\sigma_1^{-1}\, d'\sigma_1\, \sigma_1^{-1})\, d'\sigma_1)^{n-1}\big) - \operatorname{Tr}'\,\big((d'\sigma_1(d'\psi\sigma_1^{-1} - \psi\sigma_1^{-1}\, d'\sigma_1\, \sigma_1^{-1}))^{n-1}\big)\Big\}\,.$$

For an arbitrary q form ω on $T^*\mathbb{R}^{n-1}$ with values in the vector space of boundary symbols on the line we have

$$\operatorname{Tr}'\,(\psi\omega) = \psi\, \operatorname{Tr}'\,\omega, \quad \operatorname{Tr}'\,(d\psi\omega) = (-1)^q\, \operatorname{Tr}'\,(\omega\, d\psi)\,.$$

Hence, in the $(n-1)^{\text{th}}$ exterior power there may occur at most one term $d\psi$. We have

$$\begin{aligned}&\big((d'\psi\sigma_1^{-1} - \psi\sigma_1^{-1}\, d'\sigma_1\sigma_1^{-1})\, d'\sigma_1\big)^{n-1}\\ &= (-1)^{n-1}\, \psi(\sigma_1^{-1}\, d'\sigma_1)^{2n-2} + (-1)^{n-2}\,(n-1)\,\psi^{n-2}\, d'\psi(\sigma_1^{-1}\, d'\sigma_1)^{2n-3}\end{aligned}$$

and

$$\begin{aligned}&\big(d'\sigma_1(d\psi\sigma_1^{-1} - \psi\sigma_1^{-1}\, d'\sigma_1\sigma_1^{-1})\big)^{n-1}\\ &= (-1)^{n-1}\, \psi^{n-1}(d'\sigma_1\sigma_1^{-1})^{2n-2} - (-1)^{n-2}\,(n-1)\,\psi^{n-2}\, d'\psi(d'\sigma_1\sigma_1^{-1})^{2n-3}\,.\end{aligned}$$

Note that

$$\begin{aligned} d'\,\operatorname{Tr}'\,(\sigma_1^{-1}\, d'\sigma_1)^{2n-3} &= -\operatorname{Tr}'\,(\sigma_1^{-1}\, d'\sigma_1)^{2n-2}\,,\\ d'\,\operatorname{Tr}'\,(d'\sigma_1\sigma_1^{-1})^{2n-3} &= \operatorname{Tr}'\,(d'\sigma_1\sigma_1^{-1})^{2n-2}\,.\end{aligned}$$

Set $\beta = \operatorname{Tr}'\,(\sigma_1^{-1}\, d'\sigma_1)^{2n-3} + \operatorname{Tr}'\,(d'\sigma_1\sigma_1^{-1})^{2n-3}$. The form β is only defined for such x', ξ' where σ_1 is invertible. After multiplication by ψ or $d'\psi$ the resulting form is defined everywhere and vanishes where σ_1 is not invertible. For (7) we thus obtain

$$-\frac{(n-1)!}{(2n-2)!\,(2\pi i)^{n-1}}\binom{N}{n-1}\int\limits_{T^*\mathbb{R}^{n-1}} (1-\psi)^{N-n+1}\,(-1)^{n-2}\,(n-1)\,\psi^{n-2}\, d\psi\beta\,. \tag{8}$$

Observe that in view of Stokes' theorem we have that

$$\int\limits_{S_r^*\mathbb{R}^{n-1}} \beta\,, \quad S_r^*\mathbb{R}^{n-1} = \{(x', \xi') \colon |x'|^2 + |\xi'|^2 = r^2\}$$

is independent of r for $c < r < c_1$ (note that σ_1 is invertible outside $S_c^*\mathbb{R}^{n-1}$). Moreover, we have

$$\binom{N}{n-1}\int_0^1 (1-s)^{N-n+1}\,(n-1)\, s^{n-2}\, ds = 1$$

and (8) is equal to

$$-\frac{(n-1)!}{(2n-2)!\,(2\pi i)^{n-1}} \int\limits_{S_c^* R^{n-1}} \beta \,. \tag{9}$$

As observed in 4.1.1.2 the trace functional Tr′ does not vanish on commutators. We have for any p form α and q form β with values in $\mathfrak{R}$

$$\mathrm{Tr}'\,\alpha\beta - (-1)^{pq}\,\mathrm{Tr}'\,\beta\alpha = -\mathrm{i}\,\mathrm{Tr}\,\Pi'\,\partial_\nu ab$$

if a and b are the PDO symbols involved in the left upper corner of α and β, respectively. Then we get

$$\begin{aligned}&\mathrm{Tr}'\,(\mathrm{d}\sigma_1\sigma_1^{-1})^{2n-3}\\ &= \mathrm{Tr}'\,(\sigma_1^{-1}\,\mathrm{d}\sigma_1)^{2n-3} - \mathrm{i}\,\mathrm{Tr}\,\Pi'\big(\sigma(A)^{-1}\,\mathrm{d}'\sigma(A)\big)^{2n-3}\,\sigma(A)^{-1}\,\partial_\nu\sigma(A)\end{aligned} \tag{10}$$

if we change in $(\mathrm{d}\sigma_1\,\sigma_1^{-1})^{2n-3} = \big((\mathrm{d}\sigma_1\,\sigma_1^{-1})^{2n-4}\,\mathrm{d}\sigma_1\big)\,\sigma_1^{-1}$ the order of the two factors. Now the assertion follows from (9) and the above equality (10). □

Remark 4. The right hand side of (3) makes sense also in the case that the interior symbol depends on x_n near the boundary. Formula (3) expresses the index in this case, too. In fact, the form integrated over $S(\overline{R}^n_+)$ vanishes near $x_n = 0$, since after stabilization $\sigma(A)$ can be extended to all $|\xi| \leq 1$ as an isomorphism (cf. 3.2.1.2, Remark 6). Hence, without changing the right hand side in (3) we may construct a homotopy $\sigma^t = \big(\sigma(A)^t,\ \sigma_{R^{n-1}}(\mathcal{A})\big)$, $\sigma(A)^t\,(x',\ x_n,\ \xi) = \sigma(\mathcal{A})\,(x',\ x_n - tx_n\varphi(x_n),\ \xi)$, $0 \leq \varphi \leq 1$, $\varphi \in C_0^\infty(\overline{R}_+)$, $\varphi = 1$ near the origin. Then $\mathrm{ind}\,\mathcal{A}_0 = \mathrm{ind}\,\mathcal{A}_1$ and hence, formula (3) holds in the general case.

4.1.2.3. Constructions for Manifolds

In this section we shall use the notions introduced in 4.1.2.1 in order to prove an index formula for elliptic boundary value problems on manifolds. We start from a given elliptic operator in $\mathfrak{G}(X, Y)$

$$\mathcal{A}:\ \begin{matrix} C^\infty(X, E) \\ \oplus \\ C^\infty(Y, J)\end{matrix} \ \to\ \begin{matrix} C^\infty(X, F) \\ \oplus \\ C^\infty(Y, G)\end{matrix}$$

(E, $F \in \mathrm{Vect}\,(X)$, J, $G \in \mathrm{Vect}\,(Y)$).

In 4.1.2.1 we have associated a family of elliptic tupels d_A on T^*X to a given elliptic PDO on a manifold without boundary. 4.1.2.1, Theorem 6 gives a relation between ind A and the Chern character ch d_A. In a similar way we proceed now for elliptic boundary value problems in $\mathfrak{G}(X, Y)$.

Let $\mathcal{A} \in \mathfrak{G}(X, Y)$ be elliptic with the principal symbol $\sigma(\mathcal{A}) = \big(\sigma(A), \sigma_Y(\mathcal{A})\big)$,

$$\sigma^0(A): \pi^*E \to \pi^*F\,,$$

$\pi\colon T^*X \to X$, E, $F \in \mathrm{Vect}\,(X)$, being an isomorphism outside a compact set in T^*X and

$$\sigma_Y^0(\mathcal{A}): \pi_Y^*E' \otimes H^+ \oplus \pi_Y^*J \to \pi_Y^*F' \otimes H^+ \oplus \pi_Y^*G\,,$$

$\pi_Y\colon T^*Y \to Y$, J, $G \in \mathrm{Vect}\,(Y)$ is an isomorphism outside a compact set in T^*Y. As in 4.1.1.1 we consider here fixed extensions of the principal symbols from the sphere bundles to the whole cotangent bundles.

Similarly as in 4.1.2.1 the interior principal symbol $\sigma^0(A)$ corresponds to an elliptic family $d_t(x, \xi)$ with compact support on T^*X with values in the $N \times N$ matrices. $d_t(x, \xi)$ is called *interior elliptic family.*

In view of 4.1.1.3, Remark 4 we can assume (if necessary after direct addition of operators the symbol of which is an isomorphism on the whole T^*X and T^*Y, respectively) that $\sigma^0(A)$ is an isomorphism on $T^*X|_U$, where U is a sufficiently small neighbourhood of the boundary Y. Hence the support of $d_t(x, \xi)$ is compact in $T^*\Omega$.

Denote by $p^{E'}(x')$, $p^{F'}(x')$, $p^J(x')$, $p^G(x')$ projector valued functions on Y defined as in 4.1.2.1 from the bundles E', F', J, $G \in \operatorname{Vect}(Y)$, $E' = E|_Y$ $F' = F|_Y$. They take values in $\operatorname{hom}(\mathbb{C}^N, \mathbb{C}^N)$. Consider

$$\begin{matrix} p^{E'}(x') \otimes 1 \\ \oplus \\ p^J(x') \end{matrix} : \begin{matrix} \mathbb{C}^N \otimes H^+ \\ \oplus \\ \mathbb{C}^N \end{matrix} \to \begin{matrix} \mathbb{C}^N \otimes H^+ \\ \oplus \\ \mathbb{C}^N \end{matrix}$$

and

$$\begin{matrix} p^{F'}(x') \otimes 1 \\ \oplus \\ p^G(x') \end{matrix} : \begin{matrix} \mathbb{C}^N \otimes H^+ \\ \oplus \\ \mathbb{C}^N \end{matrix} \to \begin{matrix} \mathbb{C}^N \otimes H^+ \\ \oplus \\ \mathbb{C}^N \end{matrix},$$

where 1 denotes the identity operator in H^+. Obviously,

$$(p^{E'}(x') \otimes 1 \oplus p^J(x'))^2 = p^{E'}(x') \otimes 1 \oplus p^J(x')$$

and

$$(p^{F'}(x') \otimes 1 \oplus p^G(x'))^2 = p^{F'}(x') \otimes 1 \oplus p^G(x')$$

for any $x' \in Y$, i.e. they are projector-valued functions.

Let $\sigma_Y^0(\mathcal{A})(x', \xi')$ be the principal boundary symbol of $\mathcal{A} \in \mathfrak{G}(X, Y)$,

$$\sigma_Y^0(\mathcal{A})(x', \xi') : \begin{matrix} \pi_Y^* E' \otimes H^+ \\ \oplus \\ \pi_Y^* J \end{matrix} \to \begin{matrix} \pi_Y^* F' \otimes H^+ \\ \oplus \\ \pi_Y^* G \end{matrix},$$

$\pi_Y \colon T^*Y \to Y$ (cf. 2.3.3.1, Theorem 3). In view of 2.2.5.1, Definition 1 we have

$$\sigma_Y^0(\mathcal{A})(x', \xi') = (\varkappa^*_{\langle\xi'\rangle^{-1}} \oplus 1)\, \sigma_Y^0(\mathcal{A})_{[0]}(x', \xi')\, (\varkappa^*_{\langle\xi'\rangle} \oplus 1),$$

where $\sigma_Y^0(\mathcal{A})_{[0]}(x', \xi') \in \mathfrak{R}^{m,d} \otimes S^m$. In other words, there is a decomposition

$$\sigma_Y^0(\mathcal{A})_{[0]}(x', \xi') = \sum_k a_k(x', \xi')\, e_k$$

converging in $\mathfrak{R}^{m,d} \otimes S^m$, $e_k \in \mathfrak{R}^{m,d}$ and $a_k \colon \pi_Y^* E' \oplus \pi_Y^* J \to \pi_Y^* F' \oplus \pi_Y^* G$. Define $a_k'(x', \xi') \colon \mathbb{C}^N \oplus \mathbb{C}^N \to \mathbb{C}^N \oplus \mathbb{C}^N$ by 4.1.2.1. (1) with local trivializations of the bundles $E' \oplus J$ and $F' \oplus G$. Set $\sigma_Y(\mathcal{A})_{[0]}(x', \xi') = \sum_k a_k'(x', \xi')\, e_k$ and

$$\sigma_Y(\mathcal{A})(x', \xi') = (\varkappa^*_{\langle\xi'\rangle^{-1}} \oplus 1)\, \sigma_Y(\mathcal{A})_{[0]}(x', \xi')\, (\varkappa^*_{\langle\xi'\rangle} \oplus 1).$$

This boundary symbol acts between trivial bundles

$$\sigma_Y(\mathcal{A})(x', \xi') : \begin{matrix} \mathbb{C}^N \otimes H^+ \\ \oplus \\ \mathbb{C}^N \end{matrix} \to \begin{matrix} \mathbb{C}^N \otimes H^+ \\ \oplus \\ \mathbb{C}^N \end{matrix}. \tag{1}$$

Let

$$\sigma_Y^0(\mathcal{R})(x', \xi') : \begin{matrix} \pi_Y^* F' \otimes H^+ \\ \oplus \\ \pi_Y^* G \end{matrix} \to \begin{matrix} \pi_Y^* E' \otimes H^+ \\ \oplus \\ \pi_Y^* J \end{matrix}$$

be the inverse of $\sigma_Y^0(\mathcal{A})(x', \xi')$ outside a compact set in T^*Y. Denote by

$$\sigma_Y(\mathcal{R})(x', \xi'): \begin{matrix} \mathbb{C}^N \otimes H^+ \\ \oplus \\ \mathbb{C}^N \end{matrix} \to \begin{matrix} \mathbb{C}^N \otimes H^+ \\ \oplus \\ \mathbb{C}^N \end{matrix}$$

the boundary symbol associated with $\sigma_Y^0(\mathcal{R})$ by the above procedure. Then

$$p^{E'} \otimes 1 \oplus p^J - \sigma_Y(\mathcal{R})(x', \xi')\, \sigma_Y(\mathcal{A})(x', \xi') = 0$$

and

$$p^{F'} \otimes 1 \oplus p^G - \sigma_Y(\mathcal{A})(x', \xi')\, \sigma_Y(\mathcal{R})(x', \xi') = 0$$

outside a compact set in T^*Y.

Set

$$d_b(x', \xi') = \big(p^{E'}(x') \otimes 1 \oplus p^J(x'), p^{F'}(x') \otimes 1 \oplus p^G(x'), \sigma_Y(\mathcal{A})(x', \xi'), \sigma_Y(\mathcal{R})(x', \xi')\big),$$

where 1 denotes the identity operator in H^+. The above considerations show that $d_b(x', \xi')$ defines an elliptic family on T^*Y with values in boundary symbols on the line of the form (1).

Consider an elliptic tupel of principal symbols in $\mathfrak{S}(X, Y)$, i.e. an elliptic family of interior principal symbols

$$d_i(x, \xi) = \big(p_i^{(1)}(x), p_i^{(2)}(x), \sigma(A)(x, \xi), \sigma(R)(x, \xi)\big),$$

$(x, \xi) \in T^*X$, with values in hom $(\mathbb{C}^N, \mathbb{C}^N)$ and an elliptic family of principal boundary symbols

$$d_b(x', \xi') = \big(p_b^{(1)}(x'), p_b^{(2)}(x'), \sigma_Y(\mathcal{A})(x', \xi'), \sigma_Y(\mathcal{R})(x', \xi')\big),$$

$(x', \xi') \in T^*Y$, with values in $\mathfrak{R}$, the boundary symbols on the line of the form (1). Note that the elements of $d_i(x, \xi)$ and $d_b(x', \xi')$ are compatible in an obvious sense.

The matrix algebra hom $(\mathbb{C}^N, \mathbb{C}^N)$ has the usual trace functional and $\operatorname{Tr} AB = \operatorname{Tr} BA$ for $A, B \in$ hom $(\mathbb{C}^N, \mathbb{C}^N)$. For a boundary symbol $\mathfrak{a} \in \mathfrak{R}^{m,d}$ on the line

$$\mathfrak{a} = \begin{pmatrix} \Pi^+\sigma(A) + \Pi'\sigma(B) & \sigma(K) \\ \Pi'\sigma(T) & \sigma(G) \end{pmatrix}$$

a trace functional is defined by

$$\operatorname{Tr}' \mathfrak{a} = \operatorname{Tr} \Pi'\sigma(B)(\nu, \nu) + \operatorname{Tr} \sigma(Q). \tag{2}$$

The calculations in 4.1.1.2 show that Tr′ does not vanish on commutators. Thus we replace the elliptic family $d_b(x', \xi')$ by an elliptic family $d_b'(x', \xi')$ taking values in $\mathfrak{S}(I, \{0\})$, $I = [0, 1)$.

Define a trace functional on $\mathfrak{S}(I, \{0\})$

$$\operatorname{Tr}_b \sigma = \int_I \operatorname{Tr} \Pi_\nu' \sigma(A)(x_n, \nu)\, dx_n + \operatorname{Tr}' \sigma_b, \tag{3}$$

$\sigma = \big(\sigma(A), \sigma_b\big) \in \mathfrak{S}(I, \{0\})$, Tr′ defined by (2). Obviously, (3) is defined if $\sigma(A)$ decreases sufficiently fast for $|\nu| \to \infty$.

Define the composition $\sigma^1 \circ \sigma^2 = \big(\sigma(A), \sigma_b\big)$ of symbols $\sigma^i = \big(\sigma(A)^i, \sigma_b^i\big) \in \mathfrak{S}(I, \{0\})$, $i = 1, 2$, by

$$\sigma(A)(x_n, \nu) = \sigma(A)^1 \circ \sigma(A)^2 = \sigma(A)^1(x_n, \nu)\, \sigma(A)^2(x_n, \nu) + \partial_\nu \sigma(A)^1(x_n, \nu)\, D_{x_n} \sigma(A)^2(x_n, \nu), \quad \sigma_b = \sigma_b^1 \sigma_b^2.$$

Define the elliptic family $d'_b(x', \xi')$ with values in $\mathfrak{S}(I, \{0\})$ in the following way. $d'_b(x', \xi')$ consists of the elliptic family of boundary symbols $d_b(x', \xi')$ and the family of interior symbols $d''(x', \xi') = \big(p^E(x'), p^F(x'), \sigma'_A(x', x_n, \xi', \nu), \sigma'_R(x', x_n, \xi', \nu)\big)$. The matrix functions $\sigma'_A(x', x_n, \xi', \nu)$ and $\sigma'_R(x', x_n, \xi', \nu)$ are determined by the values of the interior symbols σ_A, σ_R involved in $d_i(x, \xi)$ on $x_n = 0$. In view of the compatibility of interior and boundary symbols they are determined by $d_b(x', \xi')$, too.

Let $\varphi \in C_0^\infty(I)$ be an arbitrary function, $\varphi(x_n) = 0$ near 1 and $\varphi(x_n) = 1$ near 0. Then we define

$$\sigma'_A(x', x_n, \xi', \nu) = \varphi(x_n)\, \sigma_A(x', 0, \xi', \nu)\,,$$

$$\sigma'_R(x', x_n, \xi', \nu) = \varphi(x_n)\, \sigma_R(x', 0, \xi', \nu)\,.$$

The compatibility of the interior and boundary symbols in $d'_b(x', \xi')$ is obvious. Moreover, for arbitrarily fixed x', ξ' the tupel $d'_b = (p_1, p_2, a, r)$ is elliptic. In fact, we have

$$\begin{aligned}\mathrm{Tr}_b\,(p_1 - ra) &= \int_I \mathrm{Sp}\,\Pi'_\nu(p^F - \sigma'_R \circ \sigma'_A)\, dx_n \\ &\quad + \mathrm{Tr}'\,\big(p^E \otimes 1 \oplus p^J - \sigma_Y(\mathcal{R})\,\sigma_Y(\mathcal{A})\big) \\ &= \mathrm{i}\,\mathrm{Sp}\,\Pi'_\nu\big(\partial_\nu \sigma_R(x', 0, \xi', \nu)\, \partial_{x_n}\sigma_A(x', 0, \xi', \nu)\big) \\ &\quad + \mathrm{Tr}'\,\big(p^E \otimes 1 \oplus p^J - \sigma_Y(\mathcal{R})\,\sigma_Y(\mathcal{A})\big)\,.\end{aligned}$$

Since σ_A and σ_R have the transmission property the right hand side is well-defined. The same argument applies to $\mathrm{Tr}_b\,(p_2 - ar)$.

The advantage of working with symbols in $\mathfrak{S}(I, \{0\})$ and the trace functional Tr_b rather than with boundary symbols and the trace functional Tr' is that Tr_b vanishes on commutators. This is the contents of the following

Lemma 1. *Let $\sigma^i = (\sigma^i_A, \sigma^i_b) \in \mathfrak{S}(I, \{0\})$, $i = 1, 2$, be given, $\sigma^i_A(x_n, \nu)$ $N \times N$ matrix functions. Assume that $\sigma^i_A(x_n, \nu)$ have the transmission property for all $x_n \in I$. Then $\mathrm{Tr}_b\, \sigma^1\sigma^2$ and $\mathrm{Tr}_b\, \sigma^2\sigma^1$ are defined and equal, i.e. $\mathrm{Tr}_b\,(\sigma^1\sigma^2 - \sigma^2\sigma^1) = 0$.*

Proof: $\mathrm{Tr}_b\, \sigma^1\sigma^2$ and $\mathrm{Tr}_b\, \sigma^2\sigma^1$ are defined in view of the transmission property. The last assertion follows in the same way as 4.1.1.2, Lemma 10. The only difference is that one has to forget the x', ξ' variables. □

Now we use the results of Fedosov quoted in 4.1.2.1. There is defined a closed differential form $\mathrm{ch}'\, d'_b(x', \xi')$ on T^*Y. Note that, in general, it has not a compact support with respect to ξ'. If P is a projector-valued function, the equality $d\big(P(dP)^{2k-1}\big) = d\big(P^2(dP)^{2k-1}\big) = dPP(dP)^{2k-1} + P(dP)^{2k}$ implies $d\big(\mathrm{Tr}\,(P(dP)^{2k-1})\big) = \mathrm{Tr}\,\big(dP\, P(dP)^{2k-1}\big) + \mathrm{Tr}\,\big(P(dP)^{2k}\big) = 2\,\mathrm{Tr}\,\big(P(dP)^{2k}\big)$. Since $(PdPdP)^k = P(dP)^{2k}$, it follows that the Chern character $\mathrm{ch}'\, d'_b$ is an exact form, but without compact support. $\mathrm{Tr}_b\,\big(P(dP)^{2n-1}\big)$ is a sum of positively homogeneous terms of order $\leqq 0$ for large $|\xi'|$. Hence the integral of $\mathrm{ch}'\, d'_b$ over T^*Y is finite.

Let $\mathcal{A} \in \mathfrak{G}(X, Y)$ be an elliptic operator of order m. Denote by $d_i(\mathcal{A})\,(x, \xi)$ the elliptic family $\big(p^E(x), p^F(x), \sigma(A)\,(x, \xi), \sigma(R)\,(x, \xi)\big)$ of $N \times N$ matrices, $(x, \xi) \in T^*X$. After stabilization we can assume that the support of $d_i(\mathcal{A})\,(x, \xi)$ is compact in $T^*\Omega$, Ω the interior of X. Let $d'_b(\mathcal{A})\,(x', \xi')$ be the elliptic family defined using the boundary symbol as above. It takes values in the algebra $\mathfrak{S}(I, \{0\})$ with the trace functional Tr_b.

Denote by $\operatorname{ch}' d_b'(\mathcal{A})(x', \xi')$ the Chern character of this elliptic family on T^*Y with the trace functional Tr_b. Let $\operatorname{ch} d_i(\mathcal{A})(x, \xi)$ be the Chern character of the elliptic family $d_i(\mathcal{A})(x, \xi)$ on T^*X defined with the usual trace functional on $N \times N$ matrices.

Theorem 2. *Let $\mathcal{A} \in \mathfrak{G}(X, Y)$ be elliptic and assume that X admits an embedding into $\mathbb{R}^N$ ($N \in \mathbb{Z}_+$ sufficiently large) with trivial normal bundle. Let $d_i(\mathcal{A})(x, \xi)$ and $d_b'(\mathcal{A})(x', \xi')$ be elliptic tupels defined as above. Then*

$$\operatorname{ind} \mathcal{A} = \int_{T^*X} \operatorname{ch} d_i(\mathcal{A})(x, \xi) + \int_{T^*Y} \operatorname{ch}' d_b'(\mathcal{A})(x', \xi') \tag{4}$$

where the orientation of T^*X is given by the form $d\xi_1 \, dx_1 \dots d\xi_n \, dx_n$ and on T^*Y the induced orientation is taken.

The index formula (4) is similar so that formula announced in Fedosov [5] for families of elliptic boundary value problems.

The proof of Theorem 2 corresponds to the proof of the Index Theorem for elliptic operators in $\mathfrak{G}$ given in 3.2.2.4 and reduces the given case to the case of operators on manifolds without boundary by stable homotopies. Sometimes it seems to be more natural to proceed in the spirit of the 'embedding proof' of the Index Theorem in the boundaryless case. An essential part of this proof should be the external multiplication of elliptic boundary value problems with elliptic PDOs on a closed manifold. This is outlined in 3.2.3.2.

Proof: First note that both terms on the right hand side of (4) are invariant under homotopies preserving the property that the support of $d_i(\mathcal{A})(x, \xi)$ is compact in $T^*\Omega$ (Ω the interior of X), and the properties of $d_b'(\mathcal{A})(x', \xi')$ (cf. 4.1.2.1, Theorem 5). Obviously,

$$\operatorname{ch} d_i(\mathcal{A} \oplus \mathcal{A}') = \operatorname{ch} d_i(\mathcal{A}) + \operatorname{ch} d_i(\mathcal{A}'),$$
$$\operatorname{ch} d_b'(\mathcal{A} \oplus \mathcal{A}') = \operatorname{ch} d_b'(\mathcal{A}) + \operatorname{ch} d_b'(\mathcal{A}')$$

and

$$\operatorname{ch} d_i(\mathcal{A}\mathcal{A}') = \operatorname{ch} d_i(\mathcal{A}) + \operatorname{ch} d_i(\mathcal{A}'),$$
$$\operatorname{ch} d_b'(\mathcal{A}\mathcal{A}') = \operatorname{ch} d_b'(\mathcal{A}) + \operatorname{ch} d_b'(\mathcal{A}').$$

Moreover, the support of the interior family and the boundary family of elliptic tupels is empty for the operators in 3.1.2.1. (14). This follows in the same way as 4.1.1.3, Remark 5. Then, in the same way as in the proof of 3.2.2.4, Theorem 1, we are reduced to the case of an elliptic boundary value problem $\mathcal{A} = r^+ A \oplus Q$. In this case the assertion follows from 4.1.2.1, Theorem 6. □

The proof of Theorem 2 shows that an arbitrary mapping assigning to elliptic elements of $\mathfrak{G}$ pairs in $H^*(T^*\Omega) \times H^*(T^*Y)$ with properties as in 3.2.2.4, Theorem 1 yields the desired cohomology classes. It would be interesting to express them for classical boundary problems of differential geometry in terms of local curvature invariants.

4.2. Non-Elliptic Boundary Value Problems

Let $r^+A : C^\infty(X, E) \to C^\infty(X, F)$ be an elliptic pseudo-differential operator on X and $\mathcal{A} \in \mathfrak{G}$ an operator with r^+A (plus a Green operator) in the left upper corner. $\mathcal{A}$ is then called a non-elliptic boundary value problem for r^+A if the Shapiro-Lopatinski condition (i.e. the bijectivity of $\sigma_Y(\mathcal{A})_{(y,\eta)}$, $(y, \eta) \in S^*Y$) is not satisfied for y in a non-empty subset $Z \subset Y$. A classical example of such a situation is the oblique derivative problem for the Laplacian (cf. 4.2.2.2). Under suitable assumptions on the character of degeneracy of the Shapiro-Lopatinski condition one can formulate additional trace and potential conditions with respect to Z so that a certain new operator connected with $\mathcal{A}$ has a parametrix. This shall be discussed in the following sections. Since formally we often have to deal with matrices of operators, the symbols r^+, r' shall be dropped (i.e. we write e.g. A, B, T instead of r^+A, $r'B$, $r'T$).

The theory of various classes of degenerate boundary value problems or related questions (e.g. interior boundary value problems) are treated for instance in BICADZE [1, 3], HÖRMANDER [3], VIŠIK [2], EGOROV/KONDRAT'EV [1], ÈSKIN [1, 2], SJÖSTRAND [1, 2], EGOROV [1], SOGA [1, 2], MELIN/SJÖSTRAND [1, 2], SCHULZE [8], PILLAT/SCHULZE [2].

4.2.1. Certain Fredholm Operators Connected with Non-Elliptic Boundary Problems

4.2.1.1. Reduction to the Boundary

A standard method for studying non-elliptic boundary problems is the reduction to the boundary (cf. also 3.2.1.3). For abbreviation, let

$$\mathcal{E} = C^\infty(X, E), \quad \mathcal{F} = C^\infty(X, F), \quad \mathcal{J} = C^\infty(Y, J), \quad \mathcal{G} = C^\infty(Y, G).$$

Analogous notations are used with indices (e.g. $\mathcal{G}_0 = C^\infty(Y, G_0)$). Let $A: \mathcal{E} \to \mathcal{F}$ be elliptic and

$$\mathcal{A} = \begin{pmatrix} a & K \\ T & Q \end{pmatrix} : \begin{matrix} \mathcal{E} \\ \oplus \\ \mathcal{J} \end{matrix} \to \begin{matrix} \mathcal{F} \\ \oplus \\ \mathcal{G} \end{matrix}, \tag{1}$$

where $a = A + B_0$ and B_0 a Green operator. Suppose once and for all that there exists an elliptic boundary value problem $\mathcal{A}_0 \in \mathfrak{G}$

$$\mathcal{A}_0 = \begin{pmatrix} a & K_0 \\ T_0 & Q_0 \end{pmatrix} : \begin{matrix} \mathcal{E} \\ \oplus \\ \mathcal{J}_0 \end{matrix} \to \begin{matrix} \mathcal{F} \\ \oplus \\ \mathcal{G}_0 \end{matrix} \tag{2}$$

(the operators a in (1) and (2) are the same).

The orders of the operators in $\mathcal{A}$ and $\mathcal{A}_0$ are not essential. Reduction of the order by multiplication of (1) from both sides by suitable elliptic elements in $\mathfrak{G}$ (cf. 3.1.2.1) does not change the behaviour of the considered operator. Thus, without loss of generality, we can assume $\operatorname{ord} K_0 = \operatorname{ord} K = \lambda + \frac{1}{2}$, $\operatorname{ord} T_0 = \operatorname{ord} T = \gamma + \frac{1}{2}$ and

$$\operatorname{ord} Q = 1 - \alpha + \gamma + \lambda = 0 \tag{3}$$

($\alpha = \operatorname{ord} a$).

Let $Z \subset Y$ be a subset and $\sigma_Y(\mathcal{A})_{(y,\eta)}$, $(y, \eta) \in S^*Y$, bijective on $y \in Y \setminus Z$. From $\mathcal{A}$ we pass to the operator

$$\tilde{\mathcal{A}} = \begin{pmatrix} a & K & K_0 \\ T & Q & 0 \\ 0 & 0 & 1 \end{pmatrix} : \begin{matrix} \mathcal{E} \\ \oplus \\ \mathcal{J} \\ \oplus \\ \mathcal{J}_0 \end{matrix} \to \begin{matrix} \mathcal{F} \\ \oplus \\ \mathcal{G} \\ \oplus \\ \mathcal{J}_0 \end{matrix} \tag{4}$$

(1 denotes the identical operator in $\mathcal{J}_0$). Obviously $\sigma_Y(\tilde{\mathcal{A}})_{(y,\eta)}$ is also bijective for all $y \in Y \setminus Z$. Using the potential K of $\mathcal{A}$ we define the elliptic boundary problem

$$\tilde{\mathcal{A}}_0 = \begin{pmatrix} a & K & K_0 \\ 0 & 1 & 0 \\ T_0 & 0 & Q_0 \end{pmatrix} : \begin{matrix} \mathcal{E} \\ \oplus \\ \mathcal{J} \\ \oplus \\ \mathcal{J}_0 \end{matrix} \to \begin{matrix} \mathcal{F} \\ \oplus \\ \mathcal{J} \\ \oplus \\ \mathcal{G}_0 . \end{matrix} \tag{5}$$

Let

$$\mathcal{P}_0 = \begin{pmatrix} p_0 & L_0 \\ S_0 & R_0 \end{pmatrix} : \begin{matrix} \mathcal{F} \\ \oplus \\ \mathcal{G}_0 \end{matrix} \to \begin{matrix} \mathcal{E} \\ \oplus \\ \mathcal{J}_0 \end{matrix} \tag{6}$$

be a parametrix of $\mathcal{A}_0$ (i.e. $\mathcal{A}_0\mathcal{P}_0 - \mathcal{I} \in \mathfrak{G}^{-\infty}$, $\mathcal{P}_0\mathcal{A}_0 - \mathcal{I} \in \mathfrak{G}^{-\infty}$, $p_0 = P + N_0$, $\sigma_P = \sigma_A^{-1}$, N_0 a certain Green operator). Then

$$\tilde{\mathcal{P}}_0 = \begin{pmatrix} p_0 & -p_0K & L_0 \\ 0 & 1 & 0 \\ S_0 & -S_0K & R_0 \end{pmatrix} : \begin{matrix} \mathcal{F} \\ \oplus \\ \mathcal{J} \\ \oplus \\ \mathcal{G}_0 \end{matrix} \to \begin{matrix} \mathcal{E} \\ \oplus \\ \mathcal{J} \\ \oplus \\ \mathcal{J}_0 \end{matrix} \tag{7}$$

is a parametrix of $\tilde{\mathcal{A}}_0$. Define the following pseudo-differential operator over Y

$$\delta = \begin{pmatrix} -Tp_0K + Q & TL_0 \\ -S_0K & R_0 \end{pmatrix} : \begin{matrix} \mathcal{J} \\ \oplus \\ \mathcal{G}_0 \end{matrix} \to \begin{matrix} \mathcal{G} \\ \oplus \\ \mathcal{J}_0 . \end{matrix} \tag{8}$$

Since ord $S_0 = -\lambda - \frac{1}{2}$, ord $L_0 = -\gamma - \frac{1}{2}$, ord $R_0 = \alpha - \lambda - \gamma - 1 = 0$, δ is an operator of order zero.

Lemma 1. *Let $(y, \eta) \in S^*Y$ and $\sigma_\delta(y, \eta)$ be the homogeneous principal symbol of δ (in the corresponding local coordinates). The following conditions are equivalent*

(i) *σ_δ is elliptic at (y, η),*

(ii) *$\sigma_Y(\mathcal{A})_{(y,\eta)}$ bijective.*

Proof: From $\mathcal{A}_0\mathcal{P}_0 - \mathcal{I} \in \mathfrak{G}^{-\infty}$ follows mod $\mathfrak{G}^{-\infty}$

$$\tilde{\mathcal{A}}\tilde{\mathcal{P}}_0 = \begin{pmatrix} 1 & 0 & 0 \\ Tp_0 & -Tp_0K + Q & TL_0 \\ S_0 & -S_0K & R_0 \end{pmatrix}.$$

A corresponding equation holds on boundary symbol level (since $\sigma_Y(\mathcal{A}_0)\,\sigma_Y(\mathcal{P}_0)$ is equal to the identity in $p^*(F^+ \oplus G_0)$). The bijectivity of $\sigma_Y(\mathcal{A})_{(y,\eta)}$ at a point is equivalent to that of $\sigma_Y(\tilde{\mathcal{A}})_{(y,\eta)}$. This is equivalent to the bijectivity of $\sigma_Y(\tilde{\mathcal{A}}\tilde{\mathcal{P}}_0)_{(y,\eta)}$ (since $\sigma_Y(\mathcal{P}_0)$ is an isomorphism). The bijectivity of $\sigma_Y(\tilde{\mathcal{A}}\tilde{\mathcal{P}}_0)_{(y,\eta)}$ is obviously equivalent to the bijectivity of $\sigma_\delta(y, \eta)$. This proves the lemma. □

The correspondence $\mathcal{A} \to \delta$ for fixed elliptic $\mathcal{A}_0$ is called *reduction to the boundary.* Lemma 1 says that δ is elliptic over $Y \setminus Z$. In the following we always assume that the order conditions on $\mathcal{A}$ and $\mathcal{A}_0$ are satisfied so that δ is an operator of order 0.

4.2.1.2. Additional Trace and Potential Conditions

In this section we show a connection between so called interior boundary value problems for the operator δ being Fredholm and certain Fredholm operators connected with $\mathcal{A}$ by adding trace and potential conditions with respect to Z.

Suppose that $\delta: \mathcal{J}_1 \to \mathcal{J}_2$ is a pseudo-differential operator over Y which is elliptic over $Y \setminus Z$. Let $\mathcal{L}, \mathcal{M}$ be Fréchet spaces and

$$\mathcal{D} = \begin{pmatrix} \delta & \varkappa \\ \tau & \varrho \end{pmatrix} : \begin{matrix} \mathcal{J}_1 \\ \oplus \\ \mathcal{L} \end{matrix} \to \begin{matrix} \mathcal{J}_2 \\ \oplus \\ \mathcal{M} \end{matrix} \tag{1}$$

be a continuous operator. Then $\mathcal{D}$ is called an *interior boundary problem* for δ with respect to Z. The meaning of this notation is that, in our concrete situations, $\mathcal{L}$, $\mathcal{M}$ are spaces of sections in bundles over Z (or over submanifolds of Z, cf. 4.2.2.1). For this reason we also use the notation "smoothing operator" instead of "compact operator" in the considered spaces. The following calculations are valid modulo smoothing operators. For abbreviation write simply equations. Let

$$\delta = \begin{pmatrix} \delta_{11} & \delta_{12} \\ \delta_{21} & \delta_{22} \end{pmatrix} = \begin{pmatrix} -Tp_0K + Q & TL_0 \\ -S_0K & R_0 \end{pmatrix} : \begin{matrix} \mathcal{J} \\ \oplus \\ \mathcal{G}_0 \end{matrix} \to \begin{matrix} \mathcal{G} \\ \oplus \\ \mathcal{J}_0 \end{matrix} \tag{2}$$

(cf. 4.2.1.1.(8)) and

$$\mathcal{D} = \begin{pmatrix} \delta_{11} & \delta_{12} & \varkappa_1 \\ \delta_{21} & \delta_{22} & \varkappa_2 \\ \tau_1 & \tau_2 & \varrho \end{pmatrix} : \begin{matrix} \mathcal{J} \\ \oplus \\ \mathcal{G}_0 \\ \oplus \\ \mathcal{L} \end{matrix} \to \begin{matrix} \mathcal{G} \\ \oplus \\ \mathcal{J}_0 \\ \oplus \\ \mathcal{M} \end{matrix} \tag{3}$$

an interior boundary problem for δ. Then we have the following

Theorem 1. *Let $\mathcal{A} \in \mathfrak{G}$ be a non-elliptic boundary problem of the form 4.2.1.1.(1) for an elliptic operator $A: \mathcal{E} \to \mathcal{F}$ and $\mathcal{A}_0 \in \mathfrak{G}$ an elliptic boundary problem of the form 4.2.1.1.(2). Suppose that for the operator δ defined by (2) there exists an interior boundary problem $\mathcal{D}$ of the form (3) which is Fredholm. Then*

$$\mathcal{B} = \begin{pmatrix} a & K & -K_0\varkappa_2 \\ T & Q & \varkappa_1 \\ \tau_2T_0 & \tau_1 & \varrho - \tau_2Q_0\varkappa_2 \end{pmatrix} : \begin{matrix} \mathcal{E} \\ \oplus \\ \mathcal{J} \\ \oplus \\ \mathcal{L} \end{matrix} \to \begin{matrix} \mathcal{F} \\ \oplus \\ \mathcal{G} \\ \oplus \\ \mathcal{M} \end{matrix} \tag{4}$$

is Fredholm and

$$\operatorname{ind} \mathcal{B} = \operatorname{ind} \mathcal{A}_0 + \operatorname{ind} \mathcal{D}\,. \tag{5}$$

Proof: The Fredholm property of $\mathcal{D}$ implies that of the operator

$$\tilde{\mathcal{C}} = \left(\begin{array}{c:c} 1 & 0 \\ \hdashline Tp_0 & \\ S_0 & \mathcal{D} \\ 0 & \end{array}\right) : \begin{matrix} \mathcal{F} \\ \oplus \\ \mathcal{J} \\ \oplus \\ \mathcal{G}_0 \\ \oplus \\ \mathcal{L} \end{matrix} \to \begin{matrix} \mathcal{F} \\ \oplus \\ \mathcal{G} \\ \oplus \\ \mathcal{J}_0 \\ \oplus \\ \mathcal{M}. \end{matrix}$$

Furthermore,

$$\begin{pmatrix} \tilde{\mathcal{A}}_0 & 0 \\ 0 & 1 \end{pmatrix} : \begin{matrix} \mathcal{E} \\ \oplus \\ \mathcal{J} \\ \oplus \\ \mathcal{J}_0 \\ \oplus \\ \mathcal{L} \end{matrix} \to \begin{matrix} \mathcal{F} \\ \oplus \\ \mathcal{J} \\ \oplus \\ \mathcal{G}_0 \\ \oplus \\ \mathcal{L} \end{matrix}$$

is a Fredholm operator. Thus also the operator

$$\tilde{\mathcal{B}} = \tilde{\mathcal{E}} \begin{pmatrix} \tilde{\mathcal{A}}_0 & 0 \\ 0 & 1 \end{pmatrix} : \begin{matrix} \mathcal{E} \\ \oplus \\ \mathcal{J} \\ \oplus \\ \mathcal{J}_0 \\ \oplus \\ \mathcal{L} \end{matrix} \to \begin{matrix} \mathcal{F} \\ \oplus \\ \mathcal{G} \\ \oplus \\ \mathcal{J}_0 \\ \oplus \\ \mathcal{M} \end{matrix} \tag{6}$$

is Fredholm and obviously

$$\operatorname{ind} \tilde{\mathcal{B}} = \operatorname{ind} \tilde{\mathcal{E}} + \operatorname{ind} \tilde{\mathcal{A}}_0 = \operatorname{ind} \mathcal{D} + \operatorname{ind} \mathcal{A}_0 . \tag{7}$$

Using $\mathcal{P}_0\mathcal{A}_0 - \mathcal{I} \in \mathfrak{G}^{-\infty}$ and (2) it follows from (6) that

$$\tilde{\mathcal{B}} = \begin{pmatrix} a & K & K_0 & 0 \\ T & Q & 0 & \varkappa_1 \\ 0 & 0 & 1 & \varkappa_2 \\ \tau_2 T_0 & \tau_1 & \tau_2 Q_0 & \varrho \end{pmatrix} \tag{8}$$

(note that the meaning of equations is here equivalence modulo smoothing operators). Changing the order of the last two rows and columns in $\tilde{\mathcal{B}}$ gives a Fredholm operator

$$\tilde{\mathcal{B}}_1 = \begin{pmatrix} a & K & 0 & K_0 \\ T & Q & \varkappa_1 & 0 \\ \tau_2 T_0 & \tau_1 & \varrho & \tau_2 Q_0 \\ 0 & 0 & \varkappa_2 & 1 \end{pmatrix} .$$

In view of 3.1.1.5, Proposition 1

$$\mathcal{B} = \begin{pmatrix} a & K & 0 \\ T & Q & \varkappa_1 \\ \tau_2 T_0 & \tau_1 & \varrho \end{pmatrix} - \begin{pmatrix} K_0 \\ 0 \\ \tau_2 Q_0 \end{pmatrix} (0, 0, \varkappa_2) ,$$

is a Fredholm operator, too. This is just the operator (4), and (5) is an immediate consequence of (7). This proves the Theorem. □

The left upper corner in (4) is equal to the given non-elliptic boundary value problem $\mathcal{A}$. The other matrix elements in (4) shall be interpreted as additional trace and potential conditions with respect to Z.

In order to calculate examples for operators $\mathcal{A}$ with degenerate Shapiro-Lopatinski condition for which interior boundary problems can be applied in the sense of Theorem 1, we want to go back from given δ and elliptic $\mathcal{A}_0$ to a corresponding $\mathcal{A}$. We restrict ourselves to a sufficiently interesting special case, namely, if

$$Q_0 = 0 . \tag{9}$$

Lemma 2. *Let $\mathcal{A}_0 \in \mathfrak{G}$ be elliptic, given in the form* 4.2.1.1. (2) *with the property* (9). *Then there exists a parametrix $\mathcal{P}_0 \in \mathfrak{G}$ of $\mathcal{A}_0$ with*

$$R_0 = 0 \tag{10}$$

(*here the notations of* 4.2.1.1. (6) *are used*).

Proof: It is clear that there are topological decompositions $\mathcal{E} = \mathcal{H} \oplus \mathcal{H}'$, $\mathcal{F} = \mathcal{T} \oplus \mathcal{T}'$ so that the operators $a|_{\mathcal{H}}: \mathcal{H} \to \mathcal{T}$, $\mathcal{A}_0: \mathcal{J}_0 \to \mathcal{T}'$, $T_0' = T_0|_{\mathcal{H}'}: \mathcal{H}' \to \mathcal{G}_0$ are Fredholm between the corresponding spaces. Denote by $p_0|_{\mathcal{T}}: \mathcal{T} \to \mathcal{H}$ a parametrix of $a|_{\mathcal{H}}$. Moreover, let $S_0': \mathcal{T}' \to \mathcal{J}_0$, $L_0: \mathcal{G}_0 \to \mathcal{H}'$ be parametrices of K_0 an T_0', respectively. Put $p_0 = p_0|_{\mathcal{T}} \circ q$, $S_0 = S_0' \circ q'$ ($q: \mathcal{T} \oplus \mathcal{T}' \to \mathcal{T}$, $q': \mathcal{T} \oplus \mathcal{T}' \to \mathcal{T}'$ the canonical projections). Then the matrix $\mathcal{P}_1$ formally given by 4.2.1.1.(6) with our p_0, L_0, S_0, and $R_0 = 0$ is a parametrix of $\mathcal{A}_0$. Since $\mathcal{A}_0$ is elliptic, $\mathcal{A}_0$ has a parametrix $\mathcal{P}_0 \in \mathfrak{G}$. If $\mathcal{P}_1$ is another parametrix of $\mathcal{A}_0$, we have $\mathcal{P}_1 \in \mathfrak{G}$ and $\mathcal{P}_0 - \mathcal{P}_1 \in \mathfrak{G}^{-\infty}$. This proves the Lemma. □

Theorem 3. *Let $\mathcal{A}_0 \in \mathfrak{G}$ be an elliptic boundary problem* (*of the form* 4.2.1.1.(2)) *with $Q_0 = 0$ and*

$$\delta = \begin{pmatrix} \delta_{11} & \delta_{12} \\ \delta_{21} & 0 \end{pmatrix} : \begin{matrix} \mathcal{J} \\ \oplus \\ \mathcal{G}_0 \end{matrix} \to \begin{matrix} \mathcal{G} \\ \oplus \\ \mathcal{J}_0 \end{matrix} \tag{11}$$

a pseudo-differential operator over Y of order 0 *being elliptic over $Y \setminus Z$* (*the bundles J, G are arbitrarily given*). *Suppose that for* (11) *there exists a Fredholm interior boundary problem $\mathcal{D}$ of the form* (3). *Moreover, let $T_1: \mathcal{E} \to \mathcal{G}$, $K_1: \mathcal{J} \to \mathcal{F}$ be arbitrary operators in $\mathfrak{G}$* (*i.e. T_1 a trace, K_1 a potential operator*) *with* $\operatorname{ord} K_1 = \operatorname{ord} K_0$, $\operatorname{ord} T_1 = \operatorname{ord} T_0$. *Let $\mathcal{A} \in \mathfrak{G}$ be an operator of the form* 4.2.1.1.(1) *with*

$$K = ap_0K_1 - K_0\delta_{21}\,, \quad T = T_1p_0a + \delta_{12}T_0\,, \quad Q = T_1p_0K_1 + \delta_{11}\,. \tag{12}$$

Then $\sigma_Y(\mathcal{A})_{(y,\eta)}$ is bijective iff $\sigma_\delta(y, \eta)$ is bijective, $(y, \eta) \in S^*Y$. The reduction to the boundary of $\mathcal{A}$ by means of $\mathcal{A}_0$ reproduces the operator (11) *so that for $\mathcal{A}$ and* (11) *Theorem* 1 *can be applied.*

Proof: The Fredholm property of $\mathcal{D}$ implies that of the operator

$$\mathcal{C} = \begin{pmatrix} 1 & 0 \\ \zeta_1 & \\ \zeta_2 & \mathcal{D} \\ \zeta_3 & \end{pmatrix} : \begin{matrix} \mathcal{F} \\ \oplus \\ \mathcal{J} \\ \oplus \\ \mathcal{G}_0 \\ \oplus \\ \mathcal{L} \end{matrix} \to \begin{matrix} \mathcal{F} \\ \oplus \\ \mathcal{G} \\ \oplus \\ \mathcal{J}_0 \\ \oplus \\ \mathcal{M} \end{matrix}$$

for arbitrary continuous operators ζ_j ($j = 1, 2, 3$). Let $\zeta_1 = T_1p_0$, $\zeta_2 = S_0$, $\zeta_3 = 0$. Using K_1 we get an elliptic boundary problem

$$\tilde{\mathcal{A}}_1 = \begin{pmatrix} q & K_1 & K_0 \\ 0 & 1 & 0 \\ T_0 & 0 & 0 \end{pmatrix} : \begin{matrix} \mathcal{E} \\ \oplus \\ \mathcal{J} \\ \oplus \\ \mathcal{J}_0 \end{matrix} \to \begin{matrix} \mathcal{F} \\ \oplus \\ \mathcal{J} \\ \oplus \\ \mathcal{G}_0\,. \end{matrix}$$

Then there follows an operator

$$\mathscr{B}_1 = \mathscr{C}\begin{pmatrix} \tilde{\mathscr{A}}_1 & 0 \\ 0 & 1 \end{pmatrix} = \left(\begin{array}{ccc:c} & \tilde{\mathscr{A}}_2 & & \begin{matrix} 0 \\ \varkappa_1 \\ \varkappa_2 \end{matrix} \\ \hdashline \tau_2 T_0 & \tau_1 & 0 & \varrho \end{array}\right) : \begin{matrix} \mathscr{E} \\ \oplus \\ \mathscr{J} \\ \oplus \\ \mathscr{J}_0 \\ \oplus \\ \mathscr{L} \end{matrix} \to \begin{matrix} \mathscr{F} \\ \oplus \\ \mathscr{G} \\ \oplus \\ \mathscr{J}_0 \\ \oplus \\ \mathscr{M} \end{matrix}$$

with

$$\tilde{\mathscr{A}}_2 = \begin{pmatrix} a & K_1 & K_0 \\ T_1 p_0 a + \delta_{12} T_0 & T_1 p_0 K_1 + \delta_{11} & 0 \\ 0 & S_0 K_1 + \delta_{21} & 1 \end{pmatrix}.$$

Here $Q_0 = 0$, $R_0 = 0$, $S_0 a = 0$, $S_0 K_0 = 1$, $p_0 K_0 = 0$ was used. Next define the operator

$$\mathscr{A} = \begin{pmatrix} a & K_1 \\ T_1 p_0 a + \delta_{12} T_0 & T_1 p_0 K_1 + \delta_{11} \end{pmatrix} - \begin{pmatrix} K_0 \\ 0 \end{pmatrix} (0, S_0 K_1 + \delta_{21}),$$

which is obviously equal to the operator $\mathscr{A}$ in the assertion of the Theorem. Application of 3.1.1.5, Proposition 1 on boundary symbol level then yields that $\mathscr{A}$ satisfies the Shapiro-Lopatinski condition exactly at those points $(y, \eta) \in S^*Y$ where the homogeneous principal symbol of δ is elliptic. In order to finish the proof we have to show that reduction to the boundary reproduces the operator δ, i e.

$$\delta_{11} = -Tp_0 K + Q, \; \delta_{12} = TL_0, \; \delta_{21} = -S_0 K \tag{13}$$

for the operators (11), (12) ($\delta_{22} = 0$ is automatically fulfilled). Using $\mathscr{A}_0 \mathscr{P}_0 - \mathscr{I} \in \mathfrak{G}^{-\infty}$ we now obtain

$$\begin{aligned} -Tp_0 K + Q &= -(T_1 p_0 a + \delta_{12} T_0)\, p_0 (a p_0 K_1 - K_0 \delta_{21}) + T_1 p_0 K_1 + \delta_{11} \\ &= -T_1 p_0 a p_0 a p_0 K_1 + T_1 p_0 K_1 + \delta_{11} \\ &= -T_1 p_0 (1 - K_0 S_0)(1 - K_0 S_0) K_1 + T_1 p_0 K_1 + \delta_{11} \\ &= -T_1 p_0 K_1 + T_1 p_0 K_1 + \delta_{11} = \delta_{11}, \\ -S_0 K &= -S_0 (a p_0 K_1 - K_0 \delta_{21}) = \delta_{21}, \\ TL_0 &= (T_1 p_0 a + \delta_{12} T_0) L_0 = \delta_{12}. \end{aligned}$$

Thus Theorem 3 is proved. □

If we suppose $J = J_0$, $G = G_0$, the operator (11) is elliptic at $(y, \eta) \in S^*Y$ iff both δ_{12} and δ_{21} are elliptic at that point. This simplifies the construction of interior boundary problems for δ.

Furthermore, note that for an arbitrary elliptic pseudo-differential operator $A: \mathscr{E} \to \mathscr{F}$ possessing an elliptic boundary problem (cf. 3.1.1.1) there always exists a Green operator B_0 so that there is an elliptic $\mathscr{A}_0 \in \mathfrak{G}$ of the form 4.2.1.1.(2) with $a = A + B_0$ and $Q_0 = 0$ (cf. the constructions in 3.1.1.2).

In 4.2.2.1 classes of Fredholm interior boundary problems shall be described. Using this, Theorem 3 can be applied to the construction of a lot of concrete examples of degeneracies of the Shapiro-Lopatinski condition. A special example will be the oblique derivative problem for the Laplacian (cf. 4.2.2.2).

4.2.1.3. Construction of a Parametrix

In this section an expression for a parametrix of the Fredholm operator $\mathcal{B}$ described in 4.2.1.2, Theorem 1 shall be constructed. Let

$$\mathcal{D} = \begin{pmatrix} \delta & \varkappa \\ \tau & \varrho \end{pmatrix}, \quad \varkappa = \begin{pmatrix} \varkappa_1 \\ \varkappa_2 \end{pmatrix}, \quad \tau = (\tau_1, \tau_2), \tag{1}$$

where $\mathcal{D}$ is a given Fredholm interior boundary problem for δ. Denote by

$$\mathcal{D}^{-1} = \begin{pmatrix} \varepsilon & \eta \\ \sigma & \chi \end{pmatrix} \tag{2}$$

a parametrix of $\mathcal{D}$. Here

$$\varepsilon = \begin{pmatrix} \varepsilon_{11} & \varepsilon_{12} \\ \varepsilon_{21} & \varepsilon_{22} \end{pmatrix}, \quad \eta = \begin{pmatrix} \eta_1 \\ \eta_2 \end{pmatrix}, \quad \sigma = (\sigma_1, \sigma_2).$$

Moreover, set

$$\zeta = \begin{pmatrix} \zeta_1 \\ \zeta_2 \end{pmatrix} \overset{\text{def}}{=} \begin{pmatrix} Tp_0 \\ S_0 \end{pmatrix}. \tag{3}$$

The starting point are the Fredholm operators $\tilde{\mathcal{B}}$, $\tilde{\mathcal{E}}$ occurring in the proof of 4.2.1.2, Theorem 1. Denoting parametrices of the operators $\tilde{\mathcal{B}}$ and $\tilde{\mathcal{E}}$ by $\tilde{\mathcal{B}}^{-1}$ and $\tilde{\mathcal{E}}^{-1}$, respectively, we obviously have

$$\tilde{\mathcal{E}}^{-1} = \begin{pmatrix} 1 & 0 & 0 \\ -\nu\zeta & \varepsilon & \eta \\ -\sigma\zeta & \sigma & \chi \end{pmatrix}, \quad \tilde{\mathcal{B}}^{-1} = \begin{pmatrix} \tilde{\mathcal{P}}_0 & 0 \\ 0 & 1 \end{pmatrix} \mathcal{E}^{-1}.$$

Using the identiy $\mathcal{D}\mathcal{D}^{-1} = \mathcal{J}$ modulo smoothing operators and the expression 4.2.1.2. (2) we obtain

$$\tilde{\mathcal{B}}^{-1} = \begin{pmatrix} \beta_0 & \beta_1 & \beta' & \beta_2 \\ -(\varepsilon\zeta)_1 & \varepsilon_{11} & \varepsilon_{12} & \eta_1 \\ \varkappa_2\sigma\zeta & -\varkappa_2\sigma_1 & 1 - \varkappa_2\sigma_2 & -\varkappa_2\chi \\ -\sigma\zeta & \sigma_1 & \sigma_2 & \chi \end{pmatrix} \tag{4}$$

with $\beta' = -p_0K\varepsilon_{12} + L_0\varepsilon_{22}$ and

$$\beta_0 = p_0 + p_0K_0(\varepsilon\zeta)_1 - L_0(\varepsilon\zeta)_2, \tag{5}$$

$$\beta_1 = -p_0K\varepsilon_{11} + L_0\varepsilon_{21}, \quad \beta_2 = -p_0K\eta_1 + L_0\eta_2, \tag{6}$$

$$(\varepsilon\zeta)_k = \varepsilon_{k_1}\zeta_1 + \varepsilon_{k_2}\zeta_2, \quad k = 1, 2. \tag{7}$$

According to 3.1.1.5, Proposition 1 a parametrix of $\mathcal{B}$ can be obtained by dropping the third row and column in $\tilde{\mathcal{B}}^{-1}$ (here the role of the operator α in 3.1.1.5, Proposition 1 is played by the identity in $\mathcal{J}_0$ and that of ω by the operator $\mathcal{B}$). Thus we have proved the following

Theorem 1. *Let $\mathcal{B}$ be the Fredholm operator* 4.2.1.2.(4). *Then*

$$\mathcal{B}^{-1} = \begin{pmatrix} \beta_0 & \beta_1 & \beta_2 \\ -(\varepsilon\zeta)_1 & \varepsilon_{11} & \eta_1 \\ -\sigma\zeta & \sigma_1 & \chi \end{pmatrix} : \begin{matrix} \mathcal{F} \\ \oplus \\ \mathcal{G} \\ \oplus \\ \mathcal{M} \end{matrix} \to \begin{matrix} \mathcal{E} \\ \oplus \\ \mathcal{J} \\ \oplus \\ \mathcal{L} \end{matrix} \tag{8}$$

is a parametrix of $\mathcal{B}$ with β_j ($j = 0, 1, 2$) given by the formulas (5), (6).

Remark 2. Suppose that the operator 4.2.1.1. (1) has the special form

$$\mathcal{A} = \begin{pmatrix} a \\ \\ T \end{pmatrix} : \mathcal{E} \to \begin{matrix} \mathcal{F} \\ \oplus \\ \mathcal{G}. \end{matrix}$$

Let

$$\mathcal{A}_0 = \begin{pmatrix} a \\ \\ T_0 \end{pmatrix} : \mathcal{E} \to \begin{matrix} \mathcal{F} \\ \oplus \\ \mathcal{G}_0 \end{matrix}$$

be elliptic and $\mathcal{P}_0 = (p_0, L_0)$ a parametrix of $\mathcal{A}_0$. Then boundary reduction gives $\delta = TL_0 \colon \mathcal{G}_0 \to \mathcal{G}$. If then

$$\mathcal{D} = \begin{pmatrix} \delta & \varkappa \\ \\ \tau & \varrho \end{pmatrix} : \begin{matrix} \mathcal{G}_0 \\ \oplus \\ \mathcal{L} \end{matrix} \to \begin{matrix} \mathcal{G} \\ \oplus \\ \mathcal{M} \end{matrix}$$

is an interior Fredholm boundary problem for δ, the operator 4.2.1.2.(4) has the form

$$\mathcal{B} = \begin{pmatrix} a & 0 \\ T & \varkappa \\ \tau T_0 & \varrho \end{pmatrix} : \begin{matrix} \mathcal{E} \\ \oplus \\ \mathcal{L} \end{matrix} \to \begin{matrix} \mathcal{F} \\ \oplus \\ \mathcal{G} \\ \oplus \\ \mathcal{M}. \end{matrix}$$

A parametrix is then

$$\mathcal{B}^{-1} = \begin{pmatrix} \beta_0 & \beta_1 & \beta_2 \\ -\sigma\zeta & \sigma & \chi \end{pmatrix} = \begin{pmatrix} p_0 - L_0\varepsilon T p_0 & L_0\varepsilon & L_0\eta \\ -\sigma T p_0 & \sigma & \chi \end{pmatrix}. \tag{9}$$

The constructions in the sections 4.2.1.2 and 4.2.1.3 have a local version in the following sense. Let $\mathcal{A}$ be an operator in $\mathfrak{G}$ with respect to a neighbourhood U in X of a point y of the boundary Y and suppose that there exists an elliptic $\mathcal{A}_0$ in U. Then, similarly as in 4.2.1.2, Theorem 1, one can construct an operator $\mathcal{B}$ possessing a parametrix $\mathcal{B}^{-1}$ with respect to U, and given by an expression of the form (8). All assertions (except those containing the index as a finite number) remain valid if the Fredholm property is replaced by the existence of a parametrix in the considered neighbourhood. The details are left to the reader.

4.2.1.4. Reduction to the Interior Boundary

In this section we shall formulate a certain analogue of the Agranovič-Dynin formula in the case of degenerate boundary problems. Put for abbreviation

$$\mathcal{V} = \mathcal{J} \oplus \mathcal{G}_0, \quad \mathcal{W} = \mathcal{G} \oplus \mathcal{J}_0$$

und let

$$\mathcal{D}_k = \begin{pmatrix} \delta & \varkappa^k \\ \\ \tau^k & \varrho^k \end{pmatrix} : \begin{matrix} \mathcal{V} \\ \oplus \\ \mathcal{L}_k \end{matrix} \to \begin{matrix} \mathcal{W} \\ \oplus \\ \mathcal{M}_k \end{matrix} \tag{1}$$

($k = 1, 2$) be Fredholm interior boundary problems for the operator 4.2.1.2.(2). Then the operators

$$\tilde{\mathcal{D}}_1 = \begin{pmatrix} \delta & \varkappa^2 & \varkappa^1 \\ 0 & 1 & 0 \\ \tau^1 & 0 & \varrho^1 \end{pmatrix} : \begin{matrix} \mathcal{V} \\ \oplus \\ \mathcal{L}_2 \\ \oplus \\ \mathcal{L}_1 \end{matrix} \to \begin{matrix} \mathcal{W} \\ \oplus \\ \mathcal{L}_2 \\ \oplus \\ \mathcal{M}_1 \end{matrix},$$

$$\tilde{\mathcal{D}}_2 = \begin{pmatrix} \delta & \varkappa^2 & \varkappa^1 \\ \tau^2 & \varrho^2 & 0 \\ 0 & 0 & 1 \end{pmatrix} : \begin{matrix} \mathcal{V} \\ \oplus \\ \mathcal{L}_2 \\ \oplus \\ \mathcal{L}_1 \end{matrix} \to \begin{matrix} \mathcal{W} \\ \oplus \\ \mathcal{M}_2 \\ \oplus \\ \mathcal{L}_1 \end{matrix}$$

are Fredholm, too, and obviously $\operatorname{ind} \mathcal{D}_k = \operatorname{ind} \tilde{\mathcal{D}}_k$ ($k = 1, 2$). If

$$\mathcal{D}_1^{-1} = \begin{pmatrix} \varepsilon^1 & \eta^1 \\ \sigma^1 & \chi^1 \end{pmatrix}$$

is a parametrix of $\mathcal{D}_1$, a parametrix of $\tilde{\mathcal{D}}_1$ is given by

$$\tilde{\mathcal{D}}_1^{-1} = \begin{pmatrix} \varepsilon^1 & -\varepsilon^1\varkappa^2 & \eta^1 \\ 0 & 1 & 0 \\ \sigma^1 & -\sigma^1\varkappa^2 & \chi^1 \end{pmatrix}.$$

Since

$$\tilde{\mathcal{D}}_2\tilde{\mathcal{D}}_1^{-1} = \begin{pmatrix} 1 & 0 & 0 \\ \tau^2\varepsilon^1 & -\tau^2\varepsilon^1\varkappa^2 + \varrho^2 & \tau^2\eta^1 \\ \sigma^1 & -\sigma^1\varkappa^2 & \chi^1 \end{pmatrix}$$

is Fredholm, the operator

$$\mathcal{S} = \begin{pmatrix} -\tau^2\varepsilon^1\varkappa^2 + \varrho^2 & \tau^2\eta^1 \\ -\sigma^1\varkappa^2 & \chi^1 \end{pmatrix} : \begin{matrix} \mathcal{L}_2 \\ \oplus \\ \mathcal{M}_1 \end{matrix} \to \begin{matrix} \mathcal{M}_2 \\ \oplus \\ \mathcal{L}_1 \end{matrix} \tag{2}$$

is Fredholm, too, and

$$\operatorname{ind} \mathcal{S} = \operatorname{ind} \mathcal{D}_2 - \operatorname{ind} \mathcal{D}_1 . \tag{3}$$

In view of 4.2.1.2.(5) we thus proved the following

Theorem 1. *Let $\mathcal{B}_k$ ($k = 1, 2$) be operators of the form 4.2.1.2.(4) with $\mathcal{A}$ independent of k and interior Fredholm boundary problems $\mathcal{D}_k$ ($k = 1, 2$) for δ. Then there exists a Fredholm operator*

$$\mathcal{S}: \mathcal{L}_2 \oplus \mathcal{M}_1 \to \mathcal{M}_2 \oplus \mathcal{L}_1 \tag{4}$$

expressed in terms of the operators contained in $\mathcal{B}_1$, $\mathcal{B}_2$ with

$$\operatorname{ind} \mathcal{B}_2 - \operatorname{ind} \mathcal{B}_1 = \operatorname{ind} \mathcal{S} . \tag{5}$$

Suppose now that we are given a Fredholm operator

$$\mathcal{B} = \begin{pmatrix} a & K & \lambda \\ T & Q & \varkappa_1 \\ \zeta & \tau_1 & \varrho \end{pmatrix} : \begin{matrix} \mathcal{E} \\ \oplus \\ \mathcal{J} \\ \oplus \\ \mathcal{L} \end{matrix} \to \begin{matrix} \mathcal{F} \\ \oplus \\ \mathcal{G} \\ \oplus \\ \mathcal{M} \end{matrix} \tag{6}$$

with $\mathcal{A}$ in the left upper corner as in 4.2.1.2, Theorem 1, but λ, ζ, $\varkappa_1$, τ_1, ϱ without any relation to interior boundary problems. Assume that $Q_0 = 0$ so that $R_0 = 0$, too (cf. 4.2.1.2, Lemma 2). Then we have the following

Remark 2. *The following conditions are equivalent:*

(i) *there exist finite dimensional subspaces* $\mathcal{R} \subset \mathcal{E}, \mathcal{T} \subset \mathcal{F}$ *with*

$$\ker T_0 \subseteqq \mathcal{R} \oplus \ker \zeta , \quad \mathcal{T} \oplus \operatorname{im} K_0 \supseteqq \operatorname{im} \lambda ;$$

(ii) *there exist operators*

$$\tau_2 : \mathcal{G} \to \mathcal{M} , \quad \varkappa_2 : \mathcal{L} \to \mathcal{J}$$

so that modulo operators with finite rank

$$\zeta = \tau_2 T_0 , \quad \lambda = - K_0 \varkappa_2 .$$

If (ii) in Remark 2 is satisfied, the operator $\mathcal{D}$, formally given by 4.2.1.2.(3), is Fredholm and (6) is the result of the formal constructions in 4.2.1.2, Theorem 1.

Thus condition (i) in Remark 2 shows whether two different Fredholm operators of the form (6) can be compared in the sense of Theorem 1. The choice of the operator (4) shall be considered as a boundary reduction with respect to Z.

In 4.2.2.1 we will return to this question and discuss situations in which Y appears as an elliptic pseudo-differential operator on Z.

4.2.2. Examples and Regularity in Sobolev Spaces

4.2.2.1. Some Classes of Interior Boundary Value Problems

The theory of interior boundary problems in the sense of the definition given in 4.2.1.2 is investigated in a number of papers, e.g. Sjöstrand [1, 2], Èskin [1, 2], Višik [2], Melin/Sjöstrand [1]. Here we can reproduce only a few results. The classes of concrete interior boundary problems lead to corresponding classes of degenerate boundary problems for elliptic pseudo-differential operators.

In Sjöstrand [1] the following case is considered. Let $\delta \in L^m(Y)$ be a scalar pseudo-differential operator with a principal symbol $\sigma_\delta \in C^\infty(T^*Y \setminus \{0\})$ being positively homogeneous of degree m. Local coordinates in T^*Y are denoted by (y, η). Put

$$\Sigma = \{(y, \eta) \in T^*Y \setminus \{0\} : \sigma_\delta(y, \eta) = 0\} .$$

Σ is an invariantly defined conic subset of $T^*Y \setminus \{0\}$. Denote by $\{f, g\}$ the Poisson bracket of the functions $f(y, \eta)$, $g(y, \eta)$ (f, g may be complex) and set

$$c(y, \eta) = \frac{1}{i} \{\sigma_\delta, \overline{\sigma}_\delta\} = -2 \{\operatorname{Re} \sigma_\delta, \operatorname{Im} \sigma_\delta\} .$$

Suppose that the following conditions are satisfied

(i) $c(\varrho) \neq 0$ for all $\varrho \in \Sigma$;

(ii) $n - 1 = \dim Y \geqq 3$ and $(\partial\sigma_\delta/\partial\eta_1, \dots, \partial\sigma_\delta/\partial\eta_{n-1})$

is for each $\varrho \in \Sigma$ proportional to a real vector.

If $\Sigma^\pm = \{\varrho \in \Sigma : c(\varrho) \gtrless 0\}$, then Σ is the disjoint union of Σ^+ and Σ^-. Suppose without loss of generality $m = 1$.

A consequence of the conditions (i), (ii) is the following proposition. Σ^+ and Σ^- are closed conic submanifolds of $T^*Y \setminus \{0\}$ of codimension 2. For every $\varrho \in \Sigma$ there exist local coordinates $y = (y_1, \dots, y_{n-1})$ in a neighbourhood U of $\pi_Y(\varrho)$ so that the component of $\Sigma \cap T^*Y|_U$ containing ϱ can be described by the equations $y_{n-1} = 0$, $\eta_{n-1} = \lambda(y, \eta')$, $\eta' = (\eta_1, \dots, \eta_{n-1})$ with a real function $\lambda \in C^\infty(\mathbb{R}^{n-1} \times (\mathbb{R}^{n-2} \setminus \{0\}))$ which is positively homogeneous of degree 1 with respect to η'.

For simplicity assume that $Z^\pm \overset{\text{def}}{=} \pi_Y(\Sigma^\pm)$ are C^∞ manifolds (in general $Z^\pm$ can be considered as images under immersions of manifolds $\tilde{Z}^\pm$ of dimension $n-2$). Put $Z = Z^+ \cup Z^-$, then Z, Z^+, Z^- are C^∞ submanifolds of Y of codimension 1.

In Sjöstrand [1] linear continuous operators

$$\varkappa: \mathcal{D}'(Z^-) \to \mathcal{D}'(Y), \quad \tau: \mathcal{D}'(Y) \to \mathcal{D}'(Z^+)$$

are constructed so that the operator

$$\mathcal{D} = \begin{pmatrix} \delta & \varkappa \\ \tau & 0 \end{pmatrix} : \begin{matrix} \mathcal{D}'(Y) \\ \oplus \\ \mathcal{D}'(Z^-) \end{matrix} \to \begin{matrix} \mathcal{D}'(Y) \\ \oplus \\ \mathcal{D}'(Z^+) \end{matrix} \tag{1}$$

has a C^∞ parametrix

$$\mathcal{D}^{-1} = \begin{pmatrix} \varepsilon & \eta \\ \sigma & 0 \end{pmatrix} : \begin{matrix} \mathcal{D}'(Y) \\ \oplus \\ \mathcal{D}'(Z^+) \end{matrix} \to \begin{matrix} \mathcal{D}'(Y) \\ \oplus \\ \mathcal{D}'(Z^-) . \end{matrix} \tag{2}$$

The operators $\varkappa, \tau, \varepsilon, \eta, \sigma$ induce for each $s \in \mathbb{R}$ continuous operators

$$\varkappa: H^s(Z^-) \to H^{s-1/2}(Y), \quad \tau: H^s(Y) \to H^{s-1/2}(Z^+) , \tag{3}$$

$$\varepsilon: H^s(Y) \to H^{s+1/2}(Y) , \tag{4}$$

$$\eta: H^s(Z^+) \to H^{s+1/4}(Y), \quad \sigma: H^s(Y) \to H^{s+1/4}(Z^-) . \tag{5}$$

Restriction to smooth functions gives continuous operators

$$\mathcal{D}: C^\infty(Y) \oplus C^\infty(Z^-) \to C^\infty(Y) \oplus C^\infty(Z^+) , \tag{6}$$

$$\mathcal{D}^{-1}: C^\infty(Y) \oplus C^\infty(Z^+) \to C^\infty(Y) \oplus C^\infty(Z^-) . \tag{7}$$

For compact Y the operator $\mathcal{D}$ is Fredholm.

The class of operators δ, i.e. the character of degeneracy of the ellipticity over Z, so that Fredholm interior boundary value problems $\mathcal{D}$ for δ exist, can be considerably generalized. Generalizations are already mentioned in Sjöstrand [1] and the unpleasant dimension condition in (ii) can be replaced by $\dim Y \geqq 2$ for other interesting classes of operators δ. Finally, the special form of the trace operators τ (discussed in Sjöstrand [1]) can be generalized so that restriction operators or suitable differentiations composed with restriction to Z^* are included (cf. Èskin [1], Juhl [1]).

For this case it is proved in Juhl [1] that if

$$\mathcal{D}_k = \begin{pmatrix} \delta \\ \tau^k \end{pmatrix} : \mathcal{V} \to \begin{matrix} \mathcal{W} \\ \oplus \\ \mathcal{M}_k \end{matrix} , \quad k = 1, 2 ,$$

are Fredholm interior boundary problems (cf. 4.2.1.4, Theorem 1), the operator

$$\mathcal{S} = \tau^2\eta^1: \mathcal{M}_1 \to \mathcal{M}_2$$

is an elliptic PDO over Z^+ so that $\operatorname{ind} \mathcal{D}_2 - \operatorname{ind}^* \mathcal{D}_1 = \operatorname{ind} \mathcal{S}$.

Next discuss a certain other class of interior boundary problems studied in MELIN/SJÖSTRAND [1]. Let Z^+, Z^- be disjoint submanifolds of Y of codimension 1 and $Z = Z^+ \cup Z^-$. Consider a scalar pseudo-differential operator $\delta \in L^1(Y)$ which is elliptic over $Y \setminus Z$. Suppose that there are tubular neighbourhoods U and V of Z^+ and Z^-, respectively, so tat σ_δ in U (V) with respect to local coordinates has the following form

$$\sigma_\delta(y, \eta) = \sum_{j=1}^{n-1} m_j(y)\, \eta_j + i\lambda(y, \eta)\,. \tag{8}$$

It is assumed that the m_j are real smooth functions invariantly representing a vector field μ and λ is a real smooth function on $T^*U \setminus 0$ ($T^*V \setminus 0$) positively homogeneous in η of degree 1. Thus δ is (except lower order terms) near Z the sum of two operators $\mu(y, D_y)$ and $\lambda(y, D_y)$. Let $U = U_+ \cap U_-$, $V = V_+ \cap V_-$ be decompositions into half open strips with $U_+ \cap U_- = Z^+$, $V_+ \cap V_- = Z^-$ (so that for instance, if $U \cong Z^+ \times (-1, 1)$, we have $U_+ \cong Z^+ \times [0, 1)$, $U_- \cong Z^+ \times (-1, 0]$).

Suppose

$$\lambda(y, \eta) \leqq 0 \quad \text{for} \quad y \in U_+, \quad \lambda(y, \eta) \geqq 0 \quad \text{for} \quad y \in U_-\,,$$
$$\lambda(y, \eta) \geqq 0 \quad \text{for} \quad y \in V_+\,, \quad \lambda(y, \eta) \leqq 0 \quad \text{for} \quad y \in V_-\,.$$

Moreover, suppose that the vector field μ never vanishes over Z and points into U_+ and V_+ over Z^+ and Z^-, respectively (i.e. particularly μ is transversal to Z). The results of MELIN/SJÖSTRAND [1] then yield the existence of a Fredholm interior boundary problem (for Y compact)

$$\mathcal{D} = \begin{pmatrix} \delta & \varkappa \\ \tau & 0 \end{pmatrix} : \begin{matrix} C^\infty(Y) \\ \oplus \\ C^\infty(Z^-) \end{matrix} \to \begin{matrix} C^\infty(Y) \\ \oplus \\ C^\infty(Z^+)\,. \end{matrix} \tag{9}$$

$\mathcal{D}$ has an extension as continuous operator

$$\mathcal{D}: H^s(Y) \oplus H^{s-1/2}(Z^-) \to H^{s-1}(Y) \oplus H^{s-1/2}(Z^+) \tag{10}$$

for all $s \in \mathbb{R}$. There is a C^∞ parametrix $\mathcal{D}^{-1}$ of (9) with a continuous extension

$$\mathcal{D}^{-1}: H^s(Y) \oplus H^s(Z^+) \to H^s(Y) + H^s(Z^-) \tag{11}$$

for all $s \in \mathbb{R}$.

A basic tool in the proofs of the mentioned properties are microlocal constructions and Fourier integral operator techniques (microlocal means in open conic neighbourhoods of points in $T^*Y \setminus \{0\}$). The results can be easily generalized to certain classes of systems of pseudo-differential operators on Y. Suppose, for instance, that δ is a system and that σ_δ (the homogeneous principal symbol of δ) admits microlocally near points (y, η) for which $\det \sigma_\delta(y, \eta)$ vanishes factorizations of the form

$$\beta_1(y, \eta) \begin{pmatrix} \sigma_{\delta_1}(y, \eta) & & 0 \\ & \ddots & \\ 0 & & \sigma_{\delta_k}(y, \eta) \end{pmatrix} \beta_2(y, \eta) \tag{12}$$

with β_1, β_2 elliptic and scalar diagonal elements σ_{δ_j} which are either elliptic or belong to one of the symbol classes described above for which interior boundary value problems exist possessing (micro-)local C^∞ parametrices. Then it is obvious how to construct a corresponding operator $\mathcal{D}$ for the given system. In such a sense it is not hard to construct classes of pseudo-differential operators $\delta: C^\infty(Y, J_1) \to C^\infty(Y, J_2)$ $(J_1, J_2 \in \operatorname{Vect}(Y))$ for which Fredholm interior boundary problems $\mathcal{D}$ exist, where the spaces $\mathcal{L}$, $\mathcal{M}$ mentioned in 4.2.1.2 have the form

$$\mathcal{L} = C^\infty(Z^-, L)\,, \quad \mathcal{M} = C^\infty(Z^+, M) \tag{13}$$

with suitable $L \in \text{Vect}\,(Z^-)$, $M \in \text{Vect}\,(Z^+)$, $Z = Z^- \cup Z^+$.

Further classes of interior boundary problems are considered in the papers mentioned at the beginning of this section. Especially in Èskin [2] the case of the ellipticity of δ degenerating in a suitable way with respect to the "space variable" is treated.

Now let

$$\mathcal{D} = \begin{pmatrix} \delta & \varkappa \\ \tau & \varrho \end{pmatrix} : \begin{matrix} C^\infty(Y, J_1) \\ \oplus \\ C^\infty(Z^-, L) \end{matrix} \to \begin{matrix} C^\infty(Y, J_2) \\ \oplus \\ C^\infty(Z^+, M) \end{matrix}$$

be a Fredholm interior boundary problem ($J_k \in \text{Vect}\,(Y)$, $L \in \text{Vect}\,(Z^-)$, $M \in \text{Vect}\,(Z^+)$) for the pseudo-differential operator δ on Y. Put

$$r = \text{ord}\,\delta\,, \quad k = \text{ord}\,\varkappa\,, \quad p = \text{ord}\,\tau\,, \quad q = \text{ord}\,\varrho$$

if the corresponding operators have continuous extensions

$$\delta\colon H^s(Y, J_1) \to H^{s-r}(Y, J_2)\,, \quad \varkappa\colon H^s(Z^-, L) \to H^{s-k}(Y, J_2), \ldots$$

for all $s \in \mathbb{R}$ (or for all $s \in \mathbb{R}$, $s \geqq s_1$ with some fixed $s_1 \in \mathbb{R}$). Similarly let

$$a = \text{ord}\,\varepsilon\,, \quad b = \text{ord}\,\sigma\,, \quad c = \text{ord}\,\eta\,, \quad d = \text{ord}\,\chi$$

be the orders of the operators occurring in a C^∞ parametrix $\mathcal{D}^{-1}$ of $\mathcal{D}$. Consider the following composition

$$\mathcal{D}_0 = \begin{pmatrix} \alpha & 0 \\ 0 & \gamma_+ \end{pmatrix} \mathcal{D} \begin{pmatrix} \beta & 0 \\ 0 & \gamma_- \end{pmatrix} = \begin{pmatrix} \alpha\delta\beta & \alpha\varkappa\gamma_- \\ \gamma_+\tau\beta & \gamma_+\varrho\gamma_- \end{pmatrix}.$$

Here

$$\alpha\colon C^\infty(Y, J_2) \to C^\infty(Y, J_2)\,, \quad \beta\colon C^\infty(Y, J_1) \to C^\infty(Y, J_1)\,,$$

$$\gamma_+\colon C^\infty(Z^+, L) \to C^\infty(Z^+, L)\,, \quad \gamma_-\colon C^\infty(Z^-, M) \to C^\infty(Z^-, M)$$

are Fredholm operators and assumed to be pseudo-differential operators with homogeneous principal symbols of order

$$u = \text{ord}\,\alpha\,, \quad v = \text{ord}\,\beta\,, \quad g_\pm = \text{ord}\,\gamma_\pm\,.$$

Denote by α^{-1}, β^{-1}, ... the corresponding parametrices and by

$$h = \text{ord}\,\alpha^{-1}\,, \quad l = \text{ord}\,\beta^{-1}\,, \quad f_\pm = \text{ord}\,\gamma_\pm^{-1}$$

the orders in the sense of continuous closures in corresponding Sobolev spaces (the existence is assumed). If α, β, $\gamma_\pm$ are elliptic, we obviously have $h = -u$, $l = -v$, $g_\pm = -f_\pm$ for the orders. Passing form $\mathcal{D}$ to $\mathcal{D}_0$ corresponds to an ordinary reduction of orders. Especially one can always choose α (or β) in such a way that $\text{ord}\,(\alpha\delta\beta) = 0$. If the orders of α, β are fixed, the orders of $\varkappa_0 = \alpha\varkappa\gamma_-$, $\tau_0 = \gamma_+\tau\beta$ can be arbitrarily modified by a suitable choice of the orders of γ_-, γ_+. For non-elliptic reducing operators one has to take into account a loss of smoothness. If e.g. α, β are the corresponding identity operators and $\gamma_\pm$ Fredholm but not elliptic, the operators in $\mathcal{D}_0^{-1}$ reflect a loss of smoothness, namely $g_\pm - f_\pm$. This consideration shows that a given pseudo-differential operator δ with degenerating ellipticity over Z admits in general different types of Fredholm interior boundary problems $\mathcal{D}_1$, $\mathcal{D}_2$, i.e. the operators in the parametrices $\mathcal{D}_1^{-1}$ and $\mathcal{D}_2^{-1}$ respectively, have a different behavior with respect to continuous extensions in Sobolev spaces.

Finally, note that there exist operators δ with degenerating ellipticity over Z, which are Fredholm in C^∞ spaces without additional trace or potential conditions with respect to Z (or subsets of Z) (cf. ÈSKIN [1]).

4.2.2.2. The Oblique Derivative Problem and Further Examples

As an interesting special case of non-elliptic boundary problems we discuss the classical oblique derivative problem for the Laplace operator. It is then clear how to construct further examples. Let us start with a more general situation. Consider an elliptic differential operator

$$A(x, D): C^\infty(X, E) \to C^\infty(X, F) \tag{1}$$

of order $m \in \mathbb{Z}_+$ satisfying the root condition (cf.3.1.1.3.) so that m is even (E, $F \in \operatorname{Vect}(X)$, X a smooth compact manifold with boundary). Let k be the fibre dimension of E and F. Assume that we are given a boundary problem

$$\mathcal{A} = \begin{pmatrix} A \\ T \end{pmatrix}: C^\infty(X, E) \to \begin{matrix} C^\infty(X, F) \\ \oplus \\ C^\infty(Y, G) \end{matrix} \tag{2}$$

(G the trivial bundle $Y \times \mathbb{C}^{km/2}$ over Y) where the trace operator T in local coordinates in $\overline{\mathbb{R}}^n_+$ has the following form

$$T_j w = \sum_{l=1}^{m} T_{jl}(x', D')\,(\gamma_{l-1} w) = h_j \tag{3}$$

$\left(1 \leqq j \leqq \frac{m}{2}\right)$ with $\gamma_l w \overset{\text{def}}{=} \left(\frac{1}{i}\frac{\partial}{\partial x_n}\right)^l w|_{x_n=0}$ and pseudo-differential operators $T_{lj}(x', D')$ in $\mathbb{R}^{n-1}_{x'}$. Without loss of generality assume that ord T_j is the same for all j (otherwise this assumption is satisfied after reduction of orders).

As usual σ_A and σ_T ($\sigma_{T_{jl}}$), respectively, denote the homogeneous principal symbols of the corresponding operators. On the half axis consider the equation

$$r^+\sigma_A(x', 0, \xi', D_n)\, u(x', x_n, \xi') = 0 \tag{4}$$

and denote by $p^*G^{(+)}$ the bundle of Cauchy data $\gamma u = (\gamma_0 u, \dots, \gamma_{m-1} u)$ induced by solutions u of (4) which are bounded on $\mathbb{R}_+$ (the fibre dimension of $G^{(+)}$ is $mk/2$).

A consequence of the considerations in 3.1.1.3 is the following

Theorem 1. *At a point $(x', \xi') \in T^*Y \setminus \{0\}$ the boundary problem (2) satisfies the Shapiro-Lopatinski condition in the sense of bijectivity of*

$$\begin{pmatrix} \Pi^+\sigma_A \\ \Pi'\sigma_T \end{pmatrix}: (p^*E^+)_{(x', \xi')} \to \left(p^*\begin{pmatrix} F^+ \\ \oplus \\ G \end{pmatrix}\right)_{(x', \xi')}$$

iff

$$\sum_{l=1}^{m} \sigma_{T_{jl}}(x', \xi')\,(\gamma_{l-1} u) = g_j \tag{5}$$

induces an isomorphism

$$(p^*G^{(+)})_{(x', \xi')} \to (p^*G)_{(x', \xi')}\,. \tag{6}$$

Consider the following commutative diagram

$$\begin{array}{ccccc} p^*G^{(+)} & \xrightarrow[\cong]{\gamma^{-1}} & \ker r^+\sigma_A & \xrightarrow{r'\sigma_T} & p^*G \\ & & \uparrow F^{-1} & \nearrow \Pi'\sigma_T & \\ & & \ker \Pi^+\sigma_A & & \end{array} . \tag{7}$$

Here, γ denotes the isomorphism $\gamma\colon \ker r^+\sigma_A \to p^*G^{(+)}$ induced by calculating the Cauchy data up to the order $m - 1$ and F denotes the Fourier transform on the real axis. $r'\sigma_T$ is the Fourier preimage of $\Pi'\sigma_T$.

Consider a second boundary problem for A

$$\mathcal{A}_0 = \begin{pmatrix} A \\ \\ T_0 \end{pmatrix} : C^\infty(X, E) \to \begin{matrix} C^\infty(X, F) \\ \oplus \\ C^\infty(Y, G) \end{matrix} . \tag{8}$$

Suppose that T_0 has a similar structure as T (cf. (3)). Contrarily to $\mathcal{A}$ assume that $\mathcal{A}_0$ is elliptic and discuss the boundary reduction of $\mathcal{A}$ with respect to $\mathcal{A}_0$. Let $\mathcal{P}_0 = (p_0, L_0)$ be a parametrix of (8). Since $\sigma_Y(\mathcal{A}_0)\,\sigma_Y(\mathcal{P}_0) = 1$, the isomorphisms $\Pi'\sigma_{T_0}\colon \ker \Pi^+\sigma_A \to p^*G$, $\sigma_{L_0}\colon p^*G \to \ker \Pi^+\sigma_A$ are inverse to each other. Because of the ellipticity of $\mathcal{A}_0$

$$\Pi'\sigma_{T_0} \circ F \circ \gamma^{-1}\colon p^*G^{(+)} \to p^*G \tag{9}$$

is an elliptic symbol with the inverse

$$\sigma_{\delta_0} = \gamma \circ F^{-1} \circ \sigma_{L_0}\colon p^*G \to p^*G^{(+)} . \tag{10}$$

In view of Theorem 1 and (7) the symbol

$$\sigma_{\delta_1} = \Pi'\sigma_T \circ F \circ \gamma^{-1}\colon p^*G^{(+)} \to p^*G \tag{11}$$

is elliptic at $(y, \eta) \in S^*Y$ iff $\sigma_Y(\mathcal{A})_{(y,\eta)}$ is bijective. Thus

$$\sigma_\delta = \sigma_{\delta_1} \circ \sigma_{\delta_0} = \Pi'\sigma_T \circ \sigma_{L_0} \tag{12}$$

is bijective at $(y, \eta) \in S^*Y$ iff $\sigma_Y(\mathcal{A})_{(y,\eta)}$ is bijective. This corresponds to 4.2.1.1, Lemma 1, because $\Pi'\sigma_T\sigma_{L_0}$ is the homogeneous principal symbol of the operator $\delta = TL_0$. Since σ_{δ_0} is elliptic, it is sufficient to consider

$$\sigma_{\delta_1} = r'\sigma_T \circ \gamma^{-1}\colon p^*G^{(+)} \to p^*G . \tag{13}$$

After this construction it is clear that σ_δ is changed by an elliptic factor $\sigma_{\delta'}$, if the elliptic $\mathcal{A}_0$ is replaced by another elliptic $\mathcal{A}_0'$.

Now let Ω be a smoothly bounded domain in $\mathbb{R}^n$, $X = \overline{\Omega}$, $Y = \partial\Omega$ and $A = \Delta$ (the Laplacian). Let l_j ($j = 1, \ldots, n$) be real smooth functions in a neighbourhood of X and $(\partial/\partial l)\, u(x) = \sum_{l=1}^{n} l_j(x)\, D_j u(x)$, $D_j = (1/i)\, \partial/\partial x_j$. The boundary value problem $\mathcal{A}$ defined by

$$\Delta u = f \quad \text{in } \Omega, \quad Tu = (\partial/\partial l)\, u|_Y = g \tag{14}$$

is called *oblique derivative problem.* Suppose that

$$\sum_{j=1}^{n} l_j^2(x) > 0 \quad \text{for all} \quad y \in Y . \tag{15}$$

Then we have

Proposition 2. *At a point $y \in Y$ the boundary problem* (14) *satisfies the Shapiro-Lopatinski condition for $n = \dim \Omega \geqq 3$ iff the vector $l(y) = (l_1(y), \dots, l_n(y))$ is not tangent to Y. In the case $n = 2$, the Shapiro-Lopatinski condition is always satisfied.*

This proposition can be proved by reduction to the boundary with respect to an elliptic boundary problem

$$\mathcal{A}_0 = \begin{pmatrix} \Delta \\ T_0 \end{pmatrix} : C^\infty(X) \to \begin{matrix} C^\infty(X) \\ \oplus \\ C^\infty(Y) \end{matrix}$$

for the Laplace operator. For T_0 one can take e.g. Neumann boundary conditions. They satisfy $\operatorname{ord} T_0 = \operatorname{ord} T$. Let $\mathcal{P}_0 = (p_0, L_0)$ be a parametrix of $\mathcal{A}_0$. In a neighbourhood U of a fixed $y \in Y$ introduce local coordinates in such a way that the interior normal to Y corresponds to the x_n-direction and $Y \cap U$ to an open set in $\mathbb{R}_x^{n-1}$. Then $l(y)$ can be written in the form $(l_1(y), \dots, l_n(y))$ $(y \in Y \cap U)$ so that $(l_1, \dots, l_{n-1})$ corresponds to the tangent components of l and l_n to the interior normal component (for abbreviation the same letters are used again). Consider

$$\sigma_{\delta_1}(y, \eta) = \sum_{j=1}^{n-1} l_j(y)\, \eta_j + i l_n(y)\, |\eta'| \tag{16}$$

$(\eta' = (\eta_1, \dots, \eta_{n-1}))$. Then we obtain the following

Proposition 3. *There exists an elliptic symbol $\sigma_{\delta_0}(y, \eta)$ on Y so that (in the considered local coordinates)*

$$\sigma_{TL_0}(y, \eta) = \sigma_{\delta_1}(y, \eta)\, \sigma_{\delta_0}(y, \eta)\,. \tag{17}$$

Proof: In view of (11), (12) it is sufficient to calculate σ_{δ_1} in the form (16). We have to calculate the bounded solutions of $(|\xi'|^2 + D_n^2)\, u(x', x_n, \xi') = 0$ on $x_n \geqq 0$. The bounded solutions are given by $u(x', x_n, \xi') = c\, e^{-x_n|\xi'|}$, $c \in \mathbb{C}$ arbitrary. Thus $\gamma_0 u = c$, $\gamma_1 u = ic|\xi'|$, i.e. $\ker r^+\sigma_A$ is parametrized by a constant. For $T(x, D)\, w = \sum_{j=1}^{n} l_j(x')\, D_j w|_{x_n=0}$ we obviously get $r'\sigma_T(x', \xi', D_n)\, u = c\left(\sum_{l=1}^{n-1} l_j(x')\, \xi_j + i l_n(x')|\xi'|\right)$ (cf. (13)). This proves the proposition. □

Proof of Proposition 2: For $n = 2$ we have $l_1^2 + l_2^2 \neq 0$, i.e. $l_1(y)\, \eta_1 + i l_2(y)\, |\eta_1| \neq 0$ for each $\eta_1 \neq 0$. Thus σ_{δ_1} is elliptic. For $n > 2$ the symbol σ_{δ_1} can vanish at exactly those points $y \in Y$ for wich $l_n = 0$. Then, there exists a nonvanishing vector $(\eta_1, \dots, \eta_{n-1})$ with $\sum_{j=1}^{n-1} l_j \eta_j = 0$ for each $(l_1, \dots, l_{n-1})$. □

Denote by Z the set of those points $y \in Y$ in which $l(y)$ is tangent to Y. Suppose that Z is a smooth submanifold of codimension 1. The symbol (16) is of the type 4.2.2.1.(8) if we suppose that $(l_1, \dots, l_{n-1})$ over Z is never tangent to Z (everything is formulated in the local coordinates mentioned above). We get a decomposition $Z = Z^+ \cup Z^-$, and here the situation considered in 4.2.2.1 has the following meaning. There exist tubular neighbourhoods U and V of Z^+ and Z^-, respectively, and decompositions $U = U_+ \cup U_-$, $V = V_+ \cup V_-$ with $Z^+ = U_+ \cap U_-$, $Z^- = V_+ \cap V_-$ with

$$l_n(y) \leqq 0 \quad \text{for} \quad y \in U_+, \quad l_n(y) \geqq 0 \quad \text{for} \quad y \in U_-,$$
$$l_n(y) \geqq 0 \quad \text{for} \quad y \in V_+, \quad l_n(y) \leqq 0 \quad \text{for} \quad y \in V_-.$$

Here $l' = (l_1, \dots, l_{n-1})$ is non zero and points to U_+ and V_+, respectively.

Thus we can apply the results of 4.2.2.1. There exists an interior boundary problem for $\delta_1 \in L'(Y)$ with σ_{δ_1} given by (16)

$$\mathcal{D}_1 = \begin{pmatrix} \delta_1 & \varkappa_1 \\ \tau_1 & 0 \end{pmatrix} : \begin{matrix} C^\infty(Y) \\ \oplus \\ C^\infty(Z^-) \end{matrix} \to \begin{matrix} C^\infty(Y) \\ \oplus \\ C^\infty(Z^+) \end{matrix}$$

being Fredholm. Then

$$\mathcal{D} = \begin{pmatrix} \delta_1 & \varkappa_1 \\ \tau_1 & 0 \end{pmatrix} \begin{pmatrix} \delta_0 & 0 \\ 0 & 1 \end{pmatrix} = \begin{pmatrix} \delta\delta_0 & \varkappa_1 \\ \tau_1\delta_0 & 0 \end{pmatrix}$$

is a Fredholm interior boundary problem for an operator δ with the homogeneous principal symbol (17) (δ_0 denotes a pseudo-differential operator on Y with the homogeneous principal symbol σ_{δ_0}). Applying 4.2.1.3, Remark 2 gives a Fredholm operator

$$\mathcal{B} = \begin{pmatrix} \Delta & 0 \\ T & \varkappa_1 \\ \tau_1\delta_0 T_0 & 0 \end{pmatrix} : \begin{matrix} C^\infty(X) \\ \oplus \\ C^\infty(Z^-) \end{matrix} \to \begin{matrix} C^\infty(X) \\ \oplus \\ C^\infty(Y) \\ \oplus \\ C^\infty(Z^+) \end{matrix} \tag{18}$$

and an expression for a parametrix, namely

$$\mathcal{B}^{-1} = \begin{pmatrix} p_0 - L_0\delta_0^{-1}\varepsilon_1 T p_0 & L_0\delta_0^{-1}\varepsilon_1 & L_0\delta_0^{-1}\eta_1 \\ -\sigma_1 T p_0 & \sigma_1 & 0 \end{pmatrix}. \tag{19}$$

Here δ_0^{-1} denotes a parametrix of δ_0 and ε_1, η_1, σ_1 are the elements of a parametrix of $\mathcal{D}_1$ (cf. the notations in 4.2.2.1.(2)).

As another example consider the biharmonic operator $A = \Delta^2$ in a smoothly bounded domain Ω in $\mathbb{R}^n$. For the boundary reduction of a boundary problem

$$\mathcal{A} = \begin{pmatrix} \Delta^2 \\ T \end{pmatrix} : C^\infty(X) \to \begin{matrix} C^\infty(X) \\ \oplus \\ C^\infty(Y, \mathbb{C}^2) \end{matrix} \tag{20}$$

consider, similarly as in the example (14), the bounded solutions on $x_n > 0$ of the equation $(|\xi'|^2 + D_n^2)^2 u(x', x_n, \xi') = 0$. It is easily seen that two linearly independent solutions are given by

$$u(x', x_n, \xi') = (c_0 - c_1|\xi'|\, x_n)\, \mathrm{e}^{-|\xi'|x_n} \tag{21}$$

with arbitrary $c_0, c_1 \in \mathbb{C}$. Then for the Cauchy data of (21) it follows that

$$\gamma_0(u) = c_0\,, \quad \gamma_1(u) = \mathrm{i}\, |\xi'|\, (c_0 + c_1)\,,$$
$$\gamma_2(u) = -\, |\xi'|^2\, (c_0 - 2c_1)\,, \quad \gamma_3(u) = -\,\mathrm{i}\, |\xi'|^3\, (c_0 + 3c_1)\,.$$

This shows that the Cauchy data of (21) form a two-dimensional trivial bundle over $T^*Y \setminus \{0\}$ parametrized by $(c_0, c_1) \in \mathbb{C}^2$.

If the components T_1, T_2 of T in (20) have the form (3) (with $\sigma_{T_{jl}} = \operatorname{ord} \sigma_{T_j} - (l-1)$ $(j = 1, 2)$), we get (5) in the form

$$\sum_{l=1}^{4} \sigma_{T_{jl}}(x', \xi')\, \gamma_{l-1}(u) = g_j\,, \quad j = 1, 2\,. \tag{22}$$

As an example assume that the boundary conditions have the form

$$T_1 w = (Qw)|_Y = h_1, \quad T_2 w = \sum_{j=1}^{n} l_j(x)\, D_j w|_Y = h_2 \tag{23}$$

with a vector field $l = (l_1, \ldots, l_n)$ as in the above oblique derivative problem. Use again local coordinates for which l_n corresponds to the interior normal component of l. The operator Q in (23) in supposed to be an elliptic pseudo-differential operator on Y of order 1 with the homogeneous principal symbol $q(x', \xi')$. Then, with the notations of (5), we have $\sigma_{T_{1l}} = 0$ vor $l = 2, 3, 4$, $\sigma_{T_{2l}} = 0$ for $l = 3, 4$ and

$$\sigma_{T_{21}}(x', \xi') = \sum_{j=1}^{n-1} l_j(x')\, \xi_j, \quad \sigma_{T_{22}}(x', \xi') = l_n(x').$$

For the map (5) we obtain with $\sigma_{T_{11}}(x', \xi') = q(x', \xi')$

$$\sigma_{T_{11}}(x', \xi')\, c_0 = g_1, \quad \sigma_{T_{21}}(x', \xi')\, c_0 + \mathrm{i}|\xi'|\, l_n(x')\, (c_0 + c_1) = g_2.$$

This can be considered as linear transformation $(c_0, c_1) \to (g_1, g_2)$ given by the matrix

$$\sigma_{\delta_1}(x', \xi') = \begin{pmatrix} \sigma_{T_{11}}(x', \xi') & 0 \\ \sigma_{T_{21}}(x', \xi') + \mathrm{i}|\xi'|\, l_n(x') & \mathrm{i}|\xi'|\, l_n(x') \end{pmatrix}. \tag{24}$$

For $|\xi'| \neq 0$ (24) is an isomorphism iff $l_n(x') \neq 0$. Thus we get

Proposition 4. *The boundary problem* (20) *with* $T = (T_1, T_2)$ *in the form* (23) *satisfies the Shapiro-Lopatinski condition at* $y \in Y$ *iff the normal component of the vector field* l *does not vanish at* y.

Since the assumption ord T_1 = ord T_2 is not essential, we also have

Proposition 4'. *The boundary problem*

$$\Delta^2 w = f, \quad w|_Y = h_1, \quad \frac{\partial}{\partial l} w|_Y = h_2 \tag{25}$$

satisfies the Shapiro-Lopatinski condition exactly at those points $y \in Y$ *where the normal component of* l *does not vanish.*

Note that the assertion in Proposition 4 and 4' holds for $n = 2$, too.

In order to connect Fredholm problems with the boundary problems (23) and (25), respectively, in the sense of the constructions in 4.2.1.2 one has to study interior boundary problems for the operator with homogeneous principal symbol (24). This is equivalent to the investigation of an operator with the homogeneous principal symbol $|\xi'|\, l_n(x')$. Vanishing of $l_n(x')$ means in this case degeneracy of the ellipticity with respect to the space variables in the sense of Èskin [2]. Thus one has to apply results of Èskin [2] and to make corresponding assumptions on the order of zeros of $l_n(x')$. The details are left to the reader. Note that, for non-elliptic Q in (23) under suitable conditions of degeneracy, one has to formulate additional conditions. The details are clear after the results of 4.2.1.2. More generally, another choice of the $\sigma_{T_{jl}}$ in (22) yields further examples and it is very instructive to look at the types of degeneracy of ellipticity of the resulting operators δ_1.

4.2.2.3. Regularity in Sobolev Spaces

In this section we consider closures of the various types of operators in Sobolev spaces. The corresponding orders imply a certain regularity of the boundary value problem $\mathcal{B}w = f$ ($\mathcal{B}$ as in 4.2.1.2.(4)). It is clear that we have to calculate the orders of the operators contained in $\mathcal{B}^{-1}$ (cf. 4.2.1.3.(8)).

Suppose as above that the operators contained in $\mathcal{A}$ and $\mathcal{A}_0$ (cf. 4.2.1.1.(1), (2)) have the following orders: ord $a = \alpha$,

$$\operatorname{ord} T = \operatorname{ord} T_0 = \gamma + \tfrac{1}{2}, \operatorname{ord} K = \operatorname{ord} K_0 = \lambda + \tfrac{1}{2}, \tag{1}$$

$$\operatorname{ord} Q = \operatorname{ord} Q_0 = 1 - \alpha + \lambda + \gamma = 0 \tag{2}$$

($\alpha \in \mathbb{Z}, \lambda, \gamma \in \mathbb{R}$). The condition (2) is not essential. After reduction of orders this case can be assumed without loss of generality. Then the operator δ (4.2.1.1.(8)) has order zero, and ord $p_0 = -\alpha$, ord $L_0 = -\gamma - \frac{1}{2}$, ord $S_0 = -\lambda - \frac{1}{2}$, ord $R_0 = -1 + \alpha - \lambda - \gamma = 0$ (cf. the notations in 4.2.1.1.(6) and the formulas 3.1.1.1.(7)). Suppose that the interior boundary problems $\mathcal{D}$ for δ are of one of the types described in 4.2.2.1 with $\mathcal{L} = C^\infty(Z^-, L)$, $\mathcal{M} = C^\infty(Z^+, M)$ and that the operators $\varkappa, \tau, \varrho$ contained in $\mathcal{D}$ have continuous closures $\varkappa: H^s(Z^-, L) \to H^{s-k}(Y, G \oplus J_0)$, $\tau: H^s(Y, J \oplus G_0) \to H^{s-p}(Z^+, M)$, $\varrho: H^s(Z^-, L) \to H^{s-q}(Z^+, M)$ for certain fixed real numbers k, p, q, and arbitrary $s \in \mathbb{R}$. Then we write $k = \operatorname{ord} \varkappa$ and so on. Similarly assume that the operators contained in $\mathcal{E} = \mathcal{D}^{-1}$ have the orders

$$a = \operatorname{ord} \varepsilon \geqq 0\,,\ b = \operatorname{ord} \sigma,\ c = \operatorname{ord} \eta,\ d = \operatorname{ord} \chi \tag{3}$$

for certain $a, b, c, d \in \mathbb{R}$ (cf. the notations in 4.2.1.3). Now an immediate consequence of the formula 4.2.1.3.(8) is that, for arbitrary $t, g, m \in \mathbb{R}$ sufficiently large, $\mathcal{B}^{-1}$ has a continuous extension

$$\mathcal{B}^{-1}: H^t(X, F) \oplus H^g(Y, G) \oplus H^m(Z^+, M) \to H^e(X, E) \oplus H^j(Y, J) \oplus H^l(Z^-, L) \tag{4}$$

with $e = \min\{t - \operatorname{ord} \beta_0, g - \operatorname{ord} \beta_1, m - \operatorname{ord} \beta_2\}$, $j = \min\{t - \operatorname{ord}(\varepsilon\zeta)_1, g - \operatorname{ord} \varepsilon_{11}, m - \operatorname{ord} \eta_1\}$, $l = \min\{1 - \operatorname{ord}(\sigma\zeta), g - \operatorname{ord} \sigma_1, m - \operatorname{ord} \chi\}$. Using 4.2.1.3. (5), (6), (7) we obtain

$$\operatorname{ord}(\varepsilon\zeta)_k = a - \lambda - \tfrac{1}{2}, \quad \operatorname{ord}(\sigma\zeta) = b - \lambda - \tfrac{1}{2}, \tag{5}$$

$$\operatorname{ord} \beta_0 = a - \alpha\,, \quad \operatorname{ord} \beta_1 = a - \gamma - \tfrac{1}{2}, \quad \operatorname{ord} \beta_2 = c - \gamma - \tfrac{1}{2}. \tag{6}$$

Thus $e = \min\{t - \alpha + a, g - a + \gamma + \frac{1}{2}, m - c + \gamma + \frac{1}{2}\}$, $j = \min\{t - a + \lambda + \frac{1}{2}, g - a, m - c\}$, $l = \min\{t - b + \lambda + \frac{1}{2}, g - b, m - d\}$ and $\mathcal{B}w = h \in H^t(X, F) \oplus H^g(Y, G) \oplus H^m(Z^+, M)$ implies $w \in H^e(X, E) \oplus H^j(Y, J) \oplus H^l(Z^-, L)$. It is reasonable to choose g and m in such a way that $t - a + \alpha = g - a + \gamma + \frac{1}{2} = m - c + \gamma + \frac{1}{2}$, i.e. $g = t + \alpha - \gamma - \frac{1}{2} = t + \lambda + \frac{1}{2}$, $m = t - a + c + \alpha - \gamma - \frac{1}{2} = t - a + c + \lambda + \frac{1}{2}$. Then $t - a + \lambda + \frac{1}{2} = g - a = m - c$, $t - b + \lambda + \frac{1}{2} = g - b$ and consequently

$$e = t - a + \alpha\,, \quad j = t - a + \lambda + \tfrac{1}{2}, \tag{7}$$

$$l = \min\{t - b + \lambda + \tfrac{1}{2}, t - a + c - d + \lambda + \tfrac{1}{2}\}\,. \tag{8}$$

Set $t = s - \alpha$ and define the spaces

$$\mathcal{H}_1^{(s)} = H^{s-\alpha}(X, E) \oplus H^{t+\lambda+1/2-\alpha}(Y, J) \oplus H^l(Z^-, L)\,, \tag{9}$$

$$\mathcal{H}_2^{(s)} = H^t(X, F) \oplus H^{s-\gamma-1/2}(Y, G) \oplus H^m(Z^+, M) \tag{10}$$

with

$$l = s - \gamma - \tfrac{1}{2} - \max\{-b, -a + c - d\}\,, \quad m = t + \lambda + \tfrac{1}{2} - a\,. \tag{11}$$

Moreover, set $\mathcal{H}_0 = H^{s_0}(X, E) \oplus H^{j_0}(Y, J) \oplus H^{l_0}(Z^-, L)$ with arbitrary fixed real numbers $s_0 < s - a$, $j_0 < t + \lambda + \frac{1}{2} - a$, $l_0 < l$. Denote the norms in the spaces $\mathcal{H}_1^{(s)}$, $\mathcal{H}_2^{(s)}$ and $\mathcal{H}_0$, respectively, by $\|\cdot\|_j$ $(j = 0, 1, 2)$, $s \in \mathbb{R}$ fixed, sufficiently large. $\mathcal{B}^{-1}\colon \mathcal{H}_2^{(s)} \to \mathcal{H}_1^{(s)}$ is continuous and $\mathcal{B}^{-1}\mathcal{B} = \mathcal{J} + \mathcal{C}$ with a smoothing operator $\mathcal{C}$. Thus there exists a constant $c > 0$ so that $\|\mathcal{B}^{-1}\mathcal{B}w\|_1 = \|w + \mathcal{C}w\|_1 \leqq c\|\mathcal{B}w\|_2$ for all smooth $w \in \mathcal{H}_1^{(s)}$. This proves the following

Theorem 1. *Let $\mathcal{A} \in \mathfrak{G}$ be a boundary problem with degenerate Shapiro-Lopatinski condition and let $\mathfrak{D}$ satisfy the assumptions above. Let $\mathcal{B}$ be the Fredholm operator 4.2.1.2.(4) connected with $\mathcal{A}$. Then $\mathcal{B}w \in \mathcal{H}_2^{(s)}$ ($s \in \mathbb{R}$ sufficiently large, and $w \in \mathcal{H}_0^{(s)}$ implies $w \in \mathcal{H}_1^{(s)}$). There exists a constant $c > 0$ so that for all $w \in C^\infty(X, E) \oplus C^\infty(Y, J) \oplus C^\infty(Z^-, L)$*

$$\|w\|_1 \leqq c(\|\mathcal{B}w\|_2 + \|w\|_0)\,. \tag{12}$$

Theorem 1 can be applied to the oblique derivative problem $\mathcal{A}$ given by 4.2.2.2.(14). The operator $\mathcal{B}$ has the form 4.2.2.2.(18). With the notations connected with $\mathcal{B}$ in 4.2.2.2 we have $\operatorname{ord}\delta_0 = -1$, $\operatorname{ord}\varepsilon_1 = 0$, $\operatorname{ord}\eta_1 = 0$, $\operatorname{ord}\sigma_1 = 0$, i.e. $a = \operatorname{ord}\varepsilon = 1$, $c = \operatorname{ord}\eta = 1$, $b = \operatorname{ord}\sigma = 0$, $d = 0$ and $\alpha = 2$, $\gamma = 1$. Thus, in this case

$$\mathcal{H}_1^{(s)} = H^{s-1}(X) \oplus H^{s-3/2}(Z^-)\,, \quad \mathcal{H}_2^{(s)} = H^{s-2}(X) \oplus H^{s-3/2}(Y) \oplus H^{s-5/2}(Z^+)\,.$$

Till now the Fredholm property of the boundary value problem $\mathcal{B}w = h$ has been established in spaces of C^∞ sections. For certain purposes it is reasonable to consider closures in suitable Banach spaces and to have a corresponding Fredholm property. In order to avoid too complicated notations we discuss first another situation possessing all the formal properties we need.

Let M_1, M_2 be compact closed C^∞ manifolds equipped with fixed Riemannian metrics. Denote by $H^s(M_i)$, $s \in \mathbb{R}$, the usual Sobolev spaces on M_i $(i = 1, 2)$. Let

$$D\colon C^\infty(M_1) \to C^\infty(M_2) \tag{13}$$

be a Fredholm operator and let E be a parametrix of D (i.e. $ED = 1 + K_1$, $DE = 1 + K_2$ with operators K_1, K_2 with C^∞ kernels). Suppose $m = \operatorname{ord} D$, $-l = \operatorname{ord} E$ for certain $m, l \in \mathbb{R}$ in the sense of existence of continuous extensions

$$D\colon H^s(M_1) \to H^{s-m}(M_2)\,, \quad E\colon H^t(M_2) \to H^{t+l}(M_1) \tag{14}$$

for all $s \geqq s_1$, $t \geqq t_1$ for certain fixed $s_1, t_1 \in \mathbb{R}$. For simplicity these extensions are denoted by the same letters. From $Du \in C^\infty(M_2)$ for $u \in H^s(M_1)$ it obviously follows that $u \in C^\infty(M_1)$, i.e. $\ker D$ (D as an operator in $H^s(M_1)$) is equal to the kernel of (13). Let $s, t \in \mathbb{R}$ be fixed and $s \leqq t + l$, $s - m \leqq t$. Then $H^{t+l}(M_1) \subseteqq H^s(M_1)$, $H^t(M_2) \subseteqq H^{s-m}(M_2)$. Set $W = H^t(M_2)$, $V = \{u \in H^s(M_1)\colon Du \in W\}$. D and E induce linear mappings

$$D\colon V \to W\,, \tag{15}$$

$$E\colon W \to V\,. \tag{16}$$

$E(W) \subseteqq W$ follows from $E(W) \subseteqq H^{t+l}(M_1) \subseteqq H^s(M_1)$, $C^\infty(M_1) \subseteqq V$ and $D(Ef) = f + K_2 f \in v$ for all $f \in W$ ($K_2 f \in C^\infty(M_1)$). The operator $DE = 1 + K_2\colon W \to W$ is Fredholm and has index zero. Set $U_1 = \ker DE$. Then there exists a finite dimensional subspace $U_0 \subset C^\infty(M_1)$ so that for $E_1 = E_{W \ominus U_1}$ the map

$$DE_1\colon W \ominus U_1 \to W \ominus U_0 \tag{17}$$

is an isomorphism. Thus $E_1: W \ominus U_1 \to V$ is injective, $\operatorname{im} E_1 \cap \ker D = \{0\}$ and

$$D_1 = D|_{\operatorname{im} E_1}: \operatorname{im} E_1 \to W \ominus U_0 \tag{18}$$

is an algebraic isomorphism. There exists a finite dimensional subspace $U \subset W$, $U \subset C^\infty(M_2)$ with $U \cap \operatorname{im} D = \{0\}$, $U \oplus \operatorname{im} D = W$. This immediately follows from $\operatorname{im} D \supseteqq \operatorname{im} DE$ and $DE = 1 + K_2$. From (18) follows that there exists a finite dimensional space $U' \subset C^\infty(M_1)$ with

$$U' \oplus \operatorname{im} E_1 = V\,. \tag{19}$$

Consider now the linear mappings $H \ominus U_1 \xrightarrow{E_1} \operatorname{im} E_1 \xrightarrow{D_1} H \ominus U_0$. Here E_1 and D_1 are algebraic isomorphisms. Then in $\operatorname{im} E_1$ there is a well defined Banach space topology for which E_1 is a topological isomorphism. Since $F \overset{\text{def}}{=} D_1 \circ E_1$ is a topological isomorphism, the operator $D_1 = F \circ E_1^{-1}$ is also a topological isomorphism with respect to the topology in $\operatorname{im} E_1$. Because of (19) one can define a Banach space structure in V (the direct sum of U' and $\operatorname{im} E_1$). Obviously (15), (16) become Fredholm operators between the corresponding Banach spaces V, W. It is easily seen that the index of (15) is equal to the index of (13).

In the case of the operators $\mathcal{B}$ and $\mathcal{B}^{-1}$ in Theorem 1 we have a quite similar situation as for D and E, respectively. Define the spaces

$$\mathcal{H}_1^k = H^{s_k}(X, E) \oplus H^{t_k}(Y, J) \oplus H^{r_k}(Z^-, L) \quad (k = 1, 4)\,,$$

$$\mathcal{H}_2^k = H^{s_k}(X, F) \oplus H^{t_k}(Y, G) \oplus H^{r_k}(Z^+, M) \quad (k = 2, 3)$$

with indices s_k, t_k, r_k sufficiently large and $s_3 \geqq s_2$, $t_3 \geqq t_2$, $r_3 \geqq r_2$, $s_4 \geqq s_1$, $t_4 \geqq t_1$, $r_4 \geqq r_1$. A suitable choice of these indices guarantees the existence of continuous extensions $\mathcal{B}: \mathcal{H}_1^1 \to \mathcal{H}_2^2$, $\mathcal{B}^{-1}: \mathcal{H}_2^3 \to \mathcal{H}_1^4$. We have $\mathcal{H}_1^1 \supseteqq \mathcal{H}_1^4$, $\mathcal{H}_2^2 \supseteqq \mathcal{H}_2^3$. Set $W = \mathcal{H}_2^3$, $V = \{w \in \mathcal{H}_1^1: \mathcal{B}w \in \mathcal{H}_2^3\}$ and V in the mentioned way equipped with a topology. Then $\mathcal{B}: V \to W$ defines a Fredholm operator with the parametrix $\mathcal{B}^{-1}: W \to V$ and the index is independent of the choice of the s_k, t_k, r_k. Moreover, there exists a smoothing operator $\mathcal{K}$ so that for $\mathcal{R} = \mathcal{B}^{-1} + \mathcal{K}$ (wich is a parametrix of $\mathcal{B}$, too) there exist topological decompositions $V = \ker \mathcal{B} \oplus \operatorname{im} \mathcal{R}$, $W = \operatorname{im} \mathcal{B} \oplus \ker \mathcal{R}$ with finite dimensional spaces $\ker \mathcal{B}$, $\ker \mathcal{R}$ of smooth sections.

4.2.3. Over- and Underdetermined Systems

4.2.3.1. Degenerate Underdetermined Boundary Problems

The methods for treating degenerate boundary value problems can be generalized to the case of over- and underdetermined systems (cf. 3.1.2.3). In this section we consider the underdetermined case. Use similar notations as in 4.2.1.1 and consider an underdetermined elliptic system

$$\mathcal{A}_0 = \begin{pmatrix} a & K_0 \\ T_0 & Q_0 \end{pmatrix} : \begin{matrix} \mathcal{E} \\ \oplus \\ \mathcal{J}_0 \end{matrix} \to \begin{matrix} \mathcal{F} \\ \oplus \\ \mathcal{G}_0\,, \end{matrix} \tag{1}$$

$\mathcal{A}_0 \in \mathfrak{G}$, $a = A + B_0$. Then we have a right parametrix $\mathcal{P}_0 \in \mathfrak{G}$ of the form

$$\mathcal{P}_0 = \begin{pmatrix} p_0 & L_0 \\ S_0 & R_0 \end{pmatrix} : \begin{matrix} \mathcal{F} \\ \oplus \\ \mathcal{G}_0 \end{matrix} \to \begin{matrix} \mathcal{E} \\ \oplus \\ \mathcal{J}_0 \end{matrix} \tag{2}$$

(cf. 3.1.2.3, Theorem 4). Consider an operator

$$\mathcal{A} = \begin{pmatrix} a & K \\ T & Q \end{pmatrix} : \begin{matrix} \mathcal{E} \\ \oplus \\ \mathcal{J} \end{matrix} \to \begin{matrix} \mathcal{F} \\ \oplus \\ \mathcal{G}, \end{matrix} \tag{3}$$

$\mathcal{A} \in \mathfrak{G}$ (with the same left upper corner as in (1)). Using the potential operator K in (3) we pass to the following underdetermined elliptic system

$$\tilde{\mathcal{A}}_0 = \begin{pmatrix} a & K & K_0 \\ 0 & 1 & 0 \\ T_0 & 0 & Q_0 \end{pmatrix} : \begin{matrix} \mathcal{E} \\ \oplus \\ \mathcal{J} \\ \oplus \\ \mathcal{J}_0 \end{matrix} \to \begin{matrix} \mathcal{F} \\ \oplus \\ \mathcal{J} \\ \oplus \\ \mathcal{G}_0 . \end{matrix} \tag{4}$$

Then

$$\tilde{\mathcal{P}}_0 = \begin{pmatrix} p_0 & -p_0K & L_0 \\ 0 & 1 & 0 \\ S_0 & -S_0K & R_0 \end{pmatrix} : \begin{matrix} \mathcal{F} \\ \oplus \\ \mathcal{J} \\ \oplus \\ \mathcal{G}_0 \end{matrix} \to \begin{matrix} \mathcal{F} \\ \oplus \\ \mathcal{J} \\ \oplus \\ \mathcal{J}_0 \end{matrix} \tag{5}$$

is a right parametrix of (4). With the operator

$$\tilde{\mathcal{A}} = \begin{pmatrix} a & K & K_0 \\ T & Q & 0 \\ 0 & 0 & 1 \end{pmatrix} : \begin{matrix} \mathcal{E} \\ \oplus \\ \mathcal{J} \\ \oplus \\ \mathcal{J}_0 \end{matrix} \to \begin{matrix} \mathcal{F} \\ \oplus \\ \mathcal{G} \\ \oplus \\ \mathcal{J}_0 \end{matrix}$$

we get

$$\tilde{\mathcal{A}}\tilde{\mathcal{P}}_0 = \begin{pmatrix} 1 & 0 & 0 \\ Tp_0 & -Tp_0K + Q & TL_0 \\ S_0 & -S_0K & R_0 \end{pmatrix}. \tag{6}$$

Define the operator

$$\delta = \begin{pmatrix} -Tp_0K + Q & TL_0 \\ -S_0K & R_0 \end{pmatrix} : \begin{matrix} \mathcal{J} \\ \oplus \\ \mathcal{G}_0 \end{matrix} \to \begin{matrix} \mathcal{G} \\ \oplus \\ \mathcal{J}_0 . \end{matrix} \tag{7}$$

Concerning the orders of the operators in $\mathcal{A}_0$, $\mathcal{A}$ we make the same assumptions as in 4.2.1.1. Then (7) is a PDO on Y of order zero.

Lemma 1. *Let* $(y, \eta) \in S^*Y$ *and* $\sigma_\delta(y, \eta)$ *be the homogeneous principal symbol of* δ. *If* σ_δ *is surjective at* (y, η), $\sigma_Y(\mathcal{A})$ *is surjective at this point.*

Proof: The assertion of Lemma 1 follows from the following more general observation. Let

$$\mathfrak{a} = \begin{pmatrix} \alpha & \varkappa \\ \tau & \varrho \end{pmatrix} : \begin{matrix} H_1 \\ \oplus \\ L_1 \end{matrix} \to \begin{matrix} H_2 \\ \oplus \\ L_2 \end{matrix}$$

be a linear continuous operator, H_j, L_j Fréchet spaces, α a Fredholm operator, and α^{-1} a parametrix of α. Then $\mathfrak{a}$ has a right parametrix iff the operator

$$\omega = \alpha - \varkappa\alpha^{-1}\tau : L_1 \to L_2$$

has a right parametrix. If α is an isomorphism and α^{-1} the inverse, $\mathfrak{a}$ is surjective iff ω is surjective. This follows by similar arguments as in 3.1.1.5, Proposition 1. The result can be applied to the boundary symbol of (6) with $\alpha = 1$ and $\varrho = \sigma_\partial$ at the point $(\dot{y}, \eta)$. Finally, $\sigma_Y(\tilde{\mathcal{A}})$ is surjective if $\sigma_A(\tilde{\mathcal{A}}\tilde{\mathcal{P}}_0)$ is surjective. Then $\sigma_Y(\mathcal{A})$ is surjective, too. □

Lemma 2. *Let $\mathcal{A}_0 \in \mathfrak{G}$ be an underdetermined elliptic system of the form* (1) *with*

$$Q_0 = 0\,. \tag{8}$$

Suppose that

$$\sigma_{K_0}\colon p^*J_0 \to p^*F^+ \quad \text{is injective} \tag{9}$$

and

$$\operatorname{im} \sigma_{K_0} \cap \operatorname{im} (\Pi^+\sigma_A + \Pi'\sigma_{B_0}) = \{0\}\,. \tag{10}$$

Let $\mathcal{P}_0$ be a right parametrix of $\mathcal{A}_0$ of the form (8). *Then there exists a topological decomposition*

$$\mathcal{E} = \mathcal{E}_0 \oplus \mathcal{E}_1 \oplus \mathcal{N} \tag{11}$$

with $\dim \mathcal{N} < \infty$ *so that modulo the corresponding negligible operators the following identities hold*

$$(p_0 a + L_0 T_0)|_{\mathcal{E}_1} = 1_{\mathcal{E}_1}\,, \quad p_0 K_0 = 0\,, \tag{12}$$

$$(S_0 a + R_0 T_0)|_{\mathcal{E}_1} = 0, \qquad S_0 K_0 = 1\,. \tag{13}$$

Proof: First discuss the situation on boundary symbol level. Set for abbreviation

$$\sigma_Y(\mathcal{A}_0) = \begin{pmatrix} \alpha_0 & \varkappa_0 \\ \tau_0 & 0 \end{pmatrix}, \quad \sigma_Y(\mathcal{P}_0) = \begin{pmatrix} \pi_0 & \lambda_0 \\ \sigma_0 & \varrho_0 \end{pmatrix}$$

and $p^*V = \operatorname{im} \varkappa_0 (= \sigma_{K_0}(p^*J_0))$. Denote by $p^*E^+ \ominus p^*V$ some complement of p^*V in p^*E^+. Then $\sigma_Y(\mathcal{A}_0)$ can be regarded as a surjective mapping

$$\begin{pmatrix} \begin{pmatrix} \alpha_0 \\ \tau_0 \end{pmatrix} & 0 \\ 0 & \varkappa_0 \end{pmatrix} : \begin{matrix} p^*E^+ \\ \oplus \\ p^*J_0 \end{matrix} \to \begin{matrix} p^*F^+ \ominus p^*V \\ \oplus \\ p^*G_0 \\ \oplus \\ p^*V \end{matrix}$$

with an isomorphism $\varkappa_0\colon p^*J_0 \to p^*V$. Formally the situation is as follows. We have a surjective mapping

$$\mathfrak{a} = \begin{pmatrix} \alpha & 0 \\ 0 & \varkappa_0 \end{pmatrix} : \begin{matrix} H_1 \\ \oplus \\ L_1 \end{matrix} \to \begin{matrix} H_2 \\ \oplus \\ L_2 \end{matrix}$$

with an isomorphism $\varkappa_0\colon L_1 \to L_2$ (H_j, L_j are bundles of Fréchet spaces, $j = 1, 2$).

$$\mathfrak{b} = \begin{pmatrix} \beta & \mu \\ \varrho_0 & \sigma_0 \end{pmatrix} : \begin{matrix} H_2 \\ \oplus \\ L_2 \end{matrix} \to \begin{matrix} H_1 \\ \oplus \\ L_1 \end{matrix}$$

is a right inverse of $\mathfrak{a}$ (the meaning of the notations is obvious). Then

$$\mathfrak{a}\mathfrak{b} = \begin{pmatrix} \alpha\beta & \alpha\mu \\ \varkappa_0\varrho_0 & \varkappa_0\sigma_0 \end{pmatrix} = \begin{pmatrix} 1 & 0 \\ 0 & 1 \end{pmatrix},$$

i.e. $\mu: L_2 \to \ker \mathfrak{a}$, $\varrho_0: H_2 \to \ker \varkappa_0$. Thus $\varrho_0 = 0$ and σ_0 is a right inverse of $\varkappa_0$. Since $\varkappa_0$ has an inverse, too, we obtain $\sigma_0 = \varkappa_0^{-1}$. Now

$$\mathfrak{b}\mathfrak{a} = \begin{pmatrix} \beta\alpha & \mu\varkappa_0 \\ 0 & 1 \end{pmatrix}.$$

Here the right lower corner $1 = \sigma_0\varkappa_0$ corresponds to the homogeneous principal symbol of S_0K_0. $\mathfrak{b}\mathfrak{a}$ is a projection parallel to $\ker \mathfrak{a}$ onto some complement of $\ker \mathfrak{a}$. Since $\ker \mathfrak{a} \subseteqq H_1 \oplus \{0\}$, we have $(\ker \mathfrak{a})^\perp = H_1' \oplus L_1$ for this complement with some subspace $H_1' \subseteqq H_1$. Now returning to the original notations we get for suitable subbundles E_j $(j = 0, 1)$ of p^*E^+ with $E_0 \oplus E_1 = p^*E^+$ that $\sigma_Y(\mathcal{A}_0) = E_0 \oplus \{0\}$, $(\ker \sigma_Y(\mathcal{A}_0))^\perp = E_1 \oplus p^*J_0$. Then it is clear that relations on boundary symbol level analogous to (12), (13) are valid.

Next prove (12), (13) on operator level. It follows from (9) that $\dim \ker K_0 < \infty$ (cf. 3.1.2.3, Corollary 3). From (10) it follows that $\dim (\operatorname{im} K_0 \cap \operatorname{im} a) < \infty$ (cf. 3.2.3.2, Theorem 2). Thus there exists a finite-dimensional subspace $\mathcal{V}_0 \subseteqq \ker \mathcal{A}_0$ and a subspace $\mathcal{E}_0 \subseteqq \mathcal{E}$ with

$$\begin{matrix} \mathcal{E}_0 \\ \oplus \\ \{0\} \end{matrix} \oplus V_0 = \ker \mathcal{A}_0 .$$

Furthermore, we have $\ker \mathcal{A}_0 \subseteqq \ker \mathcal{P}_0\mathcal{A}_0$ and there exists a finite-dimensional subspace $\mathcal{W}_0 \subseteqq \ker P_0\mathcal{A}_0$ with $\mathcal{W}_0 \oplus \ker \mathcal{A}_0 = \ker \mathcal{P}_0\mathcal{A}_0$. The kernel of the operator $\mathcal{I} - \mathcal{P}_0\mathcal{A}_0$ is a complement of

$$\mathcal{W}_0 \oplus \ker \mathcal{A}_0 = \begin{matrix} \mathcal{E}_0 \\ \oplus \\ \{0\} \end{matrix} \oplus \mathcal{V}_0 \oplus \mathcal{W}_0 .$$

Thus there exists a finite-dimensional subspace $\mathcal{V}_1 \subset \mathcal{E} \oplus \mathcal{J}_0$ and a subspace $\mathcal{E}_1 \subseteqq \mathcal{E}$ with

$$\ker (\mathcal{I} - \mathcal{P}_0\mathcal{A}_0) \oplus \mathcal{V}_1 = \mathcal{E}_1 \oplus \mathcal{J}_0 ,$$

and $\mathcal{E} = \mathcal{E}_0 \oplus \mathcal{E}_1 \oplus \mathcal{N}$ for some finite dimensional subspace $\mathcal{N} \subset \mathcal{E}$. The operator $\mathcal{P}_0\mathcal{A}_0$ acts as identity over $\mathcal{E}_1 \oplus \mathcal{J}_0$ modulo $\mathfrak{G}^{-\infty}$. This proves (12), (13). □

Set

$$\delta_{11} = -Tp_0K + Q,\ \delta_{12} = TL_0,\ \delta_{21} = -S_0K,\ \delta_{22} = R_0$$

and consider an interior boundary problem

$$\mathcal{D} = \begin{pmatrix} \delta_{11} & \delta_{12} & \varkappa_1 \\ \delta_{21} & \delta_{22} & \varkappa_2 \\ \tau_1 & \tau_2 & \varrho \end{pmatrix} : \begin{matrix} \mathcal{J} \\ \oplus \\ \mathcal{G}_0 \\ \oplus \\ \mathcal{L} \end{matrix} \to \begin{matrix} \mathcal{G} \\ \oplus \\ \mathcal{J}_0 \\ \oplus \\ \mathcal{M} \end{matrix} \tag{14}$$

with suitable Fréchet spaces $\mathcal{L}$, $\mathcal{M}$ and operators $\varkappa_j, \tau_j, \varrho$ $(j = 1, 2)$. Suppose that (14) is a Fredholm operator. Then the operator

$$\tilde{\mathcal{E}} = \begin{pmatrix} 1 & 0 & 0 & 0 \\ Tp_0 & \delta_{11} & \delta_{12} & \varkappa_1 \\ S_0 & \delta_{21} & \delta_{22} & \varkappa_2 \\ 0 & \tau_1 & \tau_2 & \varrho \end{pmatrix} : \begin{matrix} \mathcal{F} \\ \oplus \\ \mathcal{J} \\ \oplus \\ \mathcal{G}_0 \\ \oplus \\ \mathcal{L} \end{matrix} \to \begin{matrix} \mathcal{F} \\ \oplus \\ \mathcal{G} \\ \oplus \\ \mathcal{J}_0 \\ \oplus \\ \mathcal{M} \end{matrix} \tag{15}$$

is Fredholm, too. Furthermore, we have the Fredholm operator

$$\begin{pmatrix} \tilde{\mathscr{A}}_0 & 0 \\ 0 & 1 \end{pmatrix} : \begin{matrix} \operatorname{im} \tilde{\mathscr{P}}_0 \\ \oplus \\ \mathscr{L} \end{matrix} \to \begin{matrix} \mathscr{F} \\ \oplus \\ \mathscr{J} \\ \oplus \\ \mathscr{G}_0 \\ \oplus \\ \mathscr{L} . \end{matrix}$$

Thus

$$\tilde{\mathscr{B}} = \tilde{\mathscr{C}} \begin{pmatrix} \tilde{\mathscr{A}}_0 & 0 \\ 0 & 1_{\mathscr{L}} \end{pmatrix} : \begin{matrix} \operatorname{im} \tilde{\mathscr{P}}_0 \\ \oplus \\ \mathscr{L} \end{matrix} \to \begin{matrix} \mathscr{F} \\ \oplus \\ \mathscr{G} \\ \oplus \\ \mathscr{J}_0 \\ \oplus \\ \mathscr{M} \end{matrix} \tag{16}$$

is a Fredholm operator. $\tilde{\mathscr{B}}$ may also be considered as operator

$$\tilde{\mathscr{B}} : \mathscr{E} \oplus \mathscr{J} \oplus \mathscr{J} \oplus \mathscr{L} \to \mathscr{F} \oplus \mathscr{G} \oplus \mathscr{J}_0 \oplus \mathscr{M} \tag{17}$$

and has the form

$$\tilde{\mathscr{B}} = \begin{pmatrix} a & K & K_0 & 0 \\ T(p_0 a + L_0 T_0) & Q & T(p_0 K_0 + L_0 Q_0) & \varkappa_1 \\ S_0 a + R_0 T_0 & 0 & S_0 K_0 + R_0 Q_0 & \varkappa_2 \\ \tau_2 T_0 & \tau_1 & \tau_2 Q_0 & \varrho \end{pmatrix} \tag{18}$$

possessing a right parametrix $\tilde{\mathscr{B}}^{-1}$ with $\operatorname{im} \tilde{\mathscr{B}}^{-1} \subseteq \operatorname{im} \tilde{\mathscr{P}}_0 \oplus \mathscr{L}$. For $Q_0 = 0$ we get by rearranging the rows and columns in $\tilde{\mathscr{B}}$ the operator

$$\begin{pmatrix} a & K & 0 & K_0 \\ T(p_0 a + L_0 T_0) & Q & \varkappa_1 & T p_0 K_0 \\ \tau_2 T_0 & \tau_1 & \varrho & 0 \\ S_0 a + R_0 T_0 & 0 & \varkappa_2 & 1 \end{pmatrix} : \begin{matrix} \mathscr{E} \\ \oplus \\ \mathscr{J} \\ \oplus \\ \mathscr{L} \\ \oplus \\ \mathscr{J}_0 \end{matrix} \to \begin{matrix} \mathscr{F} \\ \oplus \\ \mathscr{G} \\ \oplus \\ \mathscr{M} \\ \oplus \\ \mathscr{J}_0 . \end{matrix} \tag{19}$$

Pass to the operator

$$\mathscr{B}_0 = \begin{pmatrix} a & K & 0 \\ T(p_0 a + L_0 T_0) & Q & \varkappa_1 \\ \tau_2 T_0 & \tau_1 & \varrho \end{pmatrix} - \begin{pmatrix} K_0 \\ T p_0 K_0 \\ 0 \end{pmatrix} (S_0 a + R_0 T_0, 0, \varkappa_2)$$

$$= \begin{pmatrix} a - K_0(S_0 a + R_0 T_0) & K & -K_0 \varkappa_2 \\ T(p_0 a + L_0 T_0) - T p_0 K_0 (S_0 a + R_0 T_0) & Q & \varkappa_1 - T p_0 K_0 \varkappa_2 \\ \tau_2 T_0 & \tau_1 & \varrho \end{pmatrix} : \begin{matrix} \mathscr{E} \\ \oplus \\ \mathscr{J} \\ \oplus \\ \mathscr{L} \end{matrix} \to \begin{matrix} \mathscr{F} \\ \oplus \\ \mathscr{G} \\ \oplus \\ \mathscr{M} . \end{matrix} \tag{20}$$

Applying the proof of Lemma 1 we obtain under the conditions of Lemma 2 that $\mathscr{B}_0$ has a right parametrix with $\operatorname{im} \mathscr{B}_0 \subseteq \mathscr{E}_1 \oplus \mathscr{J} \oplus \mathscr{L}$.

Theorem 3. *Let $\mathcal{A} \in \mathfrak{G}$ be an operator of the form* (3), *$\sigma_Y(\mathcal{A})$ surjective and $\sigma_Y(\mathcal{A})$ surjective over $Y \setminus Z$. Suppose that we are given an underdetermined elliptic system with the properties* (8), (9), (10) *and in such a way that the operator* (7) *is elliptic over $Y \setminus Z$ and admits a Fredholm problem $\mathcal{D}$ of the form* (14). *Then the operator*

$$\mathcal{B} = \begin{pmatrix} a & K & -K_0\varkappa_2 \\ T & Q & \varkappa_1 \\ \tau_2 T_0 & \tau_1 & \varrho \end{pmatrix} : \begin{matrix} \mathcal{E} \\ \oplus \\ \mathcal{J} \\ \oplus \\ \mathcal{L} \end{matrix} \to \begin{matrix} \mathcal{F} \\ \oplus \\ \mathcal{G} \\ \oplus \\ \mathcal{M} \end{matrix} \tag{21}$$

has a right parametrix $\mathcal{B}^{-1}$ with $\operatorname{im} \mathcal{B}^{-1} \subseteqq \mathcal{E}_1 \oplus \mathcal{J} \oplus \mathcal{L}$ ($\mathcal{E}_1$ *is the space defined in Lemma* 2).

Proof: Obviously, we can replace $\operatorname{im} \tilde{\mathcal{P}}_0$ by $\mathcal{E}_1 \oplus \mathcal{J} \oplus \mathcal{J}_0$. Then it is clear that by $\tilde{\mathcal{B}}$ a Fredholm operator $\mathcal{E}_1 \oplus \mathcal{J} \oplus \mathcal{L} \oplus \mathcal{J}_0 \to \mathcal{F} \oplus \mathcal{G} \oplus \mathcal{M} \oplus \mathcal{J}_0$ is induced and (20) defines a Fredholm operator $\mathcal{E}_1 \oplus \mathcal{J} \oplus \mathcal{L} \to \mathcal{F} \oplus \mathcal{G} \oplus \mathcal{M}$ (cf. 3.1.1.5). Now (12), (13) show that $\mathcal{B}_0$ and $\mathcal{B}$ are equal over $\mathcal{E}_1 \oplus \mathcal{J} \oplus \mathcal{L}$ modulo smoothing operators. The restriction of $\mathcal{B}_0$ to $\mathcal{E}_1 \oplus \mathcal{J} \oplus \mathcal{L}$ has a parametrix. It can be regarded as a right parametrix of $\mathcal{B}_0$ as well as of $\mathcal{B}$. This proves the Theorem. □

Theorem 4. *Let $\mathcal{A}_0 \in \mathfrak{G}$ be an underdetermined elliptic system with the properties* (8), (9), (10) *and let $\mathcal{P}_0 \in \mathfrak{G}$ be a right parametrix of $\mathcal{A}_0$ of the form* (2). *Let*

$$\delta = \begin{pmatrix} \delta_{11} & \delta_{12} \\ \delta_{21} & R_0 \end{pmatrix} : \begin{matrix} \mathcal{J} \\ \oplus \\ \mathcal{G}_0 \end{matrix} \to \begin{matrix} \mathcal{G} \\ \oplus \\ \mathcal{J}_0 \end{matrix} \tag{22}$$

be a pseudo-differential operator over X of order 0 *which is elliptic over $Y \setminus Z$. Suppose that* (14) *is a Fredholm operator with the given operator* (22) *in the left upper corner. Consider arbitrary operators $T_1 : \mathcal{E} \to \mathcal{G}$, $K_1 : \mathcal{J} \to \mathcal{F}$ in $\mathfrak{G}$ (i.e. T_1 of trace and K_1 of potential type) and set*

$$K = ap_0K_1 - K_0\delta_{21}\,, \quad T = T_1p_0a + \delta_{12}T_0\,, \quad Q = T_1p_0K_1 + \delta_{11}\,. \tag{23}$$

Then the operator $\mathcal{A}$ of the form (3) *with K, T, Q given by* (23) (*and a as in $\mathcal{A}_0$*) *has the property that $\sigma_Y(\mathcal{A})$ is surjective over $Y \setminus Z$ and the operator $\mathcal{B}$ given by* (21) *has a right parametrix $\mathcal{B}^{-1}$ with* $\operatorname{im} \mathcal{B}^{-1} \subseteqq \mathcal{E}_1 \oplus \mathcal{J} + \mathcal{L}$.

The proof of Theorem 4 is quite similar to the proof 4.2.1.2, Theorem 3 and shall be omitted. More details are given in Pillat/Schulze [2].

Further results of 4.2.1 can be carried over to the underdetermined case, in particular an explicit expression of a right parametrix. Details are left to the reader. From Theorem 4 and the results in 4.2.2.1 there follow many concrete examples of boundary problems for underdetermined systems for which the right Shapiro-Lopatinski condition degenerates and for which the method of reduction to the boundary is applicable.

4.2.3.2. The Overdetermined Case

Let now $\mathcal{A}_0 \in \mathfrak{G}$ be an overdetermined elliptic system with a left parametrix $\mathcal{P}_0 \in \mathfrak{G}$ (cf. 3.1.2.3, Theorem 2). Formally we use here the same notations as in 4.2.3.1.(1), (2). Consider an arbitrary operator $\mathcal{A} \in \mathfrak{G}$ (given in the form 4.2.3.1.(3)) with the left upper corner a as in $\mathcal{A}_0$. We pass to the overdetermined elliptic operator

$$\tilde{\mathcal{A}}_0 = \begin{pmatrix} a & 0 & K_0 \\ T & 1 & 0 \\ T_0 & 0 & Q_0 \end{pmatrix} : \begin{matrix} \mathcal{E} \\ \oplus \\ \mathcal{G} \\ \oplus \\ \mathcal{J}_0 \end{matrix} \to \begin{matrix} \mathcal{F} \\ \oplus \\ \mathcal{G} \\ \oplus \\ \mathcal{G}_0 \end{matrix} \tag{1}$$

with the trace operator T contained in $\mathcal{A}$. Then

$$\tilde{\mathcal{P}}_0 = \begin{pmatrix} p_0 & 0 & L_0 \\ -Tp_0 & 1 & -TL_0 \\ S_0 & 0 & R_0 \end{pmatrix} : \begin{matrix} \mathcal{F} \\ \oplus \\ \mathcal{G} \\ \oplus \\ \mathcal{G}_0 \end{matrix} \to \begin{matrix} \mathcal{E} \\ \oplus \\ \mathcal{G} \\ \oplus \\ \mathcal{J}_0 \end{matrix} \tag{2}$$

is a left parametrix of $\tilde{\mathcal{A}}_0$. Furthermore, put

$$\tilde{\mathcal{A}} = \begin{pmatrix} a & K & 0 \\ T & Q & 0 \\ T_0 & 0 & 1 \end{pmatrix} : \begin{matrix} \mathcal{E} \\ \oplus \\ \mathcal{J} \\ \oplus \\ \mathcal{G}_0 \end{matrix} \to \begin{matrix} \mathcal{F} \\ \oplus \\ \mathcal{G} \\ \oplus \\ \mathcal{G}_0 . \end{matrix} \tag{3}$$

Then

$$\tilde{\mathcal{P}}_0\tilde{\mathcal{A}} = \begin{pmatrix} 1 & p_0K & L_0 \\ 0 & -Tp_0K + Q & -TL_0 \\ 0 & S_0K & R_0 \end{pmatrix} : \begin{matrix} \mathcal{E} \\ \oplus \\ \mathcal{J} \\ \oplus \\ \mathcal{G}_0 \end{matrix} \to \begin{matrix} \mathcal{E} \\ \oplus \\ \mathcal{G} \\ \oplus \\ \mathcal{J}_0 \end{matrix} \tag{4}$$

and

$$\delta = \begin{pmatrix} \delta_{11} & \delta_{12} \\ \delta_{21} & \delta_{22} \end{pmatrix} = \begin{pmatrix} -Tp_0K + Q & -TL_0 \\ S_0K & R_0 \end{pmatrix} : \begin{matrix} \mathcal{J} \\ \oplus \\ \mathcal{G}_0 \end{matrix} \to \begin{matrix} \mathcal{G} \\ \oplus \\ \mathcal{J}_0 \end{matrix} \tag{5}$$

is a pseudo-differential operator over Y. Concerning the orders we make the same assumptions here as in the previous section so that δ has order zero.

Suppose again that $\mathcal{P}_0$ is found in such a way that (5) is elliptic over $Y \setminus Z$ for some subset Z of the boundary and that there exists an interior boundary problem

$$\mathcal{D} = \begin{pmatrix} \delta_{11} & \delta_{12} & \varkappa_1 \\ \delta_{21} & \delta_{22} & \varkappa_2 \\ \tau_1 & \tau_2 & \varrho \end{pmatrix} : \begin{matrix} \mathcal{J} \\ \oplus \\ \mathcal{G}_0 \\ \oplus \\ \mathcal{L} \end{matrix} \to \begin{matrix} \mathcal{G} \\ \oplus \\ \mathcal{J}_0 \\ \oplus \\ \mathcal{M} \end{matrix} \tag{6}$$

which defines a Fredholm operator. $\mathcal{L}$, $\mathcal{M}$ are suitable Fréchet spaces and τ_j, $\varkappa_j$, ϱ $(j = 1, 2)$ a continuous operators. Then the operator

$$\tilde{\mathcal{C}} = \begin{pmatrix} 1 & p_0K & L_0 & 0 \\ 0 & \delta_{11} & \delta_{12} & \varkappa_1 \\ 0 & \delta_{21} & \delta_{22} & \varkappa_2 \\ 0 & \tau_1 & \tau_2 & \varrho \end{pmatrix} : \begin{matrix} \mathcal{E} \\ \oplus \\ \mathcal{J} \\ \oplus \\ \mathcal{G}_0 \\ \oplus \\ \mathcal{L} \end{matrix} \to \begin{matrix} \mathcal{E} \\ \oplus \\ \mathcal{G} \\ \oplus \\ \mathcal{J}_0 \\ \oplus \\ \mathcal{M} \end{matrix}$$

is Fredholm, too. Next consider the operator

$$\tilde{\mathscr{B}} = \begin{pmatrix} \tilde{\mathscr{A}}_0 & 0 \\ 0 & 1_{\mathscr{M}} \end{pmatrix} \tilde{\mathscr{C}} = \begin{pmatrix} a & (ap_0 + K_0S_0)\,K & aL_0 + K_0R_0 & K_0\varkappa_2 \\ T & Q & 0 & \varkappa_1 \\ T_0 & (T_0p_0 + Q_0S_0)\,K & T_0L_0 + Q_0R_0 & Q_0\varkappa_2 \\ 0 & \tau_1 & \tau_2 & \varrho \end{pmatrix} : \begin{matrix} \mathscr{E} \\ \oplus \\ \mathscr{J} \\ \oplus \\ \mathscr{G}_0 \\ \oplus \\ \mathscr{L} \end{matrix} \to \begin{matrix} \mathscr{F} \\ \oplus \\ \mathscr{G} \\ \oplus \\ \mathscr{G}_0 \\ \oplus \\ \mathscr{M}. \end{matrix} \tag{7}$$

Since by $\tilde{\mathscr{B}}$ a Fredholm operator $\tilde{\mathscr{B}}: \mathscr{E} \oplus \mathscr{J} \oplus \mathscr{G}_0 \oplus \mathscr{L} \to (\mathrm{im}\, \tilde{\mathscr{A}}_0) \oplus \mathscr{M}$ is induced, $\tilde{\mathscr{B}}$ has a left parametrix $\tilde{\mathscr{B}}^{-1}$.

Lemma 1. *Let $\mathscr{A}_0 \in \mathfrak{G}$ be an overdetermined elliptic system of the form* 4.2.3.1.(1) *with*

$$Q_0 = 0\,. \tag{8}$$

Suppose that

$$\ker\,(\Pi^+\sigma_A + \Pi'\sigma_{B_0}) \cap \ker \Pi'\sigma_{T_0} = \{0\} \tag{9}$$

and that $\Pi'\sigma_{T_0}: p^*E^+ \to p^*G_0$ induces an isomorphism

$$\Pi'\sigma_{T_0}: \ker\,(\Pi^+\sigma_A + \Pi'\sigma_{B_0}) \xrightarrow{\cong} p^*G_0\,. \tag{10}$$

Let $\mathscr{P}_0 \in \mathfrak{G}$ be a left parametrix of $\mathscr{A}_0$ (use the same notations as in 4.2.3.1.(2)*). Then there exists a topological decomposition $\mathscr{F} = \mathscr{F}_0 \oplus \mathscr{F}_1 \oplus \mathscr{N}$ with $\dim \mathscr{N} < \infty$ so that*

$$\mathscr{A}_0: \mathscr{E} \oplus \mathscr{J}_0 \to \mathscr{F}_1 \oplus \mathscr{G}_0 \tag{11}$$

defines a Fredholm operator and that modulo the corresponding negligible operators the following identities hold

$$(ap_0 + K_0S_0)|_{\mathscr{F}_1} = 1_{\mathscr{F}_1}\,, \quad aL_0 + K_0R_0 = 0\,, \tag{12}$$

$$T_0p_0|_{\mathscr{F}_1} = 0\,, \quad T_0L_0 = 1\,. \tag{13}$$

The proof of Lemma 1 is in a sense dual to that of 4.2.3.1, Lemma 2 and left to the reader.

Note that an analogous remark in the overdetermined case holds as in 4.2.3.1, Lemma 1, but we do not use it explicitly. We can assume that $\mathscr{F}_1 \oplus \mathscr{G}_0 = \mathrm{im}\, \mathscr{A}_0\mathscr{P}_0$, especially $\mathrm{im}\, K_0 \subseteqq \mathscr{F}_1$. Then $\tilde{\mathscr{B}}$ defines a Fredholm operator

$$\tilde{\mathscr{B}}: \mathscr{E} \oplus \mathscr{J} \oplus \mathscr{G}_0 \oplus \mathscr{L} \to \mathscr{F}_1 \oplus \mathscr{G} \oplus \mathscr{G}_0 \oplus \mathscr{M}\,.$$

Concerning K we introduce the following additional assumption

$$\mathrm{im}\, K \subseteqq \mathscr{F}_1\,. \tag{14}$$

Then

$$\tilde{\mathscr{B}} = \begin{pmatrix} a & K & 0 & K_0\varkappa_2 \\ T & Q & 0 & \varkappa_1 \\ T_0 & 0 & 1 & 0 \\ 0 & \tau_1 & \tau_2 & \varrho \end{pmatrix} : \begin{matrix} \mathscr{E} \\ \oplus \\ \mathscr{J} \\ \oplus \\ \mathscr{G}_0 \\ \oplus \\ \mathscr{L} \end{matrix} \to \begin{matrix} \mathscr{F}_1 \\ \oplus \\ \mathscr{G} \\ \oplus \\ \mathscr{G}_0 \\ \oplus \\ \mathscr{M} \end{matrix} \tag{15}$$

is a Fredholm operator and obviously

$$\mathcal{B} = \begin{pmatrix} a & K & K_0\varkappa_2 \\ T & Q & \varkappa_1 \\ -\tau_2 T_0 & \tau_1 & \varrho \end{pmatrix} : \begin{matrix} \mathcal{E} \\ \oplus \\ \mathcal{J} \\ \oplus \\ \mathcal{L} \end{matrix} \to \begin{matrix} \mathcal{F}_1 \\ \oplus \\ \mathcal{G} \\ \oplus \\ \mathcal{M} \end{matrix}, \tag{16}$$

too (cf. 3.1.1.5). Thus we get the following

Theorem 2. *Let $\mathcal{A} \in \mathfrak{G}$ be an operator of the form* 4.2.3.1.(3) *satisfying the conditions of Lemma* 1 *and* (14). *Suppose that for* (5) *there exists a Fredholm operator* (6). *Then the operator*

$$\mathcal{B} : \mathcal{E} \oplus \mathcal{J} \oplus \mathcal{L} \to \mathcal{F} \oplus \mathcal{G} \oplus \mathcal{M} \tag{17}$$

given by the matrix (16) *has a left parametrix $\mathcal{B}^{-1}$ with* $\ker \mathcal{B}^{-1} \subseteqq \mathcal{F}_1 \oplus \mathcal{G} \oplus \mathcal{M}$.

Note that the existence of a left parametrix $\mathcal{B}^{-1}$ of $\mathcal{B}$ yields $\dim \mathcal{B} < \infty$. It also follows that $\dim(\operatorname{im} \mathcal{B} \cap \ker \mathcal{B}^{-1}) < \infty$.

A characterization of the class of situations considered in Theorem 2 gives

Theorem 3. *Let $\mathcal{A}_0 \in \mathfrak{G}$ be an overdetermined elliptic system satisfying the assumptions of Lemma* 1 *and let $\mathcal{P}_0 \in \mathfrak{G}$ be a left parametrix of $\mathcal{A}_0$. Let*

$$\delta = \begin{pmatrix} \delta_{11} & \delta_{12} \\ \delta_{21} & R_0 \end{pmatrix} : \begin{matrix} \mathcal{J} \\ \oplus \\ \mathcal{G}_0 \end{matrix} \to \begin{matrix} \mathcal{G} \\ \oplus \\ \mathcal{J}_0 \end{matrix} \tag{18}$$

be a PDO over Y of order O being elliptic over $Y \setminus Z$. Suppose that (6) *is a Fredholm operator with* (18) *in the left upper corner. Consider arbitrary operators $T_1 : \mathcal{E} \to \mathcal{G}$, $K_1 : \mathcal{J} \to \mathcal{F}$ in $\mathfrak{G}$ (i.e. T_1 a trace operator, K_1 a potential operator) and suppose that*

$$\operatorname{im} K_1 \subseteqq \mathcal{F}_1 . \tag{19}$$

Moreover, define the operators

$$K = ap_0K_1 + K_0\delta_{21}, \quad T = T_1p_0a - \delta_{12}T_0, \quad Q = T_1p_0K_1 + \delta_{11}. \tag{20}$$

Then the operator $\mathcal{A}$ of the form 4.2.3.1.(3) *with K, T, Q given by* (20) *(and a as in $\mathcal{A}_0$) has the property $\sigma_Y(\mathcal{A})$ injective over $Y \setminus Z$,*

$$\operatorname{im} K \subseteqq F_1 \tag{21}$$

and the operator $\mathcal{B}$ mentioned in Theorem 2 *has a left parametrix $\mathcal{B}^{-1}$ with* $\ker \mathcal{B}^{-1} \subseteqq \mathcal{F}_1 \oplus \mathcal{G} \oplus \mathcal{M}$.

Proof: With the given operator δ in (18) define the Fredholm operator

$$\tilde{\mathcal{C}} = \begin{pmatrix} 1 & \zeta_1 & \zeta_2 & \zeta_3 \\ 0 & \delta_{11} & \delta_{12} & \varkappa_1 \\ 0 & \delta_{21} & \delta_{22} & \varkappa_2 \\ 0 & \tau_1 & \tau_2 & \varrho \end{pmatrix} : \begin{matrix} \mathcal{E} \\ \oplus \\ \mathcal{J} \\ \oplus \\ \mathcal{G}_0 \\ \oplus \\ \mathcal{L} \end{matrix} \to \begin{matrix} \mathcal{E} \\ \oplus \\ \mathcal{G} \\ \oplus \\ \mathcal{J}_0 \\ \oplus \\ \mathcal{M} . \end{matrix} \tag{22}$$

We put $\zeta_1 = p_0K_1, \zeta_2 = L_0, \zeta_3 = 0$. The operator

$$\tilde{\mathcal{A}}_1 = \begin{pmatrix} a & 0 & K_0 \\ T_1 & 1 & 0 \\ T_0 & 0 & 0 \end{pmatrix} : \begin{matrix} \mathcal{E} \\ \oplus \\ \mathcal{G} \\ \oplus \\ \mathcal{J}_0 \end{matrix} \to \begin{matrix} \mathcal{F} \\ \oplus \\ \mathcal{G} \\ \oplus \\ \mathcal{G}_0 \end{matrix} \tag{23}$$

is overdetermined elliptic and we have modulo the corresponding negligible operators

$$\mathcal{B}_1 = \begin{pmatrix} \tilde{\mathcal{A}}_1 & 0 \\ 0 & 1_{\mathcal{M}} \end{pmatrix} \tilde{\mathcal{E}} = \left(\begin{array}{ccc|c} & \tilde{\mathcal{A}}_2 & & K_0\varkappa_2 \\ & & & \varkappa_1 \\ & & & 0 \\ \hline 0 & \tau_1 & \tau_2 & \varrho \end{array}\right)$$

with

$$\tilde{\mathcal{A}}_2 = \begin{pmatrix} a & ap_0K_1 + K_0\delta_{21} & 0 \\ T_1 & T_1p_0K_1 + \delta_{11} & T_1L_0 + \delta_{12} \\ T_0 & 0 & 1 \end{pmatrix} : \begin{matrix} \mathcal{E} \\ \oplus \\ \mathcal{J} \\ \oplus \\ \mathcal{G}_0 \end{matrix} \to \begin{matrix} \mathcal{F} \\ \oplus \\ \mathcal{G} \\ \oplus \\ \mathcal{G}_0 . \end{matrix}$$

Moreover, define $\mathcal{A}$ in the form as in the Theorem. Then injectivity of $\sigma_Y(\mathcal{A})$ over $Y \setminus Z$ is a consequence of trivial arguments about operators in matrix form similar to those in 3.1.1.5. Using (12), (13) and $\mathcal{P}_0\mathcal{A}_0 - \mathcal{J} \in \mathfrak{G}^{-\infty}$ we obtain

$$\begin{aligned} -Tp_0K + Q &= -(T_1p_0a - \delta_{12}T_0)\, p_0(ap_0K_1 + K_0\delta_{21}) + T_1p_0K_1 + \delta_{11} \\ &= -T_1p_0ap_0ap_0K_1 + \delta_{12}T_0p_0ap_0K_1 + T_1p_0K_1 + \delta_{11} \\ &= -T_1(1 - L_0T_0)(1 - L_0T_0)\, p_0K_1 + \delta_{12}T_0(1 - L_0T_0)\, p_0K_1 \\ &\quad + T_1p_0K_1 + \delta_{11} = \delta_{11}, \end{aligned}$$

$$-TL_0 = -(T_1p_0a - \delta_{12}T_0)\, L_0 = T_1p_0K_0R_0 + \delta_{12} = \delta_{12}$$

$$S_0K = S_0(ap_0K_1 + K_0\delta_{21}) = \delta_{21} - R_0T_0p_0K_1 = \delta_{21}.$$

Since $\operatorname{im} K_0 \subseteqq \mathcal{F}_1$, $\operatorname{im} K_1 \subseteqq \mathcal{F}_1$, $\operatorname{im}(ap_0K_1) = \operatorname{im}(1 - K_0S_0)\, K_1$, we finally get $\operatorname{im} K \subseteqq \mathcal{F}_1$. Thus Theorem 3 is proved. □

Now let $\mathcal{D}^{-1}$ be a parametrix of $\mathcal{D}$ and use for its elements the notations of 4.2.1.3. Set

$$\begin{aligned} &\zeta = (\zeta_1, \zeta_2) = (p_0K, L_0), \\ &(\zeta\varepsilon)_k = \zeta_1\varepsilon_{1k} + \zeta_2\varepsilon_{2k}, \quad k = 1, 2, \\ &\beta_0 = p_0 + (\zeta\varepsilon)_1 Tp_0 - (\zeta\varepsilon)_2 S_0, \\ &\beta_1 = -\varepsilon_{11}Tp_0 + \varepsilon_{12}S_0,\ \beta_2 = -\sigma_1Tp_0 + \sigma_2S_0. \end{aligned}$$

Theorem 4. *Let $\mathcal{B}$ be the operator in Theorem 2. Then*

$$\mathcal{B}^{-1} = \begin{pmatrix} \beta_0 & -(\zeta\varepsilon)_1 & -\zeta\eta \\ \beta_1 & \varepsilon_{11} & \eta_1 \\ \beta_2 & \sigma_1 & \chi \end{pmatrix} : \begin{matrix} \mathcal{F} \\ \oplus \\ \mathcal{G} \\ \oplus \\ \mathcal{M} \end{matrix} \to \begin{matrix} \mathcal{E} \\ \oplus \\ \mathcal{J} \\ \oplus \\ \mathcal{L} \end{matrix} \tag{24}$$

is a left parametrix of $\mathcal{B}$.

The proof is similar to that of 4.2.1.3, Theorem 1 and left to the reader. More details are given in PILLAT/SCHULZE [2].

Finally, we have in the overdetermined case a theorem similar to 4.2.2.3, Theorem 1. Since it is nearly verbatum the same here, we do not formulate it explicitly. Note only that the Sobolev space regularity follows again from the orders in (24) (cf. PILLAT/SCHULZE [2]).

Using Theorem 3 one can construct examples of overdetermined boundary problems the left Shapiro-Lopatinski condition of which degenerates on a subset Z of the boundary and so that Theorem 2 is applicable. For $\mathcal{A}_0$ one can take e.g. the system 3.1.2.3. (2), (3) and for interior boundary problems $\mathcal{D}$ operators as considered in 4.2.2.1.

4.3. Discussions of Further Problems

In this section we shall give some additional results related to the theory of boundary value problems presented in this book and discuss further working directions. The exposition is not self-contained and some knowledge is supposed about Lie groups, equivariant K theory, parabolic equations and other types of problems in partial differential equations.

4.3.1. G-Invariant Operators

The classical Atiyah-Singer index theorem (cf. 1.2.4.2) has a natural generalization to elliptic PDOs which are invariant under the action of a compact Lie group. The results in this section about operators in $\mathfrak{G}$ are proved in HÖPPNER [1], [2].

Let G be a compact Lie group and M a closed compact C^∞ manifold. Assume that M is a G manifold, i.e. acts on M as a group of diffeomorphisms. For simplicity denote by $g\colon M \to M$ the diffeomorphism corresponding to $g \in G$. A (smooth complex) vector bundle E over M is called a *G-bundle* if G acts on E as a group of diffeomorphisms where $\pi g = g\pi$ $(\pi\colon E \to M)$ and $g\colon E_x \to E_{gx}$ linear for each $x \in M$. By the action of G on E an action of G on $C^\infty(M, E)$ is induced by $(gs)(x) = gs(g^{-1}x)$, $s \in C^\infty(M, E)$. Let $A \in L^m_{\mathrm{cl}}(M; E, F)$ be a PDO (E, F are G-bundles on M). Then A is called *G-invariant* if the diagram

$$\begin{array}{ccc} C^\infty(M, E) & \xrightarrow{A} & C^\infty(M, F) \\ \uparrow g & & \uparrow g \\ C^\infty(M, E) & \xrightarrow{A} & C^\infty(M. F) \end{array}$$

commutes for all $g \in G$.

If A is G-invariant, the homogeneous principal symbol $\sigma_A\colon \pi^*E \to \pi^*F$ is a G-invariant homomorphism of G-bundles. Similarly as in the case $G = \{e\}$ discussed in 1.2.4.2 we get a difference element $d(\sigma_A) \in K_G(TM)$, where $K_G(\cdot)$ denotes the K_G ring of the space in the brackets (cf. ATIYAH/SINGER [2]). Denote by $R(G)$ the representation ring of G. Per def. any element of $R(G)$ can be represented by an equivalence class of pairs of G-modules (V, W). If $A \in L^m_{\mathrm{cl}}(M; E, F)$ is G-invariant and elliptic, ker A and coker A are in a natural way G-modules and the element in $R(G)$ represented by (ker A, coker A) is called *analytical G-index of A* (G-ind$_a$ A). Moreover, we have a version of the topological index for the K_G-ring, namely a homomorphism G-ind$_t$:

$K_G(TM) \to R(G)$ (cf. ATIYAH/SINGER [2]). The index theorem for elliptic G-invariant PDOs A says

$$G\text{-ind}_a A = G\text{-ind}_t\, d(\sigma_A)\,. \tag{1}$$

(1) is proved in ATIYAH/SINGER [2].

G-invariant operators arise in connection with several questions in the geometry. For instance, the Laplace operator on a Riemannian manifold commutes with isometries and the $\bar\partial$ operator on a complex manifold with biholomorphic transformations. Geometric invariants arise as G-indices of corresponding elliptic operators, e.g. the Lefschetz number in the case of the action of the isometry group.

Now let X be a compact smooth G-manifold with boundary Y. Moreover, let $E_i \in \mathrm{Vect}_G(X)$, $J_i \in \mathrm{Vect}_G(Y)$ $(i = 1, 2)$ (the index G means G-bundles). Choose a Riemannian metric on X so that G acts as group of isometries. Such a choice is always possible if we are given a G-action on X. Let

$$\mathcal{A}\colon C^\infty(X, E_1) \oplus C^\infty(Y, J_1) \to C^\infty(X, E_2) \oplus C^\infty(Y, J_2) \tag{2}$$

be an operator in $\mathfrak{G}$. The natural action of G on sections in the corresponding G bundles gives rise to an action on operators in $\mathfrak{G}$ if we define

$$(g\mathcal{A})\,(u \oplus v) = g\mathcal{A}(g^{-1}(u \oplus v))\,, \tag{3}$$

$u \in C^\infty(X, E_1)$, $v \in C^\infty(Y, J_1)$, $g^{-1}(u \oplus v) = g^{-1}\,u \oplus g^{-1}\,v$.

Let

$$\mathcal{A}\colon H^{s_1}(X, E_1) \oplus H^{t_1}(Y, J_1) \to H^{s_2}(X, E_2) \oplus H^{t_2}(Y, J_2) \tag{4}$$

be a Sobolev space closure of (2). Then the map

$$G \to \mathcal{L}\big(H^{s_1}(X, E_1) \oplus H^{t_1}(Y, J_1),\ H^{s_2}(X, E_2) \oplus H^{t_2}(Y, J_2)\big) \tag{5}$$

given by $g \to g\mathcal{A}$ is continuous with respect to the norm topology in $\mathcal{L}(...)$.

The bundles π^*E_k, $p^*(E_k^+ \oplus J_k)$, $k = 1,2$, are G-bundles in a natural way, where H^+ remains fixed in $E_k^+ = E_k' \otimes H^+$ under the G-action. For given $\mathcal{A} \in \mathfrak{G}$ denote by $g\sigma_\Omega(\mathcal{A})$ and $g\sigma_Y(\mathcal{A})$ the compositions

$$\pi^*E_1 \xrightarrow{g^{-1}} \pi^*E_1 \xrightarrow{\sigma_\Omega(\mathcal{A})} \pi^*E_2 \xrightarrow{g} \pi^*E_2$$

and

$$p^*(E_1^+ \oplus J_1) \xrightarrow{g^{-1}} p^*(E_1^+ \oplus J_1) \xrightarrow{\sigma_Y(\mathcal{A})} p^*(E_2^+ \oplus J_2) \xrightarrow{g} p^*(E_2^+ \oplus J_2)\,,$$

respectively. Then $\sigma_\Omega(g\mathcal{A}) = g\sigma_\Omega(\mathcal{A})$, $\sigma_Y(g\mathcal{A}) = g\sigma_Y(\mathcal{A})$.

For a given G-invariant pair of interior and boundary symbols $(\sigma_\Omega, \sigma_Y)$, where σ_Ω is compatible with σ_Y, there exists a G-invariant $\mathcal{A} \in \mathfrak{G}$ with $\sigma_\Omega = \sigma_\Omega(\mathcal{A})$, $\sigma_Y = \sigma_Y(\mathcal{A})$. The index theory of G-invariant operators $\mathcal{A} \in \mathfrak{G}$ is quite similar as that of usual elliptic operators $\mathcal{A} \in \mathfrak{G}$. An essential point is the relation (cf. 3.2.2.1)

$$\beta(\mathrm{ind}_{SY}\, \Pi^+\sigma) = d_2(\sigma) \tag{6}$$

for G-invariant $\sigma \in \mathfrak{A}^{(0)}$ with $d_2(\sigma) \in K_G(\mathbb{R}^2 \times SY)$ and β is the Bott isomorphism of the equivariant K-theory. (6) can be proved by using the analytical proof of the Bott periodicity theorem of K_G-theory given in SADOWSKI [1]. In SADOWSKI [1] there are used Toeplitz operators in the ball or the disc, respectively, defined by means of the Bergman projector.

This class is not quite identical with the classical Toeplitz or Wiener-Hopf operators and admits the consideration of boundary symbols. One of the consequences of the K-theoretic interpretations is a G-invariant version of 3.2.1.1, Remark 6 (cf.

3.1.1.1, Theorem 12). Let $r^+A: C^\infty(X, E_1) \to C^\infty(X, E_2)$ be an elliptic G-invariant PDO with the transmission property. Then there exists a G-invariant elliptic $\mathcal{A} \in \mathfrak{G}$ with r^+A in the left upper corner iff there is a stabilization of σ_A to G-invariant elliptic symbol possessing a G-invariant extension as an isomorphism to ∂BX, where BX denotes the unit ball bundle induced by TX.

Let $\mathcal{A} \in \mathfrak{G}$ be elliptic and G-invariant. Then one has an analytical G-index $G\text{-ind}_a\, \mathcal{A} \in R(G)$. One can introduce an Abelian group $\mathrm{Ell}_G(X, Y)$ of equivalence classes of elliptic G-invariant operators $\mathcal{A} \in \mathfrak{G}$ (cf. 3.2.1.1). Equivalent operators have the same analytical G-index. Then one has an index theorem in the following form. There exists a group homomorphism

$$d\colon \mathrm{Ell}_G(X, Y) \to K_G(T\Omega)$$

so that

$$G\text{-ind}_a[\mathcal{A}] = G\text{-ind}_t\, d[\mathcal{A}]\,.$$

Here $[\mathcal{A}]$ denotes the class in $\mathrm{Ell}_G(X, Y)$ represented by $\mathcal{A}$ and $G\text{-ind}_t$ the topological G-index (cf. Atiyah/Singer [2]).

To give an example, let X be a compact Riemannian G-manifold with boundary Y and invariant Riemannian metric. Then the Neumann problem for the Laplace operator is invariant with respect to isometries. The Lefschetz number (in the de Rham cohomology of X) can be interpreted as G-index of some elliptic invariant operator $\mathcal{A} \in \mathfrak{G}$ derived from the Neumann problem. Details are given in Höppner [2].

4.3.2. Operators without Transmission Property

An essential assumption in the whole theory of operators in $\mathfrak{G}$ is the transmission property of the pseudo-differential operator. Of course, it can be asked whether there is a Fredholm theory of boundary problems in suitable function spaces for arbitrary elliptic PDOs (a certain nice topological behaviour similar to 3.2.1.2, Remark 6 has to be supposed). Fredholm boundary problems in this sense are studied in the papers Višik/Èskin [1, 2, 4, 5, 6] and in the monograph Èskin [3]. The main part of these papers consists of explicit local calculations making essential use of singular integral equations. This is of course the hard part of the theory, since the corresponding assertions in domains or in manifolds are then simple consequences. Nevertheless, the theory of elliptic boundary problems for operators without the transmission property is not yet satisfactory from the point of view of boundary symbolic calculus. A lot of questions which have been solved for operators in the class $\mathfrak{G}$ are not clear for the general operators and it can be a program to fill this gap.

In this section we shall briefly discuss some essential points of the theory of operators without the transmission property. First consider operators in the half space $\mathbb{R}^n_+$. Let $\alpha \in \mathbb{R}$ and denote by $O'^{,\alpha}(\mathbb{R}^n)$ the set of all functions $\sigma(\xi)$ defined almost everywhere in $\mathbb{R}^n$, continuous for $|\xi'| \neq 0$ and positively homogeneous of order α in ξ. Let $O^\alpha(\mathbb{R}^n)$ be the subset of functions $\sigma \in C^\infty(\mathbb{R}^n \setminus \{0\})$ positively homogeneous of order α.

If $\sigma(\xi)$ is some given function defined almost everywhere in $\mathbb{R}^n$, measurable, and satisfying an estimate $|\sigma(\xi)| \leq c(1 + |\xi|)^m$ with some $m \in \mathbb{R}$ and a constant c, we can define an operator

$$\sigma(D)\, u = (2\pi)^{-n} \int e^{ix\xi}\, \sigma(\xi)\, \hat{u}(\xi)\, d\xi\,, \tag{1}$$

$u \in C_0^\infty(\mathbb{R}^n)$. For $\sigma(\xi) \in O^\alpha(\mathbb{R}^n)$, $\alpha \leqq -n$, we have to do something near $\xi = 0$. Since excision has some disadvantages here and since the operators will be considered in fixed Sobolev spaces, define

$$\sigma(D)\, u = (2\pi)^{-n} \int e^{ix\xi}\, \sigma\left((1 + |\xi'|)\frac{\xi'}{|\xi'|}, \nu\right) \hat{u}(\xi)\, d\xi\,. \tag{2}$$

Then we have a continuous extension $\sigma(D)\colon H^s(\mathbb{R}^n) \to H^{s-\alpha}(\mathbb{R}^n)$. Restricting to the subspace $H_0^s(\overline{\mathbb{R}}_n^+)$ of $H^s(\mathbb{R}^n)$ and composing with $r^+\colon H^{s-\alpha}(\mathbb{R}^n) \to H^{s-\alpha}(\mathbb{R}_+^n)$ we get a continuous map

$$r^+\sigma(D)\colon H_0^s(\overline{\mathbb{R}}_+^n) \to H^{s-\alpha}(\mathbb{R}_+^n)\,, \tag{3}$$

$s \in \mathbb{R}$. Now, for $s > 1/2$, there is as usual a continuous operator

$$r'\colon H^s(\mathbb{R}_+^n) \to H^{s-1/2}(\mathbb{R}^{n-1})$$

so that for $\sigma_T(\xi) \in O'^{,\gamma}(\mathbb{R}^n)$ we can define a trace operator

$$r'T \overset{\text{def}}{=} r'\sigma_T(D)\colon H_0^s(\overline{\mathbb{R}}_n^+) \to H^{s-\gamma-1/2}(\mathbb{R}^{n-1}) \tag{4}$$

for $s - \gamma > 1/2$. Finally, let $\sigma_K(\xi) \in O'^{,\lambda}(\mathbb{R}^n)$ and suppose for simplicity that $\sigma_K(\xi', \nu)$ belongs to H^+ with respect to $\nu \in \mathbb{R}$ for every $\xi' \neq 0$. Then we have a continuous map

$$K \overset{\text{def}}{=} \sigma_K(D)\colon H^{t+\lambda+1/2}(\mathbb{R}^{n-1}) \to H^t(\mathbb{R}_+^n) \tag{5}$$

for each $t \in \mathbb{R}$.

Consider the case of matrices and let $\sigma_A(\xi) \in O^\alpha(\mathbb{R}^n) \otimes \hom(\mathbb{C}^k, \mathbb{C}^k)$,

$$\sigma_T(\xi) \in O'^{,\gamma}(\mathbb{R}^n) \otimes \hom(\mathbb{C}^k, \mathbb{C}^{j'})\,, \quad \sigma_K(\xi) \in O'^{,\lambda}(\mathbb{R}^n) \otimes \hom(\mathbb{C}^j, \mathbb{C}^k)$$

(with the assumption above for the matrix elements) and $\sigma_Q(\xi') \in O^{1-\alpha+\lambda+\gamma}(\mathbb{R}^{n-1}) \otimes \hom(\mathbb{C}^{j,j'})$. Then, for $s - \gamma > 1/2$, we have a continuous operator

$$\mathcal{A} = \begin{pmatrix} r^+A & K \\ r'T & Q \end{pmatrix} \colon \begin{matrix} H_0^s(\overline{\mathbb{R}}_+^n) \otimes \mathbb{C}^k \\ \oplus \\ H^{t+\lambda+1/2}(\mathbb{R}^{n-1}) \otimes \mathbb{C}^j \end{matrix} \to \begin{matrix} H^t(\mathbb{R}_+^n) \otimes \mathbb{C}^k \\ \oplus \\ H^{s-\gamma-1/2}(\mathbb{R}^{n-1}) \otimes \mathbb{C}^{j'} \end{matrix} \tag{6}$$

$(t = s - \alpha)$. For each $\eta = \xi'/|\xi'| \in S^{n-2}$ we also have a continuous operator

$$\begin{pmatrix} r^+\sigma_A(\eta, D_n)_{(s)} & \sigma_K(\eta, D_n) \\ r'\sigma_T(\eta, D_n) & \sigma_Q(\eta) \end{pmatrix} \colon \begin{matrix} H_0^s(\overline{\mathbb{R}}_+) \otimes \mathbb{C}^k \\ \oplus \\ \mathbb{C}^j \end{matrix} \to \begin{matrix} H^t(\mathbb{R}_+) \otimes \mathbb{C}^k \\ \oplus \\ \mathbb{C}^{j'}\,. \end{matrix} \tag{7}$$

Note that we could take more general symbols. Important is only the growth at infinity. In this sense we can define operators connected with the symbols

$$\lambda_\pm(\xi) = |\xi'| \pm i\nu \pm i\,.$$

The powers $(\lambda_\pm(\xi))^q$ can be defined for all $q \in \mathbb{R}$, taking the principal branch of the logarithm function. Then it turns out that the operator $r^+\Lambda_+^s \overset{\text{def}}{=} r^+(\lambda_+^s)(D)$

$$r^+\Lambda_+^s\colon H_0^p(\overline{\mathbb{R}}_+^n) \to H_0^{p-s}(\overline{\mathbb{R}}_+^n) \tag{8}$$

is an isomorphism for each $p \in \mathbb{R}$. Denote by $l\colon H^p(\mathbb{R}_+^n) \to H^p(\mathbb{R}^n)$ some continuous operator with $r^+lu = u$ for all $u \in H^p(\mathbb{R}_+^n)$, $p \in \mathbb{R}$ fixed. Such an operator exists (cf. the extension operator used in 2.3.2.3). Then one can prove that

$$r^+\{(2\pi)^{-n} \textstyle\int e^{ix\xi} \lambda_-^{-t}(\xi)\, \widehat{lu}(\xi)\, d\xi\}\,, \quad t \in \mathbb{R}\,, \tag{9}$$

is independent of the choice of l and that (9) defines an isomorphism

$$r^+\Lambda_-^{-t}\colon H^p(\mathbb{R}^n_+) \to H^{p+t}(\mathbb{R}^n_+) \tag{10}$$

for each $p \in \mathbb{R}$ (cf. 2.3.2.4). In particular, for $p = 0$ we have the isomorphisms

$$r^+\Lambda_+^s\colon H_0^s(\overline{\mathbb{R}}^n_+) \to H^0(\mathbb{R}^n_+)\,, \quad r^+\Lambda_-^{-t}\colon H^0(\mathbb{R}^n_+) \to H^t(\mathbb{R}^n_+)$$

and composition with (3) for $\sigma = \sigma_A \in O^\alpha(\mathbb{R}^n)$ gives a continuous map

$$r^+A_0\colon H^0(\mathbb{R}^n_+) \to H^0(\mathbb{R}^n_+) \tag{11}$$

with

$$\sigma_{A_s}(\xi) = l_-^{s-\alpha}(\xi)\,\sigma_A(\xi)\,l_+^{-s}(\xi) \tag{12}$$

(note that in the definition of (11) the operator r^+A in the middle is still defined by the convention (2)). Since we are mainly interested in the homogeneous principal part, we shall look at

$$(|\xi'| - i\nu)^{s-\alpha}\,\sigma_A(\xi)\,(|\xi'| + i\nu)^{-s}\,. \tag{13}$$

Using isomorphisms of the type (8), (10) and the corresponding isomorphisms $\Lambda^r\colon H^s(\mathbb{R}^{n-1}) \to H^{s-r}(\mathbb{R}^{n-1})$, s, $r \in \mathbb{R}$ (cf. 1.2.4.2, Theorem 5), we can reduce the orders in (6) and get an operator in L^2 spaces.

Let $\sigma_A(\xi) \in O^{(0)}(\mathbb{R}^n) \otimes \hom(\mathbb{C}^k, \mathbb{C}^k)$ be elliptic. Then, by $\nu \to \sigma_A(\eta, \nu)$, $\eta \in S^{n-2}$ fixed, we get a smooth map

$$\mathbb{R} \to \mathrm{GL}\,(k, \mathbb{C}) \tag{14}$$

with well defined limits

$$\sigma_\pm = \lim_{\nu \to \pm\infty} \sigma_A(\eta, \nu) \in \mathrm{GL}\,(k, \mathbb{C})\,,$$

$\sigma_\pm = \sigma_A(0, \pm 1)$. If σ_A has the transmission property, the image of (14) is a closed curve in the space of non-degenerate matrices. Otherwise this curve is not closed and we have a non-degenerate matrix $\sigma_+^{-1}\sigma_-$. Denote by $a_1, \dots, a_k$ the eigenvalues of $\sigma_+^{-1}\sigma_-$ and introduce the condition

$$a_j \notin \mathbb{R}_-, \quad j = 1, \dots, k\,, \tag{15}$$

where $\mathbb{R}_-$ denotes the real negative half axis in the complex plane.

If $\sigma_A \in O^\alpha(\mathbb{R}^n) \otimes \hom(\mathbb{C}^k, \mathbb{C}^k)$ is elliptic and $s \in \mathbb{R}$ fixed, introduce the same condition related to the matrix function (13), namely

$$a_j^{(s)} \notin \mathbb{R}_-\,, \quad j = 1, \dots, k\,, \tag{16}$$

where $a_j^{(s)}$ denotes the eigenvalues of $(\sigma_{(s)+})^{-1}\,\sigma_{(s)-}$ with $\sigma_{(s)\pm}$ as limits for $\nu \to \pm\infty$ of (13).

Theorem 1. *Let $\sigma_A \in O^\alpha(\mathbb{R}^n) \otimes \hom(\mathbb{C}^k, \mathbb{C}^k)$ be elliptic, $s \in \mathbb{R}$, and suppose that* (16) *is satisfied. Then*

$$r^+\sigma_A(\eta, D_n)_{(s)}\colon H_0^s(\overline{\mathbb{R}}_+) \otimes \mathbb{C}^k \to H^{s-\alpha}(\mathbb{R}_+) \otimes \mathbb{C}^k \tag{17}$$

is a Fredholm operator for each $\eta \in S^{n-2}$.

Note that the index of (17) depends on s. In particular, for $\alpha = 0$ we have a family of Fredholm operators

$$r^+\sigma_A(\eta, D_n)_{(0)}\colon H^0(\mathbb{R}_+) \otimes \mathbb{C}^k \to H^0(\mathbb{R}_+) \otimes \mathbb{C}^k\,. \tag{18}$$

Taking the Fourier image we get a Fredholm family

$$\Pi^+(\sigma_A)_{(0)}: V^+ \otimes \mathbb{C}^k \to V^+ \otimes \mathbb{C}^k, \tag{19}$$

$\eta \in S^{n-2}$. It is obvious that (7) plays the role of a boundary symbol.

It can be proved that if the symbol $\sigma \in O^{(0)}(\mathbb{R}^n) \otimes \hom(\mathbb{C}^k, \mathbb{C}^k)$ satisfies the condition (15), there exists a homotopy through elliptic symbols of order O respecting (16), $(\sigma_t)_{0 \le t \le 1}$ with $\sigma_0 = \sigma$, where σ_1 has the transmission property so that $\sigma_{1+} = \sigma_{1-}$. Homotopies preserve the index. Therefore, the local version of 3.2.1.2, Theorem 1 can be applied to $\sigma = \sigma_0$, which means that $\operatorname{ind} \Pi^+(\sigma)_0 = [\mathbb{C}^l]$ for some $l \in \mathbb{Z}$ is equivalent to the existence of a homotopy respecting (16) for a certain N

$$\sigma \oplus 1_N \simeq \begin{pmatrix} z^{-l} & 0 & \dots & 0 \\ 0 & 1 & \dots & 0 \\ \vdots & \vdots & & \vdots \\ 0 & 0 & \dots & 1 \end{pmatrix}$$

through elliptic symbols of order zero with $z = (|\xi'| - i\nu)(|\xi'| + i\nu)^{-1}$.

For $\sigma_{A_0}(\xi', \nu) = (|\xi'| - i\nu)^{s-\alpha} \sigma_A(\xi', \nu) (|\xi'| + i\nu)^{-s}$ we get

$$\begin{aligned} \Pi^+\sigma_A(\xi', \nu) &= \Pi^+(|\xi'| - i\nu)^{-s+\alpha} \sigma_{A_0}(\xi', \nu) (|\xi'| + i\nu)^s \\ &= \Pi^+(|\xi'| - i\nu)^\alpha \Pi^+\sigma_{A_0}(\xi', \nu) z^{-s}. \end{aligned}$$

Since $\Pi^+(|\xi'| - i\nu)^\alpha$ is an isomorphism, $\operatorname{ind} \Pi^+\sigma_A(\eta, \nu)_{(s)} = \operatorname{ind} \Pi^+(\sigma_{A_0}(\eta, \nu) z^{-s})_{(0)}$. After the mentioned homotopy applied to $\sigma_{A_0} \oplus 1$ for $s \in \mathbb{Z}$ we get

$$\operatorname{ind} \Pi^+(\sigma_{A_0}(\eta, \nu) z^{-s})_{(0)} = \operatorname{ind} \Pi^+((z^{-l} \oplus 1) z^{-s})_{(0)}.$$

Theorem 2. *Let $\mathcal{A}$ be given in the form* (6) *with elliptic $\sigma_A \in O^\alpha(\mathbb{R}^n) \otimes \hom(\mathbb{C}^k, \mathbb{C}^k)$ satisfying the condition* (16). *If* (7) *is an isomorphism for all $\eta \in S^{n-2}$, then the operator* (6) *is an isomorphism.*

The situation discussed up to this point corresponds to operators and symbols depending on x with freezing coefficients. One can define reasonable symbol classes in the half space $\sigma_A(x, \xi)$, $\sigma_T(x', \xi)$, $\sigma_K(x', \xi)$, $\sigma_Q(x', \xi')$ with a similar meaning as for the class $\mathfrak{G}$. The half space can be regarded as a local coordinate system of a boundary neighbourhood of a manifold.

Let X be a compact C^∞ manifold with boundary Y and $E, F \in \operatorname{Vect}(X)$, $J, G \in \operatorname{Vect}(Y)$. Then operators similar to (6) can be invariantly defined

$$\mathcal{A} = \begin{pmatrix} r^+A & K \\ r'T & Q \end{pmatrix}: \begin{matrix} H_0^s(X, E) \\ \oplus \\ H^{t+\lambda+1/2}(Y, J) \end{matrix} \to \begin{matrix} H^t(X, F) \\ \oplus \\ H^{s-\gamma-1/2}(Y, G) \end{matrix}, \tag{20}$$

$t = s - \alpha$, $\alpha = \operatorname{ord} \sigma_A$, if suitable conditions to the local representations of the symbols are fulfilled and if the coordinate neighbourhoods form a covering of X. The boundary symbol of (20) has then the form

$$\begin{pmatrix} r^+\sigma_A(y, \eta, D_n)_{(s)} & \sigma_K(y, \eta, D_n) \\ r'\sigma_T(y, \eta, D_n) & \sigma_Q(y, \eta) \end{pmatrix}: \begin{matrix} p^*(H_0^s(\overline{\mathbb{R}}_+) \otimes E') \\ \oplus \\ p^*J \end{matrix} \to \begin{matrix} p^*(H^t(\mathbb{R}_+) \otimes F') \\ \oplus \\ p^*G \end{matrix}, \tag{21}$$

$p: S^*Y \to Y$. If $\sigma_A: \pi^*E \to \pi^*F$ is elliptic and $\alpha = \operatorname{ord} \sigma_A$, the condition analogous to (16) is defined with respect to local coordinates and trivializations of the bundles. Then we have

Theorem 2'. *Let $\sigma_A: \pi^*E \to \pi^*E$ be elliptic, $\alpha = \operatorname{ord} \sigma_A$, and assume that the condition analogous to* (16) *is satisfied. If* (21) *is an isomorphism,* (20) *is a Fredholm operator.*

If (21) *is an isomorphism, we have*

$$\operatorname{ind}_{S^*Y} r^+\sigma_A(y, \eta, D_n)_{(s)} = [p^*G] - [p^*J], \tag{22}$$

i.e. the index element of the Fredholm family $r^+\sigma_A(y, \eta, D_n)_{(s)}$ belongs to $p^*K(Y)$.

Theorem 3. *For a given elliptic symbol $\sigma_A: \pi^*E \to \pi^*F$ satisfying the analogue of the condition* (16) *and a given PDO r^+A belonging to σ_A there exists an operator of type* (20) *for which* (21) *is an isomorphism iff $\operatorname{ind}_{S^*Y} r^+\sigma_A(y, \eta, D_n)_{(s)} \in p^*K(Y)$.*

Finally, remark that in Èskin [3] some elements of a certain boundary symbolic calculus are discussed. We do not go into details here. In Èskin [3] an essential role is played by the Mellin transform and certain new types of operators on the half axis which are not compact in $L^2(\mathbb{R}_+)$ and arise in connection with compositions $r^+\sigma_1(\eta, D_n) \circ r^+\sigma_1(\eta, D_n)$ and the calculation of parametrices of operators of the form $r^+\sigma(\eta, D_n)$.

In Rempel/Schulze [2] is given a boundary symbolic calculus including potential and trace symbols generalizing the results in 2.1.2 to operators without the transmission property. Moreover, results on the operator level are given, in particular, a parametrix construction for elliptic boundary problems being compatible with the corresponding interior and boundary symbolic calculus.

4.3.3. Sobolev Type Problems

In this section we shall consider Fredholm boundary value problems for elliptic pseudo-differential operators r^+A on a smooth compact manifold X with boundary Y containing additional trace conditions with respect to smooth closed submanifolds $Z_j \subset X \setminus Y$ of codimension l_j, $1 \leq l_j \leq n = \dim X$, $j = 1, \dots, q$. Such problems are said to be of *Sobolev type* because of the results in Sobolev [1] concerning the polyharmonic operator in domains with boundary components of higher codimension. Problems of this type for general elliptic differential and pseudo-differential operators and general trace conditions have been investigated in a number of papers. Here we refer to the approach in the papers Sternin [1, 2, 3]. In Novikov/Sternin [1], Sternin [2] also topological and K-theoretic aspects are discussed.

Ellipticity, regularity and parametrix constructions are studied in local coordinates. Then everything can be globally defined using a covering with coordinate neighbourhoods and a partition of unity. Therefore, in order to describe the typical behaviour of Sobolev type problems it is sufficient to consider the case $q = 1$, i.e. a submanifold $Z \subset X \setminus Y$ of codimension l. For the same reason Y can be supposed to be empty, since near Y the usual elliptic theory works either in the sense of the class $\mathfrak{G}$ or in the sense of 4.3.2.

We shall do this except in the following example. Let $\Omega \in \mathbb{R}^n$ be an open set and $\partial\Omega = \bigcup_{j=1}^{q} Z_j$ where $Z_0 = \partial\bar{\Omega}$ is a smooth manifold of dimension $n-1$ and $Z_j \subset \bar{\Omega} \setminus Z_0$ disjoint smooth manifolds of codimension l_j, $j = 1, \dots, q$ (some of these manifolds may be empty). Suppose for simplicity that the normal bundles of the Z_j in Ω are trivial. Consider the polyharmonic equation

$$\Delta^m u = 0 . \tag{1}$$

Denote by w_j the field of normal frame to Z_j and pose the following boundary conditions

$$\frac{\partial^{\varkappa_j}}{\partial w_j^{\varkappa_j}} u|_{Z_j} = g_j^{\varkappa_j}, \quad |\varkappa_j| \leqq m - \left[\frac{l_j}{2}\right] - 1 , \tag{2}$$

$j = 0, \dots, q$. Here the boundary values $g_j^{\varkappa_j}$ are supposed to be admissible, i.e. the existence of some $u \in H^m(\Omega)$ is required so that (2) holds. Under these assumptions unique solvability of (1), (2) was established in SOBOLEV [1] by means of variational methods. The following result is given in SLOBODECKIĬ [1].

Theorem 1. *The problem* (1), (2) *has a unique solution* $u \in H^m(\Omega)$ *iff* $g_j^{\varkappa_j} \in H^{m-|\varkappa_j|-l_j/2}(Z_j)$, $j = 0, 1, \dots, q$. *There are constants* $c_1, c_2 > 0$ *so that*

$$c_1 \sum_{j,\varkappa_j} \|g_j^{\varkappa_j}\|_{H^{m-|\varkappa_j|-l_j/2}(Z_j)} \leqq \|u\|_{H^m(\Omega)} \leqq c_2 \sum_{j,\varkappa_j} \|g_j^{\varkappa_j}\|_{H^{m-|\varkappa_j|-l_j/2}(Z_j)}$$

for all solutions $u \in H^m(\Omega)$ *of* (1), (2).

Now let X be a smooth compact closed C^∞ manifold of dimension n and Z a smooth closed compact submanifold of codimension 1. Then, for $s \in \mathbb{R}$, $s > 1/2$, the restriction operator

$$r' : H^s(X) \to H^{s-1/2}(Z)$$

is continuous. If T is a differential operator on X with ord $T = \gamma$, we have a continuous map $T: H^s(X) \to H^{s-\gamma}(X)$, $s \in \mathbb{R}$.

By means of local coordinates $x = (z, w) \in \mathbb{R}^{n-l} \times \mathbb{R}^l$ near points of Z we can define the transversal order $\mathrm{ord}_t\ T$ of T. Namely, if locally $\sigma(T)$ is given by $\sum_{|\alpha|+|\beta|\leqq\gamma} b_{\alpha\beta}(z, w)\,\zeta^\alpha\nu^\beta$, $\mathrm{ord}_t\ T$ is defined as maximal degree in the covariables ν which are dual to $\mathcal{D}$. Then the composition

$$r'T: H^s(X) \to H^{s-\gamma-1/2}(Z)$$

is continuous for $s \in \mathbb{R}$, $s - \mathrm{ord}_t\ T - 1/2 > 0$.

Let $s \in \mathbb{R}$, $\alpha \in \mathbb{Z}_+$, and

$$r(s) = \begin{cases} [\alpha - s - l/2] & \text{if } s + l/2 \notin \mathbb{Z} , \\ \alpha - s - l/2 - 1 & \text{if } s + l/2 \in \mathbb{Z} . \end{cases}$$

If we are given a tupel $r'T_\varkappa$, $|\varkappa| \leqq r(s)$, of trace operators, we have a continuous map

$$(r'T_\varkappa)_{|\varkappa|\leqq r(s)|} : H^s(X) \to \bigoplus_{|\varkappa|\leqq r(s)} H^{s-\gamma_\varkappa-l/2}(Z) \tag{3}$$

for $s > \mathrm{ord}_t\ T_\varkappa + 1/2$ $(|\varkappa| \leqq r(s))$.

Let A be a differential operator on X, $\alpha = \mathrm{ord}\,A$. Then $A: H^s(X) \to H^{s-\alpha}(X)$ is continuous. Set $t = s - \alpha$ and denote by $H_Z^t(X)$ the subspace of those $u \in H^t(X)$ with supp $u \subseteq Z$. Then A induces a continuous operator $A: H^s(X) \to H^t(X)/H_Z^t(X)$.

Consider the operator

$$\mathcal{A} = \begin{pmatrix} A \\ \\ r'T \end{pmatrix} : H^s(X) \to \begin{matrix} H^t(X)/H^t_Z(X) \\ \oplus \\ \bigoplus_{|\varkappa| \leqq r(s)} H^{s-\gamma\varkappa-l/2}(Z) \end{matrix} \tag{4}$$

Here, $r'T = (r'T_\varkappa)_{|\varkappa|} \leqq r(s)$ and $s > \operatorname{ord}_t T_\varkappa + {}^1/_2$.

Denote by $\sigma_A(z, w, \zeta, \nu)$ and $\sigma_{T_\varkappa}(z, w, \zeta, \nu)$ the homogeneous principal symbols in local coordinates $(z, w) \in \mathbb{R}^{n-l} \times \mathbb{R}^l$ near a point of Z of the operators A and $T_\varkappa$, respectively.

Condition 2 (ellipticity).

(i) A is elliptic;

(ii) the system

$$\sigma_A(z, 0, \zeta, D_w)\,\varphi(w) = 0\,, \tag{5}$$

$$r'\sigma_{T_\varkappa}(z, 0, \zeta, D_w)\,\varphi(w) = 0\,, \quad |\varkappa| \leqq r(s) \tag{6}$$

has in $H^s(\mathbb{R}^l)$ only the trivial solution $\varphi \equiv 0$ for all $(z, \zeta) \in \mathbb{R}^{n-l} \times (\mathbb{R}^{n-l} \setminus \{0\})$.

The operator $\mathcal{A}$ is called *elliptic* if the Condition 2 is satisfied. (ii) is called *ellipticity* of the boundary conditions.

Remark 3. Let $\sigma(D_w) = \sum_{|\mu| \leqq \alpha} a_\mu D_w^\mu$ be an elliptic differential operator in $\mathbb{R}^l$ with constant coefficients of order α and suppose

$$\sigma(\nu) = \sum_{|\mu| \leqq \alpha} a_\mu \nu^\mu \neq 0 \quad \text{for all} \quad \nu \in \mathbb{R}^l\,.$$

Then the space

$$\{\varphi \in H^s(\mathbb{R}^l) : \sigma(D)\,\varphi = 0 \quad \text{in} \quad \mathbb{R}^l \setminus \{0\}\} \tag{7}$$

is finite dimensional and $\{D_w^\mu E : |\eta| \leqq r(s)\}$ is a base in this space for some fundamental solution E of $\sigma(D)$.

It can be proved that for $s \geqq \alpha - l/2$ the space (7) consists only of $\varphi \equiv 0$. Note that the problems under consideration are closely connected with the question of removability of singularities of solutions of elliptic equations.

Remark 3 can be applied to $\sigma(D) = \sigma_A(z, 0, \zeta, D_w)$, $z \in \mathbb{R}^{n-l}$, $\zeta \in \mathbb{R}^{n-l} \setminus \{0\}$. Each non-trivial solution of (5) in $H^s(\mathbb{R}^l)$ has the form

$$\varphi = \sum_{|\varrho| \leqq r(s)} c_\varrho D^\varrho E(z, \zeta, w)\,,$$

where E is a fundamental solution of (5). Then Fourier transform of (6) gives

$$\int \sum_{|\varrho| \leqq r(s)} c_\varrho \nu^\varrho \sigma_{T_\varkappa}(z, 0, \zeta, \nu)\,\sigma_A^{-1}(z, 0, \zeta, \nu)\,d\nu = 0\,,$$

$|\varkappa| \leqq r(s)$. Thus ellipticity of the boundary conditions in $\mathcal{A}$ means that the matrix of elements

$$\int \nu^\varrho \sigma_{T_\varkappa}(z, 0, \zeta, \nu)\,\sigma_A^{-1}(z, 0, \zeta, \nu)\,d\nu\,, \quad |\varkappa|, |\varrho| \leqq r(s)\,, \tag{8}$$

is non-singular for all $\zeta \neq 0$.

Theorem 4. *The following conditions are equivalent:*

(i) *$\mathcal{A}$ is elliptic in the sense of Condition* 2;

(ii) *$\mathcal{A}$ defines a Fredholm operator* (4).

Remark 5. It can be proved that for $s \geqq \alpha - n/2$ there exists no Fredholm operator (4) with the given elliptic A of order α. Thus one is limited to s sufficiently small. The most left half interval of values s being admissible in this sense is $\alpha/2 - l/2 + [l/2] \leqq s < \alpha/2 - l/2 + [l/2] + 1$.

Sobolev type problems can be studied for PDOs, too. Let $E, F \in \mathrm{Vect}\,(X)$ and

$$A: H^s(X, E) \to H^t(X, F) \tag{9}$$

be a PDO, $\alpha = s - t = \mathrm{ord}\, A$. Consider operators of the type

$$\mathcal{A} = \begin{pmatrix} A & K(r')^* \\ r'T & Q \end{pmatrix} : \begin{matrix} H^s(X, E) \\ \oplus \\ H^{t+\lambda+l/2}(Z, J) \end{matrix} \to \begin{matrix} H^t(X, F) \\ \oplus \\ H^{s-\gamma-l/2}(Z, G) \end{matrix} \tag{10}$$

$(J, G \in \mathrm{Vect}\,(Z))$, Q is a PDO on Z of order $l + \lambda + \gamma - \alpha$, K, T are PDOs on X of the form $T_U \omega$ and ωK_U, respectively, where ω is a C^∞ function on X supported by some open tubular neighbourhood U of Z,

$$T_U: H^s_{\mathrm{comp}}(U, E|_U) \to H^{s-\gamma}_{\mathrm{loc}}(U, G^U)\,,$$

$$K_U: H^{t+\lambda}_{\mathrm{comp}}(U, J^U) \to H^t_{\mathrm{loc}}(U, F|_U)$$

are PDOs of order γ and λ, respectively. J^U, G^U denote the pull backs to U of J, G with respect to a projection $U \to Z$, r' denotes the usual restriction and $(r')^*$ the extension of distributions obtained as adjoint of r'. When $m + l/2 < 0$, then $(r')^*: H^{m+l/2}(Z, J) \to H^m(U, J^U)$ is continuous. Thus (10) is continuous when

$$s - \mathrm{ord}_t\, T - \frac{l}{2} > 0\,, \quad t + \lambda + \frac{l}{2} < 0\,. \tag{11}$$

ord_t denotes the transversal order of T as above. The transversal order is a generalization of the type of a trace operator in the class $\mathfrak{G}$.

It is easily seen that the composition $r'T \cdot P \cdot K(r')^*$ is a PDO on Z for any PDO P on X. Let A in (10) be elliptic and P a parametrix of A. Then

$$r'T \circ P \circ K(r')^*: H^{t+\lambda+l/2}(Z, J) \to H^{s-\gamma-l/2}(Z, G)$$

is continuous.

Definition 6. The operator (10) is called *elliptic* if

(i) A is elliptic;

(ii) the operator

$$Q - r'T \circ P \circ K(r')^* \tag{12}$$

is elliptic on Y, where P is a parametrix of A.

Note that ellipticity refers to the homogeneous principal symbols of order α in (i) and $l + \lambda + \gamma - \alpha$ in (ii), respectively.

Since the left upper corner in (10) is Fredholm, 3.1.1.5, Proposition 1 can be applied. Thus we get

Theorem 7. *Let A be elliptic. Then the following conditions are equivalent:*

(i) *the operator* (12) *is elliptic;*

(ii) *the operator* (10) *is Fredholm.*

In the proof (ii) $\Rightarrow$ (i) one has to use 1.2.4.2, Theorem 3. Note that by 3.1.1.5.(5) a parametrix of (10) is given. Moreover, one can apply the index formula 3.1.1.5.(7). This means here

$$\operatorname{ind} \mathcal{A} = \operatorname{ind} A + \left(\operatorname{ind} Q - r'T \circ P \circ K(r')^*\right) . \tag{13}$$

Now return to the operator (4) and suppose $t + |\varrho| + l/2 < 0$, $|\varrho| \leqq r(s)$. Then

$$D_w^{\varrho}(r')^*\colon H^{t+|\varrho|+l/2}(Z) \to H^t(X)$$

is continuous and we can pass to the continuous operator

$$\begin{pmatrix} A & K(r')^* \\ r'T & 0 \end{pmatrix} : \begin{matrix} H^s(X) \\ \oplus \\ \bigoplus\limits_{|\varrho| \leqq r(s)} H^{t+|\varrho|+l/2}(Z) \end{matrix} \to \begin{matrix} H^t(X) \\ \oplus \\ \bigoplus\limits_{|\varkappa| \leqq r(s)} H^{s-\gamma\varkappa-l/2}(Z) \end{matrix} , \tag{14}$$

where K just consists of the vector $(D_w^{\varrho})_{|\varrho| \leqq r(s)}$.

Multiplication of (13) from the left by $\mathcal{R}_1 = 1 \oplus \left(\bigoplus\limits_{|\varkappa| \leqq r(s)} R^{(s-\gamma\varkappa-1/2)} \right)$ and from the right by $\mathcal{R}_2 = 1 \oplus \left(\bigoplus\limits_{|\varrho| \leqq r(s)} R^{(-t-|\varrho|-1/2)} \right)$, where R^k is an elliptic PDO on Z of order k defining an isomorphism $H^s(Z) \to H^{s-k}(Z)$, $s \in \mathbb{R}$, yields

$$\mathcal{R}_1 \mathcal{A} \mathcal{R}_2 = \begin{pmatrix} A & K_0(r')^* \\ r'T_0 & 0 \end{pmatrix} : \begin{matrix} H^s(X) \\ \oplus \\ \bigoplus\limits_{|\varrho| \leqq r(s)} H^0(Z) \end{matrix} \to \begin{matrix} H^t(X) \\ \oplus \\ \bigoplus\limits_{|\varkappa| \leqq r(s)} H^0(Z) \end{matrix} . \tag{15}$$

It can be assumed that $\sigma_{R^k}(z, \zeta) = 1$ for $|\zeta| = 1$ (for some fixed Riemannian metric on Z).

Ellipticity of (15) in the sense of Definition 6 means that A is elliptic and that the PDO

$$r'T_0 \circ P \circ K_0(r')^*\colon \bigoplus_{|\varrho| \leqq r(s)} H^0(Z) \to \bigoplus_{|\varkappa| \leqq r(s)} H^0(Z) \tag{16}$$

on Z is elliptic. The symbol of (16) for $|\zeta| = 1$ is given by (8). Thus the ellipticity of (4) is equivalent to the ellipticity of (15) or equivalently the ellipticity of A and the Fredholm property of the operator

$$r'T \circ P \circ K(r')^*\colon \bigoplus_{|\varrho| \leqq r(s)} H^{t+|\varrho|+l/2}(Z) \to \bigoplus_{|\varkappa| \leqq r(s)} H^{s-\gamma\varkappa-l/2}(Z) .$$

Thus Definition 6 is a natural generalization of Definition 2. In view of (12) the index of (4) is given by $\operatorname{ind} A + \operatorname{ind} (r'T \circ P \circ K(r')^*)$.

The reduction of orders by means of $\mathcal{R}_1$, $\mathcal{R}_2$ shows that it is sufficient to consider unified orders of the trace and potential operators in the case that they have various components. Thus the form of (10) is justified.

A discussion of the K-theoretic results in STERNIN [3] is not possible in this book because this would be an extra program. Moreover, the exposition STERNIN [2] makes it obvious that there are sufficiently many points which are not quite clear and that the whole theory of Sobolev type problems should be written down more carefully.

4.3.4. Interior Boundary Problems and Degenerate Operators

In 4.2.2.1 classes of interior boundary value problems were discussed. Here we only make some additional remarks. Interior boundary problems for operators on a closed compact manifold Y which are elliptic over $Y \setminus Z$ for a subset Z are interesting not only from the point of view of degenerate boundary problems in X with $Y = \partial X$. The investigation of interior boundary problems also gives contributions to the question of solvability and regularity of boundary problems for operators which are elliptic in the domain with degeneracy of the ellipticity at the boundary. Problems of this type are discussed in a plenty of papers, but mainly without potential conditions (cf. VIŠIK/GRUŠIN [1]). Problems containing also potential conditions are considered e.g by VIŠIK/GRUŠIN [3]. The calculus of degenerate problems in this sense is in no direction complete from the point of view of index questions, homotopy classification or a calculus in the sense of an algebra of operators. Troubles come also from the fact that degenerate problems such as interior boundary problems usually are far from a beautiful elliptic theory. The arising types of operators are "general" and one always has to select certain subclasses. Questions concerning the index are investigated e.g. by MARČENKO [1], JUHL [1], ASADA [1]. Sometimes, for very special problems, it is known that the corresponding problem is uniquely solvable (cf. e.g. BICADZE [1], [2], KIBIREV [1]). It would also be interesting to know something about the solvability of interior boundary problems over Y when Z has higher codimension (cf. 4.3.3). In this direction only special results are known, e.g. in the case when Z consists of a finite number of isolated points (cf. BICADZE [3], BAOUENDI/SJÖSTRAND [1], [2], RADKEVIČ [1], ELSCHNER/LORENZ [1]).

Degeneracies at isolated points on curves arise automatically by reduction to the boundary of problems with degenerating Shapiro-Lopatinski condition in two-dimensional domains. Then the idea of additional trace and potential conditions does not work and it is reasonable to consider the problems in adapted function spaces. Results and informations on the index in this case are given e.g. by ELSCHNER [1].

In higher dimensional cases one may expect that the theory has to refer to special Sobolev spaces from the beginning (cf. 4.3.3).

4.3.5. Transmission and Mixed Problems

Consider a compact C^∞ manifold X without boundary which is divided into two submanifolds X_1, X_2 of X with boundary, $X = X_1 \cup X_2$. Set $Y = X_1 \cap X_2$. Consider elliptic differential operators A_i over X_i (or, more generally, elliptic systems in the sense of Douglis-Nirenberg),

$$A_1\colon C^\infty(X_1, \mathbb{C}^{k_1}) \to C^\infty(X_1, \mathbb{C}^{k_1}),$$
$$A_2\colon C^\infty(X_2, \mathbb{C}^{k_2}) \to C^\infty(X_2, \mathbb{C}^{k_2}).$$

This pair of operators can be regarded as one operator with discontinuity along the manifold Y. Let T_1, T_2 be (systems of) differential operators defined near Y. Pose the problem to determine functions u_1, u_2 in suitable function spaces satisfying

$$\begin{aligned} &A_1u_1 = f_1 \quad \text{in } X_1, \qquad A_2u_2 = f_2 \quad \text{in } X_2, \\ &r'T_1u_1 - r'T_2u_2 = g \quad \text{in } Y, \end{aligned} \tag{1}$$

for given $f_1 \in C^\infty(X_1, \mathbb{C}^{k_1})$, $f_2 \in C^\infty(X_2, \mathbb{C}^{k_2})$, $g \in C^\infty(Y, \mathbb{C}^{k_3})$ (or in other suitable function spaces). (1) is called a *transmission problem*. A classical example for transmission problems is the well-known Riemann-Hilbert problem. Transmission problems arise in applied sciences (cf. PICONE [1]). For classical results cf. SCHECHTER [1], OLEĬNIK [1] and the case of Douglis-Nirenberg elliptic systems in ROĬTBERG/ŠEFTEL'

[1]. Using the theory of boundary problems for elliptic pseudo-differential operators the transmission problem is treated in Myškis [1]. Topological aspects of similar problems are discussed in Booss [1].

We shall show the relation between transmission problems for PDOs with the transmission property and the theory of boundary problems in the class $\mathfrak{G}$. In particular there is a natural notion of ellipticity of transmission conditions. Ellipticity is equivalent to the Fredholm property in adapted Sobolev spaces. A simple corollary are a-priori estimates for the solutions. The index theory developed in 3.2 can be applied to transmission problems with small modifications.

Now we describe the class $\mathfrak{G}(X_1, X_2; Y)$ (X_1, X_2, Y compact C^∞ manifolds as above) being an algebra which contains the parametrices of elliptic transmission problems in $\mathfrak{G}(X_1, X_2; Y)$. Operators in $\mathfrak{G}(X_1, X_2; Y)$ are the natural generalization of the problem (1).

Note that the considerations in Myškis [1] are slightly more general, since there in addition an elliptic PDO on $X = X_1 \cup X_2$ is involved (without discontinuity).

Let $U \subset X$ be an open neighbourhood of Y, $U \cong Y \times (-1, 1)$. Set $U_i = U \cap X_i$, $i = 1, 2$. Then there is a diffeomorphism $j: U_1 \xrightarrow{\cong} U_2$ induced by $x_n \mapsto -x_n$, $x_n \in [0, 1)$. Let $E_i, F_i \in \text{Vect}(X_i)$, $J, G \in \text{Vect}(Y)$. Then the pull back j^* defines a bijection

$$1 \oplus j^* \oplus 1 : \begin{matrix} C^\infty(U_1, E_1|_{U_1}) \\ \oplus \\ C^\infty(U_2, E_2|_{U_2}) \\ \oplus \\ C^\infty(Y, J) \end{matrix} \to \begin{matrix} C^\infty(U_1, E_1|_{U_1}) \\ \oplus \\ C^\infty(U_1, j^*E_2|_{U_1}) \\ \oplus \\ C^\infty(Y, J) \end{matrix}$$

and similarly for C_0^∞ instead of C^∞.

Denote by $\mathfrak{G}(U, Y)$ the class of operators in $\mathfrak{G}$ with respect to the manifold U with boundary Y. Non-compactness of U plays no role, since only local properties are used. In a similar sense use the notation $\mathfrak{R}(U, Y)$ for the space of boundary symbols and the notations for classes of Green, trace and potential operators.

Let $A_i \in \text{Op}(\mathfrak{A})(X_i)$, $i = 1, 2$, be given. An operator

$$\mathcal{A} = \begin{pmatrix} A_1 + B_{11} & B_{12} & K_{13} \\ B_{21} & A_2 + B_{22} & K_{23} \\ T_{31} & T_{32} & A_3 \end{pmatrix} : \begin{matrix} C^\infty(X_1, E_1) \\ \oplus \\ C^\infty(X_2, E_2) \\ \oplus \\ C^\infty(Y, J) \end{matrix} \to \begin{matrix} C^\infty(X_1, F_1) \\ \oplus \\ C^\infty(X_2, F_2) \\ \oplus \\ C^\infty(Y, G) \end{matrix} \tag{2}$$

is called *transmission problem in the class* $\mathfrak{G}(X_1, X_2; Y)$ if near Y the operator

$$\mathcal{A}_1 \overset{\text{def}}{=} (1 \oplus j^+ \oplus 1)\, \mathcal{A}(1 \oplus j^{*-1} \oplus 1) \tag{3}$$

belongs to $\mathfrak{G}(U_1, Y)$. Equivalently, $\mathcal{A} \in \mathfrak{G}(X_1, X_2; Y)$ iff $B_{11} \in \text{Op}(\mathfrak{B})(U_1)$, $B_{22} \in \text{Op}(\mathfrak{B})(U_2)$, $A_3 \in L_{cl}(Y)$, $K_{13} \in \text{Op}(\mathfrak{K})(Y, U_1)$, $K_{23} \in \text{Op}(\mathfrak{K})(Y, U_2)$, $T_{31} \in \text{Op}(\mathfrak{T})(U_1, Y)$, $T_{32} \in \text{Op}(\mathfrak{T})(U_2, Y)$ and for the Green operators B_{12} and B_{21} we have $B_{12}j^{*-1} \in \text{Op}(\mathfrak{B})(U_1)$, $j^* B_{21} \in \text{Op}(\mathfrak{B})(U_1)$. On boundary symbolic level there is a similar characterization of the considered classes of Green, potential and trace operators. Note that for $\mathcal{A} \in \mathfrak{G}(X_1, X_2; Y)$ the operator $\mathcal{A}_1 \in \mathfrak{G}(U_1, Y)$ has a PDO in the left upper corner $A_1 \oplus j^* A_2 j^{*-1}$. Thus the transmission problem (2) is equivalent to a usual boundary value problem in the class $\mathfrak{G}$ for the PDO $A_1 \oplus j^* A_2 j^{*-1}$.

The definition $j^*u(t) = u(-t)$, $t \in \mathbb{R}$, for any function u on $\mathbb{R}$ implies $Fj^*u = \overline{Fu}$. The diagram

$$\begin{array}{ccc} \mathfrak{S}(\overline{\mathbb{R}}_-) & \xrightarrow{F} & H_0^- \\ j^* \downarrow & & \downarrow \\ \mathfrak{S}(\overline{\mathbb{R}}_+) & \xrightarrow{F} & H^+ \end{array} \tag{4}$$

induces an isomorphism $H_0^- \cong H^+$ given by $h(\nu) \mapsto h(-\nu)$, $h \in H_0^-$.

Consider the principal boundary symbol of $\mathcal{A}_1$

$$\sigma_Y(\mathcal{A}_1)(x', \xi') = \begin{pmatrix} \Pi^+\sigma_{A_1} + \Pi'\sigma_{B_{11}} & \Pi'\sigma_{B_{12}}^{(1)} & \sigma_{K_{12}} \\ \Pi'\sigma_{B_{21}}^{(1)} & \Pi^+\sigma_{A_2} + \Pi'\sigma_{B_{22}}^{(1)} & \sigma_{K_{22}}^{(1)} \\ \Pi'\sigma_{T_{21}} & \Pi'\sigma_{T_{22}}^{(1)} & \sigma_{A_3} \end{pmatrix} : \begin{matrix} H^+ \otimes \mathbb{C}^{k_1} \\ \oplus \\ H^+ \otimes \mathbb{C}^{k_2} \\ \oplus \\ \mathbb{C}^{k_3} \end{matrix} \to \begin{matrix} H^+ \otimes \mathbb{C}^{k_1} \\ \oplus \\ H^+ \otimes \mathbb{C}^{k_2} \\ \oplus \\ \mathbb{C}^{k_4} \end{matrix}$$

where $\Pi^+\sigma_{A_1}$, $\Pi'\sigma_{B_{11}}$, $\sigma_{K_{12}}$, $\Pi'\sigma_{T_{21}}$ are the usual boundary symbols of operators in $\mathfrak{G}(X_1, Y)$.

By (4) a boundary symbol

$$\sigma_Y(\mathcal{A})(x', \xi') = \begin{pmatrix} \Pi^+\sigma_{A_1} + \Pi'\sigma_{B_{11}} & \Pi'\sigma_{B_{12}} & \sigma_{K_{12}} \\ \Pi'\sigma_{B_{21}} & \Pi_0^-\sigma_{A_2} + \Pi'\sigma_{B_{22}} & \sigma_{K_{22}} \\ \Pi'\sigma_{T_{21}} & \Pi'\sigma_{T_{22}} & \sigma_{A_3} \end{pmatrix} : \begin{matrix} H^+ \otimes \mathbb{C}^{k_1} \\ \oplus \\ H_0^- \otimes \mathbb{C}^{k_2} \\ \oplus \\ \mathbb{C}^{k_3} \end{matrix} \to \begin{matrix} H^+ \otimes \mathbb{C}^{k_1} \\ \oplus \\ H_0^- \otimes \mathbb{C}^{k_2} \\ \oplus \\ \mathbb{C}^{k_4} \end{matrix} \tag{5}$$

is induced, where $\sigma_{T_{22}}(\nu) = \sigma_{T_{22}}^{(1)}(-\nu)$, $\sigma_{K_{22}}(\nu) = \sigma_{K_{22}}^{(1)}(-\nu)$, $\sigma_{B_{12}}(\nu, \tau) = \sigma_{B_{12}}^{(1)}(\nu, -\tau)$, $\sigma_{B_{21}}(\nu, \tau) = \sigma_{B_{21}}^{(1)}(-\nu, \tau)$, $\sigma_{B_{22}}(\nu, \tau) = \sigma_{B_{22}}^{(1)}(-\nu, -\tau)$. Denote the space of all these boundary symbols by $\mathfrak{R}(U_1, U_2; Y)$. The isomorphism $H_0^- \cong H^+$ yields an embedding $\mathfrak{R}(U_1, U_2; Y) \to \mathfrak{R}(U_1, Y)$ with the property that the composition in $\mathfrak{R}(U_1, U_2; Y)$ corresponds to the composition in $\mathfrak{R}(U_1, Y)$. Obviously, since $\mathfrak{R}(U_1, Y)$ is an algebra, $\mathfrak{R}(U_1, U_2; Y)$ is an algebra, too, i.e. if the composition of boundary symbols as mappings (5) is defined, it is again in the class of considered boundary symbols. Moreover, any bijective mapping (5) has its inverse in the class $\mathfrak{R}(U_1, U_2; Y)$.

Theorem 1. *Let $\mathcal{A}_i \in \mathfrak{G}(X_1, X_2; Y)$ and let $\mathcal{A}_1\mathcal{A}_2$ be defined. Then $\mathcal{A}_1\mathcal{A}_2 \in \mathfrak{G}(X_1, X_2; Y)$ and the principal symbol of $\mathcal{A}_1\mathcal{A}_2$ is equal to the composition of the principal symbols.*

A transmission problem $\mathcal{A} \in \mathfrak{G}(X_1, X_2; Y)$ is called *elliptic* if the interior principal symbols σ_{A_1} and σ_{A_2} are elliptic and the principal boundary symbol $\sigma_Y(\mathcal{A})$ as a mapping (5) is bijective. Then $\mathcal{A} \in \mathfrak{G}(X_1, X_2; Y)$ is elliptic iff A_1 and A_2 are elliptic and $\mathcal{A}_1 \in \mathfrak{G}(U_1, Y)$ is an elliptic boundary value problem for $A_1 \oplus j^* A_2 j^{*-1}$.

Since the operators involved in the matrix of operators (2) are essentially in the usual classes of Green, potential and trace operators, we have the same definition of the order (cf. the definition of $\mathfrak{G}^{m,d}$ in 2.3.2.1). According to 2.3.3.3, Theorem 1 we get

Theorem 2. *Every* $\mathcal{A} \in \mathfrak{G}^{m,d}(X_1, X_2; Y)$ *defines a continuous mapping*

$$\mathcal{A}: \begin{matrix} H^s(X_1, E_1) \\ \oplus \\ H^s(X_2, E_2) \\ \oplus \\ H^{s+1/2}(Y, J) \end{matrix} \to \begin{matrix} H^{s-m}(X_1, F_1) \\ \oplus \\ H^{s-m}(X_2, F_2) \\ \oplus \\ H^{s-m+1/2}(Y, G) \end{matrix} \tag{6}$$

for any $s > d - {}^1/_2$.

Similar extensions in general Sobolev spaces $H^{s,p}$ *and Hölder spaces are valid.*

From 3.1.1.1, Theorem 2′ and 3.1.1.1, Theorem 7 we get

Theorem 3. *The following assertions are equivalent:*

(i) $\mathcal{A} \in \mathfrak{G}(X_1, X_2; Y)$ *is elliptic;*

(ii) *the extension of* $\mathcal{A}$ *to Sobolev spaces* (6) *is a Fredholm operator.*

If one of the two equivalent conditions is satisfied, the index is independent of s, ker $\mathcal{A}$ *and* coker $\mathcal{A}$ *can be represented by smooth functions and an a-priori estimate as in* 3.1.1.1, *Corollary* 3 *holds.*

If $\mathcal{A} \in \mathfrak{G}(X_1, X_2; Y)$ is *elliptic a parametrix of* $\mathcal{A}$ *can be found in the class of transmission problems* $\mathfrak{G}(X_1, X_2, Y)$.

Consider a simple example. Let

$$\mathcal{A} = \begin{pmatrix} \Delta^2 & 0 \\ 0 & \Delta \\ r'D_n^2 & r' \\ r'D_n^3 & r'D_n \\ r'D_n & 0 \end{pmatrix} \in \mathfrak{G}(X_1, X_2; Y),$$

where Δ is the Laplacian with respect to a Riemannian metric in X_2 and Δ^2 the square of the Laplacian with respect to a Riemannian metric in X_1, and D_n denotes the normal derivative to Y. In other words, we look for $u_1 \in C^\infty(X_1)$, $u_2 \in C^\infty(X_2)$ such that

$$\begin{aligned} \Delta^2 u_1 &= f_1 \quad \text{in } X_1, \\ \Delta u_2 &= f_2 \quad \text{in } X_2, \end{aligned} \tag{7}$$

$r'D_n^2 u_1 + r'u_2 = g_1$, $r'D_n^3 u_1 + r'D_n u_2 = g_2$, $r'D_n u_1 = g_3$ on Y for given $f_1 \in C^\infty(X_1)$, $f_2 \in C^\infty(X_2)$, $g_1, g_2, g_3 \in C^\infty(Y)$. It is easily checked that $\mathcal{A}$ is elliptic. According to Theorem 2 and Theorem 3 the operator $\mathcal{A}$ defines a Fredholm mapping

$$\mathcal{A}: H^s(X_1) \oplus H^{s-2}(X_2) \to H^{s-4}(X_1) \oplus H^{s-4}(X_2) \oplus H^{s-5/2}(Y) \\ \oplus H^{s-7/2}(Y) \oplus H^{s-3/2}(Y).$$

The considerations in 6.1.1.3, Corollary 4 imply that ind $\mathcal{A} = 0$. In other words, the problem (7) has a finite dimensional space of solutions for all right hand sides satisfying a finite number of conditions (equal to the dimension of the kernel).

We turn now to mixed problems. Let X be a compact C^∞ manifold with boundary Y divided into Y_1 and Y_2, two submanifolds of codimension 0 with boundary, $Y = Y_1 \cup Y_2$. Set $Z = Y_1 \cap Y_2$. Let A be an elliptic differential operator (or an elliptic system) on X. Assume that there are given elliptic boundary conditions $r'T_1$ on Y_1

and $r'T_2$ on Y_2 (or one boundary condition with discontinuity along Z). Then

$$\begin{aligned} &Au = f \quad \text{on } X\,, \\ &r'T_1u = g_1 \quad \text{in } Y_1\,, \qquad r'T_2u = g_2 \quad \text{in } Y_2 \end{aligned} \tag{8}$$

is called a *mixed problem for the elliptic operator A*.

Mixed problems have been investigated in PEETRE [1], SHAMIR [1], ZAREMBA [1] and more recently in VIŠIK/ÈSKIN [7], ÈSKIN [3].

In the case of a scalar operator of order 2 in ESKIN [3] there are described function spaces in which the problem (8) is Fredholm. For higher order operators and for systems, in general, one has to pose additional conditions on the jump manifold Z of the boundary conditions. Then the Fredholm property is obtained in Sobolev spaces with weights.

Mixed problems can be formulated in a natural way for PDOs, too, where it is supposed that the PDO in consideration has the transmission property not only with respect to Y. This case is investigated in REMPEL/SCHULZE [5]. There a class of mixed problems $\mathfrak{G}(X, Y, Z)$ is described in the sense of the class $\mathfrak{G}(X, Y)$ of operators of Boutet de Monvel type. An operator $\mathcal{M} \in \mathfrak{G}(X, Y, Z)$ without any jump in the boundary conditions corresponds to some element in $\mathfrak{G}(X, Y)$, i.e. $\mathfrak{G}(X, Y) \subset \mathfrak{G}(X, Y, Z)$. In $\mathfrak{G}(X, Y, Z)$ one has also a composition. Let $\mathcal{M} \in \mathfrak{G}(X, Y, Z)$ be some mixed problem for an elliptic PDO A. Let $\mathcal{A}_0 \in \mathfrak{G}(X, Y)$ be an elliptic boundary problem for A and $\mathcal{P}_0 \in \mathfrak{G}(X, Y)$ a parametrix. Then one can define a reduction of $\mathcal{M}$ to Y by means of $\mathcal{P}_0$, following the constructions in 4.2.1. This is in fact possible and the resulting object on Y is then a transmission problem $\mathcal{T}$. Ellipticity of $\mathcal{M}$ is then per def. the ellipticity of $\mathcal{T}$. Knowing a parametrix of $\mathcal{T}$ one can construct a parametrix auf $\mathcal{M}$ following the formal way of 4.2.1.3 and one obtains a-priori estimates in the usual way. This approach gives an additional motivation for the investigation of transmission problems.

4.3.6. Parabolic Boundary Problems

The theory of elliptic boundary value problems containing trace and potential conditions can be generalized to parabolic pseudo-differential operators, either in the sense of the class $\mathfrak{G}$ when the operators have the transmission property with respect to the corresponding boundary or in the sense of the more general operators discussed in 4.3.2. There are some papers in which boundary problems of this type are investigated (cf. VIŠIK/ÈSKIN [3], ČAN ZUI CHO/ÈSKIN [1]). The results in the literature are not so complete as for the elliptic case and it is of course desirable to have a complete analogue, for instance, to the theory of elliptic operators in $\mathfrak{G}$. This program will be fulfilled in a forthcoming paper (cf. SCHULZE [9]) a part of which we formulate here in this section. Results are also announced by GRUBB [5].

Let X be a smooth compact manifold with boundary Y, $n = \dim X$, and put $\Omega = \operatorname{int} X$. Moreover let X_0 be a closed interval on the real axis $\mathbb{R}^1$. Set $\Omega_0 = \operatorname{int} X_0$. Let U be an open neighbourhood in X of a point $y \in Y$ diffeomorphic to $\overline{\mathbb{R}}^n_+ = \{x \in \mathbb{R}^n : x_n \geqq 0\}$. Then near $(x_0, y) \in X_0 \times X$, we use local charts of the form

$$\tilde{\varkappa}: X_0 \times U \to X_0 \times \overline{\mathbb{R}}^n_+\,, \quad \tilde{\varkappa}(x_0, x) = \big(x_0, \varkappa(x_0, x)\big)\,. \tag{1}$$

A similar convention is used when U is an interior open neighbourhood $\cong \mathbb{R}^n$.

If a local coordinate system is fixed, denote the corresponding points again by (x_0, x), where $x_0 \in \mathbb{R}^1$, $x = (x_1, \dots, x_n) \in \mathbb{R}^n$ and set as usual $x' = (x_1, \dots, x_{n-1})$. The dual coordinates are denoted by (ξ_0, ξ), $\xi = (\xi', \nu)$, $\xi' = (\xi_1, \dots, \xi_{n-1})$. For abbreviation set $\tilde{x} = (x_0, x)$, $\tilde{\xi} = (\xi_0, \xi)$. It is our aim to consider PDOs on $X_0 \times X$ with an anisotropic behaviour. We begin with corresponding definitions in local coordinates. Let h be a fixed even integer. A function $p(\tilde{x}, \tilde{\xi}) \in C^\infty(\mathbb{R}^{n+1} \times (\mathbb{R}^{n+1} \setminus \{0\}))$ is called *quasi-homogeneous of degree m* if

$$p(\tilde{x}, t^h\xi_0, t\xi) = t^m p(\tilde{x}, \tilde{\xi}), \quad t > 0, \tag{2}$$

for all $(\tilde{x}, \tilde{\xi}) \in \mathbb{R}^{n+1} \times (\mathbb{R}^{n+1} \setminus \{0\})$.

For instance, the function

$$\varrho_0(\tilde{\xi}) = (\xi_0^2 + |\xi|^{2h})^{1/2h}$$

is quasi-homogeneous of degree 1. It can be regarded a sa quasi-homogeneous module of the covector $\tilde{\xi}$ and satisfies the inequality $\varrho_0(\tilde{\xi}_1 + \tilde{\xi}_2) \leqq \varrho_0(\tilde{\xi}_1) + \varrho_0(\tilde{\xi}_2)$ (cf. Grušin, Šananin [1]). It is more convenient for us to use the module

$$\varrho(\tilde{\xi}) = |\xi_0|^{1/h} + |\xi| . \tag{3}$$

Then there are constants $a, b > 0$ with $a\varrho(\tilde{\xi}) \leqq \varrho_0(\tilde{\xi}) \leqq b\varrho(\tilde{\xi})$.

Denote by ord p the degree of quasi-homogeneity of the function p with respect to $\tilde{\xi}$. If ord $p = m$, we have

$$\operatorname{ord} \frac{\partial}{\partial \xi_k} p = \begin{cases} m - h & \text{when } k = 0, \\ m - 1 & \text{when } 1 \leqq k \leqq n. \end{cases}$$

Let $\Omega \subseteqq \mathbb{R}^n$ be an open set and $\Omega_0 \subseteqq \mathbb{R}^1$ an open interval. Denote by $S_{\mathrm{cl}}^{m,h}(\Omega_0 \times \Omega)$ the set of all functions $a(\tilde{x}, \tilde{\xi}) \in C^\infty(\Omega_0 \times \Omega \times \mathbb{R}^{n+1})$ for which there is a sequence $a_{m-j}(\tilde{x}, \tilde{\xi}) \in C^\infty(\Omega_0 \times \Omega \times (\mathbb{R}^{n+1} \setminus \{0\}))$ of quasi-homogeneous functions of degree $m - j$ with

$$a \sim \sum_{j=0}^{\infty} a_{m-j} .$$

The latter asymptotic sum means that for arbitrary multi-indices α, β, for each compact set $K \subset \Omega_0 \times \Omega$ and for each integer $N \geqq 1$ there exists a constant $c = c(\alpha, \beta, K, N)$ with

$$\Big|D_{\tilde{x}}^\alpha D_{\tilde{\xi}}^\beta [a(\tilde{x}, \tilde{\xi}) - \sum_{j=0}^{N} a_{m-j}(\tilde{x}, \tilde{\xi})]\Big| \leqq c\varrho(\tilde{\xi})^{m-N-\tilde{h}\beta} \tag{4}$$

for all $\tilde{x} \in K$, $\tilde{\xi} \in \mathbb{R}^{n+1} \setminus \{0\}$ with $\varrho(\tilde{\xi}) \geqq 1$. Here $\tilde{h} = (h, 1, \dots, 1)$ and $\tilde{h}\beta = h\beta_0 + \beta_1 + \dots + \beta_n$. In this section asymptotic sums are understood in this sense.

For a given sequence a_{m-j} of quasi-homogeneous functions of degree $m - j$, $j \to \infty$, there exists an a with $a \sim \sum_j a_{m-j}$, i.e. $a \in S_{\mathrm{cl}}^{m,h}(\Omega_0 \times \Omega)$, and a is uniquely determined modulo $S^{-\infty}$. $S_{\mathrm{cl}}^{m,h}(\Omega_0 \times \Omega)$ is a subspace of $S^{m,h}(\Omega_0 \times \Omega)$ defined as the set of all functions $a(\tilde{x}, \tilde{\xi}) \in C^\infty(\Omega_0 \times \Omega \times \mathbb{R}^{n+1})$ satisfying the estimates

$$|D_{\tilde{x}}^\alpha D_{\tilde{\xi}}^\beta a(\tilde{x}, \tilde{\xi})| \leqq c\big(1 + \varrho(\tilde{\xi})\big)^{m-\tilde{h}\beta} \tag{5}$$

for arbitrary multi-indices α, β and $\tilde{x} \in K$, $K \subset \Omega_0 \times \Omega$ compact with constants $c = c(\alpha, \beta, K)$. It can be proved that for a given sequence $a_{m_j} \in S^{m_j,h}(\Omega_0 \times \Omega)$ with

$m_j \to -\infty$ for $j \to \infty$ there is an $a \in S^{m,h}(\Omega_0 \times \Omega)$, $m = \max\{m_j\}$, so that $a \sim \sum_j a_{m_j}$ and a is uniquely determined modulo $S^{-\infty}$. The asymptotic sum refers to the function $\varrho(\tilde{\xi})$.

By $L^{m,h}(\Omega_0 \times \Omega)$ we denote the space of operators $A = A_0 + C$, where A_0 has the form

$$(A_0 u)(\tilde{x}) = (2\pi)^{-(n+1)} \int e^{i\tilde{x}\tilde{\xi}} a(\tilde{x}, \tilde{\xi})\, \hat{u}(\tilde{\xi})\, d\tilde{\xi}, \tag{6}$$

$a \in S^{m,h}(\Omega_0 \times \Omega)$, and C is an operator with C^∞ kernel. By $L_{\text{cl}}^{m,h}(\Omega_0 \times \Omega)$ we denote the subset of classical operators, i.e. for which a belongs to $S_{\text{cl}}^{m,h}(\Omega_0 \times \Omega)$.

Consider a diffeomorphism $\tilde{\varkappa}: \Omega_0 \times \Omega^1 \to \Omega_0 \times \Omega^2$ of the form $\tilde{\varkappa}(x_0, x) = (x_0, \varkappa(\tilde{x}))$, Ω^j open domains in $\mathbb{R}^n$, $j = 1, 2$. The coordinates in $T^*\Omega^1$ and $T^*\Omega^2$ are denoted by $(\tilde{x}, \tilde{\xi})$ and $(\tilde{y}, \tilde{\eta})$, respectively. The following result is given in Hunt/Piriou [1]. For $A \in L^{m,h}(\Omega_0 \times \Omega^1)$ we obtain $\tilde{\varkappa}_* A \overset{\text{def}}{=} (\tilde{\varkappa}^*)^{-1} \circ A \circ \tilde{\varkappa}^* \in L^{m,h}(\Omega_0 \times \Omega^2)$. Then the complete symbol $b \in S^{m,h}(\Omega_0 \times \Omega^2)$ of $\tilde{\varkappa}_* A$ has the following asymptotic expansion

$$b(\tilde{y}, \tilde{\eta}) \sim \sum_\alpha \frac{1}{\alpha!} a^{(\alpha)}(\tilde{x}, \tilde{\xi})\, (D_{\tilde{z}}\, e^{il(\tilde{z}, \tilde{x})\tilde{\eta}})_{\tilde{z}=\tilde{x}},$$

where $\tilde{\xi} = {}^t d\tilde{\varkappa}\, \tilde{\eta}$, $\tilde{y} = \tilde{\varkappa}(\tilde{x})$ and $l(\tilde{z}, \tilde{x}) = \tilde{\varkappa}(\tilde{z}) - \tilde{\varkappa}(\tilde{x}) - d\tilde{\varkappa}(\tilde{x})\,(\tilde{z} - \tilde{x})$. Note that $A \in L_{\text{cl}}^{m,h}(\Omega_0 \times \Omega^1)$ implies $\tilde{\varkappa}_* A \in L_{\text{cl}}^{m,h}(\Omega_0 \times \Omega^2)$.

With respect to compositions we also have a rule analogous to the usual one. Let $A_i \in L^{m_i,h}(\Omega_0 \times \Omega)$ $(i = 1, 2)$, and either A_1 or A_2 be properly supported. Then $A = A_2 A_1 \in L^{m_1+m_2,h}(\Omega_0 \times \Omega)$ and (with obvious notations)

$$a(\tilde{x}, \tilde{\xi}) \sim \sum_\alpha \frac{1}{\alpha!} (\partial_{\tilde{\xi}}^\alpha a_2(\tilde{x}, \tilde{\xi}))\, D_{\tilde{x}}^\alpha a_1(\tilde{x}, \tilde{\xi}). \tag{7}$$

Note that the difference to 1.2.3.3.(7) is that the asymptotic sum has a modified meaning.

Finally, taking adjoint or transposed operators satisfies precisely the same formal rules as in 1.2.3.3 (cf. Hunt/Piriou [1]). Therefore we do not formulate it here explicitly. The set $L_{\text{cl}}^{m,h}$ of classical operators is closed under these operations. In the following restrict the consideration to classical symbols and operators.

Denote by $L_V^{m,h}(\Omega_0 \times \Omega)$ the set of all operators $A \in L_{\text{cl}}^{m,h}(\Omega_0 \times \Omega)$ for which

$$u \in C_0^\infty(\Omega_0 \times \Omega), \quad u = 0 \quad \text{for } x_0 < t$$

implies

$$Au = 0 \quad \text{when } x_0 < t$$

for all $t \in \Omega_0$. The distributional kernel $K_A(x_0, x, y_0, y)$ of an operator in $L_V^{m,h}(\Omega_0 \times \Omega)$ vanishes for $x_0 < y_0$. Operators of this type are called *Volterra operators*. Let $S_V^{m,h}(\Omega_0 \times \Omega)$ be the set of all symbols $a \in S_{\text{cl}}^{m,h}(\Omega_0 \times \Omega)$ for which the operator (6) belongs to $L_V^{m,h}(\Omega_0 \times \Omega)$. Set $\mathbb{C}^- = \{\zeta_0 \in \mathbb{C}: \zeta_0 = \xi_0 + i\eta_0,\ \eta_0 < 0\}$, $\varrho(\zeta_0, \xi) = |\zeta_0|^{1/h} + |\xi|$. The following assertions are proved in Piriou [1, 2].

Lemma 1. *Let $a \in S_{\text{cl}}^{m,h}(\Omega_0 \times \Omega)$ and assume that there exists an analytic extension of $a(\tilde{x}, \xi_0, \xi)$ as function $a(\tilde{x}, \zeta_0, \xi)$ into $\mathbb{C}^-$ which is continuous in $\overline{\mathbb{C}^-}$ and suppose that there are constants c, μ with*

$$|a(\tilde{x}, \zeta_0, \xi)| \leq c(1 + \varrho(\zeta_0, \xi))^\mu$$

for $\zeta_0 \in \overline{\mathbb{C}^-}$, $\xi \in \mathbb{R}^n$. Then the operator A defined by (6) *belongs to $L_V^{m,h}(\Omega_0 \times \Omega)$.*

Denote by $S_V^{(m),h}(\Omega_0 \times \Omega)$ the set of all functions $a(\tilde{x}, \tilde{\xi}) \in C^\infty(\Omega_0 \times \Omega \times (\mathbb{R}^{n+1} \setminus \{0\}))$ which are quasi-homogeneous in $\tilde{\xi}$ of degree m possessing an extension $a(\tilde{x}, \zeta_0, \xi) \in C^\infty(\Omega_0 \times \Omega \times ((\overline{\mathbb{C}^-} \times \mathbb{R}^n) \setminus \{0\}))$ which is analytic for $\zeta_0 \in \mathbb{C}^-$ when the other arguments (x_0, x, ξ) are fixed. Note that the extension is again quasi-homogeneous, i.e.

$$a(\tilde{x}, t^h\zeta_0, t\xi) = t^m a(\tilde{x}, \zeta_0, \xi)$$

for $t > 0$, $\zeta_0 \in \overline{\mathbb{C}^-}$.

Proposition 2. *Let* $a(\tilde{x}, \tilde{\xi}) \in S_{\mathrm{cl}}^{m,h}(\Omega_0 \times \Omega)$ *and*

$$a \sim \sum_{j=0}^{\infty} a_{m-j}\,, \tag{8}$$

where a_{m-j} *is quasi-homogeneous of degree* $m - j$*. Then, from*

$$A \in L_V^{m,h}(\Omega_0 \times \Omega) \tag{9}$$

(A *defined by* (6)), *it follows that*

$$a_{m-j} \in S_V^{(m-j),h}(\Omega_0 \times \Omega)\,, \quad j = 0, 1, 2, \ldots\,. \tag{10}$$

Conversely, if we are given a sequence (10), *there exists an* $a \in S_{\mathrm{cl}}^{m,h}(\Omega_0 \times \Omega)$ *with* (8) *and that* (9) *holds for the corresponding operator.*

We can also consider matrices of symbols and operators. Denote by $L_{\mathrm{cl}}^{m,h}(\Omega_0 \times \Omega, \mathbb{C}^k, \mathbb{C}^l)$ the set of $l \times k$ matrices of operators in $L_{\mathrm{cl}}^{m,h}(\Omega_0 \times \Omega)$. In a similar sense the notations $L_V^{m,h}(\Omega_0 \times \Omega, \mathbb{C}^k, \mathbb{C}^l)$, $S_{\mathrm{cl}}^{m,h}(\Omega_0 \times \Omega, \mathbb{C}^k, \mathbb{C}^l)$ and so on are used.

Definition 3. An element $\sigma \in S_V^{(m),h}(\Omega_0 \times \Omega, \mathbb{C}^k, \mathbb{C}^k)$ is called *parabolic* if the extension $\sigma(\tilde{x}, \zeta_0, \xi)$ is invertible for all $(\tilde{x}, \zeta_0, \xi) \in \Omega_0 \times \Omega \times ((\overline{\mathbb{C}^-} \times \mathbb{R}^n) \setminus \{0\})$. An operator $A \in L_V^{m,h}(\Omega_0 \times \Omega)$ is called *parabolic* if its quasi-homogeneous principal symbol σ_A of order m has this property.

Obviously, if σ is parabolic, σ^{-1} belongs to $S_V^{(-m),h}(\Omega_0 \times \Omega, \mathbb{C}^k, \mathbb{C}^k)$ and it is parabolic, too. For example, if A is an operator with complete symbol $(1 + i\xi_0 + |\xi|^h)^{m/h}$, we have $\sigma_A = (i\xi_0 + |\xi|^h)^{m/h}$ and A is parabolic. Another example of a parabolic operator is

$$A = \frac{\partial}{\partial x_0} + B(x_0, x, D_x)\,,$$

where for each x_0 the differential operator B is elliptic and has a real principal symbol.

Consider the space $H^{s,h}(\mathbb{R}^{n+1})$ of all $u \in \mathcal{S}'(\mathbb{R}^{n+1})$ for which the norm

$$\|u\|_{s,h} = \int (1 + \varrho(\tilde{\xi}))^{2s}\, |\hat{u}(\tilde{\xi})|^2\, d\tilde{\xi}$$

is finite, $s \in \mathbb{R}$. Use also the notations $H_{\mathrm{comp}}^{s,h}(\mathbb{R}^{n+1})$, $H_{\mathrm{loc}}^{s,h}(\mathbb{R}^{n+1})$ with an obvious meaning. Spaces of this type have a lot of properties analogous to those of usual Sobolev spaces. Mention only that $u \in H^{s,h}(\mathbb{R}_{\tilde{y}}^{n+1})$, supp u compact, implies $\tilde{\varkappa}^* u \in H^{s,h}(\mathbb{R}_{\tilde{x}}^{n+1})$ when $\tilde{\varkappa}\colon \mathbb{R}_{\tilde{x}}^{n+1} \to \mathbb{R}_{\tilde{y}}^{n+1}$ is a diffeomorphism of the form $\tilde{\varkappa}(\tilde{x}) = (x_0, \varkappa(\tilde{x}))$. This enables us to define spaces of the type $H^{s,h}$ when the underlying set is a manifold of the form $\mathbb{R} \times M$, M a manifold of dimension n.

Denote by $H_{(0)}^{s,h}(\mathbb{R}^{n+1})$ the space of all $u \in H^{s,h}(\mathbb{R}^{n+1})$ with supp $u \subseteq [0, \infty) \times \mathbb{R}^n$. In an analogous sense the notations $H_{(0),\mathrm{comp}}^{s,h}(\mathbb{R}^{n+1})$, $H_{(0),\mathrm{loc}}^{s,h}(\mathbb{R}^{n+1})$ and so on will be used.

If X is a compact smooth manifold with boundary, the space $H^{s,h}(\mathbb{R} \times X)$ can also be defined in an obvious way as well as $H^{s,h}_{(0)}(\mathbb{R} \times X)$.

Proposition 4. *Let $A \in L^{m,h}_{\mathrm{cl}}(\Omega_0 \times \Omega)$. Then A induces a continuous operator*

$$A: H^{s,h}_{\mathrm{comp}}(\Omega_0 \times \Omega) \to H^{s-m,h}_{\mathrm{loc}}(\Omega_0 \times \Omega), \quad s \in \mathbb{R}.$$

Let $\Omega_0 = (-\varepsilon, t)$ for some $\varepsilon > 0$ and $0 < t \leqq \infty$. Then we have

Proposition 5. *Let $A \in L^{m,h}_{\bar{V}}(\Omega_0 \times \Omega)$. Then A induces a continuous operator*

$$A: H^{s,h}_{(0),\mathrm{comp}}(\Omega_0 \times \Omega) \to H^{s-m,h}_{(0),\mathrm{loc}}(\Omega_0 \times \Omega), \quad s \in \mathbb{R}. \tag{11}$$

Now pass to a first result about parabolic operators. Take for simplicity scalar operators.

Proposition 6. *Let $A \in L^{m,h}_{\bar{V}}(\Omega_0 \times \Omega)$ be parabolic. Then there exists a parabolic operator $P \in L^{-m,h}_{\bar{V}}(\Omega_0 \times \Omega)$ with the property $AP - I \in L^{-\infty}_{\bar{V}}$, $PA - I \in L^{-\infty}_{\bar{V}}$.*

It is clear how to derive a local a-priori estimate for the parabolic operator A, which refers to the anisotropic Sobolev spaces in consideration. We do not formulate it explicitly here because below we shall consider more interesting situations with existence of inverses.

Consider now the global situation as at the beginning. Let W be an open neighbouring manifold of X of the same dimension and consider an atlas on $\Omega_0 \times W$ extending the given atlas for $\Omega_0 \times X$ in a natural way such that for any coordinate neighbourhood U in W the subset $\Omega_0 \times (X \cap U)$ is transformed into $\Omega_0 \times \overline{\mathbb{R}}^n_+$, $\Omega_0 \times (U \setminus (X \cap U))$ into $\Omega_0 \times \mathbb{R}^n_-$ and $\Omega_0 \times (U \cap Y)$ into $\Omega_0 \times \mathbb{R}^{n-1}_{x'}$. Then we can define the classes of operators $L^{m,h}(\Omega_0 \times W, E, F)$, $L^{m,h}_{\bar{V}}(\Omega_0 \times W, E, F)$ in an obvious way. Here E, F are vector bundles over $\Omega_0 \times W$ (the restrictions of E, F to $\Omega_0 \times X$ are denoted below by the same letters). Similarly the classes $L^{m,h}(\Omega_0 \times Y, J, G)$, $L^{m,h}_{\bar{V}}(\Omega_0 \times Y, J, G)$ can be defined, where Y is a closed compact manifold and $J, G \in \mathrm{Vect}\,(\Omega_0 \times Y)$. The anisotropic Sobolev spaces $H^{s,h}(\Omega_0 \times W, E)$, $H^{s,h}_{(0)}(\Omega_0 \times W, E)$, $E \in \mathrm{Vect}\,(\Omega_0 \times W)$, and so on can be easily defined. From now on assume that

$$\Omega_0 = (-\varepsilon, \infty), \quad \varepsilon > 0. \tag{12}$$

Distributional sections in E can be multiplied by scalar functions. By $H^{s,h,\delta}(\Omega_0 \times W, E)$, $\delta > 0$ fixed, we denote the space of all $u \in \mathcal{D}'(\Omega_0 \times W, E)$ with $\mathrm{e}^{-x_0\delta} u \in H^{s,h}(\Omega_0 \times W, E)$. Similarly define the subspace $H^{s,h,\delta}_{(0)}(\Omega_0 \times W, E)$. By

$$\|u\|_{s,h,\delta} \overset{\mathrm{def}}{=} \|\mathrm{e}^{-x_0\delta} u\|_{s,h} \tag{13}$$

we get a norm on $H^{s,h,\delta}(\Omega_0 \times W, E)$. Use also $H^{s,h,\delta}(\Omega_0 \times X, E)$, $H^{s,h,\delta}_{(0)}(\Omega_0 \times X, E)$, $E \in \mathrm{Vect}\,(\Omega_0 \times X)$, $H^{s,h,\delta}(\Omega_0 \times Y, G)$, $H^{s,h,\delta}_{(0)}(\Omega_0 \times Y, G)$, $G \in \mathrm{Vect}\,(\Omega_0 \times Y)$ and so on with analogous meaning.

An operator $A \in L^{m,h}_{\bar{V}}(\Omega_0 \times W, E, F)$ is called *parabolic* if for each local chart $\chi: \Omega_0 \times U \to \Omega_0 \times \mathbb{R}^n$ (with E, F trivial over $\Omega_0 \times U$) the operator $A|_{\Omega_0 \times U}$ in local coordinates is parabolic. Similarly we define parabolicity of operators in $L^{m,h}_{\bar{V}}(\Omega_0 \times Y, J, G)$.

Theorem 7. *Let $\Omega_0 = (-\varepsilon, \infty)$, $\varepsilon > 0$, and $A \in L^{m,h}_{\bar{V}}(\Omega_0 \times Y, J, G)$ be a parabolic operator. Then A induces an isomorphism*

$$A: H^{s,h,\delta}_{(0)}(\Omega_0 \times Y, J) \to H^{s-m,h,\delta}_{(0)}(\Omega_0 \times Y, G) \tag{14}$$

for any $s \in \mathbb{R}$ and $\delta > 0$ fixed and sufficiently large. $P = A^{-1}$ belongs to $L_{\bar{V}}^{-m,h}(\Omega_0 \times Y, G, J)$ and P is parabolic.

At this point it is clear which type of results can be expected for parabolic boundary problems.

Let X be a compact smooth manifold with boundary Y and $\Omega_0 = (-\varepsilon, \infty)$, $\varepsilon > 0$ fixed. Consider operators of the form

$$\mathcal{A} = \begin{pmatrix} r^+A + r'B & K \\ r'T & Q \end{pmatrix} : \begin{matrix} H_{(0)}^{s,h,\delta}(\Omega_0 \times X, E) \\ \oplus \\ H_{(0)}^{t+\lambda+1/2,h,\delta}(\Omega_0 \times Y, J) \end{matrix} \to \begin{matrix} H_{(0)}^{t,h,\delta}(\Omega_0 \times X, F) \\ \oplus \\ H_{(0)}^{s-\gamma-1/2,h,\delta}(\Omega_0 \times Y, G) . \end{matrix} \tag{15}$$

Here $E, F \in \operatorname{Vect}(\Omega_0 \times X)$, $J, G \in \operatorname{Vect}(\Omega_0 \times Y)$, $\delta > 0$, $s > 0$ sufficiently large, $t = s - \alpha$, $\alpha = \operatorname{ord} \sigma_A \in \mathbb{Z}$, $\alpha - 1 = \operatorname{ord} \sigma_B$, $\lambda = \operatorname{ord} \sigma_K \in \mathbb{R}$, $\gamma = \operatorname{ord} \sigma_T \in \mathbb{R}$, $1 - \alpha + \lambda + \gamma = \operatorname{ord} \sigma_Q$. The operators r^+A, $r'B$, ... belong to classes of Volterra operators defined by means of suitable symbol classes. More precisely we have

$$r^+A \in (\operatorname{Op} \mathfrak{A})_V^{\alpha,h}(\Omega_0 \times X, E, F) , \tag{16}$$

$$r'B \in (\operatorname{Op} \mathfrak{B})_V^{\alpha-1,h}(\Omega_0 \times X, E, F) , \tag{17}$$

$$r'T \in (\operatorname{Op} \mathfrak{T})_V^{\gamma,h} (\Omega_0 \times X, \Omega_0 \times Y, E, G) , \tag{18}$$

$$K \in (\operatorname{Op} \mathfrak{K})_V^{\lambda,h} (\Omega_0 \times Y, \Omega_0 \times X, J, F) , \tag{19}$$

$$Q \in L_V^{1-\alpha+\lambda+\gamma,h}(\Omega_0 \times Y, J, G) . \tag{20}$$

The classes $(\operatorname{Op} \mathfrak{A})_V^{\alpha,h}(...)$, ... in (16), ..., (20), will be described below. The quasi-homogeneous principal symbols σ_A, σ_B, ... belong to symbol classes $\mathfrak{A}_V^{(\alpha),h}$, $\mathfrak{B}_V^{(\alpha-1),h}$, .. to be defined, too.

First consider $\mathfrak{A}_V^{(\alpha),h}$. For simplicity we restrict ourselves to scalar symbols and to a local coordinate system $\Omega_0 \times \mathbb{R}^n$. The generalization to matrix valued symbols is then trivial and the invariance of the typical properties with respect to coordinate diffeomorphisms admits to pass to the global situation.

A function $\sigma_A(\tilde{x}, \tilde{\xi}) \in C^\infty(\Omega_0 \times \mathbb{R}^n \times (\mathbb{R}^{n+1} \setminus \{0\}))$ belongs to $\mathfrak{A}_V^{(\alpha),h}$ over $\Omega_0 \times \mathbb{R}^n$ if it has the following properties:

(i) σ_A is quasi-homogeneous of degree α with respect to $\tilde{\xi}$,

(ii) σ_A has an extension

$$\sigma_A(\tilde{x}, \zeta_0, \xi) \in C^\infty(\Omega_0 \times \mathbb{R}^n \times ((\overline{\mathbb{C}}^- \times \mathbb{R}^n) \setminus \{0\})) \tag{21}$$

which is analytic for $\zeta_0 \in \overline{\mathbb{C}}^-$ when the other arguments $(\tilde{x}, \xi)$ are fixed,

(iii) σ_A has the transmission property with respect to x_n, i.e.

$$\partial_{x_n}^k \partial_{\xi'}^\beta \partial_{\xi_0}^l \sigma_A(x_0, x', x_n, \xi_0, \xi', 1) = e^{i\pi(\alpha-|\beta|-hl)} \partial_{x_n}^k \partial_{\xi'}^\beta \partial_{\xi_0}^l \sigma_A(x_0, x', x_n, \xi_0, \xi', -1)$$

at $x_n = 0$, $\xi_0 = 0$, $\xi' = 0$ for all $k, l \in \mathbb{Z}_+$ and all multi-indices β, $(x_0, x') \in \Omega_0 \times \mathbb{R}^{n-1}$.

The condition (iii) can be equivalently expressed by the property

(iii)′ $\partial_{x_n}^k \sigma_A(x_0, x', 0, \xi_0, \xi', \nu)$ defines a function in $C^\infty(\Omega_0 \times \mathbb{R}^{n-1} \times ((\mathbb{R}_{\xi_0} \times \mathbb{R}_{\xi'}^{n-1}) \setminus \{0\}), H)$ for all $k \in \mathbb{Z}_+$.

A function $\sigma_B(x_0, x, \xi_0, \xi', \nu, \tau) \in C^\infty(\Omega_0 \times \mathbb{R}^n \times ((\mathbb{R}_{\xi_0} \times \mathbb{R}_{\xi'}^{n-1}) \setminus \{0\}) \times \mathbb{R}_{\nu,\tau}^2)$ belongs to $\mathfrak{B}_V^{(\alpha-1),h}$ over $\Omega_0 \times \mathbb{R}^n$ if it has the following properties:

(i) σ_B is quasi-homogeneous of degree $\alpha - 1$ with respect to (ξ_0, ξ', ν, τ), i.e. in this case

$$\sigma_B(\tilde{x}, t^h\xi_0, t\xi', t\nu, t\tau) = t^{\alpha-1}\sigma_B(\tilde{x}, \xi_0, \xi', \nu, \tau) , \quad t > 0 ,$$

(ii) σ_B has an extension

$$\sigma_B(\tilde{x}, \zeta_0, \xi', \nu, \tau) \in C^\infty(\Omega_0 \times \mathbb{R}^n \times ((\overline{\mathbb{C}}^- \times \mathbb{R}^{n-1}_{\xi'}) \setminus \{0\}) \times \mathbb{R}^2_{\nu,\tau})$$

which is analytic for $\zeta_0 \in \overline{\mathbb{C}}^-$ when the other arguments (x, ξ', ν, τ) are fixed,

(iii) $\partial^k_{x_n}\sigma_B(x_0, x', 0, \xi_0, \xi', \nu, \tau)$ defines a function in $C^\infty(\Omega_0 \times \mathbb{R}^{n-1} \times ((\mathbb{R}_{\xi_0} \times \mathbb{R}^{n-1}_{\xi'}) \setminus \{0\}), H^+_\nu \hat{\otimes} H^-_\tau)$ for all $k \in \mathbb{Z}_+$.

A function $\sigma_T(x_0, x, \xi_0, \xi', \nu) \in C^\infty(\Omega_0 \times \mathbb{R}^n \times ((\mathbb{R}_{\xi_0} \times \mathbb{R}^{n-1}_{\xi'}) \setminus \{0\}) \times \mathbb{R}_\nu)$ belongs to $\mathfrak{T}^{(\gamma),h}_V$ over $\Omega_0 \times \mathbb{R}^n$ if it has the following properties

(i) σ_T is quasi-homogeneous of degree $\gamma \in \mathbb{R}$ with respect to $\tilde{\xi}$,

(ii) σ_T has an extension

$$\sigma_T(\tilde{x}, \zeta_0, \xi', \nu) \in C^\infty(\Omega_0 \times \mathbb{R}^n \times ((\overline{\mathbb{C}}^- \times \mathbb{R}^{n-1}_{\xi'}) \setminus \{0\}) \times \mathbb{R}_\nu)$$

which is analytic for $\zeta_0 \in \overline{\mathbb{C}}^-$ when the other arguments are fixed,

(iii) $\partial^k_{x_n}\sigma_T(x_0, x', 0, \xi_0, \xi', \nu)$ defines a function in $C^\infty(\Omega_0 \times \mathbb{R}^{n-1} \times ((\mathbb{R}_{\xi_0} \times \mathbb{R}^{n-1}_{\xi'}) \setminus \{0\}), H^-_\nu)$ for all $k \in \mathbb{Z}_+$.

A function $\sigma_K(x_0, x, \xi_0, \xi', \nu) \in C^\infty(\Omega_0 \times \mathbb{R}^n \times ((\mathbb{R}_{\xi_0} \times \mathbb{R}^{n-1}_{\xi'}) \setminus \{0\}) \times \mathbb{R}_\nu)$ belongs to $\mathfrak{K}^{(\lambda),h}_V$ over $\Omega_0 \times \mathbb{R}^n$ if it has the following properties:

(i) σ_K is quasi-homogeneous of degree $\lambda \in \mathbb{R}$ with respect to $\tilde{\xi}$,

(ii) σ_K has an extension

$$\sigma_K(\tilde{x}, \zeta_0, \xi', \nu) \in C^\infty(\Omega_0 \times \mathbb{R}^n \times ((\overline{\mathbb{C}}^- \times \mathbb{R}^{n-1}_{\xi'}) \setminus \{0\}) \times \mathbb{R}_\nu)$$

which is analytic for $\zeta_0 \in \overline{\mathbb{C}}^-$ when the other arguments $(\tilde{x}, \xi', \nu)$ are fixed,

(iii) $\partial^k_{x_n}\sigma_K(x_0, x', 0, \xi_0, \xi', \nu)$ defines a function in $C^\infty(\Omega_0 \times \mathbb{R}^{n-1} \times ((\mathbb{R}_{\xi_0} \times \mathbb{R}^{n-1}_{\xi'}) \setminus \{0\}), H^+_\nu)$ for all $k \in \mathbb{Z}_+$.

Note that, similar as in the elliptic theory, it is sufficient in the calculus to consider symbols $\sigma_B, \sigma_T, \sigma_K$ which are independent of x_n. The elements in $\mathfrak{B}^{(\beta),h}_V, \mathfrak{T}^{(\gamma),h}_V, \mathfrak{K}^{(\lambda),h}_V$ are called *Green, trace* and *potential symbols* (of the corresponding homogeneities) *of Volterra type*.

Now we can speak about asymptotic expansions of symbols over $\Omega_0 \times \mathbb{R}^n$

$$a \sim \sum_{j=0}^{\infty} a_{\alpha-j}, \quad a_{\alpha-j} \in \mathfrak{A}^{(\alpha-j),h}_V,$$

$$b \sim \sum_{j=0}^{\infty} b_{\beta-j}, \quad b_{\beta-j} \in \mathfrak{B}^{(\beta-j),h}_V,$$

$$t \sim \sum_{j=0}^{\infty} t_{\gamma-j}, \quad t_{\gamma-j} \in \mathfrak{T}^{(\gamma-j),h}_V,$$

$$k \sim \sum_{j=0}^{\infty} k_{\lambda-j}, \quad k_{\lambda-j} \in \mathfrak{K}^{(\tau-j),h}_V.$$

The excision in the last three cases refers to (ξ_0, ξ').

Thus we obtain symbol spaces $\mathfrak{A}^{\alpha,h}_V(\Omega_0 \times \mathbb{R}^n)$, $\mathfrak{B}^{\beta,h}_V(\Omega_0 \times \mathbb{R}^n)$, $\mathfrak{T}^{\gamma,h}_V(\Omega_0 \times \mathbb{R}^n, \Omega_0 \times \mathbb{R}^{n-1})$, $\mathfrak{K}^{\lambda,h}_V(\Omega_0 \times \mathbb{R}^{n-1}, \Omega_0 \times \mathbb{R}^n)$. Defining operators in a similar manner as in the elliptic theory we get per def. the classes $(\mathrm{Op}\,\mathfrak{A})^{\alpha,h}_V(\Omega_0 \times \mathbb{R}^n)$, $(\mathrm{Op}\,\mathfrak{B})^{\beta,k}_V(\Omega_0 \times \mathbb{R}^n)$, $(\mathrm{Op}\,\mathfrak{T})^{\gamma,h}_V(\Omega_0 \times \mathbb{R}^n, \Omega_0 \times \mathbb{R}^{n-1})$, $(\mathrm{Op}\,\mathfrak{K})^{\lambda,h}_V(\Omega_0 \times \mathbb{R}^{n-1}, \Omega_0 \times \mathbb{R}^n)$ (cf. the corresponding versions of Proposition 2).

It turns out that for

$$r^+A \in (\operatorname{Op}\mathfrak{A})_V^{\alpha,h}(\Omega_0 \times \mathbb{R}^n), \quad r'B \in (\operatorname{Op}\mathfrak{B})_V^{\alpha-1,h}(\Omega_0 \times \mathbb{R}^n),$$
$$r'T \in (\operatorname{Op}\mathfrak{T})_V^{\gamma,h}(\Omega_0 \times \mathbb{R}^n, \Omega_0 \times \mathbb{R}^{n-1}), \quad K \in (\operatorname{Op}\mathfrak{K})_V^{\lambda,h}(\Omega_0 \times \mathbb{R}^{n-1}, \Omega_0 \times \mathbb{R}^n),$$
$$Q \in L_V^{1-\alpha+\lambda+\gamma}(\Omega_0 \times \mathbb{R}^{n-1})$$

the operator

$$\begin{pmatrix} r^+A + r'B & K \\ r'T & Q \end{pmatrix} : \begin{matrix} H^{s,h}_{(0),\mathrm{comp}}(\Omega_0 \times \mathbb{R}^n) \\ \oplus \\ H^{t+\lambda+1/2,h}_{(0),\mathrm{comp}}(\Omega_0 \times \mathbb{R}^{n-1}) \end{matrix} \to \begin{matrix} H^{t,h}_{(0),\mathrm{loc}}(\Omega_0 \times \mathbb{R}^n) \\ \oplus \\ H^{s-\gamma-1/2,h}_{(0),\mathrm{loc}}(\Omega_0 \times \mathbb{R}^{n-1}) \end{matrix} \tag{22}$$

is continuous, $t = s = \alpha$, $s \in \mathbb{R}$ sufficiently large.

As already mentioned, the definitions and properties generalize to manifolds without any troubles.

Let S^*Y be the cosphere bundle of Y with respect to a fixed Riemannian metric and consider $\Omega_0 \times S^*Y \times \overline{\mathbb{C}}^-$. Denote by

$$p_0 \colon \Omega_0 \times S^*Y \times \overline{\mathbb{C}}^- \to \Omega_0 \times Y \times \overline{\mathbb{C}}^-$$

the projection induced by $p \colon S^*Y \to Y$. For $E \in \operatorname{Vect}(\Omega_0 \times X)$ define $E' = E|_{\Omega_0 \times Y}$ and let E_0 be the pull pack of E' to $\Omega_0 \times Y \times \overline{\mathbb{C}}^-$ with respect to the projection $\Omega_0 \times Y \times \overline{\mathbb{C}}^- \to \Omega_0 \times Y$.

Proposition 8. *Let $\sigma_A \in \mathfrak{A}_V^{(\alpha),h}(\Omega_0 \times X, E, F)$ be parabolic, E, $F \in \operatorname{Vect}(\Omega_0 \times X)$. Then $g \to \Pi^+\sigma_A(x_0, x', 0, \zeta_0, \xi', \nu)\, g(\nu)$ induces a map*

$$\Pi^+\sigma_A \colon p_0^*(H^+ \otimes E_0) \to p_0^*(H^+ \otimes F_0) \tag{23}$$

which is Fredholm between the corresponding fibres over all $(x_0, x', \zeta_0, \xi') \in \Omega_0 \times S^*Y \times \overline{\mathbb{C}}^-$.

Now, similar to the elliptic theory for given $\sigma_B \in \mathfrak{B}_V^{(\alpha-1),h}(\Omega_0 \times X, E, F)$, $\sigma_T \in \mathfrak{T}_V^{(\gamma),h}(\Omega_0 \times X, \Omega_0 \times Y, E, G)$, $\sigma_K \in \mathfrak{K}_V^{(\lambda),h}(\Omega_0 \times Y, \Omega_0 \times X, J, F)$, we can define corresponding boundary symbols

$$\Pi'\sigma_B \colon p_0^*(H^+ \otimes E_0) \to p_0^*(H^+ \otimes F_0),$$
$$\Pi'\sigma_T \colon p_0^*(H^+ \otimes E_0) \to p_0^*G_0,$$
$$\sigma_K \colon \quad p_0^*J_0 \to p_0^*(H^+ \otimes F_0).$$

Consider an operator $\mathcal{A}$ of the form as on the left hand side of (15) with (16), (17), (18), (19), (20). Denote by $\mathfrak{G}_V$ the class of such operators. We have well-defined principal symbols σ_A, σ_B, σ_T, σ_K, σ_Q of the corresponding quasi-homogeneities. The family of operators

$$\sigma_{\Omega_0 \times Y}(\mathcal{A}) = \begin{pmatrix} \Pi^+\sigma_A + \Pi'\sigma_B & \sigma_K \\ \Pi'\sigma_T & \sigma_Q \end{pmatrix} : \begin{matrix} p_0^*(H^+ \otimes E_0) \\ \oplus \\ p_0^*J_0 \end{matrix} \to \begin{matrix} p_0^*(H^+ \otimes F_0) \\ \oplus \\ p_0^*G_0 \end{matrix} \tag{24}$$

is called the *boundary symbol* of $\mathcal{A}$.

Denote by $T^*_{\overline{\mathbb{C}}^-}(\Omega_0 \times X)$ the bundle defined by $T^*_{\overline{\mathbb{C}}^-}\Omega_0 \times T^*X$, where $T^*_{\overline{\mathbb{C}}^-}\Omega_0 = \{(x_0, \zeta_0) \colon x_0 \in \Omega_0, \zeta_0 \in \overline{\mathbb{C}}^-\}$. Let $\pi_1 \colon T^*_{\overline{\mathbb{C}}^-}(\Omega_0 \times X) \setminus \{0\} \to \Omega_0 \times X$ be the canonical projection. Denote by $\sigma_{1,A}$ the extension of $\sigma_A \in \mathfrak{A}_V^{(\alpha),h}(\Omega_0 \times X, E, F)$ with respect to

ζ_0 into $\overline{\mathbb{C}}^-$. Thus we get a bundle morphism

$$\sigma_{1,A}: \pi_1^* E \to \pi_1^* F\,. \tag{25}$$

If σ_A is the quasi-homogeneous principal symbol of r^+A in the operator (15), set also $\sigma_{\Omega_0\times\Omega}(\mathcal{A}) = \sigma_{1,A}$, $\Omega = \operatorname{int} X$, and denote it as the interior symbol of $\mathcal{A}$. The definition of interior and boundary symbols of operators in $\mathfrak{S}_V$ is compatible with composition, i.e.

$$\sigma_{\Omega_0\times Y}(\mathcal{A}_1\mathcal{A}_2) = \sigma_{\Omega_0\times Y}(\mathcal{A}_1)\,\sigma_{\Omega_0\times Y}(\mathcal{A}_2)\,,\quad \sigma_{\Omega_0\times\Omega}(\mathcal{A}_1\mathcal{A}_2) = \sigma_{\Omega_0\times\Omega}(\mathcal{A}_1)\,\sigma_{\Omega_0\times\Omega}(\mathcal{A}_2)\,.$$

Parabolicity of $\sigma_A \in \mathfrak{A}_V^{(\alpha),h}(\Omega_0 \times X, E, F)$ is equivalent to the property that (25) is an isomorphism.

Definition 9. An operator $\mathfrak{A}$ of the form of the left hand side of (15) with (16), (17), (18), (19), (20) is called *parabolic* if

(i) the interior symbol $\sigma_{\Omega_0\times\Omega}(\mathcal{A})$ is parabolic;

(ii) the boundary symbol (24) is an isomorphism.

A parabolic operator A is also called a *parabolic boundary value problem* and the operators r^+B, $r'T$, K, Q are called the *parabolic boundary conditions*. The second condition in Definition 9 is an obvious generalization of the Shapiro-Lopatinski condition from the elliptic theory.

Now formulate a generalization of Theorem 7.

Theorem 10. *If $\mathcal{A} \in \mathfrak{S}_V$ is parabolic in the sense of Definition* 9, (15) *is an isomorphism for $\delta > 0$, $s > 0$ sufficiently large. $\mathcal{P} = \mathcal{A}^{-1}$ is parabolic, i.e. it belongs to the class $\mathfrak{S}_V$ and $\sigma_{\Omega_0\times\Omega}(\mathcal{P}) = \sigma_{\Omega_0\times\Omega}(\mathcal{A})^{-1}$, $\sigma_{\Omega_0\times Y}(\mathcal{P}) = \sigma_{\Omega_0\times Y}(\mathcal{A})^{-1}$.*

It is clear that many other constructions and theorems of the elliptic theory can be generalized to the parabolic case (cf. Schulze [11]). We do not discuss this here and make some concluding remarks about other aspects of the theory.

First remark that there is an analogue in the parabolic case of the theory of elliptic boundary problems for operators without the transmission property, cf. 4.3.2. Results are formulated in Čan Zui Cho/Èskin [1].

One also wants to consider domains of the form $\Omega_{0,t} \times \Omega$ with $\Omega_0 = (-\varepsilon, t)$ and some finite $t > 0$. This is possible because our operators have the property that e.g for $u_j \in H_{(0)}^{s,h}(\Omega_{0,t+t'} \times X)$, $j = 1, 2$, $t' > 0$, we have

$$r^+Au_1|_{\Omega_{0,t}\times X} = r^+Au_2|_{\Omega_{0,t}\times X}$$

when $(u_1 - u_2)|_{\Omega_{0,t}\times X} = 0$ and similarly for the other Volterra operators contained in $\mathcal{A}$. Thus

$$\mathcal{A}:\ \begin{matrix} H_{(0)}^{s,h}(\Omega_{0,t}\times X, E) \\ \oplus \\ H_{(0)}^{t+\lambda+1/2,h}(\Omega_{0,t}\times Y, J) \end{matrix} \to \begin{matrix} H_{(0)}^{t,h}(\Omega_{0,t}\times X, F) \\ \oplus \\ H_{(0)}^{s-\gamma-1/2,h}(\Omega_{0,t}\times Y, G) \end{matrix} \tag{26}$$

is well-defined by first extending the functions as elements in the corresponding Sobolev spaces over $\Omega_{0,t+t'} \times X$ and $\Omega_{0,t+t'} \times Y$, respectively, then applying $\mathcal{A}$ and finally restricting the result to the original domain. Then, from Theorem 10, it follows that (26) is an isomorphism if $\mathcal{A}$ is parabolic.

A further extension of the theory concerns non-cylindric domains. If standard assumptions on the geometry of the boundary are satisfied (cf. Agranovič/Višik

[1]), we get a generalization of the theory of parabolic boundary problems which is easily formulated and left to the reader.

Consider a parabolic operator of the form $\mathcal{A} = \begin{pmatrix} r^+A \\ r'T \end{pmatrix}$ and assume that in local coordinates the complete symbol $a(\tilde{x}, \xi_0, \xi)$ of r^+A is a polynomial in ξ_0. Moreover, let $h > 0$. Then one can investigate the mixed initial-boundary problem

$$r^+Au = f\,, \quad r'Tu = g\,, \tag{27}$$

$$\left(\frac{\partial}{\partial x_0}\right)^j u|_{x_0=0} = l_j\,, \quad j = 0, \dots, \frac{\alpha}{h} - 1\,, \tag{28}$$

$\alpha = \operatorname{ord} \sigma_A$. Under a compatibility condition of the right hand sides in (27), (28) (cf. Agranovič/Višik [1]) this problem is uniquely solvable. The solution of the problem can be reduced to the case $l_j = 0$ for all j.

Finally note that the theory of parabolic boundary problems can be generalized in an obvious way to one-sided parabolicity (left or right) in the sense of 3.1.2.3. Then there exist the corresponding one-sided inverses (left and right, respectively).

4.3.7. Historical Remarks and Comments to the Literature

In this section we give some comments and additional informations about important results in the literature which have contributed to the various aspects of the theory presented in this book. It is of course not possible here to be complete in any direction. One could write a long history about the remarkable achievements in the calculus and the applications of the theories introduced in 1.1.3, 1.2.2, 1.2.3 and 1.2.4. This mainly concerns the Atiyah-Singer index theorem for elliptic pseudo-differential operators on closed compact manifolds which could be the subject of an extra monograph as well as the K-theory or the theory of pseudo-differential and Fourier integral operators. These theories can already be regarded as to be classical.

As an introduction to vector bundles and K-theory can be used Hirzebruch [1], Atiyah [2], Husemoller [1]. The classical aspects of the theory of pseudo-differential operators (PDOs) are presented in Hörmander [6], Friedrichs [1], Taylor [1], Gilkey [1], Šubin [1]. It is well known that the singular integral operators belong to the history in the development of the PDO's as operators of order zero. Many properties and basic facts have been first discovered in this special case. In this connection one should mention Noether [1], Mihlin [1], Calderon/Zygmund [1], Gohberg [1], Kreĭn [1], Simonenko [1], Seeley [2] and many other authors (detailed references are given in Michlin/Prössdorf [1]). The exposition about PDOs and Fourier integral operators in 1.2.2, 1.2.3 essentially follows the paper of Hörmander [6]. The systematic use of the Fourier transform and asymptotic expansions,. i.e. the proper calculus of PDOs, began with the papers of Kohn/Nirenberg [1], Hörmander [2] and was later enriched by essential ideas of geometric optics and Lagrange analysis leading to the theory of Fourier integral operators, cf. Hörmander [6], Duistermaat/Hörmander [1], Duistermaat [1], and the theory of asymptotic solutions, cf. Maslov [1], Maslov/Fedorjuk [1], Leray [1].

An important aspect was the possibility for studying operators on manifolds and to have an invariant symbolic calculus, cf. Seeley [2], Hörmander [2]. Together with the concept of ellipticity of PDOs this gave rise to a connection between

K-theory and index theory of PDOs. The expression of the index of an elliptic PDO on a closed compact manifold in terms of the symbol is a consequence of the Atiyah-Singer index theorem. The original papers of ATIYAH/SINGER [1], [2] are systematic and detailed expositions. The seminar of PALAIS [1] also contains a proof of the index theorem. Later other approaches and specific applications have been developed, e.g. the heat equation method, cf. ATIYAH/BOTT/PATODI [1], GILKEY [1]. A big bibliography and many comments are given in BOOSS [2]. An analogue of the index theorem for super manifolds was proved by REMPEL/SCHMITT [1].

In the present book only some few important points of the classical index theory are reproduced and we refer to the literature mentioned above. For understanding the motivation of the large and complicated constructions on symbols in the case of boundary problems it is advisable first to look at the simpler case of closed compact manifolds.

The Chapters 2 and 3 are devoted to the theory of general elliptic boundary value problems (BVPs) for elliptic PDOs with the transmission property. To the tradition of the theory of elliptic BVPs belong the papers of LOPATINSKIĬ [1], AGMON/DOUGLIS/NIRENBERG [1], SOLONNIKOV [1]. The aspect of PDOs was included by AGRANOVIČ [1], AGRANOVIČ/DYNIN [1], DYNIN [1], [2], VIŠIK/ÈSKIN [1], [2], [4], [5], [6], ÈSKIN [3], BOUTET DE MONVEL [1], [3], [4]. A part of the Chapters 2, 3 contains results of VIŠIK, ÈSKIN and of BOUTET DE MONVEL published in the papers mentioned above completed by results of the authors which are partially first published in the book.

General elliptic BVPs for PDOs with and without the transmission property have been first investigated by VIŠIK and ÈSKIN. In the case of PDOs there occur boundary and coboundary (potential) conditions satisfying an analogue of the Shapiro-Lopatinski condition. In the papers of VIŠIK/ÈSKIN there are constructed parametrices in $H^{s,2}$-spaces and this obtained a-priori estimates. An essential tool are singular integral operators and factorizations of symbols. Besides rather explicit constructions in the half space for x-independent symbols there are obtained corresponding results on manifolds. The method here is "freezing of coefficients".

By BOUTET DE MONVEL [1], [3], [4] an important progress was made in the case of operators with the transmission property by introducing the boundary symbols and establishing the algebra property of his class $\mathfrak{G}$. In connection with compositions and parametrix constructions the Green operators were introduced and so explained the nature of parametrices as elements in $\mathfrak{G}$.

Moreover BOUTET DE MONVEL proved an analogue of the Atiyah-Singer index theorem for elliptic operators $\mathcal{A} \in \mathfrak{G}$ by a difference construction and reducing to the case of manifolds without boundary. Earlier for differential boundary problems the index theorem was proved by ATIYAH/BOTT [2], cf. also the article of ATIYAH in the book PALAIS [1] and ATIYAH [1]. There also was essentially used the analytical proof of ATIYAH [2] of the Bott periodicity theorem in order to explain the connection between difference element and index element of an elliptic symbol near the boundary, cf. Section 3.2.2.1. A further nice aspect of BOUTET DE MONVEL's theory is the homotopy classification of elliptic BVPs in $\mathfrak{G}$, cf. Section 3.2. One of the important technical points is the construction of the operators Λ_E^- which admit reduction of orders. Thus the theory also includes systems in sense of Douglis-Nirenberg ellipticity, cf. Section 3.1.2.1. By reduction of orders and using adjoints in the class $\mathfrak{G}$ one is able to treat over- and underdetermined problems, cf. Section 3.1.2.3. In the Chapter 3 are given also results belonging to the authors such as $H^{s,p}$- and Hölder estimates for elliptic operators, in the class $\mathfrak{G}$, cf. Section 3.1.1.4, the clutching construction for elliptic BVPs given in Section 3.2.1.3, the contents of Section 3.1.2.2 and the results on complexes in Section 3.2.3.2., cf. PILLAT/SCHULZE [1]. The results in Section 2.3.4 essentially belong to GRUBB/GEYMONAT [1] and HÖPPNER [1] except some improvements by the authors. The boundary symbolic calculus is based on classical results on Wiener-Hopf operators, cf. Section 2.1. In Section 3.1.1.2 we give a new variant for the inversion of an elliptic boundary symbol which can be generalized to operators without the transmission property, cf. the remarks in Section 4.3.2.

The main part of Chapter 2 concerns the study of the transmission property of PDOs and of related classes of operators (trace, potential, Green operators). Instead of the notation potential operator BOUTET DE MONVEL [4] introduced the notation Poisson operator: other authors speak of coboundary operators. A generalization of the transmission property to FOURIER distributions is investigated by HIRSCHOWITZ/PIRIOU [1]. The notion of hypoellipticity of PDOs with the transmission property up to the boundary was studied by DE GOSSON [1].

In this book we do not treat spectral properties of elliptic BVPs of the class $\mathfrak{G}$. For this we refer to the papers of SEELEY [6], [7], GRUBB [1], [2], GRUBB/GEYMONAT [1], DUISTERMAAT/GUILLEMIN [1], IVRIĬ [1] and of REMPEL/SCHULZE [3], [4]. Further the algebraic considerations of CORDES [1] and related papers are not touched in this book.

The Chapter 4 is devoted to analytical index formulas foı elliptic operators of the class $\mathfrak{G}$, to non-elliptic boundary problems and to a discussion of further working directions. In Section 4.1 there are generalized corresponding results of FEDOSOV [1], [3], [4] concerning elliptic operators on closed compact manifolds and elliptic BVPs for differential operators. Section 4.1 is an exposition of the papers of REMPEL [1], [3]. The new feature in the proof of analytical index formulas for operators in $\mathfrak{G}$ is the use of complete boundary symbols in a similar manner as in FEDOSOV's proof for closed compact manifolds. This also considerably simplifies the proof for BVPs in the cases originally considered by FEDOSOV [4].

The results in Section 4.2 are an exposition of the papers of SCHULZE [10] and PILLAT/SCHULZE [2]. The constructions are based on the reduction to the boundary of the considered problem. In special cases this method was used by AGRANOVIČ [1], AGRANOVIČ/DYNIN [1], HÖRMANDER [3], SEELEY [3], BOUTET DE MONVEL [4], MELIN/SJÖSTRAND [2] and other authors. Degenerate elliptic boundary problems in various special cases were considered by BICADZE [1], [2], HÖRMANDER [3], EGOROV/KONDRAT'EV [1], ÈSKIN [1], [2], MAZ'JA/PANEJAH [1], MELIN/SJÖSTRAND [1], [2] and many other authors. In Section 4.2 we systematically develop for operators in $\mathfrak{G}$ the idea of additional trace and potential conditions with respect to the subset of the boundary where the Shapiro-Lopatinski condition degenerates and obtain parametrices and a-priori estimates with a typical loss of smoothness compared to the elliptic case, depending on a corresponding behaviour of the operator obtained after reducing to the boundary. Here one is reduced to the study of so-called interior boundary value problems, cf. the remarks in Section 4.3.4. The methods in Section 4.2 especially enable the study of over- and underdetermined systems with degenerate one-sided Shapiro-Lopatinski condition and the study of the case of degenerated (one-sided and two-sided) Douglis-Nirenberg ellipticity in a similar mannei as in Section 3.1.2.1 by reducing of orders. In Section 4.2.2.2 we return to the classical oblique derivative problem and also discuss the case of the bipotential operator.

Bibliography

Agmon, S.

[1] Maximum theorems for solutions of higher order elliptic equations, Bull. Amer. Math. Soc. **66** (1960), 77—80.

[2] Lectures on elliptic boundary value problems, Van Nostrand, Princeton (New Jersey)-New York-Toronto-London **1965**.

Agmon, S., Douglis, A., Nirenberg, L.

[1] Estimates near the boundary for solutions of elliptic partial differential equations satisfying general boundary conditions, Comm. Pure Appl. Math. (I) **12** (1959), 623—727, (II) **17** (1964), 35—62.

Agranovič, M. S. (Агранович, М. С.)

[1] Эллиптические сингулярные интегро-дифференциальные операторы, УМН **20**, 5 (1965), 3—120.

Agranovič, M. S., Dynin, A. S. (Агранович, М. С., Дынин, А. С.)

[1] Общие краевые задачи для эллиптических систем в многомерной области, ДАН СССР **146** (1962), 511—514.

Agranovič, M. S., Višik, M. I. (Агранович, М С., Вишик, М. И.)

[1] Эллиптические задачи с параметром и параболические задачи общего вида, УМН **19** (1964), 53—161.

Asada, A.

[1] Indexes of some degenerate operators, J. Fac. Sci. Shinshu Univ. **13**, 1 (1978), 1—25.

Atiyah, M. F.

[1] La formule d'indice pour les variétés à bord, Séminaire Cartan-Schwartz 16. année, fasc. 2, 1963/64.

[2] K-theory, Harvard University, Cambridge, Mass., 1965 (Russ. transl. "Mir", Moscow 1967).

[3] Bott periodicity and the index of elliptic operators, Quart. J. Math. **19** (1968), 113—140.

[4] Elliptic operators and compact groups, Lecture Notes in Math. **401**, Springer-Verlag, Berlin-Heidelberg-New York **1974**.

Atiyah, M. F., Bott, R.

[1] On the periodicity theorem for complex vector bundles, Acta Math. **112** (1964), 229—247.

[2] The index problem for manifolds with boundary, Coll. Differential Analysis, Tata Institute Bombay, Oxford University Press, Oxford 1964, 175—186.

[3] A Lefschnetz fixed point formula for elliptic complexes I, Ann. of Math. **86** (1967), 374—407.

Atiyah, M., Bott, R., Patodi, V. K.

[1] On the heat equation and the index theorem, Invent. math. **19** (1973), 279—330. (Errata Invent. math. **28** (1975), 277—280.)

Atiyah, M. F., Singer, I. M.

[1] The index of elliptic operators on compact manifolds, Bull. Amer. Math. Soc. **69** (1963), 422—433.

[2] The index of elliptic operators, Ann. of Math. (I) **87** (1968), 484—530, (III) **87** (1968), 546—604, (IV) **93** (1971), 119—138.

Atkinson, F. W. (Аткинсон, Ф. В.)
[1] Нормальная разрешимость линейных уравнений в нормированных пространствах, Мат. сб. **28 (70)** (1951), 3—14.

Baouendi, M. S., Sjöstrand, J.
[1] Non-hypoellipticité d'opérateurs elliptiques singuliers, Sém. Goulaouic-Schwartz 1975—76, Exposé 24.
[2] Régularité analytique pour des opérateurs elliptiques singuliers en un point, Arkiv för Mat. **14** (1976), 9—33.

Beals, R.
[1] Non-local boundary value problems for elliptic operators, Amer. J. Math. **87** (1965), 215—262.
[2] L^p and Hölder estimates for pseudo-differential operators, Ann. Inst. Fourier **19**, 3 (1979), 239—260.

Bers, L., John, F., Schechter, M.
[1] Partial differential equations, Interscience Publishers, New York-London-Sydney 1964.

Bicadze, A. V. (Бицадзе, А. В.)
[1] Об однородной задаче наклонной производной для гармонических функций в трехмерных областях, ДАН СССР **148**, 4 (1963), 749—752.
[2] Задача наклонной производной с полиномиальными коэффициентами, ДАН СССР **157**, 6 (1964), 1273—1276.
[3] Краевые задачи для эллиптических уравнений второго порядка, „Наука", Москва 1966.

Bojarski, B. (Боярский, Б.)
[1] On the index problem for systems of singular integral equations, Bull. Acad. Polon. Sci. Sér. Sci. Math. Astronom. Phys. **11** (1963), 653—655.
[2] Абстрактная задача линейного сопряжения и фредгольмовы пары подпространств, Дифф. и инт. уравн., Краевые задачи, Тбилиси 1979, 45—60.

Booss, B.
[1] Elliptische Topologie von Transmissionsproblemen, Bonner Math. Schriften Nr. 58, Bonn 1972.
[2] Topologie und Analysis — Eine Einführung in die Atiyah-Singer-Index-Formel, Hochschultext, Springer-Verlag, Berlin-Heidelberg-New York 1977.

Booss, B., Rempel, S.
[1] Cutting and pasting of elliptic operators, Preprint, P-Math 02/82, Berlin 1982; Math. Nachr. (to appear).

Booss, B., Schulze, B.-W.
[1] Index theory and elliptic boundary value problems — remarks and open questions, Proc. Banach Semester "Partial Diff. Equ." Warszawa 1978 (to appear).

Boutet de Monvel, L.
[1] Comportement d'un opérateur pseudo différentiel sur une variété à bord, J. Anal. Math. **17** (1966), 241—304.
[2] Pseudo-differential operators and analytic functions, C.I.M.E. "Pseudo-differential operators" (1968), 37—84.
[3] Opérateurs pseudo différentiels analytiques et problèmes aux limites elliptiques, Ann. Inst. Fourier **19** (1969), 163—268.
[4] Boundary problems for pseudo-differential operators, Acta Math. **126** (1971), 11—51.
[5] On the index of Toeplitz operators of several complex variables, Inv. math. **50**, 3 (1979), 249—272.

Breuer, M.
[1] Fredholm theories in von Neumann algebras, Math. Ann. (I) **178** (1968), 243—254, (II) **180** (1969), 313—325.

Bryčkov, Ju. A., Prudnikov, A. P. (Брычков, Ю. А., Прудников, А. П.)
[1] Интегральные преобразования обобщенных функций, „Наука", Москва 1977.

CALDERON, A. P.
[1] Intermediate spaces and interpolation, the complex method, Studia Mathematica **24** (1964), 113–190.
[2] Singular integrals, Bull. Amer. Math. Soc. **72** (1966), 427–465.
[3] The analytic calculation of the index of elliptic equations, Proc. Nat. Acad. Sci. **57** (1967), 1193–1194.

CALDERON, A. P., ZYGMUND, A.
[1] On the existence of certain singular integrals, Acta Math. **88** (1952), 85–139.
[2] On singular integrals, Amer. J. Math. **78** (1956), 289–309.

ČAN ZUI CHO, ÈSKIN, G. I. (Чан Зуи Хо, Эскин, Г. И.)
[1] Краевые задачи для параболических систем псевдо-дифференциальных уравнений, ДАН СССР **198** (1971), 50–53.

COIFMAN, R., MEYER, Y.
[1] Au delà des opérateurs pseudo-différentiels, Astérisque **57** (1978), 2–184.

CORDES, H. O.
[1] Elliptic pseudo-differential operators — an abstract theory, Lecture Notes in Math. **756**, Springer-Verlag, Berlin-Heidelberg-New York 1979.

CORDES, H. O., HERMANN, E. A.
[1] Pseudo-differential operators on a half-line, J. Math. Mech. **18** (1969), 893–908.

CORDES, H. O., WILLIAMS, D. A.
[1] An algebra of pseudo-differential operators with non-smooth symbol, Univ. Bonn, Preprint **168** (1977).

DIKANSKIĬ, A. S. (Диканский, А. С.)
[1] Сопряженные задачи к эллиптическим псевдодифференциальным краевым задачам, ДАН СССР **200**, 5 (1971), 1020–1023.
[2] Сопряженные краевые задачи к эллиптическим дифференциальным и псевдодифференциальным краевым задачам в ограниченной области, Мат. сб. **91** (133) (1973), 62–77.

DUISTERMAAT, J. J.
[1] Fourier integral operators, Lecture Notes of the Courant Institute, New York 1973.

DUISTERMAAT, J. J., GUILLEMIN, V. W.
[1] The spectrum of positive elliptic operators and periodic geodesics, Invent. math. **29** (1975), 39–79.

DUISTERMAAT, J. J., HÖRMANDER, L.
[1] Fourier integral operators II, Acta Math. **128** (1972), 183–269.

DUISTERMAAT, J. J., SINGER, I. M.
[1] Order preserving isomorphisms between algebras of pseudodifferential operators, Comm. Pure Appl. Math. **29** (1976), 1–9.

DUNFORD, N., SCHWARTZ, J. T.
[1] Linear operators, part I, II. General theory, Interscience Publishers, New York-London-Sydney 1958, 1963.

DYNIN, A. S. (Дынин, А. С.)
[1] Многомерные эллиптические краевые задачи с одной неизвестной функцией, ДАН СССР **141** (1961), 285–287.
[2] Эллиптические интегро-дифференциальные краевые задачи, УМН **18**, 2 (1963), 183–184.
[3] К теории псевдодифференциальных операторов на многообразии с краем, ДАН СССР **186**, 2 (1969), 251–253.
[4] Эллиптические краевые задачи для псевдодифференциальных комплексов, Функц. анализ и прилож. **6**, 1 (1972), 75–76.

EDWARDS, R. E.
[1] Functional analysis, Holt Rinehart Winston, New York 1965 (Russ. transl. "Mir", Moscow 1971).

Egorov, Ju. V. (Егоров, Ю. В.)
[1] Субэллиптические операторы, УМН **30**, 2 (1975), 57—114.

Egorov, Ju. V., Kondrat'ev, V. A. (Егоров, Ю В., Кондратьев, В. А.)
[1] О задаче с косой производной, Мат. сб. **78** (**120**) (1969), 148—176.

Elschner, J.
[1] Über entartete Randwert-Probleme für elliptische Differentialoperatoren zweiter Ordnung in der Ebene, Report ZIMM, R-04/80, Berlin 1980.

Elschner, J., Lorenz, M.
[1] On a class of elliptic differential operators degenerating at one point, Proc. of the Banach Semester "Partial Diff. Equ." 1978, Warszawa (to appear).

Èskin, G. I. (Эскин, Г. И.)
[1] Вырождающеся эллиптические псевдодифференциальные уравнения главного типа, Мат. сб. **82** (**124**), 4 (8) (1970), 585—628.
[2] Эллиптические псевдодифференциальные операторы с вырождением первого порядка по пространственным переменным, Труды Моск. Мат. Общ. **25** (1971), 83—118.
[3] Краевые задачи для эллиптических псевдодифференциальных уравнений, ;,Наука", Москва 1973.

Ezin, J. P.
[1] Estimées de Schauder pour des opérateurs elliptiques, Astérisque **58** (1978), 23—34.

Fedorjuk, M. V. (Федорюк, М. В.)
[1] Метод стационарной фазы и псевдодифференциальные операторы, УМН **26**, 1 (1971), 67—112.

Fedosov, B. V. (Федосов, Б. В.)
[1] Об индексе эллиптической системы на многообразии, Функц. анализ **4**, 4 (1970), 57—67.
[2] Непосредственное доказательство формулы для индекса эллиптической системы в эвклидовом пространстве, Функц. анализ **4**, 4 (1970), 83—84.
[3] Аналитические формулы индекса эллиптических операторов, Труды Моск. мат. общ. **30** (1974), 159—241.
[4] Аналитическая формула индекса эллиптической граничной задачи, Мат. сб. (I) **93** (**135**), 1 (1974), 62—89; (II) **95** (**137**), 4 (1974), 525—550; (III) **101** (**143**), 3 (1976), 380—401.
[5] Теорема периодичности в алгебре символов, Мат. сб. **105**, 3 (1978), 431—462.
[6] Об индексе эллиптических семейств на многообразии с краем, ДАН **248**, 5 (1979), 1066—1069.

Fedosov, B. V., Šubin, M. A. (Федосов, Б. В., Шубин, М. А.)
[1] Индекс случайных эллиптических операторов, Мат. сб. (I) **106** (148), 1 (1978), 108—140; (II) **106** (**148**), 3 (1978), 455—483.

Friedrich, T.
[1] Vorlesungen über K-Theorie, Teubner-Texte zur Mathematik, Leipzig 1978.

Friedrichs, K. O.
[1] Pseudo-differential operators. An introduction. Lecture Notes, Courant Institute, New York 1970.

Gårding, L.
[1] Transformations de Fourier des distributions homogènes, Bull. Soc. Math. France **89** (1961), 381—428.

Gel'fand, I. M. (Гельфанд, И. М.)
[1] Об эллиптических уравнениях, УМН **15**, 3 (1960), 121—132.

Gelfand, I. M., Schilow, G. I.
[1] Verallgemeinerte Funktionen (Distributionen) I, VEB Deutscher Verlag der Wissenschaften, Berlin 1967.

Gibbons, Greg
[1] Opérateurs pseudo-différentiels et espaces de Besov, C. R. Acad. Sci. Paris **286**, 20 (1978), 895—897.

GILKEY, P. B.
[1] The index theorem and the heat equation, Math. Lecture Series 4, Publish or Perish, Boston, Mass., 1974.
[2] The boundary integrand in the formula for the signature and Euler characteristic of a Riemannian manifold with boundary, Adv. in Math. 15 (1975), 334–360.

GOHBERG, I. C. (Гохберг, И. Ц.)
[1] Некоторые вопросы теории многомерных сингулярных интергальных уравнений, Изв. Молдавского филиала АН СССР 10 (76) (1960), 39—50.

GOHBERG, I. C., FEL'DMAN, I. A. (Гохберг, И. Ц., Фельдман, И. А.)
[1] Уравнения в свертках и проекционные методы их решения, ,,Наука", Москва 1971 (Germ. transl. Akademie-Verlag, Berlin 1974).

GOHBERG, I. C., KREĬN, M. G. (Гохберг, И. Ц., Крейн, М. Г.)
[1] Системы интегральных уравнений на полупрямой с ядрами, зависящими от разности аргументов, УМН 13, 2 (1958), 3—72.

GOHBERG, I. C., KRUPNIK, N. JA. (Гохберг, И. Ц., Крупник, Н. Я.)
[1] Об алгебре, порожденной одномерными сингулярными интегралными операторами с кусочно-непрерывными коэффициентами, Функц. анализ и прилож. 4, 3 (1970), 26—36.
[2] Введение в теорию одномерных сингулярных интегральных операторов, ,,Штиинца", Кишинёв 1973.

DE GOSSON, M.
[1] Hypoellipticité partielle à la frontière des opérateurs pseudo-différentiels de transmission. Thesis, Université de Nice 1978.

GREINER, P.
[1] An asymptotic expansion for the heat equation, Arch. Ration. Mech. and Anal. 41, 1 (1971), 163–218.

GRUBB, G.
[1] Spectral asymptotics for Douglis-Nirenberg elliptic and pseudo-differential boundary problems, Comm. Part. Diff. Equ. 2 (1977), 1071–1150.
[2] Remainder estimates for eigenvalues and kernels of pseudo-differential elliptic systems, Math. Scand. 43 (1978), 275–307.
[3] On pseudo-differential boundary problems, I. The basic calculus, Københavns Universitet Matematisk Institut, Preprint 2 (1979).
[4] On pseudo-differential boundary problems I B, I 6 Supplement to Chapter I. Corrections, Københavns Universitet Matematisk Institut, Preprint 1 (1980).
[5] On pseudo-differential boundary problems II A, Københavns Universitet Matematisk Institut, Preprint 2 (1980).

GRUBB, G., GEYMONAT, G.
[1] The essential spectrum of elliptic systems of mixed order, Math. Ann. 227 (1977), 247–276.

GRUŠIN, V. V. (Грушин, В. В.)
[1] Гипоэллиптические дифференциальные уравнения и псевдодифференциальные операторы с операторнозначными символами, Мат. сб. 88, 4 (1972), 504—521.

GRUŠIN, V. V., ŠANANIN, N. A. (Грушин, В. В., Шананин, Н. А.)
[1] Некоторые теоремы об особенностях решений дифференциальных уравнений со взвешенными главными символами, Мат. сб. 103 (145), 1 (1977), 37—51.

GUILLEMIN, V., STERNBERG, S.
[1] Geometric asymptotics, Math. Surveys 14, Amer. Math. Soc., Providence, Rhode Islands, 1977.

HIRSCHOWITZ, A., PIRIOU, A.
[1] La propriété de transmission pour les distributions de Fourier, application aux lacunes, Sém. Goulaouic-Schwartz 1976–77.

[2] Propriétés de transmission pour les distributions intégrales de Fourier, Comm. Part. Diff. Equ. **4**, 2 (1979), 113–217.

Hirzebruch, F.
[1] Topological methods in algebraic geometry, Springer-Verlag, New York 1966 (Russ. transl. "Mir", Moscow 1973).

Hirzebruch, F., Zagier, D.
[1] The Atiyah-Singer theorem and elementary number theory, Math. Lecture Series **3**, Publish or Perish, Boston, Mass., 1974.

Höppner, W.
[1] Einige Abschätzungen für die Symbolrechnung von Pseudodifferentialoperatoren im Halbraum (I), Math. Nachr. **90** (1979), 267–278; (II) Math. Nachr. **93** (1980), 33–52.
[2] Über gruppeninvariante Randwertprobleme, Report ZIMM, R-02/80, Berlin 1980.

Hörmander, L.
[1] Linear partial differential operators, Springer-Verlag, Berlin-Göttingen-Heidelberg 1963 (Russ. transl. "Mir", Moscow 1965).
[2] Pseudo-differential operators, Comm. Pure Appl. Math. **18**, 3 (1965), 501–517.
[3] Pseudo-differential operators and non-elliptic boundary problems, Ann. of Math. **83**, 1 (1966), 129–209.
[4] Pseudo-differential operators and hypoelliptic equations, Amer. Math. Soc., Symp. on Singular Integral Operators (1966), 138–183.
[5] On the index of pseudo-differential operators, Kolloqu. Ell. Diff.gln. II, Akademie-Verlag, Berlin 1971, 127–146 (Russ. transl. Сб. Математика, **14**:4 (1970), 78–97).
[6] Fourier-integral operators I, Acta Math. **127** (1971), 79–183.
[7] The Weyl calculus of pseudo-differential operators, Comm. Pure Appl. Math. **32**, 3 (1979), 360–444.

Hunt, C., Piriou, A.
[1] Opérateurs pseudo-différentiels anisotropes d'ordre variable, C. R. Acad. Sci. Paris **268**, Sér. A (1969), 28–31.
[2] Majorations L^2 et inégalité sous-elliptique pour les opérateurs pseudo-différentiels anisotropes d'ordre variable, C. R. Acad. Sci. Paris **268**, Sér. A (1969), 214–217.

Husemoller, D.
[1] Fibre bundles, Mc Graw Hill, New York 1966 (Russ. transl. "Mir", Moscow 1970).

Ivrii, V. Ja. (Иврий, В. Я.)
[1] О втором члене спектральной асимптотики для операторов Лапласа-Бельтрами на многообразиях с краем. Функц. анализ и его прилож. **14**, 2 (1980), 25—34.

Jänich, K.
[1] Vektorraumbündel und der Raum der Fredholm-Operatoren, Math. Ann. **161** (1965), 129–142.

John, F.
[1] Plane waves and spherical means applied to partial differential equations, Interscience Publ., New York 1955.

Juhl, A.
[1] Über die Lösbarkeit, die Regularität und den Index innerer Randwert-Probleme für entartete elliptische Pseudodifferentialoperatoren. Dissertation, Halle, Martin-Luther-Universität 1980 (cf. Seminar Analysis 1981/82, Berlin 1982).

Kato, T.
[1] Perturbation theory for linear operators, Springer-Verlag, Berlin-Göttingen-Heidelberg-New York 1966.

Kibirev, V. V. (Кибирев, В. В.)
[1] К задаче о наклонной производной с линейнымм коэффициентами для гармонических функции, Дифф. Уравн. **16**, 1 (1980), 80—85.

Kohn, J. J., Nirenberg, L.
[1] On the algebra of pseudo-differential operators, Comm. Pure Appl. Math. **18** (1965), 269–305.

Krasovskiĭ, Ju. P. (Красовский, Ю. П.)
[1] Оценки роста производных решений однородных эллиптических уравнений вблизи границы, ДАН СССР **184**, 3 (1969), 534—537.

Kreĭn, M. G. (Крейн, М. Г.)
[1] Интегральные уравнения на полупрямой с ядрами зависящими от разности аргументов, УМН **13**, 5 (1958), 3—120.

Krupnik, N. Ja. (Крупник, Н. Я.)
[1] О многомерных сингулярных уравнениях, УМН **20**, 6 (126) (1965), 119—123.

Kuiper, N.
[1] Contractibility of the unitary group in Hilbert space, Topology **3** (1964), 19—30.

Leray, J.
[1] Analyse Lagrangienne et mécanique quantique. Sém. sur les equ. aux dériv. part., Collège de France 1976—77.

Lions, J. L., Magenes, E.
[1] Problèmes aux limites non homogènes et applications, vol. 1, Dunod, Paris 1968 (Russ. transl. "Mir", Moscow 1971).

Lopatinskiĭ, Ja. B. (Лопатинский, Я. Б.)
[1] Об одном способе приведения граничных задач для системы дифференциальных уравнений эллиптического типа к регулярным уравнениям, Укр. мат. ж. **5** (1953), 123—151.

Luke, G.
[1] Pseudo-differential operators on Hilbert bundles, J. of Diff. Equ. **12** (1972), 566—589.

Marčenko, A. V. (Марченко, А. В.)
[1] Об индексе краевых задач для вырождающихся эллиптических операторов, Мат. сб. **87** (**129**), 4 (1972), 594—612.

Maslov, V. P. (Маслов, В. П.)
[1] Теория возмущений и асимптотические методы. Издательство МГУ, Москва 1965.

Maslov, V. P., Fedorjuk, M. V. (Маслов, В. П., Федорюк, М. В.)
[1] Квазиклассическое приближение для уравнений квантовой механики, ,,Наука", Москва 1976.

Matumoto, T.
[1] Equivariant K-theory and Fredholm operators, J. Fac. Sci. Univ. Tokyo, Sec. I A, **18** (1971—72), 109—126.

Maurin, K.
[1] Methods of Hilbert spaces, Polish Scientific Publishers, Warszawa 1967.

Maz'ja, V. G. (Мазья, В. Г.)
[1] О вырождающейся задаче с косой производной, Мат. сб. **87** (129), 3 (1972), 415—454.

Maz'ja, V. G., Panejah, B. P. (Мазья, В. Г., Панеях, Б. П.)
[1] Вырождающиеся эллиптические псевдодифференциальные операторы и задача с косой производной, Труды Моск. мат. общ. **31** (1974), 237—295.

Melin, A.
[1] Lower bounds for pseudo-differential operators, Ark. Mat. **9** (1971), 117—140.

Melin, A., Sjöstrand, J.
[1] Fourier integral operators with complex phase functions and parametrix for an interior boundary value problem, Comm. Part. Diff. Equ. **1** (1976), 313—400.
[2] A calculus for Fourier integral operators in domains with boundary and applications to the oblique derivative problem, Comm. Part. Diff. Equ. **2** (1977), 857—935.

Meyer, Y.
[1] Nouvelles estimations pour les opérateurs pseudo-différentiels, Sem. Goulaouic-Schwartz 1977—1978, Exp. 10.

Mihlin, S. G. (Михлин, С. Г.)
[1] Многомерные сингулярные интегралы и интегралные уравнения. „Физматгиз", Москва 1962

Michlin, S. G., Prössdorf, S.
[1] Singuläre Integraloperatoren, Akademie-Verlag, Berlin 1980.

Milnor, J.
[1] Morse theory, Princeton University Press, Princeton, New Jersey, 1963 (Russ. transl. "Mir", Moscow 1965).

Miranda, C.
[1] Partial differential equations of elliptic type, Springer-Verlag, Berlin-Heidelberg-New York 1970.

Mizohata, S.
[1] The theory of partial differential equations, Cambridge University Press, Cambridge 1973.

Mushelišvili, N. I. (Мусхелишвили, Н. И.)
[1] Некоторые основные задачи математической теории упругости, „Наука", Москва 1966.
[2] Сингулярные интегральные уравнения, „Наука", Москва 1968 (German transl. Akademie-Verlag, Berlin 1965).

Myškis, P. A. (Мышкис, П. А.)
[1] Об алгебре, порожденной парными псевдодифференциальными операторами на многообразии, УМН **31**, 4 (1976), 269—270.

Narasimhan, R.
[1] Analysis on Real and Complex Manifolds, Masson and Cie., Paris 1973.

Nirenberg, L.
[1] Pseudo-differential operators. Global Analysis, Proc. Symp. Pure Math. **16** (1970), 149—167.

Noether, F.
[1] Über eine Klasse singulärer Integralgleichungen, Math. Ann. **82** (1921), 42—63.

Novikov, S. P., Sternin, B. Ju. (Новиков, С. П., Стернин, Б. Ю.)
[1] Эллиптические операторы и подмногообразия, ДАН СССР **171**, 3 (1966), 525—528.

Oleĭnik, O. A. (Олейник, О. А.)
[1] Краевые задачи для линейных уравнений эллиптического и параболического типа с разрывными коэффициентами, Изв. АН СССР, сер. мат. **25** (1961), 3—20.

Palais, R. S.
[1] Seminar on the Atiyah-Singer index theorem, Princeton University Press, Princeton, New Jersey, 1965 (Russ. transl. "Mir", Moscow 1970).

Peetre, J.
[1] Mixed problems for higher order elliptic equations in two variables, Ann. Sc. Norm. Sup. Pisa (I) **15** (1961), 337—353; (II) **17** (1963), 1—12.

Picone, M.
[1] Sur un problème nouveau de la théorie mathématique classique de l'élasticité. In Second Colloque sur les équations aux dérivées partielles, Bruxelles 1954, Librairie Universitaire, Louvain 1954.

Pillat, U., Schulze, B.-W.
[1] Elliptische Randwert-Probleme für Komplexe von Pseudodifferentialoperatoren, Math. Nachr. **94** (1980), 173—210.
[2] Some classes of non-elliptic boundary value problems for pseudodifferential operators. II. Overdetermined and underdetermined systems, Comm. Partial Diff. Equ. **6**, 4 (1981), 373—398.

Piriou, A.
[1] Une classe d'opérateurs pseudodifférentiels du type de Volterra, Ann. Inst. Fourier, Grenoble **20**, 1 (1970), 77—94.

[2] Problèmes aux limites généraux pour des opérateurs différentiels paraboliques dans un domaine borne, Ann. Inst. Fourier, Grenoble **21**, 1 (1971), 59—78.

PRÖSSDORF, S.

[1] Einige Klassen singulärer Gleichungen, Akademie-Verlag, Berlin 1974 (Engl. transl. North-Holland Publishing Comp., Amsterdam, New York, Oxford 1978).

RADKEVIČ, E. V. (Радкевич, Е. В.)

[1] Теорема об устранимости особенности для решения одного класса линейных уравнений с частными производными, УМН **32**, 4 (1977), 265—266.

REMPEL, S. (Ремпель, Ш.)

[1] Eine analytische Indexformel für elliptische Randwert-Probleme für Pseudodifferentialoperatoren auf Mannigfaltigkeiten mit Rand. Dissertation, Preprint ZIMM, R-06/78, Berlin 1978.

[2] An analytical index formula for elliptic pseudo-differential boundary problems in the half-space, Banach Centre Publ. „Partial Differential Equations", Warszawa 1978.

[3] An analytical index formula for elliptic pseudo-differential boundary problems I, Math. Nachr. **94** (1980), 243—275.

[4] О ζ-функции общих эллиптических краевых задач для псевдодифференциальных операторов, Докл. Болг. АН **32**, 3 (1979), 281—283.

[5] Transmission and mixed boundary problems for elliptic pseudo-differential operators, Preprint ZIMM, Berlin (to appear).

REMPEL, S., SCHULZE, B.-W.

[1] Elliptische Operatoren auf berandeten Mannigfaltigkeiten (Operatoren auf der Halbachse), Preprint ZIMM R-02/78, Berlin 1978.

[2] Parametrices and boundary symbolic calculus for elliptic boundary problems without transmission property, Math. Nachr. **105** (1982), 45—149.

[3] Complex powers for pseudo-differential boundary problems. I, II Math. Nachr. (to appear).

[4] Complex powers for pseudo-differential boundary problems without the transmission property, Seminar Analysis 1981/82, IMath, Berlin 1982.

[5] A theory of pseudo-differential boundary value problems with discontinuous conditions, Preprint, IMath, Berlin (to appear).

REMPEL, S., SCHMITT, T.

[1] Pseudo-differential operators and the index theorem on supermanifolds, Seminar Analysis 1981/82, IMath, Berlin 1982.

ROBERTSON, A. P., ROBERTSON, W. J.

[1] Topological vector spaces, Cambridge University Press, Cambridge 1964 (Russ. transl. "Mir", Moscow 1967).

ROĬTBERG, JA. A., ŠEFTEL', Z. G. (Ройтберг, Я. А., Шефтель, З. Г.)

[1] Общие граничные задачи для эллиптических уравнений с разрывными коэффициентами, ДАН СССР **148**, 5 (1963), 1034—1037.

SADOWSKI, M.

[1] Bott periodicity theorem for the KU_G-theory and Wiener-Hopf complexes, Bull. Acad. Pol. Sci., Ser., Sci. Math., Astr. et Phys. **23** (1975), 299—304.

SCHAEFER, H. H.

[1] Topological vector spaces, Springer-Verlag, Berlin-Heidelberg-New York 1971 (Russ. transl. "Mir", Moscow 1971).

SCHECHTER, M.

[1] General boundary value problems for elliptic partial differential equations, Comm. Pure Appl. Math. **12** (1959), 457—486.

[2] A generalization of the problem of transmission, Ann. Sc. Norm. Sup. Pisa **14** (1960), 207 - 236.

[3] On L^p-estimates and regularity, I. Amer. J. Math. **85**, 1 (1963), 1—13; II. Math. Scand. **13**, 1 (1963), 47—69; III. Ricerche Mat. **13** (1964), 192—206.

SCHULZE, B.-W. (Шульце, Б.-В.)

[1] Abschätzungen in Normen gleichmäßiger Konvergenz für elliptische Randwert-Probleme, Math. Nachr. 67 (1975), 303—315.

[2] Elliptic operators on manifolds with boundary, Contr. „Globale Analysis" Ludwigsfelde, Berlin 1976.

[3] Adjoints of elliptic boundary value problems, Banach Centre Publ. "Partial Differential Equations", Warszawa 1978 (to appear).

[4] О переопределенных и недоопределенных эллиптических задач, Дифф. Уравн. 15, 2 (1979), 323—328.

[5] Multiplicative properties of elliptic boundary value problems, Proc. III. Romanian-Finnish Seminar on Complex Analysis, Lecture Notes in Math. 743, Springer-Verlag, Berlin-Heidelberg-New York 1979, 581—588.

[6] Adjungierte elliptische Randwertprobleme und Anwendungen auf über- und unterbestimmte Systeme, Math. Nachr. 89 (1979), 225—245.

[7] Über eine Verheftung elliptischer Randwert-Probleme, Math. Nachr. 93 (1979), 205—222.

[8] Some classes of non-elliptic boundary value problems for elliptic pseudo-differential operators I, Math. Nachr. 95 (1980), 327—350.

[9] Parabolic boundary value problems for pseudo-differential operators, Preprint ZIMM, Berlin 1983 (to appear).

SCHULZE, B.-W., WILDENHAIN, G.

[1] Methoden der Potentialtheorie für elliptische Differentialgleichungen beliebiger Ordnung, Akademie-Verlag, Berlin 1977.

SCHWARTZ, L.

[1] Distributions à valeurs vectorielles I, Ann. Inst. Fourier 7 (1957), 1—141; II, Ann. Inst. Fourier 8 (1958), 1—209.

[2] Théorie des Distributions. Vol. I, II, Hermann, Paris 1957, 1959.

SCHWARTZ, J. T.

[1] Differential geometry and topology, Notes on Math. and its Appl., New York 1968 (Russ. transl. "Mir" Moscow 1970).

SEELEY, R.

[1] Singular integrals on compact manifolds, Amer. J. Math. 81 (1959), 658—690.

[2] Integro-differential operators on vector bundles, Trans. Amer. Math. Soc. 117 (1965), 167—204.

[3] Singular integrals and boundary value problems, Amer. J. Math. 88 (1966), 781—809.

[4] Complex powers of an elliptic operator, Proc. Symp. Pure Math. 10 (1967), 288—307.

[5] Topics in pseudo-differential operators, C. I. M. E. "Pseudodifferential operators" (1968), 167—306.

[6] Analytic extensions of the trace associated with elliptic boundary problems, Amer. J. Math. 91 (1969), 963—983.

[7] The resolvent of an elliptic boundary value problem, Amer. J. Math. 91 (1969), 889—920.

SEGAL, G.

[1] Fredholm complexes, Quarterly J. Math. 27 (1970), 385—402.

SHAMIR, E.

[1] Mixed boundary value problems for elliptic equations in the plane, the L_p-theory, Ann. Sc. Norm. Sup. Pisa 17 (1963), 117—139.

SIMONENKO, I. B. (Симоненко, И. Б.)

[1] Новый общий метод исследования линейных операторных интегральных уравнений, Изв. АН СССР, сер. матем. I. 29, 3 (1965), 567—586, II. 29, 4 (1965), 757—782.

SJÖSTRAND, J.

[1] Operators of principal type with interior boundary conditions, Acta Math. 130 (1973), 1—51.

[2] Parametrices for pseudodifferential operators with multiple characteristics, Arkiv för Matematik 12 (1974), 85—130.

SLOBODECKIĬ, L. N. (Слободецкий, Л. Н.)

[1] Обобщенные пространства С. Л. Соболева и их приложения к краевым задачам для дифференциальных уравнений в частных производных, Уч. зап. Ленингр. гос. пед. ин-та им. А. И. Герцена **197** (1958), 54—112.

SOBOLEW, S. L.

[1] Einige Anwendungen der Funktionalanalysis auf Gleichungen der mathematischen Physik, Akademie-Verlag, Berlin 1964.

SOGA, H.

[1] Boundary value problems with oblique derivative, Publ. Research Institute for Math. Sc., Kyoto Univ. **10**, 3 (1975), 619—668.

[2] Degenerate elliptic systems of pseudodifferential equations and noncoercive boundary value problems, Osaka J. Math. **14** (1977), 537—555.

SOLONNIKOV, V. A. (Солонников, В. А.)

[1] Об общих краевых задачах для систем, эллиптических в смысле А. Даглиса-Л. Ниренберга, I. Изв. АН СССР **28** 1964, 665—706; II. Тр. МИАН СССР **92** 1966, 233—297.

[2] Переопределенные эллиптические краевые задачи, Зап. научн. Сем. ЛОМИ **21** (1971), 112—158.

SPANIER, E. H.

[1] Algebraic topology, McGraw-Hill, New York 1966 (Russ. transl. "Mir", Moscow 1971).

SPENCER, D. C.

[1] Uncoupled overdetermined systems, Proc. Symp. Pure Math. **16**, "Global Analysis" III. AMS, Providence 1970, 211—220.

STERNIN, B. JU. (Стернин, Б. Ю.)

[1] Эллиптические и параболические уравнения на многообразиях с границей, состоящей из компонентов различной размерности, Тр. Моск. мат. общ. **15** (1966), 346—382.

[2] Топологические аспекты проблемы С. Л. Соболева, Моск. инст. электронного машиностроения, Москва 1971.

[3] Относительная эллиптическая теория и проблема С. Л. Соболева, ДАН СССР **230**, 2 (1976), 287—290.

ŠUBIN, M. A. (Шубин, М. А.)

[1] Псевдодифференциальные операторы и спектральная теория, ,,Наука", Москва 1978.

SWEENEY, W. J.

[1] Estimates for elliptic complexes of higher order, J. Diff. Equ. **10**, 1 (1971), 112—140.

[2] Coerciveness in the Neumann problem, J. Diff. Geometry **6** (1972), 375—393.

TAIRA, K.

[1] On non-homogeneous boundary value problems for elliptic differential operators, Kōdai Math. Sem. Rep. **25** (1973), 337—356.

[2] On a degenerate oblique derivative problem with interior boundary conditions, Proc. Japan Acad. Sci. **52** (1976), 484—487.

[3] On a degenerate oblique derivative problem of Egorov and Kondrat'ev, J. Fac. Sc. Univ. Tokyo **23**, 2 (1976), 383—391.

[4] On some non-coercive boundary value problems for the Laplacian, J. Fac. Sc. Univ. Tokyo **23**, 2 (1976), 343—367.

TAYLOR, M.

[1] Pseudo differential operators. Princeton University Press, Princeton, New Jersey, 1981.

TREVES, F.

[1] Topological vector spaces, distributions and kernels, Academic-Press, New York, London 1967.

[2] Introduction to pseudodifferential and Fourier integral operators, Plenum Press, New York, London 1980.

Triebel, H.
[1] Interpolation Theory, Function Spaces, Differential Operators, VEB Deutscher Verlag der Wissenschaften, Berlin 1978.

Višik, M. I., Èskin, G. I. (Вишик, М. И., Эскин, Г. И.)
[1] Краевые задачи для общих сингулярных уравнений в ограниченной области, ДАН СССР **155**, 1 (1964), 24—27.
[2] Уравнения в свертках в ограниченной области, УМН **20**, 3 (1965), 89—152.
[3] Параболические уравнения в свертках в ограниченной области, Мат. сб. **71**, 2 (1966), 162—190.
[4] Эллиптические уравнения в свертках в ограниченной области и их приложения, УМН **22**, 1 (1967), 15—76.
[5] Уравнения в свертках переменного порядка, Тр. Моск. мат. общ. **16** (1967), 25—50.
[6] Нормально разрешимые задачи для эллиптических систем уравнений в свертках, Мат сб. **74**, 3 (1967), 326—356.
[7] Пространства Соболева-Слободецкого переменного порядка и их приложения к смешанными краевыми задачами, Сиб. мат. журн. **9**, 5 (1968), 973—997.

Višik, M. I., Grušin, V. V. (Вишик, М. И., Грушин, В. В.)
[1] Об одном классе вырождающихся эллиптических уравнений высших порядков, Мат. сб. **79** (121), 1 (1969), 3—36.
[2] Эллиптические псевдодифференциальные операторы на замкнутом многообразии, вырождающиеся на подмногообразии, ДАН СССР **189**, 1 (1969), 16—19.
[3] Лекции по вырождающимся эллиптическим задачам, АН Украинской ССР, институт мат., седьмая летняя мат. школа, Кацивели 1969, Киев 1970.

Vladimirov, V. S. (Владимиров, В. С.)
[1] Уравнения математической физики, ,,Наука'', Москва 1971.

Volevič, L. P. (Волевич, Л. П.)
[1] Разрешимость краевых задач для обших эллиптических систем, Мат. сб. **68** (110) (1965), 373—416.

Volevič, L. P., Gindikin, S. G. (Волевич, Л. П., Гиндикин, С. Г.)
[1] Уравнение Винера-Хопфа в обобщенных функциях, Тр. Моск. мат. общ. **35** (1976), 165—214.
[2] Уравнение Винера-Хопфа в обобщенных функциях гладких на полупрямой, Тр. Моск. мат. общ. **38** (1979), 29—74.
[3] Уравнение Винера-Хопфа для гладких и обобщенных функций в полупространстве, Тр. Моск. мат. общ. **40** (1979), 241—295.

Wells, R. O.
[1] Differential Analysis on Complex Manifolds, Englewoods Cliffs, New Jersey, 1973 (Russ. transl. "Mir", Moscow 1976).

Whitehead, G. W.
[1] Recent advances in homotopy theory, AMS Publ. 1970 (Russ. transl. "Mir", Moscow 1974).

Wiener, N., Hopf, E.
[1] Über eine Klasse singulärer Integralgleichungen, Sitzungsberichte Preußische Akademie, Math. Phys. Kl. (1931), 696—706.

Zaremba, S.
[1] Sur un problème toujours possible compenant à titre de cas particuliers, le problème de Dirichlet et celui de Neumann, J. math. pures appl. **6** (1927), 127—163.

Index

Bei Fragen zur Produktsicherheit wenden Sie sich bitte an:
If you have any questions regarding product safety,
please contact:

Walter de Gruyter GmbH
Genthiner Straße 13
10785 Berlin
productsafety@degruyterbrill.com